Lecture Notes in Electrical Engineering

Volume 650

The book series *Lecture Notes in Electrical Engineering* (LNEE) publishes the latest developments in Electrical Engineering - quickly, informally and in high quality. While original research reported in proceedings and monographs has traditionally formed the core of LNEE, we also encourage authors to submit books devoted to supporting student education and professional training in the various fields and applications areas of electrical engineering. The series cover classical and emerging topics concerning:

- Communication Engineering, Information Theory and Networks
- Electronics Engineering and Microelectronics
- Signal, Image and Speech Processing
- Wireless and Mobile Communication
- Circuits and Systems
- Energy Systems, Power Electronics and Electrical Machines
- Electro-optical Engineering
- Instrumentation Engineering
- Avionics Engineering
- Control Systems
- Internet-of-Things and Cybersecurity
- Biomedical Devices, MEMS and NEMS

For general information about this book series, comments or suggestions, please contact leontina. dicecco@springer.com.

To submit a proposal or request further information, please contact the Publishing Editor in your country:

China

Jasmine Dou, Associate Editor (jasmine.dou@springer.com)

India, Japan, Rest of Asia

Swati Meherishi, Executive Editor (Swati.Meherishi@springer.com)

Southeast Asia, Australia, New Zealand

Ramesh Nath Premnath, Editor (ramesh.premnath@springernature.com)

USA, Canada:

Michael Luby, Senior Editor (michael.luby@springer.com)

All other Countries:

Leontina Di Cecco, Senior Editor (leontina.dicecco@springer.com)

**** Indexing: The books of this series are submitted to ISI Proceedings, EI-Compendex, SCOPUS, MetaPress, Web of Science and Springerlink ****

More information about this series at http://www.springer.com/series/7818

Jiadong Sun · Changfeng Yang · Jun Xie
Editors

China Satellite Navigation Conference (CSNC) 2020 Proceedings: Volume I

Springer

Editors
Jiadong Sun
China Aerospace Science
and Technology Corporation
Beijing, Beijing, China

Changfeng Yang
China Satellite Navigation
Engineering Center
Beijing, Beijing, China

Jun Xie
China Academy of Space Technology
Beijing, Beijing, China

ISSN 1876-1100 ISSN 1876-1119 (electronic)
Lecture Notes in Electrical Engineering
ISBN 978-981-15-3706-6 ISBN 978-981-15-3707-3 (eBook)
https://doi.org/10.1007/978-981-15-3707-3

This Springer imprint is published by the registered company Springer Nature Singapore Pte Ltd.
The registered company address is: 152 Beach Road, #21-01/04 Gateway East, Singapore 189721, Singapore

Preface

BeiDou Navigation Satellite System (BDS) is China's global navigation satellite system which has been developed independently. BDS is similar in principle to the global positioning system (GPS) and compatible with other global satellite navigation systems (GNSS) worldwide. The BDS will provide highly reliable and precise positioning, navigation and timing (PNT) services as well as short-message communication for all users under all-weather, all-time and worldwide conditions.

China Satellite Navigation Conference (CSNC) is an open platform for academic exchanges in the field of satellite navigation. It aims to encourage technological innovation, accelerate GNSS engineering and boost the development of the satellite navigation industry in China and in the world.

The 11th China Satellite Navigation Conference (CSNC 2020) is held during November 22–25, 2020, Chengdu, China. The theme of CSNC2020 is "GNSS, New Global Era," including technical seminars, academic exchanges, forums, exhibitions and lectures. The main topics are as followed:

Conference Topics

S01 Satellite Navigation Applications
S02 Navigation and Location-based Service
S03 Satellite Navigation Signal and Signal Processing
S04 Satellite Orbit and System Error Processing
S05 Spatial Frames and Precise Positioning
S06 Time Primary Standard and Precision Time Service
S07 Satellite Navigation Augmentation Technology
S08 Test and Assessment Technology
S09 User Terminal Technology
S10 PNT System and Multi-source Fusion Navigation
S11 Anti-interference and Anti-spoofing Technology
S12 Policies, Regulations, Standards and Intellectual Properties
S13 Technologies for Navigation of Autonomous Systems

The proceedings (Lecture Notes in Electrical Engineering) have 201 papers in thirteen topics of the conference, which were selected through a strict peer review process from 493 papers presented at CSNC2020. In addition, another 219 papers were selected as the electronic proceedings of CSNC2020, which are also indexed by "China Proceedings of Conferences Full-text Database (CPCD)" of CNKI and Wan Fang Data.

We thank the contribution of each author and extend our gratitude to 278 referees and 57 session chairmen who are listed as members of the editorial board. The assistance of CNSC2020's organizing committees and Springer editorial office is highly appreciated.

Editorial Board

Contents

Satellite Navigation Application

High-Speed Railway Track Comprehensive Measurement System Based on GNSS/INS Multi-sensor

Qi Li, Zhengdong Bai(✉), Bobo Chen, Haohao Xin, Yuhang Cheng, and Qiang Zhang

Department of Civil Engineering, Tsinghua University, Beijing, China
zbai@tsinghua.edu.cn

Abstract. Track measurement is an importantly basic and preliminary work to ensure high-speed, safe and smooth operation of trains. In the stages of design and construction, the high-speed railway track measurement at least includes the layout and measurement of all control networks, the subgrade deformation monitoring, and the track irregularity measurement. However, there are several problems such as long retest period, high labor cost, and low detection efficiency, which cannot fully meet the needs of large-scale high-speed railway tracks rapid detection in the stages of operation and maintenance. So, a set of track comprehensive measurement methods of high-speed railway based on GNSS/INS multi-sensor are proposed, and the corresponding system is developed. It has been applied to the actual track fine adjustment engineering, and compared with the Amberg track geometry measurement system based on total station coupled with level, and the results show that the proposed system has realized the integration of track subgrade deformation monitoring and track geometry absolute measurement and relative measurement, which fully meets the requirements of the *Code for Engineering Survey of High-Speed Railway*, and significantly improves the measurement efficiency.

Keywords: Engineering survey · Deformation monitoring · Track irregularity · Measurement system

1 Introduction

By December 2019, the operating mileage of China's high-speed rail has exceeded 35 000 km, which is more than 2/3 of the world's total. China's railway network covers a wide range, with complex geological, geographical and climatic conditions along the line. Every day more than 5000 high-speed trains run at a top speed of more than 250 km/h carrying about 10 million people [1]. China's high-speed rail operation is so busy that only the skylight time (about 0:00–4:00) can repair and maintain the track. And the design and construction standard of China's high-speed railway is stricter than those of Japan, France, Germany, and Europe [2]. Furthermore, it has been more than 11 years since China's first high-speed railway "Beijing-Tianjin Intercity" operated in August 2008. During this period, China's high-speed railway has entered the long-term safe and stable operation and maintenance stage from the large-scale design and

J. Sun et al. (Eds.): CSNC 2020, LNEE 650, pp. 3–14, 2020.
https://doi.org/10.1007/978-981-15-3707-3_1

construction stage. Therefore, it is necessary to establish and improve the high-speed railway track measurement standards to meet the needs of operation and maintenance stage [3, 4].

However, the current high-speed railway track measurement standard in China is mainly based on the design and construction stage. And according to it, several high-speed railway inspection trains and track geometry measurement instruments have been researched and developed [5, 6]. It at least includes the layout and measurement of all control networks, the subgrade deformation monitoring, and the track irregularity measurement. However, there are several problems such as long retest period, high labor cost, and low detection efficiency [7], which cannot fully meet the needs of large-scale high-speed railway tracks rapid detection in the stages of operation and maintenance. So, a set of track comprehensive measurement methods of high-speed railway based on GNSS/INS multi-sensor are proposed, and the corresponding system is developed (Fig. 1). It has been applied to the actual track fine adjustment engineering, and compared with the Amberg track geometry measurement system based on total station coupled with level [8], and the results show that the proposed system has realized the integration of track subgrade deformation monitoring and track geometry absolute measurement and relative measurement with 2 mm lateral deviation accuracy and 2 mm vertical deviation accuracy of track, and 1 mm horizontal accuracy and 1.5 mm vertical accuracy of deformation points, which fully meets the requirements of the *Code for Engineering Survey of High-Speed Railway*, and significantly improves the measurement efficiency.

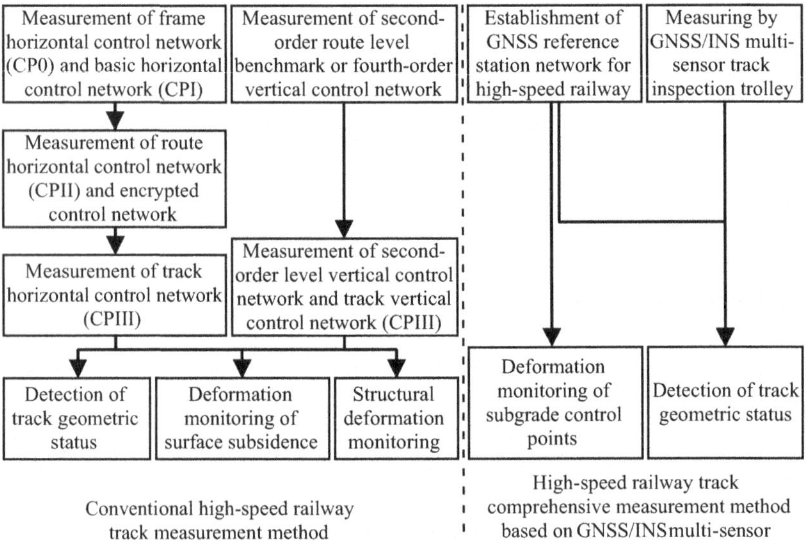

Fig. 1. Comparison of conventional and novel high-speed railway track measurement methods

The following includes the proposed system configuration (in Sect. 2), operation mode (in Sect. 3), processing flow (in Sect. 4), experimental environment (in Sect. 5), data analysis (in Sect. 6), and conclusions and prospects (in Sect. 7), which can provide a reference to the establishment of high-speed railway track measurement standard in the stages of operation and maintenance.

2 System Configuration

The proposed system mainly includes GNSS reference station network, GNSS/INS multi-sensor track inspection trolley, data centre and communication system (as shown in Fig. 2).

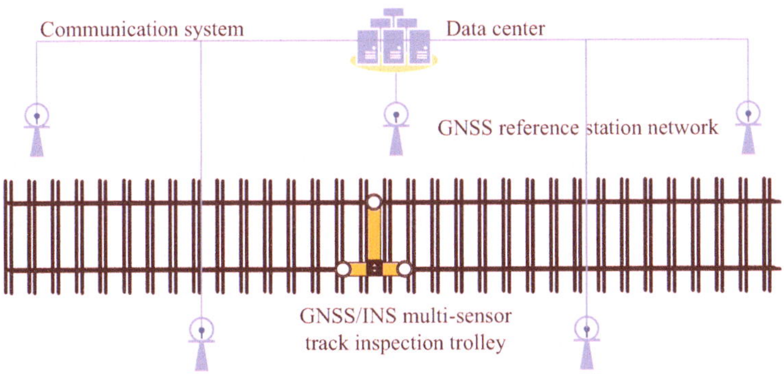

Fig. 2. Compositions of novel high-speed railway track comprehensive measurement system

2.1 GNSS Reference Station Network

GNSS reference stations are arranged in a zigzag shape along the sides of the track at a distance of 1–2 km, and are located in the open, stable and safe observation area. Simultaneously, more than 3 control points of CPII or CPIII are jointly measured to establish the relation between the GNSS reference station network and conventional high-speed railway engineering control network. GNSS reference station network can not only be used to monitor the change of the track subgrade deformation control points in real time, but also provide the position datum for the track geometric status measurement [9].

2.2 GNSS/INS Multi-sensor Track Inspection Trolley

The track inspection trolley is shown in Fig. 3. The T-shaped trolley is equipped with three-antenna GNSS receiver, fiber-optic-gyro (FOG) INS, odometer, gauge measuring sensor, sleeper recognizer sensor, inclinometer, thermometer and tablet. And it realizes the functions including multi-sensor precise installation, unified power supply, unified

Fig. 3. Appearance of GNSS/INS multi-sensor track inspection trolley

time datum, balanced vehicle weight, and kinematic data collection. The track inspection trolley can be used to measure the outer and the inner geometric parameters of the track.

Considering the special environment along the line and the weak satellite signal in some section, the development of the three-antenna GNSS receiver is mainly to increase the redundancy, continuity and reliability of observation data. Meanwhile, the receiver has been designed specifically for electromagnetic compatibility, water resistance, seismic resistance, and thermal insulation. The geometric positions of the three GNSS antennas are strictly calibrated on the T-shaped trolley, which can be used for observation preprocessing and results self-checking of the GNSS receiver. So, it is more accurate and stable than the traditional single-antenna GNSS receiver [10]. The three-antenna GNSS receiver has 440 channels, which can receive the satellite signals of the BDS B1/B2/B3, GPS L1/L2/L5, and GLONASS L1/L2. The FOG has a heading accuracy of 0.05°, a pitch accuracy of 0.02° and a roll accuracy of 0.02° [11], and the accelerometer bias is 10 μg. GNSS/INS integrated system can use GNSS position observations as update information to restrain the INS error accumulation, which can be used to provide position and attitude observations to measure the track irregularities [12, 13].

2.3 Data Centre

Data centre is used for remote setting and monitoring of GNSS reference station network. Observation data is regularly obtained from the GNSS reference station network and nearby IGS reference stations every day to calculate the accurate three-dimensional coordinates of the track subgrade deformation control points by post-processing, which can be used to analyze the track deformation trend and provide accurate position reference.

2.4 Communication System

Communication system mainly includes wired communication (i.e., fiber-optic communication) transmitted between GNSS reference station network and data centre, and wireless communication transmitted between track inspection trolley and data centre. All the observation data of multi-sensor are transmitted to the data centre for centralized processing.

3 Operation Mode

The main operating mode of the proposed system is that the GNSS reference stations are observed synchronously to monitor the trend of track subgrade deformation in real time. And during the maintenance time, the track inspection trolley is loaded on track to measure the track geometric status by the semi-kinematic (i.e., stop-and-go) mode. The details are as follows:

First of all, GNSS reference station network and nearby IGS reference stations shall be jointly observed. The number of known stations for joint observation shall not be less than 2, and the effective observation period shall not be less than 8 h a day. Data centre regularly receives, processes and calculates the observation data of GNSS reference station network and IGS reference stations to obtain the precise three-dimensional coordinate of the track subgrade deformation control points, which is used to compare with the known coordinate of the control points to analyze the trend of the track subgrade deformation.

Then, the track inspection trolley is loaded on track for observation during the maintenance time. Firstly, the track inspection trolley is pushed to the starting point and statically observes for 5 min, which is used to calibrate the attitude of the FOG INS and calculate the coordinate of the stop point. Then, the track inspection trolley is pushed about 150 m to reach the first stop point and also statically observes for 5 min, calibrates the attitude of the FOG INS again and calculates the coordinate of the first stop point, and so on until it reaches the end point and statically observes for 5 min.

Finally, according to the above observation results, the final track operation and maintenance plan for high-speed railway is analyzed and formed.

4 Processing Flow

Data centre centrally processes and analyzes the observation data of multi-sensor from the track inspection trolley, GNSS reference station network, and nearby IGS reference stations. The main processing flow is shown in Fig. 4.

- **Step 1:** When processing the observation data of GNSS reference station network and IGS reference stations, the observation data of IGS reference stations are taken as the constraint points to carry out the control network three-dimensional constraint adjustment to obtain the precise coordinates of the control points and determine the track coordinate system.
- **Step 2:** At the stop points of the track inspection trolley, the statically synchronous observation data of the three-antenna GNSS receiver and the nearby GNSS reference stations are conducted static baseline network adjustment with the GNSS short baseline distance and INS attitude constraints to calculate the precise coordinates and attitude of each stop point (Fig. 5).

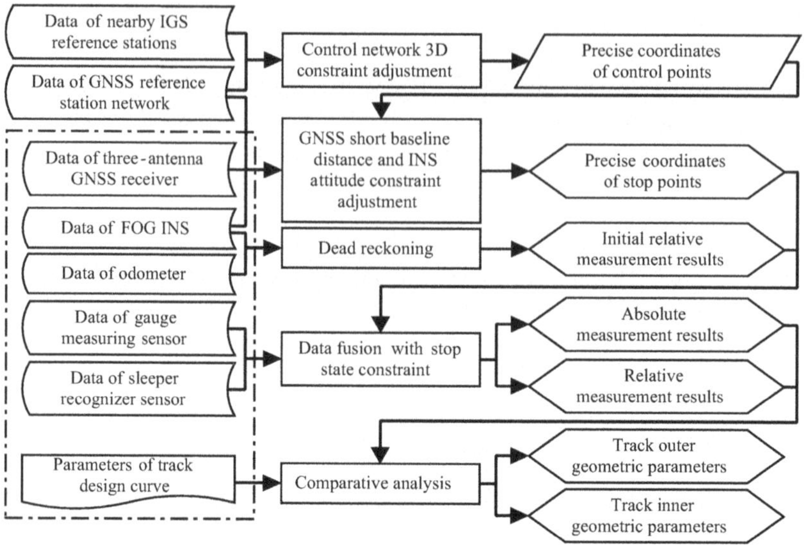

Fig. 4. Processing flow chart of high-speed railway track comprehensive measurement data

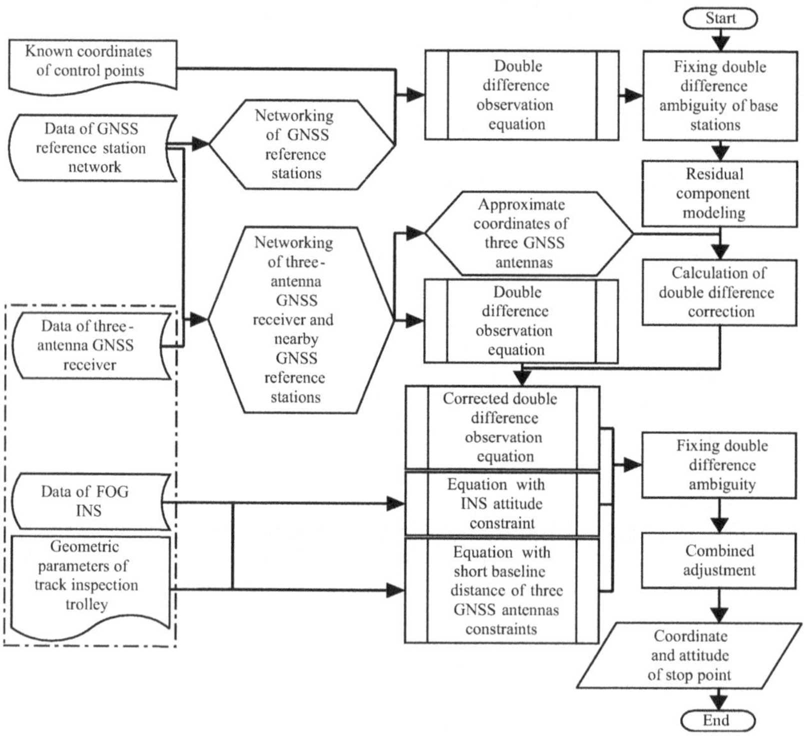

Fig. 5. Algorithm of multi-base station multi-antenna GNSS/INS integration with GNSS antenna distance and IMU attitude constraints

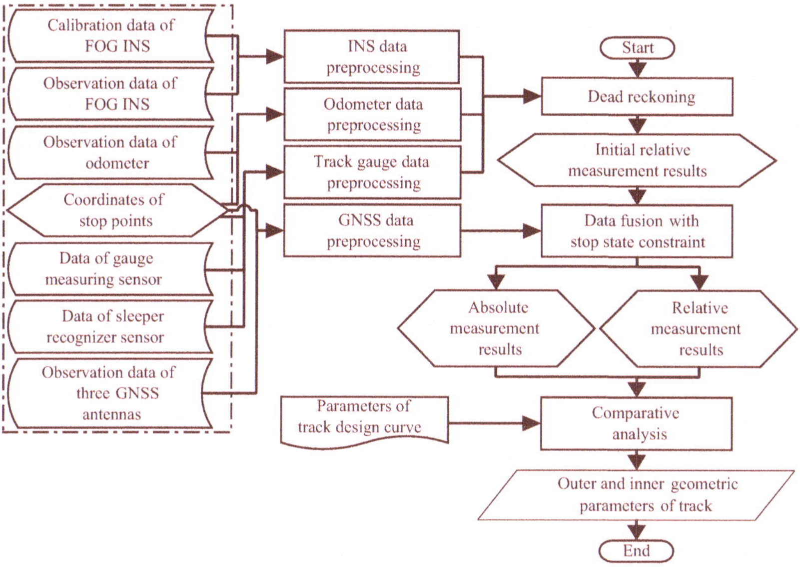

Fig. 6. Algorithm of GNSS/INS multi-sensor kinematic data fusion with stop state constraint

- **Step 3:** Observation data of INS/odometer by Dead Reckoning algorithm is processed with the precise coordinates and attitude of each stop point as constraints to calculate the absolute coordinates of the track centerline, left and right track, which are used to compared with the parameters of the track design curve to obtain all the parameters of the track geometric status (Fig. 6)

5 Experimental Environment

From August 10 to 30, 2018, the proposed system was applied to the test section of the "Shijiazhuang-Jinan High-Speed Railway", and it is a ballasted track with a design speed of 250 km/h. 6 GNSS reference stations were set up in the test section, of which 3 GNSS reference stations coincided with the existing CPIII control points. Synchronous observation was carried out from 11:00 to 21:00 on August 20. The track irregularities were measured from 15:00 to 17:00. The sampling frequency of three-antenna GNSS receiver was set to 1 Hz, and the FOG INS was set to 100 Hz.

At the same time, the Amberg GRP1000 track inspection trolley based on the Leica TCA2003 total station coupled with the Trimble DiNi high precision digital level was used to measure the same test section. The Leica TCA2003 total station has a nominal attitude accuracy of 0.5″ and a distance accuracy of 1 mm + 1 ppm, and the Trimble DiNi03 level has a nominal accuracy of 0.3 mm/km.

6 Data Analysis

Figure 7 shows the comparison of the lateral deviation and the vertical deviation between the measured value and the designed value of the same test section by two track measurement systems, and Fig. 8 shows the statistics of the difference between the lateral deviations and the difference between the vertical deviations of the two measurements. The results show that the deviations of the two kinds of track measurement systems are less than 2 mm. It can be seen that the proposed system has good consistency with the Amberg system [14, 15].

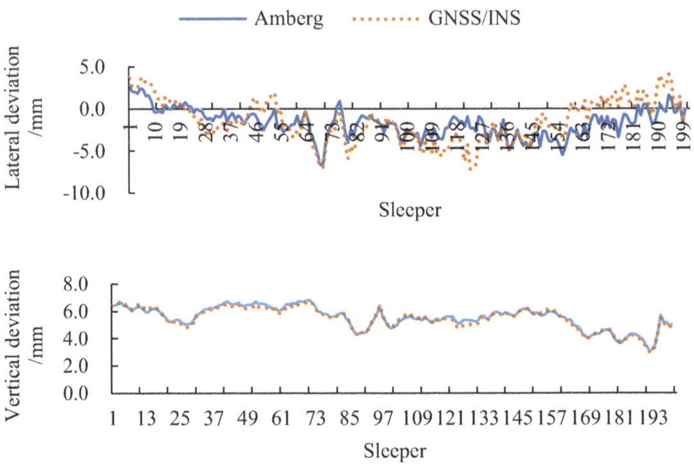

Fig. 7. Comparison of lateral deviation and vertical deviation by two track measurement systems

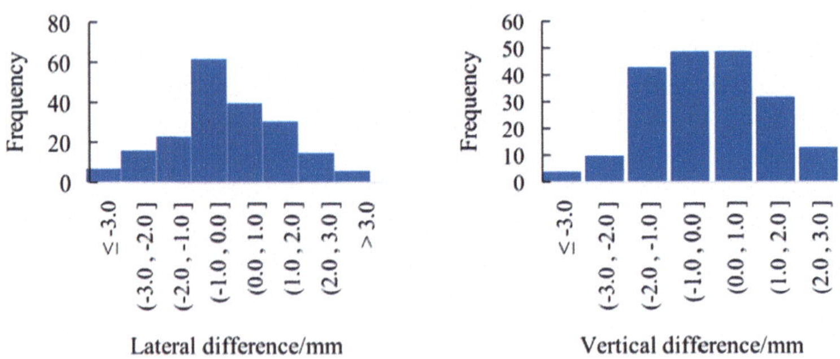

Fig. 8. Statistics of difference between lateral deviations and difference between vertical deviations of two measurements

Table 1 shows the comparison of track geometric parameters before and after track adjustment measured by the proposed system. It can be seen that the track inner geometric parameters after the track adjustment are significantly reduced. Especially the track quality index (TQI) is reduced from the maximum value of 12.44 mm before the track adjustment to 6.23 mm after the track adjustment. The track geometry measurement performance of the proposed system can meet the requirements of the track fine adjustment measurement [16].

Table 1. Comparison of partial measured results of track geometry parameters before and after track adjustment

Measured position /km	Mark position /km	Distance /m	Track gauge /mm	Gauge change rate /‰	Cross level /mm	Twist /mm	Right vertical irregularities /mm	Right track alignment /mm	Right versine /mm	Left vertical irregularities /mm	Left track alignment /mm	Left versine /mm	TQI /mm
296.2	296.2	200	0.17	0.19	0.85	0.53	1.00	0.52	0.61	0.93	0.53	0.60	5.93
296.4	296.4	200	0.24	0.20	0.7	0.51	1.26	0.75	1.25	1.25	0.70	1.17	8.03
296.6	296.6	200	0.16	0.15	0.92	0.67	1.87	0.65	0.94	1.90	0.65	0.85	8.75
296.8	296.8	200	0.14	0.13	0.85	0.67	1.22	0.81	1.25	1.01	0.80	1.19	8.07
297.0	297.0	200	0.17	0.17	1.18	0.80	2.31	1.16	1.49	2.60	1.11	1.44	12.44

a. Before track adjustment

Measured position /km	Mark position /km	Distance /m	Track gauge /mm	Gauge change rate /‰	Cross level /mm	Twist /mm	Right vertical irregularities /mm	Right track alignment /mm	Right versine /mm	Left vertical irregularities /mm	Left track alignment /mm	Left versine /mm	TQI /mm
296.2	296.2	200	0.25	0.23	0.47	0.39	0.85	0.61	0.59	0.84	0.56	0.57	5.35
296.4	296.4	200	0.43	0.30	0.38	0.38	0.94	0.61	0.63	0.87	0.51	0.59	5.65
296.6	296.6	200	0.36	0.26	0.27	0.31	0.92	0.51	0.57	0.93	0.48	0.58	5.19
296.8	296.8	200	0.29	0.23	0.35	0.44	0.94	0.62	0.72	1.06	0.61	0.63	5.88
297.0	297.0	200	0.33	0.26	0.74.	0.55	0.94	0.63	0.70	0.89	0.53	0.66	6.23

b. After track adjustment

Figure 9 shows the changes of 6 GNSS reference stations in the test section from August 20 to 30, 2018. Observation data of GNSS reference stations was processed every 2 days, and the measurement accuracy is 0.51 mm in the north-south direction, 0.74 mm in the east-west direction and 1.46 mm in the vertical direction. It can be proven that the static positioning accuracy of the proposed system can meet the accuracy requirement of the track subgrade deformation monitoring [17].

The proposed system has been successively applied to several track fine adjustment projects, including "Jinan-Qingdao High-Speed Railway", "Beijing-Shanghai High-Speed Railway", and "Shanghai-Kunming High-Speed Railway". The analysis results from several demonstration applications show that the proposed system has stable performance and accurate accuracy with the absolute accuracy of mm and the relative accuracy of sub mm, and the average measurement speed of about 1 km/h, which has significantly improved the measurement efficiency for high-speed railway track.

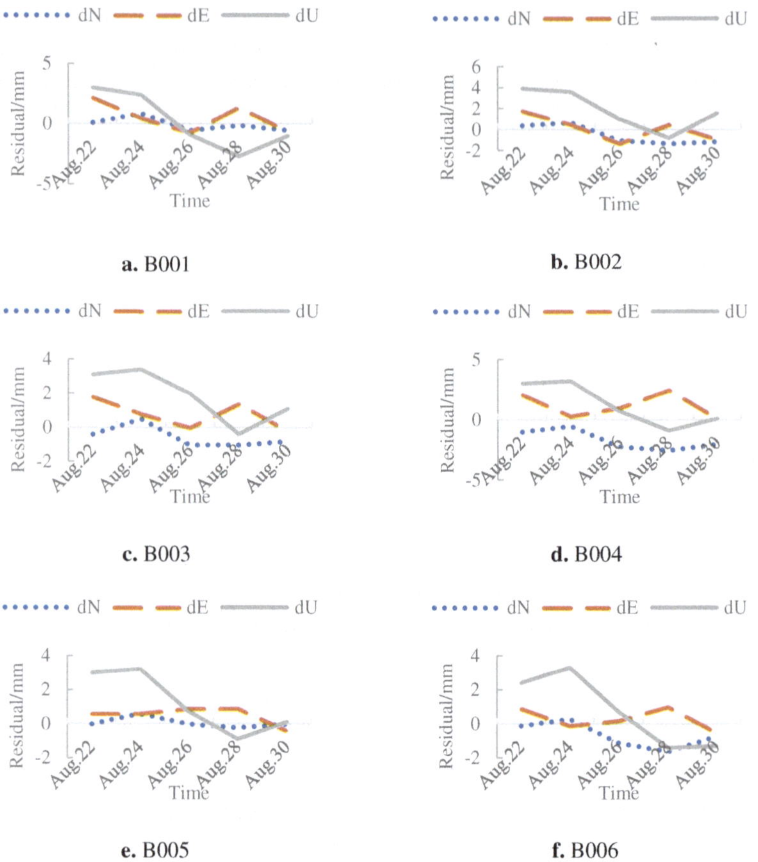

a. B001 **b.** B002 **c.** B003 **d.** B004 **e.** B005 **f.** B006

Fig. 9. Statistics of changes in GNSS reference stations

7 Conclusions and Prospects

Currently, China's high-speed railway has entered the stage of long-term, safe and stable operation and maintenance. Strict and rapid detection of the long-term service and large-scale high-speed railway tracks is an importantly basic and pioneering work to ensure the high-speed, safe and smooth operation of trains. So, a set of track comprehensive measurement methods of high-speed railway based on GNSS/INS multi-sensor are proposed, and the corresponding system is developed. It has been applied to the actual track fine adjustment engineering, and compared with the Amberg track geometry measurement system based on total station coupled with level, and the results show that the proposed system has realized the integration of track subgrade deformation monitoring and track geometry absolute measurement and relative measurement with 2 mm lateral deviation accuracy and 2 mm vertical deviation accuracy of track, and 1 mm horizontal accuracy and 1.5 mm vertical accuracy of deformation points, which fully meets the requirements of the *Code for Engineering Survey of High-Speed Railway*, and significantly improves the measurement efficiency.

Next, the proposed system will be further improved on the following aspects to provide a reference to the establishment of high-speed railway track measurement standards in the operation and maintenance stages [18]: On the one hand, the distance between the GNSS reference stations is about 1–2 km. Considering the needs of track subgrade deformation monitoring and regional surface subsidence deformation monitoring, properly expanding the distance between the GNSS reference stations can significantly reduce the construction cost of the GNSS reference station network, on the premise of ensuring the same measurement accuracy. On the other hand, the track inspection trolley adopts the "stop-and-go" mode to measure the track irregularities currently. If the observation time of the stop points is reduced or the kinematic observation distance is increased, or even the whole kinematic observation mode is adopted, the measurement efficiency will be further improved.

Acknowledgements. This work was supported in part by the National Key Research and Development Program of China (No. 2017YFB0504202).

References

1. Lawrence, M.B., Bullock, R.G., Liu, Z.: China's High-Speed Rail Development (English). International Development in Focus. World Bank Group, Washington (2019)
2. Tian, G., Gao, J., Zhai, W.: Comparative analysis of track irregularity management standards for high-speed railways. J. Chin. Railw. Soc. **37**(03), 64–71 (2015). https://doi.org/10.3969/j.issn.1001-8361.2015.03.011
3. Zhai, W., Zhao, C.: Frontiers and challenges of sciences and technologies in modern railway engineering. J. Southwest Jiaotong Univ. **51**(02), 209–226 (2016). https://doi.org/10.3969/j.issn.0258-2724.2016.02.001
4. Li, G., Fan, B.: The development of precise engineering surveying technology. Acta Geod. Cartogr. Sin. **46**(10), 1742–1751 (2017). https://doi.org/10.11947/j.AGCS.2017.20170313
5. Zhong, C., Li, H., Li, P., et al.: Review of high-speed comprehensive inspection trains. Chin. Railw. **06**, 89–93 (2013). https://doi.org/10.19549/j.issn.1001-683x.2013.06.021
6. Chen, Q.: Research on the railway track geometry surveying technology based on aided INS. Wuhan University, Wuhan (2016)
7. Ministry of Railways of the People's Republic of China: Code for Engineering Survey of High-Speed Railway: TB10601—2009. Standards Press of China, Beijing (2009)
8. Li, Q., Bai, Z., Li, Q., et al.: Geometric design of an outdoor three-dimensional kinematic verification field for a position and orientation system. J. Tsinghua Univ. (Sci. Technol.) **59** (11), 895–901 (2019). https://doi.org/10.16511/j.cnki.qhdxxb.2019.22.026
9. Jiang, W.: Challenges and opportunities of GNSS reference station network. Acta Geod. Cartogr. Sin. **46**(10), 1379–1388 (2017). https://doi.org/10.11947/j.AGCS.2017.20170424
10. Dong, D., Chen, W., Cai, M., et al.: Multi-antenna synchronized global navigation satellite system receiver and its advantages in high-precision positioning applications. Front. Earth Sci. **10**(4), 772–783 (2016). https://doi.org/10.1007/s11707-016-0559-2
11. Li, Q., Bai, Z., Zhao, S., et al.: Performance evaluation of the allan variance method for ring laser gyroscope noise analyses. J. Tsinghua Univ. (Sci. Technol.) **59**(11), 887–894. https://doi.org/10.16511/j.cnki.qhdxxb.2019.22.009

12. Naser, E.S., Ahmed, Y.: Inertial sensors technologies for navigation applications: state of the art and future trends. Satell. Navig. **1**, 1–21 (2020). https://doi.org/10.1186/s43020-019-0001-5
13. Wu, Y., He, C., Liu, G.: On inertial navigation and attitude initialization in polar areas. Satell. Navig. **1**, 1–6 (2020). https://doi.org/10.1186/s43020-019-0002-4
14. Chen, Q., Niu, X., Zuo, L., et al.: A railway track geometry measuring trolley system based on aided INS. Sensors **18**(538), 1–26 (2018). https://doi.org/10.3390/s18020538
15. Gao, Z., Ge, M., Li, Y., et al.: Railway irregularity measuring using Rauch–Tung–Striebel smoothed multi-sensors fusion system: quad-GNSS PPP, IMU, odometer, and track gauge. GPS Solute **22**(2), 36 (2018). https://doi.org/10.1007/s10291-018-0702-5
16. Ministry of Railways of the People's Republic of China: Inspecting Instrument for Railway Track: TB/T3147-2012. China Railway Publishing House, Beijing (2012)
17. Xiao, Y., Jiang, W., Chen, H., et al.: Research and realization of deformation monitoring algorithm with millimeter level precision based on BeiDou navigation satellite system. Acta Geod. Cartogr. Sin. **45**(1), 16–21 (2016). https://doi.org/10.11947/j.AGCS.2016.20140649
18. Li, Q., Mao, Q.: Progress on dynamic and precise engineering surveying for pavement and track. Acta Geod. Cartogr. Sin. **46**(10), 1734–1741 (2017). https://doi.org/10.11947/j.AGCS.2017.20170323

Research on Application of Improved S Transform in High Frequency GNSS Data Processing Results During Earthquake

Jingyu Shen, Yun Shi(✉), Kangkang Wu, Lu Zhang, Fan Tian, and Kan Zhao

College of Geomatics, Xi'an University of Science and Technology, Xi'an 710054, China
s1435973211@foxmail.com, 15364619064@163.com

Abstract. The high-frequency GNSS data at the time of earthquake was obtained by dynamic single-epoch positioning solution, and the GNSS displacement results were obtained. However, the seismic-time GNSS displacement result after dynamic single-epoch solution contains a lot of noise. It has a great impact on the analysis and research of the displacement during the earthquake, in order to minimize the noise impact in the results and improve the accuracy of dynamic positioning. Traditionally, the S transform denoising method is used to denoise the processing results, but the denoising effect is affected because its time resolution is limited.

This paper improves the S-transform method and improves its time-frequency resolution. The principle of S-transform and improved S-transform is introduced. The improved S-transform is used to denoise the high-frequency GNSS single-epoch dynamic positioning results after earthquake processing by GAMIT/TRACK, which greatly removes the noise of displacement results and improves The analysis and research of the dynamic positioning accuracy and the high-frequency GNSS displacement results during the earthquake are made to more accurately determine the time when the GNSS seismic wave arrives at the observation station. Finally, based on the characteristics of high-frequency GNSS acquisition of seismic instantaneous deformation information, this paper proposes an improved S-transform method combined with a trend term denoising method, which can minimize the effect of positioning errors of long-period low-frequency displacement signals, which is more conducive to the in-depth study of GNSS Seismology.

Keywords: High frequency GNSS · Dynamic positioning · Improved S transform · Improved S transform combined with trend term

1 Introduction

In recent years, with the advent of high-frequency GNSS receivers, high-frequency GNSS measurement technology has been widely used in various fields. As one of the main observation methods of modern geodesy, high-frequency GNSS technology has the advantages of being more efficient, fast, all-weather, and high-precision compared to traditional measurement and monitoring technologies.

J. Sun et al. (Eds.): CSNC 2020, LNEE 650, pp. 15–26, 2020.
https://doi.org/10.1007/978-981-15-3707-3_2

High-frequency GNSS technology refers to the use of single-epoch dynamic positioning to estimate the coordinates of GNSS stations and obtain high-resolution instantaneous dynamic displacement results of observation stations. This technology has developed rapidly in recent years, and has been widely used in wind vibration monitoring of buildings and acquisition of transient crustal deformation of earthquakes. Especially when high-frequency GNSS technology is used to study the transient crustal deformation and displacement of an earthquake, it is not necessary to obtain the inclination of the ground through integration and the maximum value of the displacement is not limited, like a seismograph, which has the advantage that the seismograph cannot match. As with other surveys, high-frequency GNSS surveys are also subject to interference from various measurement errors. During the data collection and data processing of GNSS dynamic surveys, there are inevitably many noise effects, resulting in a lot of uncertain random noise in the received GNSS data. These noises affect the accuracy and reliability of the measurement results to a certain extent. Therefore, in order to further improve the accuracy of dynamic positioning, it is necessary to post-process the high-frequency GNSS displacement results. Traditional high-frequency GNSS dynamic displacement result processing methods include commonly used time-frequency analysis methods such as short-time Fourier transform, wavelet transform, Wigner-Ville distribution, etc. [1–4]. In 1996, the American geophysicist Stockwell [2] combined short-time Fourier transform (STFT) and wavelet transform (WT), and first proposed a time-shifted Fourier transform method, the S-transform. Because of its high time-frequency resolution, it has received widespread attention [5–7]. The S transform is the inheritance and development of the short-time Fourier transform and continuous wavelet transform. It combines the advantages of the two transforms and makes up for their shortcomings to a certain extent. However, the S transform has the problem that the resolution of time and frequency cannot be improved at the same time, and the Gaussian window of the standard S transform is fixed with the change of frequency, resulting in a lower time resolution of the time spectrum [8], therefore, improving the S transform denoising method has become a new research.

This article introduces the principle of improved S-transform, and uses improved S-transform to denoise the high-frequency GNSS displacement results which were processed by GAMIT/TRACK. This method greatly removes the displacement signal noise and improves the dynamic positioning accuracy and earthquake. Analysis and research of time-frequency GNSS displacement results make the determination of the arrival time of GNSS seismic waves more accurate. Finally, based on the characteristics of GNSS acquired seismic wave signals, this paper proposes an improved S-variation and trend term denoising method, which can minimize the influence of positioning errors of long-period low-frequency signals. It is more conducive to the in-depth study of GNSS seismology.

2 Research Area Overview

2.1 Data Acquisition

According to the US Geological Survey (USGS), on February 27, 2010, at 06:34:14 UTC, Chile's second largest city, Concepcion, experienced a magnitude 8.8 earthquake on the Richter scale. Five major earthquakes. The epicenter was located in the sea area (35.909°S, 72.733°W) near the Maule, 320 km southwest of the Chilean capital Santiago, and the epicenter depth was about 35 km. The data used in the example is GNSS data for a total of 1 h during the earthquake. The sampling frequency is 1 Hz. High-frequency GNSS monitoring data from 5 stations near the epicenter were selected for calculation and research (.gsfc.nasa.gov/highrate/). Figure 1 below shows the distribution of seismic stations in Chile in 2010.

Distribution of seismic stations in Chile in 2010.

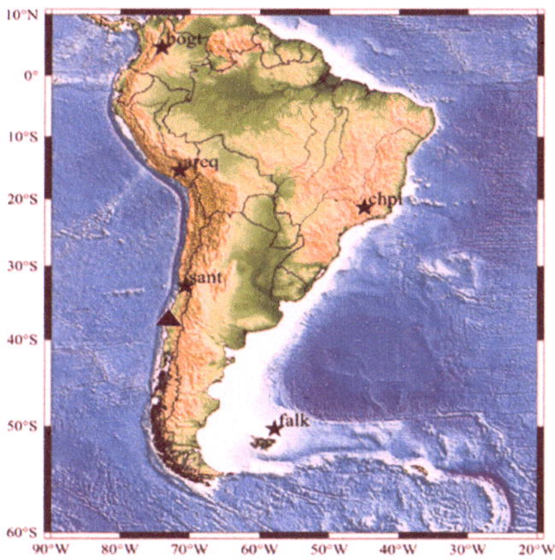

Fig. 1. Distribution of GNSS stations

2.2 Data Processing

For the 5 GNSS observation stations (areq, bogt, chpi, sant, falk) around the Chile earthquake, when processing high-frequency data at the time of the earthquake, the bogt GNSS observation station that is far from the source is selected as the reference station, and the other stations are Observation station uses the Track module of GAMIT software to calculate high-frequency data during the earthquake. The high-frequency GNSS observation data with a total of 3600 epochs from 06:00 to 07:00 are selected for single-epoch positioning processing. There is a lot of noise in the displacement result sequence data. In order to obtain the optimal high-frequency GNSS displacement result, an improved S-transform denoising method is used to denoise the high-frequency GNSS displacement result sequence, which is beneficial to the high-frequency GNSS positioning result in earthquake Application Analysis in Science.

3 Improved S-Transform Denoising Method and Calculation Example

3.1 S-Transform Theory

Based on previous studies on video analysis, Stockwell [2] proposed the S transform. The S transform combines the advantages of the short-time Fourier transform and the wavelet transform, and also overcomes their shortcomings. It uses a frequency-dependent variable Gaussian window function, and the basic wavelet does not have to meet the admissibility conditions. The S transform of the continuous signal h(t) in the time domain is defined as follows:

$$S(\tau,f) = \int_{-\infty}^{+\infty} h(t) \cdot \frac{|f|}{\sqrt{2\pi}} e^{\frac{-f(\tau-t)^2}{2}} e^{-i2\pi ft} dt \quad (1)$$

In the formula: h(t) is the signal time function, f is the frequency, and t is the time, which controls the position of the Gaussian function on the time axis. Is an exponential function provided by the Fourier transform to the oscillating part of the S transform, which does not change with the normalized Gaussian window as it moves over time.

In the S transform, the Gaussian window function and the basic wavelet are defined as:

$$g_f(t) = \frac{|f|}{\sqrt{2\pi}} e^{(-\frac{t^2 f^2}{2})} \quad (2)$$

$$\omega_f(t) = \frac{|f|}{\sqrt{2\pi}} e^{(-\frac{t^2 f^2}{2} - i2\pi ft)} = g_f(t) e^{-i2\pi ft} \quad (3)$$

It can be known from Eq. (2) that the S-transform uses a Gaussian window function with a variable width, and its time window width varies inversely with the frequency f. At low frequencies, the time window is wider to obtain higher frequency resolution; at high frequencies, the time window is narrower, so higher time resolution can be obtained.

3.2 Improved S-Transform Theory

Since the window function of the S transform changes with frequency with a fixed trend, the time-frequency spectrum obtained is unique for non-stationary signals and the time-frequency resolution is limited. Therefore, the Gaussian window needs to be improved [9–11]. In order to improve the time-frequency resolution of the S transform on the signal, a tuning parameter δ is added to the Gaussian window, where δ is a first-order linear equation of frequency:

$$\delta(f) = mf + n \quad (4)$$

Among them, m and n are constants that jointly control the width of the Gaussian window. Both m and n parameters are positive numbers, that is, the adjustment parameter δ increases linearly with the increase of the frequency f.

The Gaussian window function becomes:

$$g_f(t) = \frac{|mf+n|}{\sqrt{2\pi}} \exp\left[-\frac{t^2|mf+n|^2}{2}\right] \quad (5)$$

Substituting Eq. (5) into Eq. (1) to get the modified S-transformed expression [12]:

$$MST(\tau,f) = \int_{-\infty}^{+\infty} h(t) \cdot \frac{|mf+n|}{\sqrt{2\pi}} e^{\frac{-(\tau-t)^2(mf+n)}{2}} e^{-i2\pi ft} dt \tag{6}$$

The improved S-transform of the signal is a two-dimensional time-frequency matrix, with rows representing time and columns representing frequency. According to Heisenberg's uncertainty principle, the time-frequency resolution cannot be improved at the same time by transformation. The values of parameters m and n can be optimized to obtain higher time-frequency resolution.

3.3 Example Analysis

3.3.1 Improved S-Transform Denoising Results

The improved S-transform denoising method is used to denoise the high-frequency GNSS displacement result sequence. The high-frequency data processing results of the areq GNSS observation station are selected for analysis, and the following results are obtained.

The following figure is the result of improved S-transform denoising.

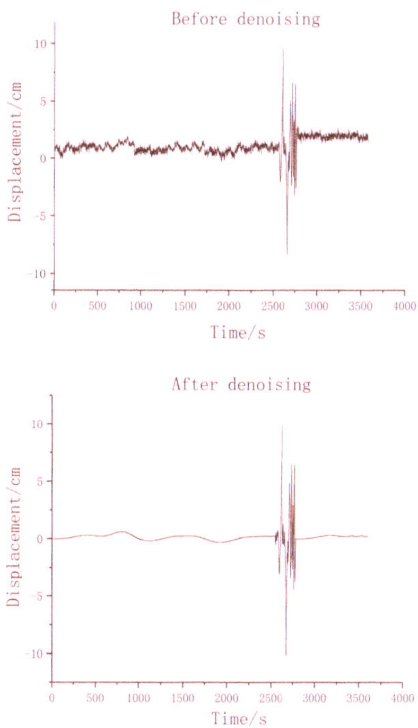

Fig. 2. Results of improved S-transform denoising in the E direction of areq station

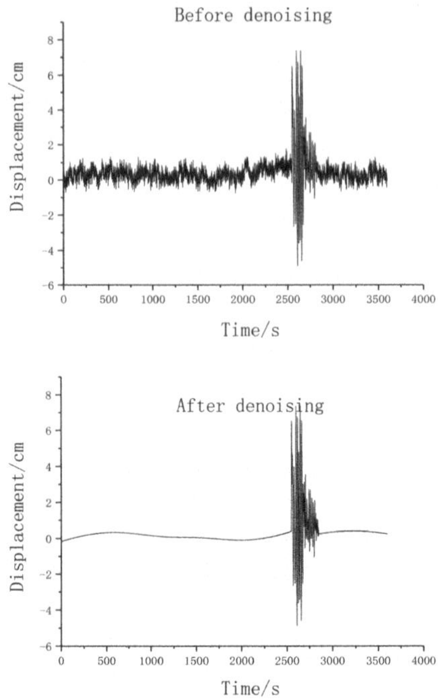

Fig. 3. Results of de-noising before and after improved S transform in N direction of areq station

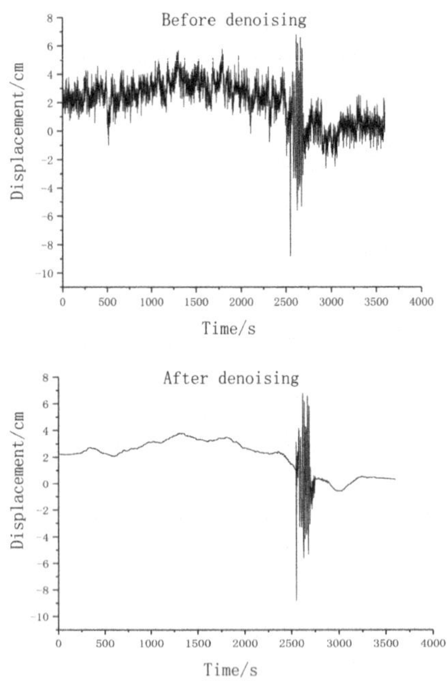

Fig. 4. Results of improved S-transform denoising in the U direction of areq station

Figures 2 and 3 are comparison diagrams of the displacement sequence results before and after adopting the improved S transform denoising method in the E and N directions of the areq GNSS observation station. From the comparison results of the upper and lower parts, it can be seen that the displacement results of the improved S transform are denoised. It is obvious that part of the influence of high-frequency signals is removed, and the sequence of displacement results changes significantly, making it easier to analyze seismic wave signals. Figure 4 shows the sequence of the displacement results before and after denoising using the improved S transform in the U direction of the areq GNSS observation station. It is obvious that the results of the U direction denoising still have a large error effect.

In order to understand the denoising effect of the improved S-transform method in detail, the RMS values of the result sequences obtained by using the S-transform denoising method and the improved S-transform denoising method are compared and analyzed.

Table 1. RMS denoising results of two denoising methods

Station	Before denoising			S transform			Improved S transform		
	E(cm)	N(cm)	U(cm)	E(cm)	N(cm)	U(cm)	E(cm)	N(cm)	U(cm)
areq	0.86	0.96	3.78	0.65	0.77	2.89	0.51	0.63	2.31
chpi	0.9	1.06	1.68	0.71	0.82	1.36	0.53	0.66	1.18
sant	0.93	0.91	4.18	0.73	0.79	2.98	0.53	0.62	2.46

It can be clearly seen from Table 1 that the improved S transform has a better denoising effect on high frequency GNSS dynamic single epoch positioning than the S transform and the improved S transform has a better effect in the U direction.

3.3.2 Determine the Arrival Time of the Seismic Wave

The sequence of displacement results in the E, N, and U directions after denoising after the improved S transform is shown in the following figure:

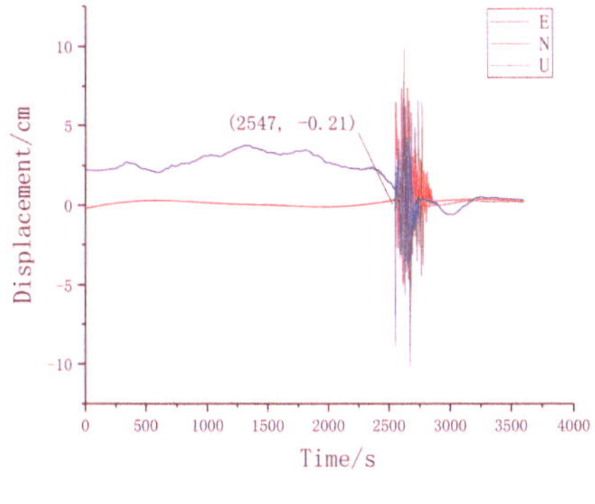

Fig. 5. Arrival time results of seismic waves at areq station

From the sequence diagram of displacement results in the E and N directions in Fig. 5, it is clear that at the 2547th epoch, the E and N displacements are abruptly changed. It can be concluded that the epoch of the seismic wave reaching the areq GNSS observation station is 2547 The time corresponding to the epoch is 6:42:27 UTC. According to the U. S. Geological Survey, the Chile earthquake occurred at 6:34:14. The Areq GNSS observation station is about 2120 km from the earthquake location. The average speed of propagation is 4.361 km/s [12], and it can be inferred that the earthquake occurred at 6:34:22, which is 8 s away from the time published by the US Geological Survey. Uniform, resulting in different transmission speeds of seismic waves in the crust.

4 Improved S Transform Method Combined with Trend Term Denoising

Combining Figs. 2, 3, 4, and 5, it can be seen that the displacement results after denoising are in two parts of the E and N directions before the earthquake (0–2547 epoch) and after the earthquake (2745–3600 epoch). The sequence is smoother, indicating that the high-frequency signal has been significantly removed, but the long-term term effects still exist in the sequence of each displacement result after denoising. The window improves the threshold of the noise signal, but the signal in its filtering window still cannot contain the noise signal with a longer period term, and the long period displacement signal will still not be eliminated by the improved S transform.

Therefore, the results of the improved S-transform denoising method are not very satisfactory and cannot satisfy the seismic analysis. This paper adopts the improved S transform combined with the trend term denoising method. The idea is to eliminate the long term trend term first, and then use the improved S transform denoising method mentioned above for denoising.

4.1 Implementation Process

The specific denoising process of the improved S transform combined with the trend term denoising method is as follows:

(1) According to the obtained displacement results, the displacement result sequence is divided into three segments, which are divided into three periods before, during and after the earthquake. The three signals are divided into L1, L2, and L3.
(2) Precisely calculate the permanent deformation S before and after the earthquake. The calculation method is to obtain the pre-seismic displacement component C1 from the GNSS static observation data for a certain time before the earthquake, and subtract the GNSS static observation data for a few minutes after the earthquake to obtain the earthquake. The rear position component is C2, then S = C2−C1.
(3) The long-term trends of the pre-seismic displacement result series (L1) and post-earthquake displacement (L3) are extracted separately. The trend terms are extracted using a composite digital filtering method of polynomial fitting, median filtering, and mean filtering [13]. After removing the pre-seismic and post-seismic trend terms, the displacement results are L1' and L3', and the displacement L2 during the earthquake remains the same as the original displacement sequence.

(4) The two pieces of data of L1' and L3' after being processed by the trend term are denoised using the improved S transform method, and the results after the improved S transform are d1 and d3.
(5) The result of d3 is reduced. The result after translation is d3', d3' = d3 + S, and finally, d1, d2, and d3' are the results after denoising.

This method needs to pay attention to the need to accurately know the time when the seismic wave signal arrives at the GNSS observation station, to ensure the integrity of the seismic wave signal at the time of the earthquake, and not to detrend the time period, so as to obtain a better sequence of seismic displacement results, which is convenient for seismology. Application Analysis

4.2 Example Analysis

The data of the AREQ station during the earthquake is also analyzed, and the denoising method using the improved S-transform and trend term correction proposed above is used to perform denoising analysis on the results of the E direction, N direction, and U direction after the GNSS data is solved.

First determine the time period of the main earthquake of the Chile earthquake. According to the processing result, the time period of the main earthquake of the Chile earthquake is (2547–2745). The entire data is divided into 3 segments using 2547 and 2545; then the pre-earthquake and earthquake are determined. The static observation result of the subsequent observation station is the permanent deformation L, and the permanent deformation of the spatial position of the station is 4.6 cm. Finally, the

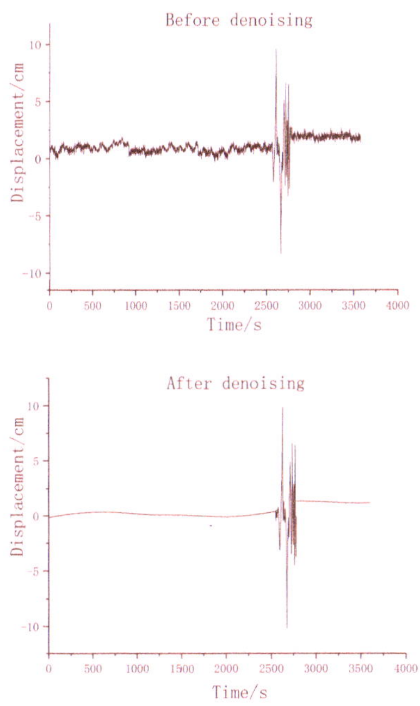

Fig. 6. Direction E of areq GNSS observation station

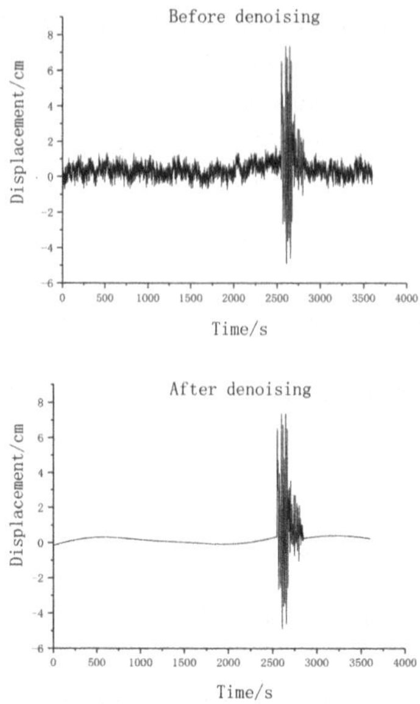

Fig. 7. N direction of areq GNSS observation station

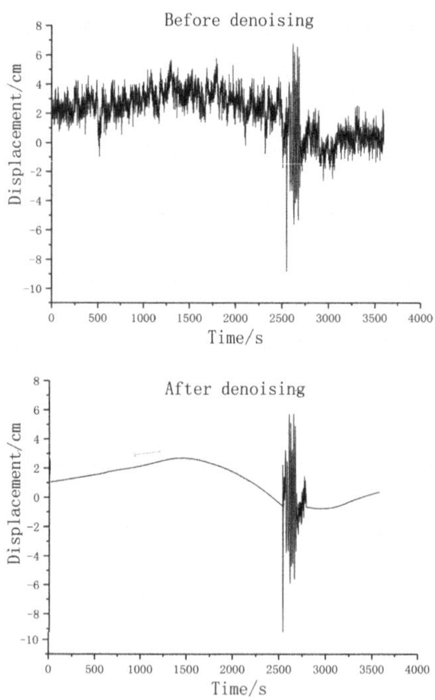

Fig. 8. U direction of areq GNSS observation station

modified S transform denoising method is used to process the data in the three directions to obtain the denoised result.

Figures 6, 7, and 8 show that the denoising results of the improved S transform combined with the trend term have greatly improved the denoising effect in three directions, and the U direction effect is the most obvious. It can be concluded that the improved S-transform combined with the trend term denoising result and the separation result are better, which is more conducive to the study and analysis of high-frequency GNSS single-epoch positioning results in seismology.

5 Conclusion

After in-depth research, an improved S transform denoising method was implemented. Applying this method to the Chile earthquake, a good denoising effect is obtained.

(1) It can be seen that the improved S-transform denoising method is suitable for denoising high-frequency GNSS single-epoch positioning results. The denoising effects in the E and N directions have been slightly improved. It is more suitable for denoising the results in the U direction. In the denoised displacement sequence results, the influence of noise is reduced.

(2) The time when the seismic wave arrived at the areq GNSS observation station is exactly 6:42:27 UTC time, which is 8 s different from the time published by the US Geological Survey, making the determination of the arrival time of the seismic wave at the high-frequency GNSS observation station more accurate.

(3) Based on the characteristics of seismic wave signals acquired from high-frequency GNSS, an improved S-transform method combined with trend term denoising is further proposed. This method can effectively eliminate long-period and low-frequency positioning errors, which is conducive to the application analysis of high-frequency GNSS data processing in seismology.

References

1. Gabor, D.: Theory of communication. J. IEEE **93**, 429–497 (1946)
2. Stockwell, R.G., Mansinha, L., Lowe, R.P.: Localization of the complex spectrum: the S transform. IEEE Trans. Signal Process. **44**(4), 998–1001 (1996)
3. Daubechies, I.: The wavelet transform, time – frequency localization and signal analysis. IEEE Trans. Inf. Theory **36**(5), 961–1005 (1990)
4. Ville, J.: Théorie et applications de la notion de signal analytique. Cables et Transm. **2**(1), 61–74 (1948)
5. Livanos, G., Ranganathan, N., Jiang, J.: Heart sound analysis using the S – transform. IEEE Comput. Cardiol. **27**, 587–590 (2000)
6. Mansinha, L., Stockwell, R.G., Lowe, R.P.: Pattern analysis with two – dimensional spectral localisation: applications of two – dimensional S transforms. Physica A **239**(1), 286–295 (1997)
7. Cui, W., Li, D., Ronghua, X.: Forward simulation of karst collapse in coal fields and application of attributes. Geophys. Geochem. Explor. **35**(5), 648–651 (2011)

8. Zhang, F., Li, C., Wanglin, X.: Improved Stockwell transform method to identify carbonate karst caves—taking Ordovician karst cave development in Ordos Basin as an example. Acta Petrolei Sinica **36**(2), 182–187 (2015)
9. Zhao, S., Zhu, G.: Time-frequency filtering denoising method of S transform. Petrol. Geophys. Prospect. **42**(4), 402–406 (2007)
10. Qi, C., Li, Y., Peng, J., et al.: An improved generalized S-transform. Petrol. Geophys. Prospect. **2**, 215–218 (2010)
11. Chen, X., He, Z.: Improved S transform and its application in seismic signal processing. Data Acquis. Process. **20**(4), 449–453 (2005)
12. Fang, R.: Research on Non-difference Precision Processing Method of High Sampling Rate GPS Data and Its Application in Seismology. Wuhan University (2010)
13. Zeng, D., Wang, X., Deng, F.: Mathematical derivation of Heisenberg's uncertainty principle in quantum mechanics and its application in wavelet analysis. J. North China Inst. Aerosp. Technol. **04**, 33–36 (2006)
14. Liu, D., Yi, J., Yan, K.: Compound Digital Filtering Algorithm and Its Application in Fan Online Monitoring System. Noise Vib. Control **03**, 50–51 (2006)
15. Li, D., Castagna, J.: Modified S-transform in time-frequency analysis of seismic data 2013 SEG Annual Meeting. Society of Exploration Geophysicists (2013)
16. Wang, T., Yang, Y., Zhong, W., et al.: Coherent unit identification using fast S transform and 2DPCA. Power Syst. Technol. **42**(10) (2018)
17. Li, M., Zhang, S., Hu, Y., et al.: Stability analysis of different GNSS satellite clocks based on high-frequency observations. Wuhan University J. (Information Science Edition) **43**(10) (2018)

Inversion of Soil Moisture by GPS-IR Combined with Wavelet Analysis and LS-SVM

Zhigang Zhang[1], Chao Ren[1,2], Yueji Liang[1,2(✉)], and Yalong Pan[1]

[1] College of Geomatics and Geoinformation, Guilin University of Technology, Guilin, China
lyjayq@glut.edu.cn

[2] Research Center of Precise Engineering Surveying, Guangxi Key Laboratory of Spatial Information and Geomatics, Guilin, China

Abstract. The research on the inversion of soil moisture by Global Navigation Satellite Signal-Interferometer and Reflectometry (GPS-IR) has become a hot research field in recent years. In the past few years, ground-based experiments using the US Panel Observation Program (PBO) have confirmed that GPS receivers primarily used for measurements can be used to measure changes in surface physical parameters. In this paper, we study the improvement of the separation model of satellite reflected signal in GPS signal to noise ratio (SNR) observation, and the inversion model of GPS-IR remote sensing soil moisture. Firstly, wavelet analysis is used to effectively separate satellite reflected signals. Further use Least squares support vector machine (LS-SVM) rolling predictive model for estimating soil moisture. Soil moisture was estimated using the GPS SNR provided by the P038 station of the PBO observation network. Comparative analysis of the feasibility and effectiveness of single- and multiple-GPS satellites for soil moisture rolling estimation. Theoretical analysis and experiments show that wavelet analysis can effectively improve the separation accuracy of satellite reflected signals. The LS-SVM rolling estimation results are highly consistent with the soil moisture verification data. This model fully exploits the advantages of LS-SVM and effectively integrates the satellites. Performance, improved the use of a single satellite for soil moisture estimation, the results are prone to abnormal jumps; the model requires less modeling data, the use of rolling can achieve long-term estimation, the estimation error is more stable.

Keywords: GPS-IR · Soil moisture · Wavelet analysis · LS-SVM · Satellite reflected signal

1 Introduction

Soil moisture is the basis for the study of surface hydrology, which affects climate and environmental change by affecting evaporation and circulation between surface water and the atmosphere, and is therefore important for global and regional climate predictions and climate system simulations [1, 2]. Some terrestrial plants and microbial status are regulated by soil moisture. Therefore, soil moisture is a key indicator of

J. Sun et al. (Eds.): CSNC 2020, LNEE 650, pp. 27–37, 2020.
https://doi.org/10.1007/978-981-15-3707-3_3

terrestrial carbon cycle [3]. Especially at the top of the soil, with strong solar radiation and moisture precipitation, the soil moisture changes very drastically [4]. Martin-Neira first demonstrated in airborne experiments in 1993 that the GPS reflected signal can achieve ocean height measurement [5]. In 2008, Larson et al. first proposed a technique based on GPS-IR remote sensing to monitor surface environmental parameters, using multipath reflection signal parameters to achieve soil moisture inversion [6]. Scholars from various countries have further extended the application prospects of GNSS remote sensing by introducing the application of GNSS remote sensing in ocean, land, atmosphere and hydrology [7, 8]. Zavorotny et al. demonstrated that the use of multipath delay phase inversion of soil moisture is more advantageous than amplitude [9]. Yan et al. based on the signal strength indicator (SSI) data in SNR, and established the relationship between SS1 phase and soil moisture for different satellite systems using multiple satellites, and extended the data acquisition method [10]. Ao et al. used the exponential function to better describe the relationship between SNR multipath delay phase and soil moisture [11]. Liang et al. further improved the accuracy and accuracy of soil moisture inversion by selecting multiple GPS satellites to form a complement [12]. Zhang et al. Used EMD to decompose the SNR signal to improve the utilization of GNSS data and the accuracy of inversion [13].

In summary, it is feasible to monitor soil moisture using GNSS reflection remote sensing. However, there are few studies that consider the separation of satellite reflected signals. Separation by low-order polynomial is inevitably subject to fixed order. The impact is not conducive to the improvement of estimation accuracy. If the SNR can be regarded as a non-stationary time series, the time-frequency multi-scale analysis is performed efficiently by wavelet transform. At the same time, soil moisture changes with time is a nonlinear event, and SNR will show a non-stationary phenomenon as soil moisture changes. Therefore, this paper proposes a satellite reflection signal separation model based on multi-scale analysis, and study the feasibility of LS-SVM model to estimate soil moisture.

2 Soil Moisture Inversion Principle

2.1 GPS-IR Principle

GPS-IR remote sensing technology realizes the monitoring of surface environmental changes through multipath reflection signals in SNR [3]. SNR is an indicator to characterize the signal quality of GPS receivers. Under the low satellite elevation angle, the multipath environment around the SNR station is more obvious, and the multipath error geometry diagram is shown in Fig. 1(a). The SNR interferogram of the PRN16 satellite on the 190th day of the P038 station is shown in Fig. 1(b).

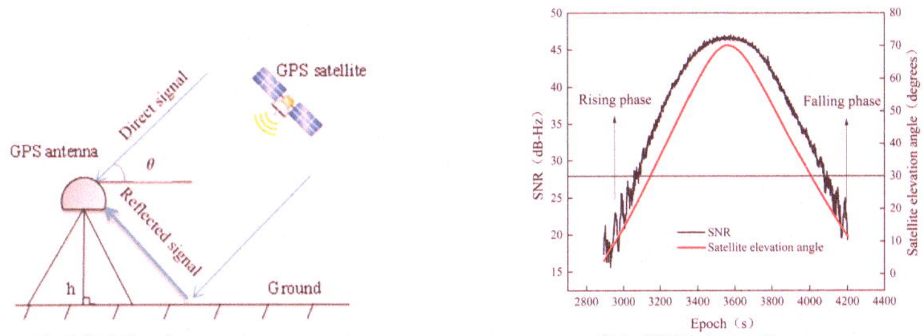

(a) Multipath error geometry (b) SNR interferogram

Fig. 1. Geometric model of ground multi-path error and SNR

It can be seen from Fig. 1(b) that the motion trajectory of GPS satellites over the antenna includes two stages: rising and falling. Below the satellite altitude angle of 30°, the GPS signal-to-noise ratio is more affected by multipath; as the satellite elevation angle increases, the multipath interference effect is reduced, the antenna gain is gradually increased, and the SNR can be expressed as:

$$SNR = SNR_{\mathrm{m}} + SNR_r \tag{1}$$

SNR_{m} denotes a direct signal component and SNR_{r} denotes a reflected signal component. The characteristics of surface environmental parameters are included in the parameters of the reflected signal components. Therefore, the effective and accurate separation of satellite reflection signals is the key factor affecting the accuracy of soil moisture inversion. It can be seen from the literature [14] that there is a sine or cosine relationship between the SNR observation and the multipath interference phase. SNR_r for:

$$SNR_r = A\cos(\frac{4\pi h}{\lambda}\sin\theta + \varphi) \tag{2}$$

In the formula, A denotes the amplitude of the multipath reflection component, h denotes the height of the GPS receiver antenna, λ represents the wavelength of the GPS carrier signal, θ represents the angle of incidence of the satellite, and φ denotes the multipath reflection component phase.

From the Eq. (2), the surface environmental characteristic parameters A and φ of the SNR multipath reflection signal can be solved. Previous studies have shown that the multipath interference phase is a strong linear relationship between surface soil moisture [15]. Therefore, this paper takes the relationship between multipath interference phase and soil moisture as the research and analysis object, and establish a multi-satellite combination inversion model.

2.2 Wavelet Analysis Principle

Wavelet analysis can effectively overcome local errors of low-order polynomials in satellite signal separation due to its good time-frequency localization. Wavelet decomposition and reconstruction essentially decomposes the original signal into multiple sets of signal components with different characteristics through different bandpass filters: high and low frequencies. Let the SNR time series be:

$$f(t) = \{x_1, x_2, x_3, \ldots x_i\}(t = 1, 2, 3, \ldots i) \quad (3)$$

f_k is the discrete sampled data of $f(t)$, $f_k = C_k^0$, then $f(t)$ multi-scale wavelet decomposition can be achieved by the Mallat algorithm [16]:

$$C_{\mathrm{n}}^{j+1} = 2^{-1/2} \sum_{k \in Z} h_{2k-n} C_k^j \quad (4)$$

$$D_{\mathrm{n}}^{j+1} = 2^{-1/2} \sum_{k \in Z} g_{2k-n} D_k^j \quad (5)$$

In the formula, C_{n}^j represents the scale factor, D_{n}^j represents the wavelet coefficient, h and g represent the low-pass and high-pass filters, j is the number of decomposition layers, and n is the number of discrete samples. Wavelet reconstruction is the inverse of wavelet decomposition, and the corresponding calculation formula is [17]:

$$C_{\mathrm{n}}^{j-1} = \sum_{n \in Z} C_n^j h_{k-2n} + \sum_{n \in Z} D_n^j g_{k-2n} \quad (6)$$

In the formula, C_{n}^{j-1} is the reconstructed signal, and the specific steps of acquiring the time series trend using wavelet decomposition and reconstruction are: decomposing the noise-containing signal into different frequency bands at a certain scale, and then detailing the frequency band in which the signal is located is set to zero, and wavelet reconstruction is performed to achieve the purpose of acquiring the trend term of the satellite direct signal.

2.3 LS-SVM Multi-star Fusion Principle

Set the GPS multipath delay phase set X to:

$$X = \begin{bmatrix} x_1^1 & x_2^1 & \ldots & x_m^1 \\ x_1^2 & x_2^2 & \ldots & x_m^2 \\ \ldots & \ldots & \ldots & \ldots \\ x_1^n & x_2^n & \ldots & x_m^n \end{bmatrix} (n = 1, 2, 3, \ldots, 32;\ m = 1, 2, 3, \ldots, i) \quad (7)$$

In the formula, n indicates the satellite number, m indicates the annual accumulation date. The corresponding soil moisture reference set Y is:

$$Y = [y_1, y_2, y_3, \cdots y_m] \quad (8)$$

X is used as the q dimension multipath delay phase input sample and Y is used as the soil moisture reference value output sample. The regression problem estimated by LS-SVM can be equivalent to the following functions:

$$\begin{cases} \min\ Q(w,e) = \frac{1}{2}\|w\|^2 + \frac{\gamma}{2}\sum_{i=1}^{l} e_i^2 \\ s.t.\ y_i = w^T\phi(x_i) + b + e_i (i = 1,\ 2,\ \cdots, l) \end{cases} \tag{9}$$

In the formula, γ denotes a regularization parameter, e_i denotes a fitting error, $\varphi(\cdot)$ denotes a kernel function, w denotes a weight coefficient vector, and b denotes a threshold. Furthermore, the regression function of the LS-SVM is:

$$y = \sum_{m=1}^{i} a_i \varphi(X_i, X) + b \tag{10}$$

In the formula, a represents the Lagrangian multiplier, X_i represents the input training sample set, and X represents the test sample set. The performance of the LS-SVM depends on the optimal selection of kernel functions, kernel parameters, and regularization parameters. Considering that the Radial Basis Function (RBF) can reflect the complexity of the model well, the generalization performance is better. Therefore, this paper chooses the kernel function of LS-SVM. At the same time, the grid search method is used to optimize the kernel parameters and regularization parameters [18].

3 Analysis of Results

This paper selects the monitoring data from the Plate boundary observation (PBO) (http://xenon.colorado.edu/portal) provided by the US Plate Boundary Observation Program to conduct experiments. The station is located in New Mexico, USA, with a peak of 256.592662°, a north latitude of 34.147255°, and an altitude of 1213. 0 m. From the Google Earth multispectral image and P038 station map in Fig. 2, it can be seen that the surrounding terrain of the station is relatively flat, open, and sparsely covered with vegetation. Conducive to soil moisture monitoring. The site GPS receivers are housed in steel triangle brackets. The receiver model is TRIMBLE NETRS, and the SCIT radome is used. The antenna model is TRM29659.00. Daily soil moisture reference values are estimated from L2C observations.

(a) Google Earth Remote Sensing Image (b) P038 Station

Fig. 2. Surround environment of station

It is known that the soil moisture information is contained in the SNR reflection signal at a low elevation angle. According to the surrounding environment of the station and multiple experimental analysis. For the P038 station, consider the station environment and experimental results comprehensively. the experiment selected the satellite elevation angle range of 5°–20°. For single satellite inversion mode, effective separation of SNR reflection components is critical. Therefore, this paper first analyzes the improvement of the separation model of 8 satellites at the P038 station. The comparative analysis of the multipath interference phase and soil moisture of each satellite is shown in Fig. 3.

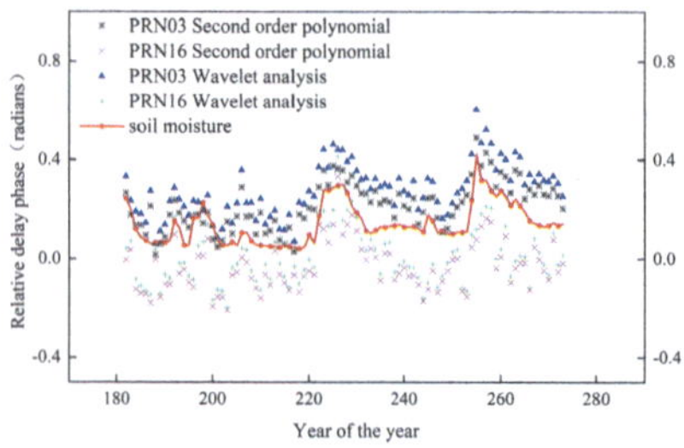

(a) PRN 03、16 Satellite multipath interference phase

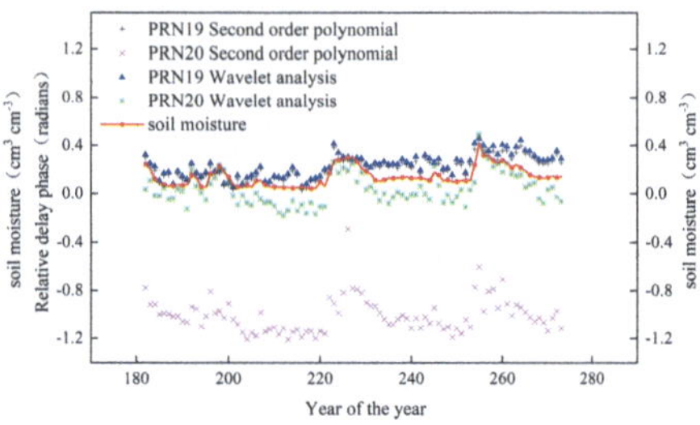

(b) PRN 19、20 Satellite multipath interference phase

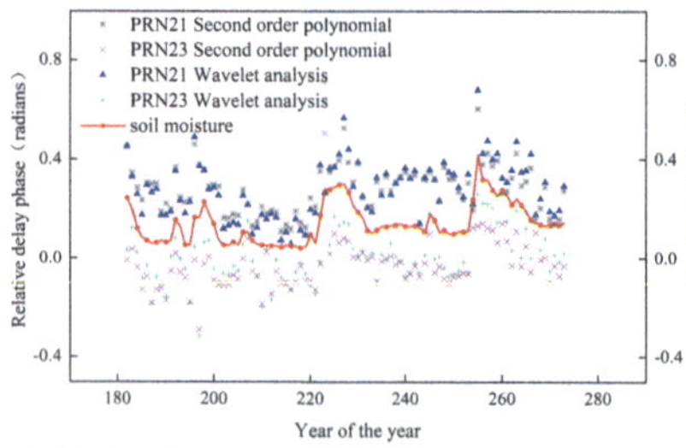

(c) PRN 21、23 Satellite multipath interference phase

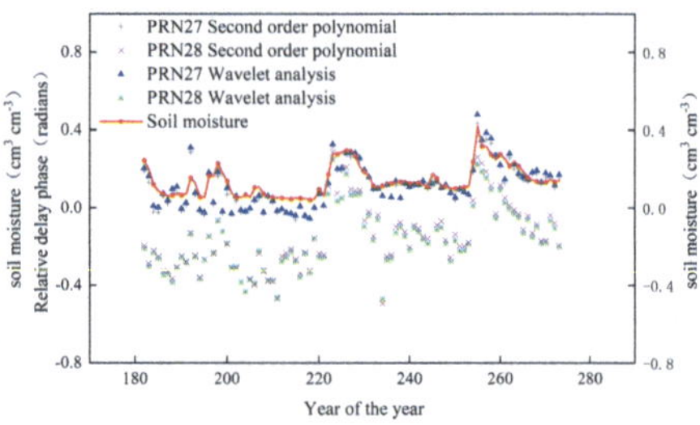

(d) PRN 27、28 Satellite multipath interference phase

Fig. 3. Multipath interference phase of each satellite

It can be seen from Fig. 3 that the multipath interference phase of each satellite can respond when the soil moisture rises or falls. The trends of different satellites are different, which is due to the satellite's own performance and the difference in azimuth of the reflection area. Further comparison shows that the multipath delay phase inversion obtained by the traditional separation satellite reflection signal method is poor, and the outliers are more, especially the PRN20 satellite. The multipath delay phase overall underestimates the soil moisture reference value; from PRN27, 28 It can also be seen that compared with wavelet analysis, the annual inversion results of some annual product days have a large reference value of soil moisture and can not reflect the trend of soil moisture. To further investigate the inversion performance of wavelet analysis, the single-star inversion results were evaluated using root mean square error (RMSE), mean absolute error (MAE), and correlation coefficients, as shown in Table 1.

Table 1. Accuracy statistics of satellite inversion results. L stands for low-order polynomial and W stands for wavelet analysis.

P038	Method	3	16	19	20	21	23	27	28	Average
R^2	L	0.687	0.474	0.599	0.609	0.526	0.506	0.782	0.717	0.612
	W	0.713	0.476	0.599	0.793	0.622	0.529	0.790	0.729	0.656
RMSE	L	0.045	0.059	0.051	0.051	0.056	0.057	0.038	0.043	0.050
	W	0.044	0.059	0.051	0.037	0.050	0.056	0.037	0.042	0.047
MAE	L	0.079	0.173	0.074	1.149	0.135	0.174	0.050	0.290	0.265
	W	0.135	0.141	0.098	0.112	0.136	0.139	0.046	0.304	0.139

It can be seen from Table 1 that the average correlation coefficient between the multipath delay phase and the soil moisture is greater than 0.6, which has a significant correlation. RMSE and MAE fluctuate between 0.044–0.059 and 0.046–1.149. The average correlation coefficient of the results estimated by wavelet analysis reached 0.656, which was 7.19% higher than the traditional method, the RMSE decreased by 0.003, and the MAE decreased by 0.126. It can be seen that wavelet analysis is feasible and effective in separating satellite reflection signals, which can effectively improve the local error of traditional method separation. However, the single satellite has the limitation of a single azimuth, and the overall accuracy of the inversion is low, which is not conducive to practical applications. Therefore, this paper considers the use of multi-star fusion to improve the spatial resolution and thus the accuracy of soil moisture retrieval.

23 schemes were set up for soil moisture inversion, as shown in Table 2. Each scheme establishes LS-SVM inversion models with different satellite number combinations. for the P038 station, a total of 60 days of 182 days to 241 days is used as the training set. Days-273 days and 32 days were used as the test set to establish the LS-SVM model for estimation. The model's rolling estimation step size is set to 1, for example: using the inversion results from days 182 to 241 as input variables and the corresponding soil moisture reference value as the output target, training the model to estimate the inversion results on day 242, and The inversion results from days 183 to 242 were used as the training set to estimate day 243, and this rolling estimation was performed until day 273. Due to space limitations, only some satellite estimation results and linear correlation analysis results are given, as shown in Figs. 4 and 5.

Table 2. Satellite combination scheme setting

Plan	Satellite combination	Plan	Satellite combination
1	PRN03	13	PRN21, 23, 28
2	PRN16	14	PRN16, 23, 28
3	PRN20	15	PRN03, 19, 20
4	PRN21	16	PRN03, 19, 23
5	PRN23	17	PRN20, 21, 23
6	PRN27	18	PRN16, 21, 23, 28
7	PRN28	19	PRN03, 20, 27, 28
8	PRN03, 19	20	PRN03, 16, 19, 20
9	PRN21, 28	21	PRN03, 16, 19, 20, 21
10	PRN20, 21	22	PRN03, 16, 19, 20, 21, 23
11	PRN03, 20	23	PRN03, 16, 19, 20, 21, 23, 28
12	PRN16, 20		

It can be seen that the correlation coefficients of the multi-satellite combinations of each scheme are 0.781, 0.829, 0.886, 0.933 and 0.940. With the increase in the number of satellite combinations, the correlation coefficient has been improved. In order to further evaluate the performance of multi-star fusion, the accuracy analysis results of each scheme are given, as shown in Fig. 6. The correlation coefficient of single star inversion fluctuates greatly, and it cannot reflect the change trend of soil moisture. Multi-star combination effectively combines the advantages of each satellite, and the correlation coefficient is significantly improved. As the number of satellite combinations increases, RMSE and MAE gradually decrease. When the number of combined satellites reaches 5 or more, RMSE is less than 0.02 and MAE is less than 0.023.

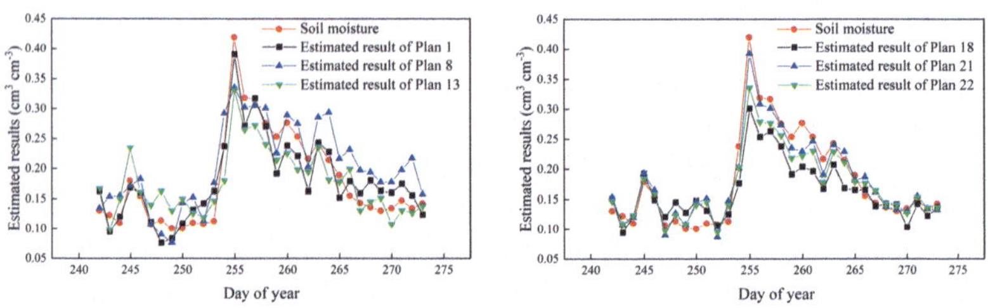

(a) Estimated results of Plan 1,8 and 13 (b) Estimated results of Plan 18,21 and 22

Fig. 4. Correlation analysis of estimation results of each program

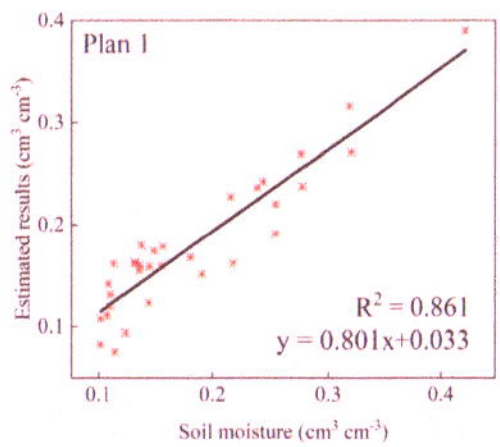

(a) Single satellite correlation

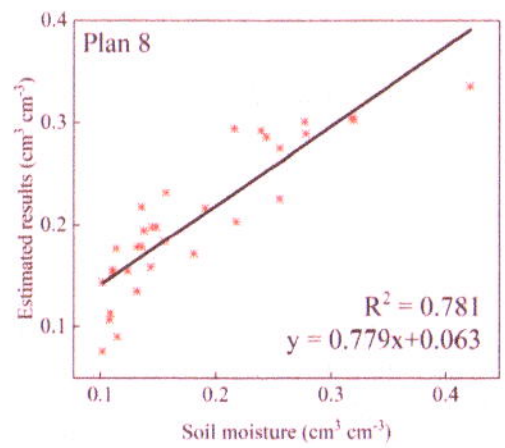

(b) Two satellite correlation

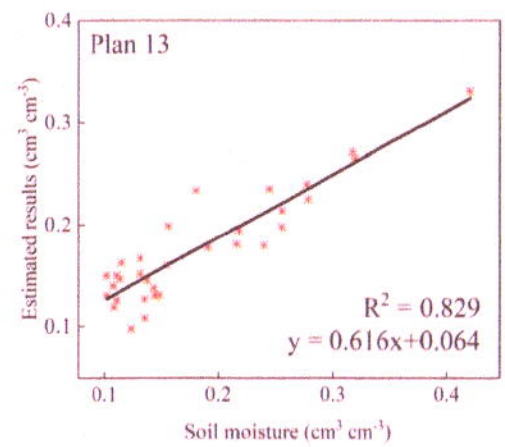

(c) Three satellite correlation

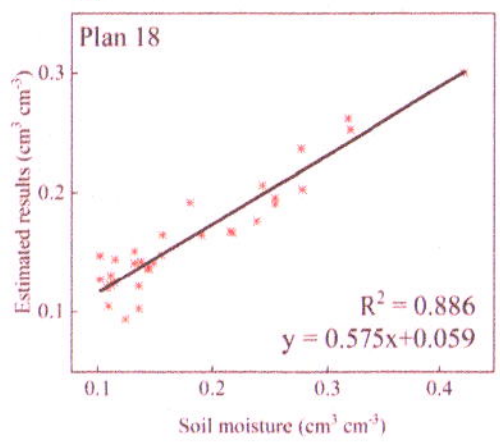

(d) Four satellite correlation

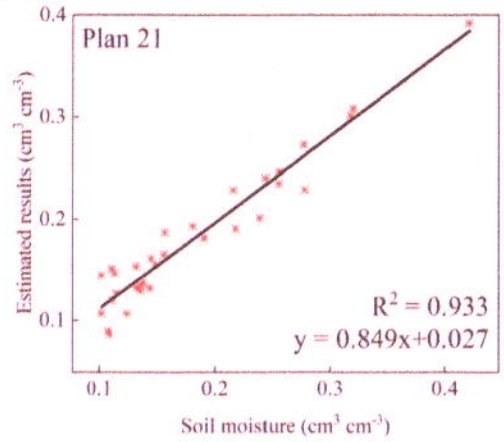

(e) Five satellite correlation

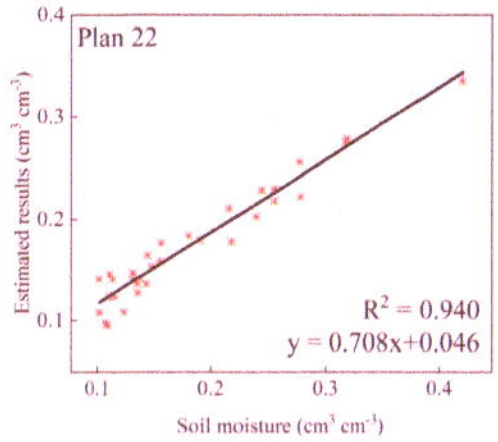

(f) Six satellite correlation

Fig. 5. Correlation analysis of estimation results of each program

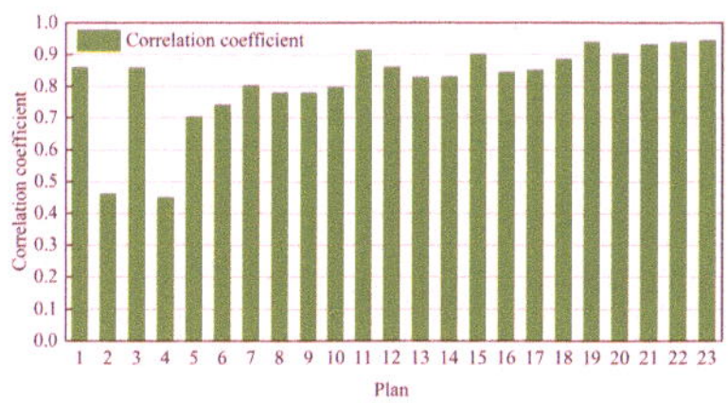

(a) Comparison of correlation coefficient of each program

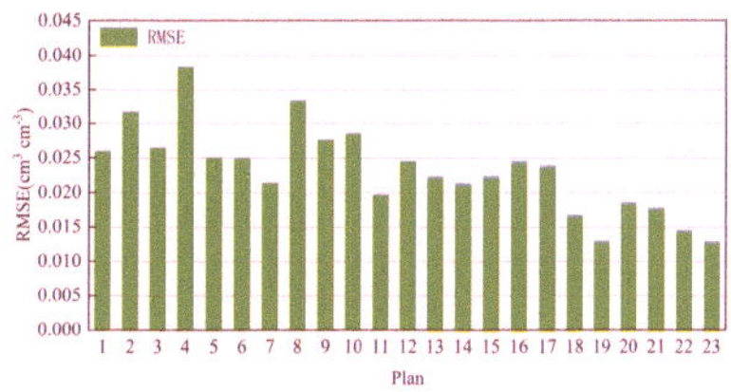

(b) RMSE comparison of each program

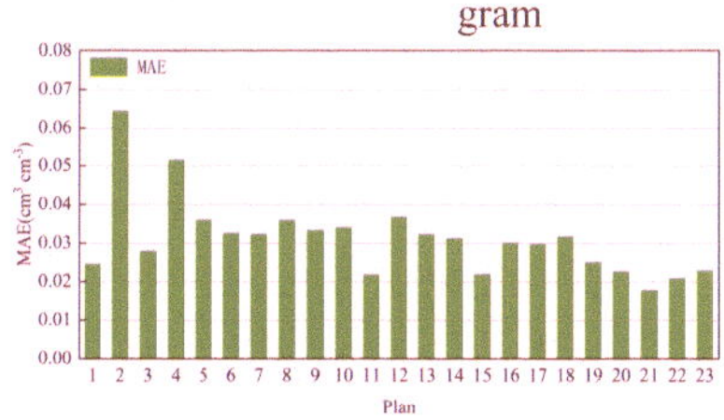

(c) MAE comparison of each program

Fig. 6. Comparative analysis of precision of each scheme

4 Conclusion

Based on the GNSS continuous operation reference station in the PBO observation network, using GPS-IR for accurate and long-term monitoring of soil moisture and its changes has important significance and application prospects. In this paper, a wavelet analysis and separation satellite reflection signal model and LS-SVM soil moisture multi-star fusion inversion model are proposed. The theoretical analysis and experiments show that: For the environment with flat terrain and single vegetation, the multipath interference phase of different satellites the response patterns to changes in soil moisture are different, and there is a linear correlation between the two; Wavelet analysis can effectively decompose the signal at multiple scales, obtain satellite reflection signals, eliminate other noise effects, and improve the inversion accuracy of single satellites, which is 7.19% higher than the traditional method; LS-SVM model can effectively integrate the effective reflection areas of each satellite to more fully reflect the soil moisture information within the effective monitoring range near the station. For the spatial and temporal resolution of soil moisture, how to make full use of multi-system data such as GPS, BDS and GLONASS to achieve soil moisture inversion with higher spatial and temporal resolution, this process needs further research on the difference of different systems and precision.

Acknowledgments. This work was supported by the National Natural Science Foundation of China (Grant Nos. 41901409); the Guangxi Young and Middle-aged Teacher Basic Ability Improvement Project (Grant No. 2018KY0247); the Guangxi Natural Science Foundation (Grant Nos. 2015GXNSFAA139230).

References

1. Liu, J., Shao, L., Zhang, X.: Advances in GNSS-R studies and key technologies. Geomatics Inform. Sci. Wuhan University **11**, 955–960 (2007)
2. Jin, S., Zhang, Q., Qian, X.: New progress and application prospects of global navigation satellite system reflectometry (GNSS+R). Acta Geodaetica Cartogr. Sin. **46**(10), 1389–1398 (2017)
3. Larson, K.M., Small, E.E., Gutmann, E.D., Bilich, A.L., Larson, C.: Use of GPS receivers as a soil moisture network for water cycle studies. Geophys. Res. Lett. **35**(24), 851–854 (2008)
4. Jung, M., Reichstein, M., Ciais, P., Seneviratne, S.I., Sheffield, J., Goulden, M.L.: Recent decline in the global land evapotranspiration trend due to limited moisture supply. Nature **467**(7318), 951–954 (2010)
5. Martin-Neira, M.: A passive reflectometry and interferometry system (PARIS): application to ocean altimetry. ESA J. **17**(4), 331–355 (1993)
6. Larson, K.M., Small, E.E., Gutmann, E., Bilich, A., Axelrad, P., Braun, J.: Using GPS multipath to measure soil moisture fluctuations: initial results. GPS Solutions **12**(3), 173–177 (2008)
7. Jin, S., Komjathy, A.: GNSS reflectometry and remote sensing: new objectives and results. Adv. Space Res. **46**(2), 111–117 (2010)
8. Zhou, W., Liu, L., Huang, L., Li, J., Chen, J., Chen, F., Xing, Y.: Monitoring snow depth based on the SNR signal of GLONASS satellites. J. Remote Sens. **22**(5), 889–899 (2018)
9. Zavorotny, V.U., Member, S., Larson, K.M., Braun, J.J.: A physical model for GPS multipath caused by land reflections: toward bare soil moisture retrievals. IEEE J. Sel. Top. Appl. Earth Obs. Remote Sens. **3**(1), 100–110 (2010)
10. Yan, S., Zhang, N., Chen, N., Gong, J.: Feasibility of using signal strength indicator data to estimate soil moisture based on GNSS interference signal analysis. Remote Sens. Lett. **9**(1), 61–70 (2018)

11. Ao, M., Zhu, J., Hu, Y., Zeng, Y., Liu, Y.: Comparative experiments on soil moisture monitoring with GPS SNR observations. Geomatics Inform. Sci. Wuhan University **40**(01), 117–120 (2015)
12. Liang, Y., Ren, C., Huang, Y., Wang, H., Lu, X., Yan, H.: Rolling estimation model of soil moisture based on multi-satellite fusion. J. Remote Sens. **23**(04), 648–660 (2019)
13. Zhang, S., Liu, K., Liu, Q., Zhang, C., Zhang, Q., Nan, Y.: Tide variation monitoring based improved GNSS-MR by empirical mode decomposition. Adv. Space Res. **63**(10), 3333–3345 (2019)
14. Chew, C.C., Small, E.E., Larson, K.M., Zavorotny, V.U.: Vegetation sensing using GPS-interferometric reflectometry: theoretical effects of canopy parameters on signal-to-noise ratio data. IEEE Trans. Geosci. Remote Sens. **53**(5), 2755–2764 (2015)
15. Chew, C., Small, E.E., Larson, K.M.: An algorithm for soil moisture estimation using GPS-interferometric reflectometry for bare and vegetated soil. GPS Solutions **20**(3), 525–537 (2016)
16. Mallat, S.: Theory for multi-resolution signal decomposition: the wavelet representation. IEEE Trans. Pattern Anal. Mach. Intell. **11**(7), 674–693 (1989)
17. Wen, L., Liu, Z.S., Ge, Y.J.: Several methods of wavelet denoising. J. Hefei Univ. Technol. (Nat. Sci.) **02**, 167–172 (2002)
18. Yang, L., Yang, S., Li, S., Zhang, R., Liu, F., Jiao, L.: Coupled compressed sensing inspired sparse spatial-spectral LSSVM for hyperspectral image classification. Knowl.-Based Syst. **79**, 80–89 (2015)

Bare Soil Freeze/Thaw Process Detection Using GNSS-R/IR Techniques: A Case Study in Alaska, USA

Xuerui Wu[1,2,3,4], Sharula[1], Xuanran Li[1], and Lei Yang[5(✉)]

[1] School of Resources, Environment and Architectural Engineering, Chifeng University, Chifeng 024000, Inner Mongolia, China
[2] Shanghai Astronomical Observatory, Chinese Academy of Sciences, Shanghai 200030, China
[3] Key Laboratory of Space Navigation and Position Technology, Shanghai 200030, China
[4] Key Laboratory of Planetary Sciences, Chinese Academy of Sciences, Shanghai 200030, China
[5] School of Information Science and Engineering, Shandong Agricultural University, Taian 271018, China
yanglei_sdau@163.com

Abstract. GNSS-IR is a new emerging remote sensing technique, and it utilizes the interference signal of direct and reflected signals for geophysical parameters retrieval. Nowadays, its application include soil moisture, snow depth, vegetation water content and sea level detection. This paper has extended its application to monitor soil freeze/thaw process. Mixed material dielectric constant model is employed to get the dielectric constant change as soil freeze/thaw process occurs. Forward GPS multipath model is used to establish the quantitative relationship between GPS multipath observables and soil freeze/thaw process. One open GPS site and a corresponding SNOTEL soil climate site are analyzed to get the relationship between multipath SNR and soil temperature (as soil freeze/thaw process occurs). To take the effect of receiver temperature into consideration, multipath data at higher elevation angles are analyzed. In-situ measurement data indicate that there is a good relationship between soil temperature and GPS multipath data and there is a bright potential for soil freeze/thaw process detection using GNSS-IR.

Keywords: GNSS-IR · Soil freeze/thaw process · Dielectric constant · Multipath

1 Introduction

Seasonal frozen soil and permanent frozen soil account for about 35% of the total land area of the earth, and they are mainly distributed in high latitudes and high altitudes. The conversion of land surface soil freeze-thaw state occurs repeatedly every season, which is closely related to human living environment. Therefore, effective monitoring of the temporal and spatial distribution of surface freezing and thawing conditions and

J. Sun et al. (Eds.): CSNC 2020, LNEE 650, pp. 38–46, 2020.
https://doi.org/10.1007/978-981-15-3707-3_4

related physical parameters is an important topic in the research fields of cryosphere, geosciences and hydrology.

Typical surface elements have strong spatial and temporal variation non-uniformity. The monitoring methods of traditional ground and meteorological stations are difficult to meet the application requirements. The development of satellite remote sensing technology provides a new method for monitoring.

Visible light and thermal infrared remote sensing are limited by weather conditions. Microwave remote sensing can be observed all day and hour, and active/passive microwave remote sensing (radar/radiometer) is one of the effective means of surface freezing and thawing monitoring [1, 2]. SMOS (Soil Moisture and Ocean Salinity) and SMAP (Soil Moisture Active and Passive) can be used for global monitoring of soil moisture and surface freeze-thaw conditions. These on-board data will increase the spatial resolution of monitoring, but time resolution (global repeat coverage every 3 days) and spatial resolution are far from meeting the scientific needs of monitoring.

Traditional GPS has entered the new era of GNSS (Global Navigation Satellite System), including USA's GPS, European's GALILEO, Indian's IRNSS, Japan's QZSS (the Quasi-Zenith Satellite System) and China's Beidou satellite navigation system. The continuous GNSS direct signal is reflected by the sea surface or the land surface, and is received by the special GNSS-R receiver/GNSS station receiver, thus it forms a new type of low cost, low power consumption, wide coverage and high spatial and temporal resolution remote sensing technique: GNSS-R (GNSS-Reflectometry)/IR (Interference-Reflectometry) [3–5]. Using this technology to remotely monitor the surface freeze-thaw state will be a useful complement to existing satellite plans in terms of space-time resolution [6–8].

No special receivers were used by GNSS-IR. Geodetic-quality receivers are employed for typical geophysical parameters detection. When the elevation angle is lower than 30°, the interference pattern of direct and reflected signals are obvious, i.e. the multipath information can be used for geophysical parameters detection [5]. The spatial resolution of GNSS-IR is about 1 km, which is just between traditional point measurement (<1 m^2) and space-borne observation (>100 km^2). This means thousands of GPS site data can be employed for soil freeze/thaw process detection in real time, which can provide data for remote sensing research of environmental science and hydrology. The establishment of soil freeze/thaw process network is of great significance for hydrological research, weather forecast and climate change monitoring. At the same time, this method can be well supplemented with L band space borne data. It is possible to use the existing global GPS net data to obtain the global soil freeze/thaw process observation network. It helps to validate and calibrate other soil freeze/thaw process detection satellite.

This paper describes in detail the theoretical basis for surface freeze-thaw monitoring using GNSS-IR, and the second part gives the theoretical basis. The principles of data analysis and influencing factors are given in the third and fourth sections respectively. Conclusions are presented in Sect. 5.

2 Theory and Fundamentals

The dielectric constant is an inherent property of the substance itself. When the dielectric constant of a substance changes, it will cause changes in its radiation characteristics or scattering characteristics, which will affect the microwave remote sensing observation [9]. This is the theoretical basis for remote sensing observation of ground parameters using microwave remote sensing. For the melted soil, it can be regarded as a mixed medium composed of solid particles, air, free water and bound water [10, 11]. The final dielectric constant is the result of the interaction of the components, i.e. Eq. 1).

$$\varepsilon^{\alpha} = V_{s}\varepsilon_{s}^{\alpha} + V_{a}\varepsilon_{a}^{\alpha} + V_{fw}\varepsilon_{fw}^{\alpha} + V_{bw}\varepsilon_{bw}^{\alpha} \tag{1}$$

where ε is the dielectric constant, the superscript α is the shape factor, V is the volumetric content, The subscripts s, a, fw, bw represent solid particles, air, free water and bound water, respectively. For frozen soil, it can be seen as a mixed medium with increased ice particle composition [12]. The final dielectric constant of frozen soil is the result of the interaction of various parts, as shown in Eq. 2.

$$\varepsilon^{\alpha} = V_{s}\varepsilon_{s}^{\alpha} + V_{a}\varepsilon_{a}^{\alpha} + V_{fw}\varepsilon_{fw}^{\alpha} + V_{bw}\varepsilon_{bw}^{\alpha} + V_{i}\varepsilon_{i}^{\alpha} \tag{2}$$

The subscript i in Eq. 2 represents ice particles.

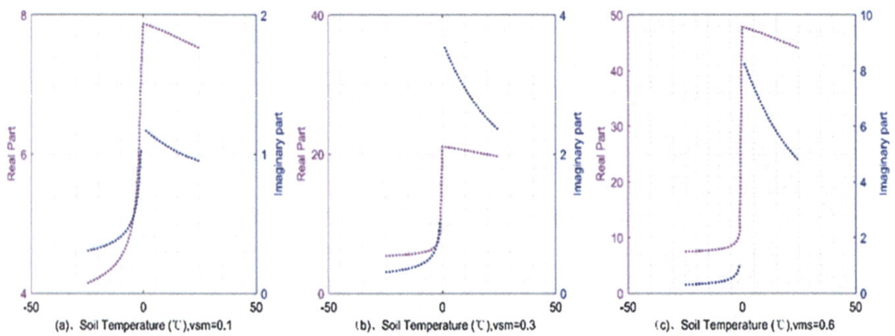

Fig. 1. Under different soil moisture (a, vsm = 0.1; b, vsm = 0.3; c, vsm = 0.6), the variations of real and imaginary parts of the dielectric constant with soil temperature.

Figure 1 shows the variation of the real and imaginary parts of the dielectric constant with soil temperature under different soil moisture. When the soil volumetric water content is 0.1, 0.3 and 0.6, as the soil temperature changes from negative temperature to positive temperature, the difference between the real part and the imaginary part is large. That is, when the soil moisture content changes, the real and imaginary parts of the dielectric constant have a large leap change when the soil freeze/thaw process occurs.

Nievinski and Larson developed a fully polarized forward GPS multipath model that can simultaneously consider GPS signal polarization, antenna and surface response [13, 14].

$$P_{\mathrm{d}} = P_d^R G_d^R W_d^2 \quad (3)$$

$$P_r = P_d^R |XSW_r|^2 \quad (4)$$

P represents the electric field energy, G represents the antenna gain, W represents the Woodward fuzzy function, subscript and r represent the direct and reflected components, respectively, and X is the surface and antenna coupling coefficient.

During the freeze-thaw transition, the large change in the dielectric constant of the soil leads to a significant change in the surface reflection characteristics, which in turn leads to significant changes in the GPS multipath observation. This is the theoretical fundamentals for surface freeze-thaw monitoring using GNSS-IR technology.

3 GPS Site Data and Meteorological Data Analysis

Feature parameter information affecting multipath observation includes soil moisture and snow thickness, Therefore, the soil moisture and meteorological information of the US NRCS (the Natural Resources Conservation Service) soil moisture and soil temperature plan were selected for analysis. SNOTEL (Snow Telemetry) records snow thickness and weather data for the western United States and Alaska. This section selects the SNOTEL site Coldfoot in Alaska (Site No. 958, Lat: 67.25; Lon: −150.18; Ele: 316.99 m) for analysis.

Within the spatial resolution of GNSS-IR, GPS stations (ID AB33, Lat: 67.251; Lon: −150.1725; Ele: 334.76 m) that meet the resolution requirements are selected in the PBO GPS network for analysis. Figures 2–3 show the changes in soil moisture and snow thickness during soil melting in 2009 and soil melting in 2012. From Figs. 2 to 3, it can be seen that both the soil moisture and the thickness of the snow have changed significantly during the freeze-thaw conversion. In order to deal with the effects of soil freezing on multipath data in the event of freezing and thawing, the effect of soil moisture and snow thickness on the results was removed during the process. In the subsequent analysis, only the soil moisture and snow thickness during the freezing or melting process were selected for analysis.

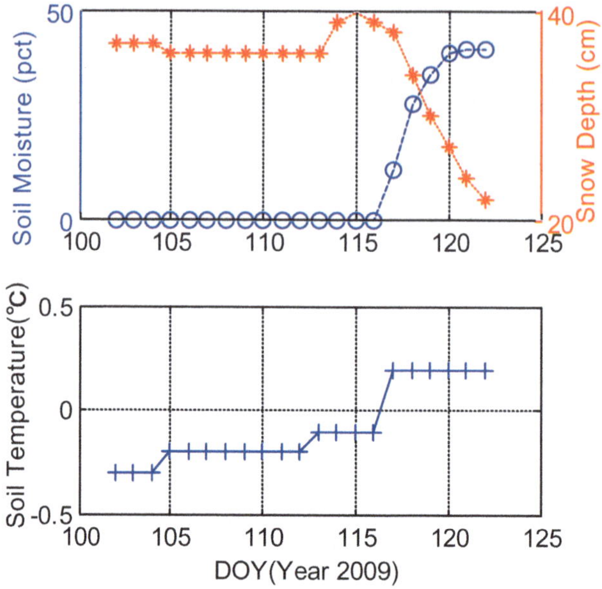

Fig. 2. Variations of soil moisture, snow depth and soil temperature in the thawing even of Year 2009.

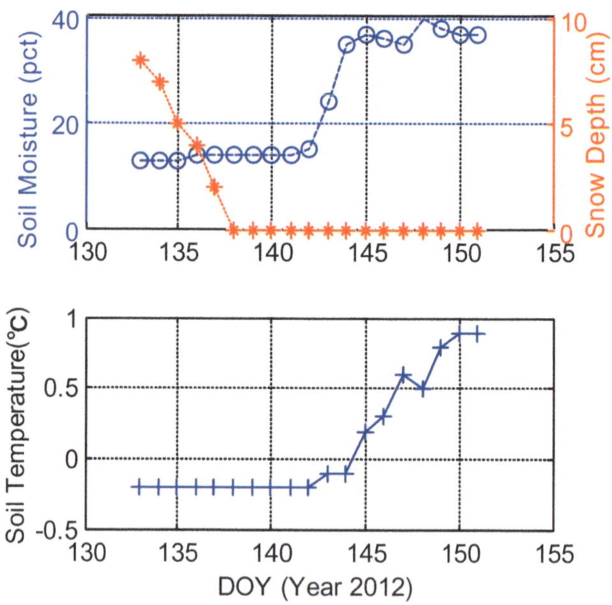

Fig. 3. Variations of soil moisture, snow depth and soil temperature in the thawing even of Year 2012.

Figure 4 shows the relationship between average SNR and soil temperature during soil melting in 2009. It can be seen from Fig. 4 that in the process of soil melting, the soil moisture is within the range of 0 and 0.1, that is, the soil moisture change is small and negligible, while the thickness of the snow remains basically unchanged. Therefore, the change in the average SNR during this process is due to the melting of the soil, and there is a positive correlation between the two parameters.

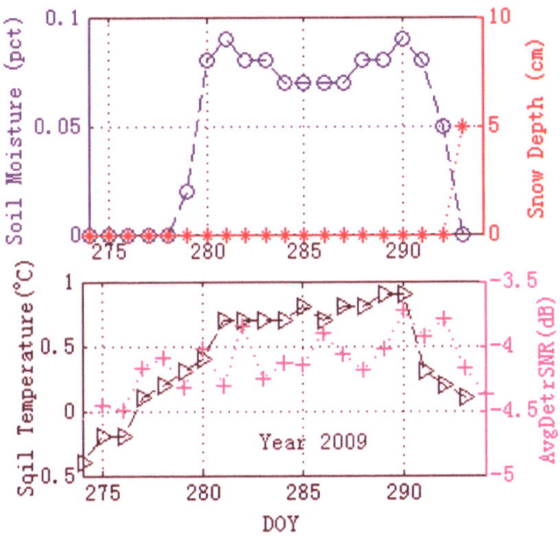

Fig. 4. Comparisons between GPS site (AB33) and SNOTEL site (ID 958) from DOY 270 (2009) to DOY 295 (2009).

Figure 5 shows the relationship between soil moisture, snow thickness and soil temperature as a function of average SNR during the soil freezing process in 2012. It can be seen from the figure that the soil moisture and snow thickness remain basically unchanged during the freezing process. The change in average SNR is due to changes in soil temperature and a negative correlation between the two parameters.

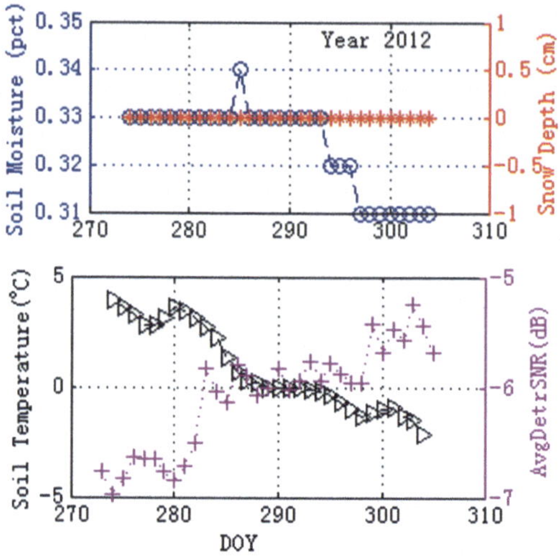

Fig. 5. Comparisons between GPS site (AB33) and SNOTEL site (ID 958) from DOY 270 (2012) to DOY 305 (2012).

4 Analysis of Influencing Factors

During the surface freeze-thaw conversion process, the receiver temperature also changes. How to eliminate the influence of receiver temperature on multipath SNR data is a key factor in analysis. Since there is no receiver temperature measurement in the measured site data, and the multipath data is only effective at low elevation angles (less than 30°). Therefore, Figs. 6–7 show the relationship between soil temperature and average SNR at high elevation angles (45°–50°) in 2009 and 2012. It can be seen from the figure that the correlation between the two parameters is poor, and the correlation coefficients between soil temperature and average SNR are 0.024 and −0.20, respectively. In the high elevation range where the multipath effect is not obvious, the non-correlation between the average SNR and the soil temperature just proves that the receiver temperature cannot affect the SNR variation, thus eliminating the influence of the receiver temperature on the multipath data.

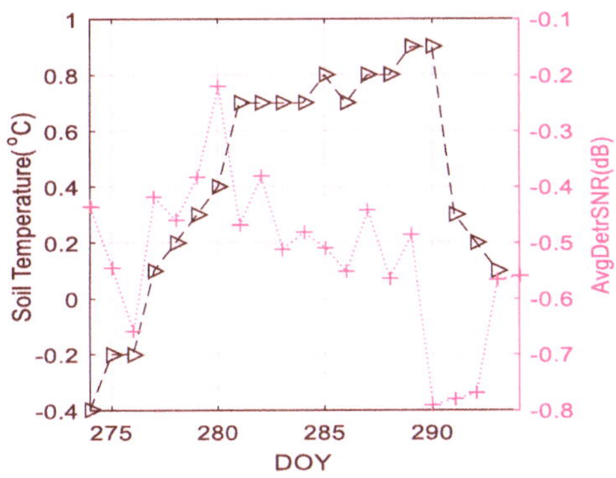

Fig. 6. Comparisons between GPS site (AB33) and SNOTEL site (ID 958) from DOY 270 (2009) to DOY 295 (2009).

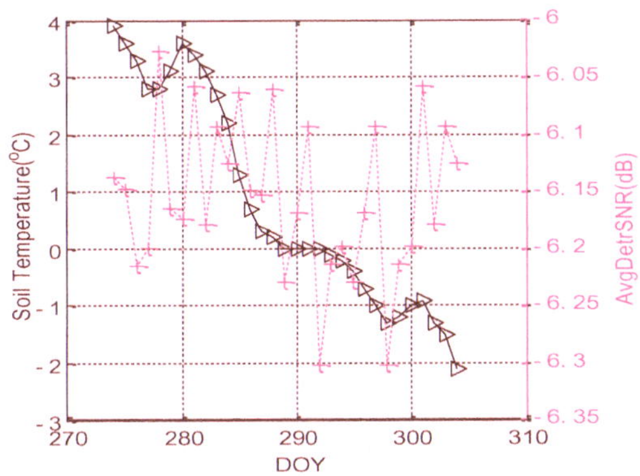

Fig. 7. Comparisons between GPS site (AB33) and SNOTEL site (ID 958) from DOY 270 (2009) to DOY 310 (2012).

5 Conclusions

GNSS-IR technology is an emerging method of remote sensing, which has attracted wide attention at home and abroad. Applications of GNSS-IR remote sensing technology include soil moisture monitoring, snow thickness, vegetation moisture and coastal tide monitoring. The use of this technology for soil freeze-thaw monitoring is an emerging application field of GNSS-IR, and the obvious change of dielectric constant is the theoretical basis for its development. Data analysis of GPS sites and soil meteorological sites in the Alaska region showed a good correlation between soil temperature and average SNR during the soil freeze/thaw process. At high elevation angles, the non-correlation analysis between the average SNR and soil temperature removes the potential for instrument temperature effects on multipath SNR. During the analysis of surface freezing and thawing characteristics in the data site, it is necessary to take into account factors such as soil moisture and snow thickness, therefore, more detailed weather station data is needed. In addition, due to the 1 km resolution of GNSS-IR, it is necessary to perform matching analysis between meteorological site data and GPS site data in the analysis. The above strict analysis conditions lead to very few site data that meet the requirements when using existing public GPS site data for analysis, this limits the long-term sequence analysis of this analytical method, therefore, it is necessary to carry out targeted GNSS-IR surface freeze-thaw monitoring experiments in the near future.

Acknowledgement. This research is supported by the Natural Science Foundation of China (NSFC) Project (Grant Nos. 31971781).

References

1. Judge, J., Galantowicz, J.F., England, A.W., Dahl, P.: Freeze/thaw classification for prairie soils using SSM/I radio brightness. IEEE Trans. Geosci. Remote Sens. **35**, 827–832 (1997)
2. Zhang, T., Armstrong, R.L., Smith, J.: Investigation of the near-surface soil freeze-thaw cycle in the contiguous United States: algorithm development and validation. J. Geophys. Res. Atmos. **108**, 1054–1058 (2003)
3. Zavorotny, V.U., Gleason, S., Cardellach, E., et al.: Tutorial on remote sensing using GNSS bistatic radar of opportunity. IEEE Geosci. & Remote. Sens. Mag. **2**(4), 8–45 (2015)
4. Cardellach, E, Fabra F, Nogués-Correig O, et al.: GNSS-R ground-based and airborne campaigns for ocean, land, ice, and snow techniques: application to the GOLD-RTR data sets. Radio Sci. **46**(6) (2016)
5. Larson, K.M.: GPS interferometric reflectometry: applications to surface soil moisture, snow depth, and vegetation water content in the western United States: GPS interferometric reflectometry. Wiley Interdisciplinary Reviews Water, 2016
6. Wu, X.R., Jin, S.G.: Can we monitor the bare soil freeze-thaw process using GNSS-R?: a simulation study. In: Proceedings of SPIE. Earth Observing Missions and Sensors: Development, Implementation, and Characterization III, vol. 9264, p. 92640I, 26 November 2014
7. Wu, X., Chang, L., Jin, S., et al.: Initial results for near surface soil freeze-thaw process detection using GPS-Interferometric Reflectometry. In: IEEE Geoscience and Remote Sensing Symposium, pp. 1989–1992 (2016)
8. Wu, X., Jin, S., Chang, L.: Monitoring bare soil freeze-thaw process using GPS-interferometric reflectometry: simulation and validation. Remote Sens. **10**(1), 14 (2017)
9. Ulaby, F.T., Moore, R.K., Fung, A.K.: Microwave Remote Sensing: Active and Passive, vol. 2. Artech House, Dedham, MA (1982)
10. Hallikainen, M.T., Ulaby, F.T., Dobson, M.C., El-Rayes, M.A.: Microwave dielectric behavior of wet soil-part 1: empirical models and experimental observations. IEEE Trans. Geosci. Remote Sens. **1**, 25–34 (1985)
11. Dobson, M.C., Ulaby, F.T., Hallikainen, M.T., El-Rayes, M.A.: Microwave dielectric behavior of wet soil-part II: dielectric mixing models. IEEE Trans. Geosci. Remote Sens. **1**, 35–46 (1985)
12. Zhang, L., Shi, J., Zhang, Z., Zhao, K.: The estimation of dielectric constant of frozen soil-water mixture at microwave bands. In: Proceedings of the 2003 IEEE International Geoscience and Remote Sensing Symposium, Toulouse, France, 21–25 July 2003, pp. 2903–2905 (2003)
13. Nievinski, F.G., Larson, K.M.: An open source GPS multipath simulator in Matlab/Octave. GPS Solut. **18**, 1–9 (2014)
14. Nievinski, F.G., Larson, K.M.: Forward modeling of GPS multipath for near-surface reflectometry and positioning applications. GPS Solut. **18**, 309–322 (2014)

High Temporal Resolution of PWV Acquisition Method and Its Preliminary Application in Yunnan

Pengfei Yang, Qingzhi Zhao(✉), and Wanqiang Yao

College of Geomatics, Xi'an University of Science and Technology, Xi'an 710054, China
ypf0323@163.com, zhaoqingzhia@163.com, sxywq@163.com

Abstract. Rapid variation of atmospheric water vapor is important to the regional hydrologic cycle and climate change. Due to the lack of high temporal resolution precipitable water vapor data, it is tough to monitor the rapid change of water vapor. This paper focuses on the hourly PWV data calculated by using the GNSS ZTD from CMONOC and meteorological parameters in ERA5 datasets from ECMWF and the application of PWV in ENSO event is also studied. This paper first verifies the pressure (P) and temperature (T) data in ERA5 datasets. Then, ZHD is calculated based on the atmospheric pressure data, and ZWD is obtained by using GNSS ZTD of CMONOC, and then PWV data of the Yunnan area is calculated based on Tm obtained by improved IGPT2w model from 2011 to 2017 and verified it. At last, the research on the abnormal daily variation of PWV during ENSO and the monitoring of ENSO events is carried out. The results show that: (1) The average RMS and bias of P/T are 3.33 hPa/1.20 K and 0.86 hPa/−0.15 K, respectively. (2) The average RMS and Bias of PWV difference from ERA5 and ERA-interim are 1.98 and 0.83 mm, respectively. (3) Based on the analysis of PWV daily variation during EI Niño Events in 2015–2016, it is found that the PWV daily variation in 2016 is significantly higher than that in 2015. (4) Combining temperature and SSTa index, a new index (ENSO Monitor Index, EMI) of ENSO events is proposed. The correlation between the index and SSTa is 0.59. Therefore, the results of this paper are considerable significance to the study of water vapor distribution and climate monitoring.

Keywords: ERA5 · PWV · ENSO · Diurnal variation

1 Introduction

Atmospheric water vapor is an important part of the troposphere, which affects the global energy, water cycle, climate change, and other related activities [1, 2]. Therefore, it is great significance to know the global high spatial and temporal distribution of atmospheric water vapor in time and accurately. Atmospheric water vapor has gradually become one of the hot spots of GNSS meteorology research at home and abroad, and atmospheric precipitable water is an important index to study atmospheric water vapor

J. Sun et al. (Eds.): CSNC 2020, LNEE 650, pp. 47–55, 2020.
https://doi.org/10.1007/978-981-15-3707-3_5

content. The precipitable water vapor (PWV) refers to the total water vapor content in the air column from the surface to the top of the troposphere, which is usually used to reflect the distribution and change of atmospheric water vapor in the troposphere. Therefore, it is great significance to quickly understand the change of atmospheric precipitable water for the analysis of water vapor distribution [3]. However, due to the lack of traditional water vapor monitoring technology, it is often difficult to monitor its distribution under high spatial-temporal resolution [4].

Since Bevis first proposed GNSS (Global Navigation Satellite System) Meteorology in the 1990s, PWV inversion based on GNSS has gradually become the main means to obtain atmospheric water vapor [5–7]. GNSS has the advantages of all-weather, high precision, high spatial-temporal resolution, and low cost, so it is widely used in PWV calculation [5, 9]. For example, Zhao, Zhang, et al. (2018) calculated the GNSS PWV in the global and china respectively [10, 11].

Meteorological data plays an important role in GNSS inversion of PWV, but the time resolution of reanalysis data or meteorological data is 6 h and 3 h, respectively [12]. The GNSS PWV with high time resolution cannot be obtained, which cannot meet the application scenario requirements of mesoscale weather prediction and monitoring. In order to solve the key problem of low time resolution of meteorological data, this paper takes the Yunnan region of china as the research area and uses the 1 h pressure and temperature data in the ERA5 datasets to interpolate in the horizontal and vertical directions. Based on ZTD (Zenith Total Delay) data provided by the crustal movement observation network of china (CMONOC) and Tm obtained from IGPT2w, an hour resolution PWV is calculated. The accuracy of PWV is evaluated by PWV in the ERA-interim datasets. Finally, the diurnal variation of atmospheric water vapor and its response to ENSO events are analyzed based on hourly PWV, and a new El Niño monitoring system is proposed ENSO Monitor Index (EMI).

2 Data and Method

2.1 Data Description

ERA-interim and ERA5 data are the 4th and 5th generation datasets in ECMWF, respectively, including the datasets of the first 2–3 months from 1979 to now. The original data can be downloaded free of charge through https://www.ecmwf.int/, and ECMWF products provide grid meteorological data such as pressure, temperature, and PWV with different temporal and spatial resolutions. Among them, ERA-interim datasets has a time resolution of 6 h, ERA5 datasets has a time resolution of 1 h. The total tropopause delay data (ZTD) can be obtained from GNSS data provided by CMONOC. In this paper, the PWV data of 2011–2017 (six years in total) are calculated by using the meteorological parameters and other data in the datasets of ZTD and ERA5 of 27 stations in Yunnan. The distribution of 27 GNSS stations is shown in Fig. 1.

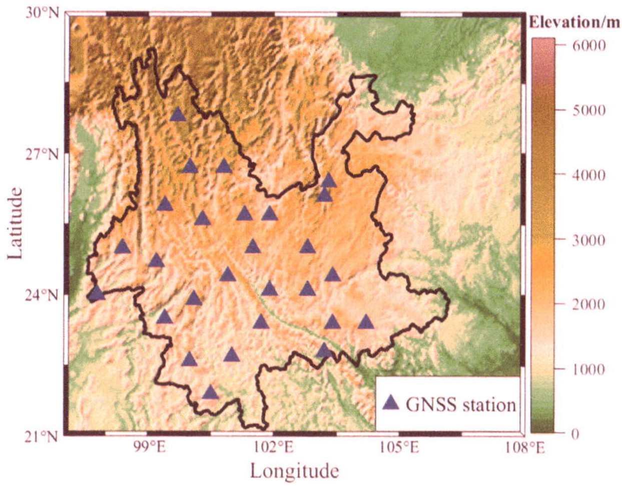

Fig. 1. Distributions of GNSS stations in Yunnan.

2.2 Study Method

In this paper, the atmospheric pressure, temperature and other meteorological parameters of GNSS station are obtained by bilinear interpolation, and the elevation is corrected according to formula (1–2), then the PWV value of GNSS station is calculated [15].

$$P = P_0(1 - 0.0000226(h - h_0))^{5.225} \tag{1}$$

$$dT/dh = -0.0065\,^{\circ}C/m \tag{2}$$

Where, P is the station pressure data (unit: hPa), P_0 is the grid point pressure data (unit: hPa), h is the station height (unit: m), h_0 is the grid height (unit: m), dT is the difference between the station temperature and the grid temperature, dh is the difference between the station height and the grid height.

According to the Saastamoinen model calculates the ZHD (Zenith Hydrostatic Delay) [17], ZWD can be obtained by using ZTD and ZHD. PWV is calculated according to Eq. (3) after obtaining *ZWD*.

$$PWV = \frac{10^6}{\left(K_2' + K_3/T_m\right) \cdot R_V \cdot \rho} \cdot ZWD \tag{3}$$

Where, ρ is the density of water vapor, the values of K_2', K_3 and R_v are 16.48 $K \cdot hPa^{-1}$, $(3.776 \pm 0.014) \times 10^5$ $K^2 \cdot hPa^{-1}$ and 461 $J \cdot kg^{-1} \cdot K^{-1}$ respectively. Tm is the weighted average temperature of the atmosphere. In this paper, we use the improved china Tm model IGPT2w proposed by Huang et al. [18].

3 Comparative Analysis of Meteorological Parameters of ECMWF

3.1 Comparison of Meteorological Parameters Between Grids

Figure 2 shows the comparison results of pressure and temperature between the ERA5 and ERA-interim grid. It can be seen from the figure that in most areas, the values of RMS and bias of pressure and temperature are small, but they are large at some grid locations. The main reason for this phenomenon is that there is a certain error between the reanalysis data and the actual value due to the influence of elevation and the data of ERA-interim. According to the calculation of RMS and bias values, the average RMS values of pressure and temperature in Yunnan are 0.47 hPa and 1.29 K respectively, and the average bias values are −2.23 hPa and 0.08 K respectively.

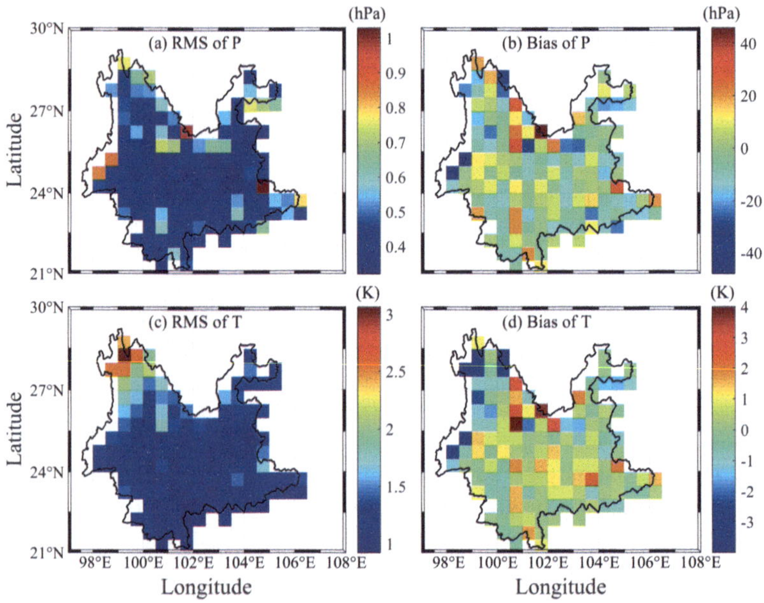

Fig. 2. RMS and bias distributions of P and T derived from ERA5 and ERA-interim at grid points over the period of 2005–2017 in Yunnan.

3.2 Comparative Analysis of Meteorological Parameters Between Stations

After removing the gross error of the data according to the percentile and three times of the mean square error method, Fig. 3 shows the comparison results of the pressure and temperature data of ERA5 and ERA-interim at 22 GNSS stations in Yunnan. It can be seen from the figure that the comparison results of pressure and temperature are generally better, and the results of pressure are worse than those of temperature. The main reason for this phenomenon is that the influence of elevation on pressure is stronger than that on temperature. However, the RMS value and bias value of pressure and temperature at some stations are relatively large. The main reason for this phenomenon is that the quality of grid raw data corresponding to these stations is relatively poor. By calculating the RMS and bias of 22 stations, we can see that the average RMS values of pressure and temperature in Yunnan are 3.33 hPa and 1.20 K, respectively, and the average bias values are 0.86 hPa and −0.15 K, respectively. It shows that the overall accuracy of ERA5 is better, which is close to the data of ERA-interim and can be better applied.

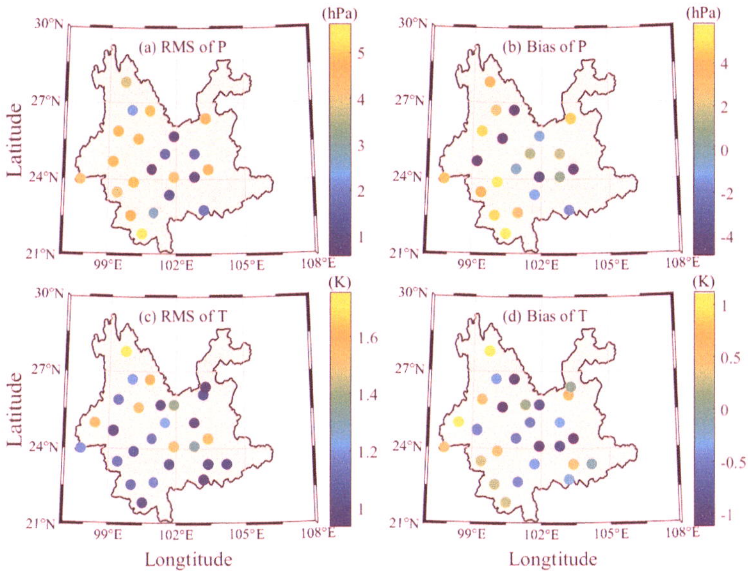

Fig. 3. RMS and bias distributions of P and T derived from ERA5 and ERA-interim at 27 GNSS stations over the period of 2005–2017 in Yunnan.

3.3 PWV Accuracy Evaluation

In the process of PWV accuracy evaluation, the PWV data of ERA-interim is interpolated to the GNSS site and the PWV calculated by the GNSS site is verified by PWV of ERA-interim. The data can be used to analyze the consistency between ERA5 PWV and ERA-interim PWV. Figure 4 shows the PWV comparison results at the GNSS site. It can be seen from the figure that the RMS value at most stations is less than 3 mm, but the RMS value at some stations reaches nearly 3.5 mm, which is mainly due to the poor pressure accuracy at this location, resulting in the low PWV accuracy at this point. The mean value of RMS and bias of PWV in Yunnan are 1.98 and 0.83 mm, respectively. The RMS mean value is higher than the theoretical value. The main reason is that there is a particular gap between the ERA-interim data and the real value for reanalysis, which results in the accuracy of the actual PWV is lower than that of the theoretical PWV. In addition, Figs. 5 and 6 show the RMS value and bias value of each site, respectively. From the RMS value statistical chart, it can be seen that the RMS values at 23 stations are less than 3 mm. Among them, the RMS value and bias value at YNYS (26.7°N, 100.8°E) stations are the largest. Through the analysis of the data at this point, it is found that the pressure data at this point is poor and at a high latitude position, resulting in poor PWV accuracy.

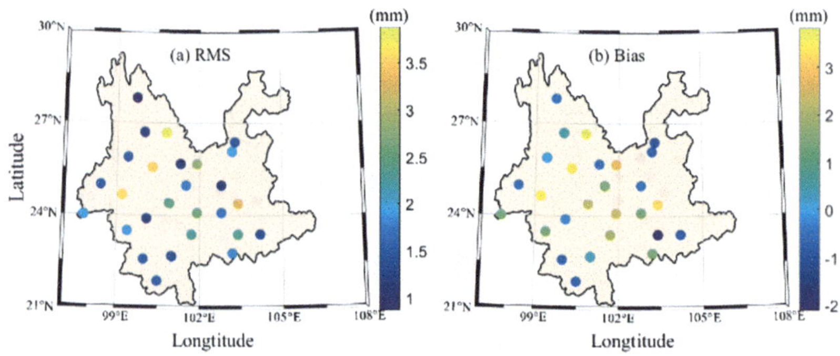

Fig. 4. RMS and bias distributions of PWV differences derived from GNSS and ERA-interim over the period of 2011–2017.

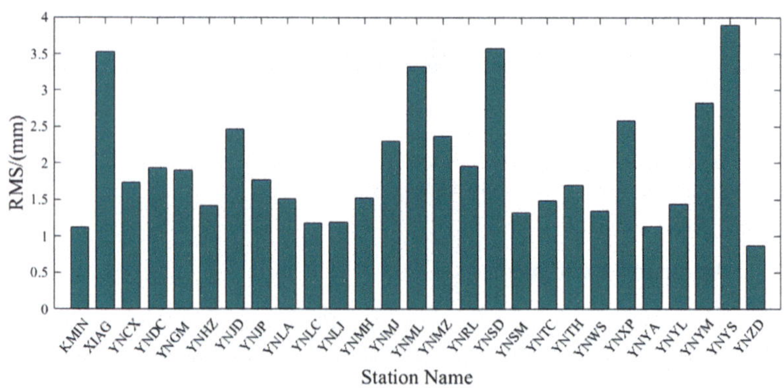

Fig. 5. RMS statistical chart of PWV differences derived from GNSS and ERA-interim at 27 stations in Yunnan.

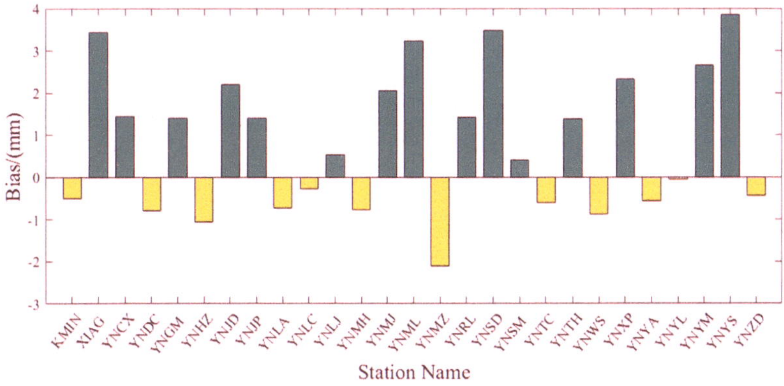

Fig. 6. Bias statistical chart of PWV differences derived from GNSS and ERA-interim at 27 stations in Yunnan.

4 Monitoring ENSO Events Based on PWV

4.1 Daily Variation of PWV

PWV data with a time resolution of 1 h can be used to study the daily change and meet the needs of climate monitoring. Therefore, this paper studies the daily variation of PWV during ENSO events, calculates the daily variation of four seasons in Yunnan, and calculates the average daily anomaly of each season by eliminating the daily average value of corresponding seasons. Among them, spring, summer, autumn, and winter correspond to March to May, June to August, September to November, and December to February, respectively. Figure 7 shows the long time series of daily variations in different seasons at YNJD and YNLA stations, respectively. It can be seen from the figure that the daily variation of PWV in each season in 2016 is significantly higher than that in 2015, of which the daily variation in autumn is the most significant. The main reason is that ENSO events occurred in this period, leading to heavy rainfall events, and then the daily variation of PWV is significantly abnormal.

4.2 PWV Response to ENSO

In order to study the application of PWV hourly data better, this paper introduces the SSTa index and temperature and studies their response to ENSO events. Taking YNTC station in Yunnan province as an example, this paper analyzes the correlation among the El Niño index SSTa, hourly PWV and temperature, and puts forward a new index, namely El Niño monitoring index (EMI), according to the response relationship between them, through which the response mechanism of PWV and ENSO events is studied. From Fig. 8, we can see the PWV and SSTa index have relatively good consistency, and the correlation coefficient is 0.59, and the significance of the p value is less than 0.05. However, the changing trend in June 2012 and August 2015 is the opposite. The main reason is that there are drought and continuous rainfall events in Yunnan during these two periods, so it may be because of this difference frequent natural disasters cause the index to change abnormally.

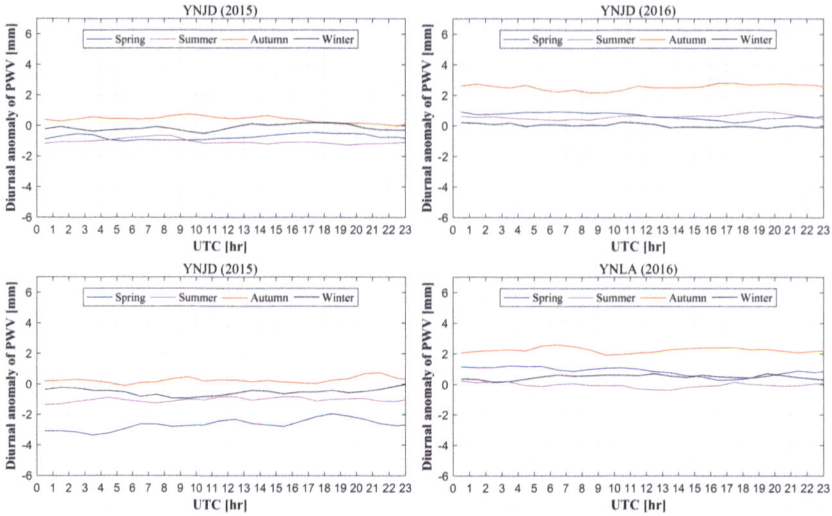

Fig. 7. Diurnal anomalies of PWV time series as a function of UTC at YNJD and YNLA station over the period of 2015–2016.

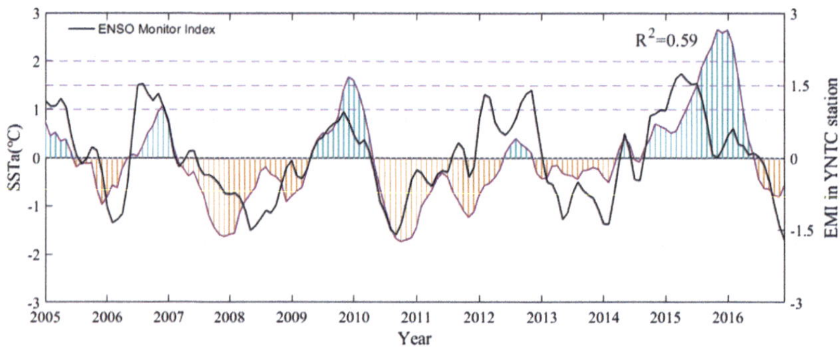

Fig. 8. Long time series correlation of PWV, temperature and SSTa index.

5 Conclusion

Aiming at the problem of low time resolution of PWV, this paper calculates hourly PWV by using the data of pressure and temperature in the fifth generation datasets ERA5 provided by ECMWF and ZTD data provided by the crustal movement observation network of china (CMONOC), and uses the data of pressure, temperature and PWV in ERA-interim datasets to calculate the pressure and temperature in ERA5 datasets and the calculated of GNSS station PWV's accuracy is evaluated. Based on the hourly resolution PWV data, the abnormal changes during ENSO events are analyzed. Finally, a new index EMI is proposed by combining the temperature and SSTa index, which reflects the temperature and the response of PWV and ENSO events. This provides a new method for ENSO event monitoring.

Acknowledgements. Thanks for the reanalysis data provided by ECMWF and ZTD data provided by CMONOC. This study was supported by the National Natural Science Foundation of China (41904036), Xi'an University of science and technology excellent youth science and Technology Fund (2018YQ3-12), Shanxi provincial key research and development program (social development field) project (201803D31224) and the open research topic of Beijing Key Laboratory of Urban Spatial Information Engineering (2019210).

References

1. Jin, S., Li, Z., Cho, J.: Integrated water vapor field and multiscale variations over China from GPS measurements. J. Appl. Meteor. Climatol. **47**(11), 3008–3015 (2008)
2. Bałdysz, Z., Nykiel, G., Figurski, M., et al.: Investigation of the 16-year and 18-year ZTD time series derived from GPS data processing. Acta Geophys. Pol. **6**, 1103–1125 (2015)
3. Wong, M.S., Jin, X., Liu, Z., et al.: Multi-sensors study of precipitable water vapour over mainland China. Int. J. Climatol. **35**(10), 3146–3159 (2015)
4. Song, S.: Sensing three dimensional water vapor structure with ground-based GPS network and the application in meteorology. Shanghai Astronomical Observatory, CAS (2004)
5. Bevis, M., Businger, S., Chiswell, S., et al.: GPS meteorology: mapping zenith wet delays onto precipitable water. J. Appl. Meteorol. **33**, 379–386 (1994)
6. Bevis, M., Businger, S., Herring, T.A., et al.: GPS meteorology: remote sensing of atmospheric water vapor using the Global Positioning System. J. Geophys. Res. Atmos. **97**, 15787–15801 (1992)
7. Rocken, C., Van Hove, T., Ware, R.H.: Near real-time GPS sensing of atmospheric water vapor. Geophys. Res. Lett. **24**, 3221–3224 (1997)
8. Basili, P., Bonafoni, S., Mattioli, V., Ciotti, P., Pierdicca, N.: Mapping the atmospheric water vapor by integrating microwave radiometer and GPS measurements. IEEE Trans. Geosci. Remote Sens. **42**(8), 1657–1665 (2004)
9. Zhao, Q., Yao, Y., Yao, W.Q., Li, Z.: Near-global GPS-derived PWV and its analysis in the El Niño event of 2014–2016. J. Atmos. Solar Terr. Phys. **179**, 69–80 (2018)
10. Zhang, W., Lou, Y., Huang, J., et al.: Multiscale variations of precipitable water over China based on 1999–2015 ground-based GPS observations and evaluations of reanalysis products. J. Clim. **31**(3), 945–962 (2018)
11. Zhang, W., Lou, Y., Haase, J.S., et al.: The use of ground-based GPS precipitable water measurements over China to assess radiosonde and ERA-interim moisture trends and errors from 1999 to 2015. J. Clim. **30**(19), 7643–7667 (2017)
12. Böehm, J., Heinkelmann, R., Schuh, H.: Short note: a global model of pressure and temperature for geodetic applications. J. Geodesy **81**(10), 679–683 (2007)
13. Wang, X., Zhang, K., Wu, S., et al.: Water vapor-weighted mean temperature and its impact on the determination of precipitable water vapor and its linear trend. J. Geophys. Res.: Atmos. **121**(2), 833–852 (2016)
14. Saastamoinen, J.: Atmospheric correction for the troposphere and stratosphere in radio ranging satellites. The Use of Artificial Satellites for Geodesy **15**, 247–251 (1972)
15. Huang, L., Liu, L., Chen, H., et al.: An improved atmospheric weighted mean temperature model and its impact on GNSS precipitable water vapor estimates for China. GPS Solutions **23**(2), 51 (2019)

Soil Moisture Inversion Based on Beidou SNR and Carrier Phase Combinations

Bo Sun[1], Lei Yang[1(✉)], Xuerui Wu[2], Chengyi Wang[1], Xiumei Guo[1], and Liguo Zhang[3]

[1] School of Information Science and Engineering, Shandong Agricultural University, Tai'an, China
yanglei_sdau@163.com
[2] Shanghai Astronomical Observatory, Shanghai, China
[3] Shandong Provincial Institute of Land Surveying and Mapping, Jinan, China

Abstract. Soil moisture is a very important variable in the study of terrestrial water and energy cycle, traditional soil moisture monitoring methods have great limitations for high resolution large area monitoring. Soil moisture monitoring based on Global Navigation Satellite System Interferometry and Reflectometry (GNSS-IR), which also known as signal-to-noise ratio (SNR) method or interference pattern technique (IPT), can be used to overcome those shortcomings, so that it receives more and more attention in recent researches. However, previous studies mostly used the observed data of Global Positioning System (GPS) to estimate soil moisture. With the maturity of the Beidou system, application of Beidou system is the focus of future research. This paper studies the soil moisture inversion methodology using the signal-to-noise ratio (SNR) data of Beidou B1, B2, and also proposes a novel method of soil moisture inversion based on geometry-free linear combinations of carrier phase on Beidou B1, B2. It also presents the soil moisture inversion model as well as the relevant signal processing flow. Moreover, an *in-situ* experiment campaign is performed for verification. The results show that the soil moisture obtained by the proposed models is significantly correlate to the ground truth data. The signal-to-noise ratio method achieves slightly higher accuracy than the carrier phase combinations method, which proves that both the two methods can realize a continuous long-term in-situ observation of soil moisture.

Keywords: Soil moisture · Global Navigation Satellite System (GNSS) · Signal-to-noise ratio (SNR) · Carrier phase combination

1 Introduction

Soil moisture, as known as soil water content, is a key parameter of soil fertility, an important factor for the growth of crops, and an important data for studying agricultural drought [1]. Accurately in-time soil moisture monitoring of cropland is of great significance for reasonable irrigation and achieving high yield of agricultural production.

J. Sun et al. (Eds.): CSNC 2020, LNEE 650, pp. 56–64, 2020.
https://doi.org/10.1007/978-981-15-3707-3_6

Comparing to traditional soil moisture measuring methods such as oven drying method, Timing-Domain Reflectometry (TDR) sensors and Frequency Domain Reflectometry (FDR) sensors, Global Navigation Satellite System Interferometry and Reflectometry (GNSS-IR) is a novel approach of soil moisture monitoring. It provides a non-contact, in-time and continuous solution for monitoring the moisture for large area of soil surface and receives more and more attention recently. With the maturity of the Beidou system [2], application of Beidou system is also the focus of future research.

GNSS-IR technology is also known as Interference Pattern Technique (IPT), which proposed by Rodriguez-Alvarez [3]. For its ability of inferring the characteristic of the reflecting plane, it has been widely applied in soil moisture [4], sea wind/wave [5, 6], snow depth [7] monitoring. Larson [8] stated that the Signal to Noise Ratio (SNR) data recorded by GNSS receiver had oscillations for the impact of multipath signal. Base on that, Chew [9] built an empirical model between phase of SNR oscillation and soil moisture, which could be used for bare soil moisture inversion. In 2016 Roussel [10] processed SNR data of GPS and GLONASS recoded by a geodetic GNSS receiver, proved that if constrained the elevation angle of satiate from 2° to 70°, single geodetic GNSS receiver can achieve high accuracy.

GNSS-IR applications by utilizing the carrier phase also become a hotspot recently. Ozeki and Heki [11] proposed a snow depth inversion model by combining GPS L1 and L2 carrier phase, which was called L4 method. Yu [12] proposed an ionospheric delay free snow depth inversion model by using linear combination of GPS L1, L2 and L5 carrier phase. Wang [13] firstly used the GPS dual frequency carrier phase data fusion algorithm in sea level monitoring application and achieved good accuracy.

This paper proposes a soil moisture inversion method by combining carrier phase of BDS B1 and B2, and make intercomparison with the result of traditional GNSS-IR method. Results show that both the two method can effectively monitor soil moisture.

2 Method

2.1 GNSS-IR Method

The concept of GNSS-IR is using a single RHCP (Right Hand Circular Polarization) antenna to receive the direct and reflected GNSS signal simultaneously. Both the direct signal the reflected signal of RHCP for low elevation angle satellites [14]. For *in-situ* application of GNSS-IR, usually the antenna height is low enough to ignore the difference of Doppler frequency shift between direct and reflected signal, so there will be an interference near the phase centre of the receiving RCHP antenna. The oscillation of SNR is more significant in condition of low elevation angle because the lower the elevation angle is the stronger the RCHP reflection component becomes. The scenario of GNSS-IR *in-situ* application is shown as Fig. 1.

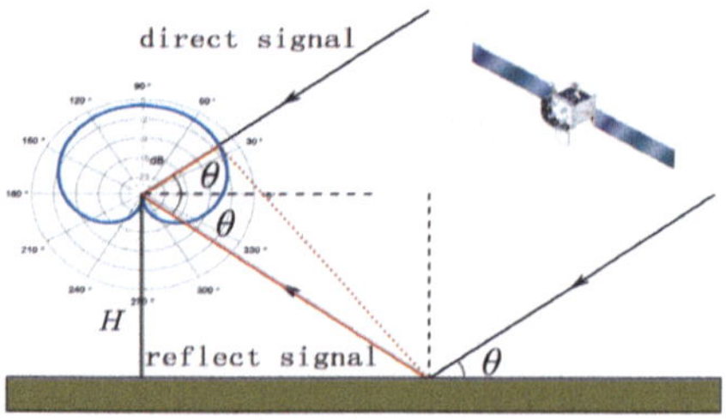

Fig. 1. Scenario of *in-situ* GNSS-IR application

Assume that the direct signal is $S_{\mathrm{d}}(t)$, then we derive reflected signal $S_{\mathrm{m}}(t)$ as Eq. (1.1) [15]:

$$\begin{cases} S_{\mathrm{d}}(t) = A_{\mathrm{d}} \sin \psi(t) \\ S_{\mathrm{m}}(t) = A_{\mathrm{m}} \sin[\psi(t) + \delta_{\varphi}(t)] \end{cases} \tag{1.1}$$

In which A_{d}, A_{m} is the amplitude of direct and reflected signal respectively, $\psi(t)$ is the phase of direct signal and $\delta_{\varphi}(t)$ is the phase difference between direct and reflected signal. $\psi(t)$ and $\delta_{\varphi}(t)$ can be derived as Eq. (1.2):

$$\begin{cases} \psi(t) = 2\pi[\varphi(t) + N] \\ \delta_{\varphi}(t) = 4\pi H \lambda^{-1} \sin \theta(t) \end{cases} \tag{1.2}$$

In which $\varphi(t)$ is the phase observable of carrier, N is the ambiguity of whole cycles, H is the antenna height, λ is the wavelength of carrier, θ is the elevation angle of GNSS satellite.

Then the SNR can be derived by Eq. (1.3) [16]:

$$SNR^2 = A_{\mathrm{d}}^2 + A_{\mathrm{m}}^2 + 2A_{\mathrm{d}}A_{\mathrm{m}} \cos \delta_{\varphi}(t) \tag{1.3}$$

A_{d}, A_{m} and $\delta_{\varphi}(t)$ all vary with elevation angle θ, which generate oscillations in SNR time series. The direct component of SNR can be deprived from by a second-order polynomial curve fitting. Then the residual of SNR time series is considered as the multipath component which can be expressed by Eq. (1.4)

$$SNR_m = A_{\mathrm{m}} \cos\left(4\pi H \lambda^{-1} \sin \theta(t)\right) \tag{1.4}$$

The oscillating frequency can by derived by Lomb-Scacle spectrum analysis. A lot of articles proved that the oscillating frequency correlated to soil moisture, and could be used for building soil moisture inversion model.

2.2 L4 Method

According to Eq. (1.1), the interfering signal received by GNSS receiver can be expressed by Eq. (1.5) [17]:

$$\begin{aligned} S(t) &= S_{\mathrm{d}}(t) + S_{\mathrm{m}}(t) \\ &= A_{\mathrm{d}} \sin \psi(t) + A_{\mathrm{m}} \sin[\psi(t) + \delta_{\varphi}(t)] \\ &= A \sin[\psi(t) + \beta(t)] \end{aligned} \tag{1.5}$$

In which A is the amplitude of interfering signal, $\beta(t)$ is the multipath error or phase error of direct signal which caused by reflected signal. A and $\beta(t)$ can be expressed as Eqs. (1.6) and (1.7) respectively.

$$A = \sqrt{A_{\mathrm{d}}^2 + A_{\mathrm{m}}^2 + 2A_{\mathrm{d}}A_{\mathrm{m}} \cos \delta_{\varphi}(t)} \tag{1.6}$$

$$\beta(t) = \arctan \frac{(\frac{A_{\mathrm{m}}}{A_{\mathrm{d}}}) sin\delta_{\varphi}(t)}{1 + (\frac{A_{\mathrm{m}}}{A_{\mathrm{d}}}) cos\delta_{\varphi}(t)} \tag{1.7}$$

For $A_{\mathrm{m}}/A_{\mathrm{d}} \ll 1$ and $\tan x = \varepsilon$, $\varepsilon \ll 1$, then $x \approx \varepsilon$, Eq. (1.7) can be simplified to Eq. (1.8):

$$\beta(t) \approx (\frac{A_{\mathrm{m}}}{A_{\mathrm{d}}}) sin\delta_{\varphi}(t) \tag{1.8}$$

Considering the impact of reflected signal, the observable of GNSS carrier phase is mainly constituted directed signal part and residual which cause by reflected signal, so the phase observable of BDS B1 and B2 can be expressed by Eqs. (1.9) and (1.10):

$$B_1 = \lambda_1 \varphi_1(t) + \lambda_1 \beta_1(t) = d + I_1 + T + \Delta + \lambda_1 \beta_1(t) \tag{1.9}$$

$$B_1 = \lambda_1 \varphi_1(t) + \lambda_1 \beta_1(t) = d + I_1 + T + \Delta + \lambda_1 \beta_1(t) \tag{1.10}$$

In which λ_1, λ_2 are wavelength of BDS B1 and B2 carrier respectively, d is the geometry distance from BDS satellite to the receiving antenna, I_1, I_2 are the errors cause by ionospheric delay respectively, T is tropospheric delay, Δ is the non-frequency related error item such as errors of receiver and satellite clock.

Then the observable L4 combined BDS dual band is expressed as (1.11):

$$L_4 = B_1 - B_2 = I_1 - I_2 + \lambda_1 \beta_1(t) - \lambda_2 \beta_2(t) \tag{1.11}$$

The L4 method eliminates all the geometric information, tropospheric delay and the non-frequency related errors in the observation of BDS measurement, leaves only the error cause by ionospheric delay and multipath. According to article [18], ionospheric delay is inversely proportional to the squared carrier frequency, then Eq. (1.11) turns to (1.12)

$$L_4 = B_1 - B_2 = \frac{40.7}{f_1^2} TEC - \frac{40.7}{f_2^2} TEC + \lambda_1 \beta_1(t) - \lambda_2 \beta_2(t) \tag{1.12}$$

In which TEC is the total electron density, f_1, f_2 are the carrier frequency of BDS B1, B2 respectively. The TEC related item can be eliminated by a high-pass filter. The time series of multipath error can be denoted from Eqs. (1.2) and (1.8):

$$\widetilde{L_4} \approx \lambda_1 (\frac{A_m}{A_d})_1 sin(4\pi H \lambda_1^{-1} \sin\theta(t)) - \lambda_2 (\frac{A_m}{A_d})_2 sin(4\pi H \lambda_2^{-1} \sin\theta(t)) \quad (1.13)$$

Comparing Eqs. (1.13) and (1.4), the similar form of equations shows that SNR method and carrier phase method are basing on similar idea. For $\widetilde{L_4}$ is a linear combination of B1 and B2 carrier phase, so there are two peaks in its power spectrum, as shown in Fig. 2(b). The power spectrum density of B2 is higher than B1, which is more convenient to for data processing, so in this paper we use B2 to build soil moisture inversion model.

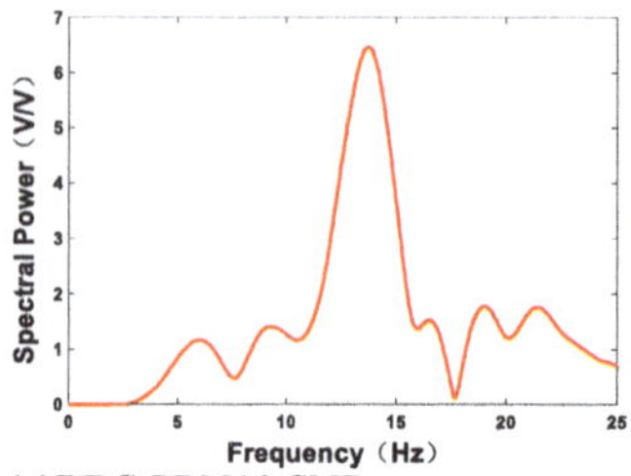

(a)BDS PRN10 SNR power spectrum

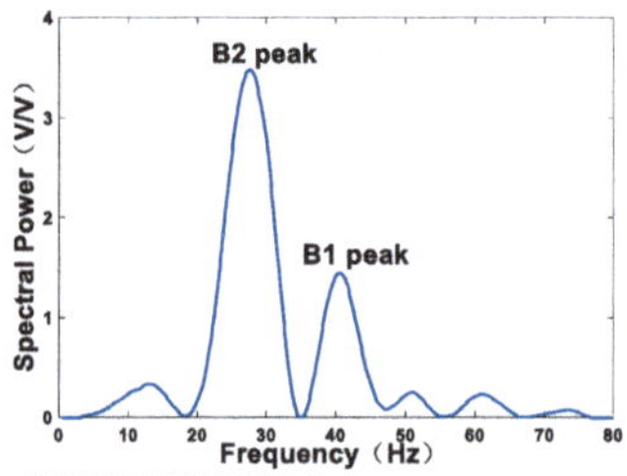

(b)BDS PRN10 L4 power spectrum

Fig. 2. Power spectrum analysis diagram

3 Experiment and Result Analysis

3.1 Field Experiment Campaign

The experiment field which locate in Chinese National Engineering Research Centre of Vegetables, Tongzhou Beiing (39°41′50.33″N, 116°41′23.19″E) is shown in Fig. 3. The experiment campaign was carried out from Sep. 10th to Nov. 9th 2018, 61 d in total. There were only three precipitations occurred during the whole experiment campaign. And there was no vegetation cover so that the experiment filed could be considered as bare soil. The direct and reflected BDS signal was collected simultaneously by Huace N72 receiver and Antcom G5Ant-52AT1 antenna. The antenna is pointing zenith with 3 m height. Three sensors are buried near the antenna for gathering soil moisture 1 t/min, two sensors are buried horizontally with 2 cm and 4 cm depth respectively and the other on is buried vertically to measuring the mean moisture of 0–6 cm depth's soil.

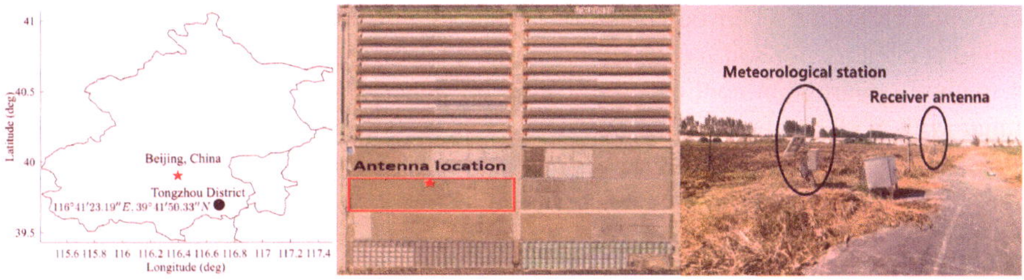

Fig. 3. Schematic diagram of experimental site

3.2 Data Processing

The data processing flow is shown in Fig. 4.

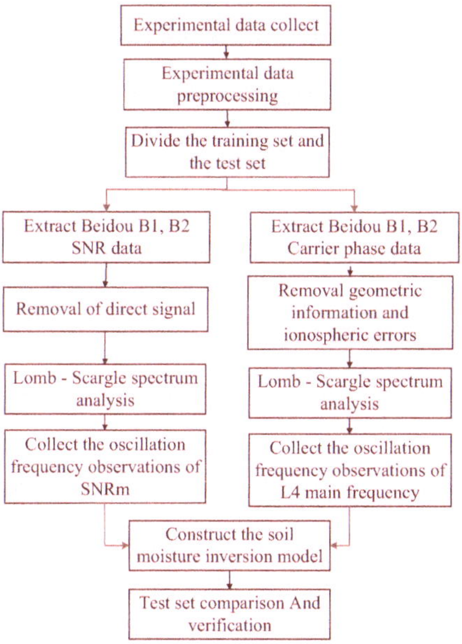

Fig. 4. Data processing flow

3.3 Result Analysis

Regarding the constraint of experiment field, only data of satellite with 90°–270° azimuth angle and 5°–30° elevation angle was processed. Then PRN9, PRN10 and PRN13 meet the criteria. The data set were divided randomly into training set and test set with a ratio of 2:1. The inversion result on test set for both models derived by univariate linear regression are shown in Fig. 5. According to Fig. 5, results of both the SNR method and L4 method are agree with the trend of soil moisture ground truth. The regression coefficient is above 0.6 between result of SNR model and ground truth, and meanwhile above 0.56 between result of L4 model and ground truth. It proves that L4 method is effective for soil moisture inversion. Comparing with SNR method, result of L4 method's regression coefficient is a little bit lower than SNR method on B1, and similar with SNR method on B2 (shown in Fig. 5(g)–(i)). The reason of L4 method doesn't achieve as high performance as it does in snow depth and sea level applications is because the reflecting plane in soil moisture application is much more complicated. The soil roughness and vegetation cover create a lot disturbance for L4 method.

- For making comparison between SNR method and L4 method, the regression coefficient(R) and root mean square error (RMSE) is shown in Table 1. The comparisons show both the two methods can effectively inverting soil moisture.

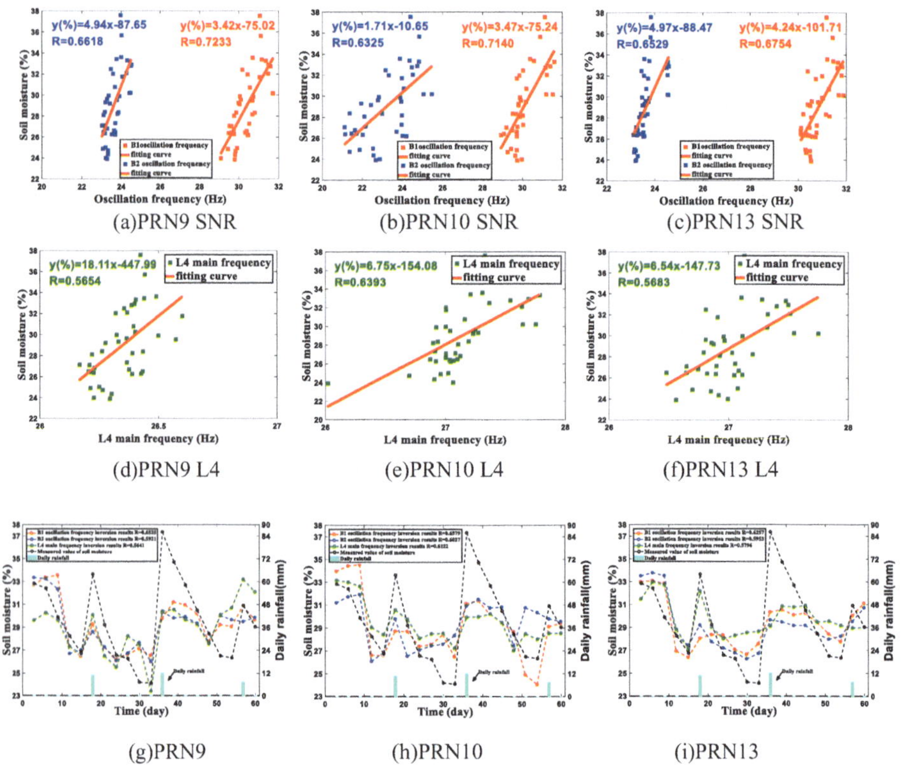

Fig. 5. (a) PRN9 SNR training set inversion model; (b) PRN10 SNR training set inversion model; (c) PRN13 SNR training set inversion model; (d) PRN9 L4 training set inversion model; (e) PRN10 L4 training set inversion model; (f) PRN13 L4 training set inversion model; (g) PRN9 test set inversion results; (h) PRN10 test set inversion results; (i) PRN13 test set inversion results;

Table 1. Comparison of soil moisture inversion results

Index	PRN9 B1	PRN9 B2	PRN9 L4	PRN10 B1	PRN10 B2	PRN10 L4	PRN13 B1	PRN13 B2	PRN13 L4
R	0.6525	0.5921	0.5641	0.6579	0.6027	0.6122	0.6257	0.5953	0.5796
RMSE (%)	2.146	2.265	2.269	1.894	2.834	1.722	1.574	2.177	2.051

4 Conclusion

This article introduced L4 method in soil moisture measurement, and make inter-comparison to traditional SNR method. Result of field experiment campaign shows:

1. For low elevation angle's data of BDS satellite, both the two methods achieved high performance, the regression coefficient are above 0.5, RMSE are lower than 3%. It proves L4 method can be used for *in-situ* soil moisture measurement.
2. For the reflecting plane is more complex, L4 method achieve lower performance in soil moisture measurement than in snow depth and sea level applications.

Overall, a lot of GNSS-R receivers haven't had ability of recording SNR data because GSI didn't add SNR data to GEONET until 2009. So L4 method is an effective alternative approach to SNR method. And it can enrich the means of GNSS-IR technology. Improving the accuracy of L4 method is a key point in future work, meanwhile with the maturity of BDS, multi-satellite multi-frequency data fusion and intercomparision by using both GPS and BDS is an important study point in future work.

Acknowledgments. The authors acknowledge the financial support of the National Natural Science Fund of China (Grant Nos. 31971781 and 41774028). The authors also would like to extend their sincere gratitude to Dr. Nazi Wang from Shandong University (Weihai) for her help in data processing.

References

1. Jackson, T.J., Schmugge, J., Engman, E.T.: Remote sensing applications to hydrology. Hydrol. Sci. J./J. Des Sci. Hydrol. **41**(4), 517–530 (1996)
2. Yang, Y., Mao, Y., Sun, B.: Basic performance and future developments of BeiDou global navigation satellite system. Satell. Navig. **1**, 1–8 (2020)
3. Rodriguez-Alvarez, N., Bosch-Lluis, X., Camps, A., et al.: Soil moisture retrieval using GNSS-R techniques: experimental results over a bare soil field. IEEE Trans. Geosci. Remote Sens. **47**(11), 3616–3624 (2009)
4. Larson, K.M., Small, E.E., Gutmann, E., et al.: Using GPS multipath to measure soil moisture fluctuations: initial results. GPS Solutions **12**(3), 173–177 (2008)
5. Santamaría-Gómez, A., Watson, C., Gravelle, M., et al.: Levelling co-located GNSS and tide gauge stations using GNSS reflectometry. J. Geodesy **89**(3), 241–258 (2015)
6. Clarizia, M.P., Ruf, C.S., Jales, P., et al.: Spaceborne GNSS-R minimum variance wind speed estimator. IEEE Trans. Geosci. Remote Sens. **52**(11), 6829–6843 (2014)
7. Jin, S.G., Najibi, N.: Sensing snow height and surface temperature variations in Greenland from GPS reflected signals. Adv. Space Res. **53**(11), 1623–1633 (2014)

8. Larson, K.M., Braun, J.J., Small, E.E., et al.: GPS multipath and its relation to near-surface soil moisture content. IEEE J. Sel. Top. Appl. Earth Obs. Remote Sens. **3**(1), 91–99 (2010)
9. Chew, C.C., Small, E.E., Larson, K.M., et al.: Vegetation sensing using GPS-interferometric reflectometry: theoretical effects of canopy parameters on signal-to-noise ratio data. IEEE Trans. Geosci. Remote Sens. **53**(5), 2755–2764 (2015)
10. Roussel, N., Frappart, F., Ramillien, G., et al.: Detection of soil moisture variations using GPS and GLONASS SNR data for elevation angles ranging from 2° to 70°. IEEE J. Sel. Top. Appl. Earth Obs. Remote Sens. **9**, 1–14 (2016)
11. Ozeki, M., Heki, K.: GPS snow depth meter with geometry-free linear combinations of carrier phases. J. Geodesy **86**(3), 209–219 (2012)
12. Kegen, Y., Ban, W., Xiaohong, Z., et al.: Snow depth estimation based on multipath phase combination of GPS triple-frequency signals. IEEE Trans. Geosci. Remote Sens. **53**(9), 5100–5109 (2015)
13. Wang, N., Xu, T., Gao, F., et al.: Sea level estimation based on GNSS dual-frequency carrier phase linear combinations and SNR. Remote Sens. **10**(3), 470 (2018)
14. Arroyo, A.A., Camps, A., Aguasca, A., et al.: Dual-polarization GNSS-R interference pattern technique for soil moisture mapping. IEEE J. Sel. Top. Appl. Earth Obs. Remote Sens. **7**(5), 1533–1544 (2014)
15. Yu, K., Li, Y., Chang, X.: Snow depth estimation based on combination of pseudorange and carrier phase of GNSS dual-frequency signals. IEEE Trans. Geosci. Remote Sens. **57**, 1–12 (2018)
16. Bilich, A., Larson, K.M.: Correction published 29 March 2008: mapping the GPS multipath environment using the signal-to-noise ratio (SNR). Radio Sci. **42**(6), 1–16 (2016)
17. Elósegui, P., Davis, J.L., Jaldehag, R.T.K., et al.: Geodesy using the global positioning system: the effects of signal scattering on estimates of site position. J. Geophys. Res. **100** (B6), 9921 (1995)
18. Kedar, S., Hajj, G.A., Wilson, B.D., et al.: The effect of the second order GPS ionospheric correction on receiver positions. Geophys. Res. Lett. **30**, 341–345 (2003)

An Improved Method of ZTD Model in Yunnan Province Based on GPT2w Model

Zheng Du, Qingzhi Zhao(✉), and Wanqiang Yao

College of Geomatics, Xi'an University of Science and Technology, Xi'an 710054, China
duzhengsss@163.com, zhaoqingzhia@163.com, sxywq@163.com

Abstract. The GPT2w model is one of the currently disclosed high-precision empirical models of tropospheric key parameters. The model can provide a variety of meteorological parameters, and the ZTD can be calculated using the model parameters. However, when using the GPT2w model to calculate ZTD directly, its accuracy is reduced, and there is a specific regional. In order to obtain higher precision ZTD parameters, this paper proposes a ZTD model that improved GPT2w(IGPT2w). The model uses the ZTD calculated by the GPT2w as the initial value of the IGPT2w and calculates the ZTD residual for the ZTD actual value and the GPT2w. The difference was fitted to obtain a ZTD residual fitting model. Finally, the IGPT2w model consists of the initial value calculated by the GPT2w and the residual estimated by ZTD residual model. Taking Yunnan, China, as an example, this paper selects 27 GNSS stations in Yunnan from 2015–2017 to establish the IGPT2w model and verify its accuracy. Firstly, MSSA and LS methods are used to analyze the characteristics of the ZTD residual and determine the periodic signals and establish a ZTD residual model considering annual, semiannual, and seasonal period, and then establish the IGPT2w in combination with the GPT2w. The IGPT2w model was established based on GNSS data from 2015–2016 and applied to ZTD estimation in 2017; The result shows that the IGPT2w estimates that the external RMS and MAE (2.4/1.8 cm) of the ZTD are lower than the GPT2w (2.8/2.3 cm). Compared with the GPT2w, the average accuracy of ZTD was improved by 13.4%. The above results show that the IGPT2w model is superior to the GPT2w model in the Yunnan province of China.

Keywords: GPT2w · Zenith Tropospheric Delay · Residual model · IGPT2w

1 Introduction

Tropospheric delay is the main error sources in the satellite navigation and positioning and an important research object in GNSS meteorology. Due to the refraction of radio signals transmitted in the troposphere, a tropospheric delay of 2–20 m is generated [1, 2]. Tropospheric delay is calculated by empirical models. Zenith Tropospheric Delay (ZTD) including Zenith Hydrostatic Delay (ZHD) and Zenith Wet Delay (ZWD). Moreover, ZWD is greatly affected by the spatiotemporal variation of atmospheric water vapour, which cannot be accurately simulated.

J. Sun et al. (Eds.): CSNC 2020, LNEE 650, pp. 65–75, 2020.
https://doi.org/10.1007/978-981-15-3707-3_7

There are three types of troposphere real-time empirical models [3]: (1) The troposphere key parameter empirical model, which estimates troposphere delay without any additional information, such as IGGtrop model [4, 5], TropGrid model [6, 7], GPT series model [8–10], ITG model [11], etc. (2) Tropospheric delay model based on measured meteorological parameters, this type of model usually combines ZHD and ZWD to express ZTD. The commonly used models include the traditional Hopfield model [12], Saastamoinen model [13], etc. In addition, there are Ifadis model [14] and AN model [15] that calculate ZWD separately. (3) Tropospheric delay model based on GNSS, which directly uses GNSS observation data to calculate ZTD and model it. Moreover, the GPT2w model is a high precision and widely used empirical model of troposphere key parameters [16].

At present, the accuracy of troposphere real-time empirical model calculation of ZTD has been significantly improved, but further improving the acquisition accuracy of ZTD is still the research focus of satellite navigation, positioning, and GNSS meteorology, especially for the region. The Yunnan region is a mountain plateau with complicated terrain and few GNSS stations. The water vapour in the Indian Ocean mainly flows into China through this region [17]. The establishment of a high-precision tropospheric real-time empirical model in this region is of vital significance to study the distribution and evolution of water vapour in China. Therefore, this paper proposes an improved GPT2w (IGPT2w) model to estimate ZTD in the Yunnan region. The IGPT2w model added ZTD residual model, and combining the GPT2w ZTD and the ZTD residuals estimated by the residual model to calculate IGPT2w ZTD. Numerical results show that the IGPT2w model is superior to the GPT2w model in Yunnan, China.

2 GPT2w Model and Acquisition of ZTD

As an empirical model of troposphere key parameters, the GPT2w model can provide 9 parameters such as pressure, temperature, and water vapor pressure by inputting ellipsoidal coordinates (latitude, longitude, and height) of GNSS stations and corresponding dates. The establishment of this model takes into account the mean value, annual and semi-annual period of parameters, and provides grid-based parameter information files with a resolution of 1° and 5°. The GPT2w model used in this paper is shown in http://ggosatm.hg.tuwien.ac.at/DELAY/SOURCE/GPT2w/. The main parameters obtained in this paper include: pressure, weighted mean temperature, water vapor pressure, and water vapor decrease factor.

The calculation of GPT2w ZTD depends on the tropospheric delay model based on measured meteorological parameters. In this paper, the output parameters of GPT2w are used to calculate ZHD and ZWD, and then the sum is added to GPT2w ZTD.

3 Improved GPT2w Model for ZTD Estimation

3.1 Experiment Introduction

The GNSS ZTD data of Yunnan were obtained from the Crustal Movement Observation Network of China (CMONOC). The observation data of GNSS stations are processed by GAMIT/GLOBK (Ver. 10.4) software to obtain hourly ZTD from 2015–2017. The Geographical distribution of 27 GNSS stations in Yunnan is shown in Fig. 1. This study compares the accuracy of GNSS ZTD with co-located sounding data. The results show that the root mean square (RMS) of GNSS ZTD is between 0–3 cm, and the mean RMS and Bias of GNSS data are about 1.9 cm and 0.1 cm respectively.

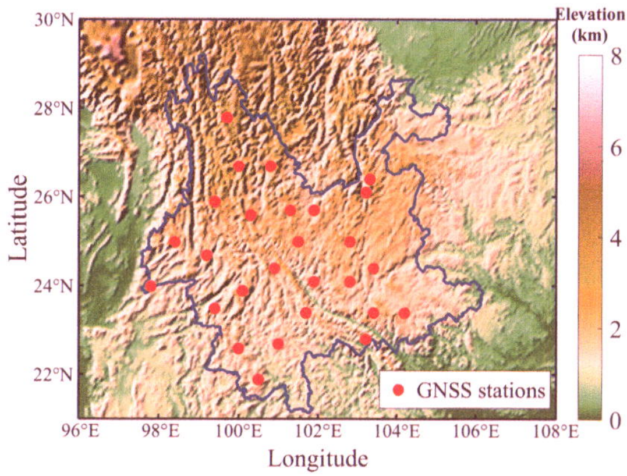

Fig. 1. Geographical distribution of GNSS stations in Yunnan

This paper first calculates the residual between GPT2w ZTD and GNSS ZTD. Multi-channel Singular Spectrum Analysis (MSSA) and Lomb-Scargle periodogram (LS) methods are used to detect the periodic signals of the residual. The residual is modeled, and this model is introduced into GPT2w to get the IGPT2w. During the experiment, the ZTD data from 2015–2016 was used to establish the IGPT2w model, and the 2017 GNSS ZTD was used to test the IGPT2w model. RMS and Mean Absolute Error (MAE) were selected for accuracy evaluation.

3.2 Time Series Analysis and Establishment of ZTD Residuals Model

3.2.1 Time Series Analysis of ZTD Residuals

In this paper, we analyze the ZTD residual and find that there are apparent periodic signals with monthly time resolution. In order to verify that there are still periodic signals in the ZTD residual with higher time resolution, this article introduces the MSSA method to decompose the hourly ZTD residual. This method is more effective than the singular spectrum analysis (SSA) in determining the periodic variation of time series [18]. Figure 2 shows the first 10 reconstructed components (RC) signals after the ZTD residual sequence of YNHZ station in Yunnan is decomposed by the MSSA method. The MSSA window selected in the experiment is 350. It can be seen from the figure that the RC2-3, RC4-5, RC6-7, and RC8-9 signals have visible periodic characteristics, and they are shown as annual, semiannual, seasonal (120), and seasonal (90) period, respectively. However, this method cannot determine the value of the periodic signal.

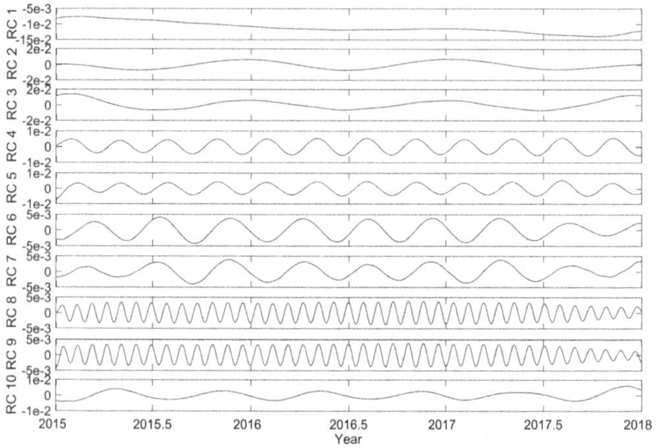

Fig. 2. RC (mm) corresponding to the 10 largest eigenvalues when the MSSA method applied to the time series of ZTD residual at YNHZ throughout 2015–2017

The LS method can handle time series with certain intermittent or uneven intervals [19], so this method is used to determine the periodic signal value of the hourly ZTD residual of YNHZ, YNMH, and YNZD stations in Yunnan. Figure 3(a, c), (d, f), and (g, i) show the power spectrum changes of the three stations using the LS method during 2015–2017, respectively. Figure 3(b, e, h) and (c, f, i) is the periodic changes of the first and second power spectrum values composed of the signal period type, respectively. It can be observed that the three stations have an annual period, YNHZ and YNMH have seasonal (120) period, while YNHZ and YNZD have seasonal (90) period. The difference is that only YNHZ exist semi-annual period. The experimental results further confirmed the existence of the periodic signal in the hourly residual of ZTD.

In this paper, the LS method is used to analyze the ZTD residual of 27 GNSS stations in Yunnan, China. The statistical results show that the ZTD residual of GNSS stations at different locations have different periodic signals. Table 1 shows the types of ZTD residual periods and the number of corresponding GNSS sites, where A, B, C, and D represent annual, semi-annual, seasonal period (120), and seasonal period (90), respectively. It can be seen from the table that the sequence of ZTD residuals of most GNSS stations has at least two types of periodic signals, and more than 50% of GNSS stations have three types of periodic signals. Figure 4 shows the distribution of the first four GNSS sites with the most period. The results show that the location distribution of GNSS sites of the ZTD residual period type has specific regional characteristics.

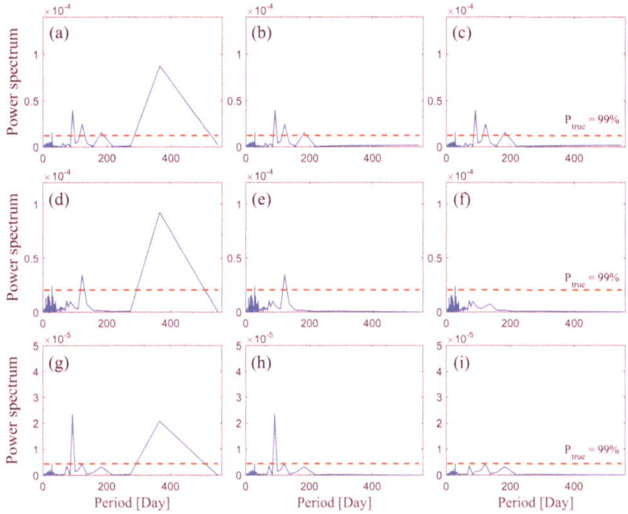

Fig. 3. Periodic variations for YNHZ, YNMH, and YNZD stations obtained using the LS method

Table 1. Number of GNSS stations and the corresponding periods

Type	Number
A + C + D	10
A + B + C + D	7
A + C	7
A + B + D	2
A + D	1

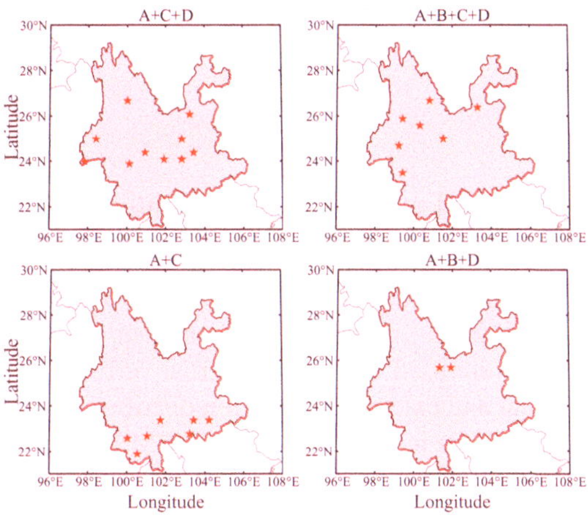

Fig. 4. First four distributions of GNSS stations with the maximum number of period

3.2.2 Establishment of ZTD Residual Model

The periodic signal in the ZTD residual sequence can be fitted by establishing the corresponding periodic model. In this paper, the periodic model of ZTD residual is determined as the combination of mean value and four periodic terms, and its expression is as follows:

$$\Delta ZTD = A_0 + A_1\cos\left(\frac{doy}{365.25}2\pi - \varphi_1\right) + A_2\cos\left(\frac{doy}{365.25}4\pi - \varphi_2\right) + \ldots \\ A_3\cos\left(\frac{doy}{365.25}6\pi - \varphi_3\right) + \ldots A_4\cos\left(\frac{doy}{365.25}8\pi - \varphi_4\right) \quad (1)$$

Where ΔZTD is ZTD residual (hourly resolution); A_0 is the mean value of ZTD residual; A_1–A_4 represents the annual, semi-annual, and seasonal amplitudes (120, 90), respectively.

The least-square method is used to estimate these 9 unknown parameters to establish the ZTD residual model. Figure 5 shows the ZTD residual model and the ZTD time series derived from GNSS, GPT2w, and IGPT2w. Figure 5(a) shows the residual fitting of the ZTD residual model to the monthly time resolution. Figure 5(b) shows that IGPT2w ZTD has the right consistency compared with GNSS ZTD.

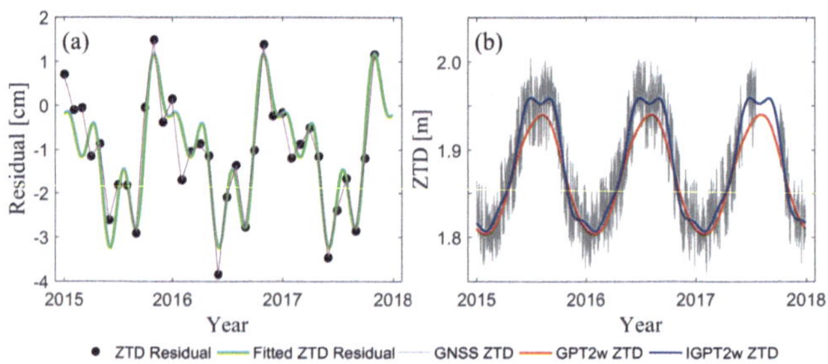

Fig. 5. ZTD residual model and the ZTD time series derived from GNSS, GPT2w, and IGPT2w

4 Model Validation

4.1 IGPT2w Model Internal and External Verification

In order to study the applicability of the IGPT2w model in Yunnan, China, this paper verifies the accuracy of the IGPT2w model. In the experiment, GNSS ZTD and GPT2w ZTD from 2015–2016 were used to establish the IGPT2w model, and GNSS ZTD data in 2017 were used to test the external verification of IGPT2w model. Figure 6 shows the internal and external verification distribution of the IGPT2w model. Figure 6(a) and (b) represent the internal and external RMS distribution of the IGPT2w model, and Fig. 6(c) and (d) represent the internal and external MAE distribution of IGPT2w model.

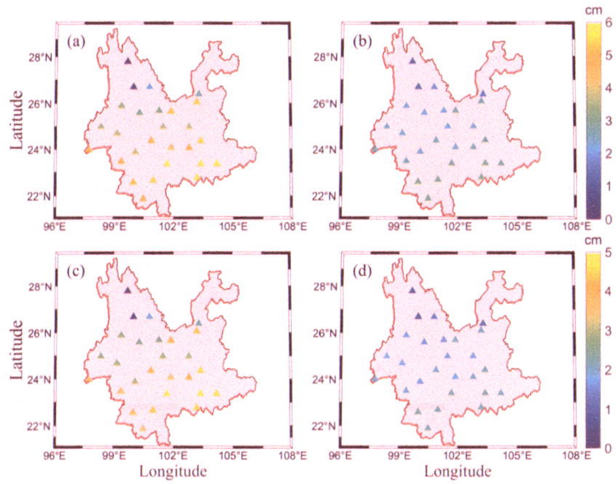

Fig. 6. Internal and external validations of the IGPT2w model in Yunnan

It can be seen in Fig. 6 that the RMS and MAE of the IGPT2w model show a downward trend from the south to the north of the Yunnan, corresponding to the distribution of atmospheric water vapor in this region. Combining the statistical results of the internal and external accuracy of the GPT2w and IGPT2w models in Table 2, we know that: the RMS (about 2.4 cm) of IGPT2w ZTD is lower than that of GPT2w ZTD (about 2.8 cm), and IGPT2w model has good prediction accuracy when GNSS ZTD is taken as a reference.

Table 2. Statistical result of internal/external accuracies of GPT2w and IGPT2w models in Yunnan when compared with the GNSS-derived ZTD (unit: cm)

Model	Type			
	Internal		External	
	RMS	MAE	RMS	MAE
GPT2w	2.81	2.28	2.77	2.27
IGPT2w	2.45	1.92	2.42	1.83

4.2 Improvement Rate of ZTD Estimation by IGPT2w

In addition, the Improvement Rate (IR) was introduced to evaluate the performance of the IGPT2w model. After the RMS of GPT2w-GNSS and IGPT2w-GNSS are calculated respectively, the IR of the IGPT2w model for ZTD estimation compared with GPT2w can be obtained (Fig. 7); the accuracy of ZTD estimates of 27 GNSS stations is improved in different degrees. The statistical results show that the mean IR of 27 GNSS stations in Yunnan is 13.4%, which shows that the accuracy of this model is better than that of the GPT2w model. The experimental results show that the IR of IGPT2w has regional characteristics, and the IR in the north of Yunnan is higher than that in the south.

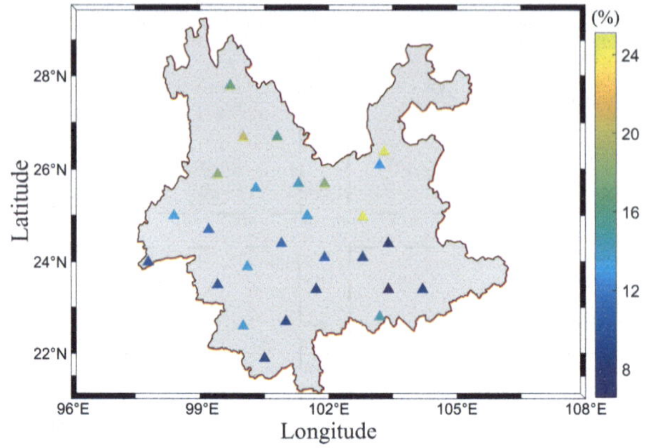

Fig. 7. Improvement rate of IGPT2w model for ZTD estimation compared with GPT2w model in Yunnan

In addition, the RMS frequency distributions of GPT2w ZTD and IGPT2w ZTD of 27 GNSS stations in Yunnan are calculated in this paper (see Fig. 8). The results show that GNSS sites with an RMS of less than 2.8 cm in the IGPT2w model account for 77.7% of all stations, while GNSS sites with an RMS of less than 2.8 cm in the GPT2w model account for only 44.4% of all stations. Moreover, the RMS of IGPT2w model in all stations in the Yunnan is less than 3 cm, which shows that the accuracy of the ZTD estimation of this model is higher than that of the GPT2w model.

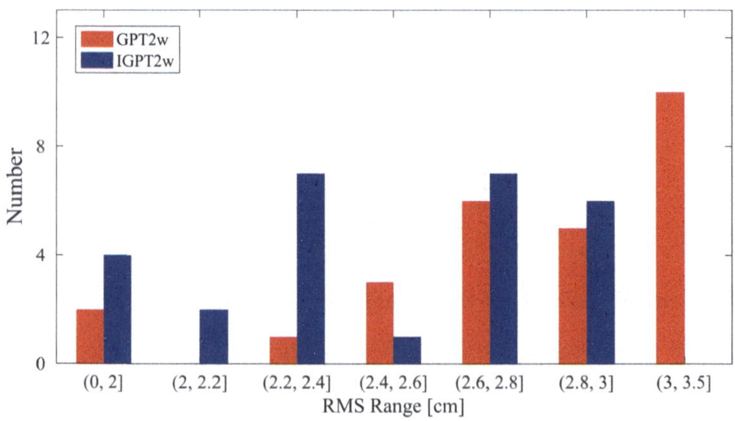

Fig. 8. Frequency distributions of RMS derived from GPT2w and IGPT2w in Yunnan

4.3 The 2D Distribution of IGPT2w ZTD in Different Seasons

In order to analyze the accuracy of the IGPT2w model in different seasons in Yunnan, this paper uses the Inverse Distance Weighted (IDW) method to interpolate data from

27 stations in the Yunnan. Figure 9 shows the ZTD residual between GNSS-GPT2w (Fig. 9(a1-4)) and GNSS-IGPT2w (Fig. 9(b1-4)) in different seasons and the 2D distribution of the IR of ZTD estimation between IGPT2w and GPT2w. This figure shows that the 2D distribution of ZTD residuals between GNSS-IGPT2w is generally smaller than that between GNSS-GPT2w, especially in summer and autumn. Moreover, the IR of IGPT2w (Fig. 9(c1-4)) is evident in summer and autumn, and the performance of IGPT2w in spring and winter is similar. The established IGPT2w is better than GPT2w in all seasons.

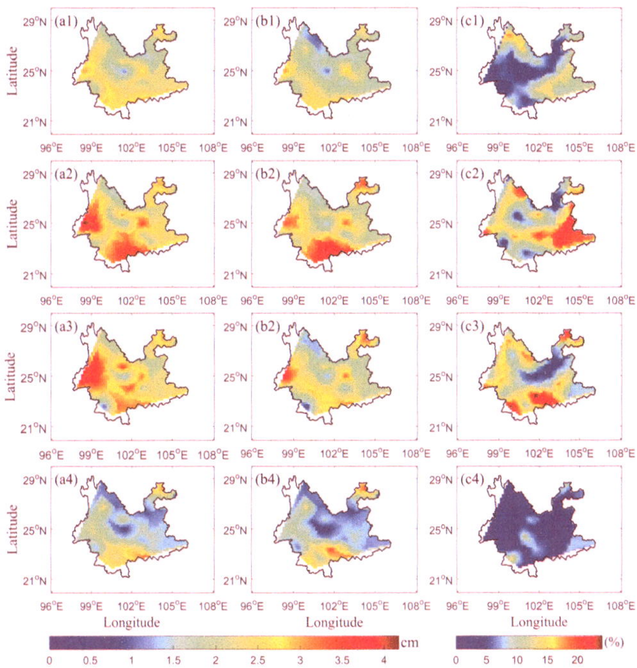

Fig. 9. 2D distributions of GPT2w & IGPT2w ZTD residuals and IR in Yunnan

5 Conclusion

This paper proposes an IGPT2w model that combines the GPT2w and the ZTD residual model to estimate the ZTD in Yunnan. The IGPT2w model was established using the GNSS data from 2015–2016 and applied to the ZTD estimation (hour resolution) in 2017. The model prediction results showed that the accuracy of the IGPT2w model was better than that of the GPT2w model, with an overall average accuracy improvement of 13.4%. In addition, since the ZTD values of summer and autumn are relatively large, the GPT2w does not describe the ZTD changes well, so the IGPT2w model mainly improves the ZTD estimation of summer and autumn, while the performance of the two models in spring and winter is similar. The ZTD residual periodic model complements the IGPT2w when the water vapour changes significantly in summer and autumn. The relationship between the regional characteristics of the ZTD residual and the atmospheric water vapour distribution needs further study.

Acknowledgments. Thanks for the relevant experimental data provided by Integrated Global Radiosonde Archive (IGRA) and CMONOC, and thanks for the GPT2w_1w model provided by GGOS. This research was supported by the National Natural Science Foundation of China (41904036), The Excellent Youth Science and Technology Fund Project of Xi'an University of Science and Technology (2018YQ3-12), the Key Research and Development Projects of Shanxi Province (201803D31224) and Beijing Key Laboratory of Urban Spatial Information Engineering open research project (2019210).

References

1. Jin, S.G., Luo, O., Ren, C.: Effects of physical correlations on long-distance GPS positioning and Zenith tropospheric delay estimates. Adv. Space Res. **46**(2), 190–195 (2010)
2. Kong, J., Yao, Y., Shan, L., et al.: The accuracy analysis of GPT2w at the antarctic area. Acta Geodaetica et Cartographica Sinica **47**(10), 1316–1325 (2018)
3. Yao, Y., Xu, X., Xu, C., et al.: Establishment of a real-time local tropospheric fusion model. Remote Sens. **11**(11), 1321 (2019)
4. Li, W., Yuan, Y.B., Ou, J.K., et al.: A new global zenith tropospheric delay model IGGtrop for GNSS applications. Chin. Sci. Bull. **57**(17), 2132–2139 (2012)
5. Li, W., Yuan, Y., Ou, J., et al.: New versions of the BDS/GNSS Zenith tropospheric delay model IGGtrop. J. Geodesy **89**(1), 73–80 (2015)
6. Krueger, E., Schueler, T,. Hein, G.W., et al.: Galileo tropospheric correction approaches developed within GSTB-V1. In: Proceedings of ENC-GNSS. p. 1619 (2004)
7. Schüler, T.: The TropGrid2 standard tropospheric correction model. GPS Solutions **18**(1), 123–131 (2014)
8. Böhm, J., Heinkelmann, R., Schuh, H.: Short note: a global model of pressure and temperature for geodetic applications. J. Geodesy **81**(10), 679–683 (2007)
9. Lagler, K., Schindelegger, M., Böhm, J., et al.: GPT2: empirical slant delay model for radio space geodetic techniques. Geophys. Res. Lett. **40**(6), 1069–1073 (2013)
10. Böhm, J., Möller, G., Schindelegger, M., et al.: Development of an improved empirical model for slant delays in the troposphere (GPT2w). GPS Solutions **19**(3), 433–441 (2015)
11. Yao, Y., Xu, C., Shi, J., et al.: ITG: a new global GNSS tropospheric correction model. Sci. Rep. **5**, 10273 (2015)
12. Hopfield, H.S.: Two-quartic tropospheric refractivity profile for correcting satellite data. J. Geophys. Res. **74**(18), 4487–4499 (1969)
13. Saastamoinen, J.: Atmospheric correction for the troposphere and stratosphere in radio ranging satellites. Use Artif. Satell. Geodesy **15**, 247–251 (1972)
14. Ifadis I. The atmospheric delay of radio waves: modelling the elevation dependence on a global scale. Licentiate Thesis, Technical Report, 38 (1986)
15. Askne, J., Nordius, H.: Estimation of tropospheric delay for microwaves from surface weather data. Radio Sci. **22**(3), 379–386 (1987)
16. Cheng, Z.M., Sheng, H.W., Shun, W.L.: Accuracy test and analysis for GPT2w model in China. Geomat. Inf. Sci Wuhan Univ. **44**(09), 1304–1311 (2019)

17. Xiangde, X., et al.: The relationship between water vapor transport features of Tibetan Plateau-monsoon "large triangle" affecting region and drought-flood abnormality of China. Acta Meteorologica Sinica **603**, 257–266 (2002)
18. Zhang, B., Liu, L., Khan, S.A., et al.: Transient variations in glacial mass near Upernavik Isstrøm (West Greenland) detected by the combined use of GPS and GRACE data. J. Geophys. Res. Solid Earth **122**(12), 10626–10642 (2017)
19. Zhao, Q., Yao, Y., Yao, W.: Studies of precipitable water vapour characteristics on a global scale. Int. J. Remote Sens. **40**(1), 72–88 (2019)

Inclusion of Side Signals on GNSS Water Vapor Tomography with a New Height Factor Model

Wenyuan Zhang[1,2](✉), Nan Ding[3], and Shubi Zhang[1,2](✉)

[1] MNR Key Laboratory of Land Environment and Disaster Monitoring, China University of Mining and Technology, Xuzhou 221116, China
zhangwy160617@163.com, zhangsbi@163.com
[2] School of Environment Science and Spatial Informatics, China University of Mining and Technology, Xuzhou 221116, China
[3] School of Geography, Geomatics and Planning, Jiangsu Normal University, Xuzhou 221116, China

Abstract. GNSS tomography has bloomed into an efficient tool for sensing the high spatiotemporal variations of tropospheric water vapor. Presently, GNSS signals passing from the top boundary are selected as effective rays to tomography system in the most studies. However, a number of side signals penetrating from the side of tomography area are eliminated, which reduces the utilization of GNSS rays and aggravates the morbidity of tomographic observation equations. In this paper, the integration of top signals and side ones for GNSS water vapor tomography system is explored and developed. The sectional slant wet delay (SWD) corresponding to part signals, regarded as the key for utilizing side rays, is accurately estimated by a new height factor model (HFM). In addition, dynamic top boundary of tomography area is analyzed and determined based on the same radiosonde data. Three experimental schemes are carried out using 31 days observation data from the Satellite Positioning Reference Station Network (SatRef) and Radiosonde in Hong Kong. The experimental results show that the average number of effective signals increased by 66.29% and the average utilization rate of GNSS signals is enhanced by 31.86% with side signals absorbed into the tomography system. Furthermore, with the proposed method, the statistics suggest that the mean RMSE is reduced from 1.59 g/m^3 (Scheme I) to 1.08 g/m^3 (Scheme III), and the accuracy is remarkably improved by 32.08%. On the other hand, compared to the present approach for modeling side rays, the improved model proposed in this paper has a better retrieval capability.

Keywords: GNSS water vapor tomography · Side signals · Height factor model (HFM) · Dynamic tropopause

1 Introduction

Tropospheric water vapor, characterized by dynamic and variable, plays a key role in global climate change and extreme meteorological disasters. With the development of Global Navigation Satellite System (GNSS), it has bloomed into an efficient atmospheric

J. Sun et al. (Eds.): CSNC 2020, LNEE 650, pp. 76–88, 2020.
https://doi.org/10.1007/978-981-15-3707-3_8

water vapor detection tool to study rainstorm events and climatic change [1]. Due to several significant advantages including observation data accessibility, all-weather availability, wide coverage, and high spatiotemporal resolution, GNSS tomography has been favored by many researchers and is developed rapidly in the past 20 years [2, 3].

In GNSS tomography model, due to the geometric constraints constrained by GNSS station networks and satellite constellations, the satellite signal is an inverted cone and is mostly concentrated in the middle and upper layer of the tomographic region. These structural geometries causes that a large number of voxels are not passed through by any ray, which results in ill-conditioned equations and a rank-deficient matrix when constructing tomographic observation equations [1]. To solve this problem, there are several approaches have been developed. It is a given that (1) horizontal and vertical constraints are added and optimized [2, 4]; (2) a priori information, which is initialized and updated by radiosonde, the European Centre for Medium-Range Weather Forecasts (ECMWF) and Atmospheric Infrared Sounder (AIRS), are employed in GNSS tomography [3, 5]; and (3) data in addition to GNSS observations (e.g., InSAR and GNSS-R) are used to add constraint information [6, 7]. Increasing the number of effective observation signals is also one of the useful methods to solve the problem of rank deficiencies. Observations from GPS, BDS, GALILEO and GLONASS have been used in the GNSS tomography model [8–10]. In addition, virtual GNSS stations and slant signals were constructed to raise the number of rays [11, 12]. These methods optimize the geometric distribution of tomographic signals and reduce the number of the blank voxel in the tomography model, which greatly improve the performance of tomography solutions.

However, restricted by the acquisition of the partial SWV content of inside signals (the part of the side ray inside the tomography area), the foregoing studies only select GNSS rays passing from the top boundary to structure the tomography system, which wastes a great deal of beneficial observation information. The Yao's method proposed in [5, 13], utilizes the top and side signals for GNSS tomography based on the factor model. However, the accuracy of the initial water vapor field is the key to computing high-precision SWV of inside signals, which was neglected in both works.

In this paper, the integration of top signals and side ones for GNSS water vapor tomography system is explored and developed. We study the quantitative relation between the partial ZWD and height, and establish the HFM based on 30 years radiosonde data, which can be used to calculate the high-precision SWV value of side signals. The proposed method was tested using 31 days observation data from the Satellite Positioning Reference Station Network (SatRef) and Radiosonde in Hong Kong. The tomographic solutions derived from proposed method is compared with those of the Yao's approach.

2 GNSS Tomography Method with Height Factor Model

In this section, a new improved GNSS tomography method based on HFM is introduced and described, which increases the utilization of GNSS rays and relieves the morbidity of tomographic observation equations. This section is structured as follows. Section 2.1 is devoted to the description of the principle for GNSS tomography.

Dynamic tropopause, the criterion for judging the availability of GNSS signals, is analyzed and determined based on 30 years radiosonde data in Sect. 2.2. Section 2.3 shows the construction process of HFM. Tomographic observation equations based on top signals and side ones as well as constraint equations are built in Sect. 2.4.

2.1 The Principle of GNSS Tomography

The tropospheric delay (ZTD) usually refers to the total delay in the zenith direction of the GNSS station, which includes the zenith hydrostatic delay (ZHD) and zenith wet delay (ZWD). The main methods of obtaining ZTD are double-difference method and precise point positioning (PPP) method. Saastamoinen model [14], empirical tropospheric delay model, is used to estimate ZHD values. Then, ZWD can be extracted from ZTD.

In general, *SWD* of GNSS signals, consisting of isotropic and anisotropic part, is calculated with the wet mapping function $mf_w(\varepsilon)$ and ZWD, as well as horizontal gradient function $mf_g(\varepsilon)$ and wet horizontal gradient *G*, as shown in Eq. (1) [2].

$$SWD = \underbrace{mf_w(\varepsilon) \cdot ZWD}_{isotropic} + \underbrace{mf_g(\varepsilon) \cdot (G_{NS} \cdot \cos(\alpha) + G_{EW} \cdot \sin(\alpha))}_{anisotropic} \tag{1}$$

The slant water vapor (SWV) is practically proportional to the SWD as follows [2].

$$SWV = SWD \cdot \Pi \tag{2}$$

where $\Pi = \frac{10^5}{R_V \cdot \left(\frac{k_3}{T_m} + k_2'\right)}$ denotes the conversion factor. k_2' and k_3 are the constant factors with values of 16.48 K/hPa and $3.75 \times 10^5\,\mathrm{K}^2/\mathrm{hPa}$ [2], and $R_V = 461.53\,\mathrm{J/(kg \cdot K)}$ represents the specific gas constants for water vapor. T_m is the weighted mean temperature.

In GNSS tomography model, the tomographic area is divided into a lot of voxels, as seen in Fig. 1. The water vapor density (WVD) of the voxels be represented by those at the voxel center, and an assumption that water vapor density is constant within a voxel is imposed on the tomographic model to facilitate the construction of tomographic equations. Accordingly, tomographic equations can be expressed by the following form [3].

$$SWV = \sum_{i=1}^{n} d_{ij} \cdot x_i \tag{3}$$

where d_{ij} denotes the distance traveled by the *j*th ray in the *i*th voxel, whose WVD is represented by x_i. *n* is the total number of voxels in tomographic model.

For the red rays puncturing from tropopause in Fig. 1(a), can be easily parameterized using Eq. (3). As far as the green side signals passing from the side face of tomographic area, however, it is the premise of equations that accurately estimate the water vapor content of the part side signals in the tomographic region, which will be detailed described in the followings.

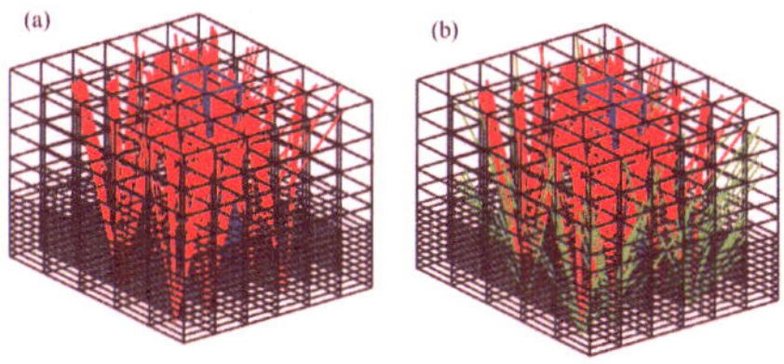

Fig. 1. 3-D distribution of GPS signals for traditional tomography model (a) and proposed one (b). Red lines represent signals passing from the top boundary of the tomographic area, while green lines denote rays penetrating from the side face of tomography region. The blue vertical columns represent the voxels where the radiosonde is located

2.2 Dynamicity of Tropopause

According to Fig. 1, signals are classified into two categories (top and side rays) in light of the positional relationship between themselves and the tropopause [15]. In consequence, it is essential to identify an appropriate tropopause to utilize more signals and to avoid unnecessary voxels.

Currently, the height of the tomography top boundary is empirically determined, e.g., 10 km [3, 6] and 15 km [2, 16]. In the work of [5], a principle that the WVD less than 0.2 g/m^3 at a certain height can be considered as the tropopause and the top boundary of the tomography model is proposed.

In this study, the idea of determining the optimal tomography top boundary based on the water vapor vertical distribution information derived from 30 years (1989–2018) of radiosonde data at Hong Kong King Park station (114.17°E, 22.31°E) is introduced. Accordingly, water vapor profiles of different months in Hong Kong are shown in Fig. 2. It was found that the distribution trend of water vapor in the atmosphere varies greatly in different seasons.

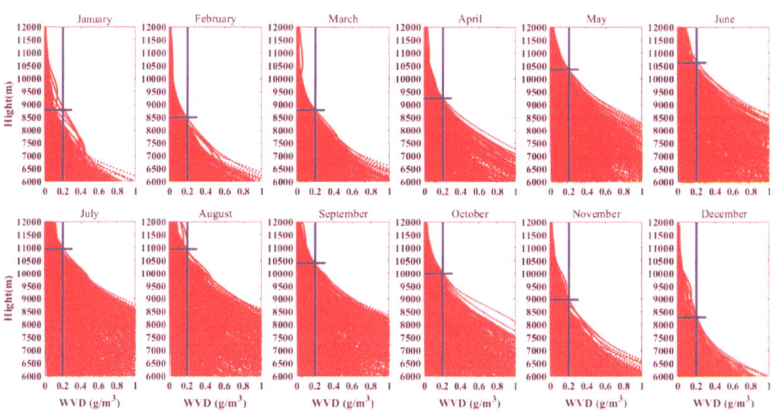

Fig. 2. The vertical distribution of water vapor in the atmosphere in different months. Blue vertical lines correspond to the principle that the WVD is 0.2 g/m^3 and blue horizontal ones describe the optimal tropopause during different months

It is evidently observed that tropopause in summer and autumn is higher than that in spring and winter. This phenomenon is related to the climate type of Hong Kong, in where extreme weather such as rainstorm and typhoon mainly occurs in summer and autumn, which results in the larger water vapor content and change scale in both seasons [17]. The mean height value of tropopause for four seasons (spring, summer, autumn, winter) are 10200 m, 10967 m, 10600 m and 9583 m, respectively. As a consequence, the tropopause changes in different seasons, even for the same region, which can be deduced from the research of this work. What's more, dynamic top boundary of tomography area, particularly in areas with complex weather conditions, should be determined from the local water vapor profiles in different months.

2.3 Height Factor Model for Side Signals

GNSS signals are reasonably separated into top rays and side ones based on the optimal tomography top boundary. However, a number of side signals penetrating from the side of tomography area were eliminated in traditional tomography model. Consequently, The HFM introduced in this paper will show the process of side signals modeling. The crucial point for modeling side rays is the estimation of the SWD of inside signals, which is lucubrated in HFM.

For the integral rays penetrating from the top boundary, the SWDs of these signals are calculated by Eq. (1). For instance, the SWD of signal *OM* shown in Fig. 3, obtained by the ray path mapping of ZWD of *OM*′, plus the gradient delay caused by the entire troposphere. Accordingly, the SWD of side signal (e.g., signal *OQ*), can be estimated by the mapping of ZWD of *OQ*′ and the gradient delay component.

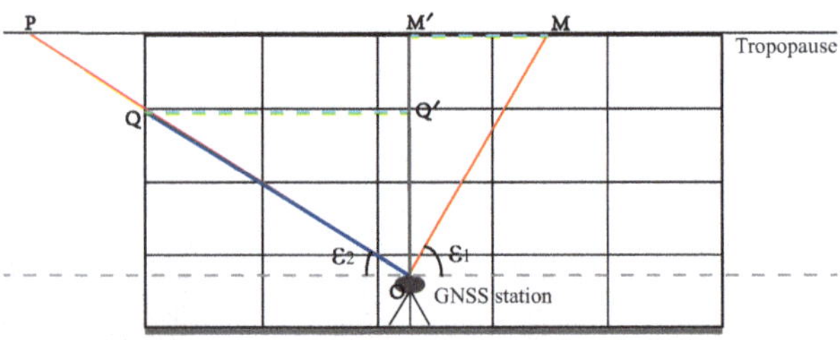

Fig. 3. Two dimensional schematic diagram of GNSS tomography to illustrate the estimation method of SWD for top signals and side ones

As mentioned in Sect. 2.1, the ZWD in the zenith direction of the GNSS station can be accurately estimated. Thus, our intention was to discuss the functional relationship between the sectional ZWD from surface to a different altitude and the total in the zenith direction, which may be of great benefit to calculate the SWV of side rays. Consequently, a height factor λ was defined to describe this relationship, as seen in Eq. (4).

$$\lambda(h) = \frac{ZWD_h}{ZWD_{zenith}} \tag{4}$$

where ZWD_h denotes the partial ZWD from surface to the height h, and ZWD_{zenith} is the total ZWD in the zenith direction of stations. Figure 4 shows the relationship between the height factor and the height based on the 30 years radiosonde data in January. Fitted images of other months, similar to Fig. 4, were not shown in this section.

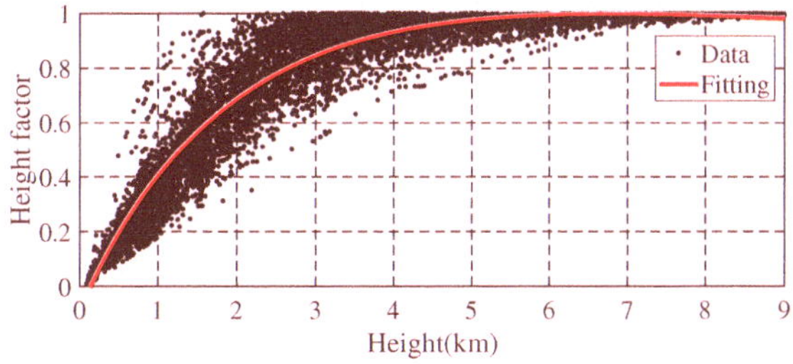

Fig. 4. Height factor change with the height of intersection between the satellite signal and the side face of the tomographic area

An Obvious finding is that the height factor has an approximate exponential relationship with height. Accordingly, the HFM was established as follows.

$$\lambda_{iso}(h) = a_1 \cdot e^{b_1 \cdot h} + a_2 \cdot e^{b_2 \cdot h} \tag{5}$$

where a_1, b_1, a_2 and b_2 represent the coefficients of HFM, which are determined by the least squares fitting. Table 1 shows the optimal coefficients as well as the Root Mean Squared Error (RMSE) and R-square of fitting results for each month.

Table 1. The optimal coefficients for HFM as well as the RMSE and R-square of fitting results for each month

Month	a_1	b_1	a_2	b_2	RMSE	R-square
January	1.093	−0.044	−0.184	−1.305	0.028	0.991
February	1.090	−0.042	−0.176	−1.351	0.020	0.995
March	1.122	−0.053	−0.220	−1.155	0.024	0.997
April	1.132	−0.060	−0.213	−1.133	0.035	0.985
May	1.117	−0.048	−0.206	−1.054	0.024	0.999
June	1.181	−0.070	−0.264	−0.911	0.024	0.993
July	1.089	−0.041	−0.153	−1.126	0.027	0.988
August	1.121	−0.051	−0.191	−1.019	0.026	0.989
September	1.088	−0.040	−0.158	−1.165	0.027	0.990
October	1.119	−0.053	−0.199	−1.130	0.033	0.986
November	1.089	−0.042	−0.172	−1.255	0.042	0.982
December	1.053	−0.026	−0.125	−1.516	0.021	0.994

2.4 Tomographic Observation Equations with the Fusion of Side Signals

The height factor λ corresponding to any side ray can be readily estimated by the HFM (5), and the SWV_{side} of it is expressed in Eq. (6). It should be noted that GNSS signals with an elevation angle below 15°, in spite of passing from the side face of tomography region, are eliminated due to their bending effect caused by atmosphere in this paper [18].

$$SWV_{side} = \Pi \cdot \lambda \cdot \left[m^{w}(e) \cdot ZWD + m_{g}^{w}(e) \cdot \left(G_{NS}^{w} \cdot \cos(\varphi) + G_{EW}^{w} \cdot \sin(\varphi) \right) \right] \tag{6}$$

The tomographic observation system with side rays can be written in the following equations.

$$\begin{bmatrix} SWV_{top} \\ SWV_{side} \end{bmatrix} = \begin{bmatrix} A_{top} \\ A_{side} \end{bmatrix} X \tag{7}$$

where SWV_{top} and SWV_{side} is the observation vector of the top signals and side ones, respectively. Both A_{top} and A_{side} represent the observation matrix including the traveled distance in each voxel. X denotes the unknown vector for the 3D tomographic model.

Although side signals with elevation angle greater than 15° are utilized, there are still some empty voxels in 3D tomographic model. In this paper, spatial constraints including horizontal and vertical constraints are introduced in the tomography system to relieve the ill-conditioned problem [2, 4, 19].

As a result, the final tomographic system equations can be written as follows.

$$\begin{pmatrix} SWV_{top} \\ SWV_{side} \\ 0 \\ 0 \end{pmatrix} = \begin{pmatrix} A_{top} \\ A_{side} \\ A_{H} \\ A_{V} \end{pmatrix} \cdot X \tag{8}$$

where A_H and A_V represent the constrained matrix of the horizontal constraints and vertical ones, respectively.

Due to the morbidity of the coefficient matrix, Algebraic Reconstruction Technique (ART), with advantages of simple iteration and fast convergence [20], is introduced to solve the equations in this work.

3 Tomographic Experiment

The GNSS data from 12 stations (red dot in Fig. 5(a)) provided by Hong Kong Satellite Positioning Reference Station Network (SatRef) and 3 IGS stations (BJFS station, GMSD station, and PIMO station) were processed using GAMIT/GLOBK (v. 10.6) in this work. The sampling rate of GNSS data is 30 s, while the time resolution of estimated ZWD and gradient delay is 5 min, which is sufficient to reflect the time variation characteristics of water vapor. The 30 years radiosonde observation data comes from the King's Park Meteorological Station (HKKP, blue dot in Fig. 5(a)). The integral area of tomography model spanned from 113.85°E to 114.39°E and 22.15°N to 22.50°N, the horizontal resolution is 0.09° in longitude and 0.07° in latitude. The experimental time is set in August 2017, that is, day of year (DOY) 213–243, when Hong Kong is in the summer, and there are more rainstorms.

As far as vertical resolution, it is nonuniform stratification that the tomography region is segmented into 15 non-uniform layers from 0 km to 11000 m is introduced in this paper [16, 19] (Fig. 5(b)). Additionally, it is noticeable that HKSC GNSS station and radiosonde are located in the same voxel columns (small blue rectangle in Fig. 5 (a)), GNSS observations from HKSC station would interfere with the validation of side signals. Therefore, all observations from HKSC station are excluded in tomography model.

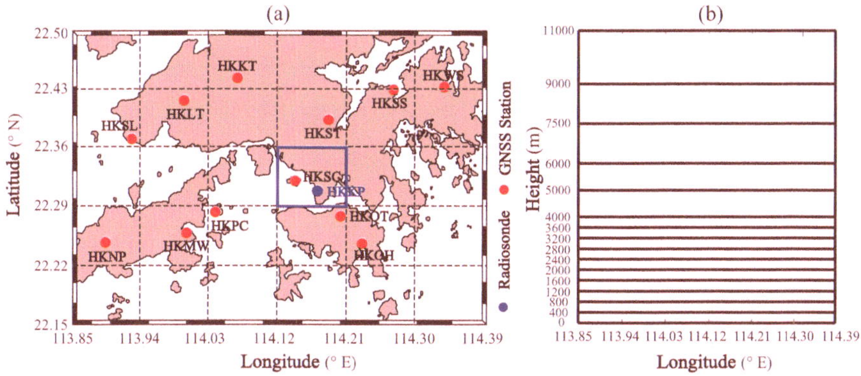

Fig. 5. (a) Plane map of the Hong Kong area obtained by Mercator projection with 12 GNSS reference stations (red dots) and radiosonde HKKP (blue dot) in Hong Kong, (b) Non-uniform vertical layer strategy for the tomographic model

In this work, three schemes are developed to assess the benefit of side rays for the tomographic solutions. The extant approach for using side rays introduced by Yao and Zhao [5], is compared to the proposed method. The three schemes are introduced and described in detail as follows.

1. *Scheme I*: Adopt the traditional tomography scheme which only consider the GNSS signals travelling the top boundary to construct the observation equations.
2. *Scheme II*: Use the Yao's method to estimate the SWV_{side} of side signals, and build the equations with both rays passing from the top and side boundary.
3. *Scheme III*: Employ the HFM to estimate the SWV_{side} of side rays, and structure the tomography system with top and side signals.

4 Results and Discussion

4.1 Contribution Analysis of the Side Signals

The average number of effective rays and the mean of signal utilization rate for traditional model and optimized one are compared in Fig. 6. With side signals absorbed into the tomography system, the average effective rays is increased by 66.29%, whereas the average utilization rate of GNSS signals is enhanced by 31.86% from 54.13% to 85.99%. In particular, the number of effective signals combining side ones is approximately twice as much as the quantity of top rays for some tomographic epoch, which enhances the stability of improved tomography model.

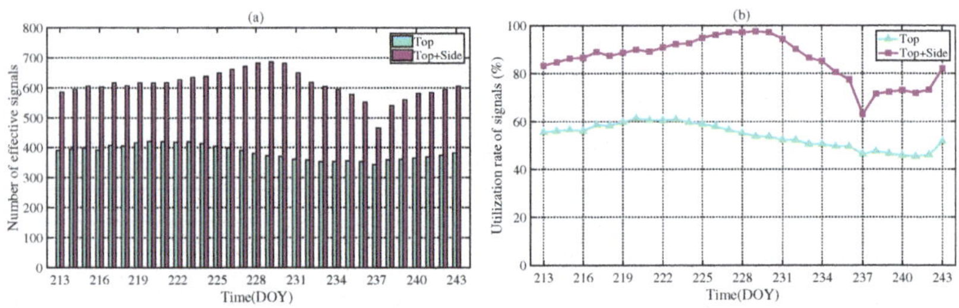

Fig. 6. The number of effective signals (a) and the utilization rate of signals (b) for only top rays and combing signals during the period of DOY 213-243, 2017

4.2 Comparison of Water Vapor Profiles

To evaluate and compare the water vapor vertical distribution derived from different methods, tomographic results, described in Fig. 7, illustrate the water vapor vertical profile at 0:00 UTC and 12:00 UTC on DOY 220 (sunny day) and DOY 234 (stormy day).

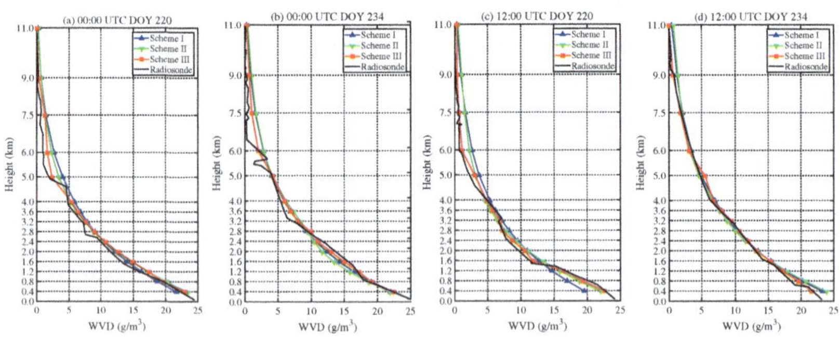

Fig. 7. Comparison of tomographic water vapor profiles for three schemes under different weather conditions using the radiosonde (black lines) as a reference

Similar water vapor profiles derived from three tomography models, also having agreement with radiosonde profiles, can be found in Fig. 7. However, with the input of side observations, both Yao's method and proposed one have a more coincidental reconstruction. In particular, the tomographic results retrieved by the model of this paper show a higher matching accuracy than those from Yao's model in both weather.

In addition, the RMSE between the three schemes and the radiosonde data are shown in Fig. 8, which refers to the 31-day period from DOY 213 to DOY 243, 2017, at 00:00 and 12:00 UTC daily. Besides, Table 2 lists the statistics including the maximum, minimum and mean values of Bias, STD and RMSE.

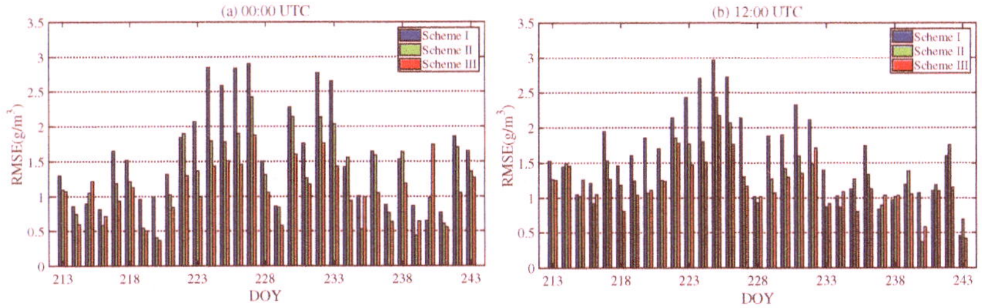

Fig. 8. Comparisons of the tomography results derived from different schemes during the 31-day period from DOY 213 to DOY 243, 2017, at UTC 00:00 (a) and UTC12:00 (b) daily

Table 2. Statistics (Bias, STD, RMSE) of the tomography results for three schemes. unit: g/m^3

Scheme	Bias			STD			RMSE		
	Max.	Min.	Mean	Max.	Min.	Mean	Max.	Min.	Mean
Scheme I	0.55	−1.26	−0.26	3.37	0.45	1.60	3.36	0.50	1.59
Scheme II	0.66	−1.33	−0.14	2.30	0.41	1.25	2.42	0.40	1.28
Scheme III	0.87	−0.55	0.16	2.00	0.29	1.07	1.98	0.28	1.08

It should be noted that Scheme II and Scheme III has a much smaller RMSE than Scheme I in a majority of time periods, and the accuracy of some epochs is increased by 50%. The mean accuracy improvement rate of Scheme II and Scheme III are 19.50% and 32.08%, respectively. Additionally, the improvement of Scheme III with respect to Scheme II has been of 0.20 and 0.18 g/m^3 in the mean RMSE and STD, respectively, which suggest that the proposed tomography model have a better reconstruction capabilities than Yao's model.

4.3 Comparison of Each Layer Tomographic Results

In the following, to quantitatively compare the accuracy of each layer tomographic results, the reference value of WVD, along the vertical columns, is interpolated using the radiosonde data, and the difference between tomographic results obtained from three schemes and radiosonde has been calculated and reported in Fig. 9.

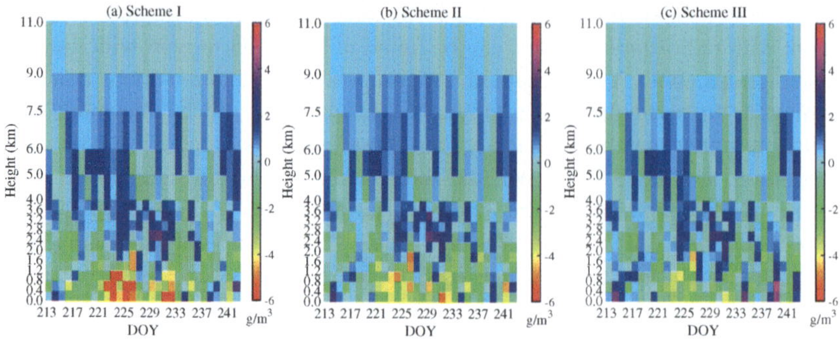

Fig. 9. The difference between tomographic results and radiosonde along the radiosonde vertical columns from DOY 213 to DOY 243, 2017

It can be observed that the differences obtained with side signals (Scheme II and Scheme III) are small than those provided by the tomography solution using top rays only (Scheme I), particularly in the lower layers from 0 to 2 km, which illustrates the benefit of side rays for tomography model. Furthermore, Scheme III shows a smaller differences, with a small improvement in the reconstruction accuracy compared to the Scheme II. A similar phenomenon has been also noticed from Fig. 10, showing the RMSE values of WVD derived from three tomography models on every horizontal layer.

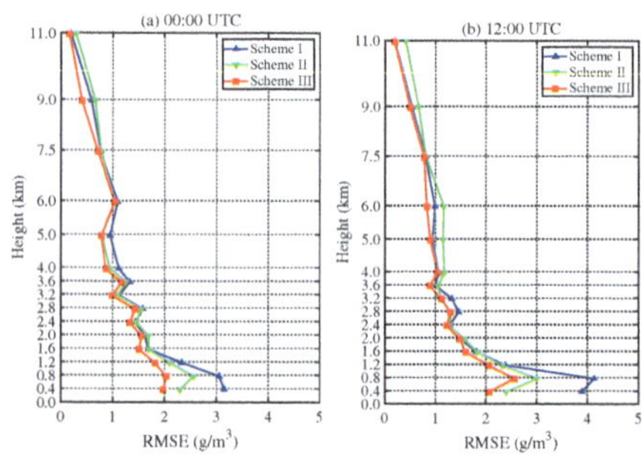

Fig. 10. RMSE at each layer of the tomographic results from three schemes at 0:00 UTC (a) and 12:00 UTC (b) during the tomographic period

It clearly stands out that the solutions of scheme II and scheme III, compared with those of scheme I, has a remarkably smaller RMSE from 0 to 2 km. In this altitude range, plenty of water vapor, probably accounting for 80% of the total, is concentrated and variable, resulting in extreme weather. Additionally, it is noticeable that the scheme III has an improvement compared to the scheme II in both 00:00 UTC and 12:00 UTC. Consequently, the optimized model proposed in this paper is helpful for analyzing the spatial distribution of water vapor near the surface.

5 Conclusion

A novel tropospheric tomography methodology based on the HFM, combining top and side signals, has been introduced and demonstrated in this paper. GNSS rays penetrating from the side of tomography area, usually eliminated in conventional tomography model, be used innovatively with the HFM. Dynamic top boundary of tomography area, as a critical factor for distinguishing top rays and side ones, is proposed and analyzed according to 30 years radiosonde data. It is evidently observed that top boundary in summer and autumn is higher than that in spring and winter. As far as the side signals, a height factor λ, representing the ratio between the partial content of zenith wet delay (ZWD) and the total ZWD, is first defined and modeled by the 30 years radiosonde data.

Three tomographic schemes based on the traditional method, Yao's approach and proposed one were implemented, and the GNSS observations during the period of August 2017 from Hong Kong SatRef were processed in all tests. Radiosonde data from HKKP were used as reference values to evaluate the tomographic performance of the three ways. With the inclusion of side signals, the average number of effective signals increased by 66.29% and the average utilization rate of GNSS signals is enhanced by 31.86%. The statistical analysis shows that the both tomography solutions derived from proposed method and Yao's one have higher accuracy than those of traditional way. The layered tomographic results obtained by the new approach suggested that there is a remarkable decline in the mean layer RMSE values respect to the conventional tomographic solutions. Furthermore, compared with the water vapor profiles derived from radiosonde data, the proposed approach showed better retrieval capabilities than that of the other two approaches, particularly in the lower layers from 0 to 2 km.

Acknowledgments. This study is supported by the National Natural Science Foundation of China (Grant Nos: 41774026 and 41904013). The authors acknowledge the support of the Survey and Mapping Office (SMO) of Lands Department, HongKong, for provision of the SatRef GNSS data and the ground-based meteorological data. The King's Park Observatory is also acknowledged for providing the high-precision radiosonde data. The GAMIT/GLOBK software is provided by the Department of Earth Atmospheric and Planetary Sciences, MIT. The authors would like to thank anonymous reviewers for the review and suggest of this paper.

References

1. Yao, Y., Zhang, S., Kong, J.: Research progress and prospect of GNSS space environment science. Acta Geodaetica Cartogr. Sin. **46**(10), 1408–1420 (2017)
2. Flores, A., Ruffini, G., Rius, A.: 4D tropospheric tomography using GPS slant wet delays. In: Annales Geophysicae (2000)
3. Benevides, P., Catalao, J., Nico, G., Miranda, P.M.A.: 4D wet refractivity estimation in the atmosphere using GNSS tomography initialized by radiosonde and AIRS measurements: results from a 1-week intensive campaign. GPS Solutions **22**, 91 (2018)
4. Song, S., Zhu, W., Ding, J., Peng, J.: 3D water-vapor tomography with Shanghai GPS network to improve forecasted moisture field. Chin. Sci. Bull. **51**, 607–614 (2006)

5. Yao, Y., Zhao, Q.: Maximally using GPS observation for water vapor tomography. IEEE Trans. Geosci. Remote Sens. **54**, 7185–7196 (2016)
6. Heublein, M., Alshawaf, F., Erdnüß, B., Zhu, X.X., Hinz, S.: Compressive sensing reconstruction of 3D wet refractivity based on GNSS and InSAR observations. J. Geodesy **93**, 197–217 (2019). https://doi.org/10.1007/s00190-018-1152-0
7. Jaberi Shafei, M., Mashhadi-Hossainali, M.: Application of the GNSS-R in tomographic sounding of the Earth atmosphere. Adv. Space Res. **62**, 71–83 (2018). https://doi.org/10.1016/j.asr.2018.04.003
8. Dong, Z., Jin, S.: 3-D water vapor tomography in Wuhan from GPS, BDS and GLONASS observations. Remote Sens. **10**, 62 (2018). https://doi.org/10.3390/rs10010062
9. Zhao, Q., Yao, Y., Cao, X., Zhou, F., Xia, P.: An optimal tropospheric tomography method based on the Multi-GNSS observations. Remote Sens. **10**, 324 (2018). https://doi.org/10.3390/rs10020234
10. Zhao, Q., Yao, Y., Cao, X., Yao, W.: Accuracy and reliability of tropospheric wet refractivity tomography with GPS, BDS, and GLONASS observations. Adv. Space Res. **63**, 2836–2847 (2019). https://doi.org/10.1016/j.asr.2018.01.021
11. Yao, Y., Zhao, Q., Luo, Y.: An approach of imposing virtual signals to sophisticate water vapor tomographic model. Geomat. Inf. Sci. Wuhan Univ. **11**, 161–167 (2017)
12. Adavi, Z., Mashhadi-Hossainali, M.: 4D tomographic reconstruction of the tropospheric wet refractivity using the concept of virtual reference station, case study: northwest of Iran. Meteorol. Atmos. Phys. **126**, 193–205 (2014)
13. Zhao, Q., Zhang, K., Yao, Y., Li, X.: A new troposphere tomography algorithm with a truncation factor model (TFM) for GNSS networks. GPS Solutions **23**, 64 (2019)
14. Saastamoinen, J.: Contributions to the theory of atmospheric refraction. J. Geodesy **105**, 279–298 (1972)
15. Chen, B.Y., Liu, Z.Z.: Voxel-optimized regional water vapor tomography and comparison with radiosonde and numerical weather model. J. Geodesy **88**, 691–703 (2014). https://doi.org/10.1007/s00190-014-0715-y
16. Perler, D., Geiger, A., Hurter, F.: 4D GPS water vapor tomography: new parameterized approaches. J. Geodesy **85**, 539–550 (2010)
17. Liu, B., Wang, Y., Lou, Z., et al.: The MODIS PWV correction based on CMONOC in Chinese mainland. Acta Geodaetica Cartogr. Sin. **48**(10), 1207–1215 (2019)
18. Möller, G., Landskron, D.: Atmospheric bending effects in GNSS tomography. Atmos. Meas. Tech. **12**, 23–34 (2019). https://doi.org/10.5194/amt-12-23-201
19. Xia, P., Cai, C., Dai, W., et al.: Three-dimensional water vapor tomography using ground-based GPS and COSMIC occultation observations. Geomat. Inf. Sci. Wuhan Univ. **38**(8), 892–896 (2013)
20. Xiaoying, W., Ziqiang, D., Enhong, Z., Fuyang, K.E., Yunchang, C., Lianchun, S.: Tropospheric wet refractivity tomography using multiplicative algebraic reconstruction technique. Adv. Space Res. **53**, 156–162 (2013). https://doi.org/10.1016/j.asr.2013.10.012

Real-Time Attitude Estimation for High-Speed UAV in High-Frequency Environmental Dithering Based on AMCF

Zebo Peng[1], Lianwu Guan[1(✉)], Xu Xu[1], Jianhui Zeng[1], Yanbin Gao[1], and Jie Yang[2]

[1] College of Automation, Harbin Engineering University, Harbin 150001, China
guanlianwu@hrbeu.edu.cn

[2] Chongqing Flagship Intelligent Technology Research Institute Co., Ltd., Chongqing 400039, People's Republic of China

Abstract. Attitude Measurement System (AMS) that comprised of the low-cost Micro-Electro-Mechanical System (MEMS) based Inertial Measurement Unit (IMU) is usually used as the backup equipment for high-speed Unmanned Aerial Vehicle (UAV) in high-frequency environmental dithering condition. However, both the large-amplitude acceleration during UAV high-speed taxiing and the high-frequency environmental dithering caused by the propeller are important reasons decreasing the real-time attitude measurement precision of the UAV. Furthermore, there is no any other aiding sensors could be used to correct the measurement errors except for the gyroscopes and accelerometers in MEMS IMU. In this paper, an Adaptive Mahony Complementary Filter (AMCF) is used to estimate the real-time attitude of oil-powered single-propeller industrial-grade UAV with low-cost MEMS AMS. Meanwhile, the AMCF based on interference acceleration compensation is proposed to compensate the external disturbance acceleration and the dynamic tuning PI parameters of AMCF. Moreover, the attitude angle is updated by the quaternion updating algorithm to improve the real-time performance and reliability of the AMS. Finally, the UAV high-speed taxiing and flight experiments are included to verify the practical measurement accuracy of low-cost MEMS AMS when it compared with the high-precision and expensive reference system. The flying experimental results demonstrated that the statistical RMS errors of AMS by low-cost MEMS IMU do not exceed 0.882° in pitch and 0.864° in roll when installed in the aviation UAV with high-speed and high-frequency dithering environments. These results not only provide powerful supports for UAV developers but also provide useful method for low-cost MEMS AMS developing and application.

Keywords: UAV attitude measurement · Micro-inertial AMS · Interference acceleration compensation · AMCF · Quaternion update

J. Sun et al. (Eds.): CSNC 2020, LNEE 650, pp. 89–98, 2020.
https://doi.org/10.1007/978-981-15-3707-3_9

1 Introduction

In general, the flight control system for the Unmanned Aerial Vehicle (UAV) is mainly focus on the feasibility, reliability and stability during its flying process [1, 2]. Usually, low-cost and low-precision Micro-Electro Mechanical System (MEMS) Inertial Measurement Unit (IMU), high-precision and expensive Inertial Navigation System (INS), airspeed meter, barometer, Global Navigation Satellite System (GNSS) and other sensors are properly installed and integrated to implement the real-time control and flexible flight of the UAV. Specifically, the MEMS IMU based Attitude Measurement System (AMS) is used as the backup equipment for high-speed UAV in high-frequency environmental dithering condition [3, 4]. While, the traditional mechanical, fibre optic or laser gyroscopes present their disadvantages in large-volume, high power consumption, expensive, and complexity in maintenance and so on when compared with the MEMS AMS. Hence, they cannot meet the requirements of UAV AMS for low-cost, miniaturization, low power consumption, maintenance-free and so on [5, 6].

Recently, the MEMS technology developed rapidly and matured gradually, and its advantages are increasingly prominent. So the IMU based multi-sensor integrated navigation system is utilized in industrial fields widely [7–9]. However, the MEMS IMU are easily to be disturbed by the environment, such as high-frequency dithering, large-amplitude acceleration and temperature variation, which are difficult for the MEMS AMS to measure the attitude accurately without aiding information in long-term applications [10]. Moreover, the gyroscope presents better dynamic characteristic in maneuvering condition and the accelerometer only provides accurate attitude in static condition, which means they present complementary property in AMS for UAV applications [11–13]. Furthermore, the UAV attitude calculation algorithm mainly includes the complementary filter and Extended Kalman Filter (EKF). When it compared with EKF, the complementary filter reveals smaller calculation consumption, less-complexity, which is mainly used in low-cost and real-time AMS [14, 15]. The literature [16] shows that Mahony Complementary Filter (MCF) has presented its superiorities in low-cost navigation application when it compared with EKF. However, the PI parameter is fixed during the overall flying process in various dynamic conditions such as taxiing, take-off, accelerating, slowing down, turning and landing. The PI parameter cannot be adjusted according to the amplitude and frequency changes of the vehicular motions. The literature [17] has also used the enhanced MCF for UAV attitude measurement, while it only uses the helicopter to verify its performance in low-dynamic condition and its attitude measurement precision cannot satisfy the requirement for the high-speed UAV. Hence, both its dynamic adaptability performance and the measurement precision are poor during the overall UAV flight process [18]. What's worse, it is difficult to separate the external disturbance acceleration that induced by the velocity change of UAV from the gravity field component, uncompensated external disturbance acceleration would also cause huge attitude measurement error especially when UAV in high-speed and high-frequency environmental dithering conditions [19].

In this paper, the AMCF based on external disturbance acceleration compensation is proposed to adjust the PI parameters automatically, and the quaternion is also used to describe the attitude angles of the UAV to ensure its real-time performance. Moreover, a low-cost and shock-absorption structured MEMS AMS is specially designed for these environments. Finally, both the high-speed taxiing and flight experiments are conducted to demonstrate the feasibility and effectiveness of the proposed algorithm.

2 Frame Definitions

Vehicular frame (b-frame): the b-frame ($OX_bY_bZ_b$) is fixed to the vehicle, and its origin is defined as the centre of gravity of the vehicle. The OX_b axis points to the right along the horizontal axis of the vehicle, the OY_b axis points to the front along the vertical axis of the vehicle, and the OZ_b axis points up along the vertical axis of the vehicle [20].

Navigation frame (n-frame): the n-frame ($OX_nY_nZ_n$) utilizes the local geographic coordinate system. The origin is defined as the center of gravity of the vehicle. The OX_n axis points to the geographic East, the OY_n axis points to the geographic North, and the OZ_n axis points to the sky perpendicular to the local ellipsoid of rotation.

To be more intuitive, the relationship among the above mentioned coordinate systems is shown in Fig. 1.

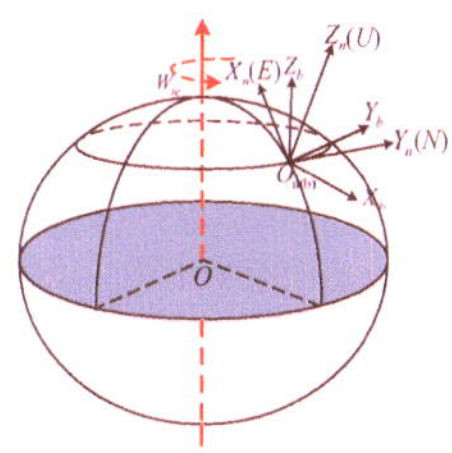

Fig. 1. The diagram of frame definitions

The orientation relationship of the b-frame with respect to the n-frame can be described by yaw Ψ, roll γ and pitch θ. Specifically, yaw is the angle between the projection line of the vertical axis on the local horizontal plane and the local geographic North direction. Pitch represents the angle between the vertical axis of the vehicle and its horizontal projection line. Roll shows the angle between the vertical axis of the vehicle and the vertical axis on the vertical plane.

3 Attitude Calculation by Quaternion

Euler angle is intuitive and can be used for the final attitude display, but when pitch is 90°, the universal joint deadlock will occur, resulting in the failure of the full attitude. In addition, the shortcomings of large-computation of Direction Cosine Matrix (DCM) cannot meet the requirements of real-time attitude output. However, the quaternion is widely used because of their advantages of fast calculation speed, high accuracy and full attitude. Therefore, in this paper, quaternion updating is used to calculate attitude angle. As the name implies, a quaternion is a number containing four elements, which can be expressed as:

$$\mathbf{q} = q_0 + q_1\boldsymbol{i} + q_2\boldsymbol{j} + q_3\boldsymbol{k} \tag{1.1}$$

Where: q_0, q_1, q_2, q_3 are real numbers, q_0 is called the real part, $\boldsymbol{q}_v = q_1\boldsymbol{i} + q_2\boldsymbol{j} + q_3\boldsymbol{k}$ is called the imaginary part.

The conversion relationship between the quaternions and the DCM is as follows:

$$\begin{aligned} C_{11} &= q_0^2 + q_1^2 - q_2^2 - q_3^2; C_{12} = 2(q_1q_2 - q_0q_3); C_{13} = 2(q_1q_3 + q_0q_2); \\ C_{21} &= 2(q_1q_2 + q_0q_3); C_{22} = q_0^2 - q_1^2 + q_2^2 - q_3^2; C_{23} = 2(q_2q_3 - q_0q_1); \\ C_{31} &= 2(q_1q_3 - q_0q_2); C_{32} = 2(q_2q_3 + q_0q_1); C_{33} = q_0^2 - q_1^2 - q_2^2 + q_3^2; \end{aligned} \tag{1.2}$$

The transformation relationship between the DCM and the attitude angles is shown as follows:

$$\begin{aligned} C_{11} &= c_\Psi c_\gamma - s_\Psi s_\theta s_\gamma; C_{12} = -s_\Psi c_\theta; C_{13} = c_\Psi s_\gamma + s_\Psi s_\theta c_\gamma; \\ C_{21} &= s_\Psi c_\gamma + c_\Psi s_\theta s_\gamma; C_{22} = c_\Psi c_\theta; C_{23} = s_\Psi s_\gamma - c_\Psi s_\theta c_\gamma; \\ C_{31} &= -c_\theta s_\gamma; C_{32} = s_\theta; C_{33} = c_\theta c_\gamma; \end{aligned} \tag{1.3}$$

Where, trig functions $s_\vartheta = \sin\vartheta, c_\vartheta = \cos\vartheta\ (\vartheta = \Psi, \gamma, \theta)$.

First calculate the DCM from the quaternion, and then calculate the attitude angle from the DCM, combined Eqs. (1.3) and (1.2), the attitude angles are calculated as:

$$\begin{aligned} \theta &= \sin^{-1}(2(q_2q_3 + q_0q_1)) \\ \gamma &= -\tan^{-1}\left(\frac{2(q_1q_3 - q_0q_2)}{q_0^2 - q_1^2 - q_2^2 + q_3^2}\right) \end{aligned} \tag{1.4}$$

4 Attitude Calculation Optimized in UAV Dynamic Condition

4.1 Disturbance Acceleration Compensation

Firstly, compare the output value of the tri-axial accelerometer $\left|\tilde{f}_{sf}^b\right|$ with the local gravity g. If $\left|\left|\tilde{f}_{sf}^b\right| - g\right| < a_1^m$ is satisfied (a_1^m is the preset acceleration threshold), there is no maneuvering acceleration. Usually, In order to reduce the impact of accelerometer measurement noise, the averages of the accelerometer in a certain fixed time period is used instead of the instantaneous value under the stable flight state of UAV.

Next, on the basis of $\left|\left|\tilde{f}_{sf}^b\right| - g\right| < a_1^m$, to judge the measured value of the horizontal accelerometer $\left|\delta\tilde{f}_{sf}^h\right|$ as follows:

(1) When $\left|\delta\widetilde{f_{sf}^{h}}\right| < a_2^m$ (a_2^m is another preset threshold), there is no maneuvering acceleration, the acceleration is only used to solve or estimate the attitude angle.
(2) When $\left|\delta\widetilde{f_{sf}^{h}}\right| \geq a_2^m$, there are two cases: one is that the attitude angle in the attitude matrix is relatively large, and the other is that the UAV has a large horizontal maneuvering acceleration. If the maneuvering acceleration occurs only in a short period of time, it is considered that there is a short-term high-acceleration maneuver. If this condition occurs for a long time, it is necessary to combine the gyro output to check whether the vehicle is performing a circle or turning movement. Otherwise, the attitude angle correction should be performed by using the horizontal acceleration.

Finally, according to the acceleration caused by the horizontal or vertical maneuvering of the UAV, they are eliminated and compensated respectively, and the compensation basis is:

$$\begin{cases} f_{sfx}^{b} = f_{sfx}^{b} \pm \beta_1 (if\ f_{sfx}^{b} > 0\ \ then-; else+) \\ f_{sfy}^{b} = f_{sfy}^{b} \pm \beta_2 (if\ f_{sfy}^{b} > 0\ \ then-; else+) \end{cases} \tag{1.5}$$

Where, β_1 and β_2 are compensation constants that are set according to the maneuvering situation of the UAV, which are jointly determined by the measurement accuracy of the used inertial sensor and the magnitude of the maneuvering acceleration of the UAV.

4.2 Adaptive Mahony Complementary Filter

The MCF is an attitude calculation algorithm in low-cost MEMS AMS. In this paper, data collected by IMU are fused. While high-pass filter is utilized to process the gyroscope measurement signals and low-pass filter is utilized to smooth the accelerometer measurement signals, and adaptive PI adjustment is added to the complementary filter to form an adaptive complementary filter. The main formula of the AMCF is shown:

$$\begin{gathered} \dot{\hat{q}} = \frac{1}{2}\hat{q} \otimes P(w+\delta) \\ \delta = k_p \cdot e + k_i \cdot \int e \cdot dt \\ \mathrm{e} = \mathrm{f} \times \mathrm{d} \end{gathered} \tag{1.6}$$

Where, $\hat{q}$ represents the quaternion that is used to calculate the attitude of AMS, $P(\cdot)$ represents the rotation vector, δ represents the innovation generated by the PI regulator, w represents the value measured by gyroscope, e represents the relative rotation error between the measured inertial vector and the predicted vector. The proportional parameter k_p of the PI regulator is used to control the cross frequency between the accelerometer and the gyroscope, and the integral parameter k_i is used to correct the error of the gyroscope.

The AMCF process is divided into two stages: prediction and correction. In the prediction stage, the angular rate measured by the tri-axial gyroscope is predicted by the quaternion update. In the correction stage, the attitude prediction of the quaternion is corrected by the measurement of the tri-axial accelerometer after the maneuvering acceleration compensation. Moreover, the parameters of the PI regulator are adjusted by the output modulus of the tri-axial accelerometer at the previous time to reflect the dynamic condition of the high-speed UAV. The specific designing process is shown in Fig. 2.

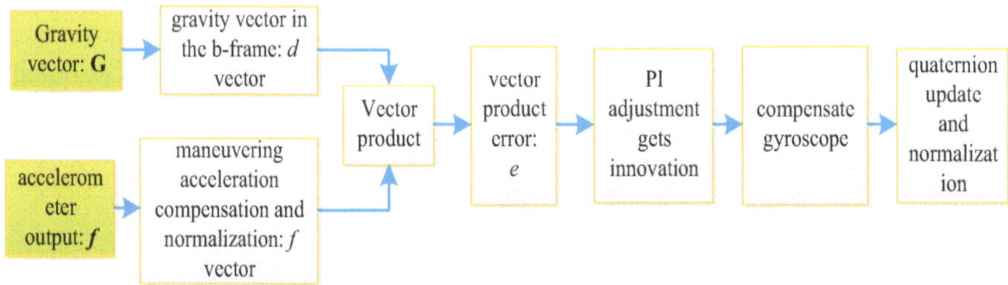

Fig. 2. The flowchart of the Mahony complementary filter

5 Experimental Verification and Results Analysis

Both UAV high-speed taxiing and UAV flight experiments are demonstrated to verify the practical performance of the low-cost MEMS AMS when it compared with the high-precision and expensive reference system.

5.1 Experiment Equipment

The experimental equipment includes the MEMS IMU based FS-VG445 from Harbin Engineering University, and the reference system from AVIC Xi'an Flight Automatic Control Institute, both systems are demonstrated and firmly installed in the oil-powered single-propeller industrial-grade UAV. The experimental equipment, the reference system and the UAV platform are shown in Fig. 3. Moreover, their performances of the key components are shown in the Table 1.

Fig. 3. Experimental equipment, the reference system and the UAV

Table 1. Key performances of FS-VG445 and AVIC INS

Equipment	Gyro bias instability	Acc bias instability	Size (cm)	Weight (g)	Attitude precision (°) (Maneuver condition)	
FS-VG445	8(°/h)	4(mg)	9.6 × 7.5 × 6.5	730		
AVIC INS	0.05(°/h)	10(ug)	32.5 × 9.8 × 15.2	5200	Pitch: 0.02	Roll: 0.02

5.2 UAV High-Speed Taxiing Experiment

Three groups of UAV high-speed taxiing experiments are conducted on the ground to test the measurement precision of the MEMS AMS, the taxiing experimental parameters of UAV are listed in Table 2.

Table 2. The taxiing parameters of experimental UAV

Taxiing phase	Maximum propeller speed	Terminal velocity of UAV	Maximum acceleration	Terminal time	Taxiing times
High-speed	5100 r/min	72 km/h	10 m/s^2	10 s	3

The attitude measurement and their errors of the FS-VG445 and AVIC INS on the high-speed UAV ground taxiing experiment are demonstrated in Figs. 4 and 5, respectively. More intuitively, the attitude errors statistical results of the experimental FS-VG445 are listed in Table 3. The statistical RMS errors of pitch and roll are not exceeds 0.174° and 0.114° respectively. Therefore, the proposed disturbance acceleration compensation technology and adaptive complementary filter algorithm are effective for UAV attitude measurement in high-speed and high-frequency environmental dithering UAV.

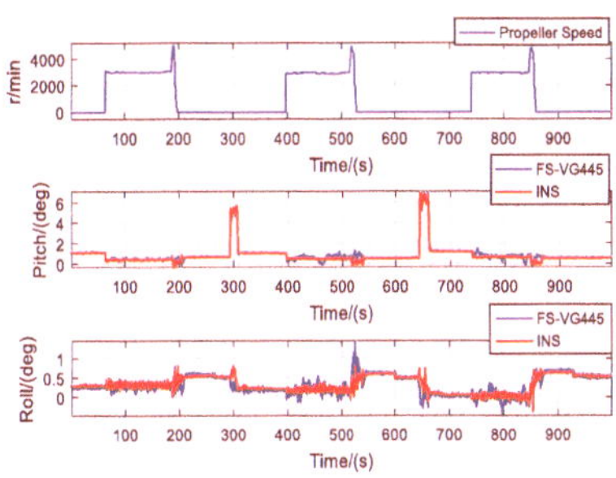

Fig. 4. Attitude of UAV in different propeller speed

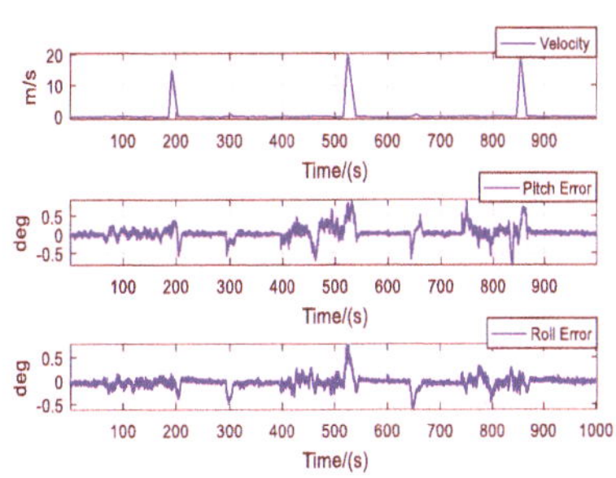

Fig. 5. Attitude errors in different velocity

Table 3. The attitude errors statistical results of the FS-VG445

Errors	Mean (°)	Maximum (°)	RMS (°)
Pitch	0.058	0.935	0.174
Roll	−0.024	0.804	0.114

5.3 UAV Flight Experiment

The flight experiment of the oil-powered single-propeller industrial-grade UAV is organized by high-speed taxiing, taking-off, turning, accelerating and straight line flying procedures to test the performance of the MEMS AMS. The overall flight time is around 33 min, and its flying trajectory is shown in the blue curve of Fig. 6.

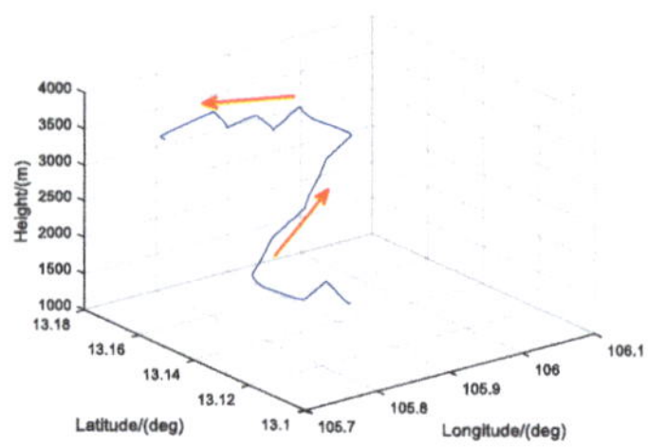

Fig. 6. The flight trajectory of the UAV

The attitude angles measured during the flight process are shown in the upper panels of the Figs. 7 and 8. In addition, the lower panels in both figures also demonstrate the measurement errors of pitch and roll when the MEMS AMS compared with the high-precision and referenced AVIC INS. Specifically, the statistical results of the MEMS AMS attitude measurement precision are listed in Table 4 and the RMS errors are occurred during the UAV turning, which reach to 0.882° and 0.864° in pitch and roll respectively.

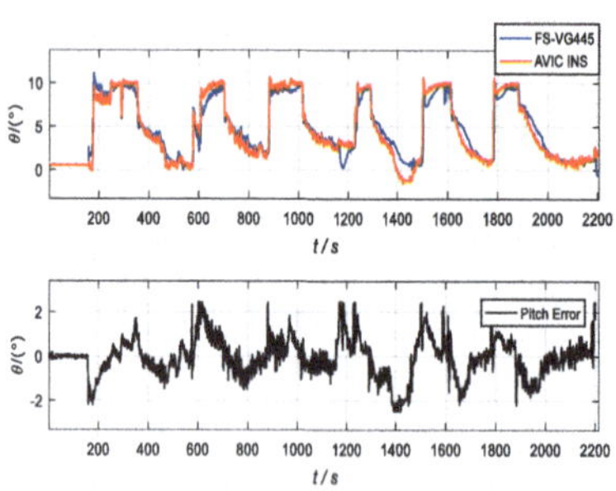

Fig. 7. UAV pitch and its error during flying

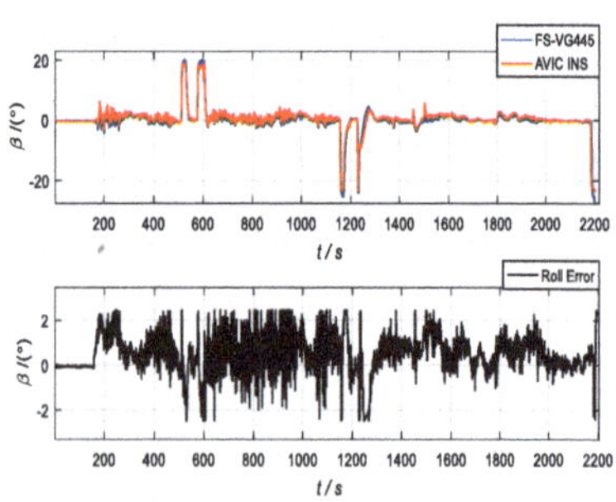

Fig. 8. UAV roll and its error during flying

Table 4. The attitude errors statistical results of the FS-VG445

Errors	Mean (°)	Maximum (°)	RMS (°)
Pitch	−0.055	2.496	0.882
Roll	0.486	2.391	0.864

6 Conclusions

The low-cost MEMS IMU constructed AMS for high-speed UAV in high-frequency environmental dithering condition is one of the challenge research works in engineering application. In this paper, an AMCF based on external disturbance acceleration compensation is proposed to estimate the real-time attitude of a high-speed aerial UAV under a high-frequency environmental dithering through a low-cost MEMS AMS. The high-speed taxiing experiment and UAV flight experiment are demonstrated. The results show that the MEMS AMS has good attitude measurement accuracy and superior performance in high-speed and high-frequency dithering environments. The paper provides an effective method for the development and application of low-cost MEMS AMS.

Acknowledgments. This work is supported by the NSFC (61803118), the Science and Technology Research Program of Chongqing Municipal Education Commission (KJZD-K201804701), and the Post Doc. Foundation of Heilongjiang Province (LBH-Z17053). The AVIC Guizhou Aircraft Co., Ltd is thanks for UAV experiment.

References

1. Wang, S., Zhen, Z., Zheng, F.: Design of autonomous flight control system for small-scale UAV. In: Proceedings of 2014 IEEE Guidance, Navigation and Control Conference, 8–10 Aug 2014, pp. 1885–1888 (2014)
2. Saeed, S., Younes, B., Cai, C., Cai, G.: A survey of hybrid unmanned aerial vehicles. Prog. Aerosp. Sci. **98**, 91–105 (2018)
3. Gross, J., Gu, Y., Rhudy, M.: Fixed-wing UAV attitude estimation using single antenna GPS signal strength measurements. Aerospace **3**(2), 14 (2016)
4. Zhang, X., Li, H., Yuan, D.: Dual redundant flight control system design for micro-miniature UAV. In: 2nd International Conference on Electrical, Computer Engineering and Electronics (ICECEE), pp. 785–791 (2015)
5. Sukkarieh, S., Gibbens, P., Grocholsky, B., Willis, K., Hugh, F.D.: A low-cost, redundant inertial measurement unit for unmanned air vehicles. Int. J. Robot. Res. **19**(11), 1089–1103 (2000)
6. Zhou, Y., Zhang, H.: A fusion attitude determination method based on quaternion for MEMS gyro/accelerometer/magnetometer. In: China Conference on Control and Decision-Making, pp. 3228–3232 (2013)
7. Qin, W., Yuan, W.Z., Chang, H.L.: Real-time and high-performance attitude and heading reference system based on MIMU/Magnetometers. Adv. Mater. Res. **6**(1), 219–223 (2009)
8. Du, X., Lan, X., Zhai, J.: Initial attitude determination of aerial platform based on MIMU. Appl. Mech. Mater. **568–570**, 964–969 (2014)

9. Liu, X.: Design and test of MEMS attitude measurement unit for fall detection. Key Eng. Mater. **483**, 465–470 (2011)
10. Chang, J., Jérôme, C., Chang, J.: Quadrotor attitude estimation with gyroscope bias reconstruction capabilities. IFAC Papers-Online **49**(5), 260–265 (2016)
11. Kada, B., Munawar, K., Shaikh, M.S.: UAV attitude estimation using nonlinear filtering and low-cost MEMS sensors. IFAC Papers-Online **49**(21), 521–528 (2016)
12. Gu, Y., Gross, J., Rhudy, B.: A fault-tolerant multiple sensor fusion approach applied to UAV attitude estimation. Int. J. Aerosp. Eng. **2016**(3), 1–12 (2016)
13. Bauer, P., Bokor, J.: Multi-mode extended Kalman filter for aircraft attitude estimation. IFAC Papers Online **44**(1), 7244–7249 (2011)
14. Kukillaya, P., Kamali, C., Saraf, A.: Evaluation of a novel attitude estimation algorithm for a high performance fighter aircraft. IFAC Papers-Online **47**(1), 26–33 (2014)
15. Zhu, R., Sun, D., Zhou, Z.: A linear fusion algorithm for attitude determination using low cost MEMS-based sensors. Measurement **40**(3), 322–328 (2007)
16. Li, N., Gao, Y., Wang, Y., Liu, Z., Guan, L., Liu, X.: A low-cost underground garage navigation switching algorithm based on Kalman filtering. Sensors **19**(8), 1861–1875 (2019)
17. Wang, M., Guan, L., Xiong, D., Gao, Y., Xu, X., Chen, X.: UAV attitude measurement based on enhanced Mahony complementary filter. In: Proceedings of 2018 IEEE International Conference on Mechatronics and Automation, Changchun, China, 5–8 August, pp. 544–550 (2018)
18. Sun, Q., You, P., Zhong, U.: Attitude estimation based on adaptive explicit complementary filter. Measur. Control Technol. **34**(4), 24–27 (2015)
19. Yan, G., Weng, J.: Strapdown inertial navigation algorithm and integrated navigation principle, ISBN 978-7-5612-6547-5, pp. 222–224. Northwestern Polytechnical University Press (2019)
20. Chang, R.H., Mu, X.D., Shen, X.W.: Attitude estimation with complementary filter. Appl. Mech. Mater. **44–47**, 3781–3784 (2011)

Modeling and Simulation of GNSS-R Signals with Ocean Currents

Bowen Li[1], Baoguo Yu[2], Lei Yang[1,3(✉)], Dongkai Yang[1], and Hua Han[2]

[1] School of Electronics and Information Engineering, Beihang University, Beijing, China
yanglei@sdau.edu.cn
[2] State Key Laboratory of Satellite Navigation System and Equipment Technology, The 54th Research Institute of China Electronics Technology Group Corporation, Shijiazhuang, Hebei, China
[3] College of Information Science and Engineering, Shandong Agricultural University, Taian, Shandong, China

Abstract. In recent years, with the development of Global Navigation Satellite System (GNSS), GNSS reflectometry (GNSS-R) has become a new method in the field of remote sensing. And the simulation of the GNSS-R research is very important. The simulation of GNSS-R signals mostly uses wind-driven wave spectrum, which ignoring other factors in the real ocean environment, such as the currents. Therefore, this paper proposes a method for modeling GNSS-R signal with considering currents. The influence of currents on the wave spectrum is simulated, and the wave spectrum model of the wind and currents is established. The Kirchhoff approximation-geometric optics is used to calculate the composite scattering coefficient. Finally, with the newly built model, the delay waveform (DW) and the Delay-Doppler maps (DDM) obtained from the simulation in the space-borne scene are analyzed. The results show that ocean currents of different speeds and directions have different effects on the simulation of GNSS reflected signals. The simulation results of the proposed model have good consistency with the theoretical waveform, and the correlation coefficient between DWs is raised to 0.9999. It is proved that the proposed model of GNSS-R signals considering the influence of currents is feasible and effective. And it made the simulated reflected signal more realistic.

Keywords: GNSS · Reflected signals · Ocean currents · Modeling · Simulation

1 Introduction

On the surface of the earth, the ocean has a huge impact on human production and life. In recent years, in the field of ocean remote sensing, using Global Navigation Satellite System-Reflectometry (GNSS-R) technology to retrieve sea surface parameters has gradually become a new trend [1]. This technology has the advantages of all-weather, all-day time, multiple signal sources, and high spatio-temporal resolution. It is widely used in the sea surface height, sea surface wind speed and significant wave height [2–4].

J. Sun et al. (Eds.): CSNC 2020, LNEE 650, pp. 99–110, 2020.
https://doi.org/10.1007/978-981-15-3707-3_10

Since Martin-Neira proposed the PARIS (Passive Reflectometry and Interferometry System) system in 1993, GNSS-R technology has developed rapidly [5]. Zavorotny and Voronovich in 2000 combined the Kirchhoff geometrical optics approximation and radar equations to derive the correlation power model (Z-V model) of GNSS (Global Navigation Satellite System) scattered signals [6]. In 2005, Thompson proposed an improved optical approximation model to simulate the scattered power of GNSS reflected signals [3]. In order to reduce the calculated amount of Delay-Doppler Maps (DDM) simulation in Z-V model, Marchan-Hernandez proposed an effective algorithm based on GNSS reflected signals correlation power convolution calculation [7]. Schiavulli proposed a panel-based scattering model in order to study the polarization characteristics of GNSS reflected signals [8]. On the basis of generating random sea surface, this model uses KA-PO model (Kirchhoff Approximation-Physical Optics) to calculate the correlation power of GNSS scattered signals and output DDMs. In 2016, Giangregorio summarized a large number of previous studies and presented a stochastic model of reflected signals correlation power based on finite time series [9]. This model can better simulate the speckle noise of scattered signals, and can describe the statistical characteristics of GNSS reflected signals more. In 2016, the Space Science Center of the Chinese Academy of Sciences introduced the research on software simulation of GNSS reflected signals, and analyzed the delay and Doppler interval of GNSS-R DDM simulation in 2018 [10, 11]. From 2017 to 2019, the GNSS-R remote sensing team at Beihang University proposed a backward fitting method, which equivalently reflects the reflected signal into multiple signals with different delays, Doppler and amplitudes [12, 13].

At present, most of the researches on GNSS-R simulation are based on the wind-driven ocean wave spectrum, without considering other factors (such as ocean currents). Taking the ocean currents as an example, the ocean currents will affect the roughness of the sea surface, and then affect the scattering of GNSS reflected signals. It will have a significant effect in the L-band [14]. Therefore, GNSS-R simulation considering the effects of ocean currents is very important.

Based on the analysis of the basic formulas for simulation of GNSS reflected signals on the sea surface, this paper proposes a GNSS-R simulation method that with the effects of ocean currents.

2 Ocean Currents Model

2.1 Wave Spectrum Model

The surface wind field causes wave motion, which affects the change of ocean surface roughness. This process can generally be regarded as a stochastic process, and the wave spectrum is often used to describe sea surface statistical characteristics with sea surface roughness. Ocean waves can be regarded as consisting of an infinite number of waves with different amplitudes, different frequencies, different directions, and disordered phases. The wave spectrum is a combination of these waves [1]. The wave spectrum is defined as the Fourier transform of the sea surface autocorrelation function, which can be expressed as [15]:

$$S(k) = FT\{\langle\zeta(r_0)\zeta(r_0 + r)\rangle\} \quad (1)$$

where FT represents the Fourier transform, <...> represents the set average, and k is the wave number vector.

There have been many ocean wave spectrum models, among which the Elfouhaily spectrum is a two-dimensional energy spectrum commonly used in ocean surface remote sensing [15]. It can be expressed as:

$$S_{\mathrm{E}}(k, \varphi) = M_{\mathrm{E}}(k) f_{\mathrm{E}}(k, \varphi) \quad (2)$$

where $M_E(k)$ is the isotropic part, $f_E(k, \varphi)$ is the corresponding bearing function.

For mature seas, one-dimensional Eifouhaily wave spectrum can be simulated under different wind speed conditions as shown in Fig. 1(a).

2.2 Ocean Currents Model

Ocean currents are different from wind waves. Ocean currents exist on the surface of various sea areas all year round, affecting the sea surface roughness on their paths. Therefore, considering the effects of ocean currents, the GNSS reflected signals can be simulated more realistically and accurately. The wave spectrum under the influence of the sea surface wind and ocean currents can be modeled as [16]:

$$S_{\text{wind + currents}}(k) = \frac{\beta}{2(1 + \frac{U_c}{c})^7} \cdot \frac{1}{k^3} \exp\left[\frac{-\alpha g^2}{k^2 \cdot U_{10}^4 \cdot (1 + \frac{U_c}{c})^4}\right] \quad (3)$$

where $\alpha = 0.74$ and $\beta = 0.81 \times 10^{-2}$, U_c is the velocity of the ocean current, c is the phase velocity, k is the wave number, g is the acceleration of gravity, and U_{10} is the average wind speed at a height of 10 m above the sea surface. Therefore, the simulation of the composite ocean wave spectrum considering the influence of ocean currents is performed. The simulation results at different ocean current velocities are shown in Fig. 1(b).

The dashed line in Fig. 1(b) indicates the simulation results of the ocean wave spectrum model (wind-driven ocean wave spectrum) under the condition of sea surface wind speed of 10 m/s and sea surface ocean current speed of 0 m/s. The solid line shows the simulation result of the composite ocean wave spectrum model under different sea surface ocean current speed conditions (a negative number indicates that the sea surface ocean current is opposite to the direction of the ocean wind). It can be seen from Fig. 1(b) that the composite ocean wave spectrum model under the influence of ocean currents and wind-driven ocean wave spectrum models have very similar energy distributions and wave trends, mainly concentrated in the range of the large-scale roughness. When the ocean current is in the same direction as the ocean wind, the ocean wave spectrum gradually changes with the increase of the ocean current speed, the descending speed of the trailing edge accelerates, the coverage area shrinks, and the energy is relatively concentrated. When the ocean current and the wind direction are reversed, the increase of the ocean current speed leads to the ocean wave spectrum and the coverage area become larger.

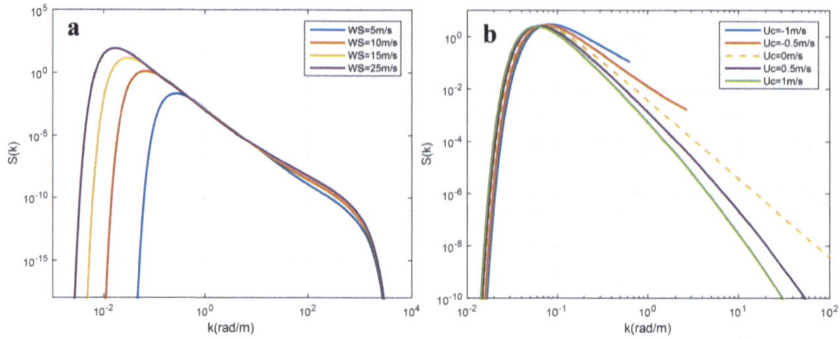

Fig. 1. Elfouhaily wave spectrum and composite wave spectrum of wind and current. (a) Elfouhaily wave spectrum of different wind speeds, (b) composite wave spectrum of different current speeds

3 Model of GNSS Reflected Signals with Effect of Ocean Currents

3.1 Establishment of Model

In the process of GNSS reflected signals processing, the reflected signals received by the receiver can be considered as the weighted sum of the scattered signals of various scattering units [17]. The GNSS reflected signals is correlated with the locally generated pseudo-random code, and the continuous correlation result is non-coherently accumulated to obtain the basic view measurement DDM of the GNSS reflected signals. The delay-Doppler two-dimensional correlation power of GNSS reflected signals can be expressed as [18]:

$$\left\langle |Y(\tau,f)|^2 \right\rangle = \frac{\lambda^2 P_t G_t T_c^2}{(4\pi)^3} \times \int \frac{G_{r\_x,y}\sigma_{p,q\_x,y}}{R^2_{t\_x,y} R^2_{r\_x,y}} \Lambda^2(\tau - \frac{R_{r\_x,y} + R_{t\_x,y}}{c}) S^2(f_{x,y} - f) dxdy \quad (4)$$

where P_t is the transmitter power, G_t is the transmitting antenna gain, λ is the navigation signal wavelength, T_c is the coherent integration time, $G_{r\_x,y}$ is the receiving antenna gain on the scattering unit $S_{x,y}$, $\sigma_{p,q\_x,y}$ is the scattering coefficient of the scattering unit $S_{x,y}$ when the p polarized incident wave corresponds to the q polarized reflected wave, $R_{t\_x,y}$ and $R_{r\_x,y}$ respectively represent the distance between the GNSS satellite and the receiving antenna to the scattering unit $S_{x,y}$, Λ is the autocorrelation function of pseudo-random code, and S is the Doppler filter function.

When considering the effects of ocean currents, the ocean currents have an effect on the roughness of the surface, resulting in changing in the scattering coefficient of the reflected surface. Therefore, accurate calculation of the scattering coefficient affected by ocean currents is a prerequisite for modeling. When the GNSS signals is incident on the ocean surface, a scattering effect occurs, and the roughness of the reflected surface will affect the scattering intensity. The strength of the scattering intensity is usually described by the scattering coefficient, which is defined as [19]:

$$\sigma_{qp} = \frac{4\pi R^2 |E_r^s|^2}{A |E_t^i|^2} \tag{5}$$

where q and p represent the polarization of the incident and reflected signals, respectively, E_r^s and E_t^i represent the incident and reflected electric fields, respectively. R is the distance between the observation point and the reflected point, and A represents the signal irradiation area. Under KA-GO (Kirchhoff approximation-geometric optics) model, it can be expressed as [19]:

$$\sigma_{KA-GO} = \pi |\Re|^2 \frac{|\mathbf{q}|^4}{q_z^4} P_{\mathrm{pdf}}\left(-\frac{q_\perp}{q_z}\right) \tag{6}$$

where $\Re$ is the Fresnel reflection coefficient, $\mathbf{q}$ is the scattering vector, q, q_z, and $q_\perp$ denote the modulus of the scattering vector $\mathbf{q}$, the modulus of the normal component of $\mathbf{q}$, and the modulus of the horizontal component of $\mathbf{q}$, respectively. P_{pdf} represents the probability distribution of the mean square slope of the sea surface and can be calculated using a wave spectrum model. Assuming a zero-mean Gaussian distribution, the Mean Square Slope (MSS) of the sea can be calculated using the wave spectrum model under the influence of ocean currents analyzed in Sect. 2.2:

$$MSS = \int_0^{K_c} k^2 S(k) dk \tag{7}$$

where k represents the wave number of the incident wave, $S(k)$ is the composite wave spectrum considering the influence of the ocean currents, K_c is the cut-off wave number of large-scale roughness, which can be obtained from $K_c = 2\pi\cos\theta/3\lambda$, θ and λ are the incident angle and the signal wavelength, respectively.

3.2 Noise Analysis

GNSS reflected signals generally consider two types of noise interference: thermal noise and speckle noise. The two types of noise are analyzed separately here. Assuming that the additive thermal noise obeys the Gaussian distribution of $(0, \sigma^2)$, for two branches the correlation power is the sum of the squares of the two branches in phase and quadrature. Since the sum of the squares of the two standard Gaussian distributions obeys a central chi-square distribution with a degree of freedom of 2, the thermal noise can be expressed as:

$$n_w(\tau, f) = n_I^2 + n_Q^2 \sim \sigma_w^2 \chi^2(2) \tag{8}$$

where $\chi^2(2)$ is the central chi-square distribution with a degree of freedom of 2 and mean of 2, σ_w^2 is the noise power after correlation [20].

Speckle noise originates from random scattering in the basic resolution unit and is often expressed as multiplicative noise. It obeys an exponential distribution with a mean of 1 and a variance of σ_s^2 [19]. Therefore, the two-dimensional delay-Doppler correlation power after noise addition can be expressed as:

$$\left\langle |Y_n(\tau,f)|^2 \right\rangle = \left\langle |Y(\tau,f)|_0^2 \right\rangle n_s(\tau,f) + n_w(\tau,f) \tag{9}$$

Calculate the delay-Doppler two-dimensional correlation power according to Eq. (9). The noise can be added to the GNSS simulation, and the obtained results can be non-coherently accumulated, which can reduce the influence of noise.

4 Simulation and Verification

According to the model of GNSS reflected signals that considering the effect of ocean currents established in Sect. 3, the model can be used to simulate and analyze the GNSS reflected signals under the effect of ocean currents.

Assuming that the GNSS reflection event is a spaceborne scene, the position and velocity of the GNSS satellite set in the simulation environment are [−11178791.991 294 m, −13160191.204988 m, 20341528.127540 m], [2523.258023 m/s, −361.5928 39 m/s, 1163.748104 m/s]. The position and velocity of the receiver mounted on the satellite is [−4069896.7033860330 m, −3583236.9637350840 m, 4527639.271758 1640 m], [−4738.0742342063 m/s, −1796.2525689964 m/s, −5654.9952013657 m/s]. The sea surface wind speed is 5 m/s, the wavelength is 400 m, the receiving antenna gain is 12 dB, the antenna 3 dB beam width is 20°, the ocean current speed is 1 m/s, and the direction is the same as the sea wind.

Simulations are performed using the proposed GNSS reflected signals model considering the effects of ocean currents. The delay waveform (DW) and DDM obtained by the simulation are shown in Figs. 2 and 3. In Fig. 3, (a)–(e) are the ocean current velocity of 1 m/s, 0.5 m/s, 0 m/s, −0.5 m/s, and −1 m/s, respectively. It can be seen from Figs. 2 and 3 that only considering the influence of the sea wind and both the influence of the sea wind and the ocean current, the DW and DDM results obtained by the simulation have similar waveform shapes and trends. The peak of the DW shifts backwards and the tail flicks. The DDM has obvious horseshoe characteristics, which are accord with the typical theoretical DW and DDM waveform laws. It can be proved that the model of GNSS reflected signals with the effect of ocean currents is feasible.

Further analysis of the numerical ranges of correlation power in Fig. 2 and Fig. 3 shows that under the same sea surface wind speed, different sea surface current will obviously affect the correlation power obtained by simulation. The data results are shown in Table 1. It can be clearly seen from Table 1 that after adding the influence factors of ocean currents, the change of sea surface roughness causes the correlation power to change. The specific manifestations are: when the wind and the ocean currents

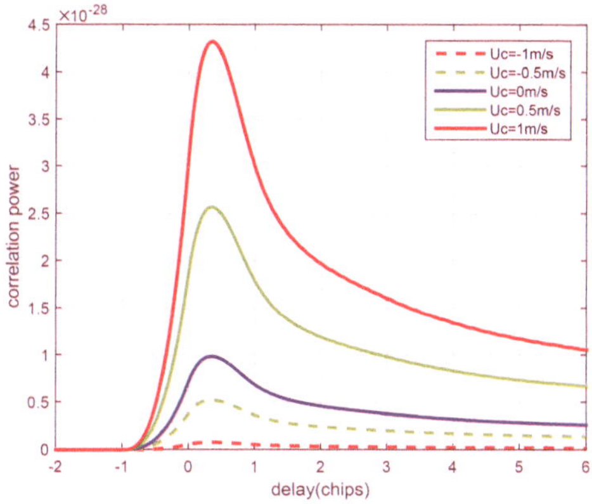

Fig. 2. Simulation results of DW, DW with the effect of currents

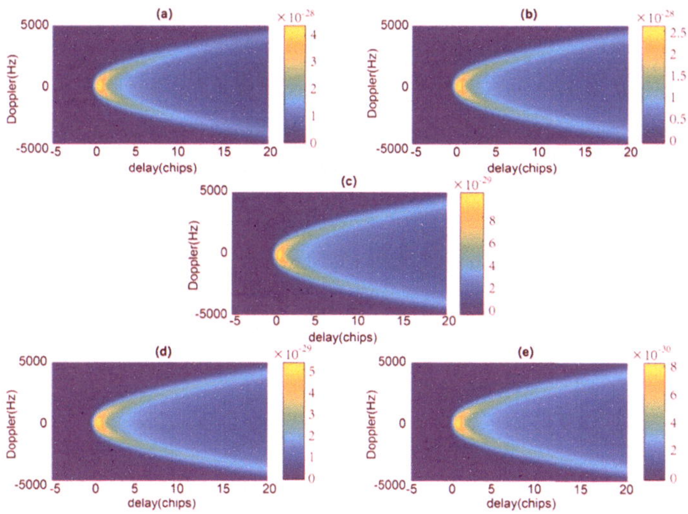

Fig. 3. Simulation results of DDM, DDM with the effect of currents

are in opposite directions, the collision between the ocean current and the wind wave causes the sea surface roughness to increase, the diffuse reflection at the reflecting surface is enhanced, the specular reflection part is reduced, and the correlation power is significantly reduced; When they are in the same direction, the sea surface tends to be gentler, the sea surface roughness becomes smaller, the diffuse reflection at the reflecting surface is reduced, the specular reflection is increased, and the correlation power is significantly increased.

At the same time, for the ocean currents in a fixed direction, as the speed of the ocean currents increases, the effect on the surface roughness becomes more significant. When the wind and the ocean currents are in opposite directions, the increase of the

Table 1. Peak correlation power at different current speeds

Speed of ocean current	−1 m/s	−0.5 m/s	0 m/s	0.5 m/s	1 m/s
Correlation power	8.1×10^{-30}	5.2×10^{-29}	9.8×10^{-29}	2.6×10^{-28}	4.3×10^{-28}

ocean currents speed will cause the surface roughness to increase, and the correlation power of the simulated reflected signals will decrease. When the ocean wind and the ocean currents are in the same direction, the increase of the ocean currents speed will cause the surface roughness to decrease. The correlation power of the reflected signals increases.

In order to further analyze the influence of ocean currents on GNSS-R simulation, the same ocean currents and different wind speeds were simulated and analyzed. The mean speed of ocean currents is 1 m/s (the ocean current is in the same direction as the wind), and the wind speed is set to range from 1 to 25 m/s. The DW obtained by simulation is shown in Fig. 4(a). It can be seen that as the wind speed changes from small to large, the correlation power of simulated DW gradually decreases. The increase of wind speed leads to the increase of sea surface roughness, the decrease of specular reflection and the increase of diffuse reflection, which significantly reduces the correlation power.

Using the DW curve at the wind speed of 25 m/s as the reference curve, the correlation coefficients between the DW curve and the reference curve under different wind speed is calculated. It can be seen from Fig. 4(b) that as the wind speed increases, the correlation coefficient between DWs gradually converges, and finally converges to 1. Therefore, although the ocean currents will affect the simulation of GNSS-R signals, the main factor affecting the surface scattering field is still the wind speed of sea surface.

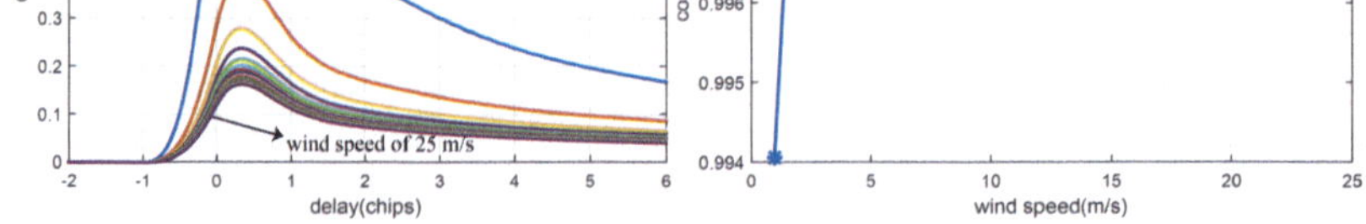

Fig. 4. Simulation results with different wind speeds. (a) DW with different wind speeds, (b) correlation coefficients of DW with different wind speeds

In the basic theoretical research of GNSS-R, the satellite elevation angle will affect the correlation power of the scattered signals. In order to analyze the influence of elevation angle on the newly established GNSS-R model with ocean currents, simulation analysis was performed on fixed speed of ocean current, fixed wind speed, and different elevation angles. The wind speed on the sea surface is set to 5 m/s, and the speed of ocean currents is 1 m/s. The simulation results are shown in Fig. 5. It can be seen from Fig. 5 that after considering the influence of the ocean current, the change of the elevation angle has little effect on the correlation power of the simulation result of the reflected signals, and its influence is mainly reflected in the expansion and extension of the delay. As the elevation angle increases, the tail of the DW of scattered signals becomes longer and the slope of the falling edge becomes lower. This is consistent with the effect of the elevation angle when the wind-driven wave spectrum model simulates the reflected signals. Satellite elevation affects the geometric relationship of GNSS-R reflection events, and is more sensitive to the size of the scattering area, but has little effect on the roughness of the reflected surface. Therefore, it has less effect on the ocean currents.

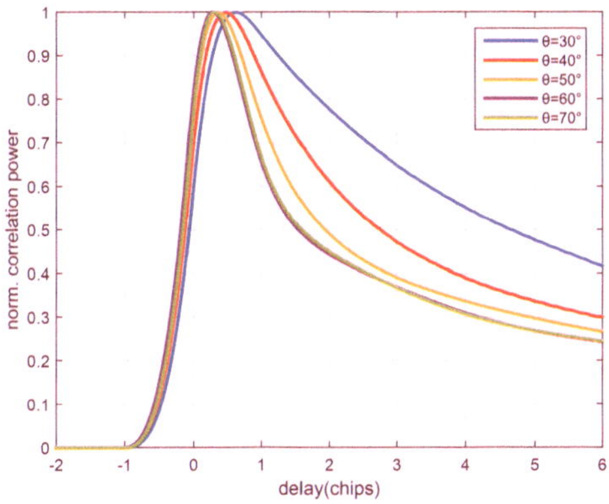

Fig. 5. Simulation results of DW with different elevations

Finally, in order to verify the accuracy of the model of GNSS reflected signals with the effect of ocean currents, the theoretical Z-V model is used as a benchmark to compare with the GNSS-R model with and without ocean currents. The comparison results are shown in Fig. 6. The solid line in Fig. 6 represents the simulated DW obtained from the theoretical Z-V model. The dashed line above the solid line represents the simulation result of the GNSS-R model without the influence of ocean currents. The dotted line below the solid line represents the simulation result with ocean currents. It can be seen from Fig. 6 that two models can all simulate the GNSS reflected signals well, and have high consistency with the theoretical Z-V model. However, it can be found that the DW with the introduction of ocean currents is closer to the DW obtained by the theoretical Z-V model, and the correlation coefficient between the two curves reaches 0.9999. Under the same simulation conditions, the correlation coefficient between the DW without considering the effect of ocean currents and the DW of the theoretical Z-V model is only 0.9997. Therefore, the model of GNSS reflected signals with effect of ocean currents is not only closer to the real scene, but also the simulation accuracy is improved.

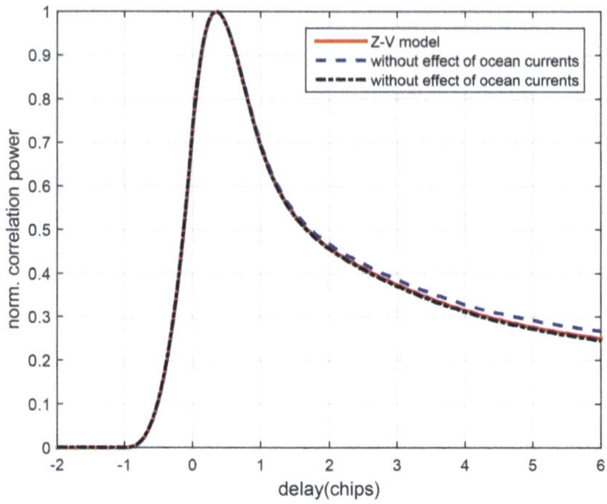

Fig. 6. DW of Z-V model, GNSS-R model without currents and GNSS-R model with currents

5 Conclusion

Based on the introduction of the statistical characteristics of the sea surface, the influence of ocean currents on GNSS reflected signals is analyzed through the wave spectrum. A method for calculating the scattering coefficient of the reflecting surface considering the influence of ocean currents by KA-GO is proposed. And a method for establishing the model of GNSS reflected signals with the influence of ocean currents is proposed. The established model is used for multi-dimensional simulation verification, and the results show that:

(1) The established model of GNSS reflected signals with the influence of ocean currents has a clear idea and clear algorithms, which is convenient for simulation testing and generation of real signals in the future.
(2) Ocean currents will cause the sea surface roughness to change. When the wind and ocean currents are in opposite directions, the sea surface roughness becomes larger; when they are in the same direction, the sea surface roughness becomes smaller. Therefore, considering the influence of ocean currents in GNSS-R remote sensing applications, the accuracy of inversion can be effectively improved.
(3) With the increase of wind speed, the influence of ocean currents in reflection events gradually weakens, and sea wind becomes the main influence factor. The elevation angle of the satellite mainly affects the size of the scattering area and is not sensitive to ocean currents.
(4) The correlation coefficient between the DWs with the effect of ocean currents and the DWs of the theoretical Z-V model is as high as 0.9999, which is better than the simulation result without considering the influence of the ocean current. The

feasibility and correctness of this modeling method are proved, which lays the foundation for the subsequent construction of model of GNSS reflected signals in complex environments with multiple influence factors and the study of the generation of real signals.

Acknowledgments. This work is supported by the National Natural Science Foundation of China (Grant 41774028, 31971781) and the New Technology Research University Cooperation Project (Grant KX162600040).

References

1. Yang, D.K., Zhang, Q.S.: GNSS Reflected Signal Processing: Fundamentals and Applications. Publishing House of Electronics Industry, Beijing (2012)
2. Cardellach, E., et al.: GNSS-R ground-based and airborne campaigns for ocean, land, ice, and snow techniques: application to the GOLD-RTR data sets. Radio Sci. **46**(5), 1–16 (2011)
3. Thompson, D.R., et al.: An improved geometrical optics model for bistatic GPS scattering from the ocean surface. IEEE Trans. Geosci. Remote Sens. **43**(12), 2810–2821 (2005)
4. Wang, F., et al.: Sea-state observation using reflected BeiDou GEO signals in frequency domain. IEEE Geosci. Remote Sens. Lett. **13**(11), 1656–1660 (2016)
5. Martin-Neira, M.: A passive reflectometry and interferometry system (PARIS): application to ocean altimetry. ESA J. **17**(4), 331–355 (1993)
6. Zavorotny, V.U., Voronovich, A.G.: Scattering of GPS signals from the ocean with wind remote sensing application. IEEE Trans. Geosci. Remote Sens. **38**(2), 951–964 (2000)
7. Marchan-Hernandez, J.F., et al.: An efficient algorithm to the simulation of delay-Doppler maps of reflected global navigation satellite system signals. IEEE Trans. Geosci. Remote Sens. **47**(8), 2733–2740 (2009)
8. Schiavulli, D., Ghavidel, A., Camps, A., et al.: A simulator for GNSS-R polarimetric observation over the ocean. In: IGARSS 2014, pp. 3802–3805 (2014)
9. Giangregorio, G., et al.: Stochastic modeling and simulation of delay-Doppler maps in GNSS-R over the ocean. IEEE Trans. Geosci. Remote Sens. **54**(4), 2056–2069 (2009)
10. Bai, W., et al.: GREEPS: an GNSS-R end-to-end performance simulator. In: IGARSS 2016, pp. 4831–4834 (2016)
11. Xia, J.M., et al.: Research on the effects of delay and Doppler intervals on GNSS-R DDM simulation. J. Remote Sens. **23**(3), 456–463 (2019)
12. Li, B.W., Zhang, B., Yang, D.K., et al.: An improved model and simulator design of GNSS ocean reflected signals. In: CPGPS 2017, pp. 157–161 (2017)
13. Li, B., et al.: A random model and simulation for generating GNSS ocean reflected signals. IEEE Geosci. Remote Sens. Lett. **16**(7), 1036–1040 (2019)
14. Camps, A., et al.: Sea surface brightness temperature at L-band: impact of surface currents. In: IGARSS 2004, pp. 3481–3484 (2004)
15. Elfouhaily, T., et al.: A unified directional spectrum for long and short wind-driven waves. J. Geophys. Res. **102**(15), 781–796 (1997)
16. Huang, N.E., et al.: Interactions between steady non-uniform currents and gravity waves with applications for current measurements. J. Phys. Oceanogr. **2**(4), 420–431 (1972)
17. Qi, Y.Q., Zhang, B., Yang, D.K., et al.: Three dimensional modeling method of GNSS sea surface reflection signal. J. Beijing Univ. Aeronaut. Astronaut. **44**(1), 125–131 (2018)

18. Park, H., Camps, A., Pascual, D.: A generic level 1 simulator for spaceborne GNSS-R missions and application to GEROS-ISS ocean reflectometry. IEEE J. Sel. Topics Appl. Earth Obser. Remote Sens. **10**(10), 4645–4659 (2017)
19. Wangfeng: Observation of sea wind speed using spaceborne global navigation satellite system-reflectometry. Beihang University, Beijing (2019)
20. Gleason, S.: Space-based GNSS scatterometry: ocean wind sensing using an empirically calibrated model. IEEE Trans. Geosci. Remote Sens. **51**(9), 4853–4863 (2013)

Coastal GNSS-R Code Delay Altimetry Using GPS L5 Signals

Xinyue Meng[1,2,3], Fan Gao[1,2,3](✉), Tianhe Xu[1,2,3], Yunqiao He[1,2,3], Ti Chu[4], and Nazi Wang[1,2,3]

[1] School of Space Science and Physics, Shandong University, Weihai, China
gaofan@sdu.edu.cn
[2] Institute of Space Science, Shandong University, Weihai, China
[3] Shandong Provincial Key Laboratory of Optical Astronomy and Solar-Terrestrial Environment, and Institute of Space Sciences, Shandong University, Weihai, China
[4] College of Marine and Space Information, China University of Petroleum, Qingdao, China

Abstract. In recent years, with the gradual completion of different GNSS constellations, the available L-band navigation signals have become more abundant. Not only can they be used for PNT (Positioning, Navigation, and Timing), but the reflected signals can also be used for retrieving a variety of geophysical parameters such as sea surface height. Coastal GNSS-R (Global Navigation Satellite System Reflectometry) code delay altimetry obtains the path delay measurements between reflected signal and corresponding direct signal based on ranging code. And the height from sea surface to antenna can be calculated according to the geometric relationship of sea surface altimetry. In order to achieve path delay with higher accuracy, we performed coastal GNSS-R altimetry experiment on a trestle using GPS L5 signals with high ranging code rate. The digital IF (intermediate frequency) data at 25 MHz sample rate were collected during the experiment. Then, self-developed software-defined receiver was used to process the IF data. As the code modulation methods of in-phase (I component) and quadrature-phase (Q component) on GPS L5 carrier are different, we tested the accuracy of I and Q component for sea surface altimetry, separately. In addition, considering different coherent integration time have a certain impact on the accuracy, 1 ms and 10 ms coherent integration time are adopted to compute the path delay based on I and Q component. The results show: for 1 ms coherent integration, the RMSE of I and Q component are 0.70 m and 0.61 m, respectively; while the coherent integration time is 10 ms, the RMSE of I and Q component are 0.54 m and 0.43 m, respectively. To sum up, increasing coherent integration time can improve the accuracy; the altimetry accuracy of the Q component signal is better than that of I component signal.

Keywords: GNSS-R · Code phase delay · Altimetry · L5

J. Sun et al. (Eds.): CSNC 2020, LNEE 650, pp. 111–120, 2020.
https://doi.org/10.1007/978-981-15-3707-3_11

1 Introduction

The concept of GNSS-R altimetry was proposed by Martin-Neira [1] of ESA in 1993. Compared with traditional tide stations and satellite radar altimetry, its main advantage is reflected in high space-time resolution and low cost [2]. It also has the ability to observe small and medium-scale ocean phenomena. In recent years, research on the use of spaceborne GNSS-R code phase altimetry has been carried out internationally, and some progress has been made.

The essential measurement of coastal GNSS-R altimetry is the path delay between direct signal and reflected signal. Now, there are mainly three methods: carrier phase delay altimetry [3, 4], SNR altimetry [5, 6], and code delay altimetry [7]. For carrier phase method, path delay measurements are derived from the continuous carrier phase observations, which are characterized by high accuracy. But, while water surface is rough, the carrier phase of the reflected signal cannot be tracked stably, which limits its application. For the SNR method, the height of antenna to water surface is achieved from phase interference information between direct signal and reflected signal at low elevation angles [8, 9]. However, due to its low elevation angle limitation, available observation arcs are few, which result in a low temporal-spatial resolution [10]. For code phase delay altimetry, path delay values are calculated based on the measurement of ranging code phase. It has high stability, but its measurement accuracy is limited by chip width.

Theoretically, if GNSS-R altimetry is performed using a ranging code with a higher code rate, the accuracy of its measurement will also be improved accordingly. Compared with C/A code, L5 code has higher power, and the code rate is ten times that of L1 C/A. Therefore, using L5 code for altimetry will theoretically improve its accuracy. In addition, as there is no data code on L5 Q component, its performance will be better than that of I component.

In order to verify the accuracy of altimetry using L5 code, we conducted a 7-day coastal GNSS-R altimetry experiment. The IF digital signal data with a sampling rate of 25 MHz was collected and self-developed GNSS-R software-defined receiver was used to obtain the code delay observations between direct and reflected signals. Then, the vertical distance from water surface to phase centre of down-looking antenna was obtained based on the geometric relationship. Finally, experimental results were compared with measurement provided by a high-precision ground-based radar altimeter.

In the following sections, the paper is arranged as follows: the second part introduces the basic principles of GNSS-R code delay altimetry; the third part describes coastal experiments and analyses the experimental results; the fourth part gives a summary of our work.

2 The Principle of Coastal GNSS-R Code Delay Altimetry

2.1 Geometric Model of Coastal GNSS-R Altimetry

As shown in Fig. 1, up-looking antenna receives direct signals, and down-looking antenna receives signals reflected by sea surface; h is vertical distance from down-looking antenna phase center to sea surface; θ is elevation angle of GNSS satellite; AB is the line connecting antennas phase center; ab is projection of antennas phase center line on the reflection path.

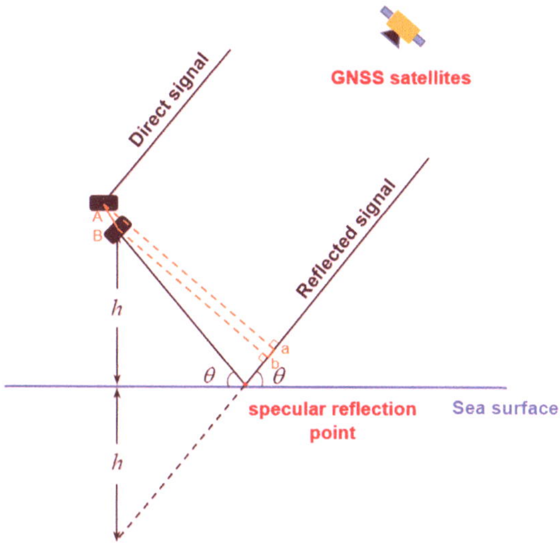

Fig. 1. Geometric relationship of GNSS-R sea surface altimetry

According to the geometric relationship in Fig. 1, the projection of $\overrightarrow{AB}$ on reflection path is:

$$\overline{ab} = \overrightarrow{AB} \cdot \overrightarrow{e_{ab}} \tag{1}$$

Calculate path delay between GNSS direct signal and reflected signal, and then calculate the height from the antenna to sea surface according to the geometric relationship among antenna, specular reflection point and GNSS satellite:

$$h = \frac{\rho_r - \rho_d - \overline{ab} + \delta_{Trop} + \delta_{Iono}}{2\sin\theta} + \varepsilon \tag{2}$$

where ρ_r is the length of reflected signal path; ρ_d is the length of direct signal path; δ_{Trop} is the error caused by troposphere; δ_{Iono} is the error caused by ionosphere. Tropospheric and ionospheric errors have little effect on the accuracy of the final calculation results, so these two errors are not considered in calculation.

2.2 Code Delay

The path delay between direct signal and reflected signal can be seen as time difference between signals reaching antenna. In order to obtain the time delay, this paper used the same local code to perform phase correlation with direct and reflected signal, and then obtained two waveform diagrams (Fig. 2). The delay between the peaks of two waveforms is called code phase delay, and corresponding distance is path delay between direct signal and reflected signal. Based on the geometric relationship between antennas' phase centres, specular reflection point, and GNSS satellite, the vertical distance from down-looking antenna to sea surface can be obtained by inversion.

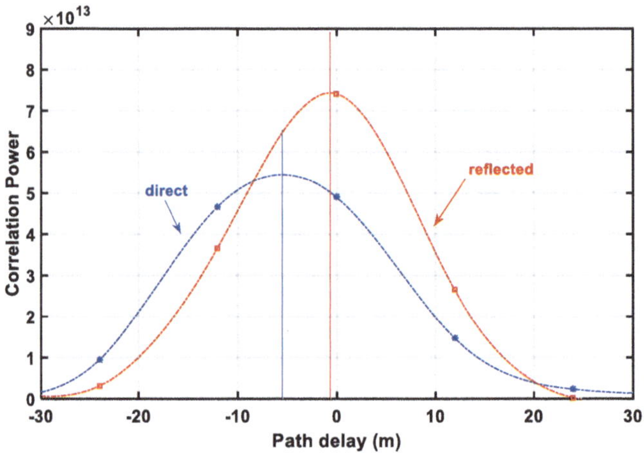

Fig. 2. Waveforms of reflected signal and direct signal

The waveforms in this paper were obtained by coherent integration of local code and received signal. The reflected signal is subject to more interference, so its SNR is low. Using long-term coherent integration time can effectively improve SNR of reflected signal, thereby improve the accuracy of code delay altimetry. Therefore, coherent integration times of 1 ms and 10 ms are used in this paper to verify the altimetry performance of L5 I component and Q component signals.

3 Coastal Experiment

3.1 Experiment Location and Instruments

We conducted a 7-day coastal GNSS-R altimetry experiment in Weihai, Shandong Province from September 17 to September 23, 2019. The location of instruments is shown by Fig. 3.

The installation of GNSS-R receiver is shown in Fig. 4. Up-looking antenna is used to receive direct signals, and down-looking antenna is used to receive reflected signals. The sampling rate of IF data is 25 MHz, and then the data is stored in a 32G SD card.

Therefore, the SD card needs to be updated manually approximately every 1 h, resulting in the experimental data discontinuous. During the experiment, due to excessive wind speed on September 18 (DOY 261), sea conditions were very bad. For safety reasons, equipment was powered off and no data was recorded. The experiment was restarted on September 19.

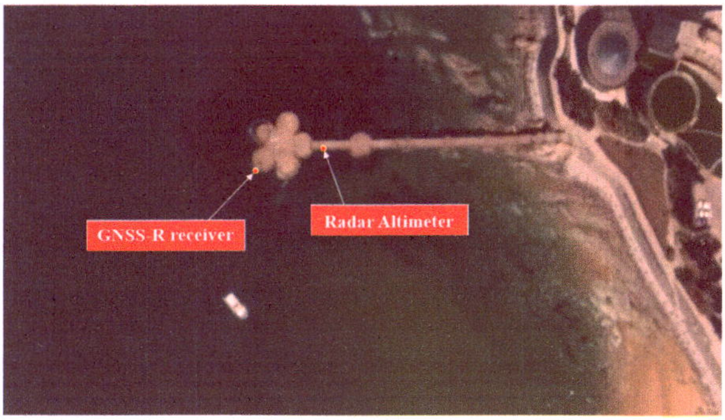

Fig. 3. Coastal GNSS-R experiment location

Fig. 4. GNSS-R receiver and anemorumbometer. GNSS-R receiver used to collect IF digital signals, and anemorumbometer used to measure wind speed and direction during experiment

The installation situation of high-frequency radar altimeter is shown in Fig. 5. It is fixed on the side of trestle with a high-strength aluminum alloy plate. It did not move during the experiment. The frequency of device is 26 GHz, and the accuracy of measurement is 3 mm. The vertical distance between phase center of radar altimeter and phase center of down-looking antenna is 1.942 m.

Fig. 5. High-frequency radar altimeter. It can measure actual height of instrument to sea surface

The anemometer was installed inside antenna support, as shown in Fig. 3, to ensure that it does not affect signal received. The accuracy of wind speed measurement is 0.1 m/s, and wind direction measurement accuracy is 0.1°.

3.2 Process of Data

In the data processing, cutoff elevation angle is 30°, and available azimuth angle ranges from 90° to 270°. The altimetry results were used double standard deviation to remove gross error, and then smooth it. Both I and Q component data were processed using this strategy.

After data processing, there are 1480 data points under the condition of 1 ms coherent integration, while a total of 1283 data points for 10 ms. Under the conditions of 1 ms and 10 ms coherent integration, heights derived from GNSS-R and radar altimeter data both have the same trend. However, none of them formed a continuous curve. The difference is that dispersion of result obtained by 10 ms coherent integration is lower (Fig. 6).

On the second day of this experiment (DOY 261), wind speed and waves at trestle was too high. The equipment was powered off and no data collection was performed. Therefore, the data at DOY 261 was interrupted. During the experiment, data collection was mainly focused on day and evening, so the data obtained by inversion did not form a continuous curve. In addition, the average number of captured satellites per epoch is only one or two, which results in no redundant observation for adjustment and affects the accuracy of inversion results to a certain extent.

Data of Q component adopted the same processing strategy as I component data. There are 1315 data points at 1 ms and 1276 at 10 ms. Similar to the comparison of I component, since experiment time is not been performed for 24 h, there is no continued curve. Inversion results of I and Q component are the same in that dispersion decreases after increasing the duration of coherent integration (Fig. 7).

Table 1 shows the accuracy indexes of I and Q component results, including root mean square error, standard deviation and mean bias. When coherent integration time is increased from 1 ms to 10 ms, root mean square error and mean bias of I component

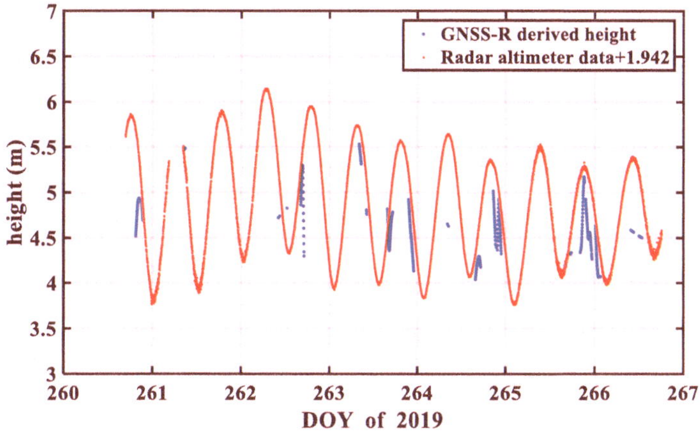

(a)

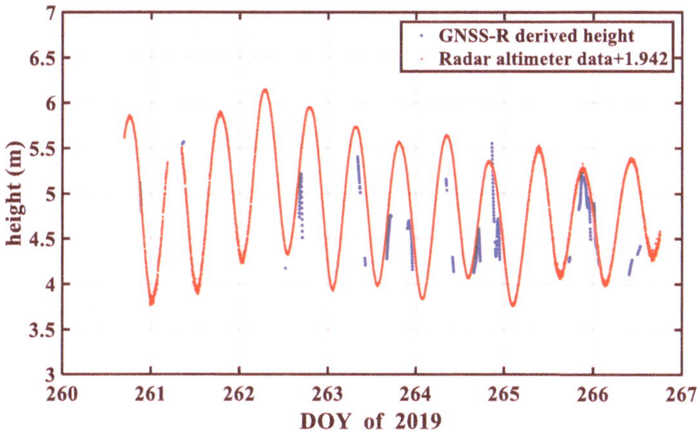

(b)

Fig. 6. Height derived by I component signal under 1 ms and 10 ms coherent integration and Radar Altimeter data. Blue scatters are the height derived by GNSS-R, and Red curve represents the observation of altimeter plus 1.942 m (the vertical distance between radar altimeter and down-looking antenna). (a) shows comparison between the result derived by I component signal under 1 ms coherent integration and the data of radar altimeter, and (b) shows comparison under the condition of 10 ms.

result are increased by 0.16 m and 0.20 m, respectively, while standard deviation values are slightly reduced. For 10 ms case, all the accuracy indexes of Q component result have been improved: root mean square error has increased by 0.18 m, standard deviation has increased by 0.10 m, and the mean bias has increased by 0.15 m.

Comparing the results of Q and I component, it can be found that root mean square error, standard deviation and mean bias of Q component data are better than those of I component for the same coherent integration time.

Fewer data points are important reasons for experimental errors. GPS L5 signal was officially broadcast in 2010, with 24 broadcast satellites planned by 2020. As of

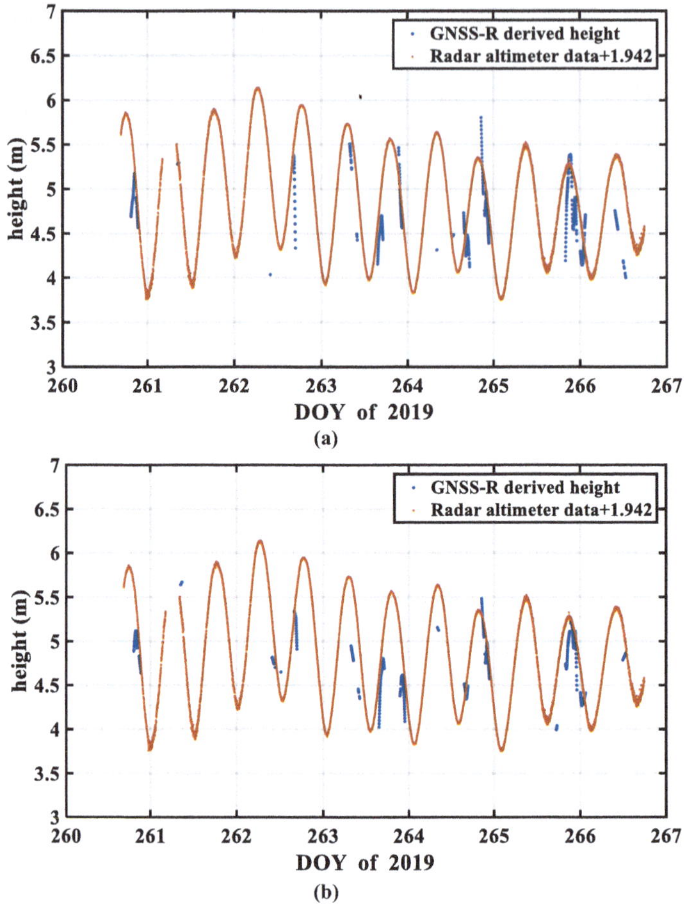

Fig. 7. Height derived by Q component signal under 1 ms and 10 ms coherent integration and Radar Altimeter data. (a) shows comparison between result derived by Q component signal under 1 ms coherent integration and data of radar altimeter, and (b) shows comparison under the condition of 10 ms.

Table 1. Root mean square error, standard deviation and mean bias of I component and Q component results

	Coherent time	Root mean square error (m)	Standard deviation (m)	Mean bias (m)
I	1 ms	0.6966	0.3359	−0.6104
	10 ms	0.5369	0.3437	−0.4126
Q	1 ms	0.6094	0.3910	−0.4675
	10 ms	0.4333	0.2909	−0.3212

October 17, 2019, a total of 12 satellites have been broadcast worldwide. During the experiment, since the average number of captured satellites per epoch was one to two, the effect of adjustment was very small during data processing. After eliminating the data with elevation angle less than 30° and error greater than double standard deviation, the final result has a total of 1200 to 1500 data points.

3.3 Conclusion

There is no previous sea surface altimetry experiment based on the L5 code has been performed. In order to verify its performance, we conducted a coastal experiment at Weihai coast, using self-developed software-defined receiver to process IF data. The same data processing strategy was used for I and Q component data. We selected data with elevation angle greater than 30°, remove gross errors, and then compare the results with radar altimeter data.

This experiment can be improved from the following aspects for better accuracy:

Firstly, increase the sampling rate of GNSS-R IF data. The sampling rate in this experiment is 25 MHz. Measurement accuracy of L5 code may be greatly improved with higher sampling rate.

Secondly, increase coherent integration time of Q component data. The I component of GPS L5 code contains data code information, and Q component does not contain it. Therefore, compared with I component, Q component can perform coherent integration for a longer time, thereby obtaining more accurate code delay information.

Thirdly, experimental data collection time can be more continuous. This experiment was carried out only during day and midnight, therefore it was not possible to obtain a continuous waveform diagram of sea level changes. Longer continuous data collection time can make the data result more complete.

Finally, the effect of wind speed and direction on code phase altimetry can be further analyzed. Both the effects of different wind speeds and whether wind direction has an impact on accuracy of measurement should be further studied.

4 Discussion

In order to test GNSS-R altimetry performance of GPS L5 code, a coastal experiment was carried out in Weihai, Shandong Province from September 17 to 23, 2019. We used dual-antenna GNSS-R device to obtain IF data with a sampling rate of 25 MHz. Then, self-developed GNSS-R software-defined receiver was used to obtain path delay between direct and reflected signals. Finally, the vertical distance from downward-looking antenna to sea surface is calculated according to the geometric relationship. And 26 GHz radar altimeter data are used as the comparison data.

In this paper, 1 ms and 10 ms coherent integration time are adopted, and I and Q component data of L5 are used for sea surface altimetry. During data processing, I and Q component adopt the same processing strategy. Data with satellite elevation greater than 30° was selected, and the inversion results with errors greater than twice standard deviation are eliminated.

The experimental results show that when coherent integration time is 10 ms, the accuracy of altimetry results of I and Q component is improved compared to 1 ms. Under the condition of same coherent integration time, the accuracy of Q component data is better than I component.

To analyze error of the experimental results, smaller amount of data is an important influencing factor. As of October 17, 2019, there were only 12 L5 code broadcast satellites, so the number of satellites captured during data collection was small, and the adjustment effect during data processing was minimal.

Based on the results of this coastal experiment, further improvements can be made: Use GNSS-R receiver with a higher IF sampling rate to compare the accuracy of altimetry results at different sampling rates; Increase coherent integration time of Q component signal to verify the improvement of its measurement accuracy; Finally, the effect of wind speed and direction on code phase altimetry can be further studied.

Acknowledgements. This research was jointly funded by the National Natural Science Foundation of China (Nos. 41604003, 41704017 and 41574013) and China Postdoc Science Foundation (2017M612274).

References

1. Martin-Neira, M., Colmenarejo, P., Ruffini, G., Serra, C.: Altimetry precision of 1 cm over a pond using the wide-lane carrier phase of GPS reflected signals. Can. J. Remote. Sens. **28**(3), 394–403 (2002)
2. Penna, N.T., Morales, M.M.A., Ian, M., Jing, G., Foden, P.R.: Sea surface height measurement using a GNSS wave glider. Geophys. Res. Lett. **45**, 5609–5616 (2018)
3. Martin-Neira, M., Colmenarejo, P., Ruffini, G., et al.: Ocean altimetry using the carrier phase of GNSS reflected signals. Cersat J. **11**(22) (2000)
4. Liu, W., Beckheinrich, J., Semmling, M., et al.: Coastal sea-level measurements based on GNSS-R phase altimetry: a case study at the Onsala space observatory, Sweden. IEEE Trans. Geosci. Remote Sens. **55**, 1–12 (2017)
5. Larson, K.M., Ray, R.D., Nievinski, F.G., et al.: The accidental tide gauge: a GPS reflection case study from Kachemak Bay, Alaska. IEEE Geosci. Remote Sens. Pap. **10**(5), 1200–1204 (2013)
6. Roggenbuck, O., Reinking, J.: Sea surface heights retrieval from ship-based measurements assisted by GNSS signal reflections. Mar. Geod. **42**, 1–24 (2019)
7. Martin-Neira, M., Caparrini, M., Font-Rosselló, J., et al.: The PARIS concept: an experimental demonstration of sea surface altimetry using GPS reflected signals. IEEE Trans. Geosci. Remote Sens. **39**(1), 142–149 (2001)
8. Anderson, K.D.: Determination of water level and tides using interferometric observations of GPS signals. J. Atmos. Ocean. Technol. **17**(8), 1118–1127 (2000)
9. Benton, C.J., Mitchell, C.N.: Isolating the multipath component in GNSS signal-to-noise data and locating reflecting objects. Radio Sci. **46**(6), 1–11 (2011)
10. Nazi, W., Tianhe, X., Fan, G., et al.: Sea level estimation based on GNSS dual-frequency carrier phase linear combinations and SNR. Remote Sens. **10**(3), 470 (2018)

An Improved Height Rate Correction Method Based on Robust Regression for Sea Level Estimation in GNSS Interferometry Reflectometry

Xiaolei Wang(✉) and Jie Wang

School of Earth Sciences and Engineering, Hohai University,
Nanjing 211100, Jiangsu, China
chd_wxl@qq.com

Abstract. Global Navigation Satellite System-Interferometry Reflectometry (GNSS-IR) sea level altimetry, which is based on analysis of GNSS signals reflected from the sea surface, has demonstrated unique advantages for sea level monitoring. The signal-to-noise ratio (SNR) of a geodetic GNSS receiver can be used to estimate the distance between antenna and sea surface called reflecting height (RH), and subsequently to retrieve sea level. The classical SNR analysis method uses the multipath frequency of the SNR to estimate RH by assuming the sea is static within a limited period. Then, the bias caused by the moving sea surface is corrected by the height variation rate $\dot{h}$ from the dynamic surface and by the elevation angle variation rate $\dot{e}$ from the dynamic satellite. However, this method cannot correct this height rate error completely because $\dot{h}$ cannot be calculated accurately from the raw RH series. Recently, a dynamic SNR method has been developed that can estimate h and its variation rate $\dot{h}$ concurrently from the variational multipath frequency based on the least square method. However, this method has no ability to avoid error, leading to many outliers in retrieval series. So, we introduced the robust regression method to improve this method to avoid errors. The different performance of the classical SNR method and the improved dynamic SNR method is firstly analyzed and compared, using the SNR data from the Kachemak Bay GPS site. The results showed the retrievals of improved dynamic method based on robust regression achieves a higher accuracy than that of classical method. And it has the potential for short-term monitoring (waves and individual days when an event such as a large storm surge might occur).

Keywords: Global Navigation Satellite System-Interferometry Reflectometry · Sea level altimetry · Non-stationary surface · Signal-to-noise ratio · Height rate correction

1 Introduction

Sea level monitoring is important to global climate analysis and human living environments [1–4]. Recently, one technique called Global Navigation Satellite System-Interferometry Reflectometry (GNSS-IR, see [5, 6]) becomes a new way to monitor sea

J. Sun et al. (Eds.): CSNC 2020, LNEE 650, pp. 121–128, 2020.
https://doi.org/10.1007/978-981-15-3707-3_12

level [7–9]. Using this technique, the vertical distance between the antenna phase center and the reflecting surface (reflector height) can be estimated from signal-to-noise (SNR) data. The reflector height ($\bar{h}$, reflector height calculated in static assumptions) is calculated in the straightforward model for GPS multipath frequency for a horizontal planar surface [10]. However, this model is not sufficiently accurate for GPS sites with large sea level fluctuations.

Larson et al. improved this classical SNR method for large tidal variations [11]. The first iteration determines $\bar{h}$ without integrating $\dot{h}$ (reflector height variation rate); then, $\dot{h}$ is estimated from this coarse $\bar{h}$ time series to correct $\bar{h}$ for dynamic height h (reflector height calculated in dynamic assumptions). Löfgren et al. improved the second iteration [12]. Larson et al. adopted a similar approach to solve height rate directly by tidal analysis [9].

Roussel et al. proposed a dynamic SNR method to estimate h and $\dot{h}$ concurrently using least square method (LSM) [13]. They chose an offshore site with a height of approximately 60 m (the top of Cordouan lighthouse, France) and an LSP computed with a 5-min moving window (the period of five minutes corresponds to classical tide gauge sampling) to determine the multipath frequency. However, GPS sites are commonly located at much lower heights, meaning the SNR has much lower frequency and the 5-min SNR time series is too short to obtain the sea surface frequency (see Section SNR analysis). The objective of the current study was to improve the dynamic SNR method for use at GNSS sites by studying the performance of both methods and discussing how best to use them for sea level and sea state estimations.

2 SNR Analysis

The classical SNR analysis method has been proposed in earlier work [10, 11] (Fig. 1).

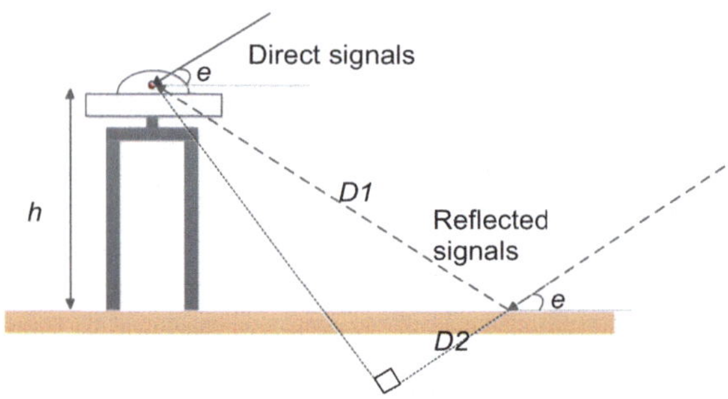

Fig. 1. Simple reflection model

At this point, the phase difference between direct and reflected signals is:

$$\phi = 2\pi D/\lambda = 4\pi h \sin(e)/\lambda \tag{1}$$

where λ is the wavelength of the GPS signal. Therefore,

$$2\pi\tilde{f} = \frac{d\phi}{dsin(e)} = 4\pi\frac{\bar{h}}{\lambda} \tag{2}$$

where $\tilde{f}$ is the sinusoid frequency. Then, $\bar{h}$ can be determined by $\bar{h} = \frac{\lambda\tilde{f}}{2}$. However, this model is not sufficiently accurate for data obtained from Kachemak Bay GPS site PBAY (AK, U.S.A.) with a large sea level fluctuation. Considering the movement of sea surface, Eq. (1) can be modified as:

$$2\pi f = \frac{d\phi}{dt} = \frac{4\pi\dot{h}}{\lambda}\sin(e) + \frac{4\pi h}{\lambda}\cos(e)\dot{e} \tag{3}$$

where $\dot{h}(=\frac{dh}{dt})$ is the height variation rate and $\dot{e}(=\frac{de}{dt})$ is the elevation angle variation rate. We thus obtain:

$$\frac{\lambda\tilde{f}}{2} = \frac{\tan(e)}{\dot{e}}\dot{h} + h \tag{4}$$

where the left-hand term of Eq. (6) is equivalent to $\bar{h}$. Therefore,

$$\bar{h} = \frac{\tan(e)}{\dot{e}}\dot{h} + h \tag{5}$$

The classical and dynamic methods use two different methods to obtain the value of h in Eq. (5): an iterative approach and the least square method.

2.1 Classical Correction Method

Larson et al. [11] stated that for a reflecting surface that is nonstationary during the measurement period, the retrieval would be biased by an amount equal to

$$\frac{tan(e)}{\dot{e}}\dot{h} \tag{6}$$

Larson et al. used a sinusoidal curve to estimated height variation rate $\dot{h}$ [9]. Traditionally, in an LSM tidal analysis, one would solve for the sine and cosine coefficients (S_i, C_i) of N independent tidal frequencies ω_i (N depends on the length of the series) with known nodal amplitude factors f_i and equilibrium phases (including nodal corrections) υ_i, as given by

$$\bar{h} = \sum\nolimits_{i=1}^{N} C_i f_i \cos(\omega_i t + \upsilon_i) + S_i f_i \sin(\omega_i t + \upsilon_i) \tag{7}$$

So, the height variation rate can be estimated. Then, the corrected h can be calculated:

$$h = \bar{h} - \frac{tan(e)}{\dot{e}}\dot{h} \tag{8}$$

Recall, the sea level is retrieved based on a constant minus the reflector height. This constant is determined by the station height and the reference frames of the GPS and sea level. So, the sea level variation rate could be estimated by $-\dot{h}$.

2.2 Dynamic Correction Method

Roussel et al. chose a 5-min moving window to tailor the SNR series, and calculated the frequency $\tilde{f}$ for every SNR series within the 5-min moving window [13]. For PBAY ($\sim$9 m above sea surface), the 5-min SNR time series is unsuitable for obtaining the sea surface frequency (as explained above); therefore, a new moving window is required. We used a moving window with width of 2-h and sampling step of 10 min to tailor RH series.

The solution presented by Roussel et al. [13] is obtained by the combination of the SNR measurements within a given period to determine h and $\dot{h}$ concurrently using classical LSM resolution. If $M = \frac{tan(e)}{\dot{e}}$, the ith RH in one window equals to

$$\bar{h}_i(t) = M_i\dot{h}(t) + h(t) \tag{9}$$

Combining all RHs in this window, a system of linear equations is obtained

$$\begin{pmatrix} \bar{h}_1(t) = M_1\dot{h}(t) + h(t) \\ \bar{h}_2(t) = M_2\dot{h}(t) + h(t) \\ \bar{h}_3(t) = M_3\dot{h}(t) + h(t) \\ \cdots \end{pmatrix} \tag{10}$$

or equivalently in terms of a matrix:

$$H = M\dot{h}(t) + h(t) = AX \tag{11}$$

with

$$H = \begin{pmatrix} \bar{h}_1(t) \\ \bar{h}_2(t) \\ \bar{h}_3(t) \\ \cdots \end{pmatrix} A = (M1), X = \begin{pmatrix} \dot{h}(t) \\ h(t) \end{pmatrix} \tag{12}$$

We used the Robust regression to solve this equation set. Please refer to [14] for detailed information. Fast convergence can be achieved due to fewer unknown parameters.

3 Result

One GPS site named PBAY, located on a small rocky island on the south side of Kachemak Bay in Cook Inlet in Alaska (U.S.A.) (see Fig. 2) is used. Its SNR sampling period is 30 s. We chose an azimuth mask of 90° to 260° to calculate the GPS tide gauge of PBAY, as in the study of [11, 15]. The Seldvia tide gauge, sited 30 km away, provided an extended time series of hourly sea level data.

We opted to use L1 and L2 signals. The SNR arcs of all satellites in azimuth mask of 90° to 260° in elevation angle of 5° to 25° are used. Their reflecting zones are shown in Fig. 2.

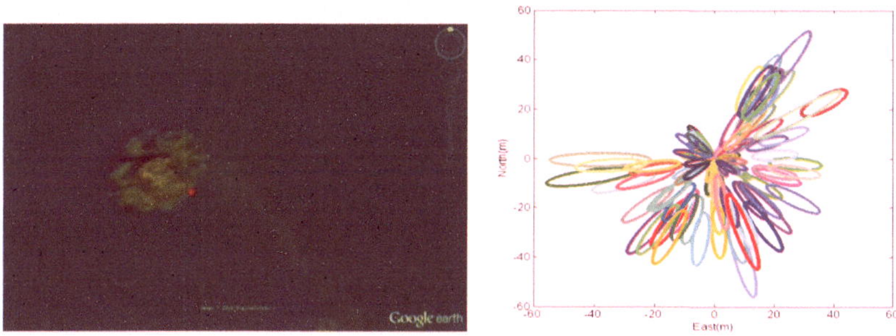

Fig. 2. (Left) Image obtained from Google Maps™. Red circle indicates the location of the PBAY GPS site. (Right) The reflecting zones for all SNRs in selected azimuths with specific elevation angles of 5°–25° for PBAY on doy 106 2012.

We first used the classical method. We calculated the tidal parameters for sea level retrievals using the tidal analysis. Thus, the sine and cosine fitting curves could be obtained, and we converted the $\bar{h}$ fitting curve to the sea level retrieval fitting curve based on its subtraction from a constant, as shown in Fig. 3.

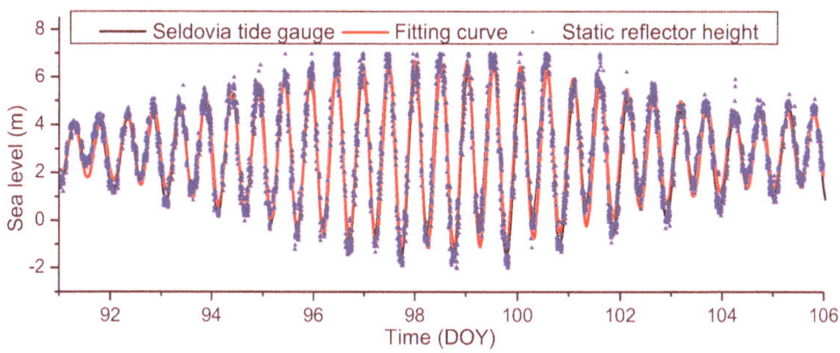

Fig. 3. PBAY sea levels estimated from $\bar{h}$ (blue circles), measured by the Seldovia tide gauge (black line), and fitted by the sea level retrieval curve (red line) from the fitting curve calculated by $\bar{h}$ (blue circles)

Figure 3 shows that the sea level retrieval fitting curve corresponds well with Sedovia tide gauge and that the fitting curve can reflect the fluctuation of sea level. Thus, $\dot{h}$ can be calculated and h can be corrected. The sea level records estimated from h are shown in Fig. 4.

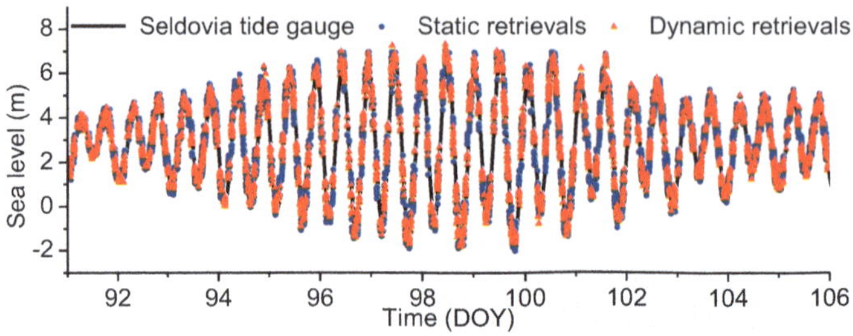

Fig. 4. PBAY sea levels estimated from $\bar{h}$ (blue circles), estimated from h (red triangles), and measured by tide gauge (black line)

Figure 4 shows retrievals from $\bar{h}$ and h, both of which convincingly reproduce the variation of tide gauge (correlations are both >0.97). The RMS error between the sea level retrievals from $\bar{h}$ (blue circles in Fig. 4) and the Seldovia tide gauge is 25.91 cm. The RMS between the retrievals and the measurements is 16.61 cm. The classical method can correct the error caused by the moving sea surface and the moving satellites, benefiting the GPS-R monitoring of tidal variations and long-term sea level changes. Using the fitting curve to fit $\bar{h}$ means unusual values of $\dot{h}$ cannot be recognized.

We subsequently used the dynamic method for determining h and $\dot{h}$. We calculated h and $\dot{h}$ concurrently, from which the sea level records estimated from h and sea level variation rates from $\dot{h}$ could be obtained, as shown in Fig. 5.

Figure 5 displays the sea level retrievals h and sea level variation rates from $\dot{h}$ derived using the dynamic SNR method. The RMS error between the sea level retrievals from h (red circles in Fig. 5) and the Seldovia tide gauge is 12.63 cm. The accuracy is improved marginally by 24% for sea level retrievals of classical correction method (red circles in Fig. 5). Most importantly, different from the $\dot{h}$ of the classical method calculated by the fitting curve, the dynamic method provides $\dot{h}$ for very short periods. The calculation of the short-term sea level variation rate means the monitoring of swell and waves is potentially achievable, enhancing the spatiotemporal monitoring of tides and waves considerably.

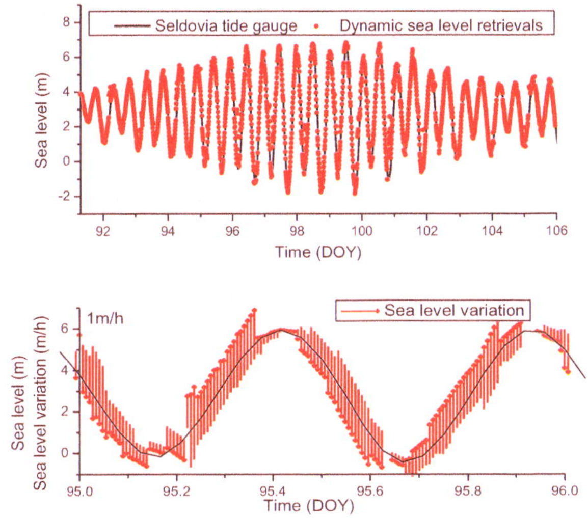

Fig. 5. (Upper) PBAY sea level records estimated from h using the dynamic method (red circles) and measured by tide gauge (black line). (Lower) Sea level variation rates obtained from $\dot{h}$ using the dynamic SNR method (red arrows) and sea level records measured by tide gauge (black line)

Acknowledgments. This work is supported by "Jiangsu Province Surveying Mapping and Geoinformation Research Project" (JSCHKY201901), Natural Science Foundation of Jiangsu Province (20195044012) and "the Fundamental Research Funds for the Central Universities" (2019B01714).

References

1. Douglas, B.C., Kearney, M.S., Leatherman, S.P.: Sea Level Rise: History and Consequences. Academic, San Diego (2001)
2. Jin, S.G., van Dam, T., Wdowinski, S.: Observing and understanding the earth system variations from space geodesy. J. Geodyn. **72**, 1–10 (2013). https://doi.org/10.1016/j.jog.2013.08.001
3. Feng, G.P., Jin, S.G., Zhang, T.Y.: Coastal sea level changes in the Europe from GPS, tide gauge, satellite altimetry and GRACE, 1993–2011. Adv. Space Res. **51**(6), 1019–1028 (2013). https://doi.org/10.1016/j.asr.2012.09.011
4. Jin, S.G., Qian, X.D., Wu, X.: Sea level change from BeiDou navigation satellite system-reflectometry (BDS-R): first results and evaluation. Glob. Planet. Change **149**, 20–25 (2017). https://doi.org/10.1016/j.gloplacha.2016.12.010
5. Jin, S.G., Feng, G.P., Gleason, S.: Remote sensing using GNSS signals: current status and future directions. Adv. Space Res. **47**(10), 1645–1653 (2011). https://doi.org/10.1016/j.asr.2011.01.036
6. Jin, S.G., Cardellach, E., Xie, F.: GNSS Remote Sensing: Theory, Methods and Applications, p. 276. Springer, Dordrecht (2014)

7. Santamaría-Gómez, A., Watson, C.: Remote leveling of tide gauges using GNSS reflectometry: case study at Spring Bay, Australia. GPS Solut. **21**(2), 451–459 (2016). https://doi.org/10.1007/s10291-016-0537-x
8. Santamaría-Gómez, A., Watson, C., Gravelle, M., King, M., Wöppelmann, G.: Levelling co-located GNSS and tide gauge stations using GNSS reflectometry. J Geod. **89**(3), 241–258 (2015). https://doi.org/10.1007/s00190-014-0784-y
9. Larson, K.M., Ray, R.D., Williams, S.D.P.: A ten-year comparison of water levels measured with a geodetic GPS receiver versus a conventional tide gauge. J Atmos. Oceanic Technol. **34**(2), 295–307 (2017). https://doi.org/10.1175/JTECH-D-16-0101.1
10. Larson, K.M., Löfgren, J.S., Haas, R.: Coastal sea level measurements using a single geodetic GPS receiver. Adv. Space Res. **51**(8), 1301–1310 (2013). https://doi.org/10.1016/j.asr.2012.04.017
11. Larson, K.M., Ray, R.D., Nievinski, F.G., Freymueller, J.T.: The accidental tide gauge: a GPS reflection case study from Kachemak Bay, Alaska. IEEE Geosci. Remote Sens. Lett. **10** (5), 1200–1204 (2013b). https://doi.org/10.1109/lgrs.2012.2236075
12. Löfgren, J.S., Haas, R., Scherneck, H.G.: Sea level time series and ocean tide analysis from multipath signals at five GPS sites in different parts of the world. J. Geodyn. **80**, 66–80 (2014)
13. Roussel, N., Ramillien, G., Frappart, F., Darrozes, J., Gay, A., Biancale, R., Striebig, N., Hanquiez, V., Bertin, X., Allain, D.: Sea level monitoring and sea state estimate using a single geodetic receiver. Remote Sens. Environ. **171**, 261–277 (2015). https://doi.org/10.1016/j.rse.2015.10.011
14. Wang, X., He, X., Zhang, Q.: Evaluation and combination of quad-constellation multi-GNSS multipath reflectometry applied to sea level retrieval. Remote Sens. Environ. **231** (2019b). https://doi.org/10.1016/j.rse.2019.111229
15. Wang, X., Zhang, Q., Zhang, S.: Sea level estimation from SNR data of geodetic receivers using wavelet analysis. GPS Solut. **23**, 6 (2019). https://doi.org/10.1007/s10291-018-0798-7

Application Research and Error Analysis of GNSS-MR Technology in Snow Depth Measurement

Zheng Li, Peng Chen(✉), Naiquan Zheng, Hang Liu, and Lixia Liu

College of Geomatics, Xi'an University of Science and Technology, Xi'an 710054, China
chenpeng0123@gmail.com

Abstract. As a part of global fresh water resources, the change of snow storage has an important impact on the earth's ecological environment, climate change and human development. Recently years, with the continuous development of Global Navigation Satellite System, GNSS Multipath Reflectometry technology has become one of the emerging means of snow depth measurement because of its all-weather, low cost, high spatial-temporal resolution and other advantages. Based on the study of the basic principle of snow depth measurement by GNSS-MR technology, this paper uses the observation data of p351 station in the PBO network of the United States to retrieve the local snow depth. And then the retrieved snow depth is compared with the measured snow depth of 490 station in the SNOTEL network of the United States. The results show that the correlation coefficient between the two is about 0.98, and the Root Mean Square Error is about 0.1 m, which has a good consistency. On this basis, this article analyzes GNSS-MR snow depth measurement error from three aspects: azimuth, elevation angle and snow depth, and finds that there is a certain quantitative relationship between the negative error generated by GNSS-MR snow depth measurement and the snow depth. Aiming at different error characteristics and error sources, the weakening methods are proposed and verified in practice, which provides a reference for the error correction of GNSS-MR technology in snow depth measurement and other environmental monitoring.

Keywords: GNSS · Snow depth · GNSS-MR · Error analysis

1 Introduction

As a part of fresh water resources, the change of snow storage has an important impact on the ecological environment, climate change and social development of the earth. Therefore, it is of great significance to obtain snow storage quickly and accurately.

Recently years, with the development of GNSS Technology, GNSS-MR (GNSS Multipath Reflectometry) technology has become a new method of surface environmental monitoring due to the advantages of its low cost, easy distribution, high spatial-temporal resolution and so on. Through the multi-path effect of GNSS signal in the process of propagation, GNSS-MR technology uses the signal-to-noise ratio (SNR) data in the observation file to retrieve the environmental parameters such as soil moisture, snow depth and tide height.

J. Sun et al. (Eds.): CSNC 2020, LNEE 650, pp. 129–140, 2020.
https://doi.org/10.1007/978-981-15-3707-3_13

Comp and Axelrad proved that the frequency and amplitude of different SNR data could build the model [1]. Bilich et al. studied how to separate the direct and reflected signal in SNR, and the relationship between the SNR of the reflected signal and the multipath effect [2]. Larson et al. proposed using GPS receiver to measure snow depth [3]. Zhang et al. studied and analyzed the feasibility and accuracy of GPS-MR snow depth detection by using single and multiple GPS satellites respectively [4]. Zhang et al. put forward the principle of plane grid, infer the reflection region according to the interference characteristics between the direct signal and the reflected signal, calculate the interference characteristic parameters, finally place the parameter in the grid of the corresponding inversion area [5]. Zhou et al. used SNR data of GLONASS satellite to measure snow depth [6]. Li et al. proposed a snow depth estimation method of GNSS single frequency signal based on the combination of pseudo range and carrier phase observation [7].

At present, GNSS-MR technology has made some progress in surface environment monitoring, but there are still some problems that need to be further studied. GNSS-MR technology is often disturbed by the surrounding environment of the station when it is used to measure the snow depth. First of all, GNSS-MR technology can measure the vertical distance between the antenna phase center and the snow surface. Due to the topographic relief, the height of the ground is lower or higher than the height of the reference point of the station, resulting in the measurement error of snow depth. Secondly, when there is less snow, the snow depth measured by GNSS-MR is higher than the actual snow depth due to the vegetation or other coverings on the surface.

As the snow reaches a certain depth, some signals will pass through the snow surface when the signal is reflected on the snow surface, which makes the snow depth measured by GNSS-MR less than the actual snow depth. Therefore, in order to improve the accuracy of GNSS-MR technology in the application of snow depth measurement, this paper uses the observation data of p351 station in PBO (Platform Boundary Observatory) network in the United States to measure the local snow depth, and analyzes the factors affecting the accuracy in the measurement process. The snow depth measured by GNSS-MR is compared with that measured at 490 station of the U.S. SNOTEL network and that obtained by the PBO H2O research group using GNSS-MR.

2 Basic Principle of Snow Depth Measurement Using GNSS-MR Technology Based on SNR

GNSS receiver can receive two kinds of composite signals. One is direct signal, which directly enters the receiver. The other is reflected signal, which enters the receiver after the surface reflection. The phase and amplitude characteristics of the signal will change after being reflected from the surface.

Figure 1 is a schematic diagram of snow depth measurement by GNSS-MR technology. H is the height of the antenna, that is, the vertical distance between the antenna phase center and the earth surface. RH is the vertical distance between the antenna phase center and the snow surface. SH is snow depth. E is the incident angle of the direct signal entering the receiver, which named elevation angle.

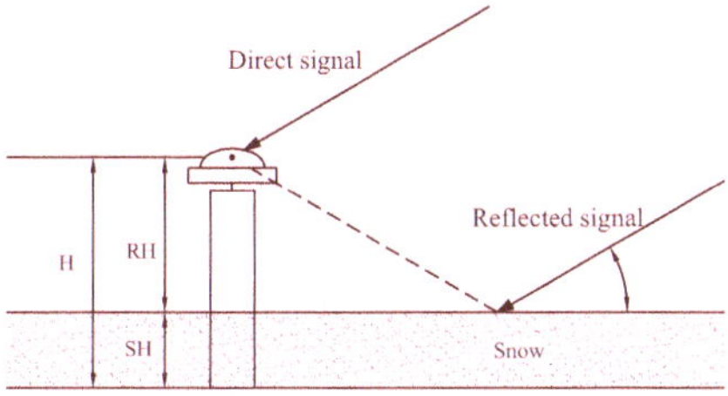

Fig. 1. Schematic diagram of snow depth measurement by GNSS-MR Technology

SNR is an observation value in the observation file of GNSS receiver. The magnitude of SNR reflects the signal strength of GNSS. Figure 2 is the analysis chart drawn by using the SNR data of PRN15 satellite at doy70 of P351 station in 2014. In the chart, the blue curve is the mixed signal SNR, and the red curve is the direct signal SNR fitted by low-order polynomial. When the elevation angle is high, the antenna gain is larger, the signal is less interfered by multipath, the SNR is higher, and the amplitude of oscillation is smaller; when the elevation angle is low, the antenna gain is small, the signal is greatly affected by multipath, the SNR is small, and the amplitude of oscillation is large.

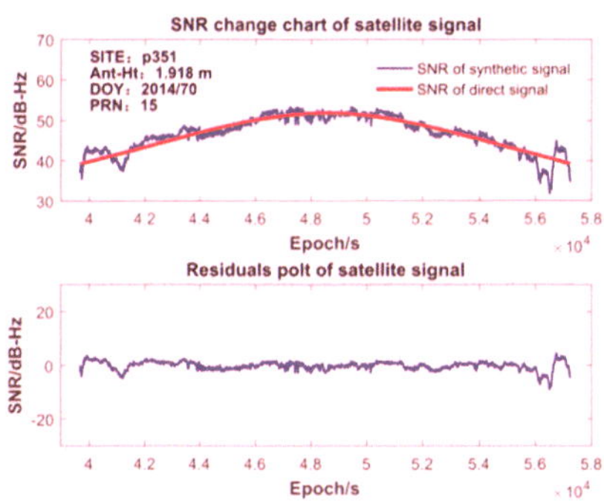

Fig. 2. SNR analysis chart of satellite signal

In SNR of synthetic signal, A_d is the amplitude of the direct signal, A_m is the amplitude of the reflected signal. The SNR in the observation file of the geodetic GNSS receiver is the SNR of the synthetic signal. The mathematical expressions of the amplitude of the synthetic signal, the amplitude of the direct signal and the amplitude of the reflected signal are as follows:

$$SNR^2 = A_c^2 = A_d^2 + A_m^2 + 2A_dA_m \cos Q \tag{2.1}$$

Where A_c is the amplitude of the composite signal; $A_d \gg A_m$; Q is the angle between the direct signal and the reflected signal.

In order to extract the SNR data of the reflected signal from the SNR data of the synthetic signal, the low order polynomial is used to fit the overall trend term of the SNR data. Because $A_d \gg A_m$, the overall trend of SNR of synthetic signal can be regarded as A_d, which is the amplitude of direct signal. The residual SNR image of satellite signal in Fig. 2 is the SNR image after removing the overall trend item. The amplitude of the reflected signal can be expressed as:

$$A_m = A\cos(\frac{4\pi h}{\lambda}\sin E + \varphi) \tag{2.2}$$

Where λ is the carrier wavelength; E is the elevation angle of satellite; H is the vertical reflection distance; if $t = \sin E, f = \frac{2h}{\lambda}$. Then formula (2.2) can be expressed as:

$$A_m = A\cos(2\pi ft + \varphi) \tag{2.3}$$

Because in formula (2.3) $f = \frac{2h}{\lambda}$, where h is the vertical reflection distance RH from the antenna phase center to the snow surface in Fig. 1, the frequency f is obtained through Lomb-Scargle spectrum analysis, and then the vertical distance RH from the antenna phase center to the snow surface layer is obtained from the frequency f. Finally, the snow depth is obtained by subtracting the antenna height from RH.

3 An Example Analysis of GNSS-MR Snow Depth Measurement

3.1 Data Introduction

In order to verify the feasibility of snow depth measurement by GNSS-MR technology, this paper uses the L1 band observation data of p351 station of PBO network in the United States to measure the local snow depth, and compares the measured snow depth of 490 station which is 1.8 km away from P351 station on SNOTEL network, analyzes the accuracy of the measurement results, and puts forward the technical scheme to improve the accuracy.

This paper measures snow depth by using the data of p351 station from the 274th day of 2011 to the 50th day of 2012, the 274th day of 2013 to the 43rd day of 2014, the 290th day of 2017 to the 50th day of 2018, then performs error analysis and accuracy improvement.

3.2 Selection of the Satellite Azimuth Angles

This section uses the observation data of p351 station from the 274th day of 2011 to the 50th day of 2012 to measure snow depth by GNSS-MR technology, and compares the results with the actual snow depth. First, use the data from the station azimuth 0° to 360° to measure snow depth by GNSS-MR. The correlation coefficient between the GNSS-MR measured result and the measured snow depth is 0.344, and the root mean square error is 0.43. The following azimuth angles are 0°–45°, 45°–90°, 90°–135°,

135°–180°, 180°–225°, 225°–270°, 270°–315°, 315°–360° to perform snow depth measurement respectively.

Figure 3 is the surrounding environment of p351 station, Fig. 3(a) is the east terrain of the station, Fig. 3(b) is the west terrain of the station, Fig. 3(c) is the south terrain of the station, and Fig. 3(d) is north terrain of the station. It can be seen from the figure that the terrain around the p351 station is obvious undulating, and the surface around the station is covered by dense vegetation. In order to avoid the influence of topographic fluctuations in different areas around the station on the GNSS-MR snow depth measurement, this article divides the area around the station by azimuth. In order to avoid accidental error affecting the results, the azimuth range cannot be divided too small, so it is divided according to 45o for each area. The azimuth angles are selected as 0°–45°, 45°–90°, 90°–135°, 135°–180°, 180°–225°, 225°–270°, 270°–315°, 315°–360° for GNSS-MR snow depth measurement, respectively.

(a) (b) (c) (d)

Fig. 3. The surrounding environment of p351 station

Table 1 shows the comparison of different azimuth angles measurement results. The azimuth angles are divided every 45° from 0° to 360°, and snow depth measurements are made using GNSS-MR technology, respectively. When the azimuth is 270° to 315°, the correlation coefficient is the smallest which is −0.340 and the root mean square error is 0.85 m; when the azimuth is 90° to 135°, the correlation

Table 1. Comparison of different azimuth angles measurement results

Azimuth angles	Correlations coefficients	Root mean square errors/m
0°–360°	**0.344**	**0.430**
0°–45°	−0.002	0.553
45°–90°	−0.282	0.775
90°–135°	0.920	0.175
135°–180°	0.258	0.475
180°–225°	0.091	0.561
225°–270°	0.035	0.593
270°–315°	−0.345	0.853
315°–360°	0.845	0.228
105°–150°	**0.983**	**0.130**

coefficient is the largest which is 0.92 and the root mean square is 0.17 m. Therefore, the choice of azimuth angle has an important impact on the accuracy of GNSS-MR snow depth measurement. In order to further improve the accuracy of snow depth measurement, the azimuth angle is selected again around 90° to 135° in the experiment of this section, and the optimal azimuth angle of p351 station is determined to be 105° to 150°.

Figure 4 is a comparison chart of GNSS-MR measured snow depth and actual measured snow depth when the azimuth angles is 105° to 150° at p351 station from 274th day of 2011 to 50th day of 2012. It can be seen from the figure that when the azimuth angles is selected from 105° to 150°, GNSS-MR technology can accurately measure the snow depth and change trend more accurately, but with the increase of snow depth, the error between the actual measured snow depth and the measured snow depth by GNSS-MR decreases at first, and then increases gradually.

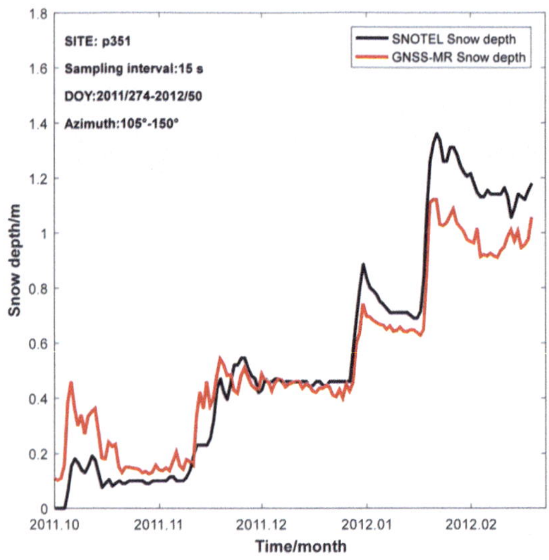

Fig. 4. Comparison between actual measured snow depth and measured snow depth by GNSS-MR from 274th day of 2011 to 50th day of 2012

3.3 Selection of the Satellite Elevation Angle

The basic principle of snow depth measurement by GNSS-MR technology is to use the reflected signal affected by multi-path effect under low elevation angles to measure snow depth, so the selection of satellite elevation angle is also an important factor to improve the accuracy of GNSS-MR snow depth measurement. In this section, the observation data of the p351 station from the 274th day of 2011 to the 50th day of 2012 are selected, and the GNSS-MR snow depth measurement is carried out with elevation angles of 5°–20°, 5°–25°, 5°–30°, 10°–25°, 10°–30°, 15°–30°.

Table 2 shows the measurement results of different elevation angles, and Fig. 5 shows the error analysis diagram of snow depth measured by GNSS-MR at different elevation angles. It can be found that when the elevation angles are 5°–20°, 5°–25°, 5°–30°, 10°–25°, the correlation coefficients are better, the root mean square errors are smaller, and GNSS-MR can reflect the actual snow depth more accurately.

When the elevation angle is 5°–25°, the correlation coefficient is the highest, which is 0.985, and the root mean square error is the smallest, which is 0.128 m; when the elevation angles are 10°–30°, 15°–30°, the correlation is poor, and the mean square root error is large. Preliminary analysis, the main reason is that as the altitude angle increases, the reflection area becomes smaller, the amount of data decreases, and when the altitude angle range is high, the signal is less affected by multipath. At this time, the GNSS-MR snow depth measurement technology cannot get the snow depth value more accurately.

Table 2. Comparison of different elevation angles measurement results

Elevation angles	Correlation coefficients	Root mean square errors/m
5°–20°	0.984	0.150
5°–25°	0.985	0.128
5°–30°	0.983	0.130
10°–25°	0.980	0.135
10°–30°	0.871	0.204
15°–30°	0.665	0.315

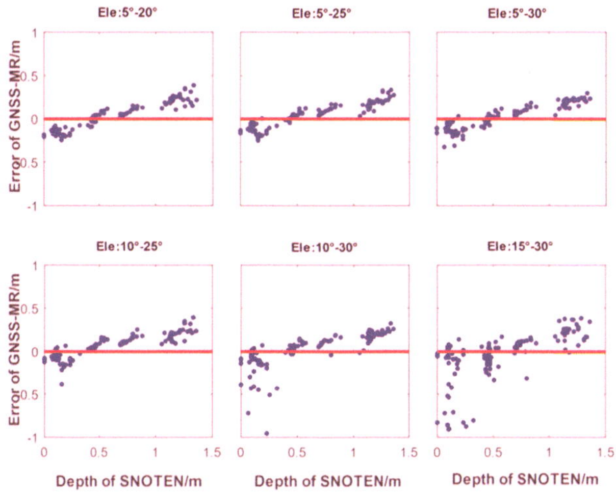

Fig. 5. Error analysis chart of GNSS-MR snow depth measurement at different elevation angles from 2011 to 2012

Figure 6 is a comparison chart of the measured snow depth, GNSS-MR measured snow depth, and PBO H2O measured snow depth from the 274th day of 2013 to the 43rd day of 2014. The correlation coefficient between PBO H2O measured snow depth and GNSS-MR measured snow depth is 0.855, and the root mean square error is 0.278 m; the correlation coefficient between GNSS-MR measured snow depth and actual measured snow depth after screening azimuth angles and elevation angle is 0.974, root mean square error is 0.118 m, indicating that the snow depth measured by GNSS-MR is better than the snow depth measured by PBO H2O.

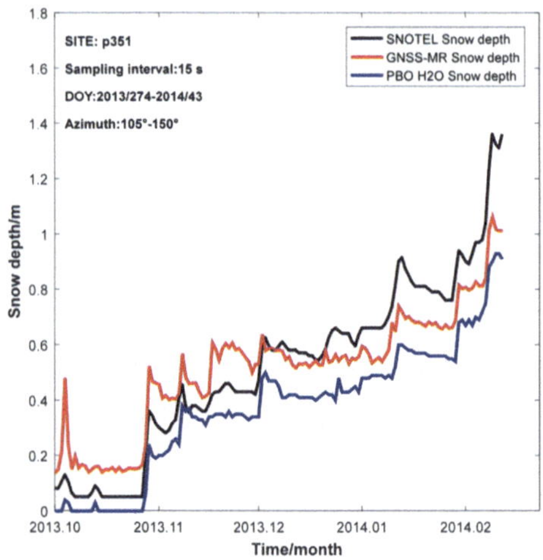

Fig. 6. Comparsion of measured snow depth from 2013 to 2014

Table 3 shows the GNSS-MR snow depth measurement results for three periods which are from the 274th day of 2011 to the 50th day of 2012, from the 274th day of 2013 to the 43th day of 2014, and from the 290th day to the 50th day of 2018. As can be seen from Table 3, the correlation coefficient between the measured GNSS-MR snow depth and the actual snow depth from day 274 of 2011 to day 50 of 2012 is 0.985, and the root mean square error is 0.128 m. Compared with the measurement accuracy in Table 1, there is a slight improvement when the azimuth angle is 105°–150° and the elevation angle is 5°–30°.

Table 3. Snow depth measurement results of GNSS-MR in different periods

Periods	Correlation coefficients	Root mean square errors/m
2011/274–2012/50	0.985	0.128
2013/274–2014/43	0.974	0.118
2017/290–2018/50	0.979	0.138

3.4 Study and Analyze the Relationship Between Snow Depth and Measurement Error

Based on the research and analysis in the previous two sections, we can basically determine that the optimal azimuth angles is 105°–150° and the optimal elevation angles is 5°–25° for p351 station. And the GNSS-MR technology can accurately measure snow depth.

Figure 7 is the correlation analysis chart of GNSS-MR measured snow depth and actual measured snow depth in three periods, the abscissa is the actual measured snow depth value, and the ordinate is the GNSS-MR measured snow depth value. It can be seen from the figure that when the actual measured snow depth is about 0.5 m, the GNSS-MR snow depth is the close to the actual measured snow depth. When the measured snow depth is less than 0.5 m, the GNSS-MR snow depth is basically greater than the measured snow depth. When the actual measured snow depth is greater than 0.5 m, the GNSS-MR snow depth is basically smaller than the actual measured snow depth. According to preliminary analysis, GNSS-MR technology has a negative error in snow depth measurement, and this negative error increases with the increase of snow depth. When the snow depth is small, the snow depth value of GNSS-MR is greater than the actual snow depth due to the influence of ground vegetation and other coverings. When the actual snow depth is about 0.5 m, the negative error is about 0.1 m, which can basically offset the error that caused by the ground coverings around the station. At this time, the error is about 0 m. As the snow depth continues to increase, negative errors begin to appear. There are two possible reasons. On the one hand, some of the signals received by GPS are reflected from the middle of the snow layer or from the soil beneath the snow, and these reflections will cause negative errors in GNSS-MR snow depth measurement. This error is greatest when the surface has fresh, low-density snow. On the other hand, the bare ground for GNSS-MR snow depth measurements may be a few centimeters higher than the ground detected by manual snow detectors. Manual detection may penetrate deeper into vegetation litter and topsoil than this reflecting surface, resulting in negative errors in GNSS-MR snow depth measurements.

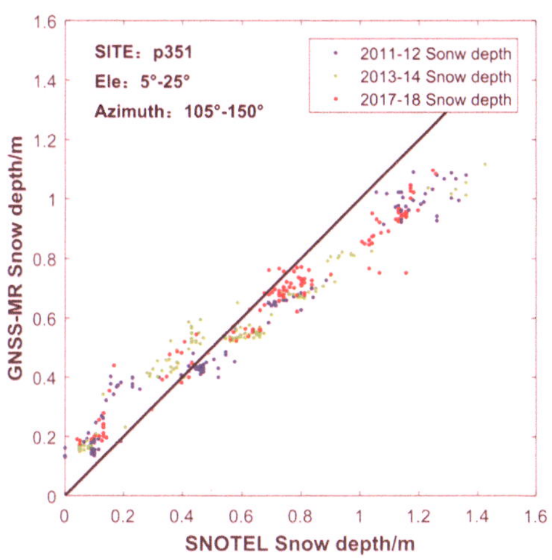

Fig. 7. Correlation analysis of snow depth measured by GNSS-MR and actual measured snow depth

Therefore, this paper uses the L1 band observation data of p351 station from the 274th day of 2011 to the 50th day of 2012 and from the 274th day of 2013 to the 43th day of 2014 to conduct a comparative study of GNSS-MR measured snow depths and measured snow depths. The snow depth values measured by GNSS-MR are sorted from small to large, and an error average value is obtained every 10 cm interval. An empirical model of snow depth and error measurement by GNSS-MR is established by calculating the average values of several errors.

The snow depth data from the 290th day of 2017 to the 50th day of 2018 are used for verification. The snow depth value measured by GNSS-MR is substituted into the above empirical model to obtain the error value predicted by the model. Then, the snow depth value measured by GNSS-MR is corrected according to the error predicted by the empirical model.

Figure 8 is a comparison chart of the actual measured snow depth, the snow depth measured by GNSS-MR, and the snow depth corrected by empirical model. The results show that the correlation coefficient between GNSS-MR measured snow depth and actual measured snow depth has decreased from 0.979 to 0.976 after correction of the empirical model, and the root mean square error has decreased from 0.131 m to 0.079 m. The feasibility of this empirical model is preliminary proved, which can better weaken the negative error of GNSS-MR when measuring snow depth.

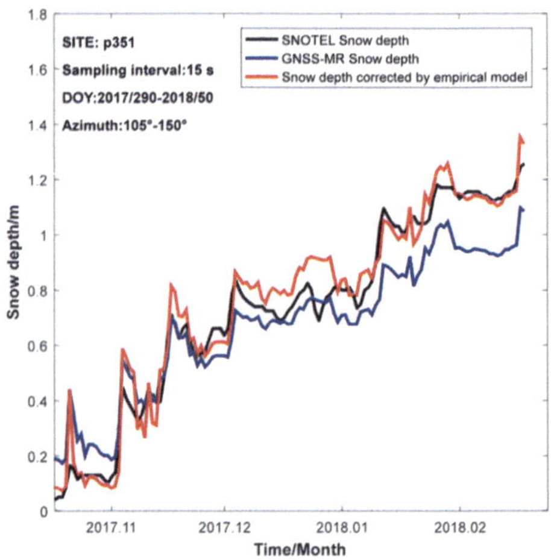

Fig. 8. Comparison of actual measured snow depth, the snow depth measured by GNSS-MR and the snow depth corrected by empirical model

4 Conclusions

Based on the study of the snow depth measurement principle of GNSS-MR technology, this paper uses the observation data of p351 station to carry out measurement experiments, and compares the snow depth measured by GNSS-MR with the actual measured snow depth to analyze the causes of the error. The weakening methods are proposed for the errors of different reasons, and the specific conclusions are as follows:

(1) Satellite observation data at different azimuth angles have an important impact on the accuracy of GNSS-MR snow depth measurement. The reason may be caused by the topographic fluctuations and ground coverings around the station. Therefore, when using GNSS-MR technology to measure snow depth, it is necessary to fully understand the terrain and landform around station.

(2) By measuring and comparing satellite observation data of different elevation angles selected at the p351 station, it can be found that the accuracy is highest when the satellite elevation angle is 5°–25°. Therefore, this article considers that the 5°–25° elevation angle can be used as the optimal range of elevation angles for GNSS-MR to measure snow depth, but it can be adjusted appropriately for different stations and the environment of the station.

(3) Through studying and analyzing the relationship between snow depth and error, it can be found that there is a negative error in snow depth measurement by GNSS-MR technology. As the snow depth increases, the negative error also increases. The empirical model obtained by analyzing the error of previous years can reduce the negative error better. However, the negative error does not appear when the snow depth is small. This may be due to the positive error caused by the ground cover is greater than the negative error in the GNSS-MR snow depth measurement. However, he quantitative relationship between negative error and snow depth still needs further research.

Acknowledgements. The authors would like to thank the PBO H2O research team for providing experimental data, and the US Department of Agriculture (USDA) Natural Resources Conservation Service Organization (NRCS) for providing measured snow depth data.

References

1. Comp, C., Axelrad, P.: Adaptive SNR-based carrier phase multipath mitigation technique. IEEE Trans. Aerosp. Electron. Syst. **34**(1), 264–276 (1998)
2. Bilich, A., Larson, K.M., Axelrad, P.: Modeling GPS phase multipath with SNR: case study from the Salar de Uyuni, Boliva. J. Geophys. Res. **113**(B4), B04401 (2008)
3. Larson, K.M., Gutmann, E.D., Zavorotny, V.U., Braun, J.J., Williams, M.W., Nievinski, F. G.: Can we measure snow depth with gps receivers? Geophys. Res. Lett. **36**(17), L17502 (2009)
4. Zhabg, S., Dai, K., Liu, K., Hou, X., Zhao, Y.: Research of GPS-MR on snow depth monitoring. Progr. Geophys. **31**(4), 1879–1884 (2016). https://doi.org/10.6038/pg20160461
5. Zhang, S., Wang, X., Zhang, Q.: Avoiding errors attributable to topography in GPS-IR snow depth retrievals. Adv. Space Res. **59**(6), 1663–1669 (2017)
6. Zhou, W., Liu, L., Huang, L., Li, J., Chen, J., Chen, F.D., Xing, Y., Liu, L.B.: Monitoring snow depth based on the SNR signal of GLONASS satellites. J. Remote Sens. **22**(5), 889–899 (2018)
7. Li, Y., Chang, X., Yu, K., Wang, S., Li, J.: Estimation of snow depth using pseudorange and carrier phase observations of GNSs single-frequency signal. GPS Solut. **23**(4), 118 (2019)

8. Small, E.E., Larson, K.M., Braun, J.J.: Sensing vegetation growth with reflected GPS signals. Geophys. Res. Lett. **37**(12) (2010)
9. Mccreight, J.L., Small, E.E., Larson, K.M.: Snow depth, density, and SWE estimates derived from GPS reflection data: validation in the western U.S. Water Resour. Res. **50**(8), 6892–6909 (2014)
10. Larson, K.M., Van Dam, T.: Measuring postglacial rebound with gps and absolute gravity. Geophys. Res. Lett. **27**(23), 3925–3928 (2000)
11. Wan, W., Larson, K.M., Small, E.E., Chew, C.C., Braun, J.J.: Using geodetic GPS receivers to measure vegetation water content. GPS Solut. **19**(2), 237–248 (2015)
12. Bilich, A., Larson, K.M.: Mapping the GPS multipath environment using the signal-to-noise ratio (SNR). Radio Sci. **42**(6) (2007)
13. Larson, K.M., Wahr, J., Kuipers Munneke, P.: Constraints on snow accumulation and firn density in greenland using GPS receivers. J. Glaciol. **61**(225), 101–114 (2015)

Tide Height Inversion and Accuracy Analysis Based on GNSS-MR Technology

Naiquan Zheng, Peng Chen(✉), Zheng Li, Yongchao Ma, and Lixia Liu

College of Geomatics, Xi'an University of Science and Technology, Xi'an 710054, China
chenpeng0123@gmail.com

Abstract. The tide height of offshore waters is an indispensable marine geographic information data for marine traffic, offshore environmental management, and marine disaster warning and forecasting. Recently years, with the advantages of low power consumption and low cost, the Global Navigation Satellite System Multipath Reflection (GNSS-MR) technology has become an emerging technical method of tide height measurement. In this article, the observation files of SC02 station in 2018 in Washington state of United States is used to retrieve the distance in vertical direction from antenna phase center to sea surface by using the change of signal-to-noise ratio (SNR) caused by the multipath effect of navigation satellite signal. Afterwards, the elevation based on International Terrestrial Reference Frame (ITRF) is converted to the elevation based on the Mean Lower Low Water (MLLW). And then the tide height of each inversion time relative to MLLW is obtained. Finally, the retrieved tide height is compared and analyzed with the tide height offered by the tide gauge station named FRIDAY HARBOR. The results show that the mean error of the two is about −12 cm, the root mean square error is about 15 cm, and the correlation coefficient is about 0.98. Overall, the retrieved tide height agrees well with the measured tide height. Hence, the tide height retrieved by GNSS-MR technology can make up for the lack of ocean observation data in the coastal waters and plays an important role in marine scientific research.

Keywords: Tide height · GNSS-MR · SNR · MLLW

1 Introduction

The Global Navigation Satellite System (GNSS) not only provides navigation and positioning, speed measurement, timing and other information for users of spatial information, but also provides L-band microwave signals that can be used for a long time, which contribute to the creation of the GNSS-MR technology [1]. The tide height in the coastal waters is an important marine geographic information data, which is of great significance to marine scientific research.

J. Sun et al. (Eds.): CSNC 2020, LNEE 650, pp. 141–152, 2020.
https://doi.org/10.1007/978-981-15-3707-3_14

At present, the tide height is mainly obtained by means of tide gauge and satellite altimeter. Recently years, with the continuous progress of GNSS-MR technology, it has become possible to measure tide height by GNSS-MR technology. In 2014, Lofgren et al. extracted the observation files of five GPS stations distributed in the northern and southern hemispheres to analyze the variations of sea level with time series, and the correlation coefficients were all between 0.89–0.99, which has a good consistency in general [2]. In the same year, Lofgren and Haas proposed to use SNR and carrier phase to retrieve the tide height respectively, and the results showed good correlation [3]. In 2015, Roussel et al. proposed a new mean which is based on the least square method (LSM) that combined the SNR of GPS and GLONASS system to retrieve the tide height. The correlation coefficient was significantly improved [4]. In 2016, Zhang et al. selected SC02 station to monitor the change of sea level, and compared it with the tide height of tide gauge. Results showed that the correlation coefficient between them was better than 0.98 [5]. In the same year, Strandberg et al. proposed to use L1 and L2 signals of GPS and GLONASS for inversion, proving that both signals can be used for tide height inversion [6]. In 2018, Wang et al. proposed a method to determine the sea area azimuth angles and elevation angles of the station, and conducted analysis on stations of PBAY, SC02 and BRST respectively [7]. In 2019, Puente and Valdés used multi-system and multi-frequency to invert the tide height in the coast of Spain, and analyzed the advantages and disadvantages of this method [8]. In the same year, Fade Chen used the SNR data of the BDS, GPS, and GLONASS of the MAYG station to generate the initial tide waveform, combined with wavelet denoising to retrieve the tide height, and the correlation coefficient was increased to 0.95 [9].

In this paper, the method and theory of GNSS-MR technology to invert the tide height is systematically introduces. The observation data of SC02 station in 2018 is selected for inversion of tide height, and compared with the tide height of FRIDAY HARBOR which is 359 m away from SC02 station, then accuracy analysis is performed. The results show that the tide height retrieved by GNSS-MR technology can be used to make up for the shortage of ocean observation data in the coastal area.

2 Theoretical of Tide Height Inversion by GNSS-MR

2.1 Fundamental

In actual, the inversion of tide height by GNSS-MR technology is to retrieve the distance in vertical direction from the antenna phase center to sea level by using the change of SNR. As we all know, the MLLW is used as the chart depth reference surface in the United States. Afterwards, the elevation based on ITRF is converted to the elevation based on the MLLW. And then the tide height of each inversion time relative to MLLW is obtained. Figure 1 shows the diagram of tide height inverted by

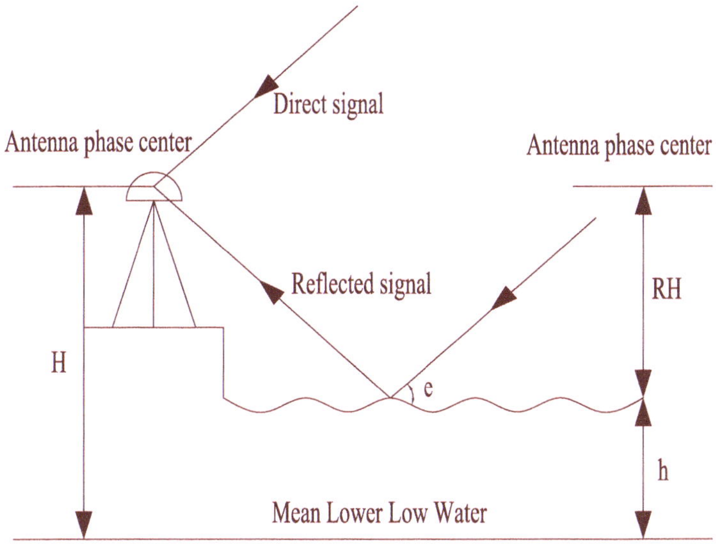

Fig. 1. Diagram of tide height inverted by GNSS-MR

GNSS-MR Technology. Where H is the height from the MLLW to the antenna phase center, RH is the vertical reflection distance from the sea surface to the antenna phase center, h is the measured tide height from the MLLW to the sea surface, and e represents the satellite elevation angle.

Figure 1 shows the SNR change of the PRN10 at SC02 station of doy335 in 2018, and the overall SNR trend of the satellite in the ascending and descending phases is shown. In the short time of satellite ascent and descent, there is a tendency that the SNR amplitude is significantly larger. Figure 2(a) extracts the original SNR data of PRN10 and obtains the direct signal through low-order polynomial fitting [10, 16]. The blue line represents the change of original SNR with time series, and the orange line reflects the overall trend of SNR after low-order polynomial fitting. Figure 2(b) is the SNR residual (δSNR) caused by multipath effect in the SNR after removing the overall trend. The blue line reflects the change of δSNR, and the orange line reflects the change of the satellite elevation angle during this period. The figure clearly shows that the variations of δSNR is closely related to the satellite elevation angle. At high elevation angles, δSNR is about ±2 dB-Hz; while at low elevation angles, δSNR amplitude is obviously larger, reaching ±5 dB-Hz. It indicates that δSNR is affected by multipath effect at low elevation angles, which provides a basis for using multipath effect to invert tide height.

In Fig. 1, the path delay caused by the multipath effect is:

$$D = 2RH\ sin\ e \tag{2.1}$$

Where D is the path delay caused by the reflected signal and the direct signal during the propagation of the satellite signal.

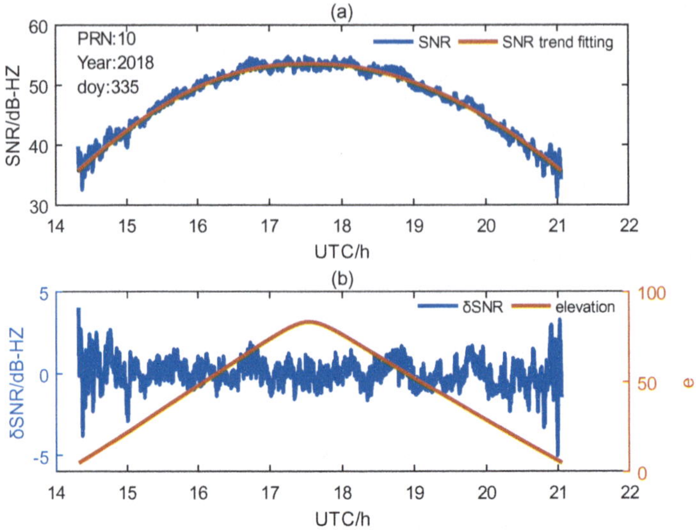

Fig. 2. SNR variations of PRN10

In Fig. 2, the phase difference of δSNR affected by multipath is [11]:

$$\varphi_i = \frac{2\pi D}{\lambda} \tag{2.2}$$

Where φ is the phase difference between the reflected signal and the direct signal, λ is the length of the band, and the signal length of the L1 band is taken as $\lambda_1 = 19.03$ cm.

Lomb-Scargle (L-S) spectral analysis is performed on the δSNR in Fig. 2, and the frequency of the δSNR is [12, 13]:

$$\omega_\varphi = 2\pi f = \frac{d\varphi_i}{d\ sin\ e} \tag{2.3}$$

From the formulas (2.1), (2.2) and (2.3), the vertical reflection distance from the sea surface to the antenna phase center is [14, 15]:

$$RH = \frac{\lambda f}{2} \tag{2.4}$$

According to Fig. 1, the tide height retrieved by GNSS-MR is:

$$GNSSMR_h = H - RH \tag{2.5}$$

Where $GNSSMR_h$ is the tide height retrieved by GNSS-MR technology.

2.2 Accuracy Analysis

In this paper, the observation data with an ebb period and a year of SC02 station in 2018 are taken as examples to invert tide height and perform accuracy analysis. The selected accuracy indicators are mean error (*MEAN*), root mean square error (*RMSE*), and correlation coefficient (*COEF*). The calculation formulas are as follows:

$$MEAN = \frac{1}{n}\sum\nolimits_{i=1}^{n}(GNSSMR_h - TG_h) \tag{2.6}$$

$$RMSE = \sqrt{\frac{1}{n}\sum\nolimits_{i=1}^{n}(GNSSMR_h - TG_h)^2} \tag{2.7}$$

Where TG_h is the tide height measured by the tide gauge.

3 An Example Analysis of Tide Height Inverted by GNSS-MR

3.1 Station Environment

In order to ensure that the satellite signal received in the experiments are reflected from sea level, the elevation angles range is selected as 5°–12° and the azimuth angles range is selected as 90°–150° when using the L1 signal of the satellite for tide height inversion.

When using the observation files of SC02 station of the GPS Continuous Operation Reference Station (CORS) located in Washington state, USA to retrieve the tide height, it involves the elevation conversion between datum of ITRF08 and MLLW. The parameters of SC02 station are listed in Table 1:

Table 1. Parameters of SC02 in 2018

Station	SC02
Latitude (°)	48.54619494
Longitude (°)	−123.00761033
Height (m)	−15.0345
Region	Contiguous United States
Reference frame	ITRF08
Reference epoch	2004.4973
Tide gauge	FRIDAY HARBOR
Distance to tide gauge (m)	359
Antenna type	TRM59800.80
H (m)	6.603
TCR (m)	0.1020

Through the above parameters, the elevation of the antenna phase center based on MLLW is obtained. The schematic is shown in Fig. 3. Then the tide height retrieved by GNSS-MR technology is calculated by formula (2.5). And it is compared with the measured tide height of FRIDAY HARBOR provided by the National Oceanic and Atmospheric Administration (NOAA) every 6 min.

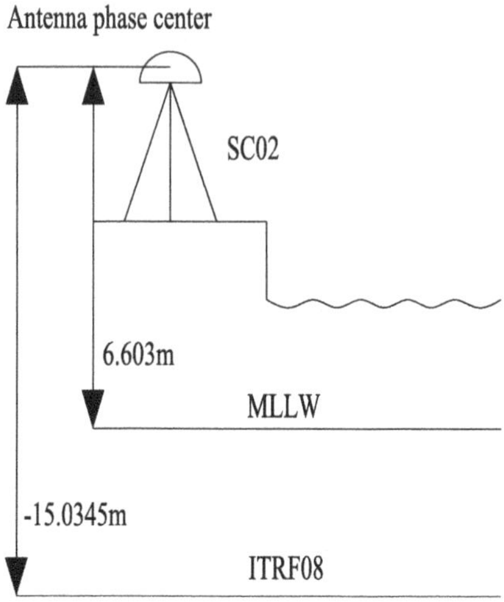

Fig. 3. The elevation conversion between datum of ITRF08 and MLLW

3.2 Analysis of the Inversion Results of an Ebb Tide Period

In order to analyze the periodicity of tide height inverted by GNSS-MR multi-system, SNR in L1 band provided by GPS and GLONASS at SC02 station during the period of doy335–350 in 2018 is extracted for tide height inversion. Table 2 takes the observation data of doy335 as an example, and gives the comparison results of the tide height retrieved by GPS and the tide height measured by tide gauge. It is found that the deviations between the two are −0.2–0.2 m, which has a good consistency. And T represents the time length of L-S spectrum analysis. Mean UTC represents the average time during the spectrum analysis period.

Table 2. Comparison of GNSS-MR and tide gauge

PRN	T/min	Mean UTC/h	RH/m	$GNSSMR_h$/m	TG_h/m	Deviations/m
01	23	4.88	5.41	1.295	1.415	−0.120
10	19	20.90	3.89	2.815	2.655	0.160
13	23	13.38	5.95	0.755	0.729	0.026
20	19	19.74	4.00	2.705	2.675	0.030
24	26	16.56	4.92	1.785	1.707	0.078
27	22	1.00	5.27	1.435	1.464	−0.029
28	20	8.69	5.25	1.460	1.340	0.120
07	18	6.08	5.25	1.455	1.507	−0.052
08	20	1.47	5.32	1.385	1.382	0.003
15	20	13.52	5.93	0.775	0.753	0.022
16	18	22.35	4.46	2.250	2.306	−0.056
23	16	3.38	5.28	1.425	1.285	0.140
26	20	21.56	4.25	2.450	2.528	−0.078
30	17	7.30	5.29	1.415	1.504	−0.089

Figure 4 shows the δSNR variations and L-S spectral analysis results of the PRN01 satellite after linearization. In Fig. 4(a), the abscissa is the sine of the elevation angles and the ordinate is δSNR. It is found that the amplitude becomes smaller and smaller with the increase of the elevation angles. In Fig. 4(b), the abscissa is the vertical reflection distance and the ordinate is the L-S spectral amplitude of δSNR. And the vertical reflection distance corresponding to the peak of the L-S spectral amplitude is the height from the antenna phase center to the instantaneous sea level, so the vertical reflection distance calculated by the PRN01 satellite at mean UTC of 4.88 h is 5.41 m.

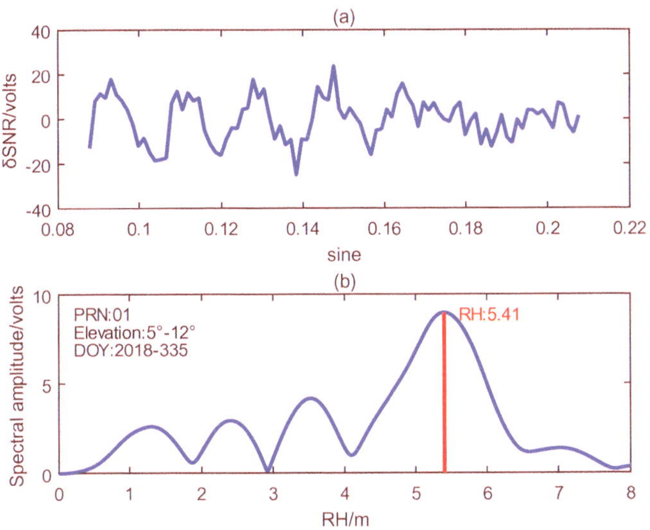

Fig. 4. SNR residual sequence and L-S spectral analysis

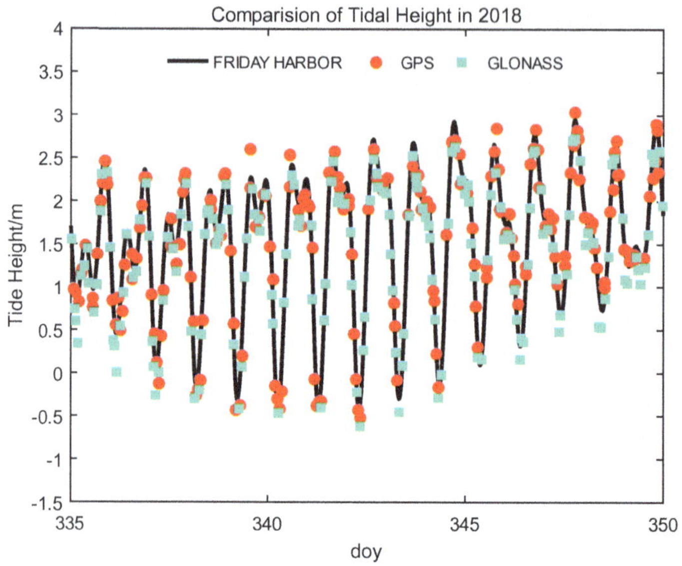

Fig. 5. Comparison of tide height between GNSS-MR and tide gauge

Figure 5 shows the comparison of tide height between GNSS-MR and FRIDAY HARBOR. In the figure, the black line represents the measured tide height of FRIDAY HARBOR, the red dots represent the 208 instantaneous tide height retrieved by the GPS satellite, and the cyan dots represent the 181 instantaneous tide height retrieved by the GLONASS satellite. It can be seen from the figure that there exists tide height retrieved by GNSS-MR at both the tide peaks and the tide valleys, indicating that this method can not only reflects the rising and falling tide trend of a day, but also reflects the variations trend of tide height during the ebb tide period. Overall, the retrieved tide height agrees well with the measured tide height.

Table 3. Inversion of tide height at SC02 station in 2018 of doy335–350 by GPS, GLONASS and GPS+GLONASS

Constellation	MEAN/m	RMSE/m	COEF
GPS	−0.042	0.154	0.984
GLONASS	−0.157	0.129	0.987
GPS+GLONASS	−0.096	0.154	0.983

Table 3 shows the comparison statistics of tide height retrieved by GPS, GLONASS and GPS+GLONASS. Results show that the MEAN of all three are negative values, and the MEAN of GPS is the smallest, which is −0.042 m. The RMSE of GLONASS is the smallest, which is 0.129 m, and the COEF is the largest, which is 0.987. Although the COEF of GPS+GLONASS dual satellite system is slightly lower than that of single satellite system, the retrieved tide height still has good consistency. With the increase of the number of retrieved instantaneous tide height, its spatial-temporal resolution is also significantly improved, which verifies the feasibility of multi-system inversion of tide height.

Figure 6 shows the correlation analysis and error deviations of tide height between GNSS-MR and FRIDAY HARBOR tide gauge. Figure 6(a)–(c) are the correlation analysis diagrams between the tide height retrieved by GPS, GLONASS and GPS +GLONASS and the tide height measured by tide gauge. It can be seen from the figure that the tide height retrieved by GPS and GPS+GLONASS is evenly distributed on both sides of the anastomosis line, while most of the tide height retrieved by GLONASS is below the anastomosis line, which makes the retrieved tide height less than the measured tide height. It may be caused by some uncertain factors of GLONASS, which needs further discovery and research. Figure 6(d)–(f) intuitively reflect the deviations between the tide height retrieved by GPS, GLONASS and GPS+GLONASS and the tide height measured by tide gauge. Most of the tide height deviations retrieved by GPS are concentrated in −0.3–0.2 m, and most of the tide height deviations retrieved by GLONASS are concentrated in −0.4–0.1 m, indicating that the tide height retrieved by GNSS-MR has a high accuracy.

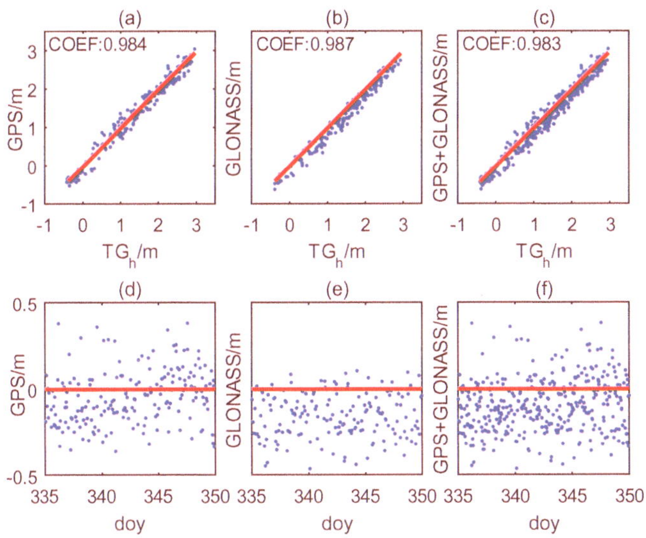

Fig. 6. Correlation analysis and error deviations of tide height between GNSS-MR and tide gauge

3.3 Analysis of the Inversion Results of a Year in 2018

In order to analyze the continuity of tide height inverted by GNSS-MR multi-system, SNR in L1 band provided by GPS and GLONASS at SC02 station in 2018 is extracted for tide height inversion. It should be noted that the long-period tide height inversion value covers the vertical movement of the GPS station, and the vertical displacement of the GPS station of SC02 station studied in this paper is small, so it can be ignored.

Table 4 shows the comparison statistics of tide height retrieved by GPS, GLONASS and GPS+GLONASS in 2018. Results show that the MEAN of all three are negative values, and the MEAN of GPS is −0.074 m, and that of GLONASS is −0.165 m. There is a certain deviation between the two systems. The RMSE of GLONASS is the smallest, which is 0.124 m, and the COEF is the largest, which is 0.987. Although the COEF of GPS+GLONASS dual system is slightly lower than that of single system, the retrieved tide height still has good consistency. With the increase of the number of retrieved instantaneous tide height, its spatial-temporal resolution is also significantly improved, indicating that the tide height retrieved by GNSS-MR multi-system can reflect the continuous change of a year, and generally can be used to describe the tide height change of a long time series.

Table 4. Inversion of tide height at SC02 station in 2018 by GPS, GLONASS and GPS +GLONASS

Constellation	MEAN/m	RMSE/m	COEF
GPS	−0.074	0.163	0.976
GLONASS	−0.165	0.124	0.987
GPS+GLONASS	−0.116	0.152	0.980

Figure 7 shows the correlation analysis and error deviations of tide height retrieved by GPS+GLONASS and the tide height measured by FRIDAY HARBOR tide gauge. Figure 7(a) are the correlation analysis diagrams of the two. The MEAN is −0.116 m, the RMSE is 0.152 m, and the COEF is 0.980, which indicates that the tide height retrieved by GNSS-MR can show the continuity of the tide height with time series. Figure 7(b) shows the variations of the errors between the two with day of year (doy). As shown in the figure, most of the errors are concentrated in −0.4–0.2 m, which is affected by GLONASS, leading to the overall value to be negative. At present, this part of errors may be caused by two reasons. On the one hand, the instantaneous tide height retrieved by GNSS-MR deviates from the measured tide height by tide gauge due to the wave fluctuations on the sea level. On the other hand, the GNSS observation stations and tide gauge are not located at the same place, which inevitably causes errors in the conversion process of elevation datum.

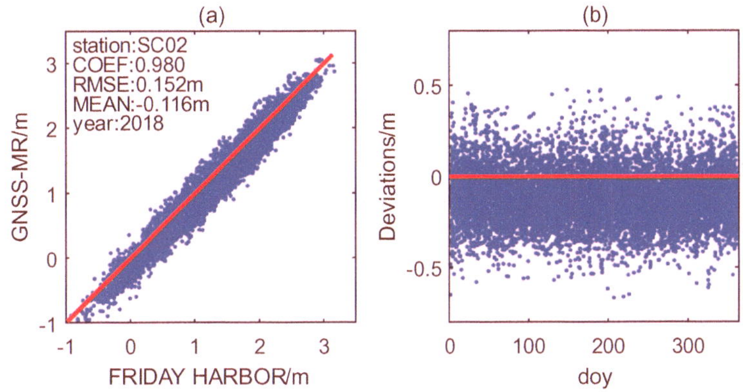

Fig. 7. Correlation analysis and error deviations of tide height between GNSS-MR and tide gauge

4 Conclusions

In this paper, the observation files of SC02 station in 2018 are taken as examples to invert the tide height and analyze the accuracy. The conclusions are as follows:

(1) From the data analysis of 2018, the instantaneous tide height retrieved by GNSS-MR technology based on SNR has obvious periodicity and continuity, which can show the dynamic change of tide height with time series. Compared with the measured tide height by tide gauge, it has good consistency in general. As a new remote sensing method, the retrieved tide height by GNSS-MR can be used to make up for the lack of ocean observation data in the coastal waters, which plays an important role in marine scientific research.

(2) Combining GPS and GLONASS dual system to perform tide height inversion, the spatial-temporal resolution is significantly improved. The MEAN between the two is about −12 cm, the RMSE is about 15 cm, and the COEF is about 0.98. In the future, the effects of integrating the four major systems of GPS, GLONASS, Galileo and BDS on the accuracy of the tide height inversion will be further discussed.

(3) At present, the GNSS-MR technology has preliminarily realized the function of the tide gauge. The next step will continue to study how to reduce the impact of wave fluctuations on the tide height and the errors caused by the no co-located between GNSS observation stations and the tide gauge.

Acknowledgements. The authors would gratefully thank SONEL for providing GNSS observation data and NOAA for providing the measured data from the tide gauge.

References

1. Liu, J., Shao, L., Zhang, X.: Advances in GNSS-R studies and key technologies. J. Wuhan Univ. Inform. Sci. Edn. **32**(11), 955–960 (2007)
2. LoFgren, J.S., Haas, R., Scherneck, H.G.: Sea level time series and ocean tide analysis from multipath signals at five GPS sites in different parts of the world. J. Geodyn. **80**, 66–80 (2014)
3. Löfgren, J.S., Haas, R.: Sea level observations using multi-system GNSS reflectometry. In: NKG, 17th General Assembly (2014)
4. Roussel, N., Ramillien, G., Frappart, F., Darrozes, J., Gay, A., Biancale, R., Allain, D.: Sea level monitoring and sea state estimate using a single geodetic receiver. Remote Sens. Environ. **171**, 261–277 (2015)
5. Zhang, S., Nan, Y., Li, Z., Zhang, Q., Dai, K., Zhao, Y.: Analysis of tide variation monitored by GNSS-MR. J. Surv. Mapp. **9**, 1042–1049 (2016)
6. Strandberg, J., Hobiger, T., Haas, R.: Sea level measurements from inverse modelling of GNSS SNR data-initial results. In: EGU General Assembly Conference Abstracts, vol. 18 (2016)
7. Wang, X., Zhang, Q., Zhang, S.: Azimuth selection for sea level measurements using geodetic GPS receivers. Adv. Space Res. **61**(6), 1546–1557 (2018)
8. Puente, V., Valdés, M.: Sea level determination in the Spanish coast using GNSS-R. In: Multidisciplinary Digital Publishing Institute Proceedings, vol. 19, No. 1, p. 11 (2019)
9. Chen, F., Liu, L., Guo, F.: Sea surface height estimation with multi-GNSS and wavelet de-noising. Sci. Rep. **9**(1), 1–10 (2019)
10. Zimmermann, F., Schmitz, B., Klingbeil, L., Kuhlmann, H.: GPS multipath analysis using Fresnel zones. Sensors **19**(1), 25 (2019)
11. Wang, X., Zhang, Q., Zhang, S.: Water levels measured with SNR using wavelet decomposition and Lomb-Scargle periodogram. GPS Solut. **22**(1), 22 (2018)
12. Wang, X., He, X., Zhang, Q.: Evaluation and combination of quad-constellation multi-GNSS multipath reflectometry applied to sea level retrieval. Remote Sens. Environ. **231**, 111229 (2019)
13. Wang, X., Zhang, Q., Zhang, S.: Sea level estimation from SNR data of geodetic receivers using wavelet analysis. GPS Solut. **23**(1), 6 (2019)
14. Song, M., He, X., Wang, X., Zhou, Y., Xu, X.: Study on the quality control for periodogram in the determination of water level using the GNSS-IR technique. Sensors **19**(20), 4524 (2019)
15. Jin, S., Qian, X., Wu, X.: Sea level change from BeiDou navigation satellite system-reflectometry (BDS-R): first results and evaluation. Glob. Planet. Change **149**, 20–25 (2017)
16. Roesler, C., Larson, K.M.: Software tools for GNSS interferometric reflectometry (GNSS-IR). GPS Solut. **22**(3), 80 (2018)

Research on Sea Surface Height Measurement Based on GNSS-IR Dual Frequency Data Fusion

Jie Wang[1], Tianhe Xu[2(✉)], Nazi Wang[2,3], Yunqiao He[2], and Fan Gao[2]

[1] Chang'an University, Xi'an, China
aibuwjl12@163.com
[2] Institute of Space Science, Shandong University, Weihai, China
thxugfz@163.com
[3] State Key Laboratory of Geo-Information Engineering, Xi'an, China

Abstract. It is an important application of GNSS in marine remote sensing to measure sea level by using Global Navigation Satellite System Interferometry Reflectometry (GNSS-IR), which provides a new direction for global sea level change research. At present, the accuracy of sea level retrieved from GNSS single-frequency observations needs to be further improved. In order to solve this problem, this paper proposed a new retrieval method based on the peak weighting scheme of integrating the respective sea level retrieved from signal-to-noise ratio (SNR) data of two single frequency L1 and L2 of the Global Positioning System (GPS). The peak weighting method takes the peak power of the spectral analysis of the observed signal-to-noise ratio as the weight, thereby integrating dual-frequency data for sea level retrieval. In order to verify the validity of the mentioned method, the continuous multi-day dual-frequency SNR observation data of two coastal GPS stations were processed and analyzed. At first, the Lomb-Scargle Periodogram (LSP) spectral analysis method was used to obtain the peak oscillation frequency. Then the peak weighting method is used to fuse the two-frequency sea level retrieval results. At last, the sea level results obtained from the single-frequency observations are compared with that of the double-frequency observations, and the proposed weighting method is verified by comparing with tide gauge. The results show that the accuracy of the retrieval results using the peak weighting method is better than that of the GPS single-frequency observations when sea condition is good, while the accuracy improvement is not obvious when the sea surface is rough.

Keywords: GNSS-IR · Sea surface height · SNR · LSP analysis · The peak weighting method

1 Introduction

As global mean sea level rises due to climate warming, monitoring and studying sea level changes are playing an increasingly important role [1]. The traditional method to monitor the change of sea level is through the observation data of the tide gauge. With

J. Sun et al. (Eds.): CSNC 2020, LNEE 650, pp. 153–165, 2020.
https://doi.org/10.1007/978-981-15-3707-3_15

the development of GNSS, a new remote sensing technology, named as Global Navigation Satellite System Interferometry Reflectometry (GNSS-IR) has been developed. This method uses GNSS reflected signals to observe sea level. The GNSS receiver receives direct satellite signals as well as signals reflected from the surrounding ground. The reflected signal carries the characteristic information of the reflection surface, which can be obtained by analyzing the variation, polarization characteristics, amplitude, phase, frequency and other parameters of the reflection signal waveform. At present, GNSS-IR is applied in many fields, such as soil moisture, vegetation water content, sea level change, snow depth and moving target detection and so on. Because it can use the abundant data of established coastal observation stations when monitoring sea level changes, GNSS-IR technology becomes one of the research hotspots in the field of ocean remote sensing.

Martin-Neira first introduced the concept of passive reflectometry and interferometry system (PARIS) in 1993 [2]. that is, the GPS receiver receives the direct signal and the ground reflected signal transmitted by the satellite together, and applied this technology to the sea surface altimetry. In later studies, Anderson considered the interference effect between the direct signal and the reflected signal in the same antenna, and used the signal-to-noise ratio (SNR) data from the sea surface to retrieve sea level [3]. Bilich used the data from two stations to verify the consistency between the results of sea level retrieval from GPS SNR data and from the tide gauge data. The root mean square (RMS) error of both stations was less than 10 cm with correlation coefficient of better than 0.97 [4]. In 2012, Larson proposed a multipath theory based on specular reflection and a new method using SNR data to monitor sea level change, which can be applied to the calibration and verification of altimeters [5]. Considering the time-varying characteristics of sea level, Löfgren proposed an extended analysis method with the change rate of sea level in 2014. The experimental results show that new method performs significantly better than the standard analysis method [6]. In 2016, Strandberg proposed a new method for retrieval of sea surface height using GNSS SNR data, which relies on the inverse model of detrended SNR (dSNR), and verified that the RMS error of this method is about 1.8 cm with worse performance than the spectral analysis [7]. In 2017, Jin et al. used three-frequency combined observations of BDS carrier phase, pseudorange and SNR data to observe the sea level with accuracy of less than 0.6 m [8]. Wang et al. adopted five different observations to observe sea level change based on GNSS carrier phase and SNR observations in 2018, and the RMS error was both less than 0.2 m [9].

At present, the sea level retrieval precision from GNSS-IR single frequency observation is not high enough to meet the precision requirements of geodetic survey. Therefore, the focus of current research on GNSS-IR is to adopt new processing methods to achieve higher accuracy sea level retrieval results. Based on the retrieval principle of sea surface height from GPS SNR data, this paper proposes a dual-frequency data fusion method to combine the SNR observations of L1 and L2 in a weighted way to improve the retrieval accuracy, and uses the measured data from two GPS observation stations to verify its validation.

2 GNSS-IR Sea Surface Height Retrieval Principle

GNSS-IR usually uses SNR data received by geodetic receivers for sea level retrieval. SNR is defined as the ratio between signal power and noise power, which is an important parameter to measure the signal quality received by the receiver. It is mainly affected by antenna gain, receiver noise and multipath effects [10].

The principle of observing sea surface height by GNSS-IR is shown in Fig. 1. The GNSS antenna installed along the coast receives both the direct satellite signal and the signal reflected by the sea surface. The direct signal and the reflected signal produce interference effects at the antenna and affect the SNR value recorded by the receiver. By analyzing the SNR value, the vertical distance h between the antenna phase center and the reflecting surface can be obtained. The geodetic height of the GPS antenna can be obtained by precise point positioning (PPP) technology or other positioning methods. The sea level change can be obtained by subtracting h from the geodetic height of the antenna.

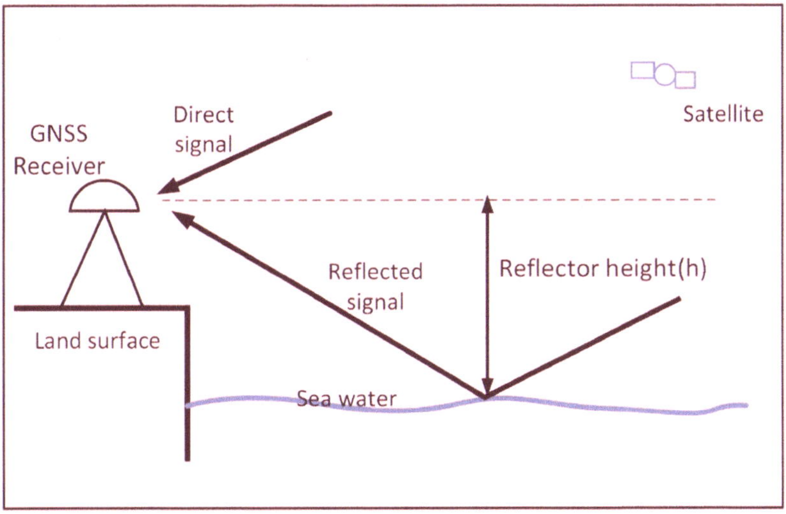

Fig. 1. GNSS-IR sea surface height retrieval

The GNSS receiver receives the combined signals of the direct signal and the reflected signals transmitted by the satellite. The amplitude of the direct signal is A_d and the amplitude of the reflected signal is A_m. Since the multipath signal is an important error source in GNSS positioning, the GNSS antenna are designed to be sensitive to the direct satellite signals and to suppress the reflected signals from the environment around the receiver, i.e., the antenna gain in the upper hemisphere has higher values, thus gives [11]

$$A_d \gg A_m \tag{1}$$

SNR can be described as

$$SNR^2 = A_c^2 = A_d^2 + A_m^2 + 2A_d A_m cos\varphi \tag{2}$$

Where A_c is the amplitude of the composite signal and φ is the phase of the reflected signal vector relative to the direct signal vector. Satellite elevation angle and azimuth can be calculated according to the satellite position provided by the satellite ephemeris file. When the altitude angle of the satellite is lower, the energy transmitted to the reflector by the incident GPS signal is lower, the energy of the reflected wave is higher, and the SNR value recorded by the receiver is more affected by the reflected wave [12]. Thus, the SNR data at the low satellite elevation angle (usually 5°–25°) are adopted. In order to obtain the SNR data from the sea surface, the corresponding azimuth should be selected. Figure 2 depicts the SNR time series in the satellite elevation angle range from 5° to 15° at Weihai coastal station of Shandong University. The SNR includes the direct signal and the reflected signal. The low-order polynomial fitting is often used to remove the direct signal to obtain the SNR residual sequence which is denoted as dSNR [5]. Figure 3 depicts the relationship between dSNR and *sine* of satellite elevation angle for GPS PRN15 at coastal station. The functional relationship can be expressed as [5]

$$\text{dSNR} = A_m = Acos(\theta + \varphi) = Acos\left(\frac{4\pi h}{\lambda} sin\epsilon + \varphi\right) \tag{3}$$

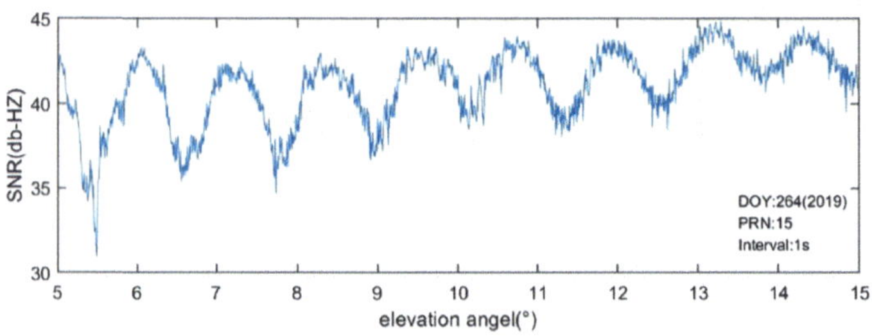

Fig. 2. SNR observation sequence of GPS PRN15 at a coastal station of Shandong university, Weihai, China

Where A is the amplitude, θ is the phase of the dSNR, φ is a phase offset, h is the vertical distance between the antenna phase center and the sea surface, λ is the GPS carrier wavelength, and ε is the satellite elevation angle. When the coastal site has small tidal variations (<7 m), the reflector is regarded as horizontal and time independent. At this time, the phase θ of dSNR can be regarded as a constant related to the satellite elevation angle. Fast Fourier Transform (FFT) is usually used to process the data with evenly spaced samples. Since dSNR is evenly sampled in time, it is considered to be unevenly sampled from the *sine* of satellite elevation angle, so Lomb-Scargle

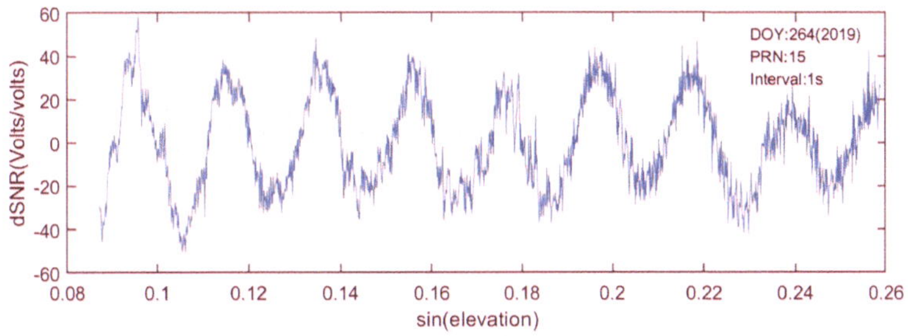

Fig. 3. dSNR observation sequence of GPS PRN 15 at a coastal station of Shandong university, Weihai, China

Periodogram (LSP) analysis method is used to perform spectral analysis for the reflected components. Figure 4 depicts the spectral analysis diagram of satellite PRN15 at the coastal station of Shandong University, Weihai, China.

After spectral analysis of the reflection component sequence by Eq. (4), the vertical distance h between the antenna phase center to the sea surface can be calculated according to the peak frequency f from the following formula.

$$\mathrm{f} = \frac{2h}{\lambda} \tag{4}$$

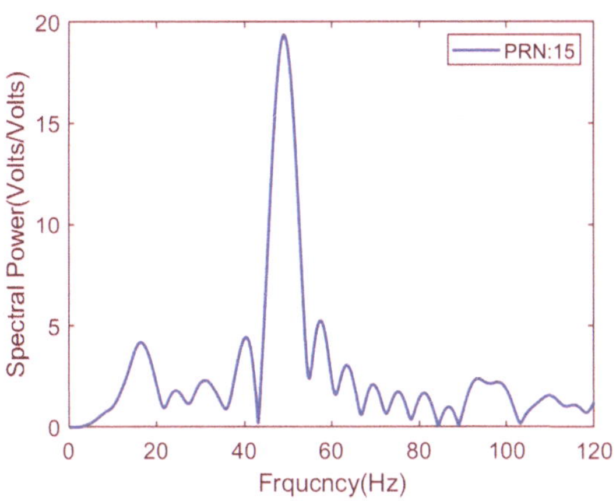

Fig. 4. Lomb-Scargle Periodogram (LSP) analysis of GPS PRN15 at a coastal station of Shandong university, Weihai, China

The flowchart of sea surface height retrieval using GPS SNR data can be summarized as Fig. 5.

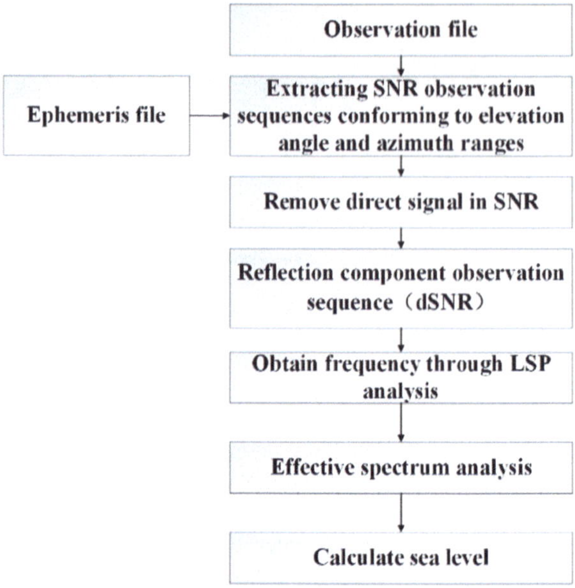

Fig. 5. Flow chart of sea surface height retrieval by GNSS-IR

3 Double-Frequency Data Fusion Based on the Peak Weighting Method

The accuracy of the above method used GPS single-frequency observations for sea level retrieval is not high. Therefore, this paper aims to use a new method to fuse the SNR data of GPS two single frequency L1 and L2 to improve the retrieval accuracy. It is known that the peak frequency and power spectral density of the LSP spectrum can be obtained as shown in Fig. 4 after performing spectral analysis on the reflection component of the signal-to-noise ratio. The RMS of the peak frequency estimation error is inversely proportional to the power spectral density at the peak frequency [13], which indicates that the greater the power spectral density at the peak frequency, the more accurate the estimated peak frequency f is, and the higher the accuracy of the sea level h can be obtained. Therefore, the sea surface height retrieval results of GPS S1 and S2 with the peak power spectral density as the weight can improve the retrieval accuracy. The weighting calculation formula can be expressed as:

$$\mathrm{h} = \frac{\sum_{i=1}^{n} P_i h_i}{\sum_{i=1}^{n} P_i} \tag{5}$$

where P_i is the peak power spectral density of the reflected signal components of GPS at different frequency after LSP analysis, and h_i is the retrieval sea level in different frequency. It can be seen from Eq. (5) that the peak weighting method gives larger weight to the retrieval height with larger signal power since the signal power reflects the overall quality of the signal. Therefore, the peak weighting method can improve the accuracy of sea surface retrieval.

4 Experimental Analysis of GNSS-IR Monitoring Sea Level Change

4.1 Experiment Description

In order to evaluate the effect of the peak weighting method on the accuracy of GNSS-IR observation results, two stations in different environments were selected in the experimental analysis. One is the SC02 station located in Friday Harbor, Washington, USA and the second station located in trestle of seaside in Shandong University, Weihai, Shandong province, China.

4.1.1 SC02 Station, Friday Harbor, USA

The SC02 station is located in Friday Harbor, Washington, USA (48.5° N, 123.0° W). It belongs to the US Plate Boundary Observation (PBO) network. The receiver of the station is a geodetic receiver Trimble NETR9 with a choke-ring antenna TRM59800.80 of a hemisphere, and the sampling rate of the receiver is 1 Hz as shown in Fig. 6. The SC02 station is built near the sea with open sea from the east to southeast. The mean vertical distance between the antenna phase center and the sea level is about 5.5 m. SNR data for GPS L1 and L2 were collected during 10 days from February 9 (day of year 40) to February 18 (day of year 49), 2014. In order to verify the retrieval accuracy of sea surface height from the SNR observations at SC02 station, the measured data with sampling rate of 1 min from Friday Harbor tide gauge 359 m away from SC02 station (see Fig. 6 right) were selected for comparison.

Fig. 6. Environment of SC02 station (left), Friday Harbor tide gauge (right)

4.1.2 Shandong University Seaside Trestle Station

The second station is located at the seaside trestle of Shandong University, Weihai, Shandong Province, China (37° 32′2.75″ N, 122° 2′44.93″ E). The receiver uses a multi-frequency GNSS receiver PolarX5 as shown in Fig. 7, which can track all visible signals. The sampling rate of the receiver is 1 Hz, and the observation data of different frequency of GPS are collected in the experiment. This station is located in a coastal area with open sea from northwest to southeast. The vertical distance from the receiver antenna phase center to the sea surface is about 4.5–6.5 m. GPS S1C and S2L were collected during four days from September 19 (day of year 262) to September 22 (day of year 265), 2019 in this experiment. During the experiment, a radar altimeter with a time resolution of 1 s and height measurement precision of 3 mm (see Fig. 7 right) was used to measure the real-time change of sea level, and the sea surface height retrieved by the station was compared and analyzed with the radar altimeter.

Fig. 7. Environment of trestle bridge station (left), co-located radar altimeter (right)

4.2 Processing and Analyzing Experimental Data

During experimental data processing, in order to select SNR data under low elevation angle which is seriously affected by multipath and obtain the data reflected from sea surface area, it is necessary to select the SNR data within the range of low satellite elevation angle and corresponding azimuth from two stations, as shown in Table 1.

Table 1. Station satellite elevation angle and azimuth

	SC02	Trestle bridge
Start/stop elevation angle (°)	5/12	5/15
Start/stop azimuth (°)	90/150	135/330

At station SC02 and trestle bridge, the sea surface height retrieved by SNR of GPS two frequency and the sea surface height retrieved by fusing dual-frequency data using peak weighting method are compared with the measured data of tide gauge or radar altimeter as shown in Figs. 8 and 9. The abscissa in the figures is the day of year (DOY), the ordinate is the sea surface height, and the sea surface height retrieved by different observations are marked with different symbols. Figures 10 and 11 depict the correlation coefficient between the retrieval results of sea level using the two methods (i.e. the retrieval results of GPS single frequency observations and the sea level using the peak weighting method) and the external data (tide gauge or radar altimeter). *R* in the figure shows the correlation coefficient between the retrieval results and the comparison data in different cases.

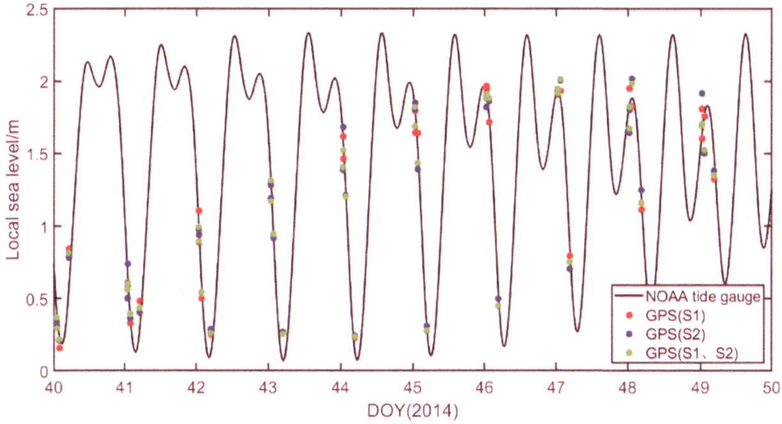

Fig. 8. Relationship between GPS sea level retrieval at SC02 station and measured data from co-located tide gauge

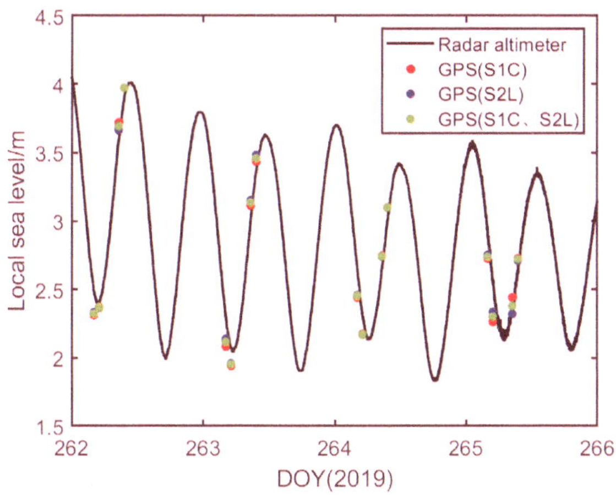

Fig. 9. Relationship between GPS sea level retrieval at trestle bridge station and measured data from co-located radar altimeter

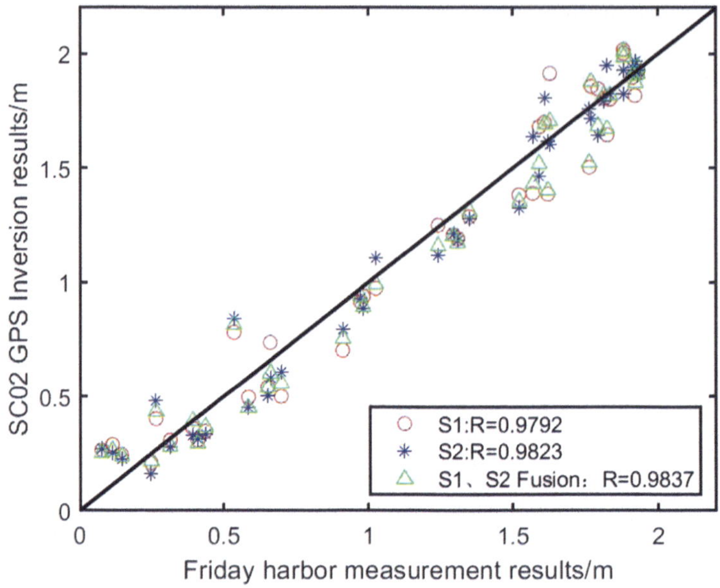

Fig. 10. Correlation analysis between GPS sea level retrieval at SC02 station and measured data from co-located tide gauge

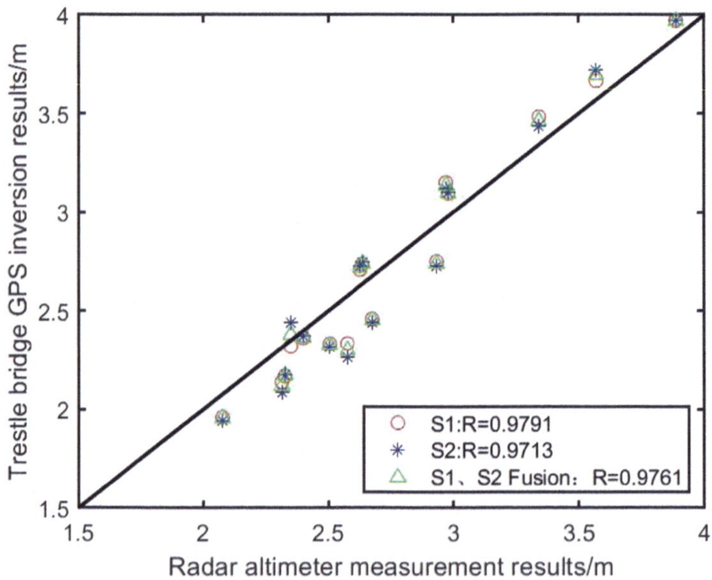

Fig. 11. Correlation analysis between GPS sea level retrieval at trestle bridge station and measured data from co-located radar altimeter

Tables 2 and 3 show the comparison of the retrieval accuracy of GPS single-frequency observations and the retrieval accuracy of dual-frequency observations fused by peak weighting method at different station.

Table 2. The precision of single frequency observation retrieval result at SC02 station and the precision of retrieval results of dual frequency data fusion by the peak weighting method

SC02 station	RMS (m)	STD (m)	Correlation coefficient
S1 process result	0.1250	0.1260	0.9792
S2 process result	0.1168	0.1178	0.9823
The peak weighting method	0.1124	0.1118	0.9837

Table 3. The precision of single frequency observation retrieval result at trestle bridge station and the precision of retrieval results of dual frequency data fusion by the peak weighting method

Weihai trestle bridge	RMS (m)	STD (m)	Correlation coefficient
S1C process result	0.1452	0.1464	0.9791
S2L process result	0.1636	0.1650	0.9713
The peak weighting method	0.1522	0.1537	0.9761

Fig. 12. Anemometer at trestle bridge station

In SC02 station, compared with the measured data of tide gauge, the precision of the retrieval result by the peak weighting method with GPS dual frequency observation is higher than that with S1 and S2 single-frequency observation, and the RMS and STD have different degrees of improvement respectively. The precision of the retrieval result of the peak weighting method fused with dual-frequency observations is 4% higher than that of S1, and 10% higher than that of S2. Compared with S2L, the result of peak weighting method at trestle is improved by 7%, however, compared with S1C, the accuracy has no improvement. The trestle station is located near the bay of the Bohai,

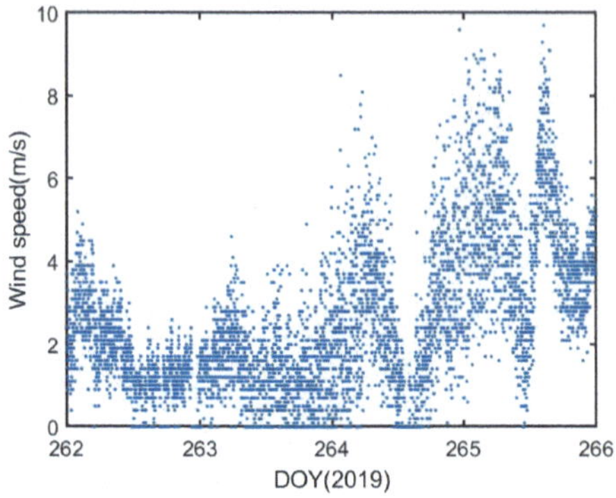

Fig. 13. The figure of sea surface wind speed at trestle bridge station

so the sea surface roughness is mainly affected by the sea wind and the surrounding topography, and the relationship with the ocean current is negligible. Considering that GNSS-IR height measurement is affected by environmental factors, as shown in Fig. 13, the sea surface wind speed measured by anemometer as shown in Fig. 12 during the four-day experiment at trestle bridge can be up to 10 m/s, while larger wind speed has larger impact on the accuracy of sea surface height measurement. Sea surface wind speed will affect the change of sea waves, which will affect the retrieval results of single frequency observations and the data of radar altimeter. Wind speed will cause the shaking of the GNSS antenna, which will change the GNSS antenna phase center correction, and the sea level will deviate from the data of the radar altimeter. Therefore, when the external environment is seriously affected, the precision of the retrieval result of the peak weighting method fused with the dual-frequency observations is not obviously improved.

5 Conclusion

The tide gauge data are often used for sea surface height observation with some disadvantages such as highly construction cost and susceptibility to vertical crustal movement, etc. GNSS-IR can use geodetic receivers with cheap in cost, easy to construct to provide abundant data using existing coastal CORS stations for sea surface height retrieval. Therefore, the sea level retrieval based on GNSS-IR can be used to supplement the observation of tide gauge to a certain extent.

Based on the principle of GNSS-IR sea level retrieval, this paper expounds the procedure of sea level retrieval using single-frequency observations of geodetic GNSS receivers, and proposes a peak weighting method to fuse the retrieval results of dual-frequency data in order to improve the accuracy of sea level retrieval. Experiments show that the accuracy of GNSS retrieval results will be improved using the proposed method to a certain extent in the case of good sea conditions, but the improvement are not obvious when the sea surface roughness is caused by large sea surface wind speed.

Acknowledgement. Thanks to the US Plate Boundary Observation (PBO) and International GNSS Service (IGS) for providing GPS observation data, and the US National Oceanic and Atmospheric Administration (NOAA) for providing tide gauge observation data. The study is funded by State Key Laboratory of Geo-information Engineering Program (SKLGIE2017-M-3-4) and National Natural Science Foundation of China (41704017, 41931076, 41874032, and 41731069).

References

1. Zhang, S., Nan, Y., Li, Z., et al.: Analysis of tide variation monitored by GNSS-MR. Acta Geodaetica Cartogr. Sin. **45**(9), 1042–1049 (2016)
2. Martin, M.: A passive reflectometry and interferometry system (PARIS): application to ocean altimetry. ESA J. **17**, 331–355 (1993)
3. Anderson, K.D.: Determination of water level and tides using interferometric observations of GPS signals. J. Atmos. Oceanic Technol. **17**(8), 1118–1127 (2000)
4. Bilich, A., Larson, K.M., Axerad, P.: Observations of signal-to-noise ratios (SNR) at geodetic GPS Site CASA: implications for phase multipath. Proc. Cent. Eur. Geodyn. Seismol. **23**, 77–83 (2004)
5. Larson, K.M., Löfgren, J.S., Haas, R.: Coastal sea level measurements using a single geodetic GPS receiver. Adv. Space Res. **51**, 1031–1310 (2013)
6. Löfgren, J.S., Haas, R., Scherneck, H.: Sea level time series and ocean tide analysis from multipath signals at five GPS sites in different parts of the world. J. Geodyn. **80**, 66–80 (2014)
7. Strandberg, J., Hobiger, T., Haas, R.: Inverse modelling of GNSS multipath for sea level measurements—initial results. In: Proceedings of the 2016 IEEE International Geoscience and Remote Sensing Symposium (IGARSS), Beijing, China, 10–15 July 2016, pp. 1867–1869 (2016)
8. Jin, S., Qian, X., Wu, X.: Sea level change from BeiDou navigation satellite system-reflectometry (BDS-R): first results and evaluation. Glob. Planet. Change **149**, 20–25 (2017)
9. Wang, N., Xu, T., Gao, F., Xu, G.: Sea level estimation based on GNSS dual-frequency carrier phase linear combinations and SNR. Remote Sens. **10**, 470 (2018). https://doi.org/10.3390/rs10030470
10. Liu, L., Fend, H., Chen, W., et al.: Study on sea level height retrieval based on GPS signal-to-noise ratio. J. Guilin Univ. Technol. **37**(4), 629–634 (2017)
11. Wu, J., Yang, R.: Using GPS receiver reflected signal to measure sea level. Geodesy Geodyn. **32**(6), 135–138 (2012)
12. Sun, X., Zhang, S., Wu, J., et al.: Mechanism analysis of GPS-IR technology based on SNR. Adv. Earth Sci. **34**(2), 156–163 (2019)
13. Li, Y., Chang, X., Yu, K., et al.: Estimation of snow depth using pseudorange and carrier phase observations of GNSS single-frequency signal. GPS Solut. **23**, 118 (2019)

Application of Fitting of Moving Quadric Surface to Height Anomaly Fitting in the Band-Shaped Area

Puyu Sun[1], Chengfa Gao[1(✉)], Xinde Zhai[2], and Yongsheng Liu[1]

[1] School of Transportation, Southeast University, Nanjing 211189, China
sunpuyu@163.com, gaochfa@163.com
[2] Changjiang Waterway Bureau, Nanjing 430010, China

Abstract. Global Navigation Satellite System (GNSS), which can provide high-precision coordinates in the global geodetic framework, is able to replace most geodetic surveying, but it is difficult to replace conventional leveling surveying due to height anomaly. The Fitting of Moving Quadric Surface (FMQS) is based on the known points in the circular region with the center of the point to be fitted and a specific radius. In this paper, the method of FMQS is used to fit the height anomaly of Nanjing section of Yangtze River and the downstream, and the results show that the fitting accuracy of the 226 points evenly distributed along the river is 1.86 cm; At the same time, the fitting of GNSS height determination using this set of data can replace the fourth order leveling that the leveling line more than 3.46 km at 95.5% confidence interval. Benchmarks data collected in the field show that the difference between the height of National Vertical Datum 1985 obtained by geodetic height fitting with Jiangsu CORS and the height of National Vertical Datum 1985 obtained by third order leveling is within 2 cm.

Keywords: Quasi-geoid · Height anomaly · Fitting of Moving Quadric Surface · Leveling surveying

1 Introductions

Global Navigation Satellite System (GNSS) can provide real-time, all-weather, high-precision three-dimensional absolute coordinates. GNSS can obtain quickly in the plane direction centimetre-level accuracy even millimetre, but it is regrettable that there is still something unsatisfactory in the accuracy of height determination. Since the geoid is an irregular surface, which cannot be simulated by an accurate surface. So, GNSS can only provide us geodetic height with high precision, instead of the normal height usually used in engineering. This has seriously affected the development of GNSS three-dimensional positioning applications. Therefore, the development of GNSS in the field of height determination is severely limited.

Since the introduction of GPS in China in the 1980s, the subject of GPS height transformation has become the research direction of many scholars. As early as 1996, Li [1] did research on GPS height determination, he used the quadric surface fitting method to compute the height anomaly applied to the GPS geodetic height. The test

J. Sun et al. (Eds.): CSNC 2020, LNEE 650, pp. 166–175, 2020.
https://doi.org/10.1007/978-981-15-3707-3_16

results showed that the root mean square error (RMSE) of the fitting is 3 cm. which can meet some of the needs of the survey at the time. Cheng [2] studied the height fitting of the linear survey area, and the experiment showed that the RMSE obtained by the cubic spline interpolation was about 3 cm. With the continuous refinement of the quasi-geoid, more and more scholars have carried out research on height fitting. Wang [3] conducted an accuracy test on the Yangtze River Jiangsu section based on the non-tidal surveying technology, and the height anomaly were fitted using piecewise surface & neural networks fitting. The test results showed that the final accuracy of the internal and external coincidence accuracy is 1.83 cm and 2.17 cm. Zhang [4] used the measured data from the Xiaoxian X014 reconstruction project to analysis the accuracy of the cubic spline fitting method. which shows that the internal conformance is 1.5 cm and the external conformance is 2.1 cm. Bai [5] verified the feasibility of height fitting based on the orthogonal matrix fitting method in the construction of a strip-shaped narrow area. The experimental results showed that the RMSE of the orthogonal function curve fitting method is less than 1.3 cm and the maximum deviation is 2.8 cm. Plane fitting, quadric fitting, and polyhedral function fitting are often used in planar surveying areas. Cubic spline functions, curve fitting, and Chebyshev fitting curve can be used in linear surveying areas such as railways. In the band-shaped areas, like the Yangtze River, methods such as Fitting of Moving Quadric Surface and fitting of moving quadric surface can be used.

A successful example of the use of quasi-geoid results is the non-tidal water depth surveying technology. The RTK is used to obtain the 3D coordinates of the water surface points in the geodetic coordinate system, and then the height anomaly correction is applied to the geodetic height to obtain the planar coordinates and normal height coordinates of the underwater points [6].

2 Height Anomaly

Quasi-geoid is an irregular surface, which cannot be expressed by a simple shape and mathematical formula, and it is difficult to be accurately determined at present. In order to facilitate the research, scientists constructed a regular ellipsoid to fit the earth. This ellipsoid is called the reference ellipsoid, and its surface is called the surface of reference ellipsoid, which is the datum of geodetic surveying.

The distance from a point on the ground along the plumb line to the surface of reference ellipsoid is called the geodetic height, and the distance to the quasi-geoid is called the normal height (Fig. 1). There are many ways to obtain the height anomaly of a point. In engineering applications, it is difficult to accurately compute the height anomaly of each point. It is only necessary to obtain an approximate value of the height anomaly within the error tolerance range, so in most cases, several known points are used to interpolate height anomaly of unknown points [7].

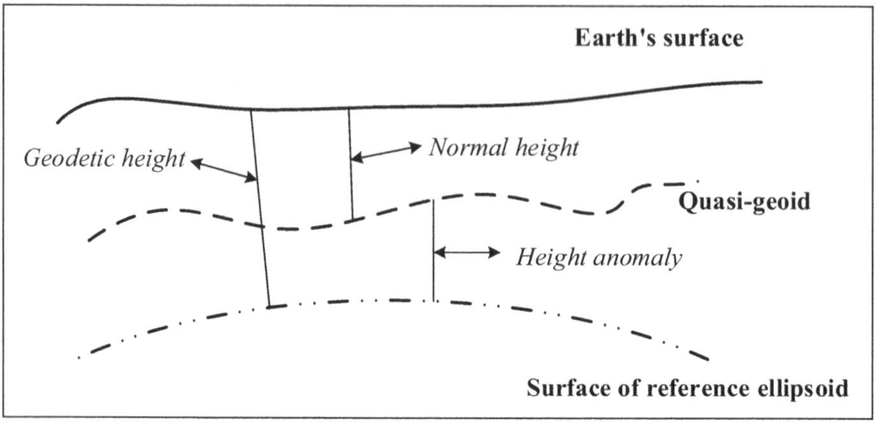

Fig. 1. Diagram of height anomaly

3 Methods of Height Anomaly Fitting

3.1 Quadric Surface Fitting

Quadric surface fitting is one of the most commonly used methods in height anomaly fitting, and Its formula is shown in formula (1).

$$\xi(x, y) = a_0 + a_1 \Delta x + a_2 \Delta y + a_3 \Delta x^2 + a_4 \Delta x \Delta y + a_5 \Delta y^2 + \varepsilon \quad (1)$$

where, $\Delta x = x - x_0$, $\Delta y = y - y_0$. (x, y) denote the geodetic coordinate. In order to prevent the value from being too large and affecting the calculation accuracy of the computer, set the relative origin (x_0, y_0). When there are more than 6 known points in the area, the error equation has an optimal solution. The value of the parameters $(a_0, a_1, a_2, a_3, a_4, a_5)$ of the model to be determined is obtained according to the least square method, and then substituted into it to obtain the model of quadric surface. When selecting the quadric surface fitting, you must ensure that the selected points are evenly distributed around and around the area, and that there are more than 6 fitting points.

According to the post-adjustment information after adjustment, the weights of the observations (known points) are gradually adjusted to appropriate values, the method of reducing the effect of small gross errors on the adjustment results is called the two-dimensional surface robust fitting. The 2D curved surface robust fitting can get better results [8].

This method can achieve high accuracy when the terrain changes gently in the surveying area and the known points are evenly distributed. However, when the area of the surveying area is too large or the height of the abnormal changes is large, the fitting accuracy of this method is not ideal, and it is not suitable for areas with insufficient known points.

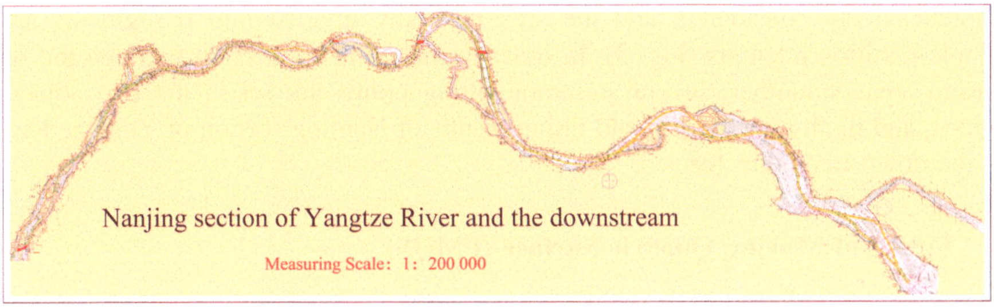

Fig. 2. Nanjing section of Yangtze River and the downstream

There are more than 300 km of Nanjing section of Yangtze River and the downstream, and which is obviously narrow and long (Fig. 2). From the analysis of the available data, it can be seen that the height anomaly increases from west to east, and the increase amplitude is greatly different, and does not show a clear linear rule [9]. The range of the height anomaly of this part of the Yangtze River from the west to the east is nearly 12 m, and the difference of height anomaly between the north and south banks can reach decimeter-level.

3.2 Fitting of Piecewise Quadric Surface (FPQS)

The Fitting of Piecewise Quadric Surface is to divide the Yangtze River into several sections, and each section is fitted separately. There is uncertainty in the division of each segment, which is mainly based on the actual terrain fluctuations of the surveying area and the distribution of control points. The segmentation work requires rich engineering experience. The difference in segmentation results has a great impact on the fitting results and fitting accuracy of the entire region [10].

Based on the principle of high overall fitting accuracy and as few areas as possible, factors such as the distribution of control points in Nanjing section of Yangtze River and the downstream, the direction of the Yangtze River, and height anomalies were

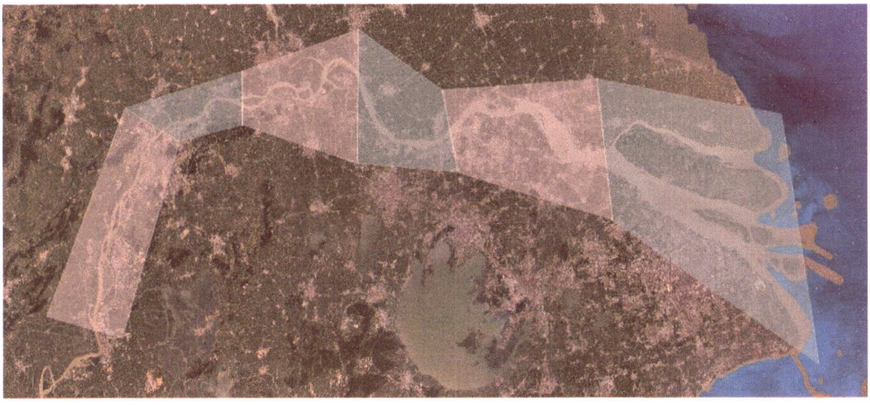

Fig. 3. Partition diagram of Nanjing section of Yangtze River and the downstream

comprehensively considered, and the area is finally divided into 6 segments after multiple partition attempts (Fig. 3). In order to make the connection between the two adjacent areas smoother, several common fitting points are set in the two adjacent sections, and finally the quasi-geoid fitting results of Nanjing section of Yangtze River and the downstream are formed.

3.3 Fitting of Moving Quadric Surface (FMQS)

Moving quadric surface means that fitting area is changing with the location change of the points. Fitting of Moving Quadric Surface is to perform surface fitting by taking all known points in a circular area with the center of the point to be fitted and a radius of R. First, the value of the radius R is given randomly to determine the initial fit area. Secondly, in this initial area, some points are randomly and uniformly selected for the calculation of model parameters. These points are called fitting points; the other points called checkpoints are used to check the model. Third, if the accuracy of the check points meets the specified conditions, the radius of the fitted area is the current value of R, otherwise, change the value of R and repeat the above steps until a fit radius that meets the conditions is selected.

According to the principle of the Fitting of Moving Quadric Surface, for many points to be fitted, it is difficult to compute the accuracy of the checkpoints and the accuracy of the fitting points during the fitting process. Only the deviations of these points to be fitted can be counted to determine the final accuracy.

Generally, the number of fitting points is less than the number of checkpoints, and the number of fitting points is greater than 8. In order to prevent the fitting area from being too large, the maximum number of fitting points is not more than 25. In Fig. 4, the middle pentagram represents the point to be fitted, and the red known points within the blue circle will be randomly and uniformly divided into fitting points and check points.

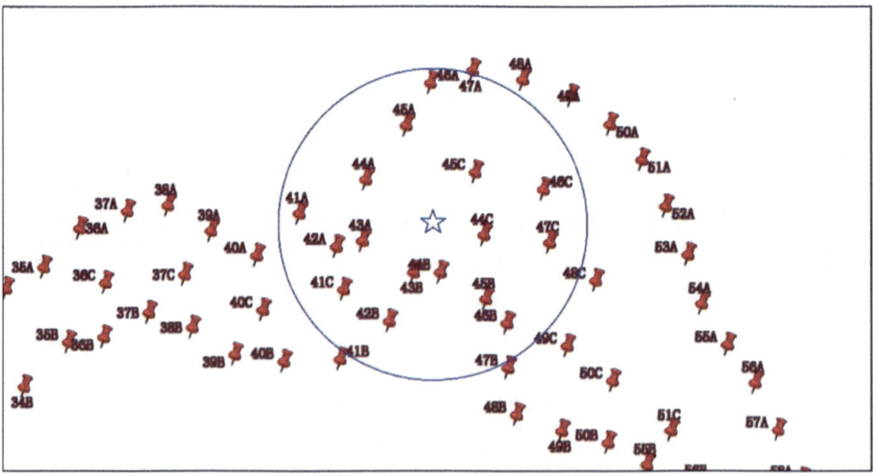

Fig. 4. Schematic diagram of Fitting of Moving Quadric Surface

4 Results and Discussion

This test is based on the assumption that the height anomalies of 226 known points are precise. These 226 points are evenly distributed along the river, and a pair of points about 5 km are distributed on both sides of the Yangtze River. If encountering the island, a few points will be added evenly.

4.1 Accuracy Analysis of FPQS

The points participating in the model parameter solution are called fitting points, and the others are called checkpoints. The accuracy of the fitting points and the checkpoints are used to determine the fitting accuracy [11].

The accuracy of the fitting points: According to the deviation of the fitting points, the fitting accuracy is calculated according to formula (2).

$$M_f = \sqrt{\frac{[vv]}{n}} \tag{2}$$

where, v is the residual value of the fitting points, which is the simulated value calculated by the model minus the precise value, and n is the number of the fitting points. Fitting accuracy is used to illustrate the model's ability to learn from samples.

The accuracy of the checkpoints: According to the deviation of the checkpoints, the accuracy of the checkpoints is calculated according to formula (3).

$$M_c = \sqrt{\frac{[vv]}{n}} \tag{3}$$

where, v is the residual value of the check point, which is the simulation value calculated by the model minus the precise value, and n is the number of the checkpoints. The accuracy of the checkpoints is used to check the accuracy of the model.

Table 1. Results of Fitting of Piecewise Quadric Surface

No	Number of the fitting points	M_f/cm	Number of the checkpoints	M_c/cm
1	16	1.26	23	1.81
2	13	1.00	21	2.40
3	17	1.90	33	2.43
4	12	2.29	22	3.18
5	16	1.77	33	1.96
6	12	2.18	21	2.39
Statistics	86	1.72	153	2.47

Nanjing section of Yangtze River and the downstream were divided into 6 segments fitted separately. The accuracy of the fitting points and checkpoints each segment are shown in Table 1. Figure 5 shows the error distribution of the Fitting of Piecewise Quadric Surface.

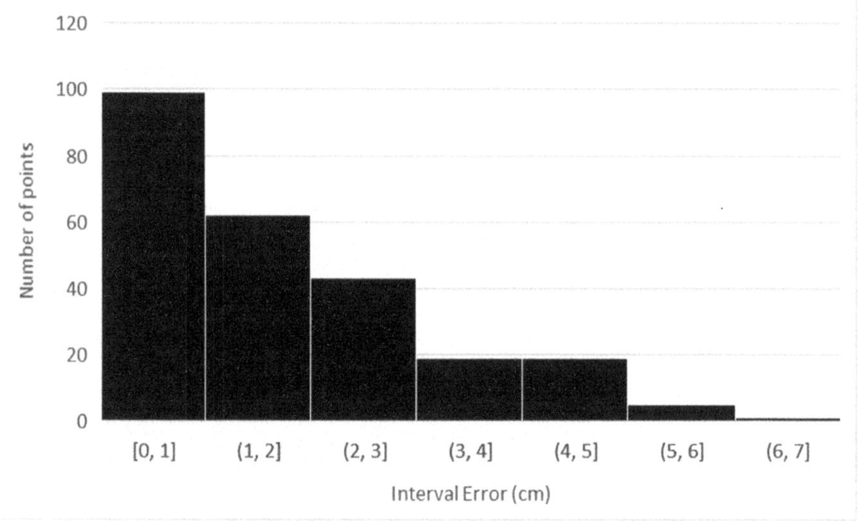

Fig. 5. Histogram of deviation statistics of FPQS

4.2 Accuracy Analysis of FMQS

The accuracy analysis of the FMQS does not distinguish between checkpoints and fitting points [12]. The calculation formula is as shown in formula (2).

$$M = \sqrt{\frac{[vv]}{n}} \tag{4}$$

where, v is the difference between the fitted value and the precise value, n is the number of sample points, in this test, it is 226, and the final RMSE is 1.86 cm.

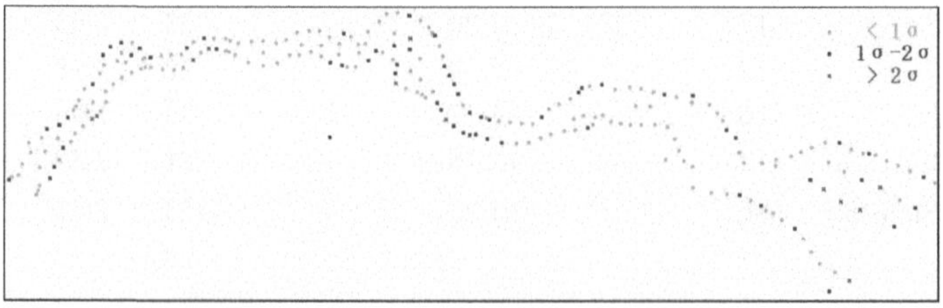

Fig. 6. Error distribution of FMQS

Figures 6 and 7 show error distributions of FMQS and histograms.

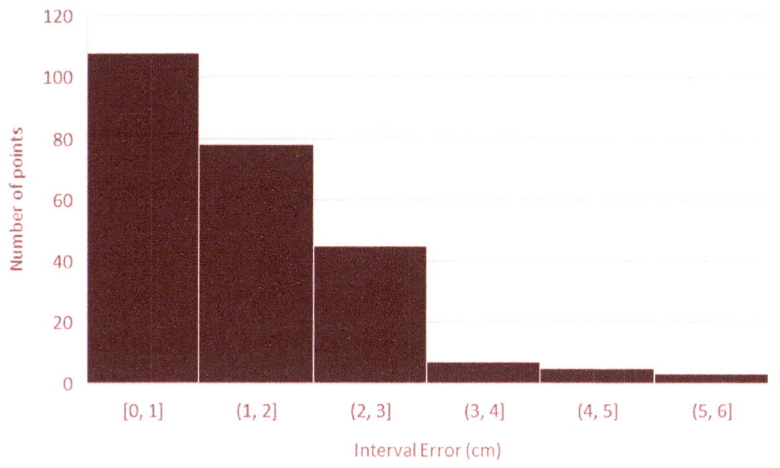

Fig. 7. Histogram of deviation statistics of FMQS

4.3 Comparison with Third Order Leveling

In order to verify the reliability of the fitting results, the geodetic coordinates of the three third order Benchmarks were obtained through Jiangsu CORS. In order to ensure the accuracy of the geodetic coordinates, 5 sets of data are measured at each point, and the results are smoothed 10 times for each set of data. Table 2 shows the geodetic coordinates and the height of National Vertical Datum 1985 of the three third order benchmarks.

Table 2. Geodetic coordinates and height of National Vertical Datum 1985 of benchmarks

Point	B (dd.mmss)	L (dd.mmss)	Geodetic height/m	Levelling height/m
NJ15	32.06321836	118.4426186	10.01	8.248
NJ16	31.56377217	118.3712471	9.95	8.584
NJ17	31.46595923	118.3027057	10.07	8.993

Table 3 shows the heights fitted by Fitting of Piecewise Quadric Surface (HFPQS) and Fitting of Moving Quadric Surface (HFMQS), and the difference between the two fitting results and the levelling height.

Table 3. HPSF, HMSF and the difference

Point	HFPQS/m	Difference/cm	HFMQS/m	Difference/cm
NJ15	8.268	−1.97	8.238	1.01
NJ16	8.567	1.67	8.569	1.47
NJ17	8.991	0.16	8.979	1.38

As can be seen from the table, the differences are all below 2 cm; and the FMQS is better than the FPQS.

4.4 Feasibility of Leveling Replacement

In China, leveling is mainly divided into four levels, of which the most commonly used are third and fourth order leveling. It can be known from specifications for the third and fourth order leveling that the third order leveling requires compliance with the leveling line closure difference tolerance of $\pm 12\sqrt{L}$ mm, and the fourth-level leveling requirement meets the leveling line closure difference tolerance of $\pm 20\sqrt{L}$ mm (where L is the length of the leveling line, The unit is km) [13].

It can be known from the leveling specification that the leveling tolerance increases with the length of the leveling line, but the GNSS measurement can be considered to be independent of distance to a certain extent, so the length of leveling line can be used as a condition for GNSS surveying to replace leveling.

In Nanjing section of Yangtze River and the downstream, the RMSE of the FMQS is 1.86 cm. Generally, twice the RMSE (95.5% confidence interval) is used as the tolerance. It can be concluded that the third order leveling can be replaced at L > 9.61 km, and fourth order leveling can be replaced at $L > 3.46$ km.

5 Conclusions

In this paper, 226 known points and 3 benchmarks were used for experiments, and the main conclusions could be obtained as follow:

First, the results of the Fitting of Moving Quadric Surface are better than the Fitting of Moving Quadric Surface, and the Fitting of Moving Quadric Surface is easier to implement.

Second, in Nanjing section of Yangtze River and the downstream, the fourth order leveling with length of a leveling line 3.46 km or more can be replaced by height determination using GNSS with a probability of 95.5% confidence interval.

Third, the distribution of the deviation in Nanjing section of Yangtze River and the downstream is uneven. As can be seen from Fig. 6, the fitting accuracy of the middle section and the last section is relatively lower than others, and the deviation of other sections are almost within 1 times the RMSE.

There are still deficiencies in this paper. For example, the benchmarks are not enough due to funding limitations, only three points were measured, and all of three points are located in the Nanjing section, which may not be representative.

References

1. Li, X.: GPS height determination application. J. Kunming Metall. Coll. **12**, 1–2 (1996)
2. Cheng, W.: Research on Several Problems of GPS height Fitting. Surveying and Mapping of Geology and Mineral Resources, 10–12 (2003)
3. Wang, L., Wang, Y., Zhang, W.: Tide-free measurement and accuracy test based on RTK in Jiangsu section of the Yangtze River. Pearl River **39**(240(04)), 62–64 (2018)
4. Zhang, T., Dai, H., Xu, J.: Application of GPS height fitting in highway surveying and mapping and taking Xiaoxian County X014 reconstruction project as an example. J. Henan Sci. Technol. **8**, 127–129 (2019)
5. Bai, J., Li, W., Li, S., Guo, Y.: Research on height fitting of satellite control network based on orthogonal matrix elevation fitting method. J. Qiqihar Univ. (Nat. Sci. Ed.) **35**(06), 49–52 (2019)
6. Gao, C., Hu, W.: The Principle and Application of Satellite Navigation and Positioning. China Communications Publishing, Beijing (2011)
7. Li, A.: Research on GPS height fitting using partition fitting method. Pioneer. with Sci. Technol. Mon. **28**(05), 93–94 (2015)
8. Hu, W.: The research of the theory and the method and application of precise GPS height Surveying. Ph.D. thesis, Hohai University, Nanjing, China (2001)
9. Jiang, C., Gai, Y., Peng, J.: Investigation and analysis of GPS height anomalies in China. Site Invest. Sci. Technol. **1**, 49–50 (2003)
10. Yang, R.: Segmentation fitting and precision analysis of GPS altimetry on high-grade highways in high mountain areas. Bull. Surveying Mapp. **12**, 38–40 (2005)
11. Liu, L.: The study and application of GPS level elevation detection of outliers fitting models. Ph.D. thesis, Liaoning Technology University, Liaoning, China (2008)
12. Wu, D., Xiong, W.: GPS elevation fitting method for sea-crossing bridge. Sci. Surv. Mapp. **42**(06), 125–129 (2017)
13. General Administration of Quality Supervision, Inspection and Quarantine of the People's Republic of China. GB/T 12898-2009 Specifications for the third and fourth order leveling. Beijing, China

Instant PPP with Low-Cost Multi-constellation Dual-Frequency GNSS Chipset

Fei Liu[1,2(✉)], Hongzhou Yang[1], and Yang Gao[1,2]

[1] Profound Positioning Inc., Suite 820, 800 6 Ave SW, Calgary, AB, Canada
fliu@profoundpositioning.com, liul9@ucalgary.ca
[2] University of Calgary, 2500 University Dr NW, Calgary, AB, Canada

Abstract. With increased measurement quality of low-cost Global Navigation Satellite System (GNSS) receivers, Precise Point Positioning (PPP) becomes feasible using a low-cost multi-constellation single-frequency GNSS receiver. Profound Instant PPP (IP3), for example, is a real-time fast convergence PPP technology that offers world-wide 50-cm circular error probability (CEP) PPP solution using a low-cost GNSS chipset. IP3 is able to output precise positioning solutions within seconds, which makes it suitable for various emerging applications such as self-driving cars, UAV and smartphones. With strong demand on higher accuracy by those applications, several low-cost dual-frequency GNSS chipset products (e.g. u-blox F9P, Broadcom BCM47755, etc.) have been released to market. The availability of GNSS observations at the second frequency can further improve the positioning accuracy with low-cost GNSS chipsets. Compared to single-frequency PPP, a dual-frequency PPP system not only increases the redundancy of the positioning system but also helps mitigate the ionospheric effect, leading to more accurate and robust positioning solutions. This paper describes the development of a dual-frequency based Instant PPP system (DF-IP3) using low-cost multi-constellation dual-frequency GNSS chipsets. This system is able to output precise positioning solutions at 30 cm CEP within a few seconds and can be applied to support a wide range of applications.

Keywords: Instant · Dual-frequency · Low-cost · PPP

1 Introduction

With public access to real-time precise satellite orbit and clock correction provided by satellite-based augmentation system (SBAS) or the international GNSS service (IGS), real-time PPP attracts increasing attention both in academic and industrial communities [1–3]. Usually, an expensive dual-frequency geodetic receiver has to be applied in PPP in order to achieve centimeter to decimeter level positioning accuracy. This is because ionosphere-free combination can be formed to mitigate the ionospheric effect [4, 5]. Such ionosphere-free combination based conventional PPP system, however, suffers a long convergence time at tens of minutes and high cost using a geodetic receiver. Single-frequency PPP has been intensively studied in recent years [6–10]. With the release of improved low-cost GNSS receivers, it is possible to develop low-cost

J. Sun et al. (Eds.): CSNC 2020, LNEE 650, pp. 176–186, 2020.
https://doi.org/10.1007/978-981-15-3707-3_17

real-time PPP systems. For low-cost GNSS devices, single-frequency receiver currently still dominates the market and single-frequency PPP provides cost-effective precise positioning solution to support sub-meter or lane-level applications (e.g. self-driving cars, UAV, etc.) [9]. For instance, Profound Instant PPP (IP3) is a real-time fast convergence single-frequency PPP technology using low-cost GNSS receivers to provide 50 cm CEP world-wide solution [11]. Recently, GNSS manufactures released low-cost dual-frequency GNSS chipsets (e.g. u-blox F9P, Broadcom BCM47755, etc.) capable of receiving dual-frequency civil signals. With the additional signals, ionosphere-free combination can be formed to further reduce the ionospheric delay, resulting in higher accuracy [1].

For low-cost dual-frequency (DF) receiver, not all satellite measurements can form ionosphere-free (IF) combination since Block IIA and IIR Global Positioning System (GPS) satellites only transmit civil signal at L1 C/A. This will lead to a smaller number of IF measurements available for position estimation. Moreover, the code measurement noise level of a low-cost receiver is higher than a geodetic receiver [12], which means the larger noise (about 3 times) in IF measurements with low-cost receiver will increase the convergence time and degrade the obtainable accuracy. To address those challenges in the development of DF PPP using low-cost receivers, we propose the integration of ionospheric-corrected single-frequency measurements for all the satellites and ionosphere-free measurements for the satellites transmitting dual-frequency civil signals. In this way, all observed satellite signals are used in position estimation and the effect of large IF code noise can be reduced. A dual-frequency Instant PPP (DF-IP3) has been developed along with a DF-IP3 evaluation kit (EVK). The EVK integrates a u-blox F9P chipset and has been tested in a number of field tests to verify the performance of the DF-IP3 system. In the kinematic field tests, the DF-IP3 EVK and a NovAtel FlexPak6 receiver were connected to a dual-frequency u-blox patch antenna placed on the vehicle roof through a signal splitter. The NovAtel receiver was applied to generate RTK solutions as the reference for accuracy evaluation. In all field tests, DF-IP3 can offer position solutions at 30 cm CEP within seconds.

2 Methodology

The GNSS observation models of the code and phase measurements from multiple constellations are given as [1]

$$\begin{aligned} P_i &= \rho + c \cdot dt_r - c \cdot dt^s + isb^M + T + \mu_i \cdot I + B_{pr,i} - B_i^{ps} + \varepsilon_{P_i} \\ L_i &= \rho + c \cdot dt_r - c \cdot dt^s + isb^M + T - \mu_i \cdot I + \lambda_i \cdot N_i + b_{lr,i} - b_i^{ls} + \varepsilon_{P_i} \end{aligned} \tag{1}$$

where P_i and L_i are the code and carrier-phase measurements of frequency i (unit: meter); ρ is the geometric distance between the satellite and receiver; c is the light speed; dt_r represents the receiver clock offset for the reference GNSS system; dt^s is the satellite clock offsets; isb^M is the inter system bias (ISB) between the current system M and the reference system (GPS in this work); T is the tropospheric delay; I is the ionospheric delay on L1 and μ_i is the ionospheric scalar factor for the ith carrier frequency f_i with value of $\mu_i = f_i^2/f_1^2$; λ_i and N_i represent the wavelength and the integer ambiguity; $B_{pr,i}$ and B_i^{ps} are the receiver and satellite code biases; similarly, $b_{lr,i}$ and b_i^{ls} are the phase biases on receiver- and satellite-end; ε_{Pi} and ε_{Li} are the measurement noise.

In estimation, the precise satellite orbit and clock offset are accessible by application of satellite orbit and clock corrections to the broadcast GNSS ephemerides via state space representation (SSR) or SBAS. Similarly, the ionospheric delay can be corrected by using SSR or SBAS corrections. The tropospheric delay is corrected by Saastamoinen model and the code bias on satellite-side is corrected by differential code bias (DCB) products. The carrier-related biases are absorbed into the ambiguities in estimation since the integer ambiguities are not fixed. Please note that not all the error sources are listed in Eq. (1). Other errors such as the Sagnac effect, special relativistic effect, Shapiro time delay, satellite/receiver antenna phase center offset (PCO) and phase center variation (PCV), phase wind-up effect, etc. have to be corrected before estimation [13–15].

Note that ionosphere advances the carrier-phase but delays the code measurements at the same frequency with the same magnitude. Therefore, the code and phase measurements can be combined to form the GRAPHIC (GRoup And Phase Ionospheric Correction) [16] to reduce the ionospheric delay, given as

$$L_G = \rho + c \cdot dt_r - c \cdot dt^s + isb^M + T + 0.5 \cdot A_G + \varepsilon_{L_G} \tag{2}$$

where L_G is the GRAPHIC measurement (unit: meter); A_G is the float ambiguity (unit: meter) absorbing the related biases; ε_{LG} is the measurement noise; the rest has the same meaning as in Eq. (1).

For low-cost GNSS receiver, L1 signals are available. After linearization of Eqs. (1) and (2) and correction of the error sources mentioned above, the L1 single-frequency observation model can be rewritten as

$$\begin{aligned} p_1 &= -e \cdot \delta r + c \cdot dt_r + isb^M + B_{pr,1} + \varepsilon_{P_1} \\ l_G &= -e \cdot \delta r + c \cdot dt_r + isb^M + 0.5A_G + \varepsilon_{L_G} \end{aligned} \tag{3}$$

where p_I and l_G denote observed-minus-computed L1 code and GRAPHIC measurement residuals; e is the unit vector from receiver to satellite; δr represents the position error.

For satellites with dual-frequency measurements available from the receiver, the IF combination can be formed to largely reduce the ionospheric delay, given as

$$\begin{aligned} P_{IF} &= \alpha P_1 + (1-\alpha)P_2 \\ L_{IF} &= \alpha L_1 + (1-\alpha)L_2 \\ \alpha &= f_1^2/(f_1^2 - f_2^2) \end{aligned} \tag{4}$$

where P_{IF} and L_{IF} are the code and phase measurements of IF combination (unit: meter). Based on the coefficient α in the equation above, the measurement noise for IF combination is around three times of L1 measurements, which makes conventional dual-frequency IF PPP inappropriate for low-cost GNSS receiver, given that low-cost receiver measurement noise is larger than geodetic receiver.

Similarly, Eq. (4) can be linearized as

$$\begin{aligned} p_{IF} &= -e \cdot \delta r + c \cdot dt_r + isb^M + \varepsilon_{P_{IF}} \\ l_{IF} &= -e \cdot \delta r + c \cdot dt_r + isb^M + A_{IF} + \varepsilon_{L_{IF}} \end{aligned} \tag{5}$$

where p_{IF}, l_{IF} denote observed-minus-computed code and phase ionospheric-free measurement residuals; A_{IF} represents the float IF ambiguity. In DF-IP3, both ionosphere-corrected observations and ionosphere-free observations will be incorporated into position determination.

3 Field Tests and Results

The DF-IP3 is featured with fast convergence within seconds, which makes it suitable for positioning and navigation for different platforms. A number of field land vehicle kinematic tests were conducted, and some results will be provided to show the positioning accuracy and convergence time of the DF-IP3 system. In the kinematic tests, a dual-frequency IP3 EVK was used with a u-blox multi-band patch antenna placed on the roof. The IP3 EVK is developed by Profound Positioning Inc. (PPI) with a u-blox F9P chipset providing the raw GNSS measurements. BeiDou is not used due to insufficient signals during the tests.

To evaluate the IP3 EVK performance, a NovAtel FlexPak6 receiver was also used to generate RTK solutions as reference with IGS UCAL station as base. The NovAtel FlexPak6 receiver and IP3 EVK were connected to the U-box patch antenna on the vehicle roof through a signal splitter. For land vehicle navigation, the horizontal positioning accuracy is more important than the vertical positioning accuracy. Therefore, in this section, only horizontal positioning results of field tests are presented. The field kinematic test setup is shown in Fig. 1 and several field test results were described in the following paragraphs.

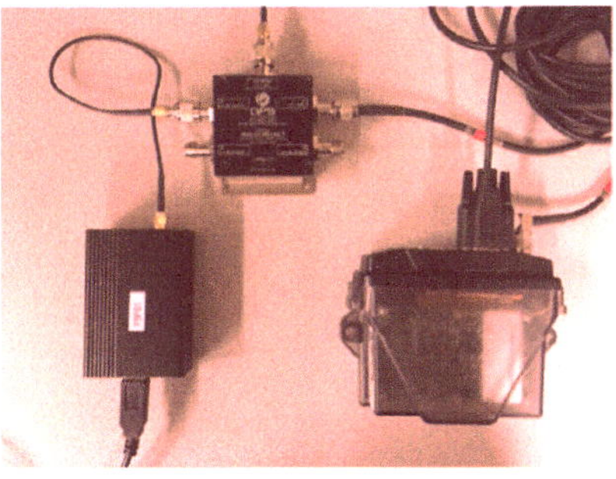

Fig. 1. Kinematic test setup

The first kinematic test lasted for about 68 min. The land vehicle was moving on Crowchild Trail NW, Shaganappi Trail, 144 Ave NW, Range Rd 24, Township Rd 263, Range Rd 25, Range Rd 24A, Range Rd 25A, Big Hill Springs Rd, Bearspaw Road and Bow Valley Trail in Calgary with tress and low constructions along the road from time to time. The first test trajectory and test environment can be seen from Fig. 2. During the test, GPS, GLONASS and Galileo satellites were applied in processing.

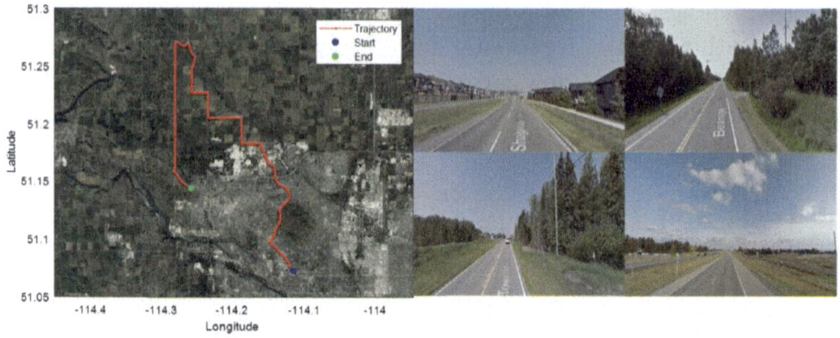

Fig. 2. The first kinematic test trajectory and test environment

Shown in Fig. 3 is the positioning errors of the DF-IP3 in the horizontal plane. The blue and green lines represent the positioning errors in the east and north. It can be seen that the DF-IP3 got converged at the beginning with horizontal positioning error less than 0.5 m. During the first test, the horizontal positioning error is less than 0.3 m in most time. The small fluctuation around 3000 s is due to re-convergence caused by signal blockage. The overall root mean square (RMS) values of positioning errors in the east and north are 0.14 m and 0.19 m respectively. It is clear that the DF-IP3 system can achieve 0.3 m CEP horizontally with fast convergence in seconds.

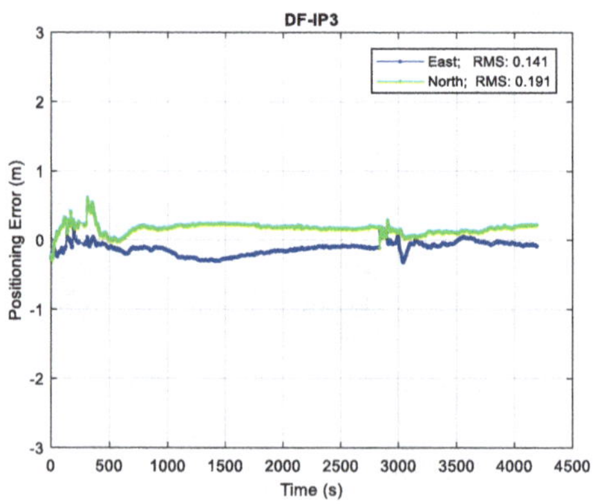

Fig. 3. Positioning errors of the DF-IP3 in the east and north directions during the first test

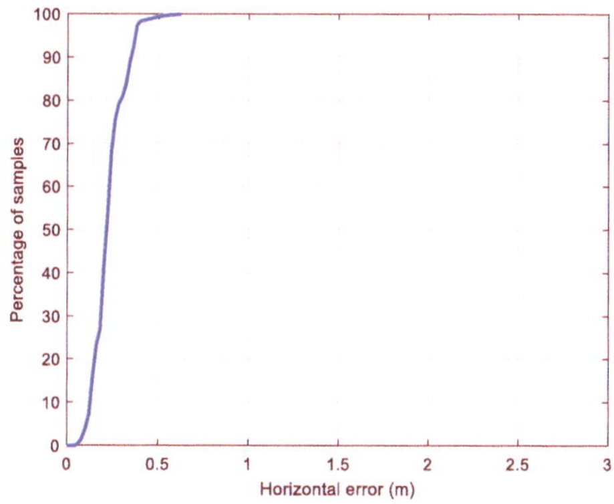

Fig. 4. Horizontal CDF of the DF-IP3 for the first test

The cumulative distribution function (CDF) of the DF-IP3 horizontal positioning errors is provided in Fig. 4. From Fig. 4, it can be seen that the horizontal CEP of the DF-IP3 is better than 0.3 m. Over 90% DF-IP3 solution errors are within 0.4 m horizontally and the maximum horizontal error is about 0.7 m.

To fully assess the convergence time of the DF-IP3 system, a post-mission simulated test for convergence time was conducted. The details of the simulated tests are as follows. The open-sky kinematic test dataset was divided into 230 15-s sessions. For each session, the processing was restarted manually, which means 230 15-s solutions can be achieved. The performance of each method (CEP value) in the first 15 s can be obtained by making statistics of the 230 sessions. For instance, the CEP of the first second is based on the first second positioning results in 230 sessions. In this way, the general performance in the first 15 s can be achieved and the convergence time can be seen directly. Figure 5 shows the horizontal CEP for the DF-IP3 in the beginning 15 s. It is clear that the DF- IP3 reached accuracy better than 0.5 m at the very beginning. Based on the above results, the first test verified the accuracy and fast convergence of the developed DF-IP3 system.

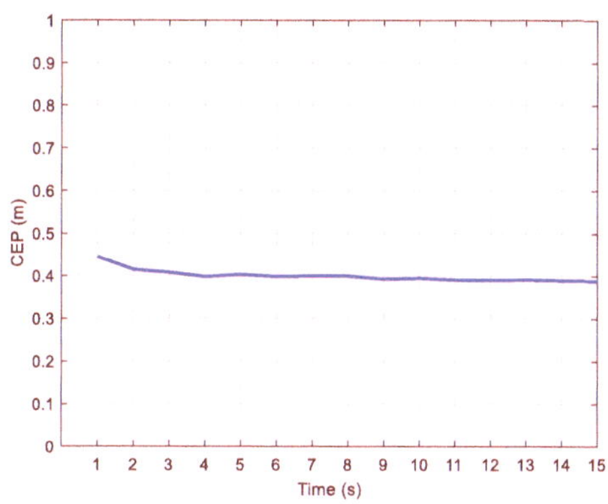

Fig. 5. Horizontal positioning errors of the DF-IP3 in the beginning 15 s for the first test

The second test was conducted in residential areas with houses along the road, which represent typical environments expected in many applications. The land vehicle was moving on Crowchild Trail NW, Shaganappi Trail, John Laurie Blvd NW, 14 St NW, Macewan Dr NW, Country Hills Blvd NW and roads in Community Hamptons in Calgary. This test lasted for about 1 h and GPS, GLONASS, Galileo satellites were applied in this test. The trajectory and test environment of the second test can be seen from Fig. 6.

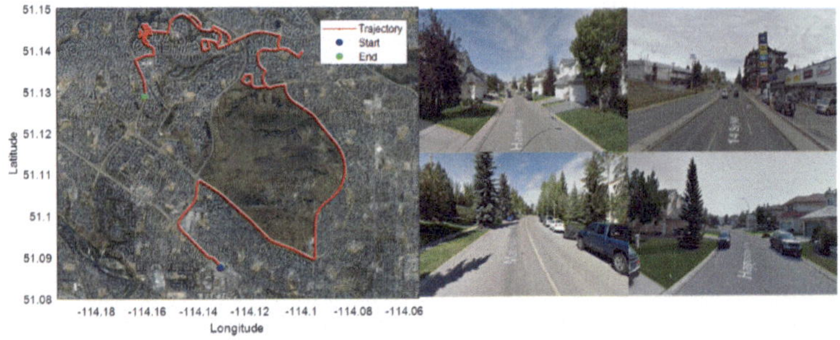

Fig. 6. The second kinematic test trajectory and test environment

In this test, the DF-IP3 had similar performance to the first test. To assess the improvement of positioning accuracy by a dual-frequency system, the results from both SF-IP3 and DF-IP3 systems are also compared in terms of positioning accuracy. Shown in Fig. 7 are the horizontal positioning results of the DF-IP3 and SF-IP3 with blue and green lines representing the positioning errors in the east and north directions, respectively. Both the DF-IP3 and the SF-IP3 got converged within seconds at the beginning of the test. In terms of positioning accuracy, the DF-IP3 positioning errors are less than the SF-IP3 in each direction. The east and north RMS values of the DF-IP3 are 0.09 m and 0.22 m while the corresponding values of the SF-IP3 are 0.11 m

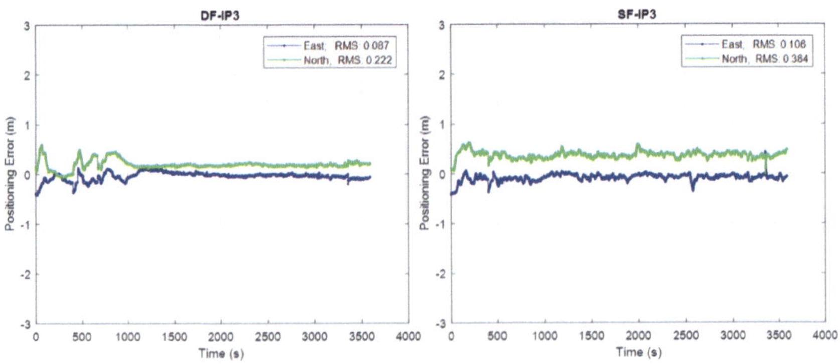

Fig. 7. Positioning errors of the DF-IP3 and SF-IP3 in the east, north directions during the second test

and 0.38 m. This indicates the dual-frequency observations can further improve the positioning accuracy of a single-frequency based system.

The horizontal CDFs of the DF-IP3 and SF-IP3 are plotted in Fig. 8 with the blue line representing the SF-IP3 and green line representing the DF-IP3. For the second test, the horizontal CEP of the DF-IP3 is better than 0.3 m as well. It can be seen that, similar to the previous test, over 90% DF-IP3 solution errors are less than 0.4 m. As comparison, for the SF-IP3, the majority of horizontal solution errors are between 0.3 m and 0.5 m. It is clear that the positioning accuracy of the DF-IP3 is improved, compared to the SF-IP3.

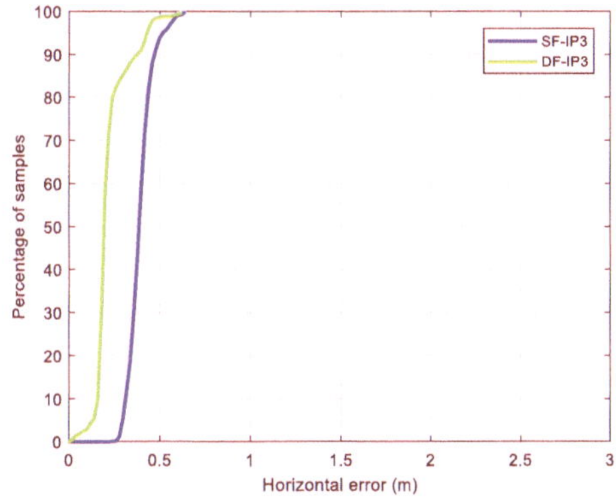

Fig. 8. Horizontal CDFs of the SF-IP3 and DF-IP3 for the second test

The third test was conducted in residential areas and sparse urban areas with trees and buildings along the road. The land vehicle was moving on roads in Community Banff Trail, Crowchild Trail NW, Shaganappi Trail, John Laurie Blvd NW, McKnight Blvd NW, McKnight Blvd NE, 68 St NE and 16 Ave NE in Calgary. This test lasted for about 50 min and GPS, GLONASS, Galileo satellites were applied during the test. The trajectory and test environment for the third test can be seen in Fig. 9.

Fig. 9. The third kinematic test trajectory and test environment

The environment of this test is more challenging than the previous two tests with more signal blockages. For performance comparison of different systems in challenging environments, the DF IF PPP solution is included in addition to the DF- and SF-IP3 solutions. The horizontal positioning errors of the SF-, DF-IP3 and the conventional DF IP PPP are plotted in blue, green and red lines respectively in Fig. 10. The results indicate that the conventional method suffered frequent re-convergence due to signal blockage and its overall accuracy is the worst with horizontal error over 1 m. Therefore, it is not suitable for precise positioning in such environment. The IP3 systems can still maintain the half a meter accuracy without re-convergence during the test. Compared to the SF-IP3 with horizontal RMS at 0.52 m, the DF-IP3 had the best solution in most time during the test with horizontal RMS at 0.39 m.

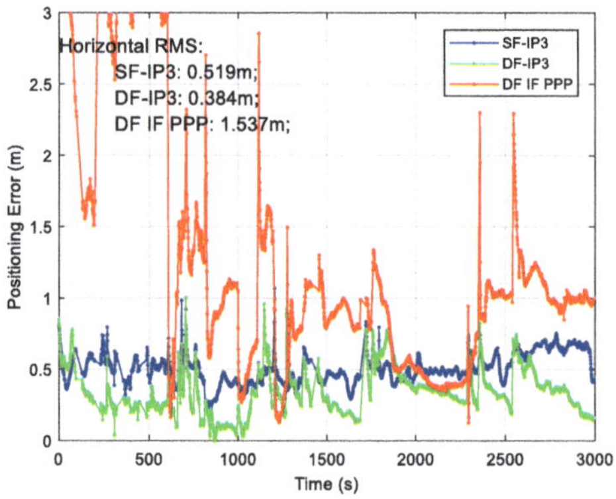

Fig. 10. Horizontal positioning errors of the SF-IP3, DF-IP3 and DF IF PPP during the third test

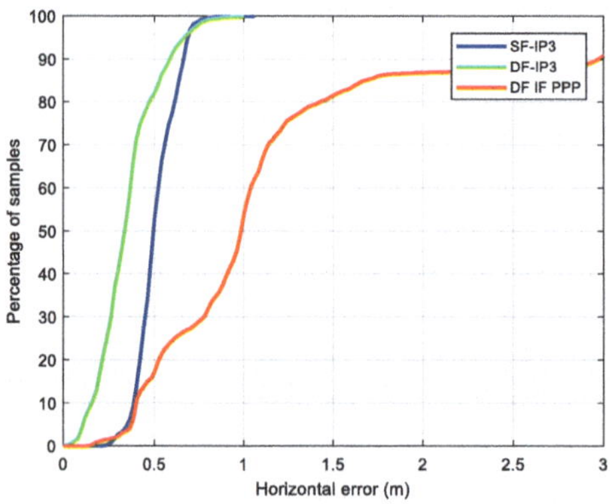

Fig. 11. Horizontal CDFs of the SF-IP3, DF-IP3 and DF IF PPP for the third test

Figure 11 shows the horizontal CDFs of the SF-IP3, DF-IP3 and the conventional PPP method. Although the performance of the DF-IP3 degraded in this test due to more challenging environment, its horizontal CEP could still reach 0.3 m. In contrast, over 80% conventional PPP solution errors are over 0.5 m horizontally and about 50% solution errors are even larger than 1 m. For the SF-IP3, it reached CEP 0.5 m, which however is not as good as the DF-IP3.

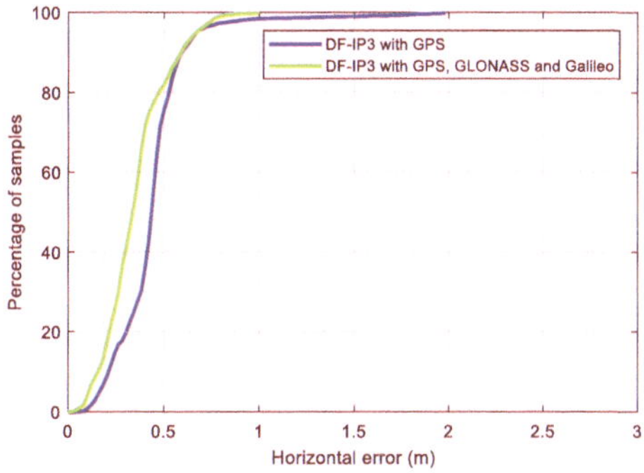

Fig. 12. Horizontal CDFs of DF-IP3 using GPS and multiple systems for the third test

In addition, in such environment, more satellites can be observed in multi-constellation positioning, compared with one system applied, resulting in improved performance. Shown in Fig. 12 is comparison of DF-IP3 CDFs using GPS and multiple systems. The horizontal errors are clearly reduced with more satellites observed.

To summarize the performance of IP3 in all the tests, the RMS values of DF-IP3 positioning error in east and north directions are listed in Table 1.

Table 1. DF-IP3 positioning RMS in three tests

	Test 1	Test 2	Test 3
East	0.141 m	0.087 m	0.158 m
North	0.191 m	0.222 m	0.350 m

4 Conclusions

The release of low-cost dual-frequency GNSS chipsets has attracted more attention to PPP applications and pushed for the development of low-cost dual-frequency PPP system. The Profound dual-frequency IP3 (DF-IP3) is a new real-time PPP system based on low-cost dual-frequency GNSS chipsets, which is capable of providing worldwide fast convergence PPP solution with accuracy at 30 cm (CEP). The system has overcome issues in current low-cost dual-frequency chipset products in precise point positioning. According to the simulated re-start post-mission test using the first kinematic dataset acquired in an open-sky environment, both the SF- and DF- IP3 can get converged within seconds. The field tests with the Profound DF-IP3 EVK (integrated with a u-blox F9P chipset) show that DF-IP3 can further enhance the positioning accuracy and stability than a single-frequency system.

References

1. Nie, Z., Liu, F., Gao, Y.: Real-time precise point positioning with a low-cost dual-frequency GNSS device. GPS Solut. **24**(1), 9 (2020)
2. Nie, Z., Zhou, P., Liu, F., Wang, Z., Gao, Y.: Evaluation of orbit, clock and ionospheric corrections from five currently available SBAS L1 services: methodology and analysis. Remote Sens. **11**(4), 411 (2019)
3. Hadas, T., Bosy, J.: IGS RTS precise orbits and clocks verification and quality degradation over time. GPS Solut. **19**(1), 93–105 (2015)
4. Bisnath, S., Gao, Y.: Current state of precise point positioning and future prospects and limitations. In: Observing Our Changing Earth, pp. 615–623. Springer, Berlin (2009)
5. Zumberge, J.F., Heflin, M.B., Jefferson, D.C., Watkins, M.M., Webb, F.H.: Precise point positioning for the efficient and robust analysis of GPS data from large networks. J. Geophys. Res. Solid Earth **102**(B3), 5005–5017 (1997). http://doi.wiley.com/10.1029/96JB03860
6. Li, L., Jia, C., Zhao, L., Cheng, J., Liu, J., Ding, J.: Real-time single frequency precise point positioning using SBAS corrections. Sensors **16**(8), 1261 (2016)
7. Li, B., Zang, N., Ge, H., Shen, Y.: Single-frequency PPP models: analytical and numerical comparison. J. Geod. **93**(12), 2499–2514 (2019)
8. Lou, Y., Zheng, F., Gu, S., Wang, C., Guo, H., Feng, Y.: Multi-GNSS precise point positioning with raw single-frequency and dual-frequency measurement models. GPS Solut. **20**(4), 849–862 (2016)
9. de Bakker, P.F., Tiberius, C.C.J.M.: Real-time multi-GNSS single-frequency precise point positioning. GPS Solut. **21**(4), 1791–1803 (2017)
10. Pan, L., Zhang, X., Liu, J., Li, X., Li, X.: Performance evaluation of single-frequency precise point positioning with GPS, GLONASS, BeiDou and Galileo. J. Navig. **70**(3), 465–482 (2017)
11. Profound Positioning Inc.: Profound-IP3 2019. http://www.profoundpositioning.com/
12. Odolinski, R., Teunissen, P.J.G.: Single-frequency, dual-GNSS versus dual-frequency, single-GNSS: a low-cost and high-grade receivers GPS-BDS RTK analysis. J. Geod. **90**(11), 1255–1278 (2016)
13. Ashby, N.: Relativity in the global positioning system. Living Rev. Relativ. **6**(1), 1 (2003)
14. Schmid, R., Steigenberger, P., Gendt, G., Ge, M., Rothacher, M.: Generation of a consistent absolute phase-center correction model for GPS receiver and satellite antennas. J. Geod. **81** (12), 781–798 (2007)
15. Wu, J.T., Wu, S.C., Hajj, G.A., Bertiger, W.I., Lichten, S.M.: Effects of antenna orientation on GPS carrier phase. In: Advances in the Astronautical Sciences, pp. 1647–1660 (1992)
16. Cai, C., Liu, Z., Luo, X.: Single-frequency ionosphere-free precise point positioning using combined GPS and GLONASS observations. J. Navig. **66**(3), 417–434 (2013)

Preliminary Research on GNSS Multipath Interpret the Process of Vegetation Growth

Jilun Peng[1], Shuangcheng Zhang[1,2(✉)], Jingjiang Zhang[3], Qi Liu[1], and Tao Wang[1]

[1] College of Geology Engineering and Geomantic, Chang'an University, Xi'an, China
shuangcheng369@vip.163.com
[2] Institute of Desert Meteorology, China Meteorological Administration, Urumqi, Xinjiang, China
[3] Institute of Urban Meteorology, CMA, Beijing, China

Abstract. Vegetation is a significant link between soil, moisture, organisms and the atmosphere. Recording vegetation changes has important significance in climate change. Traditional remote sensing has the drawback with low temporal resolution in monitoring vegetation, while GNSS multipath can provide a more complete vegetation growth process. Firstly, the multipath characteristics of GNSS are analyzed, and then basic principles and processes of multipath interpret vegetation growth are introduced in detail. Finally, GNSS continuous tracking stations P476 and P542 deployed in California are used for analysis. The preliminary results show that, at P476, the correlation coefficient between multipath and Normalized Difference Vegetation Index (NDVI) is 0.64 (When the lag time is considered, the correlation can reach 0.88), and similar results are obtained at P542. The results show that GNSS multipath interpret vegetation growth process is effective. In phenology, GNSS multipath estimated phenology variables have later growth season start time and peak time, and shorter season length, mainly affected by lag time.

Keywords: GNSS · Multipath · NDVI · Vegetation growth · Phenology

1 Introduction

Vegetation is an indicator of the ecological environment. Accurate and comprehensive understanding of vegetation conditions are important for the assessment of ecological environment. Of course, testing climate change models, monitoring surface phenology, and verifying ecological hydrological hypotheses also all need to measure vegetation growth [1]. At present, the methods for studying vegetation are roughly divided into two categories: one is optical remote sensing; the other is microwave remote sensing. NDVI is the most widely used indicator for measuring vegetation, and is usually used for vegetation cover change, drought monitoring, and land use change [2]. There are also

J. Sun et al. (Eds.): CSNC 2020, LNEE 650, pp. 187–195, 2020.
https://doi.org/10.1007/978-981-15-3707-3_18

some shortcomings of NDVI for measuring vegetation growth, such as soil background effects, cloud cover effects, and low temporal resolution. Unlike optical remote sensing, microwave remote sensing is not affected by cloud cover or time and can be used to monitor plant phenology, vegetation water content and estimate biomass. However, the biggest difficulty is how to remove the effects of soil moisture and surface roughness. With the continuous development of GNSS, many scholars have carried out research on the mechanism and related suppression algorithms of multipath errors, and obtained valuable results [3–6]. Subsequently, it's found that multipath signals can also be used as useful information for the inversion of surface environmental parameters, becoming a new remote sensing technology [7].

Multipath signals for surface environmental monitoring has achieved good development in Europe and the United States, and relevant research has been carried out in China. Vegetation growth is a hot spot in the study of GNSS reflection signals. Alvarez and Egido et al. [8, 9] found significant power changes in GNSS reflected signals for different soil moisture and vegetation conditions. Chew et al. [10] proposed a one-dimensional planar layered model of simulated signal-to-noise ratio data response to soil moisture and vegetation change. The simulations indicate that the amplitude of SNR oscillations may be used to estimate vegetation amount when vegetation wet weight is below 1.5 kg/m2. Wan et al. [11] used GPS SNR observations to interpret the water content of three plant types: desert grass, wheat and alfalfa. By comparing with the measured data, the validity of GPS-MR technology for vegetation water content is verified. Wu et al. [12] developed a GPS multipath simulator based on the radiative transfer equation model, which explicitly links vegetation parameters with multipath observations. Small et al. [13] proposed multipath interpret vegetation growth. It was verified by experiments that the water content of vegetation was inversely proportional to the magnitude of multipath, and similar results were obtained by PBO network site. Larson et al. [14] introduced the effects of vegetation water content for multipath in detail, and proposed Normalized Microwave Reflection Index (NMRI) to interpret vegetation water content, and obtained consistent results with NDVI. Subsequently, Small et al. [15] verified the effectiveness of NMRI interpret vegetation water content in the grassland of Montana, and obtained consistent results at four stations. Evans et al. [16] compared NMRI and NDVI in different regions in detail, and found that the lag time between the two varies from region to region. Domestic and international research on multipath interpret vegetation growth is in infancy, and it is necessary to conduct in-depth research on key issues in order to promote its faster and better application in practice.

At present, there are few studies on vegetation growth based on multipath in China. In this paper, the theoretical system involved in this technology is discussed in detail, and the data of grass and shrub area are used for analysis, in order to lay the foundation for further research.

2 Theoretical System of Multipath Interpret Vegetation Growth

Multipath is the interference between direct microwave signals from the satellites and those reflected before reaching the antenna, typically by the ground (Fig. 1). Signal energy is absorbed by ground objects and the antenna has a suppression effect on the reflected signal, thus the signal strength is reduced. Multipath has become the main source of error for high-precision positioning, mainly related to the structure of the reflective surface and dielectric parameters. The multipath is related to the reflection ability (reflection coefficient) of the ground objects, and the ground objects becomes a medium for signal propagation. Therefore, multipath changes in GNSS can be used to interpret surface environment parameters. Among them, the pseudorange multipath error can be extracted according to the code tracking loop discriminator equation or the carrier and code combination [17].

$$M_1 = \frac{\alpha D \cos \varphi_1}{1 + \alpha D \cos \varphi_1} \tag{1}$$

$$MP_1 = P_1 - \frac{f_1^2 + f_2^2}{f_1^2 - f_2^2}\lambda_1\phi_1 + \frac{2f_2^2}{f_1^2 - f_2^2}\lambda_2\phi_2 = M_1 + C + \varepsilon_1 \tag{2}$$

M_1 is pseudorange multipath on the L_1 frequency; $\alpha = A_m/A_d$ is the reflection coefficient, A_d and A_m are the amplitude of the direct and reflected signal, respectively; D and φ_1 are the path delay and phase difference between reflected and direct signal; P_1 is the pseudorange observation; ϕ_1 and ϕ_2 are carrier observations; f_1 and f_2 are the frequencies; λ_1 and λ_2 are the wavelengths; C is the deviation caused by the ambiguity; ε_1 is the measurement error. When the cycle slip is fixed, C is a constant, and M_1 can be obtained by the moving average method.

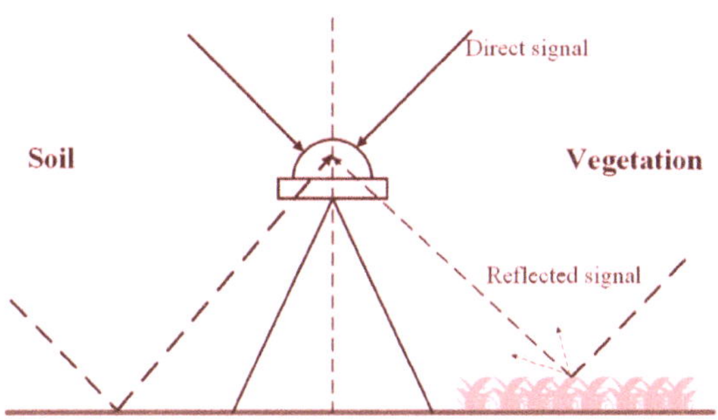

Fig. 1. Diagram of multipath interpretation of vegetation growth

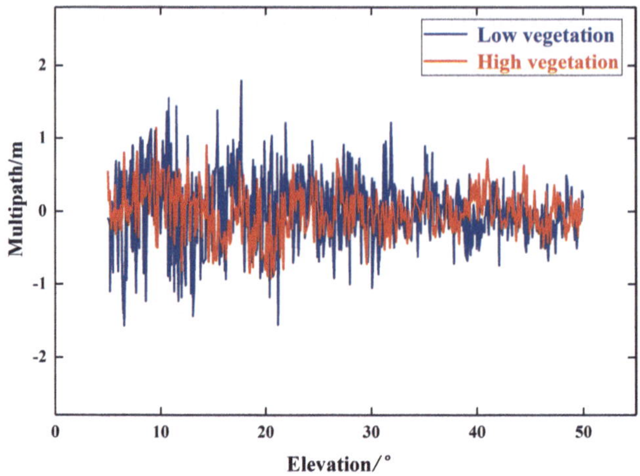

Fig. 2. Effects of multipath amplitude vegetation growth

In Fig. 1, we can know that the signal received by the GNSS receiver is a composite signal. When the signal passes through the vegetation, the signal will produce more diffuse reflection and absorption [18], and there is a smaller specular scattering cross section relative to the bare soil, resulting in a decrease in the amplitude of the reflected signal (A_m), and thus the amplitude of the multipath (M_1) is reduced. Of course, M_2 also has the same characteristics, so it will not be described again. Figure 2 shows the multipath amplitude variation of the P476 station at different stages of vegetation growth. The blue curve represents the multipath at low vegetation period; the red curve represents the multipath at high vegetation period. It can be seen that the multipath amplitude is affected by the growth of the vegetation. That is, as the vegetation grows, the multipath amplitude continuously decreases.

Thus, the multipath amplitude variation characteristics can be used to interpret the vegetation growth process. The average of all satellite multipath RMS (MP_1rms) can be used as an indicator to measure reflected signal, which is not only sensitive to the vegetation changes around the station, but also can weaken the influence of random errors.

At the same time, in order to interpret the vegetation growth process between different stations, it is necessary to eliminate the influence of the path delay inconsistency caused by the terrain. That is, normalizing the MP_1rms [15].

$$\text{NMRI} = \frac{-(MP_1rms - \max(MP_1rms))}{\max(MP_1rms)} \tag{3}$$

MP_1rms is the average of all satellite multipath RMS; $\max(MP_1rms)$ (the maximum MP_1rms) is based on the average of the largest 5% daily MP_1rms values, which can effectively avoid the error caused by the single maximum value abnormality.

In summary, the corresponding data processing flow is drawn, as shown in Fig. 3. First, the NMRI and NDVI sequence were extracted, and the correlation was calculated. T statistics were used to verify whether the correlation was significant, which is used to verify the validity of NMRI interpret vegetation growth.

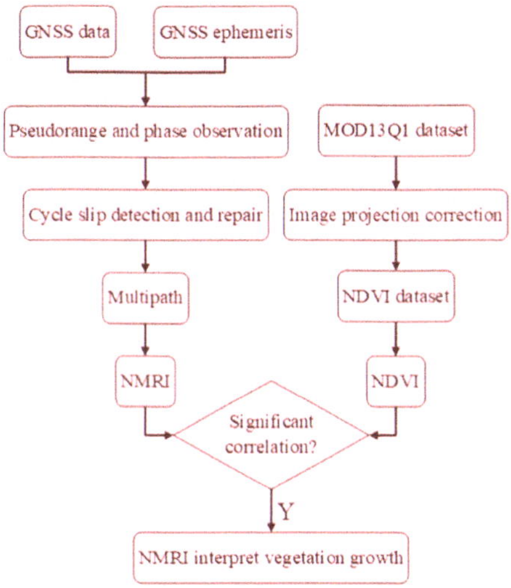

Fig. 3. Flow chart for multipath interpret vegetation growth

3 Result of GNSS Multipath Interpret Vegetation Growth

3.1 Data Sources

NDVI is derived from the MOD13Q1 dataset, which is processed by moderate-resolution imaging spectrometer data carried by the Terra satellite. Time resolution is 16 days and spatial resolution is 250 m. NDVI calculation formula is as follows:

$$\mathrm{NDVI} = \frac{\rho_{\mathrm{NIR}} - \rho_{\mathrm{RED}}}{\rho_{\mathrm{NIR}} + \rho_{\mathrm{RED}}} \tag{4}$$

ρ_{NIR} and ρ_{RED} are reflectances of the near-infrared and red-light bands, respectively. NDVI (ratio of red and near-infrared band reflectance) is sensitive to the amount of plants and linearly related to plant density. It is the best indicator for measuring plant growth and spatial distribution density.

GNSS data is the observation data of the GPS continuous operation tracking stations P476 and P542 (2015–2018) deployed in California, USA. The station environment is shown in the following figure.

Fig. 4. The surrounding environment of the P476 and P542 station

Figure 4 (left) shows the vegetation in the north direction of the P476 station. It can be seen that the vegetation types around the station is single (short shrubs), which effectively avoids errors caused by spatial heterogeneity. At the same time, P476 is part of the PBO network. It is equipped with a TRIMBLE NETRS dual-frequency geodetic receiver and antenna is TRM41249.0 with a fairing (SCIT) choke. Long-term observation data is available and data quality is excellent. Different from P476 station, the environment of P542 station is grassland, and the seasonal variation of grass is more obvious.

3.2 Results and Analysis

3.2.1 Correlation Analysis

Time series of NMRI and NDVI sequences are plotted (Fig. 5). It can be seen from the figure that there is a good consistency. In order to further calculate the correlation between NMRI and NDVI, the correlation test was carried out by taking P476 station as an example. The results are shown in Fig. 6.

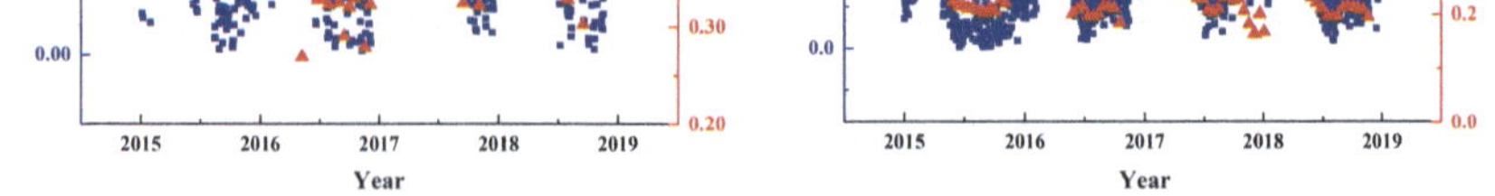

Fig. 5. Time series of NMRI and NDVI at P476 and P542 station

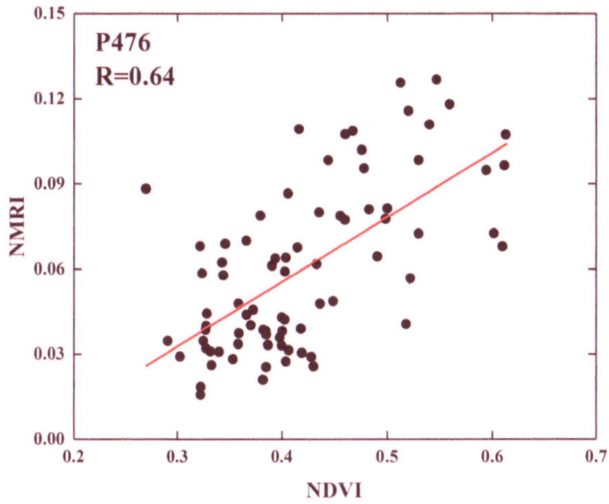

Fig. 6. Correlation analysis of P476 station

Figure 5 are the time series of NMRI and NDVI at P476 and P542 stations in 2015–2018, respectively. Red triangles represent NDVI and blue squares represent NMRI. It can be seen that with the growth of vegetation, NDVI gradually increases, and NMRI increases. Once the vegetation begins to senesce, NDVI begins to decrease, and NMRI decreases. The trends of NDVI and NMRI are basically the same, and have strong annual periodicity. NDVI and NMRI are linked through vegetation growth or vegetation water content. The greening of leaves and increase in photosynthetic activity yields higher NDVI. The effect on NMRI is only apparent after the increased photosynthetic activity yields additional plant growth or vegetation water content. Therefore, NMRI lags behind NDVI for some time.

Figure 6 is a correlation analysis of NDVI and NMRI at the P476 station. According to the calculation, the correlation coefficient is 0.64. When considering the effect of lag time, the correlation coefficient can reach 0.88. Finally, T statistic ($t = 7.34$) is greater than the threshold ($tcrit = 1.66$), indicating that the above correlation coefficient is significant. It can be seen that NMRI has good consistency with NDVI, and high temporal resolution NMRI can be used to interpret the vegetation growth, which has a good complement to NDVI.

3.2.2 Comparing Phenology Metrics Estimated by NMRI and NDVI

In this work, the threshold is used to extract the phenology metrics [19], and the start time and end time of the growing season are determined by the threshold. The data of two stations in 2017 are selected, and the differences between the phenology metrics estimated by NMRI and NDVI are compared and analyzed. The results are shown in Table 1.

Table 1. NMRI and NDVI estimate phenology metrics

Phenology metrics	P476		P542	
	NDVI	NMRI	NDVI	NMRI
Start of growing season (doy)	−16	7	−54	3
Season length (days)	209	171	177	155
Timing of peak growth (doy)	65	86	49	108

As can be seen from Table 1, at P476 station, NMRI estimated start and peak time of growing season are approximately 20 days behind NDVI, and NMRI estimated season length is shorter. Similar results were obtained from P542 station, but the lag time was different from P476 station. The results are consistent with the above-mentioned lag analysis results. NDVI and NMRI are linked through vegetation growth, so NMRI estimated phenology metrics have a late growth season start time and peak time, and a shorter season length.

4 Conclusions

Vegetation plays an important role in maintaining soil and water, regulating the atmosphere, and slowing the rise of greenhouse gas. In this work, the basic theory of multipath interpret vegetation growth is given in detail. The effectiveness of method is verified by data, and the preliminary results are obtained. The preliminary research in this work also have key issues such as denoising, spatial heterogeneity, which need to be further studied and resolved. The advantage of multipath interpret vegetation growth is that daily multipath data can provide high-resolution seasonal fluctuation description without being affected by clouds, and more complete records of plant growth and senesce processes. It has a good complement to the vegetation index such as NDVI.

Multipath interpret vegetation growth process has become a new remote sensing method. The advantage of method is that based on the multi-frequency multi-mode GNSS system realizes near real-time monitoring, which further compensates for the shortcomings of space-time resolution of the conventional instrument and satellite radar monitoring surface environment. It also provides a data source for data fusion, which plays an important role in GNSS monitoring of global environmental change.

Acknowledgements. We gratefully acknowledge the provision of data, equipment, and engineering services by the Plate Boundary Observatory operated by UNAVCO for Earth Scope. The authors thank Carolyn Roesler and Kristine M. Larson for providing Software Tools for GNSS Interferometric Reflectometry (GNSS-IR) by (https://www.ngs.noaa.gov/gps-toolbox/GNSS-IR.htm). This work was supported by Institute of Desert Meteorology in China Meteorological Administration (Ground-based GNSS monitoring snow depth in Altay, Sqj2017002) and. Thanks to anonymous reviewers for providing valuable comments and suggestions in this paper!

References

1. Nemani, R.R., Keeling, C.D., Hashimoto, H., et al.: Climate-driven increases in global terrestrial net primary production from 1982 to 1999. Science **300**(5625), 1560–1563 (2003)
2. Wan, X.: Change detection for landuse in city based on NDVI. Geosp. Inf. **7**(4), 111–113 (2009)
3. Xia, L.: Theoretical Study on GPS Observations Multi-path Effects and Numerical Results. Wuhan University, Wuhan (2001)
4. Yuan, L., Huang, D., Ding, X., et al.: On the influence of signal multipath effects in GPS carrier phase surveying. Acta Geod. Cartogr. Sin. **33**(3), 210–215 (2004)
5. Huang, S., Li, P., Yang, B., et al.: Study on the characteristics of multipath effects in GPS dynamic deformation monitoring. Geomat. Inf. Sci. Wuhan University **30**(10), 870–875 (2005)
6. Dai, W., Ding, X., Zhu, J.: Study on multipath effect in structure health monitoring using GPS. J. Geodesy Geodyn. **28**(1), 65–70 (2008)
7. Larson, K.M., Gutmann, E.D., Zavorotny, V.U., et al.: Can we measure snow depth with GPS receivers? Geophys. Res. Lett. **36**(17), 1–5 (2009)
8. Alvarez, N.R., Bosch, X., Camps, A., et al.: Vegetation water content estimation using GNSS measurements. IEEE Geosci. Remote Sens. Lett. **9**(2), 282–286 (2012)
9. Egido, A., Caparrini, M., Ruffini, G., et al.: Global navigation satellite systems reflectometry as a remote sensing tool for agriculture. Remote Sens. **4**(8), 2356–2372 (2012)
10. Chew, C.C., Small, E.E., Larson, K.M., et al.: Vegetation sensing using GPS-interferometric reflectometry: theoretical effects of canopy parameters on signal-to-noise ratio data. IEEE Trans. Geosci. Remote Sens. **53**(5), 2755–2764 (2015)
11. Wan, W., Larson, K.M., Small, E.E., et al.: Using geodetic GPS receivers to measure vegetation water content. GPS Solutions **19**(2), 237–248 (2015)
12. Wu, X., Jin, S., Xia, J.: A forward GPS multipath simulator based on the vegetation radiative transfer equation model. Sensors **17**(6), 1291 (2017)
13. Small, E.E., Larson, K.M., Braun, J.J.: Sensing vegetation growth with reflected GPS signals. Geophys. Res. Lett. **37**(12), 1–5 (2010)
14. Larson, K.M., Small, E.E.: Normalized microwave reflection index: a vegetation measurement derived from GPS data. IEEE J. Sel. Top. Appl. Earth Obs. Remote Sens. **7**(5), 1501–1511 (2014)
15. Small, E.E., Larson, K.M., Smith, W.K.: Normalized microwave reflection index: validation of vegetation water content estimates at Montana Grasslands. IEEE J. Sel. Top. Appl. Earth Obs. Remote Sens. **7**(5), 1512–1521 (2014)
16. Evans, S.G., Small, E.E., Larson, K.M.: Comparison of vegetation phenology in the Western United States from reflected GPS microwave signals and NDVI. Int. J. Remote Sens. **35**(9), 2996–3017 (2014)
17. Axelrad, P., Larson, K.M., Jones, B.A.: Use of the correct satellite repeat period to characterize and reduce site-specific multipath errors. In: Proceedings ION-GNSS 2005, Institute of Navigation Long Beach, CA, USA, 13–16 September 2005
18. Katzberg, S.J., Torres, O., Grant, M.S., et al.: Utilizing calibrated GPS reflected signals to estimate soil reflectivity and dielectric constant: results from SMEX02. Remote Sens. Environ. **100**(1), 17–28 (2006)
19. Jones, M.O., Kimball, J.S., Jones, L.A., et al.: Satellite passive microwave detection of North America start of season. Remote Sens. Environ. **123**, 324–333 (2012)

Calibration and Error Analysis of the BF-1 Demonstration GNSS-R Satellites

Bei Wan(✉), Xinliang Niu, Cheng Jing, Bochi Lei, and Chong Han

Chinese Academy of Space Technology (Xi'an), Xi'an, China
wanbeiddd123@163.com

Abstract. The reflection signal of Global Navigation Satellite System can be applied to the research of Remote Sensing Applications (GNSS-Reflectometry, GNSS-R), which shows such advantages as all-time, all-weather, light weight, small volume and low cost. As the first generation of GNSS-R microwave remote sensing satellite in China, BF-1 demonstration satellite is capable to produce high spatial and temporal resolution wind speed observation results, which plays an significant complementary role in the measurement of global wind speed, with the continuous observation of typhoon in particular. In this paper, an introduction is made of the calibration process from the metadata of payload output to the normalized bistatic scattering cross section (NBRCS) of wind speed retrieval. The entire process involves two steps. Firstly, the metadata of payload is calibrated using the calibration load and temperature information, and then it is calibrated according to the geometry of transmitter and receiver. The errors introduced in the process of calibration are analyzed by formula derivation, based on which the total error of the calibration result is obtained as 0.74 dB. This term demonstrates the accuracy and validity of the data processing results regarding BF-1 test satellite, in addition to providing a reference for the design of the next-generation GNSS-R microwave remote sensing satellite.

Keywords: GNSS-R · BF-1 · Calibration · Error analysis

1 Introduction and Background

GNSS-R technology relies on the reflected signals of GNSS satellites for the inversion of the relevant physical characteristics to the reflecting surface. Not only is it applicable to sea surface wind speed, average sea surface height, soil moisture, etc., it also exhibits such advantages as high space-time resolution, all-weather and low cost [1].

On June 5, 2019, China utilized the Long March 11 launch vehicle to conduct its first-ever maritime launch test, which led to a complete success [2]. As the main load of this emission, the BF-1 demonstration GNSS-R satellites are tasked with the first spaceborne GNSS-R test for China. It is mainly purposed to observe the surface wind speed with high spatiotemporal resolution, which provides an effective supplement for traditional active remote sensing on the time resolution of observation.

With reliance on the sensitivity of the power of the navigation reflection signal to the roughness of the sea surface, the GNSS-R sea surface wind speed detection technology has the capability of retrieving the sea surface wind speed. However, the

J. Sun et al. (Eds.): CSNC 2020, LNEE 650, pp. 196–205, 2020.
https://doi.org/10.1007/978-981-15-3707-3_19

delay-Doppler maps (DDM) of the reflected signal calculated by the receiver is not only related to the sea surface wind speed, but also to the equivalent isotropic radiated power (EIRP) of the navigation satellite, the receiver antenna gain, the receiver temperature, and the geometry of the transmission and reception, which makes it necessary to calibrate these additional factors for the enhanced correlations and consistency of observations with sea surface wind speeds.

In the process of BF-1 data pre-processing, power calibration and geometry calibration were conducted successively.

2 Power Calibration

The purpose of power calibration is to eliminate the effects of receiver temperature and channel differences on DDM.

2.1 Hardware Structure and Working Method

At present, the power calibration of the GNSS-R device is carried out with the addition of a switch to enable the signal source to switch between the antenna and the calibration load, as shown in Fig. 1.

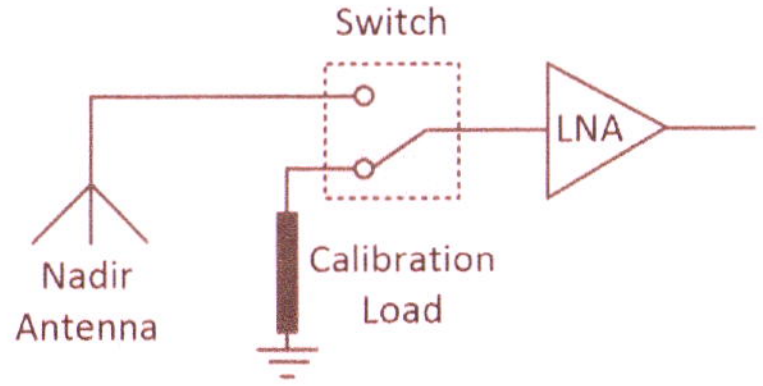

Fig. 1. LNA design with calibration load and switch

When the switch is turned on the antenna side, the traditional DDM image is received. When the switch is shifts to the calibration load, the absolute power calibration is conducted in conjunction with the temperature measurement of the calibration load.

In general, power calibration is interspersed in the process of conventional DDM image collection. The BF-1 switches to the calibration load once per minute, the entire switching process of which takes 3 s. In the first second, it switches from the nadir antenna to the calibration load, then remain on the calibration load for one second, and in the third second it switches back from the calibration load to the nadir antenna.

2.2 DDM Value

The pixel value of a single output DDM can be obtained using the following formula [3]:

$$C = G(P_a + P_r + P_g) \tag{1}$$

where C is the DDM image pixel value, G denotes the total gain, P_a is the antenna thermal noise power, and P_r is the GNSS-R device thermal noise power value. It is worth noting that P_g is the value of the reflected power on the surface of the earth, that is, the direct observation measurement as required for GNSS-R remote sensing.

In addition, C includes the reflection signal and noise signal of the sea surface C_N, which is defined as the GNSS-R device noise when the signal does not reach the surface of the earth, and is expressed as follows:

$$C_N = G(P_a + P_r) \tag{2}$$

In each DDM image C_N are the pixel values before the specular point chip. In theory, these pixel values only include the thermal noise of the GNSS-R device and the antenna.

2.3 Total Gain and Power Value After Calibration

The calculation of the total gain G plays a crucial role in the process of power calibration, which needs to be performed with the assistance of the calibration load. After obtaining the total gain G, the post-calibration power value can be obtained directly by combining of formulas (1) and (2).

2.3.1 Thermal Noise of Calibration Load and Receiver Device

In design, the BF-1 calibration load is a 50 Ω resistor, which is sufficient to drown the GNSS signal. Through the switch, the signal source is switched from the nadir antenna to the calibration load at the preset time, so that the noise DDM generated by the calibration load is received. It can be expressed as follows:

$$C_B = G(P_b + P_r) \tag{3}$$

where C_B is the noise DDM image obtained by the receiver when the switch is pointed to the calibration load, P_b is the thermal noise power value of the calibration load, and P_r is the thermal noise power value of the GNSS-R device. Based on the actual temperature of the calibration load, the thermal noise power value of the calibration load can be obtained as follows:

$$P_b = kT_bB_W \tag{4}$$

where P_b is the thermal noise power value of the calibration load, k refers to the Boltzmann constant, B_W is the signal bandwidth (usually 1000 Hz, corresponding to the coherent integration time T_i = 1 ms), and T_b is the actual measurement value of the calibration load temperature.

It is necessary to obtain the thermal noise power of the GNSS-R device over the course of routine power calibration using only the calibration load. This power value can be determined by the noise figure of the GNSS-R device, as shown below:

$$P_r = kT_rB_W = k[(NF - 1)290]B_W \tag{5}$$

where NF represents the noise figure of the GNSS-R device, which is directly associated with the physical temperature of the GNSS-R device. A corresponding lookup table of the noise figure with temperature can be obtained through ground temperature cycle experiments. During orbital operation, the noise figure is determined directly by taking reading of the temperature as measured by the calibrated load, and the noise power of the GNSS-R device is further calculated using the lookup table.

2.3.2 Power Value After Calibration

The total gain can be directly obtained using formula (3):

$$G = \frac{C_B}{P_b + P_r} \tag{6}$$

Substituted into formulas (1) and (2), the power value of the reflected signal after calibration can be obtained as follows:

$$P_g = \frac{(C - C_N)}{G} = \frac{(C - C_N)(P_b + P_r)}{C_B} \tag{7}$$

where P_g is the calibrated power value, which represents the absolute power value of the GNSS-R signal at the corresponding delay and Doppler.

3 Geometry Calibration

The power calibration is successful in converting the DDM pixel counts obtained by the GNSS-R receiver into an absolute power value, based on which the power information generated by the reflection of the earth's surface can be acquired. Nevertheless, for the generation of the geophysical parameters, it is necessary to eliminate interference in addition to the influence of sea surface roughness at each pixel. In doing so, the same value can be obtained for each DDM pixel under the identical conditions of marine environment.

3.1 Influencing Factors

The forward scattered power model of the GNSS-R sea surface reflection signal and its equivalent simplified form are presented in the following formula [3]:

$$P^{g}_{\hat{\tau},\hat{f}} = \iint_{A} \frac{P^{T}\lambda^{2}G^{T}_{x,y}\sigma^{0}_{x,y}G^{R}_{x,y}\Lambda^{2}_{\hat{\tau};x,y}S^{2}_{\hat{f};x,y}}{(4\pi)^{3}\left(R^{R}_{x,y}\right)^{2}\left(R^{T}_{x,y}\right)^{2}L}dxdy = \frac{P^{T}\lambda^{2}\bar{G}^{T}_{\hat{\tau},\hat{f}}\bar{\sigma}^{0}_{\hat{\tau},\hat{f}}\bar{G}^{R}_{\hat{\tau},\hat{f}}\bar{A}_{\hat{\tau},\hat{f}}}{(4\pi)^{3}\left(\bar{R}^{R}_{\hat{\tau},\hat{f}}\right)^{2}\left(\bar{R}^{T}_{\hat{\tau},\hat{f}}\right)^{2}\bar{L}} \tag{8}$$

where $P^{g}_{\hat{\tau},\hat{f}}$ indicates the power of the reflected signal under relative delay $\hat{\tau}$ and relative Doppler $\hat{f}$, P^{T} is the transmitter power, λ is the corresponding GNSS signal wavelength, $\bar{G}^{T}_{\hat{\tau},\hat{f}}$ and $\bar{G}^{R}_{\hat{\tau},\hat{f}}$ is the equivalent antenna gain for transmission and reception, $\bar{\sigma}^{0}_{\hat{\tau},\hat{f}}$ represents the equivalent normalized cross section area for scattering, $\bar{A}_{\hat{\tau},\hat{f}}$ indicates equivalent scattering area, $\bar{R}^{R}_{\hat{\tau},\hat{f}}$ and $\bar{R}^{T}_{\hat{\tau},\hat{f}}$ refer to the equivalent distance of the transmitting and receiving paths, respectively, and $\bar{L}$ means the atmospheric attenuation factors.

In general, the observation measurement is purposed to select the signal near the specular point (SP). According to the aforementioned formula, the calculation formula of the normalized cross section area can be obtained as follows:

$$\bar{\sigma}^{0}_{\hat{\tau},\hat{f}} = \frac{(4\pi)^{3}P^{g}_{\hat{\tau},\hat{f}}L_{SP}\left(R^{R}_{SP}\right)^{2}\left(R^{T}_{SP}\right)^{2}}{P^{T}\lambda^{2}G^{T}_{SP}G^{R}_{SP}\bar{A}_{\hat{\tau},\hat{f}}} \tag{9}$$

where $\bar{A}_{\hat{\tau},\hat{f}}$ refers to the equivalent scattering area:

$$\bar{A}_{\hat{\tau},\hat{f}} = \iint_{A} \Lambda^{2}_{\hat{\tau};x,y}S^{2}_{\hat{f};x,y}dxdy \tag{10}$$

That is to say, for the output results $P^{g}_{\hat{\tau},\hat{f}}$ of the receiver at different moments and geometric configurations, there is a necessity to calibrate the transmitter power P^{T} and gain factors G^{T}_{SP}, the reception gain factors G^{R}_{SP}, distance attenuation factors R^{R}_{SP} and R^{T}_{SP}, and equivalent scattering areas $\bar{A}_{\hat{\tau},\hat{f}}$. Following geometry calibration, the normalized bistatic scattering cross section is obtained.

3.2 Calibration Effect

Direct modeling is conducted using the BF-1 spaceborne data obtained in October 2019 and the Tropical Atmosphere Ocean Array Buoy data National Data Buoy Center (NDBC), with the space and time matching interval set to within 100 km and 10 min, respectively. Figure 2 shows the model established by the observation using the raw DDM, while Fig. 3 presents the model constructed by the NBRCS obtained after calibration.

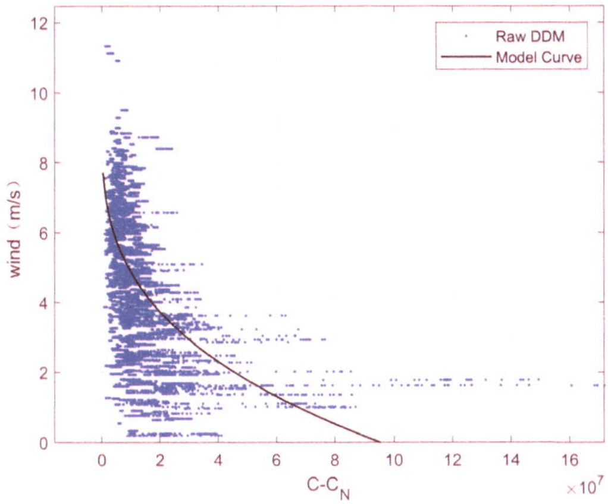

Fig. 2. Modeling with Raw DDM

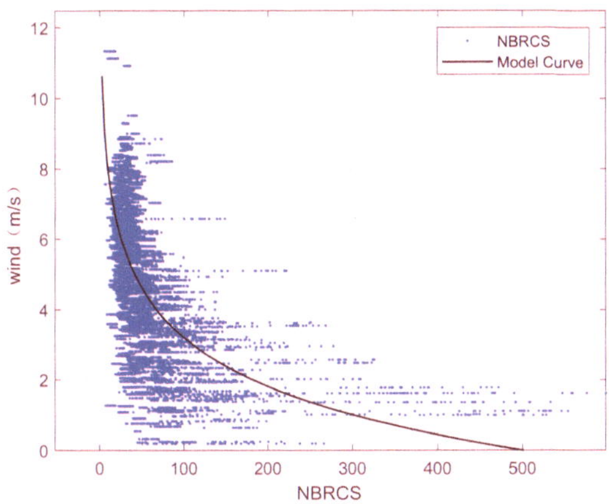

Fig. 3. Modeling with NBRCS

As shown in Fig. 2, the correlation between observations and wind speed is relatively low, with many observations sample values concentrated in tiny section. In Fig. 3, the observations demonstrate a more noticeable relation to wind speed, with the model curve being more reasonable. The former model shows a RMSE of 1.6942 m/s and a correlation coefficient of 0.5170, while the latter model shows a RMSE of 1.5597 m/s and a correlation coefficient of 0.6157. These results are considered qualified for spaceborne GNSS-R wind retrieval. In this paper, only the simplest and direct modeling method is applied. Though the more complex approach to modeling can produce more desirable retrieval results, this is not the focus of this paper.

It can be seen that the correlation between the power measurement and the geometry calibration observation NBRCS and the wind speed improves, and the modeling error is less significant, indicating the validity of the calibration.

4 Error Analysis

It is assumed that the error factors associated with the two steps of power calibration and geometry calibration as mentioned above are independent of each other. The total error of BF-1 is the Root of Sum of Squares (RSS) [4]:

$$RSS = \left\{ \sum_{i=1}^{n} [E(p_i)]^2 \right\}^{\frac{1}{2}} \tag{11}$$

4.1 Error Analysis of Power Calibration

By calculating the partial derivatives of the error terms in the expression of the calibration result, and taking into account the $1-\sigma$ error of each term, the formula for calculating each error can be expressed as follows:

$$E(p_i) = \left| \frac{\partial P_g}{\partial p_i} \right| \Delta p_i \tag{12}$$

the quantities of individual error are defined as follows: $p_1 = C$, $p_2 = C_N$, $p_3 = C_B$, $p_4 = P_b$ and $p_5 = P_r$, so that it can be obtain that:

$$E(C) = \frac{P_b + P_r}{C_B} \Delta C \tag{13}$$

$$E(C_N) = \frac{P_b + P_r}{C_B} \Delta C_N \tag{14}$$

$$E(C_B) = \frac{C - C_N}{C_B^2} (P_b + P_r) \Delta C_B \tag{15}$$

$$E(P_b) = \frac{C - C_N}{C_B} \Delta P_b \tag{16}$$

$$E(P_r) = \frac{C - C_N}{C_B} \Delta P_r \tag{17}$$

The typical values of C_N, C_B, P_b and P_r can be obtained from the spaceborne data. Due to the stable working state of the satellite, the values of these variables will show no significant change much, as a result of which the above-mentioned values can be brought into the partial derivative formula to obtain a more accurate result. In addition, C represents a matrix with the same specifications of DDM, and varies at different wind speeds. Therefore, in order to calculate these error terms, consideration is given to using the value near the specular point under common wind speeds.

ΔC, ΔC_N and ΔC_B are mainly related to the design parameters of receiver. ΔP_b is primarily determined by the accuracy of temperature measurement for the receiver and considered according to the measurement error of 1 °C. ΔP_r is mainly determined by the measurement error of the low-noise amplifier noise figure curve (Table 1).

Table 1. Power calibration error terms (dB)

Error term	$E(C)$	$E(C_N)$	$E(C_B)$
Error	0.18	0.08	0.009
Error term	$E(P_b)$	$E(P_r)$	**RSS**
Error	0.017	0.25	0.32

4.2 Error Analysis of Geometry Calibration

In order for the geometry calibration error to be calculated, the NBRCS value at the specular point is the analysis target, that is:

$$\sigma_{SP}^{0} = \frac{(4\pi)^3 P_{SP}^{g} L_{SP} \left(R_{SP}^{R}\right)^2 \left(R_{SP}^{T}\right)^2}{P^T \lambda^2 G_{SP}^{T} G_{SP}^{R} A_{SP}} \tag{18}$$

The error analysis formula is expressed as follows:

$$E(p_i) = \left|\frac{\partial \sigma_{SP}^{0}}{\partial p_i}\right| \Delta p_i \tag{19}$$

the quantities of individual error are defined as follows: P_{SP}^{g}, L_{SP}, $R_{SP}^{R} R_{SP}^{T}$, $P^T G_{SP}^{T}$, G_{SP}^{R}, and A_{SP}, for which it can be obtained that:

$$E\left(P_{SP}^{g}\right) = \frac{(4\pi)^3 L_{SP} \left(R_{SP}^{R}\right)^2 \left(R_{SP}^{T}\right)^2}{P^T \lambda^2 G_{SP}^{T} G_{SP}^{R} A_{SP}} \Delta P_{SP}^{g} \tag{20}$$

$$E(L_{SP}) = \frac{(4\pi)^3 P_{SP}^{g} \left(R_{SP}^{R}\right)^2 \left(R_{SP}^{T}\right)^2}{P^T \lambda^2 G_{SP}^{T} G_{SP}^{R} A_{SP}} \Delta L_{SP} \tag{21}$$

$$E\left(R_{SP}^{R} R_{SP}^{T}\right) = \frac{2(4\pi)^3 P_{SP}^{g} L_{SP} R_{SP}^{R} R_{SP}^{T}}{P^T \lambda^2 G_{SP}^{T} G_{SP}^{R} A_{SP}} \Delta\left(R_{SP}^{R} R_{SP}^{T}\right) \tag{22}$$

$$E\left(P^T G_{SP}^{T}\right) = \frac{(4\pi)^3 P_{SP}^{g} L_{SP} \left(R_{SP}^{R}\right)^2 \left(R_{SP}^{T}\right)^2}{\lambda^2 \left(P^T G_{SP}^{T}\right)^2 G_{SP}^{R} A_{SP}} \Delta\left(P^T G_{SP}^{T}\right) \tag{23}$$

$$E\left(G_{SP}^{R}\right) = \frac{(4\pi)^3 P_{SP}^{g} L_{SP} \left(R_{SP}^{R}\right)^2 \left(R_{SP}^{T}\right)^2}{P^T \lambda^2 G_{SP}^{T} \left(G_{SP}^{R}\right)^2 A_{SP}} \Delta G_{SP}^{R} \tag{24}$$

$$E(A_{SP}) = \frac{(4\pi)^3 P_{SP}^g L_{SP} \left(R_{SP}^R\right)^2 \left(R_{SP}^T\right)^2}{P^T \lambda^2 G_{SP}^T G_{SP}^R (A_{SP})^2} \Delta A_{SP} \tag{25}$$

Herein, $E\left(P_{SP}^g\right)$ is set to the previous error result of power calibration, that is, 0.32 dB. The atmospheric attenuation is taken into consideration for the design of the navigation system, and the L band is used to ensure the attenuation factor quite small [5]. The position accuracy of GNSS satellites is in no excess of 1 m [3], and that of the receivers is within 10 m. Considering that the calculation error of the specular point of the earth's ellipsoid model and sea level fluctuations is of such magnitude as kilometer, it is negligible compared to the values of R_{SP}^R and R_{SP}^T since the magnitude is merely 100 m after conversion to $\Delta R_{SP}^R + \Delta R_{SP}^T$. The navigation satellite transmitter power P^T and antenna gain G_{SP}^T are considered here, the uncertainty of which can be obtained from published data [3]. $E\left(G_{SP}^R\right)$ can be estimated from the satellite platform's attitude angle accuracy and antenna pattern lookup table. Due to the transmitter and the high accuracy of the position and velocity of the receiver and the specular point, $E(A_{SP})$ is mainly determined by the calculation error of the simulation parameters. Through a large number of simulations and comparisons, the value of $E(A_{SP})$ can be estimated (Table 2).

Table 2. Geometry calibration error terms (dB)

Error term	$E\left(P_{SP}^g\right)$	$E(L_{SP})$	$E\left(R_{SP}^R R_{SP}^R\right)$	$E\left(P_T G_{SP}^T\right)$
Error	0.32	0.04	0.005	0.40
Error term	$E\left(G_{SP}^R\right)$	$E(A_{SP})$	Margin	**RSS**
Error	0.47	0.06	0.25	0.74

By analyzing the above-mentioned error terms one by one, the final calibration error RSS is calculated to be 0.74 dB, which is better than the design target 1 dB and ensures the wind speed retrieval accuracy. In addition, the error introduced by $E\left(P_T G_{SP}^T\right)$ and $E\left(G_{SP}^R\right)$ is shown to be most significant, which requires their optimization to be given prime consideration in the design of next-generation system.

5 Conclusions

Following power calibration and geometry calibration, BF-1 can eliminate the effects on the DDM by temperature, channel differences, GNSS satellite and BF-1 satellite position and speed, GNSS satellite transmission power, and antenna pattern, so as to make the data characteristics associated only with the surface wind speed. As reveled by the results of error analysis, the error introduced by power calibration is 0.32 dB, while the error after geometry calibration is 0.74 dB. The calibration is validated as effective, since the error is insignificant, and the performance of the overall data product of BF-1 improves. It is thus believed to provide a better input and basis for subsequent use and wind speed retrieval.

References

1. Yu, K., Rizos, C., Burrage, D., Dempster, A.G., Zhang, K., Markgraf, M.: An overview of GNSS remote sensing. EURASIP J. Adv. Sig. Process. **2014**(1), 134 (2014)
2. Tongyu, L., Yakun, P.: Summary of China's first maritime launch technology test. Aerosp. China **2019**(6), 2 (2019)
3. Gleason, S., Ruf, C.S., Clarizia, M.P., O'Brien, A.J.: Calibration and unwrapping of the normalized scattering cross section for the cyclone global navigation satellite system. IEEE Trans. Geosci. Remote Sens. **54**(5), 2495–2509 (2016)
4. Janssen, M.A., Ruf, C.S., Keihm, S.J.: TOPEX/Poseidon Microwave Radiometer (TMR). II. Antenna pattern correction and brightness temperature algorithm. IEEE Trans. Geosci. Remote Sens. **33**(1), 138–146 (1995)
5. Ulaby, F.T., et al.: Microwave Radar and Radiometric Remote Sensing. University of Michigan Press, Ann Arbor (2014)

Application and Technology of Bufeng-1 GNSS-R Demonstration Satellites on Sea Surface Wind Speed Detection

Xinliang Niu[1(✉)], Feng Lu[2], Yuanhua Liu[3], Cheng Jing[1], and Bei Wan[1]

[1] Chinese Academy of Space Technology (Xi'an), Xi'an, China
xlniu1983@qq.com
[2] National Satellite Meteorological Centre, Beijing, China
lufeng@cma.gov.cn
[3] Xi'an University of Posts and Telecommunications, Xi'an, China
yuanhliu@163.com

Abstract. GNSS-R microwave remote sensing technology has increasingly become an important branch of satellite navigation applications. In June 2019, China successfully achieved the deployment of BuFeng-1 A/B satellites by Chinese first time sea platform launch. BuFeng-1 is mainly to conduct GNSS-R on-orbit tests of sea wind field detection and its applications. In the processing method, the power self-calibration method is introduced to eliminate the influence of the GNSS-R load receiving link changes by the space environment. Then, the GNSS-R observation geometric change by the GNSS-R geometry is corrected. Finally, DDMs with power self-calibration and geometric correction can directly show an agreement with the sea surface wind speeds. On-orbit results show that the detection accuracy is of 2.63 m/s up at 30 m/s of wind speeds by European Centre for Medium-Range Weather Forecasts (ECMWF) and 1.97 m/s at 20 m/s compared with ASCAT observations, which are accorded to the pre-design requirements of 2 m/s or 10%.

Keywords: GNSS-R · Microwave remote sensing · Sea surface wind speed · Power self-calibration · Geometric correction

1 Introduction

The concept of Global Navigation Satellite System-Reflectometry (GNSS-R) microwave remote sensing originated in the 1990s [1, 2]. After that, ESA and NASA successively launched GNSS-R satellites and constellations such as UK-DMC, TDS-1, and CYGNSS, which greatly accelerate the development of GNSS-R microwave remote sensing technology [3]. Now, it has become an important branch of GNSS applications. Compared with traditional microwave remote sensing technology, GNSS-R can have an equivalent performance with lower mass, lower power consumption, and low costly, which is very suitable for microsatellites networking. With the advantage of spatial-temporal resolution, GNSS-R can be a good composition and supplement of existing satellite remote sensing systems.

J. Sun et al. (Eds.): CSNC 2020, LNEE 650, pp. 206–213, 2020.
https://doi.org/10.1007/978-981-15-3707-3_20

GNSS-R sea surface wind field detection is realized by receiving and processing the direct signals of the existing satellite navigation systems and the sea surface reflected signals. The basic principles are as follows: sea surface winds causes the change of sea surface roughness, and then varies the geometric power distribution of GNSS signals on the ocean surface. Therefore, delay Doppler map (DDM) correspondingly changes by the winds. In another word, the power variation in the certain delay-Doppler window of DDM can be the sensitive observable to sense the sea surface wind speed.

In the next section, the overview of the Bufeng-1 demonstration satellite is introduced. The Sect. 3 gives the design of GNSS-R sea surface wind field. The Sect. 4 tests Bufeng-1 preliminary on-orbit technical indicators, which gives a good start for future GNSS-R microwave remote sensing technology and applications of China.

2 Overview of the Demonstration Satellites

2.1 Overview of the Satellites

At 12:06 on June 5, 2019, the Bufeng-1 on-orbit demonstration satellites were successfully deployed in orbit by Chinese first time sea platform launch [4]. It is the first generation of Chinese GNSS-R microwave remote sensing satellite, whose main client is China Meteorological Administration. The DFH co. ltd. and the China Academy of Space Technology Xi'an Branch are the united research and development team (Fig. 1).

Fig. 1. Bufeng-1 sea platform launch

2.2 Twin Satellites Networking

The Bufeng-1 demonstration twin satellites operate at altitude of 579 km with inclination of 45°. The orbital phase between the two satellites is 45°. An uniform global coverage between 45° S and 45° N latitudes can be finished in 4 days (Fig. 2).

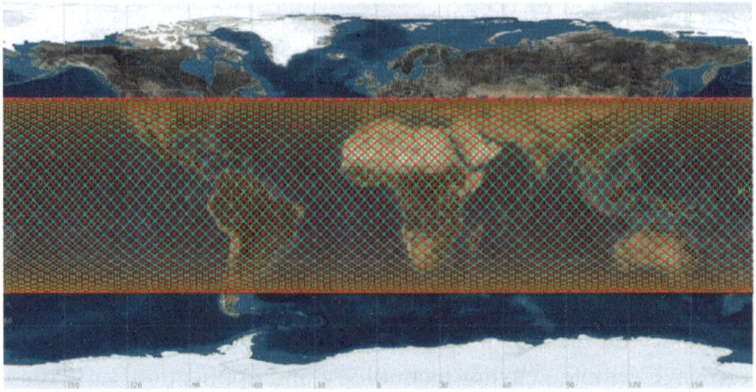

Fig. 2. Trajectories of Bufeng-1 twin satellites

In addition, Bufeng-1 satellites receive two types of GNSS signals, GPS L1C/A and BDS B1I. Thus, by taking the advantage of Beidou Satellite Navigation System (BDS) the spatial coverage of the Asia-Pacific region is significantly improved.

3 Design of GNSS-R Winds Detection

3.1 Power Self-Calibration

The overall power of the signal received by the GNSS-R payload of the Bufeng-1 demonstration satellites is as follows,

$$C = G\left(P_g + P_a + P_r\right) \tag{1}$$

where C is the overall power, G is receiving antenna gain, P_g is the power of reflected signal, P_a and P_r are the thermal noise powers of the receiving antenna and the instrument, respectively.

The receiving gain and thermal noise power mentioned above change with the temperature of the space environment. Therefore, the power self-calibration design of "3s calibration + 57s antenna in every minute" is introduced (Fig. 3).

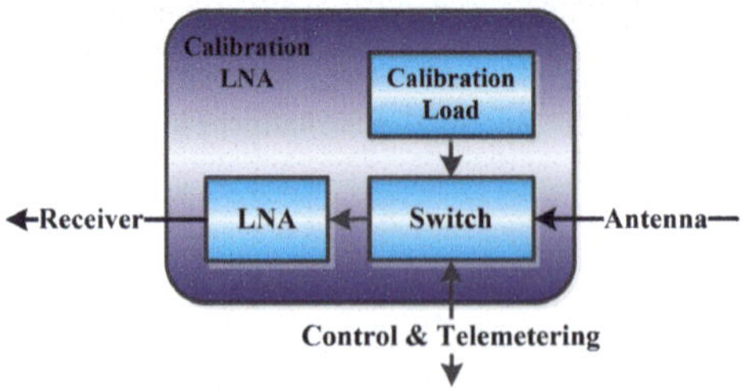

Fig. 3. Design of power self-calibration

When the microwave switch switches to the calibration load, the overall signal power [5] becomes as follows:

$$C_B = G(P_B + P_r) \tag{2}$$

where C_B is the overall power of noise power by the load, P_B and P_r are still the thermal noise powers of the receiving antenna and the instrument, respectively.

Combining the Eqs. (1) and (2), the P_g can be rewritten as follows,

$$P_g = \frac{C - G(P_a + P_r)}{G} = \frac{(C - C_N)(P_B + P_r)}{C_B} \tag{3}$$

$$P_B = kT_B B_w \tag{4}$$

$$P_r = kT_r B_w \tag{5}$$

where C_N is a noise statistic numeric result of DDM, P_B and P_r can be inversed by the LUT by prelaunch calibration and on-orbit telemetered temperatures, k is the Boltzmann constant, and B_w is the equivalent bandwidth of the instrument.

Power self-calibration design of Bufeng-1 can largely mitigate the effects of receiver link in different temperatures in the space environment (Fig. 4).

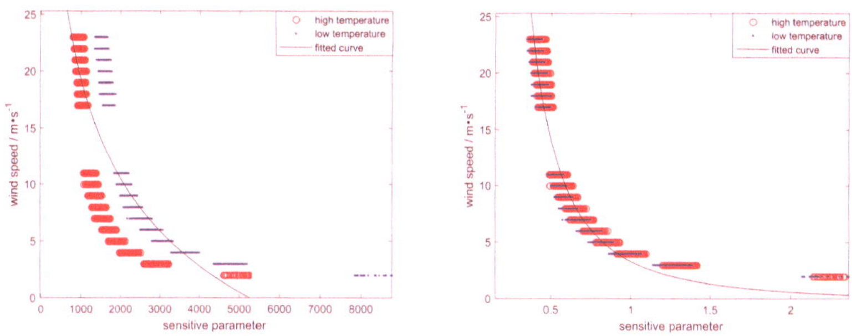

Fig. 4. Results before and after power self-calibration

3.2 Geometric Correction and Wind Speed Inversion

The relationship between Bufeng-1 DDMs after power self-calibration and sea surface scattering coefficient is as the equation below,

$$\sigma_0(\vec{\rho}) = \frac{P_g(4\pi)^3 R_T^2(\vec{\rho}) R_R^2(\vec{\rho})}{P_T G_T(\vec{\rho}) G_R(\vec{\rho}) \lambda^2} \times \frac{1}{\iint_A \Lambda^2[\delta_\tau(\vec{\rho})] S^2[\delta_f(\vec{\rho})]\, dxdy} \tag{6}$$

where $\sigma_0(\vec{\rho})$ is the sea surface scattering coefficient, $R_T(\vec{\rho})$ and $R_r(\vec{\rho})$ are ranges of transmitter to specular point and receiver to specular point, respectively, P_T is the

GNSS transmit power, $G_T(\vec{\rho})$ is the transmit antenna gain to the specular point, $G_R(\vec{\rho})$ is the receiver antenna gain to the specular point, λ is the wavelength of GNSS signal, $\Lambda^2[\delta_\tau(\vec{\rho})]$ and $S^2[\delta_f(\vec{\rho})]$ are delay correlation function and Doppler spread function, respectively, and $\iint\limits_A (\cdot)dxdy$ describes the geometric distribution (Fig. 5).

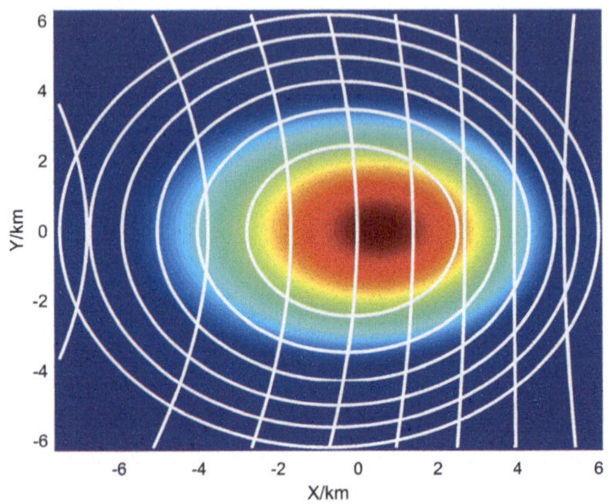

Fig. 5. Geometric distribution of delay and Doppler

Basically, the power of the DDM after power self-calibration will vary with the change of GNSS-R observation geometry. Here, geometric correction is performed by formula (6) to obtain how the sea surface reflection coefficient change, thereby realizing the sea surface wind speed retrieval (Fig. 6).

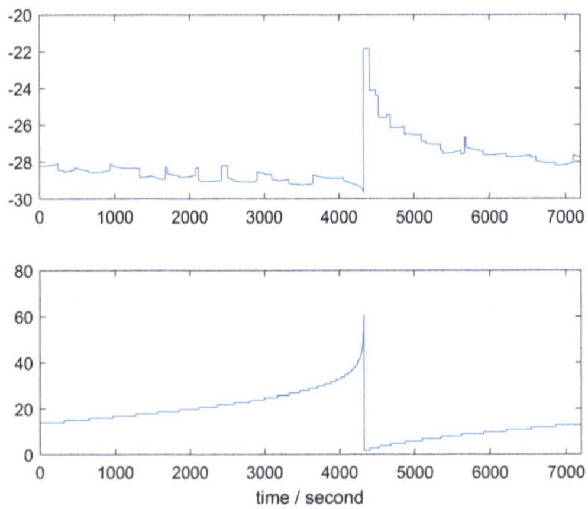

Fig. 6. Performance of power self-calibration

One should note that the Fresnel reflection coefficient [6] is also needed to be taken into consideration that it will vary as the elevation angle increases.

4 On-Orbit Testing

4.1 DDM Collection

4.1.1 The Chinese First Spaceborne Sea Surface DDM

Bufeng-1 demonstration satellite GNSS-R payload successfully obtained Chinese first satellite GPS-R sea surface DDM, and the resolution surpassed the current missions (Fig. 7).

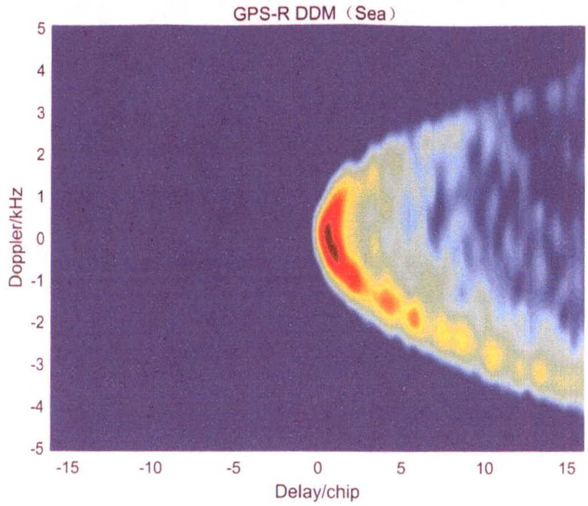

Fig. 7. GPS-R DDM by Bufeng-1 mission

4.1.2 The World's First Spaceborne Sea Surface BDS-R DDM

At the same time, the world's first sea surface BDS-R DDM was also mapped by Fufeng-1 mission, which supplements the gap of spaceborne GNSS technology (Fig. 8).

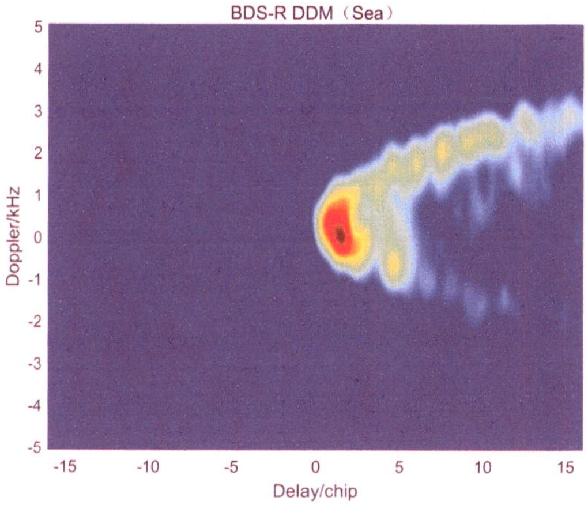

Fig. 8. BDS-R DDM by Bufeng-1 mission

4.1.3 The World's First Spaceborne Inland BDS-R DDM

Moreover, Bufeng-1 successfully measured the world's first inland BDS-R DDM, which shows a great difference compared with sea surface DDMs (Fig. 9).

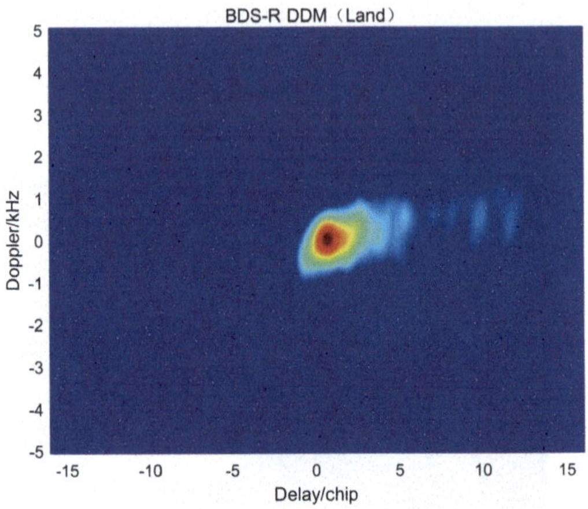

Fig. 9. Inland BDS-R DDM by Bufeng-1 mission

4.2 Performances on Wind Observation

4.2.1 ECWMF Trained Results

After trained by ECMWF dataset, the Bufeng-1 GMF was tested a performance of 2.63 m/s RMSE at wind speed up to 30 m/s (Fig. 10).

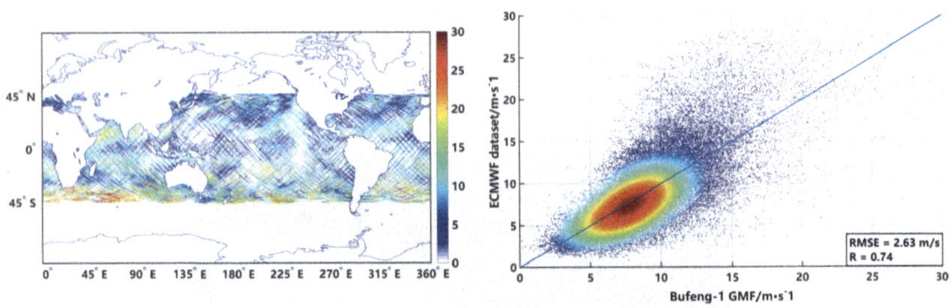

Fig. 10. Performance of Bufeng-1 GMF by ECMWF training

4.2.2 ASCAT Trained Results

The on-orbit data was also trained by the ASCAT observations, and the GMF has the RMSE of 1.97 m/s at wind speed of 20 m/s (Fig. 11).

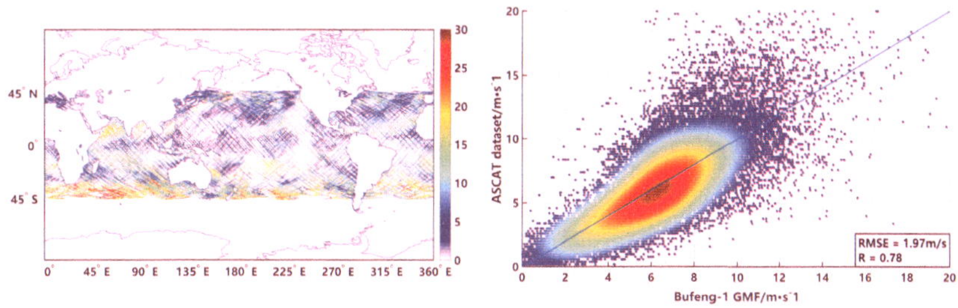

Fig. 11. Performance of Bufeng-1 GMF by ASCAT training

5 Conclusion

As the first-generation GNSS-R microwave remote sensing technology demonstration satellites in China, Bufeng-1 mission has achieved global winds observation by building the relationship between the DDMs after power self-calibration and geometric correction and the sea surface wind speed. The preliminary results of on-orbit testing show that it meets the predetermined design requirements of 2 m/s or 10%, which gives a good start for the future GNSS-R microwave remote sensing technology and application in China.

References

1. Martin-Neira, M.: A passive reflectometry and interferometry system (Paris): application to ocean altimetry. ESA J. **17**(4), 331–355 (1993)
2. Auber, J.C.: Characterization of multipath on land and sea at GPS frequencies. In: Proceedings of ION GPS 1994, pp. 1155–1171 (1994)
3. Bei, W.: Research on key technologies of GNSS-R sea surface wind field detection. M.A. thesis. Xi'an: Chinese Academy of Space Technology (2017)
4. Tongyu, L., Yakun, P.: Summary of China's first maritime launch Technology test. Aerosp. China **2019**(6), 2 (2019)
5. Gleason, S., Ruf, C.S., Clarizia, M.P., O'Brien, A.J.: Calibration and unwrapping of the normalized scattering cross section for the cyclone global navigation satellite system. IEEE Trans. Geosci. Remote Sens. **54**(5), 2495–2509 (2016)
6. Jing, C., Yang, X., Ma, W., Yu, Y., Dong, D., Li, Z., Xu, C.: Retrieval of sea surface winds under hurricane conditions from GNSS-R observations. Acta Oceanol. Sin. **35**(9), 91–97 (2016)

Study on the Correlation Between GNSS Vertical Time Series and the Space-Time Distribution of Groundwater in California

Xiaoguang Pang[1(✉)], Yong Luo[2], Shaodong Jing[3], Shuangcheng Zhang[1,4], and Kunchao Lei[2]

[1] College of Geology Engineering, Chang'an University, Xi'an 710054, China
895530083@qq.com
[2] Beijing Institute of Hydrogeology and Engineering Geology, Beijing 100195, China
[3] Sinopec Petroleum Engineering Corporation, Dongying 257000, China
[4] Institute of Desert Meteorology in China Meteorological Administration, Urumqi, Xinjiang, China

Abstract. The agriculture in California mainly depends on irrigation, but it is one of the regions that suffer serious groundwater loss in the world. The authors obtain the four-year data of 48 survey stations uniformly distributed by PBO network in California and the vertical time series of each station based on GAMIT/GLOBK software. Based on the time series, the contour map of surface elevation variation in California from 2014 to 2017 can be obtained, and it is found that there is an obvious subsidence funnel in southern California. In order to analyze the cause of subsidence funnel, the authors introduce the burial depth data, which shows that the groundwater depth in this area decreases 25.65 cm annually in the past four years. The groundwater loss caused by agricultural irrigation is the main reason for the formation of settlement funnel in this area, while rainfall can effectively alleviate the exploitation of groundwater for agriculture. Through analyzing the TRMM rainfall data of this area, the authors find that the cumulative rainfall in northern California from 2014 to 2017 is significantly higher than that in the south. The maximum reaches about 6000 mm, but the minimum rainfall occurs in the south, less than 1000 mm. According to the GPS vertical time series, TRMM rainfall data and groundwater depth analysis, there is a high correlation between the P566 survey station with severe surface subsidence and the groundwater depth data of nearby wells. In addition, the rainfall data shows that the rainfall in southern California is significantly lower than that in the north. The experimental results indicate that the time series of the surface elevation change acquired by the GPS can effectively reflect the groundwater timing change; the ground surface settlement monitored by the GPS is highly correlated with the measured groundwater depth data, which proves that the surface settlement of the southern California is mainly caused by the exploitation of the underground water; the loss of groundwater in the southern region can not be effectively alleviated due to inadequate rainfall, which further demonstrates that the surface settlement in California is mainly caused by the excessive exploitation of groundwater.

Keywords: GPS · TRMM · Groundwater depth

J. Sun et al. (Eds.): CSNC 2020, LNEE 650, pp. 214–221, 2020.
https://doi.org/10.1007/978-981-15-3707-3_21

1 Introduction

With the rapid development of Global Navigation Satellite System (GNSS) and the great improvements on its positioning accuracy, GNSS technology has become one of the major research means in geoscience field based on the massive observational data acquired by global and regional GNSS stations. The coordinate time series not only reflects the changing trend of the station, but also reveals its non-linear changes. Specifically, the trend changes mainly reflect the tectonic movements of the area, and the nonlinear motion of the station, especially in elevation direction, is mostly caused by environmental loads such as atmospheric loads, non-tidal ocean loads, and hydrological loads. Researchers at home and abroad have conducted extensive research on the coordinate displacements caused by surface loads, and have made a series of achievements. Their researches confirmed that atmospheric loads, non-tidal ocean loads, and hydrological loads are strongly correlated with the time series of elevation directions of the stations, and the contributions of each station to different regions vary (Farrell 1972; He et al. 2014; Jiang et al. 2014).

Land water consists of surface water, soil water, surface runoff, canopy water, and groundwater, etc. The movement of land water causes changes in surface load, which in turn leads to the surface deformation. Traditional methods of monitoring land water changes include oven drying and weighing, sensors, etc., through which only changes of shallow groundwater within 2 m under the surface can be obtained and researchers cannot effectively monitor changes in deep confined water. A great number of domestic and foreign studies have been carried out on the use of GNSS elevation time series in analyzing the changes of groundwater storage. The results show that the seasonal and long-term variation trends of land water and groundwater storage can be inversion through GNSS elevation time series (He et al. 2014; Jiang et al. 2014; Jiang et al. 2013).

This paper collects the data of 48 PBO stations in California, calculates and analyzes the elevation time series of each station. The results show that the vertical time series of the stations fluctuates in different degrees, including annual cycle, semi-annual cycle and long-term change trend (Wang et al. 2018; Zhang and Cheng 2014; Zhang et al. 2017). Obtain the change trend of the station, and verify the validity of the GNSS elevation time series inversion of groundwater time series changes with the measured well data. Furthermore, the outliners in time series are also analyzed based on TRMM rainfall data.

2 Data and Methods

2.1 GPS Data Processing

This article uses the GNSS observation data provided by the United States PBO Observation Network (https://www.unavco.org). In order to obtain the overall elevation time series of California, the authors evenly selected the data from 48 PBO observational stations in California, and the time span is from January 2014 to December 2017. Its spatial distribution is shown in Fig. 1. Five IGS (International GNSS Service) stations including BREW, KOKV, QUIN, PIE1 and GOLD are combined for calculation in order to present the precise coordinates and time series of each station.

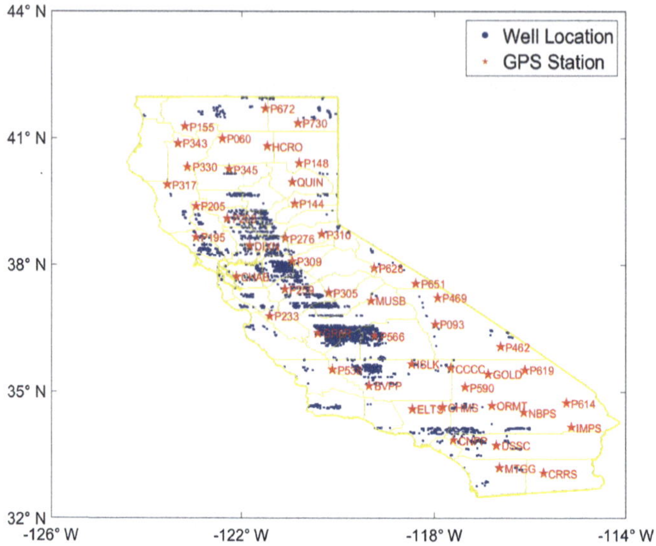

Fig. 1. GNSS station and water well spatial distribution

This paper combines 5 IGS stations as reference stations to collect the single-day results whose adjustment can be calculated based on GAMIT/GLOBK. Firstly, GAMIT is adopted to obtain the single-day results of the data, the position of the station and the parameters of satellite orbit, etc. The data processing parameters are set at the satellite height cut-off angle of 15°; the influence of ionospheric refraction is eliminated by LC observations; the phase center deviation of the satellite and receiver antennas is corrected by the absolute antenna phase center; the sampling interval is 30 s; the ITRF2008 reference frame is used; the double-difference model is adopted; solid tides and polar tides are deducted and the ocean tides is deducted by FES2004; the tropospheric dry component provided by Global Mapping Function (GMF) and global pressure and temperature (GPT) models is taken as a priori value in tropospheric model; the satellites clock difference model is corrected based on the clock difference parameter in the broadcast ephemeris. After the baseline calculation, the single-day results without reference to the station are obtained, and then the network adjustment of GLOBK is used to obtain the time series and speed of each station (Wang et al. 2008; Zhang et al. 2012).

The trend term is mostly non-linear. Based on the four-year trend of each station, a contour map of the cumulative change in California from January 2014 to December 2017 can be drawn. Figure 2 is the contour map with significant regional characteristics. It can be seen that the ground near P566 in the central valley has severely subsided, and the maximum variation in four years reaches −52.29 mm.

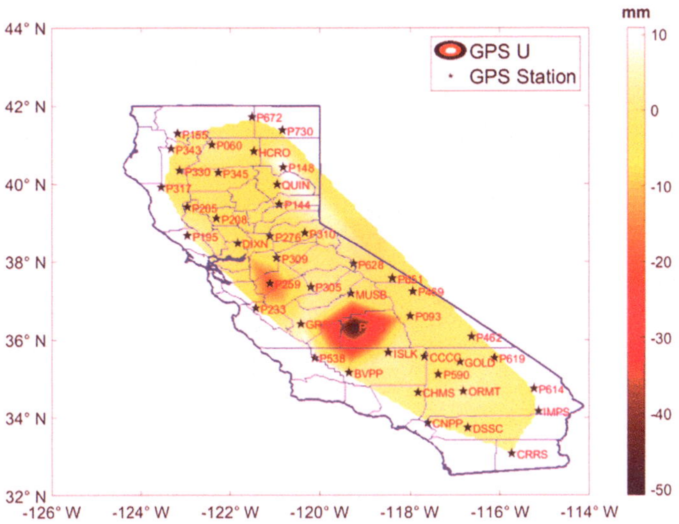

Fig. 2. Spatial distribution of surface elevation changes in California over four years based on vertical time series of 48 GPS stations

2.2 Measured Well Data and Rainfall Data

In order to verify the reliability of GNSS vertical time series inversion on groundwater time series, GNSS vertical time series and measured well data are compared and analyzed. This paper collects the statewide well location and measured data of well water level published by the California government official website (https://data.ca.gov/), as shown in Fig. 1. The earliest time span of the well data is from 1900 to now, and the observation frequency is once a week. Three kinds of results are obtained, which include the Water Surface Elevation (WSE) indicated by referring to the North American vertical datum in 1988, the depth below the reference point or the distance from the reference point to the water surface (RPE_WSE), the depth below ground or distance from ground to water (GSE_WSE). Since the original unit of measurement is feet, the authors have converted it into standard unit *cm*.

Directing at the GNSS vertical time series and the abnormal value of groundwater level, this paper uses the rainfall data of Tropical Rainfall Measurement Mission (TRMM, https://pmm.nasa.gov/) published by NASA for analysis and verification. TRMM is a satellite specially used to measure rainfall in tropical and subtropical regions. This paper selects the data of TRMM-3B43, which has a spatial resolution of $0.25^{\circ} \times 0.25^{\circ}$, a time resolution of a month and a coverage range of 50°S–50°N. The data format is hdf and the unit is mm/hr which can be converted to monthly rainfall by the formula mm/month = mm/hr $\times$ 24 (hr) $\times$ N (days). Figure 3 shows the spatial distribution of cumulative rainfall in California from 2014 to 2017.

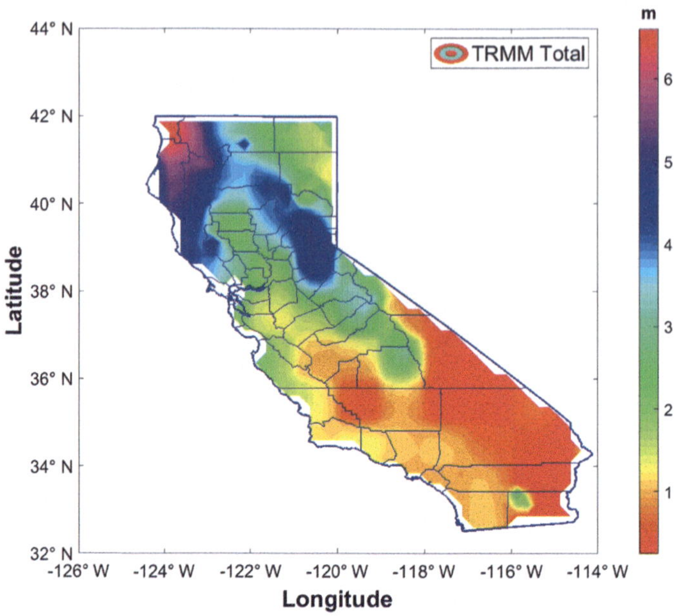

Fig. 3. Spatial distribution of cumulative rainfall in California over four years

3 Results

Figure 2 shows that the settlement P566 stations are obvious, and it's compared with the nearby No. 374527N1209768W001 and No. 362985N1192201W001 wells for time series comparison and analysis. As shown in Fig. 2, the ground settlement of the P566 station is about 52.29 mm over four years, and the average annual settlement rate is about 13.1 mm/a. The water level of the well dropped by about 76.5 cm in four years, and the average annual decline rate is about 25.65 cm/a. A long-term consistency trend is reflected between the change in ground elevation based on the vertical time series of GPS data and the change in shallow groundwater level measured by surface water wells, which means the time consistency is relatively good and indicates that the groundwater losses account for the major contribution in land subsidence. This paper relies on the TRMM rainfall data to analyze the relationship between the spatial - temporal changes of precipitation and the groundwater level and the resulting ground subsidence over the four years in California. Figure 3 shows the spatial distribution of cumulative precipitation in California from 2014 to 2017. The precipitation in the south is significantly less than that in the north. There appear abnormal rainfall near the P566 stations, which is in good agreement with the contour map of GPS ground subsidence. The two stations are analyzed.

With GPS vertical time series, rainfall data, and groundwater level changes combined for analysis, Fig. 4(a) shows that the GPS vertical time series of the P566 station and the water level of the well 374527N1209768W001 continued to decline, and the rainfall data show obvious seasonal changes. From 2014 to 2016, the cyclical changes in rainfall did not significantly improve the groundwater level and ground subsidence. The direct cause was excessive extraction of underground confined water, and the rainfall did not alleviate the excessive exploitation of groundwater, which caused continuous ground subsidence. From 2017 to 2018, after a month of heavy precipitation, the groundwater level and surface subsidence slowed down, indicating that precipitation can alleviate groundwater extraction and, consequently, ground subsidence slowed down.

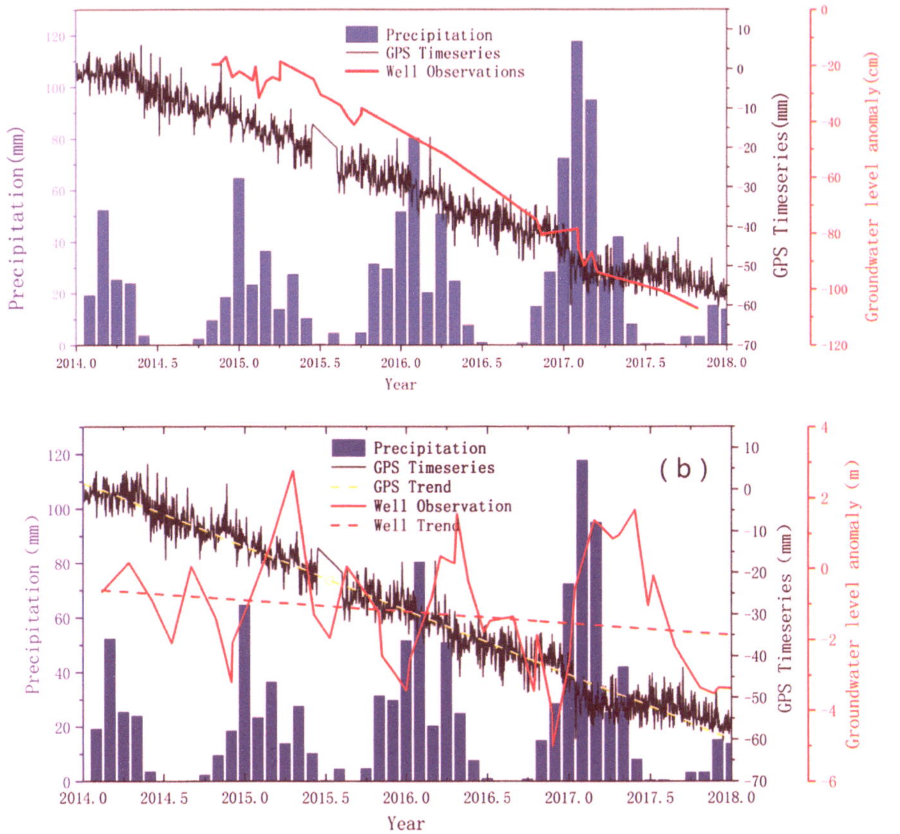

Fig. 4. The blue histogram in the figure is the monthly rainfall of the GPS station, the red is the time series of groundwater level changes, and the black is the GPS time series (a) is the vertical time series of the P566 station. (b) Comparison analysis of vertical time series, rainfall data of P259 station and wells 362985N1192201W001

In order to verify the reliability between the above GPS station and the water level of the well, a comparison was made on the well No. 361980N1193446W001 near the P566 station, and it was found that the water level of the well was more volatile, but the peak rainfall and the peak water level were about 2 months Delay, mild type is better; and GPS and time series show a downward trend. This shows that groundwater extraction plays a major role in ground subsidence. The above-mentioned fluctuations in the tree sperm water level are presumed to be due to the shallow water level of the well and the susceptibility to the surface environment.

4 Conclusion

This article analyzes the changes in surface elevation, groundwater level, and temporal - spatial changes of rainfall data in California from 2014 to 2017. Starting from the relationship between groundwater changes and ground subsidence, the author analyzes the impact on ground of groundwater loss. Through the analysis of the experimental results, we found that:

1. Agricultural irrigation in the southern central valley plains of California caused excessive extraction of groundwater. During the period of 2014–2017, the P566 station dropped by about 52.29 mm, and the water level of the well 374527N1209768W001 dropped by about 76.5 cm in four years. Good consistency between the two, indicating that excessive groundwater extraction is the main factors of land subsidence.
2. The vertical time series, precipitation, and groundwater levels of the two selected stations agree well, and the consistency is high, which indicates that the use of GPS stations can effectively retrieve groundwater loss.

This paper studies the changes in groundwater reserves in California by combining GPS, TRMM, and Well data. In the analysis of groundwater changes in California, GPS observations are used, and there is a problem of limited surface settlement area and points reflected by a single station; only the impact of hydrological loading is analyzed, and geophysical structures are not considered; station analysis at a single point This is not conducive to medium- and long-scale research. Consideration should be given to joint GRACE-Follow on data to study and analyze regional land water storage changes and groundwater storage changes, extending from time series to spatial analysis.

Acknowledgments. We gratefully acknowledge the provision of data, equipment, and engineering services by the Plate Boundary Observatory operated by UNAVCO for Earth Scope. Thanks to the GAMIT/GLOBK software provided by MIT in the United States; thanks to the measured well data provided by DWR; and the TRMM rainfall data provided by NASA. The pictures in this article were drawn by MATLAB and ORIGIN software. This work was supported by Institute of Desert Meteorology in China Meteorological Administration (Ground-based GNSS monitoring snow depth in Altay, Sqj2017002). And thanks to anonymous reviewers for providing valuable comments and suggestions in this paper!

References

Farrell, W.E.: Deformation of the earth by surface loads. Rev. Geophys. Space Phys. **10**(3), 761–797 (1972)

He, X., Hua, X., Zhou, S.: Analysis of the influence mechanism of abnormal periodic signals in GPS time series. Geomat. Surv. Mapp. **39**(02), 22–26 (2014)

Jiang, W., Xia, C., Li, Z., Guo, Q., Zhang, S.: Impact of environmental load on the time series of regional GPS reference stations. J. Surv. Mapp. **43**(12), 1217–1223 (2014)

Jiang, W., Li, Z., Liu, H., Zhao, Q.: Analysis on the causes of non-linear variation of coordinate time series of IGS base stations in China. Chin. J. Geophys. **56**(07), 2228–2237 (2013)

Wang, M., Shen, Z., Gan, W., Liao, H., Li, T., Ren, J., Qiao, X., Wang, Q., Yang, Y., Kato, Z., Li, P.: GPS continuous monitoring of the Xianshuihe fault deformation field dynamic evolution. Chin. Sci. (Ser. D: Earth Sci.) **38**(05), 575–581 (2008)

Wang, X., Zhang, Q., Zhang, S.: Periodic oscillation analysis of GPS water vapor time series based on EMD and WD joint algorithm. Wuhan Univ. J. (Inf. Sci. Ed.) **43**(04), 620–628 (2018)

Zhang, H., Cheng, P.: Recognition and extraction of noise in GPS elevation time series based on EEMD. Geodesy Geodyn. **34**(02), 79–83 (2014)

Zhang, L., Jin, S., Zhang, T.: Using GPS and satellite gravity observation to estimate the seasonal variations of surface deformation. Geodesy Geodyn. **32**(02), 32–38 (2012)

Zhang, S., He, Y., Li, Z., Hou, X., Qu, W., Nan, Y.: EMD for GPS time series noise reduction analysis. J. Geodetics Geodyn. **37**(12), 1248–1252 (2017)

Suspension Cable Bridge Deflection Determination Using Kinematic PPP with High-Rate GPS Satellite Clock Corrections

Xu Tang[1(✉)], Fei Guo[2], Craig Matthew Hancock[3], and Huib de Ligt[3]

[1] School of Remote Sensing and Geomatics Engineering, Nanjing University of Information Science and Technology, Nanjing 210044, China
Xu.Tang@hotmail.com
[2] Wuhan University, 129 Luoyu Road, Wuhan 430079, China
[3] University of Nottingham Ningbo China, 199 Taikang East Road, Ningbo 315100, China

Abstract. Many previous studies have utilized the advantages of GPS for structural health monitoring however high data rate (higher than 1 Hz) is required for structures' vibration detection using GPS. This paper investigates the advantages of utilizing high-rate satellite clock in a kinematic PPP processing strategy specifically applied to the detection of bridge displacement. Global distributed IGS stations high rate observations were used for GPS satellite clock estimations. The Uncalibrated Phase Delay (UPD) and the Zenith Total Delay (ZTD) parameters were firstly estimated using PPP in static mode. The GPS satellite clock product with 1 Hz sampling rate were successfully estimated by applying the UPD and tropospheric delays as compensations to the carrier phase. The deformation of the suspension cables (20 Hz datasets) and towers (10 Hz datasets) were estimated using both the kinematic PPP, using the 1 s, 30 s, 300 s clocks, and the DD data processing method. The 1 s GPS satellite clock gave the most precise and stable solution when compared with the 30 s and 300 s clocks obtained from IGS. Correlation coefficient of PPP solution and DD solution reflects bridge displacement waveform using 1 s PPP is more consistency and accuracy. The 1 s PPP is an alternative method for suspension cable bridge deflection determination, particular in the scenario that DD is impossible due to failure of reference station.

Keywords: Satellite clock correction · Bridge displacement monitoring · PPP

1 Introduction

GNSS has been well used for the measurement of the deformation of bridges during the past few decades [1, 2]. The use of low cost accelerometers can improve upon a GNSS-only sampling rate for the extraction of high vibration structure frequency bands [3]. Many previous studies were processed using the GPS only double differenced (DD) method. Sub-centimeters accuracy for structure monitoring can be achieved due

J. Sun et al. (Eds.): CSNC 2020, LNEE 650, pp. 222–231, 2020.
https://doi.org/10.1007/978-981-15-3707-3_22

to majority errors are eliminated or reduced. DD GPS applications has the advantage of high accuracy. However, DD applications are limited by scenarios that compel the need for long distance between reference station and surveying point, power outages or natural disasters. Precise Point Positioning (PPP) is a viable option in this specific situation. PPP precision and performance significantly rely on the quality and configuration of products, which is offered by the International GNSS service (IGS). Kinematic PPP has been successfully applied to the suspension bridge displacement detection in the vertical component, but failed in extracting the suspension cable displacement in the longitudinal and lateral components [4, 5] with relatively small movements. GPS PPP is a prominent tool for retrieving the seismic displacement waveforms which movements could be up to a few decimeters, or more, during an earthquake event of a short period [6]. Therefore, PPP is suitable for detecting large movements that occur over a short period, but is less suitable for detecting small movements over longer periods [7]. PPP displacement monitoring accuracy is degraded by the unmodeled corrections, measurement noise or using IGS products improperly.

Bridge oscillating displacement is a long term and continuous movement, which is quite different from seismic movement that normally happens during a short period. Apart from that, each bridge has its own characteristic movements at different points. Movements on the bridge deck are usually significant bigger than the ones of the supporting towers. Extracting bridge movements by kinematic PPP has more challenges than the deriving seismic movement. 1 Hz GPS satellite clock corrections will be computed, then applied for the kinematic PPP for the bridge deflection detection. This will solve the failure of small movements detection in the component of longitudinal and lateral at bridge suspension cable survey points, as well as the ones of supporting towers [4].

The performance of high-rate kinematic PPP for bridge monitoring is assessed against the true bridge deflection estimated by the double differenced method. The manuscript is organized as follows: first, data processing strategy of 1 s interval satellite clock corrections estimation will be described then the quality of satellites clock correction is assessed. Secondly, the bridge experiment is described and analysed for assessing the kinematic PPP performance in high-rate real life oscillating applications using the advantage of high-rate satellite clock corrections. Finally, the manuscript is concluded by giving evidence of the advantage and disadvantage of high-rate kinematic PPP, as well as its challenges in structural monitoring applications.

2 High-Rate GPS Satellite Clock Correction Estimation

This section describes how high-rate satellite clock corrections are estimated using the 1-Hz observations from global distributed stations. The receiver clock offset at each station is computed using SPP (Standard Point Positioning). Then, ZTDs, UPDs and positions of the IGS stations, which are stable or change slowly over time, are computed by the static PPP with either 30 s or 300 s interval between the neighboring epochs (depending on the computer configuration). In this step, the receivers' clocks are taken as unknown parameters for estimation in the PPP procedure. Linear interpolation is applied for generating the 1 s-interval ZTDs, UPDs and positions parameters. The 1 s-interval ZTDs, UPDs, positions parameters (estimated by PPP) and receiver clock offsets (estimated by SPP) are then used as ionosphere-free observation corrections for the 1 Hz satellite clock correction estimation. Figure 1 gives the technical process for generating GPS satellite clock corrections.

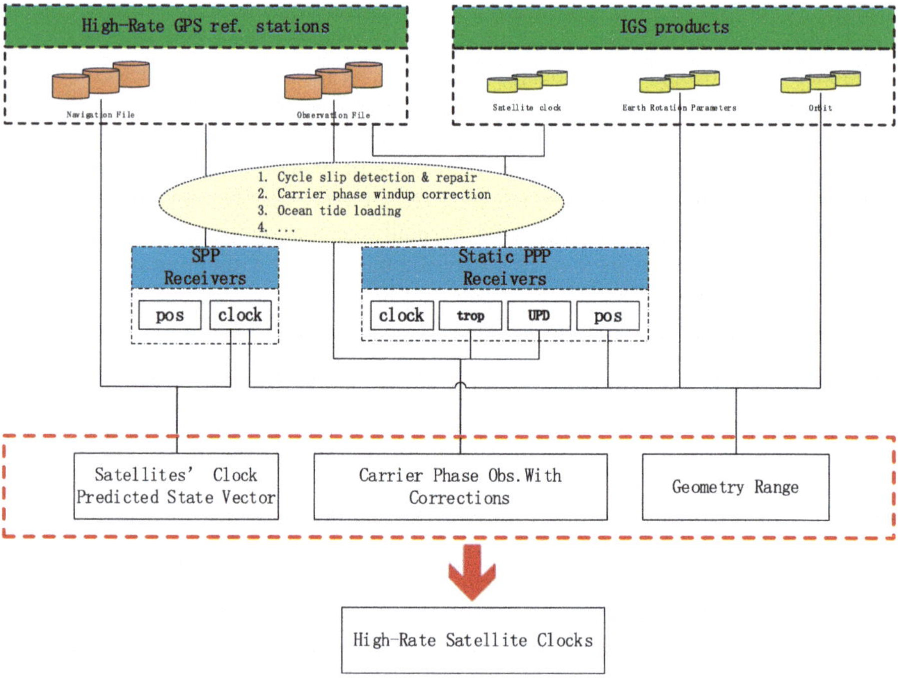

Fig. 1. Flowchart of generation 1 Hz GPS satellite clock corrections, including 3 major steps: 1. SPP; 2. static PPP; 3. clock correction estimation.

UPDs, ZTDs and position coordinates of global distributed stations can be estimated and modeled by the static PPP. Then, the GPS satellite clock correction observation equation can be written as,

$$\varphi_{IF,i}^{j} = \Delta T^{j} + \Delta T_{i} + \rho_{i}^{j} + M_{i}^{j} \cdot \delta_{ZTD,i} + UPD_{i}^{j} \tag{1}$$

Where, φ_{IF} is the Ionosphere-free (IF) linear combination carrier phase measurement, j and i denote the jth satellite and i^{th} station, respectively. ΔT with subscript and superscript denote GPS satellite clock correction and receiver clock offset, respectively. ρ denotes geometry range between GPS satellite and receiver. M denotes mapping function coefficient. δ_{ZTD} is the zenith total delay, can be precisely estimated by the static

PPP. *UPD* denotes the satellite Uncalibrated Phase Delay. Static PPP offers the reference station precise position coordinate, UPDs and ZTD with low rate. The high rate of these corrections applied in high-rate GPS satellite clock correction estimation is precisely provided by the linear interpolation, due to their features of stable and gradual movement.

$$\begin{bmatrix} \boldsymbol{u}_1 \\ \vdots \\ \boldsymbol{u}_i \\ \vdots \\ \boldsymbol{u}_n \end{bmatrix}_{(m_1+\cdots+m_i+\cdots+m_n)\times 32} \cdot \begin{bmatrix} \Delta T^1 \\ \vdots \\ \Delta T^j \\ \vdots \\ \Delta T^{32} \end{bmatrix}_{32\times 1} = \begin{bmatrix} \varphi_{1,IF} \\ \vdots \\ \varphi_{i,IF} \\ \vdots \\ \varphi_{n,IF} \end{bmatrix}_{(m_1+\cdots+m_i+\cdots+m_n)\times 1} \quad (2)$$

With the assumption of 32 GPS satellites tracked, $\boldsymbol{u}_i$ in the left hand of Eq. 2 is the satellite clock correction equation coefficient matrix with dimension $m_i \times 32$, contains

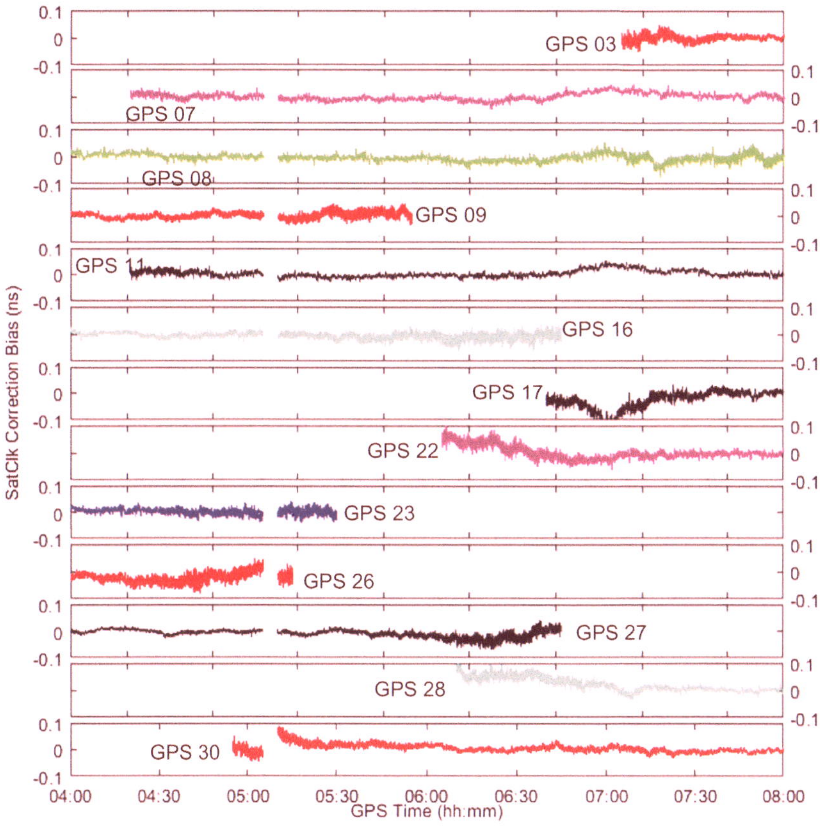

Fig. 2. 13 GPS satellite clock correction biases estimated by using only one GPS tracking station

only one "1" and all others are zeros at each row. Right hand of Eq. 2 $\varphi_{i,IF}$ denotes the IF carrier phase measurements vector of the tracked satellite at the ith station. Satellite clock corrections can be estimated by Eq. 2 with the observations corrected by ZTDs and UPDs.

The satellites clock correction changes are quite small, and cannot be illustrated properly in the plots, due to their big-scale changes during the short period. The difference between the estimated satellite clock correction and the linear interpolation time series from the satellite clock production, defined as satellite clock correction bias, is introduced for clearly illustrating the details of satellite clock correction. Figure 2 shows the GPS satellites' clock correction biases, which are computed using only one of Continuously Operating Reference Stations (CORS) of the roof of The Sir Peter Mansfield Building(PMB)at UNNC. The GNSS receiver's sampling rate is configured as 10 Hz. 13 GPS satellites were tracked over these 4 h. The GPS raw data have a logging gap between 5:00 and 5:30, due to the GPS signal lost over that period. All panels of Fig. 2 show the correction biases has relatival bigger changes at beginning (GPS 03, GPS 17, GPS 22, GPS 28 and GPS 30, etc.) or end of time series (GPS 09, GPS 16, GPS 23, GPS 26 and GPS 27, etc.). The feature of the clock correction biases' time series is due to the bigger observation errors of GPS satellites which start to go into the view from low elevation to high, or get out of the view from high to low elevation. The slowly changes in the satellite clock correction time series is due to the limited number of tracking station. Data from Global distributed IGS station allow the better satellite clock correction estimation equation geometry and the redundant observations. Precise IGS stations' positions, ZTDs and UPDs of global distributed IGS stations are, respectively, estimated before they are used for satellite clock correction estimation. In order to reduce the computation load in the static PPP, a batch of dataset with no more than 15 stations, 555 unknown parameters, was processed. Figure 3 shows the GPS PRN 28 clock correction, estimated using the only one UNNC reference station dataset and the world-wide distributed IGS stations datasets. This improvement is significant, from 1×10^{-1} ns to 2×10^{-2} ns, in other words, improvement from 3.0 cm to 6 mm on the range of GPS PRN 28's signal. Figure 4 top panel, satellites clock bias error time series implies the stabilities of satellites' clocks are very different. The general range of bias error difference is from $\pm 3 \times 10^{-2}$ ns over half an hour, from 4:30 to 05:00 on 15th June 2016. Bottom panel of Fig. 4 shows the standard deviation (std) of each satellite's clock bias error. GPS PRN 06 has the smallest std over this short period, 2.2×10^{-3} ns. Oppositely, PRN 01 has the biggest std, 8.2×10^{-3} ns.

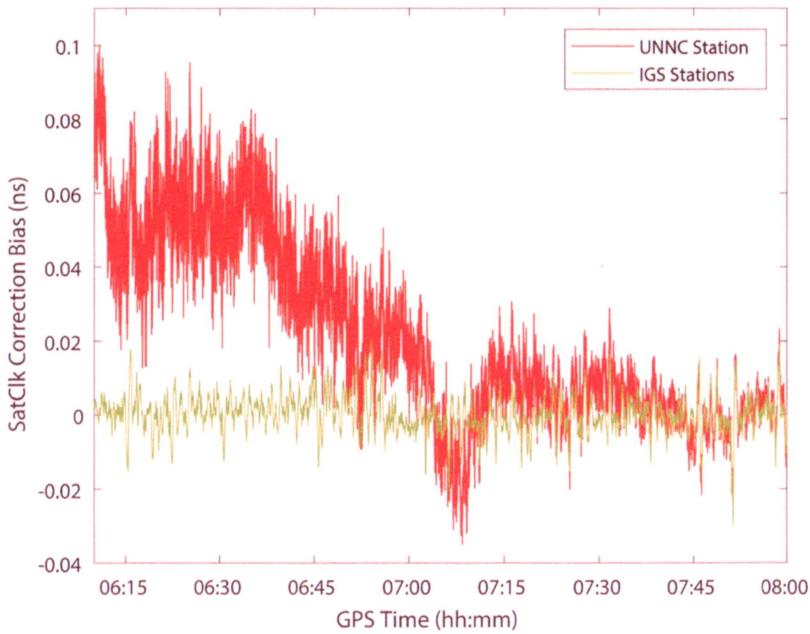

Fig. 3. Comparison of GPS PRN 28 clock corrections, which are estimated by the observation from UNNC only one reference station (red line) and the global distribution IGS stations (green line).

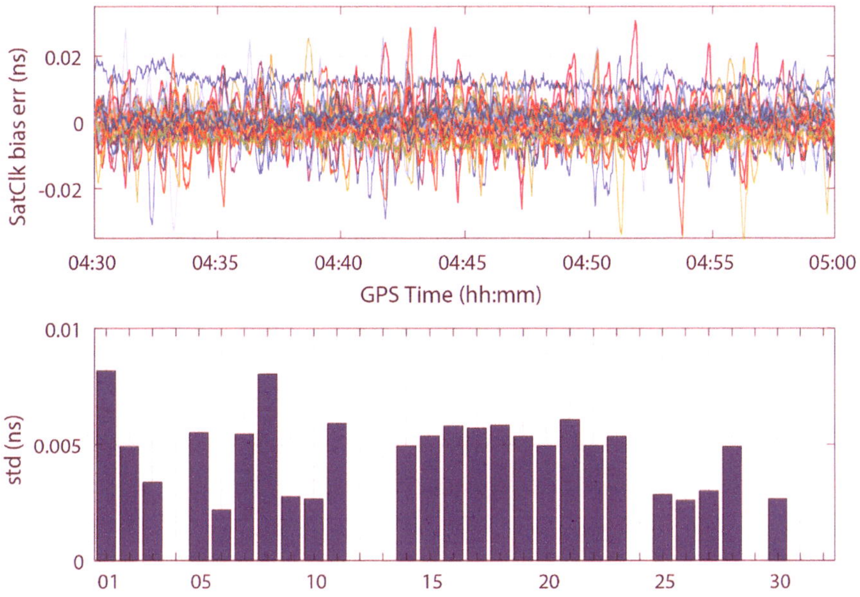

Fig. 4. 25 GPS satellites clock bias errors estimated from global distributed IGS stations (top panel); and their standard deviation (bottom panel).

3 High-Rate Kinematic PPP Experiment

The main span of Severn Bridge between the two towers is 988 m. 9 GNSS chock ring antennas were mounted on the bridge in this experiment. 5 of them were mounted on the suspending cables with the receivers' sampling rate 20 Hz, while the left 4 were mounted on the towers with the receivers' sampling rate 10 Hz. The orange dots show the location of antenna on bridge. The mask-off angel was set at 15° for reducing the random noise.

Weigh in motion (WIM) sensors is equipped on the welsh side of the Severn bridge. These sensors allow analysis of the bridge deflections in conjunction with the temperature and loading changes. Figure 5 shows the locations of the GNSS antennas and the location of WIM sensor. The WIM sensors are placed on the highway to the west of the Severn bridge, at 1.522 km away from the Beachley support towers. The speed, direction, and weight of vehicles can be recorded by the WIM sensors, when the vehicles pass the bridge. There is no congested traffic between 16:00 and 17:00 11th March 2010. With the assumption that each vehicle crosses the bridge at a constant speed which is recorded by WIM sensors, the whole load of traffic on the main-span can be estimated.

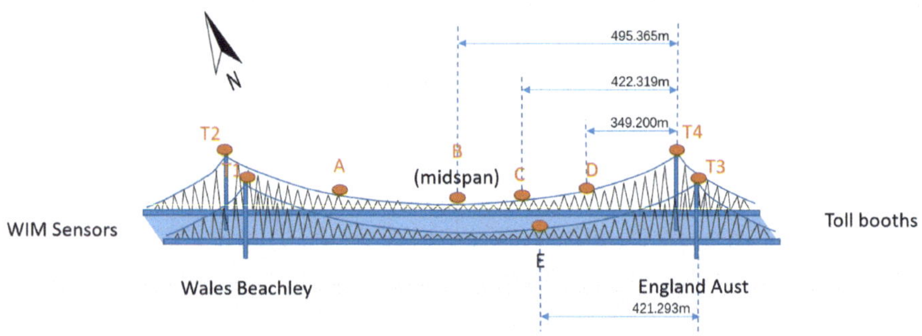

Fig. 5. Surveying points' positions on the Severn Bridge, UK.

Double differenced method is well established for the structure monitoring, treated as the truth deflection of Severn bridge at the vertical component. Figure 6 top panel shows the bridge time series of suspension cable movements at the vertical direction using PPP with different types of time-interval satellites clock corrections, as well as that of DD time series. The black, red and green waveforms in Fig. 6 are the 1 s, 30 s and 300 s PPP time series of surveying point A, respectively. The observable movements of 1 s and 30 s PPP are matched with DD results very well, which implies that both PPP can be used for extracting the big movements. Figure 6 bottom panel shows the differences between the DD and 1 s, 30 s and 300 s PPP time series, respectively. The differences between DD and PPPs reflect that 1 s PPP has small biases, compared with the 30 s and 300 s PPP results. This is evidence between 16:05 and 16:13. The big bias between 16:13 and 16:17 reflects there is a kind of gradually movements in 1 s PPP time series. Black waveform is much more smooth than red waveform in Fig. 6

bottom panel implies that 1 s PPP is better to extract bridge small movements than that of 30 s PPP. The correlation coefficient between DD and 300 s PPP time series is 0.689, while the correlation coefficients 1 s and 30 s PPP time series and that of DD are 0.966 and 0.931, respectively. This reflects the 1 s PPP has the best performance, comparing with 30 s and 300 s PPP. 300 s PPP cannot use for extracting the suspension cable bridge deflection properly.

Figure 7 shows the 4 cable and 4 tower surveying points deflections time series in the vertical and longitudinal components, which are estimating by the PPP kinematic model with 1 Hz satellites' clocks corrections, precise ephemeris and tropospheric corrections which is derived from reference station, approximate 2 km far away from the middle of bridge main span. The consistency of the WIM time series (bottom panel in Fig. 7) compared with the bridge cable surveying point movements in the vertical component implies the bridge movements are largely due to the traffic loading. The corresponding movements at the tower surveying point in the longitudinal component (top panel of Fig. 7) proves that high-rate kinematic PPP is qualified to monitoring the small movements of big structure, such as cable suspension bridge. Surveying point E is located at the opposite of surveying point C. Comparing the location C and E allows analysis the bridge rotation. There is a heavy traffic load at 16:15 at the north side of main-span, and light traffic load at the south side of the bridge. All 3 cable surveying points have relative bigger movements in the order of A, B and C. This means the heavy traffic cross over the bridge in the order of A, B and C points. The heavy weight

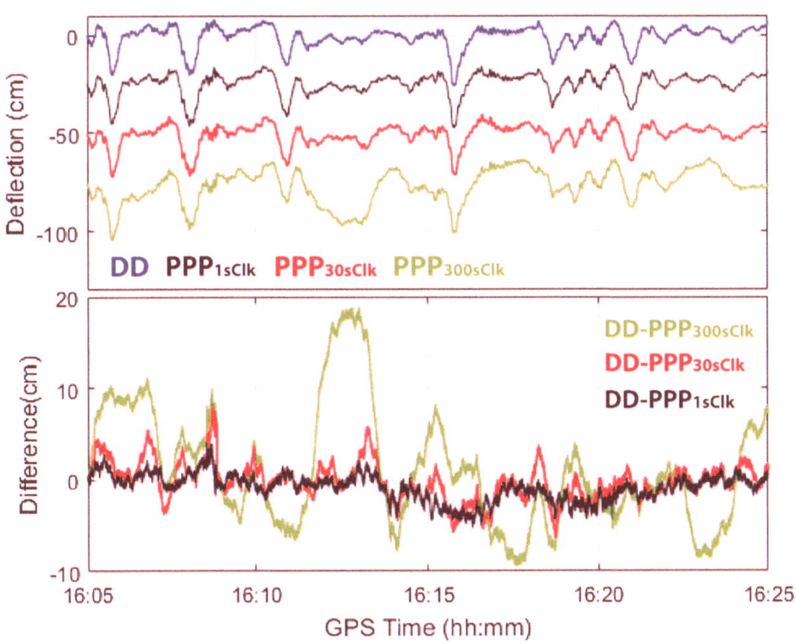

Fig. 6. PPP time series of surveying point A using 1 s, 30 s and 300 s intervals satellite clock corrections, as well as the DD time series.

load at approximate 16:11 cross the bridge from the south side of bridge, then the big movements at the 3 surveying points in the order of C, B and A. C and E big movements occurred at the same time. This implies that the Severn bridge both north and south suspension cables' deflections are due to either the north or south side traffic load. The biggest weight load is at 16:18, but the suspension cable surveying points is smaller than the period will less weight load (approx. 16:16 and 16:21). The heavy weight load is contributed from only one side of bridge at 16:16 and 16:21, while the heavy weight load is contributed from both sides at 16:18. This indicates that weight load asymmetry of can introduce bigger cable deflections, but less when the weight load comes from the both sides.

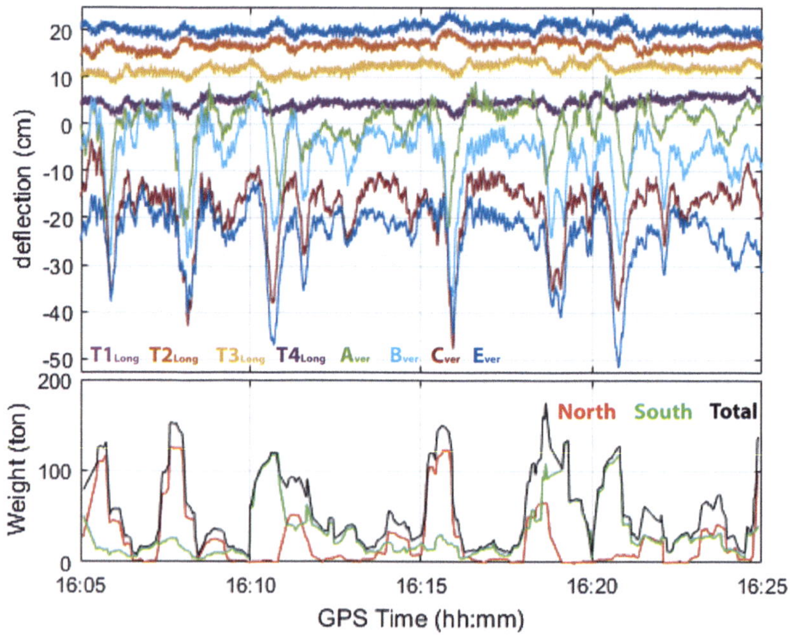

Fig. 7. High-rate kinematic PPP movement comparison of cable and tower surveying points (top panel). Bridge weight load at North and South sides of bridge main-span, as well as the total bridge weight load (bottom panel).

4 Conclusion

PPP was used for retrieving earthquake movements in the past few years. Earthquake movements tend to be large and only occur for short time periods. Structural movements however tend to be relatively small with some long-term features. This paper introduced the kinematic PPP for the bridge deflection detection, with the aid of high-rate satellites clock corrections. The procedure of satellites clock correction estimation was explained in detail, then dataset from one of the UNNC CORS station on roof of PMB was used for estimating the satellite clock offset. Clock corrections solutions are generally larger when the satellites are at low elevation. This study utilized 96 worldwide distributed IGS reference stations for improving the GPS satellite clock corrections. The numerical statistics and figures reveal the evident improvement by using a large number of GPS tracking stations. Validation of high-rate PPP displacement monitoring was performed using a real-life dataset from the Severn Bridge, UK. The bridge experiment shows the kinematic PPP with high-rate satellites' clock correction can identify small movements, which is not possible when applying the satellite clock products downloaded directly from the IGS websites. The kinematic PPP offers an alternative approach for the precise structure deflection detection, while the DD method is not possible or failed because of large baseline lengths, power outages or natural disasters.

Acknowledgment. The work in this paper is supported by Young Scientist program of Natural Science Foundation of China (NSFC) with a project code No. 41704024, general programme of NSFC with project code No. 41774034, as well as the Ningbo science and technology bureau commonweal programme (project 20181JCGY020386). Authors are grateful to the anonymous reviewers, whose comments significantly improved this manuscript.

References

1. Ashkenazi, V., Roberts, G.W.: Experimental monitoring of the Humber bridge using GPS. In: Proceedings of the Institution of Civil Engineers-Civil Engineering, vol. 120, no. 4, pp. 177–182, November 1997 (in English)
2. Roberts, G.W., Brown, C.J., Tang, X., Meng, X., Ogundipe, O.: A tale of five bridges; the use of GNSS for monitoring the deflections of bridges. J. Appl. Geodesy **8**(4), 241–264 (2014)
3. Meng, X., Roberts, G.W., Cosser, E., Dodson, A.H., Barnes, J., Rizos, C.: Real-time bridge deflection and vibration monitoring using an integrated GPS/Accelerometer/Pseudolite system. In: Proceedings of 11th International Symposium on Deformation Measurements, International Federation Surveyors (FIG), Commission 6-Engineering Surveys, Working Group, vol. 6, pp. 25–28 (2003)
4. Tang, X., Roberts, G.W., Li, X., Hancock, C.M.: Real-time kinematic PPP GPS for structure monitoring applied on the Severn Suspension Bridge, UK. In: Advances in Space Research (2017)
5. Tang, X., Li, X.: Differenced measurements between satellites applied on RTK PPP for structure monitoring. In: 2017 Proceedings of China Satellite Navigation Conference (CSNC) (2017). (in English)
6. Chen, K., Ge, M., Babeyko, A., Li, X., Diao, F., Tu, R.: Retrieving real-time co-seismic displacements using GPS/GLONASS: a preliminary report from the September 2015 Mw 8.3 Illapel earthquake in Chile. Geophys. J. Int. **206**, 941–953 (2016)
7. Moschas, F., Avallone, A., Saltogianni, V., Stiros, S.C.: Strong motion displacement waveforms using 10-Hz precise point positioning GPS: an assessment based on free oscillation experiments. Earthq. Eng. Struct. Dyn. **43**(12), 1853–1866 (2014). (in English)

Radiosonde-Based New Spatiotemporal Modelling for the Construction of Temperature Profiles for GNSS Applications

Longjiang Li, Zhen Shen, Qimin He, Mofeng Wan, Kefei Zhang, and Suqin Wu(✉)

School of Environment Science and Spatial Information, China University of Mining and Technology, Xuzhou, China
suqin_wu@cumt.edu.cn

Abstract. In order to construct temperature profiles over a GNSS station of interest, a series of spatiotemporal models for determining the ratio of the zenith dry delay (ZDD) for 40 selected pressure levels were established using a harmonic function and 18 years radiosonde data from 92 stations distributed in East Asia. The ZDD over a site is defined by the integral of the temperature profile and the dry pressure profile over the site. In the case that neither the temperature profile nor the dry pressure profile is available, a method for the ZDD obtained from precipitation water vapor (PWV) and zenith hydrostatic delay (ZHD) is derived in this study. Then, the ZDD value at a dry pressure level, named ZDD-lev, was obtained by multiplying the above ZDD with the ratio of the ZDD interpolated from the above mentioned spatiotemporal model for the level. The ZDD-lev and the dry pressure were used to determine the temperature of the level. The temperature profile was composed of the temperature values at all the dry pressure levels. The temperature profiles constructed at 10 IGS stations during a 8-year-period were evaluated using the observed ones as the references. The results indicated that the mean root-mean-square error (RMSE) and mean bias of the temperature profiles at 10 IGS stations was 3.24 K and 0.5 K, respectively.

Keywords: Zenith dry delay · GNSS · Models · Temperature profile · Precipitation water vapour

1 Introduction

Atmospheric temperature is one of the most important elements in the troposphere and plays a critical role in climatology, weather forecast, positioning etc. Nowadays, radiosonde is routinely used for accurate measurements of temperature profiles. However, it has the disadvantages of low spatiotemporal resolution, high cost etc. (Gui et al. 2017). For those sites rather than radiosonde stations, their temperature profiles can be obtained from empirical models, e.g. the linear model ($T_h = T_s + B \times \delta_h$) (Wang et al. 2016; Saastamoinen 1972; Xia et al. 2019) and MSISs. In the above linear model, which is most commonly used for the lower atmosphere, the coefficient B is used as a constant, which is in fact not true in reality, thus it has low accuracy.

J. Sun et al. (Eds.): CSNC 2020, LNEE 650, pp. 232–239, 2020.
https://doi.org/10.1007/978-981-15-3707-3_23

The MSISs, which can be used to obtain temperature over the sea surface level, perform well in the upper atmosphere, i.e. the thermosphere, while they perform poorly in the lower atmosphere (Picone et al. 2002; Hedin 1991; Batten et al. 1987).

As a new technique to remotely sense the atmosphere, the global navigation satellite systems (GNSS) have been found more and more applications that are not only limited to positioning, navigation, and timing, but also to meteorology and climatology (Wang et al. 2018; Rohm et al. 2014; Yao et al. 2017; Zhang et al. 2015) due to their high accuracy, high resolution and low cost. However, the atmospheric parameters retrieved from ground-based GNSS signals such as the zenith tropospheric delays are the total delays, rather than profiles, over the GNSS stations. For the best use of GNSS data, this study is mainly focused on the use of GNSS-derived atmospheric parameters to construct temperature profiles for the GNSS stations.

In this study, first, the method to use the zenith hydrostatic delay (ZHD) and precipitation water vapour (PWV) to determine the zenith dry delay (ZDD) over a GNSS station was derived. The ZHD was obtained from the Saastamoinen model (Wang et al. 2017; Bevis et al. 1992; Jiang et al. 2016) and PWV was converted from the GNSS-derived zenith wet delay. Secondly, spatiotemporal models for determining the ratio of the ZDD at each of the 40 selected dry pressure levels (ZDD-lev) to the ZDD were developed based on radiosonde data, and then the ratio and the ZDD from ZHD and PWV were used to construct temperature profiles for the GNSS stations. Thirdly, the temperature profiles constructed at 10 selected international GNSS service (IGS) stations were evaluated by comparing them against the observed ones.

2 Data and Methodologies

2.1 Data

The two types of data used in this study were radiosonde and GNSS. The radiosonde data during the 18-year-period of 2000–2017 at 92 selected stations (denote the red triangles) for constructing spatiotemporal models are shown in Fig. 1. The GNSS data during the 8 years period of 2010–2017 at the 10 IGS stations denoted by the black squares were used to evaluate the temperature profiles constructed from the above mentioned models.

The aforementioned selected radiosonde data were from the Integrated Global Radiosonde Archive (IGRA), a high-quality radiosonde dataset developed by the American National Climatic Data Center (NCDC). Their temporal resolution is usually twice per day (at 00:00 and 12:00 UTC) and all the radiosonde stations included in the dataset are inhomogeneously distributed.

The GNSS data at the above 10 selected IGS stations were from Crustal Dynamics Data Information System (CDDIS), which included raw GNSS observations and some products. These raw GNSS observations were used to estimate the zenith total delays over the stations and the software package Bernese v5.2 was used to process the GNSS data.

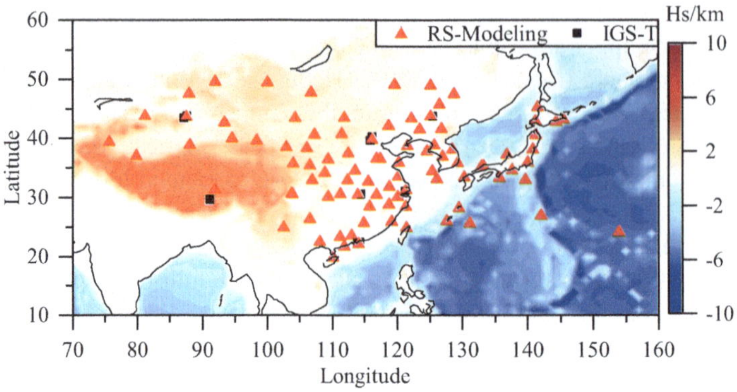

Fig. 1. Distribution of radiosonde stations (triangles) and IGS stations (squares) selected in this study

2.2 Methodologies

2.2.1 Investigating Relationship Among ZDD, ZHD and PWV

ZDD is defined as (Davis et al. 1985; Bevis 1994):

$$ZDD = 10^{-6} \int_{H_s}^{H_a} k_1 (\frac{P_d}{T}) Z_d^{-1} dh \tag{1}$$

where H_s(m) and H_a(m) are the heights of the site and the tropopause respectively; k_1 is an empirical constant; P_d(hPa) is the partial dry pressure; T(K) is the partial temperature; Z_d^{-1} is the inverse compressibilities for dry gas (Owens 1967).

ZHD is defined as (Davis et al. 1985):

$$ZHD = 10^{-6} (\int_{H_s}^{H_a} k_1 \frac{P_d}{T} Z_d^{-1} dh + \int_{H_s}^{H_a} k_1 \frac{R_d}{R_w} \frac{e}{T} Z_w^{-1} dh) \tag{2}$$

It can be seen from Eq. (4) that the first term on the right-hand side is ZDD, and the second term is the effect of water vapor. Subtracting Eq. (1) from Eq. (4) lead:

$$\begin{aligned} ZHD - ZDD &= 10^{-6} k_1 R_d \int_{H_s}^{H_a} \frac{\mathrm{e}}{R_w T} Z_w^{-1} dh = 10^{-6} k_1 R_d \int_{H_s}^{H_a} \rho_w dh \\ &= 10^{-6} k_1 R_d \rho \cdot PWV \end{aligned} \tag{3}$$

Then,

$$ZDD = ZHD - 10^{-6} k_1 R_d \rho \cdot PWV \tag{4}$$

Formula (4) indicates the relationship among ZHD, ZDD and PWV, thus it can be used to obtain ZDD from ZHD and PWV, especially in the case that the temperature and dry pressure profiles over a site are not available, e.g. over GNSS stations, formula (1) cannot be used.

ZHD can be obtained from a standard empirical model such as the most widely used Saastamoinen model and PWV can be obtained from a conversion of the troposphere delay estimated from GNSS data.

2.2.2 Establishing Models for the Ratio of ZDD-lev to ZDD

The radiosonde data at 40 pre-determined dry pressure levels from the aforementioned 92 stations were used to establish 40 spatiotemporal models. For a specific dry pressure level, only those radiosonde data that were at this level were used to establish the model for the level. The modelling was for the ratio of the ZDD-lev (which is the ZDD at the level) to the ZDD (which is the total value over the site). The procedure for the establishment of this model is as follows.

1. *Determining the ratio of ZDD-lev to ZDD.*
   (a) For a specific pre-determined dry pressure level at a radiosonde station (lat, lon, h) and time (t), if the temperature and dry pressure measurements are available, the ZDD-lev over the station at the time was obtained using formula (1) in which the value of 1 m for the thickness of the level (denoted by dh) was adopted in this study;
   (b) Using formula (4), together with the ZHD from the Saastamoinen model and PWV from the radiosonde data to obtain the ZDD over the station at the time;
   (c) Calculating the ratio of the ZDD-lev and the ZDD;
   (d) Repeating the above three steps to obtain the ratios at all the pre-determined levels over all radiosonde stations at all times.

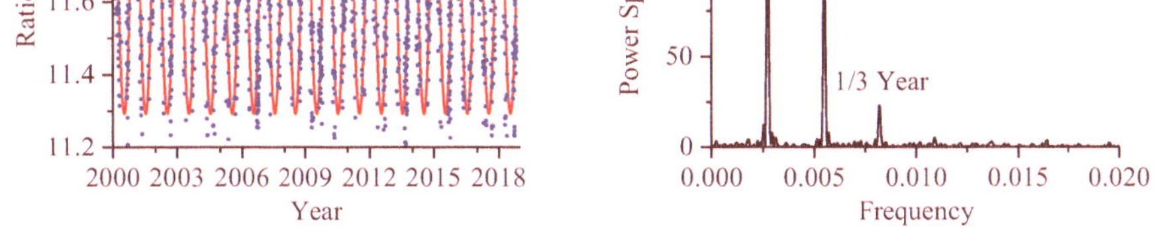

Fig. 2. Time series of the ratio of ZDD-lev to ZDD (ID: CHM00058027, Dry Pressure: 995 hPa)

2. *Determining fitting functions for ratio time series.*
   (a) Investigating the trend of the ratio time series at a pre-determined dry level over a station in the 19-year period 2000–2018, as shown by the blue points in Fig. 2 (a) for an example, from which a periodic trend can be seen;

(b) Using the Fourier transform to extract the periodic terms (see Fig. 2(b)) from the result shown in Fig. 2(a), then only the two terms that are significant were selected for the annual and semiannual terms.

(c) Using the two selected periodic terms to develop a harmonic function that fits the ratio time series at the level over the station (see the red curve in Fig. 2(a) for the fitting result of the example), which is:

$$\begin{aligned} Ratio = a_0 + a_1 \times \cos\left(\frac{2\pi \times doy}{365.25}\right) + a_2 \times \sin\left(\frac{2\pi \times doy}{365.25}\right) \\ + a_3 \times \cos\left(\frac{2\pi \times 2 \times doy}{365.25}\right) + a_4 \times \sin\left(\frac{2\pi \times 2 \times doy}{365.25}\right) \end{aligned} \quad (5)$$

where a_i (i = 0, 1, 2, 3, 4) is the amplitude (or coefficient) to be estimated; *doy* is the day of year.

(d) Repeating the above three steps to obtain fitting functions for all pre-determined levels over all stations.

3. *Determining spatial fitting functions.*

(a) Extracting the values for the same amplitude (from formula (5)) estimated from the previous step from all the 92 stations at a pre-determined dry pressure level, then developing a spatial fitting function for the level, expressed as:

$$a_i = b_0 + b_1 \times \varphi + b_2 \times \lambda + b_3 \times h \quad (6)$$

where a_i is one of the five amplitudes expressed in formula (5); b_j (j = 0, 1, 2, 3) is the coefficients to be estimated; φ, λ and h are the latitude, longitude and height of the station.

(b) Repeating the above step to obtain the spatial fitting models for all the 40 pre-determined levels.

2.2.3 Constructing Temperature Profiles

The procedure of constructing temperature profiles over a GNSS station is as follows.

(1) Using the spatiotemporal models established in the last session to predict the 40 ratios of the ZDD-levs to ZDD for all the 40 pre-determined levels over the station for any time.

(2) Using formula (4), together with the ZHD calculated from the Saastamoinen model and PWV from a conversion of the ZTD estimated from the GNSS data of the station to obtain the ZDD over the station.

(3) Multiplying the above ratios with the ZDD to obtain the ZDD-levs for the 40 pre-determined levels.

(4) Dividing the above ZDD-levs by the thickness of the level (dh = 1 m) to obtain the dry refractive indexes for all the 40 levels.

(5) Using the above dry refractive indexes and the formula below to obtain the temperature values for all the 40 levels, and all these temperature values are the components of the temperature profile for the GNSS station.

$$T_i = 77.604\frac{Pd_i}{R_i} \tag{7}$$

where i denotes the i-th level; T_i, Pd_i and R_i are the temperature, dry pressure and dry refractive index at the i-th level respectively.

3 Results

For the assessment of the above established models, the data for a 8-year period during 2010–2017 at 10 IGS stations in East Asia were used for testing. Due to the unavailability of a reference for validating the model predicted ratio values, the temperature values obtained from the ratios for all the 40 pre-determined dry pressure levels were used to evaluate the models performance. The reference was the radiosonde data at those radiosonde stations that were nearest to the selected IGS stations.

Table 1 shows the statistical results of all the 40 levels during the while period of time at each of the stations. The RMSE value indicates the overall accuracy of the temperature results. It is noted that the largest bias and largest RMSE occurring at the LHAZ station is most likely due to the large distance between the station and its nearest radiosonde station (about 223 km). At the other 9 stations, the biases were in the range from −0.74 K to 1.8 K; the RMSEs were in the range from 2.51 K to 3.67 K. Among the 9 stations, the worst result was at the GUAO station, which had the large distance between the station and the nearest radiosonde station (50 km); the best result was from the HKWS station with the smallest RMSE value of 2.3 K.

Table 1. Statistics of temperature at 10 IGS stations

Name	Latitude	Longitude	Height (m)	Distance (km)	Bias (K)	STD (K)	RMSE (K)
BJFS	39.61	115.89	61.00	49.18	1.2	3.5	3.67
BJNM	40.25	116.22	100.00	35.05	0.6	3.4	3.42
CHAN	43.79	125.44	235.00	21.87	1.0	3.34	3.49
GUAO	43.47	87.18	2092.00	49.55	1.8	4.23	4.62
HKSL	22.37	113.93	49.00	25.24	1.1	2.27	2.51
HKWS	22.43	114.34	41.00	20.88	0.38	2.26	2.29
JFNG	30.52	114.49	60.00	43.25	−0.74	2.86	2.95
LHAZ	29.66	91.10	3649.00	222.94	−5.87	3.34	6.75
SHAO	31.10	121.20	9.00	42.49	−0.61	3.00	3.06
URUM	43.81	87.60	868.00	3.03	−0.23	3.12	3.13

Note: 'Distance' is the distance between IGS station and the nearest radiosonde station, unit: km.

4 Conclusion

In order to investigate temperature profiles over GNSS stations, a series of spatiotemporal models for determining the ratio of ZDD for 40 selected dry pressure levels were established using a harmonic function and radiosonde data from 18 years and 92 stations distributed in East Asia. In addition, a new formula for the determination of ZDD is derived using both PWV and ZHD in the case that neither the temperature profile nor the dry pressure profile is available. ZDD-lev was in such a way that multiplies the above approximate ZDD with the ratio of the ZDD interpolated from the model for the level. The ZDD-lev and the dry pressure were used to determine the temperature at the level. The temperature profile was composed of the temperature values at all the dry pressure levels. Finally, the temperature profiles constructed from the models are evaluated. Test results for temperature from 10 IGS stations during a 8-year-period showed that the mean RMSE of the differences between the temperature profiles resulting from the new models and observed ones were 3.24 K.

This study proposed a new approach of constructing temperature profiles at GNSS stations. We believe that the new approach can be applied into other models or reanalysis datasets to improve their performance. One of the main follow-on benefits is that GNSS can be used to increase the density of temperature profiles due to the ever-increased availability of dense GNSS stations.

Acknowledgements. This work was funded by the State Key Program of National Natural Science Foundation of China (No. 41730109). And this project also thanks to IGRA for provision of radiosonde data, ECMWF for provision of ERA-Interim dataset, and CDDIS for IGS data and products.

References

Batten, S., et al.: A numerical data base for VAX and personal computers for the storage, reconstruction and display of global thermospheric and ionospheric models. Planet Space Sci. **35**(9), 1167–1179 (1987)

Bevis, M., et al.: GPS meteorology: remote sensing of atmospheric water vapor using the Global Positioning System. J. Geophys. Res. Atmos. **97**(D14), 15787–15801 (1992)

Bevis, M.: GPS meteorology: mapping zenith wet delays onto precipitable water. J. Appl. Meteorol. **33**(3), 379–386 (1994)

Davis, J.L., et al.: Geodesy by radio interferometry: effects of atmospheric modeling errors on estimates of baseline length. Radio Sci. **20**(6), 1593–1607 (1985)

Gui, K., et al.: Evaluation of radiosonde, MODIS-NIR-Clear, and AERONET precipitable water vapor using IGS ground-based GPS measurements over China. Atmos. Res. **197**, 461–473 (2017)

Hedin, A.E.: Extension of the MSIS Thermosphere Model into the middle and lower atmosphere. J. Geophys. Res. Space Phys. **96**(A2), 1159–1172 (1991)

Jiang, P., et al.: Retrieving precipitable water vapor data using GPS zenith delays and global reanalysis data in China. Remote Sens. Basel **8**(5), 389 (2016)

Owens, J.C.: Optical refractive index of air: dependence on pressure, temperature and composition. Appl. Opt. **6**(1), 51–59 (1967)

Picone, J.M., et al.: NRLMSISE-00 empirical model of the atmosphere: statistical comparisons and scientific issues. J. Geophys. Res. Space Phys. **107**(A12), 11–15 (2002)

Rohm, W., et al.: Ground-based GNSS ZTD/IWV estimation system for numerical weather prediction in challenging weather conditions. Atmos. Res. **138**, 414–426 (2014)

Saastamoinen, J.: Atmospheric correction for the troposphere and stratosphere in radio ranging satellites. Use Artif. Satell. Geodesy **15**(6), 247–251 (1972)

Wang, X., et al.: Water vapor-weighted mean temperature and its impact on the determination of precipitable water vapor and its linear trend. J. Geophys. Res. Atmos. **121**(2), 833–852 (2016)

Wang, X., et al.: Determination of zenith hydrostatic delay and its impact on GNSS-derived integrated water vapor. Atmos. Meas. Tech. **10**(8), 2807–2820 (2017)

Wang, X., et al.: The correlation between GNSS-derived precipitable water vapor and sea surface temperature and its responses to El Niño-Southern Oscillation. Remote Sens. Environ. **216**, 1–12 (2018)

Xia, P., et al.: Retrieval of tropospheric refractivity profiles using slant tropospheric delays derived from a single ground-based Global Navigation Satellite System station. Earth Space Sci. **6**(7), 1081–1097 (2019)

Yao, Y., et al.: Establishing a method of short-term rainfall forecasting based on GNSS-derived PWV and its application. Sci. Rep. U.K. **7**(1), 12465 (2017)

Zhang, K., et al.: Capturing the signature of severe weather events in Australia using GPS measurements. IEEE J-STARS **8**(4), 1839–1847 (2015)

Analysis of Temporal and Spatial Variation of Crustal Strain Around Longmenshan Fault Based on GNSS Observation

Huijuan Liu[1(✉)], Xianchun Chen[1], Caiya Yue[2], and Qiang Yang[3]

[1] Sichuan No. 1 Surveying and Mapping Engineering Institute, Chengdu, Sichuan, China
474253346@qq.com
[2] China University of Mining and Technology (Beijing), Beijing, China
[3] Chinese Academy of Surveying and Mapping, Beijing, China

Abstract. The Longmenshan fault zone is located in the transition zone between the eastern margin of Qinghai-Xizang Plateau and Sichuan Basin. It is an important part of the north-south seismic belt in China, and the tectonic activity is frequent, and the 2008 Wenchuan earthquake and the 2013 Ya'an earthquake are all on the tectonic fault. In this paper, the coordinates' time series and velocity fields in the region with high precision are obtained based on GAMIT/GLOBK. The strain parameters in the region are calculated by using the strain model and the high-precision dynamic model of the fault zone and the surrounding crust are constructed, and the time-space distribution of the abnormal information of the fault zone is extracted and analyzed with the reference of the earthquake such as Wenchuan and Ya'an. The study in this paper can be used as a reference for the analysis of crustal movement characteristics and the assessment of geological hazards in Sichuan-Yunnan region.

Keywords: Longmenshan fault · Motorial block · Strain rates · Tectonic movement · CORS data

1 Introduction

The Sichuan-Yunnan region is located on the southeastern edge of the Qinghai-Tibet Plateau, which is the marginal zone where the Eurasian and Indian plates interact. It belongs to the north-south earthquake zone of China [1, 2]. The western boundary of this region is driven by the squeezing force of the plateau potential energy flowing southeast and the northeast inflow of the Indian plate, while its eastern boundary is blocked by the stable South China block, which makes the region's geological structure complicated and the tectonic activity strong. The frequency of earthquakes is high and large. It is the most prominent region of crustal deformation in China [3]. Therefore, in recent years, this region has become a hotspot for geological researchers to analyze the evolution characteristics of the Qinghai-Tibet Plateau, seismic crustal strain models, and crustal deformation in the north-south seismic zone of China [4].

The strong tectonic deformation activity in the Sichuan-Yunnan region has been in long-term so faults of various sizes, different occurrences, and different activity rates

J. Sun et al. (Eds.): CSNC 2020, LNEE 650, pp. 240–250, 2020.
https://doi.org/10.1007/978-981-15-3707-3_24

have developed in the region [5]. The compressional displacement of these fault zones caused intense seismic activity in the area; The complexity and multiplicity of long-term tectonic movement characteristics also made the region prone to sudden geological disasters and severe disasters [6]. Based on this characteristic, domestic and foreign experts have studied the characteristics of fault activity in the region using various techniques and methods, such as leveling, GNSS velocity field, Insar, and gravity measurement. For example, Chen and others considered that the deformation of the Sichuan-Yunnan block mainly exhibits the clockwise rotation and non-uniform sliding of the block boundary fault zone [7]. Guo Xiaohu and Wei Dong ping et al. established a linear elastic block model consisting of 24 secondary blocks in the Sichuan-Yunnan region, and the long-term motion speed of each block and the current activity rate of each major boundary fault are obtained by inversion [8]. Yue et al. used the long-term GPS velocity field information to invert the activity rates of the main fault zones in the Sichuan-Yunnan region, and they found that relatively active fault zones are mainly distributed in several deep faults [9]. Wang and Guo et al. divided the Sichuan-Yunnan region into more detailed secondary plots based on geological and satellite observations, and they obtained the long-term motion speed of each block and the current trend of major boundary faults by inversion [8, 10]. A large number of studies have shown that the relative motion of the fault zone in the Sichuan-Yunnan region is the direct cause of earthquakes in the region.

There are many fault zones in the Sichuan-Yunnan region, and the Longmenshan fault is one of the most active fault zones which is also the dividing line between the Bayankla block and the South China block. For the secondary block division in the Sichuan-Yunnan region, the region is the dividing line between the Longmenshan block and the Chengdu block [11]. Both the 2008 Wenchuan earthquake (depth 14 km, magnitude 8.0, longitude 103.4, latitude 31.0) and the 2013 Ya'an earthquake (depth 13 km, magnitude 7.0, longitude 103.0, latitude 30.3) occurred on the fault zone. These two connected earthquakes have led scholars to launch a new round of research on the characteristics of fault motion and the strain characteristics of surrounding blocks based on geodetic observation techniques and geological observations [12–14], and fruitful results have been obtained.

In this paper the strain parameters of the region are calculated by using the strain model built on a large amount of GNSS data from the Crustal Movement Observation Network of China (CMONOC) and Sichuan CORS network. Also the high-precision crustal deformation dynamic model of the fault zone and the surrounding area are constructed. Eventually the time-space distribution of the abnormal information of the fault zone is extracted and analyzed with the reference of the Wenchuan and Ya'an earthquake.

2 Research Area and GNSS Data

The Longmenshan fault is located at the southeast of the Bayankla block, the northwestern of the South China block, and the northeast section of the Sichuan-Yunnan diamond block, including the southwest section (fault 6, as shown in Fig. 1) and the northeast section (fault 7, as shown in Fig. 1) of the Longmenshan fault. The fault associated with a length of about 500 k and a width of 40–50 km has a northeast-southwest trend. Its maximum elevation is about 6,500 m and its terrain elevation difference is up to 5000 m [15]. For the secondary block division scheme in the Sichuan-Yunnan region, the fault mainly involves three secondary blocks, namely the Longmen block in the northwest direction, the Chengdu block and the asbestos block in the southeast direction. Besides, it is intersected by the fault of Xianshuihe-Xiaojiang (fault 1 2 3 4 5, as shown in Fig. 1). For a long time, there has been great controversy about the special undulating terrain uplift mechanism in the Longmenshan region, which can be summarized as a brittle crustal thickening model and a lower crustal flow model. In recent years, the Wenchuan and the Ya'an earthquake occurred on the fault caused the region to be taken seriously again by researchers.

The GPS data used in this study comes from CMONOC and Sichuan CORS network. The CMONOC GPS data period is from 2001 to 2016, and the Sichuan CORS data period is from 2010 to 2015. Data processing can be divided into 5 steps:

Step 1: filter the stations to remove the stations with short periods and poor data; Step 2: obtain single-day relaxation solutions for station and satellite orbits; Step 3: estimate the position of each station using GLOBS to synthesize all the single-day relaxation solutions, taking into account the coseismic displacement of the station, and in conjunction with IGS stations in and around China; Step 4: the obtained station position is used as the prior coordinates required for the GAMIT solution. After the single-day relaxation solution is obtained, the GLOBK adjustment is used to obtain the position of each station and the velocity field under the framework of ITRF2008, which is on the premise of removing the influence of coseismic displacement. Step 5, In order to accurately reflect the characteristics of horizontal crustal movement in the Sichuan-Yunnan region, 15 stable reference stations in the Eurasian plate be used and the optimal Euler rotation vector be solved between ITRF2008 and the Eurasian frame. Thus, the GNSS horizontal motion velocity field is calculated for the Eurasian frame (Fig. 1).

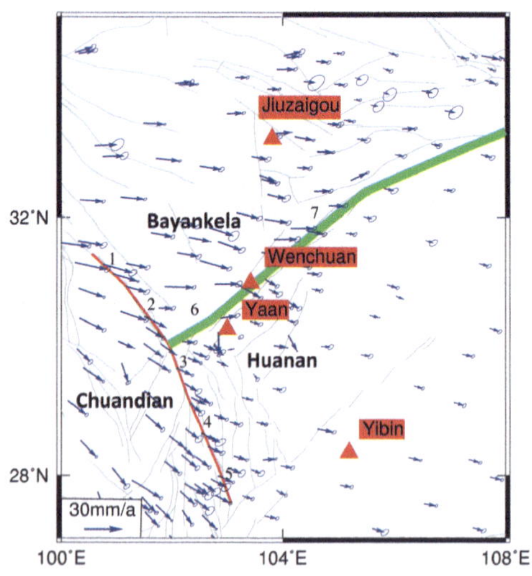

Fig. 1. GPS station velocity field under the Eurasian framework from 2001 to 2016

During the long-term velocity field calculations, the GNSS station may be affected by co-seismic displacement, strong solar magnetic storms, and GNSS station antenna replacement, which may cause the coordinate time series to jump. In order to obtain an accurate velocity field, the GNSS stations with a significant impact on jumps are eliminated, and the stations with a small impact are corrected using a function fitting method.

3 Block Model

The premise of using Euler's law to study plate motion is based these assumptions: (1) the tectonic block is a rigid block which can transfer stress over a long distance; (2) there is no obvious plastic deformation inside; and (3) the size of the earth does not change; thus, the area of plate expansion proliferation and crush extinction is equal [9]. However, a new model must be established to calculate the block internal strain since the block will deform under the external force, which is known as elastoplastic deformation [16]. Considering the rigid motion and elastoplastic strain of the block in combination with the overall displacement characteristics, a more accurate translation-rotation-strain model is established [11].

$$\boldsymbol{V}_s = \boldsymbol{R}_t \bullet \boldsymbol{\Omega}$$

$$\boldsymbol{R}_t = \begin{bmatrix} -\sin\lambda & \cos\lambda & 0 & -r\cos\lambda\sin\varphi & -r\sin\lambda\sin\varphi & -r\cos\varphi & r(\lambda-\lambda_0) & 0 & \frac{1}{2}r(\varphi-\varphi_0) \\ -\sin\varphi\cos\lambda & -\sin\varphi\sin\lambda & \cos\varphi & r\sin\lambda & r\cos\lambda & 0 & 0 & r(\varphi-\varphi_0) & \frac{1}{2}r(\lambda-\lambda_0)\cos\varphi \end{bmatrix}$$

$$\boldsymbol{\Omega} = \begin{bmatrix} \Delta x & \Delta y & \Delta z & \omega_x & \omega_y & \omega_z & \varepsilon_e & \varepsilon_n & \gamma_{en} \end{bmatrix}^{\mathrm{T}}$$

$$\boldsymbol{V}_s = \begin{bmatrix} v_e & v_n \end{bmatrix}^{\mathrm{T}}$$

Where $\boldsymbol{R}_t$ is the design matrix; Ω represents translation, rotation, and strain parameters; Vs represents North-south and east-west velocity at any point of the block (λ, φ), r represents the radius of the earth, (λ_0, φ_0) is the geometric center coordinate of the block, $(\Delta x, \Delta y, \Delta z)$ represents the overall displacement in the x, y, z directions of the block; ω_x, ω_y, ω_z are the Euler rotation vector; ε_e, ε_n, γ_{en} are the east and north principal strains and shear strains relative to the geometric center of the block, respectively.

When the eastward and northward principal strains and the two directions of shear strain are known, the principal strain direction θ and the area expansion rate Δ can be calculated by the following formula:

$$\Delta = \varepsilon_e + \varepsilon_n$$

$$\tan 2\theta = -\frac{2\gamma_{en}}{\varepsilon_e - \varepsilon_n}$$

The key to calculating the block strain parameters is to obtain high-precision velocity field information and high-precision coordinates of GPS stations. The calculation of the strain parameters of the block has no bearing on the selected reference frame. In this paper, the crustal strain information is calculated based on the velocity field under the Eurasian frame.

4 Analysis of Crustal Strain Around Longmenshan Fault

Since the 2008 Wenchuan earthquake, the 2013 Lushan earthquake also occurred on the Longmenshan fault zone in less than 5 years, which was associated with a magnitude above 7.0. In this paper, the long-term velocity field information of the region was obtained based on GAMIT/GLOBK by using the long-term land state network data in the attachment block of the Longmenshan fault zone and the Sichuan CORS data. And the staged strain was calculated in this region by using displacement- rotation-strain model.

4.1 Construction of Grid Velocity Field Model

In this paper, the GNSS data, including CMONOC and Sichuan CORS, is from 2001 to 2016. All the CMONOC data are derived from the regular observation data and a few continuously operating reference stations before 2009. After 2009, with the construction of the CMONOC, the station distribution density and the number of continuously operating reference stations keep growing. Given the observation time of GNSS stations, the observation data collected in this paper is divided into 6 phases: 2001–2004, 2004–2007, 2007–2009, 2009–2011, 2011–2013, and 2013–2016. The velocity field is first obtained based on the ITRF2008 reference frame, and then the optimal fitting Euler vector between the ITRF and the Eurasian framework is solved by using the 15 stable reference stations in the Eurasian plate. Finally, the horizontal velocity field under the ITRF can be reduced to the Eurasian framework. According to statistics, the average daily errors in the N and E directions of the stations are 2.0 mm and 2.3 mm, and the annual errors are 1.3 mm/a.

A $0.5^\circ \times 0.5^\circ$ grid interpolation (Fig. 2) is constructed by using the block model. The GNSS station velocity, which is inverted by the block model, is used to be compared to the measured velocity, which is to analyze the accuracy of the grid velocity field model through the unbiased and effective estimated of differences [17, 18]. The analysis results show that the unbiased estimate of the model is 0.36 mm/a, and its the efficient estimate is 2.86 mm/a, which indicates that the grid velocity field model established in this paper can describe the actual crustal movement and strain more accurately, and can meet the requirements of large-scale block movement and strain research.

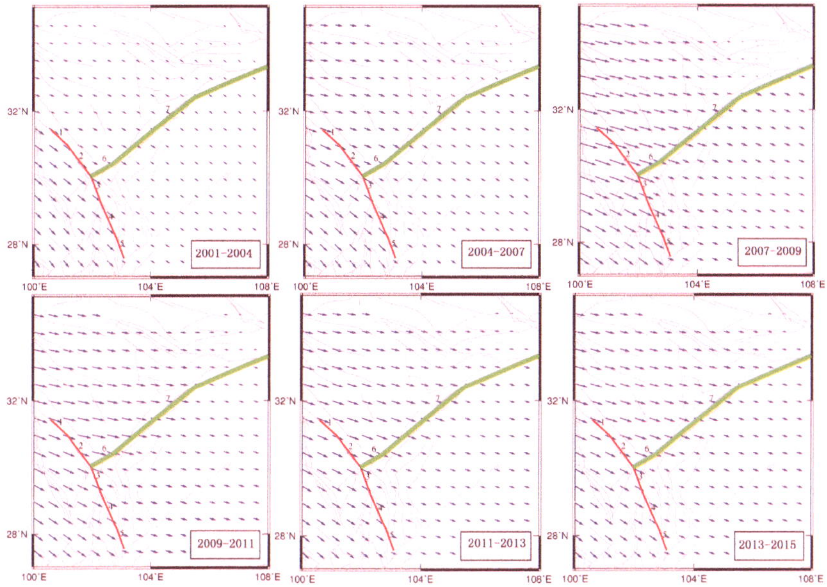

Fig. 2. Staged grid velocity field of the Longmenshan fault and adjacent blocks

The grid velocity field can reflect the overall characteristics of the crustal movement in the region. According to the grid velocity field analysis under several periods of the Eurasian framework, it can be seen that the block around the Longmenshan fault has a decreasing velocity field from north to south. The movement direction gradually changed from the weak movement trend of east-south movement in the far north to the south-east movement in the south most. It shows that the south China plate blocks the crustal movement from west to east in the east. The east movement is gradually attenuated, turning to the overall characteristics of south movement.

According to the velocity field differences in each period, the magnitude of the velocity field movement in the region during the period 2004–2007 is significantly higher than that during 2001–2004, and the velocity has increased from the original 9.6 mm/a to 15.3 mm/a in the northwestern of the Longmenshan fault. The difference in the size of the velocity field shows that the crustal movement increased horizontally during 2004–2007 in the study region. The reason why this crustal movement became active may be related to the 2004 Sumatra earthquake in Indonesia (depth 46 km, magnitude 8.7, longitude 97.1, latitude 2.4), since the earthquake stimulated crustal activity in the region. The overall velocity magnitude of the velocity field of 2007–2009 is also significantly larger than that during 2004–2007. For the period 2007–2009, the horizontal movement of the northwestern crust of the Longmenshan fault is about 24 mm/a, and the horizontal crustal movement speed in the southeast is about 8.1 mm/a. Compared with the period 2004–2007, the active crustal movement during 2007–2009 was related to the 2008 Wenchuan earthquake which occurred in the Longmenshan fault zone, since the earthquake made the crustal movement more active around the Longmen fault. Analysis of the velocity field during 2009–2011 shows that the velocity field during this period is smaller than that during 2007–2009, but larger than the velocity field during 2001–2004 and 2004–2007. The reason for the movement characteristics may be due to the fact that the Wenchuan Earthquakes continued to adjust the crustal movement in the region. Compared with the period 2009–2011, the speed during 2011–2013 increased slightly, especially in the northwest of the Longmenshan fault, and the crustal movement is not apparent in the southeast of the Longmen fault. This phenomenon may be related to the 2010 Yushu earthquake in the northwest of the Longmenshan fault (depth 14 km, magnitude 7.1, longitude 98.1, and latitude 31.1). During the periods of 2013–2015 and 2011–2013, the overall movement trend did not change much even though some regions were adjusted, which is mainly concentrated at the junction of Longmenshan fault and Xianshuihe-Xiaojiang fault.

4.2 Dynamic Characteristics of Regional Crustal Strain Rate Field

The crustal motion rate obtained by GNSS direct observation varies with the reference frame and is not unique. However, the crustal strain tensor calculated based on the velocity field has nothing to do with the selection of the reference frame, but it can reflect the crustal deformation characteristics and Main force direction in more detail. In this paper, based on the velocity field of the continuous distribution in the Eurasian framework, the least-squares configuration method is used through the partial derivative relationship between strain rate and displacement velocity to calculate the strain rate field during 6 phase: 2001–2004, 2004–2007, 2007–2009, 2009–2011, 2011–2013 and 2013–2016, as shown in Fig. 3.

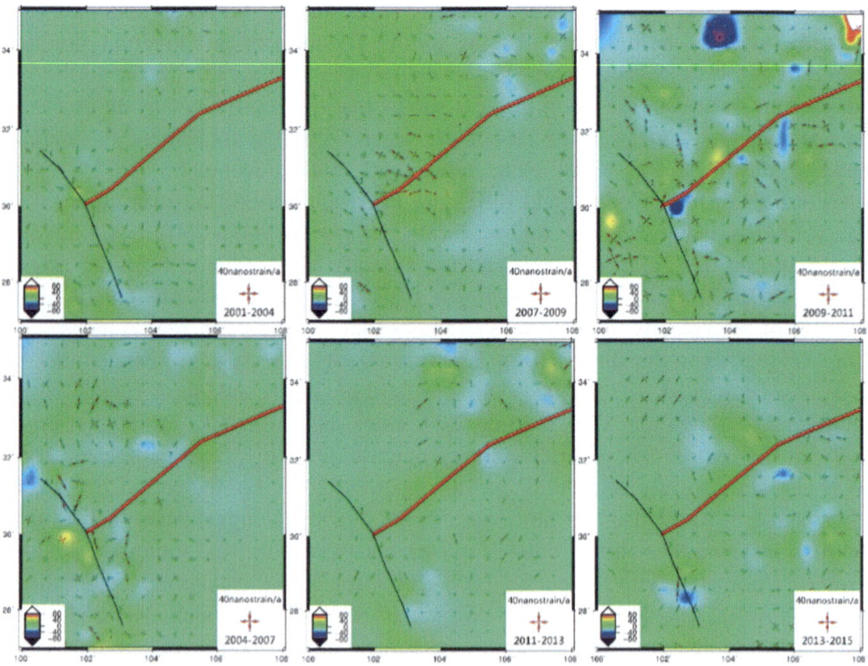

Fig. 3. Staged crustal strain rate field around the Longmenshan fault

Figure 3 reflects the continuously changing principal strain rate field and area expansion rate of the crust around the Longmenshan fault. As can be seen, the main strain rate field and the area expansion rate show different forms with time. In general, the crustal strain rate field showed a trend of strong south and weak north during 2001–2004 and 2009–2011, and the trend of strain rate field became weak south and strong north in other periods. And since 2011, the inner crustal strain rate field has become smaller gradually in the entire region.

From the analysis of the crustal surface compression and tension characteristics, it can be known that the crustal is mainly dominated by tension around the Longmenshan fault, and is accompanied by a small amount of compression deformation. According to the analysis of the crustal principal strain rate around the Longmenshan fault, from 2001 to 2004, the crustal strain rate showed strong south and weak north, and the region with significant strain rate was mainly distributed in the southwest of the Longmenshan fault. During 2004–2007, the trend of crustal strain rate activity in the area gradually evolved to the north, and the strain gradually increased, which may be affected by the adjustment of the 2004 Sumatra earthquake. However, from 2004 to 2007, the crust associated with a small strain rate and severe crustal compression mainly became strongly compressed on both sides of the Longmenshan fault, which created specific conditions for the 2008 Wenchuan earthquake. Since the period of 2007–2009 spanned the Wenchuan earthquake, the high strain area during this time period was mainly distributed in the crust near the source. And during 2009–2011, the high-value area of strain rate gradually expanded typically. The maximum flat maximum principal strain rate is about 55 nanostrian/a, which indicates that the Wenchuan earthquake has a more significant impact on the adjustment of crustal deformation in the entire study region, and increased the degree of crustal activity in the region. For 2011, the strain rate is gradually decreasing, and the typical maximum principal strain rate is gradually reduced from the original 50–60 nanostrian/a to 30–40 nanostrian/a, and finally reduced to about 15 nanostrian/a before the Lushan earthquake, such as 2009–2011 and 2011–2013. However, after the Lushan earthquake in 2013, crustal activity in the study region became slightly active, which mainly distributed in the northwest and southeast of the Longmenshan fault, but the degree of crustal deformation was small. Its strain rate values were about 20 nanostrian/a and 50 nanostrian/a. According to the changes in the strain field between 2011–2013 and 2013–2015, we can see that the crustal adjustment effect of the Wenchuan earthquake has gradually receded. Although earthquakes such as Lushan and Yushu occurred in this region during this period, these earthquakes have a far-reaching effect on the crust but not as evident as the Wenchuan earthquake.

In Fig. 3, the continuous background color mode is the surface expansion strain rate. The characteristics of a slight compression and tension present in the entire study region. The typical magnitude is between -40 to 40 nanostrian/a, and the change was not significant over a period. However, the crustal surface expansion value changed significantly in some areas. The period of large negative crustal surface expansion is during 2004–2007, 2009–2011, and 2013–2015, which may be due to the Sumatra earthquake and the Wenchuan earthquake, Lijiang earthquake in Yunnan (2012, depth 11 km, magnitude 5.7) and Shaotong earthquake in Yunnan (2012, depth 10 km, magnitude 5.7). And the Wenchuan earthquake is believed to have the most severe impact. The positive and negative surface expansion values can reflect the change in crustal thickness. Thus the positive and negative values of the surface expansion value and the distribution changes in the region have been analyzed, and the analysis shows that the crust is undergoing thinning and thickening in the long-term and it is only happening in a small number of areas.

4.3 Analysis of Crustal Strain Characteristics Before and After Strong Earthquakes

Based on the staging grid velocity field motion characteristics and strain rate field variation characteristics, it can be seen that the 2008 Wenchuan earthquake has a significant effect on crustal movement in the study region. Therefore, a detailed analysis of the characteristics of the secondary blocks near the source before and after the Wenchuan earthquake is of great significance for understanding the near-source source crust plate adjustment of the Wenchuan earthquake. Based on Guo Xiaohu [8] and others' subdivision scheme [8], this paper studies the strain characteristics of Chengdu, Longmen, and Shimian blocks near the Longmenshan fault, as shown in Fig. 4.

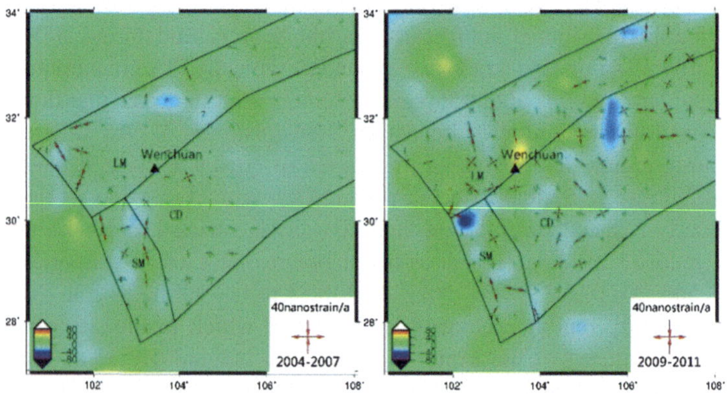

Fig. 4. Changes in crustal surface expansion and strain rate fields of Chengdu, Asbestos and Longmen blocks before and after the earthquake

"Wenchuan" in the picture is the location where the Wenchuan earthquake occurred. It is located in the Longmenshan fault zone at the junction of the Chengdu block and the Longmen block. From the analysis of the figure, it can be concluded that the values of the surface expansion and the strain rate of the three blocks before the earthquake were tiny, and the typical values were between 10 nanostrain/a and 25 nanostrain/a. These tiny values indicate that the crust was in a stable state at this time. Since the compression and tension value is very small, further judgment and analysis can be performed by considering both pressure and tension on the crust. After the earthquake, the strain rate field of the Longmen block and the Chengdu block became very active, but the Shimian block showed relatively small change. The reason may be that fewer stations located in the block, leading to reduced accuracy of the required strain field. The Longmen block mainly exhibits intense shear deformation, which can be identified by a typical value of about 30–50 nanostrain/a. The Chengdu block shows the strain characteristics of simultaneous tensile and shear strain, since its maximum value can reach 55 nanostrain/a. But the strain in the southeast of the Chengdu block is small, which may be related to the blocking effect of the stable South China block. Therefore, it can be concluded that the strain field on the crust surface was small before the earthquake. However after the earthquake, the strain rate field and the surface expansion of the crust changed significantly, and the crust surface stress was reactivated, which may result from the fact that the earthquake redistributes the stress field around the crust.

5 Conclusion

In this paper, multiple periods of GNSS data were used to establish 6-phase velocity field models from 2001–2004, 2004–2007, 2007–2009, 2009–2011, 2011–2013 and 2013–2016, and then the strain rate field is calculated based on the velocity field model, which can more intuitively reflect the spatiotemporal continuous dynamic changes of crustal deformation around the Longmenshan fault.

Based on multi-period GNSS velocity field analysis, the results show that inherited movements around the Longmenshan fault dominate the horizontal movement of the crustal. And the movement trend is consistent and maintained near-horizontal movement in the north. The south-east movement was shown in the south, and the magnitude of movement was active in the west and weak in the east. Moreover, the change in the magnitude of the velocity field corresponds to the occurrence of an earthquake. Specifically, the crustal movement speed increases after the earthquake.

According to the analysis of the multi-period main strain rate field and the characteristics of area expansion rate change in the region, it showed a trend of strong south and weak north during 2001–2004 and 2009–2011, and the trend of strain rate field became weak south and strong north in other periods. And since 2011, the inner crustal strain rate field has become smaller gradually in the entire region. The Wenchuan earthquake and the Lushan earthquake that occurred on the fault had a more significant impact on the adjustment of crustal deformation in the whole study region, and both increased the degree of crustal activity in the region. However, according to the magnitude of the main strain rate field, it can be seen that the Wenchuan earthquake has a far more significant effect on the crustal adjustment than the Lushan earthquake in the region. The period of large negative crustal surface expansion is during 2004–2007, 2009–2011, and 2013–2015 in the region. According to the analysis of the positive and negative values of the surface expansion and the distribution rule, it is known that the crust is undergoing thinning and thickening in a long-term and it is only happening in a small number of areas.

References

1. Dang, Y.M., Yang, Q., Xue, S.Q., Yue, C.Y., Liu, Z.Q.: GNSS monitoring dynamic variation characteristics of crustal deformation in the Sichuan-Yunnan region. J. Geodesy Geodyn. **39**(02), 111–116+177 (2019)
2. Yue, C.Y., Dang, Y.M., Yang, Q., Li, C.H.: Study of active trend of main fault in Sichuan-Yunnan region before and after Wenchuan earthquake. Sci. Surv. Mapp. **41**(12), 101–105 (2016)
3. Fang, Y., Jiang, Z.S., et al.: Analysis of fault deformation model and strain accumulation in Sichuan-Yunnan area. J. Geodesy Geodyn. **26**(2), 48–52 (2006)
4. Yue, C.Y., et al.: Crustal deformation characteristics of Sichuan-Yunnan region in China on the constraint of multi-periods of GPS velocity fields. Adv. Space Res. **61**(8), 2180–2189 (2018). https://doi.org/10.1016/j.asr.2018.01.039
5. Tang, W.Q., Chen, Z.L., Liu, Y.P., et al.: Present-day tectonics activity in the intersection area of the Xianshuihe fault and Longmenshan fault on the eastern margin of the Qinghai-Tibet Plateau. Geol. Bull. China **24**(12), 1169–1172 (2005)
6. Qiao, X.J., Wang, Q., Du, R.L.: Characteristics of current crustal deformation of active blocks in the Sichuan-Yunnan region. Chin. J. Geophys. (in Chinese) **47**(5), 805–811 (2004)
7. Chen, Z.L., Zhang, Y.Y., Shen, F., et al.: GPS monitoring for crustal movement in Southwest China. Chin. Sci. Bull. **44**(8), 851–854 (1999)
8. Guo, X.H., Wei, D.P., Zhang, K.L.: GPS-constrained estimate method of present-day slip rate along major faults of Sichuan-Yunnan region in China. J. Graduate Univ. Chin. Acad. Sci. **30**(1), 74–82 (2013)
9. Yue, C.Y., Dang, Y.M., Yang, Q., Liu, Z.Q., Zou, B.: Analysis of the current activity in Sichuan-Yunnan region and its sub blocks of main faults. J. Geodesy Geodyn. **37**(02), 176–181 (2017)
10. Wang, Y.Z., Wang, E.N., Shen, Z.K., et al.: GPS-constrained Inversion of present-day slip rates along major faults of the Sichuan-Yunnan region in China. Sci. China (Ser. D): Geosci. **38**(5), 582–597 (2008)
11. Dang, Y.M., Yang, Q., Liang, S.M., et al.: Block movement and strain characteristics effected by earthquake in Sichuan-Yunnan region. Acta Geodaetica et Cartographica Sin. **47** (05), 559–566 (2018)
12. Liu, X., Sun, D.Y., Ma, J., et al.: Present-day deformation and stress state of Longmenshan fault from GPS results comparative research on active faults in Sichuan-Yunnan region. Chin. J. Geophys. (in Chinese) **57**(4), 1092–1100 (2014)
13. Feng, G.C., Hetland, E.A., Ding, X.L., et al.: Coseismic fault slip of the 2008 Mw7.9 Wenchuan earthquake estimated from InSAR and GPS measurements. Geophys. Res. Lett. **37**(1), Lol302 (2010)
14. Zhang, P.Z., Xu, X.W., Wen, X.Z., et al.: Slip rates and recurrence intervals of the Longmen Shan active fault zone and tectonic implications for the mechanism of the May 12 Wenchuan earthquake, 2008, Sichuan. China. Chin. J. Geophys. **51**(4), 1066–1073 (2008)
15. Xia, T.T.: 3D modeling and gravity inversion of crustal structure in Longmenshan fault zone. Inst. Crustal Dyn. China Earthq. Adm. **06**, 1–7 (2019)
16. Qu, W., Zhang, Q., Zhang, D.J., et al.: Analysis of deformation-strain characteristics of the northeast margin of Qinghai-Tibet block using the RELSM model based on the GPS velocity field. Prog. Geophys. (in Chinese) **24**(1), 67–74 (2009)
17. Ding, K.H., Xu, C.J., Zou, R., Hu, Y.J.: Crustal movement and strain model of active blocks analyzed by GPS in Sichuan-Yunnan region. Geomat. Inf. Sci. Wuhan Univ. **38**(7), 822–827 (2013)
18. Xu, C.J., Wen, Y.M.: Identification of models for crustal movement and strain of blocks. J. Geodesy Geodyn. **23**(3), 50–55 (2003)

Design and Research of Missile-Borne High Dynamic Satellite Navigation Device

Feiping Lu and Wen Xue(✉)

Beidou Engineering R&D Center, Shanghai Electric Control Research Institute, Shanghai 200092, China
xuewen1217@sina.cn

Abstract. A design scheme of the missile-borne high dynamic satellite navigation device is given for the requirements of the trajectory correction projectile for the measurement system, analyzing the dynamic stress characteristics of the trajectory correction projectile. In order to solve the problem of the signal's lock-lose of the trajectory ascending section, the strategy of the carrier tracking loop is improved. Finally, the effectiveness of the improved strategy is verified by the simulation scenario test.

Keywords: High dynamics · Carrier tracking · Dynamic stress · Missile-borne

1 Introduction

With the development of navigation, communication, optoelectronics, microwave, microelectronics and information processing technologies, modern war has come into the high-tech war based on informatization. Informatization, intelligence and unmanned have become the trend of weapon equipment development. Guns play a key role in suppressing enemy firepower and killing the enemy's living force in the war. Modern war has an increasing demand of the accuracy of gun strikes. As a new projectile which is different from the traditional projectile and missiles, meanwhile, the smart projectile has the characteristics of high strike accuracy, low combat risk, less incidental damage, and low cost which can be installed in large numbers. At present, the smart projectile has become an important factor that dominates the war and determines the outcome of the war. Also it is in an advantageous position in the international arms trade competition.

The trajectory correction projectile is a typical representative of smart projectile and can be divided into one-dimensional trajectory correction projectile and two-dimensional trajectory correction projectile due to different correction methods. A high-reliability, high-precision and low-cost measurement system is required for measuring real-time elements during projectile flight. A high dynamic satellite navigation device can be designed with a dedicated integrated chip, with low power consumption, small size and low cost. It can achieve positioning accuracy better than 10 m in all weather and all day time, and is widely used in trajectory correction projectile.

Compared with the common environment, the missile-borne environment has two characteristics of high dynamics and high spin. The navigation signal has a larger

J. Sun et al. (Eds.): CSNC 2020, LNEE 650, pp. 251–261, 2020.
https://doi.org/10.1007/978-981-15-3707-3_25

Doppler frequency shift and Doppler frequency change rate. It requires a special design aiming at the high dynamic tracking loop of the missile-borne to meet the requirement of full positioning during the course of the projectile. Literature [1] analyzed the application requirements of the rotary guided projectile, modeled and decomposed the flat and rotary motions during the carrier rotation process, and the changes of Doppler, phase and amplitude during the carrier spin process. Literature [2] proposed a method of INS-assisted receiver code loop for high dynamic conditions, which provided compensation for the code loop. After verification by the software receiver, this method could improve the receiver's tracking performance for the chirped dynamic signal. Literature [3] proposed a method for measuring the missile's attitude by using a magnetic sensor to compensate for the carrier Doppler frequency shift and phase offset caused by the rotation in view of the high-speed rotation characteristics of the missile-borne receiver.

A missile-borne high dynamic satellite navigation device is designed, using a combination of actual measurement and theoretical analysis to analyze the dynamic stress characteristics of the missile-borne environment. The carrier tracking loop of the first 30 s of the projectile is a third-order phase locked loop (PLL) which is aided by a second-order frequency locked loop (FLL). It switches to third-order PLL after 30 s. The simulation scenario test proves that the strategy can meet the requirements of global positioning.

2 Missile-Borne High Dynamic Satellite Navigation Device Solution

Aiming at the trajectory correction projectile, a missile-borne high dynamic satellite navigation device is designed to meet the requirements of a real-time measurement system which has the characteristics of low-power, miniaturization and low-cost trajectory. The device is composed of a high dynamic receiver and two sidewall antennas.

2.1 High Dynamic Receiver

The block diagram of the high dynamic receiver is shown in Fig. 1. Two GPS L1 frequency point antennas are connected to the receiver, signal combining in the receiver, and via SAW filtered into the RF chip. The RF chip outputs analog IF to the baseband chip after mixing, IF filtering and amplification. After sampling by the built-in 12-bit high-speed ADC in the baseband chip, the signal capture tracking and PVT calculation are completed, the external interaction is performed through the RS232 serial port, and the 1 PPS time signal is output. The RF chip uses XN112-1 and the baseband chip uses BP2015. The core chips are all special integrated chips, the number of chips is greatly reduced, the diameter of the receiver is less than 28 mm, and the power consumption is less than 1 W, which can meet the requirements of low power consumption, miniaturization and low cost.

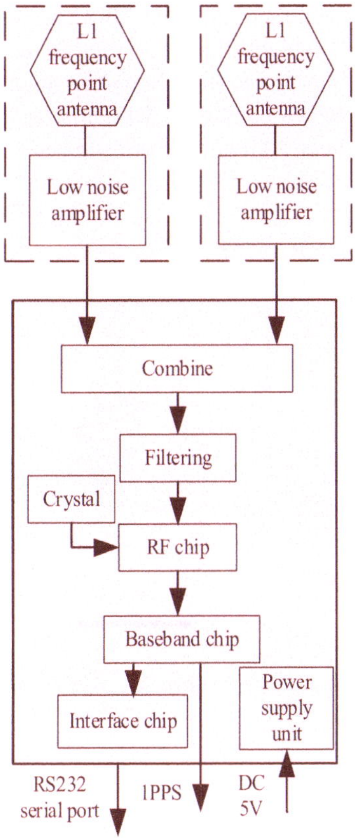

Fig. 1. High dynamic receiver block diagram

The receiver and antennas are compactly mounted on the head of the projectile. The antennas are very close to the receiver. Two side-wall microstrip antennas are used to get an omnidirectional pattern. The electromagnetic interference radiated by the receiver itself is received by the antennas, thereby affecting the sensitivity of the satellite navigation device. In order to improve the electromagnetic compatibility of the system, a filter circuit combining EMI capacitors and magnetic beads is designed on the serial port and the power supply to reduce the in-band electromagnetic interference generated by the serial port and the power supply. In addition, the PCB is designed as a 10-layer board, and each layer of the signal has a complete reference ground plane. The ground plane adopts a unified ground scheme, and the analog area and the digital area are virtually divided to avoid cross-zones.

2.2 Two Side-Wall Antennas

The trajectory correction projectile has the characteristic of spins. Two side-wall microstrip antennas are used to solve the problem of satellite signal reception due to the spin. The Fig. 2 shows that the two antennas are mounted on the conical surface of the warhead, and are distributed symmetrically at 180°. The combined pattern of the two antennas is shown in Fig. 3, which allows continuous reception of satellite signals during the entire rotation of the projectile. Microstrip antennas are active antennas to prevent additional insertion losses caused by RF cables. The passive part is made of composite material and can bear 20,000 g of transmission impact.

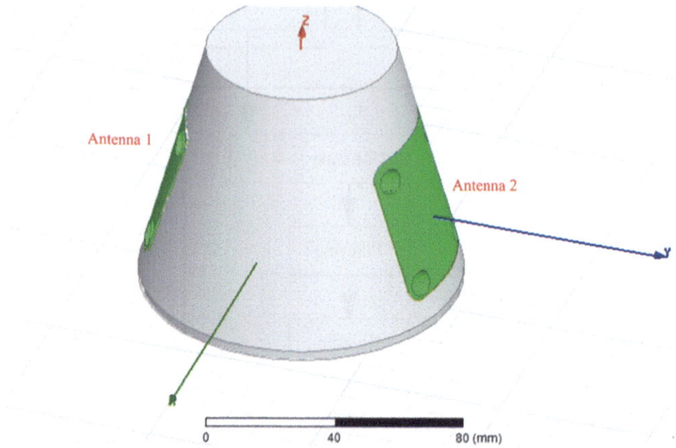

Fig. 2. Installation diagram of double-lobed sidewall antenna

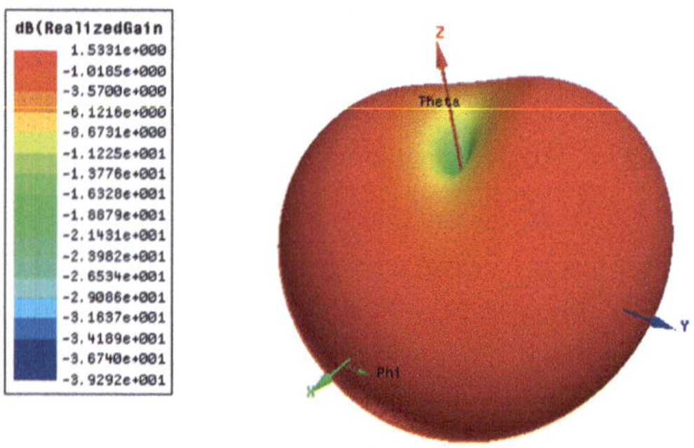

Fig. 3. Dual antenna synthetic pattern

3 Dynamic Stress of Missile-Borne Environment

Compared with ordinary wearable and vehicle-mounted environments, the missile-borne environment has two characteristics of high dynamics and high spin, which will cause a large Doppler change of the signal and have a certain impact on the normal signal reception. It is difficult to keep a stable acquisition and tracking of the signal [4]. Harsh requirements on the performance of missile-borne satellite navigation receivers are put forward.

3.1 High Dynamic

High dynamic refers to the extremely fast movement of the carrier, which is accompanied by large acceleration and jerk. This dynamic stress will produce a large Doppler effect. The Doppler effect is the phenomenon that the relative frequency between the satellite and the receiver causes the frequency of the received signal to shift from the frequency of the transmitted signal. The Doppler frequency shift produced by the Doppler effect is an important observational measure of the satellite receiver. The Doppler frequency shift and its change rate are related to the search range and the frequency accuracy of the tracking loop during the acquisition process.

The mathematical expressions of the signal's Doppler frequency shift and its change rate are [5]:

$$f_d = \frac{v_d f_L}{c} \tag{1}$$

$$\delta f_d = \frac{dv_d}{dt}\frac{f_L}{c} \tag{2}$$

Among them, f_d, δf_d are the Doppler frequency shift and Doppler frequency change rate respectively; v_d is the radial relative speed of the satellite and the receiver; f_L represents the carrier frequency; c is the signal propagation speed, which means the speed of light.

Considering only the impact of satellite motion, taking GPS as an example, the maximum speed of the satellite relative to the ground is about 929 m/s [6], the acceleration is about 0.178 m/s^2, and the C/A code modulates the L1 frequency (f_{L1} = 1575.42 MHz) At this time, the maximum carrier Doppler frequency shift caused by satellite motion and its change rate are:

$$f_d = \frac{v_d f_{L1}}{c} = \frac{929 \times 1575.42 \times 10^6}{3 \times 10^8} = 4.9\,\text{kHz} \tag{3}$$

$$\delta f_d = \frac{dv_d}{dt}\frac{f_L}{c} = \frac{0.178 \times 1575.42 \times 10^6}{3 \times 10^8} = 0.936\,\text{Hz/s} \tag{4}$$

It can be seen that in the case of low-speed carrier movement, the maximum Doppler frequency shift range is within ±5 kHz, and the tracking program frequency accuracy of the tracking loop can meet the requirements within 1 Hz. The Doppler effect caused by satellite motion has a small effect to the tracking loop. If in a highly dynamic scenario, the effect of the Doppler effects caused by the carrier dynamic on reception must be considered. Therefore, the trajectory correction projectile of dynamic characteristic in the specific application environment should be in-depth analyzed. In this method, it could improve the dynamic adaptability of the conventional high dynamic receiver and avoid the reduction of tracking performance because receiver's dynamic adaptation range is too wide or too narrow. In order to analyze the dynamic stress of the trajectory correction projectile during flight, the trajectory of the projectile is tracked by the radar throughout, and the trajectory curve obtained by filtering the radar measurement error is shown in Fig. 4. The variation of flight speed, acceleration and jerk with flight time is shown in Fig. 5. As can be seen from Figs. 4 and 5, the projectile flight time is about 110 s and the flight distance is about 20.8 km, and the jerk of the whole process is small. The initial speed and acceleration are relatively large. The maximum speed is 730 m/s (the corresponding f_d is 3.8 kHz) and the maximum acceleration is 1.6 g (the corresponding δf_d is 82.3 Hz/s). As for the projectile flying, the speed and acceleration gradually decrease. The speed gradually increased after reaching the highest point. It can be seen that compared with low-speed motion carriers, the loop design is suitable for a wider range.

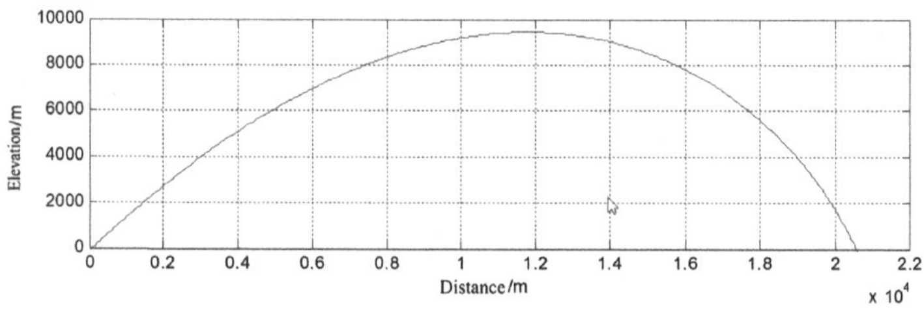

Fig. 4. Curve: measured trajectory

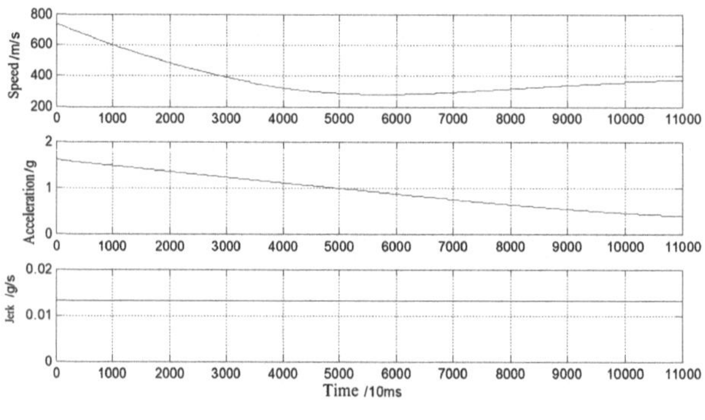

Fig. 5. Curve: trajectory speed, acceleration, jerk

3.2 High Spin

In addition to the high dynamic characteristics of the trajectory correction projectile, it also performs continuous high-speed rolling around the rotation axis during flight, which means the characteristics of high spin. High spin helps the projectile to be more stable during flight and ensures the projectile to run in a very stable state. The projectile is spun in the gun barrel, and is in a high spin state at the moment of emission. The rotating speed of different types of trajectory correction projectile is between several and hundreds per second, and then gradually decreases during all flight.

The phase center of two side-wall microstrip antennas and the geometric center of the projectile are not the same, which will also cause periodic deviations in the carrier frequency and phase of the satellite signal received by the receiver. The rotational angular velocity ω of the trajectory correction projectile during its flight can be decomposed into three axial velocity components ω_x, ω_y, and ω_z. In order to ensure the stability of the projectile during flight, there are mainly rotations along the center axis of the projectile, while the rotation components in the other two axis directions are very small. Although the rotation axis of the projectile and its own central axis may be deflected by the influence of air resistance and other factors, this effect is weak. It can be assumed that the rotation axis of the carrier coincides with its central axis in a short period of time [1]. In order to facilitate the calculation, the projectile coordinate system shown in Fig. 6 is established. The OX axis coincides with the center axis of the projectile and points to the forward direction of the projectile; the OY axis points to the composite phase center of the double-lobed antenna; The OZ axis and the other two axes form a right-handed rectangular coordinate system, L1 represents the direction of incident electromagnetic waves, L2 represents the projection of L1 on the OY-OZ rotation plane of the missile body; β represents the angle between L1 and projection L2, and α represents the L2 and OY Angle; point C represents the composite phase center of the two-lobed antenna (OC = R); point O represents the geometric center of the two-lobed antenna. Because the electromagnetic wave signals emitted by satellites meet the far-field conditions, they can be regarded as parallel signals on the missile body. The projection of OC on L1 is the distance d between the phase center and geometric center of the dual-lobe antenna and the satellite.

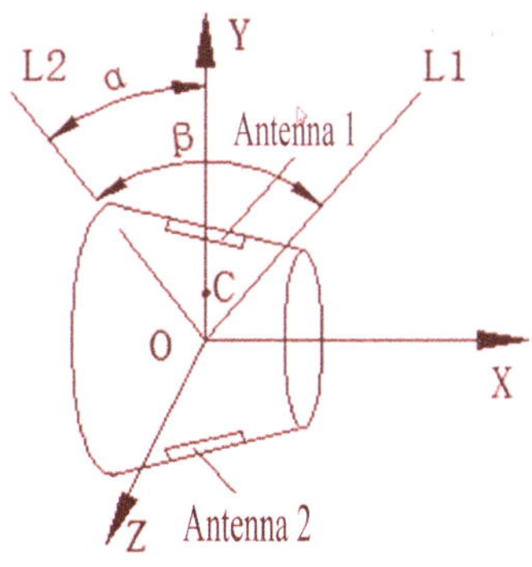

Fig. 6. Missile spin coordinate system

According to Fig. 6, the unit vector on L1 is $\overrightarrow{r} = (\sin\beta \quad \cos\beta\cos\alpha \quad \cos\beta\sin\alpha)$, Using the vector quantity product method, we can get the projection of $\overrightarrow{OC}$ on L1

$$d = R\cos\beta\cos\alpha,\ \alpha = 2\pi\omega_x t + \theta \tag{5}$$

In the formula, ω_x is the angular velocity of the body rotating around the X axis, and θ is the initial phase.

Regardless of the satellite's motion, the first, second, and third derivatives of Eq. (5) can be obtained to obtain the relative radial velocity, acceleration, and jerk of the antenna phase center and the satellite caused by the spin. Taking the rotation speed as 10 r/s and R = 5 mm, the maximum radial relative velocity, acceleration and jerk between the phase center of the two-lobed antenna and the satellite caused by spin are 0.3 m/s, 2 g, and 126.6 g/s, respectively. This dynamic stress is far greater than the characteristics of the translational state. The high spin causes the carrier Doppler of the signal received by the receiver to change more drastically. In engineering, in order to eliminate the effects of high spin, multiple circularly polarized microstrip antennas are usually conformed to the outer surface of a cylindrical or conical shell, so that the phase center of the loop antenna is stabilized on the body's rotation axis. The antenna still can not completely eliminate signal distortion, and the phase centers of different direction of arrival (DOA) do not coincide. Trajectory correction bombs are difficult to design and debug into an ideal conformal antenna due to cost, volume, and installation. The antenna phase center must be offset from the rotation axis. Therefore, for high-spin trajectory correction projectiles, high-spin is the key factor affecting signal acquisition and tracking.

4 Carrier Tracking Loop Strategy

The trajectory correction projectile has a short flight time. High dynamics and high spins cause the velocity, acceleration and jerk of the projectile to change drastically, which causes the received signal carrier Doppler to change drastically, furtherly affecting both the carrier tracking loop and the code tracking loop. The PN code frequency is much smaller than the carrier frequency, for GPS L1 frequency signals, PN code frequency is only 1/1540 of the carrier's which can be eliminated by carrier assistance. The dynamic stress change has a relatively small impact on the code tracking loop, so the design of the high dynamic receiver's tracking loop mainly focuses on the optimization of the carrier tracking loop [7]. Carrier tracking is to further precise carrier Doppler frequency estimates and code phase estimates which are obtained at the signal acquisition stage in tracking stage. GPS signal is refined from the carrier signal. Only the signals are tracked stably, the measured values of various GPS signals can be obtained, and the navigation message data can be obtained by processing the signals to complete the PVT.

The drastic carrier Doppler changes increase the difficulty of receiver signal carrier tracking [8]. The characteristics of the receiver carrier tracking loop are determined by the discriminator and filter of the carrier loop. The carrier tracking loop commonly uses PLL and FLL. PLL controls the loop through the phase error. In the coherent demodulation mode, the frequency phase of the locally generated signal must be the same as the frequency phase of the input signal carrier. FLL is a carrier tracking loop controlled by frequency error. In non-coherent demodulation mode, the frequency of the locally generated signal must be the same as the frequency of the input signal carrier. Compared with FLL, PLL has better measurement error accuracy, but has lower adaptability to dynamic stress.

Since the third-order PLL is not sensitive to acceleration stress, it can withstand a certain jerk stress, and has good tracking accuracy. Therefore, the receiver carrier tracking loop often uses a third-order PLL. The third-order PLL phase error is composed of thermal noise, dynamic stress, receiver vibration, and Allan deviation error [9], among which thermal noise and dynamic stress error are the main influencing factors. Thermal noise error can be expressed by the following formula:

$$\sigma_t = \frac{360}{2\pi}\sqrt{\frac{B_n}{C/N_0}\left(1 + \frac{1}{2TC/N_0}\right)} \tag{6}$$

Where σ_t is 1σ thermal noise; B_n is the carrier loop noise bandwidth, Hz; C/N_0 is the carrier-to-noise ratio; T is the pre-integration time. The third-order PLL dynamic stress error can be expressed by the following formula:

$$\theta_e = 0.4828\frac{dR^3/dt^3}{B_n^3} \tag{7}$$

Where dR^3/dt^3 is the relative jerk between the carrier and the satellite, deg/sec^3.

It can be known from Eqs. (6) and (7) that the phase error increases with the increase of the loop noise bandwidth, and the third-order PLL is sensitive to jerk. Trajectory correction projectiles have the characteristics of high spin in the ascending section after the projectile exits. The jerk stress caused by spin brings the total measurement error of the third-order PLL tracking loop to exceed the loop working threshold, which will cause lose-lock. In the initial sample test of a certain project, the receiver immediately filled the satellite ephemeris and should be positioned within 10 s, but quick positioning failed, and it was only positioned after more than 20 s. In order to adapt to the dynamic stress of the rising section, it is necessary to further increase the loop noise bandwidth, but at some time the thermal noise error increases, and the carrier loop measurement accuracy deteriorates with the widening of the noise bandwidth [10], which is not desirable.

In order to improve the dynamic performance and ensure the signal not to lose-lock in the rising section, the carrier tracking loop is improved by adopting a strategy. In the first 30 s of the rising section, a second-order FLL is used to assist a 3rd-order PLL. FLL has worse accuracy than PLL, but its dynamic stress adaptability is stronger than PLL. In order to tolerate the dynamic stress of the first 30 s and shorten the pre-detection integration time, the carrier tracking loop uses FLL as a discriminator. The noise bandwidth of the carrier loop filter is widened and the loop measurement accuracy is reduced. As the projectile speed decreases, it switches to the PLL after 30 s to ensure measurement accuracy.

5 Simulation Scene Test

The projectile is at a standstill before launch. At the moment of launch, the phase center disturbance caused by 10 r/s high spin and 5 mm phase center shift is added to the trajectory measured by radar as shown in Fig. 4. Simulated trajectory is generated and imported into the satellite navigation signal simulator, and the positioning performance of the high dynamic receiver is tested. Figure 7 shows the whole process positioning curve of the third-order PLL with 10 Hz loop noise bandwidth, the upward section cannot be continuously positioned. Figure 8 shows the whole process positioning curve of the improved strategy that second-order FLL with 3 Hz loop noise bandwidth assisting third-order PLL at the beginning, and adopting third-order PLL with 10 Hz loop noise bandwidth 30 s later, the upward section can be continuously positioned.

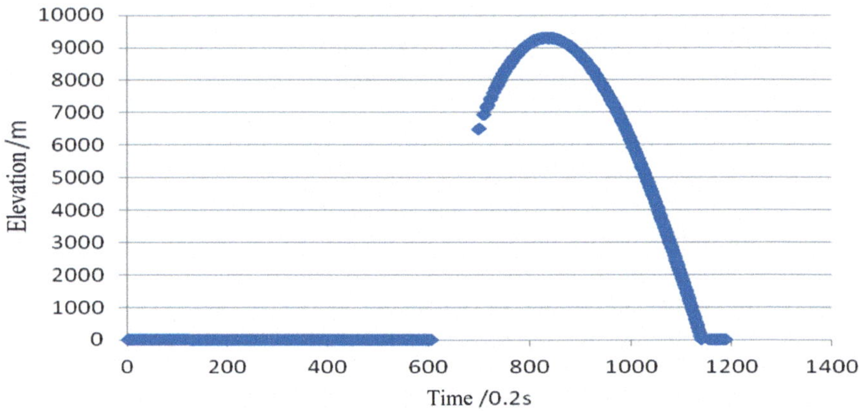

Fig. 7 Positioning results before improvement

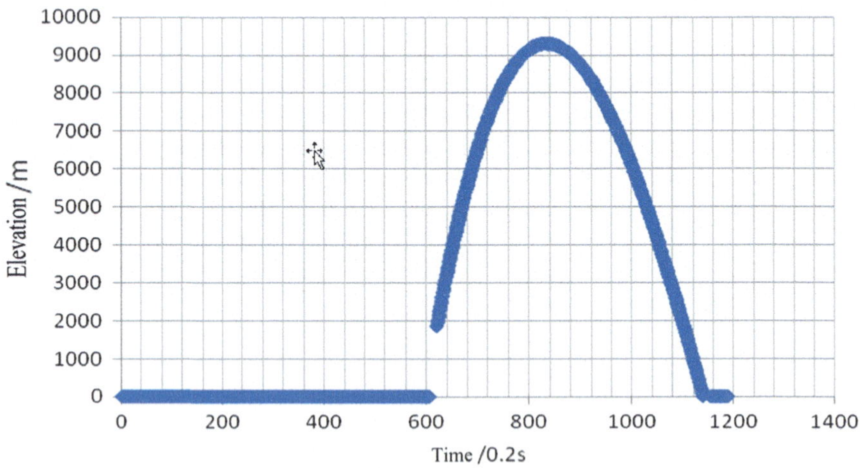

Fig. 8 Positioning results after improvement

In Fig. 8, there is a discontinuity in positioning for about 2 s at the moment of launch, which is caused by the super-acceleration of 20000 g at the moment of launch. The lose-lock time is short and does not affect guidance control.

6 Conclusion

Aiming at the requirement of low-power consumption, miniaturization and low cost real-time measurement system for trajectory data, the high dynamic satellite navigation device of trajectory correction projectile is designed, also the high dynamic and high spin characteristics of flight course are analyzed. The spin of the projectile causes the relative acceleration and jerk between the antenna phase center and the satellite to be much higher than its translation dynamic characteristics, which makes Doppler frequency shift change dramatically, resulting in carrier tracking loop loses lock at the upward section of flight course. In order to solve the problem of lose-lock tracking loop, a solution strategy is given, which uses second-order FLL to assist third-order PLL at the beginning of upward section, and then changes to third-order PLL, which not only maintains the positioning continuity in the upward section, but also ensures the subsequent positioning accuracy. Finally, the effectiveness of the strategy is verified by simulation experiments.

References

1. Qu, J., Xu, Z., Hui, O.: Modeling of high dynamic rotating GNSSS signal. Guidance Fuze **29** (1), 38–42 (2018)
2. Zhang, L., Chang, Q.: INS assisting GPS receiver code loop in high dynamic environment. Comput. Eng. Appl. **53**(S2), 234–238 (2018)
3. Zhu, J., Wu, P., Bo, Y.: Novel assisted carrier tracking loop based on magnetic sensors for the projectile receiver. J. Chin. Inert. Technol. **25**(5), 650–655 (2017)
4. Yang, R.: Study on flying trajectory and control system design for rolling guided projectile. Nanjing University of Science & Technology, Nanjing (2012)
5. Kaplan, E.D., Hegarty, C.J.: Understanding GPS: Principles and Applications, 2nd edn. Artech House Mobile Communications, Boston (2005)
6. Tui, J.: Fundamentals of Global Positioning System Receivers-A Software Approach. Wiley, Hoboken (2000)
7. Virball, V.G., Michalson, W.R., Levin, P.L., et al.: A GPS integrity channel based fault detection and exclusion algorithm using maximum solution separation. In: Position Location and Navigation Symposium, pp. 747–754 (1994)
8. Pan, X., Nie, Y.: Improved high dynamic GPS tracking algorithm for precise attack weapons. Acta Armamentarii **32**(1), 70–74 (2011)
9. Wang, W.: Research on tracking technique of satellite navigation signals in high dynamic applications. Harbin Institute of Technology, Harbin (2013)
10. Jin, T., Ren, J.X.: Stability analysis of GPS carrier tracking loops by phase margin approach. GPS Solut. **17**(3), 423–431 (2013)

Navigation and Location-Based Service

Detecting Community Structure of Urban Hotspot Regions

Rui Chen(✉), Mingjian Chen, Wanli Li, and Naikun Guo

Information Engineering University, Zhengzhou 450000, China
chenrui2408@outlook.com

Abstract. The travel behavior of residents is influenced by environment factor such as urban transportation system and administrative division. In turn, users equipped with navigation devices act as sensors detecting the environmental dynamics. The long-term accumulation of navigation big data contains massive valuable spatio-temporal information. We propose to detect the spatio-temporal distribution and community structure of urban hotspot regions from navigation big data. A framework including data preprocessing, hotspot region detection, urban spatial discretization, and community structure detection is designed in this work. Hotspot regions are detected by kernel density estimation and density-based clustering on origin-destination (OD) points of navigation trajectories. The hotspot regions are discretized into Voronoi polygon grids based on the spatial distribution of OD points. Finally, we analyze the complex network formed by hotspot region grids and employ Louvain algorithm to detect the community structure of hotspot region network. This framework is implemented on the taxi dataset of Chengdu. The experimental results reveal the spatio-temporal distribution and community structure of urban hotspot regions in different periods of weekdays and weekends. The urban hotspot regions and the community structure are influenced by inherent geographical environment and dynamically evolve with time. The spatio-temporal characteristics of urban hotspot regions are proved to be the result of coaction of environment and human activities. Findings of this work could provide decision-making support for transportation system optimization, city layout plan, and smart city construction etc.

Keywords: Navigation big data · Hotspot region · Clustering · Community detection

1 Introduction

With the rapid development of Global Navigation Satellite System (GNSS), e.g., the third generation of BeiDou System has already been providing global service, the application cost of GNSS gets lower and the application fields expand wider. By means of portable low-cost navigation devices, users are able to obtain their position and time information anytime and anywhere. Users equipped with navigation devices can be regarded as mobile sensors placed in the environment [1]. The long-term accumulation of navigation big data contains massive valuable spatio-temporal information. These data are important resources for the study of environment spatio-temporal dynamics [2].

J. Sun et al. (Eds.): CSNC 2020, LNEE 650, pp. 265–278, 2020.
https://doi.org/10.1007/978-981-15-3707-3_26

Plenty of work are dedicated to mining navigation big data produced in travel activities of residents to explore urban dynamics. E.g., the crowded zones and crowded roads can be detected by mining the float car data [3]. Trajectory data of taxi can be used to mine the spatio-temporal patterns of residents' travel behaviour and recognize different function zones of city [4]. The hotspot regions of city have both the static geographical and spatio-temporal dynamic characteristics, since they are influenced by both the static layout of city and human activities. In addition, studies have shown that there also exist associations among hotspot regions [5]. Investigation about the underlying characteristics of hotspot regions are extensively concerned, such as the study on relation network among hotspot regions. Researches based on complex network theory are able to reveal some valuable knowledge, e.g., separate out the function-independent zones in city [6] and analyse the interactive network of transportation system [7].

Community structure refers to sub-network composed of tightly related objects in complex network system [8]. Community structure of hotspot regions can be embodied in the travel patterns of residents. While there are intrinsic mechanisms such as the city layout which lead to this specific structure. Comprehending the knowledge about hotspot region community structure is helpful for transportation system optimization, reasonable city layout and the construction of smart city etc. In this work, we design a framework for detecting hotspot region community structure based on navigation big data by utilizing community detection algorithm. This framework is implemented on real taxi navigation data to transform raw data into valuable knowledge.

2 Method

Navigation data contains multidimensional information including time, position, and user status etc. Arranging each records in chronological order can make up navigation trajectories. The origin-destination (OD) points are the most important waypoints of navigation trajectories. There are massive residents' travel activities occurring in the regions where OD points gather and these regions can be regarded as hotspot regions. There exists tight association among hotspot regions which are connected by plenty of OD pairs. Further, community structure will be composed of these tightly related hotspot regions. Framework of the community structure detection designed in this work includes four parts: data preprocessing, hotspot region detection, spatial discretization, and community structure detection.

2.1 Data Preprocessing

The purpose of data preprocessing is obtaining useful data from raw data that contain outliers, missing data, and redundancy data. There are three main steps of data preprocessing: data selection, data cleaning, and OD points extraction.

(1) **Data selection.** According to the position and time attributes, select the data that are located in the study region and time range.
(2) **Data cleaning.** Firstly, set a threshold of normal value for every field based on domain knowledge to exclude outliers. Then handle the missing data and redundancy data by interpolating and dropping duplicates.
(3) **OD points extraction.** OD points are the start and end of trips, which equal to stay points detected by calculating the transition speed. In addition, they also can be recognized according to the switch of recorded mobility status.

2.2 Hotspot Region Detection

It is essential to obtain the spatial distribution of hotspot regions before analyzing the community structure. In this work, KDE (kernel density estimation) [9] and DBSCAN (density-based spatial clustering of applications with noise) [10] algorithm are employed to detect the distribution of hotspot regions.

KDE utilizes density function to calculate the density of unit area in the neighborhood of OD points. Then the distribution of discrete OD points is expressed in continuous space using raster value. The density function is defined as follows:

$$P = \frac{1}{r^2}\sum_{i=1}^{n}\frac{3(1-\frac{s_i^2}{r^2})^2}{\pi}, \tag{1}$$

where r denotes the search radius, s_i is the distance between the i-th OD point in the raster and the center of the raster, n denotes the number of OD points located in the raster. The raster value P forms a smooth surface and equals to 0 at the edge of neighborhood.

The most prominent advantage of DBSCAN algorithm is the ability to deal with irregular distributed OD points. There are two key parameters that need to be adjusted: the neighborhood radius *Eps* and the minimum number of points in the neighborhood of core objects *Minpts*. After DBSCAN clustering, densely distributed OD points will be recognized as clusters while the sparse points will be regarded as noise and be excluded.

2.3 Spatial Discretization

There are several widely used strategies of spatial discretization including even grids, road networks and referring to POIs (points of interest) etc. Given that the even grids often lead to sparse issue and the OD points are located on roads, these strategies are not suitable for the spatial discretization of hotspot regions.

In this work, we adopt a spatial discretization strategy using Voronoi polygons. This strategy is based on the density distribution of OD points obtained by DBSCAN clustering, as presented in Fig. 1. DBSCAN clustering will generate several separated clusters at the edge of city. These clusters are categorized into individual hotspot grids. Whereas in the downtown, OD points are densely distributed over a wide range. Thus, we implement K-means algorithm [11] on the cluster in the downtown to divided it into smaller clusters. Then, we obtain the center points of all the K-means clusters as the reference points to generate Voronoi polygon girds. The distance from OD points to the center point of the same grid is closer than to the center points of any other grids. This spatial discretization strategy is competent to recognize the hotspot regions where OD points gather and the generated grids match the distribution characteristics of OD points well.

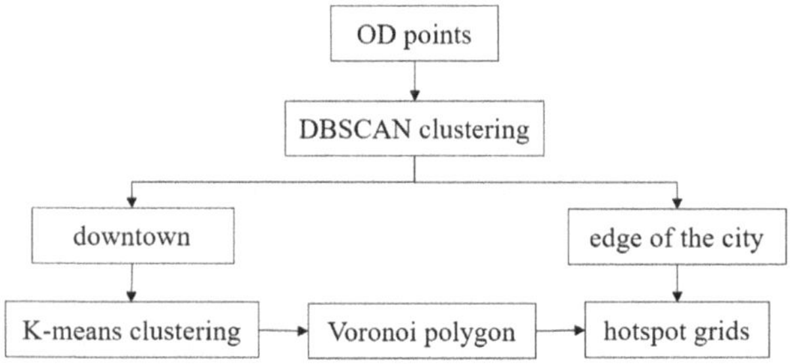

Fig. 1. Framework of urban spatial discretization strategy

2.4 Community Structure Detection

The formation of hotspot region community structure is related to both the objective city spatial layout and the travel activities of residents. Thus, the latent knowledge about community structure of hotspot region network can be mined from navigation big data generated in travel activities. We employ Louvain algorithm [12] based on graph theory to analyze the complex network that is composed of hotspot region grids.

The discretized hotspot region grids are regarded as nodes v. These nodes are connected by edges e due to the travel of residents back and forth among them. Ultimately, all the hotspot region grids form an undirected weight graph $G(v, e)$. The nodes are tightly associated when there are a large amount of trips connecting them. The association is quantified as the weight of edge e. The weight of edge e_{ij} between node v_i and v_j is defined as A_{ij}:

$$A_{ij} = N\{O \in v_i \cap D \in v_j\} + N\{O \in v_j \cap D \in v_i\}, \tag{2}$$

where O denotes the origin of a trip and D denotes the destination. N is the number of trips. Assume there are totally n hotspot region grids, we can construct a weight matrix $\mathbf{V}$ with the size of $n \times n$:

$$\mathbf{V} = \begin{bmatrix} A_{11} & \cdots & \cdots & A_{1j} & \cdots \\ \vdots & & & \vdots & \\ A_{i1} & \cdots & \cdots & A_{ij} & \cdots \\ \vdots & & & \vdots & \end{bmatrix}_{n \times n}. \tag{3}$$

According to the definition, $\mathbf{V}$ is a symmetric matrix where $A_{ij} = A_{ji}$. $\mathbf{V}$ is also named as the adjacent matrix of graph $G(v, e)$.

The key factor of Louvain algorithm is the calculation of modularity. In the process of graph dividing, the modularity is regarded as the objective function to be optimized. The community structure is obtained from complex network system when the optimization process convergence. The modularity is defined as:

$$Q = \frac{1}{2m}\sum\nolimits_{i,j}\left[A_{ij} - \frac{k_i k_j}{2m}\right]\delta(C_i, C_j), \tag{4}$$

where $m = \frac{1}{2}\sum_{i,j} A_{ij}$ is the sum of weight of the whole graph. $k_i = \sum_j A_{ij}$ denotes the sum of weight of edges connected to node v_i. C_i represents the community that v_i belongs to, δ is the Kronecker function:

$$\delta(C_i, C_j) = \begin{cases} 1, & C_i = C_j \\ 0, & C_i \neq C_j \end{cases}. \tag{5}$$

The process of community structure detection includes two stages: (1) Treat each node in the graph as an individual community. Take a node out from its original community and put it into another community. Perform this operation on every node in the graph and calculate the increment of modularity. Then put node v_i to the community which could lead to the largest increment of modularity. This operation will be iterated until the increment becomes 0. The increment of modularity are defined as follows when node v_i are put into community C:

$$\begin{aligned} \Delta Q = & \left[\frac{\sum in + 2k_{i,in}}{2m} - \left(\frac{\sum tot + k_i}{2m}\right)^2\right] \\ & - \left[\frac{\sum in}{2m} - \left(\frac{\sum tot}{2m}\right)^2 - \left(\frac{k_i}{2m}\right)^2\right] \end{aligned}, \tag{6}$$

where $\sum in$ represents the sum of weight of community C, $k_{i,in}$ represents the sum of weight of all the edges connecting node v_i and community C, $\sum tot$ denotes the sum of weight of all the edges between all other nodes in the graph and community C. (2) Treat the communities obtained in the first stage as nodes and construct a new graph. Then repeat the operation of the first stage until all the communities no longer merge.

3 Experimental Results

The trajectories of working taxis are able to express the characteristics of residents' travel behavior, since taxis provide customized service for passengers and the routes of taxis are not fixed. Thus, taxi navigation data are employed for the experiment in this work. The dataset covers historical records of over 13000 taxis in Chengdu in one week from 4 August 2014 to 10 August 2014. The dataset contains data of 5 weekdays and 2 rest days on weekends. The rectangular study region in Chengdu covers the districts of Jinniu, Qingyang, Wuhou, Jinjiang, Chenghua, and a part of Shuangliu.

Firstly, we perform data selection according to the study region and time range. Given that there are few people going out at the time after midnight, we exclude the data within the period from 0 am to 6 am. After that, we perform data cleaning and OD points extraction. OD points are extracted by referring to the switch of "passenger status".

3.1 Distribution of Hotspot Regions

The temporal distribution of average number of trips on weekdays and weekends is counted, as depicted in Fig. 2. There are obviously different temporal patterns between weekdays and weekends, respectively. There are three evident peak periods on weekdays, whereas on weekends, the number of trips is almost steady. There are the most trips during the period from 20 pm to 23 pm, indicating that the night activities are various for city residents on both weekdays and weekends.

According to the temporal patterns of number of trips, it is also essential to analyze the temporal characteristics of both hotspot region and its community structure. In the following study, we differentiate two modes of weekdays and weekends and divide one day into 4 periods: 7 am to 11 am, 11 am to 15 pm, 15 pm to 19 pm, and 19 pm to 23 pm.

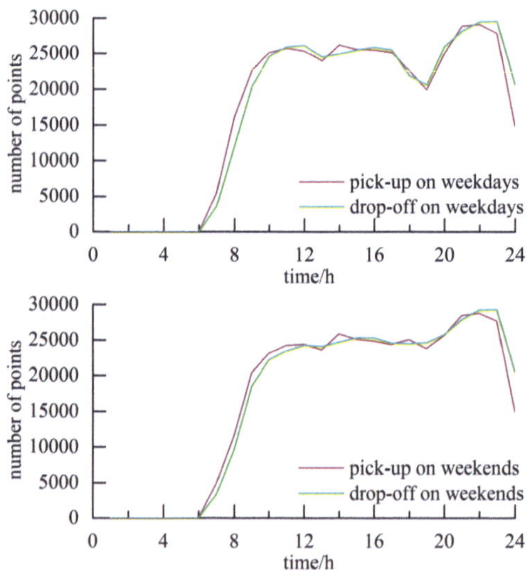

Fig. 2. Temporal distribution of average number of trips

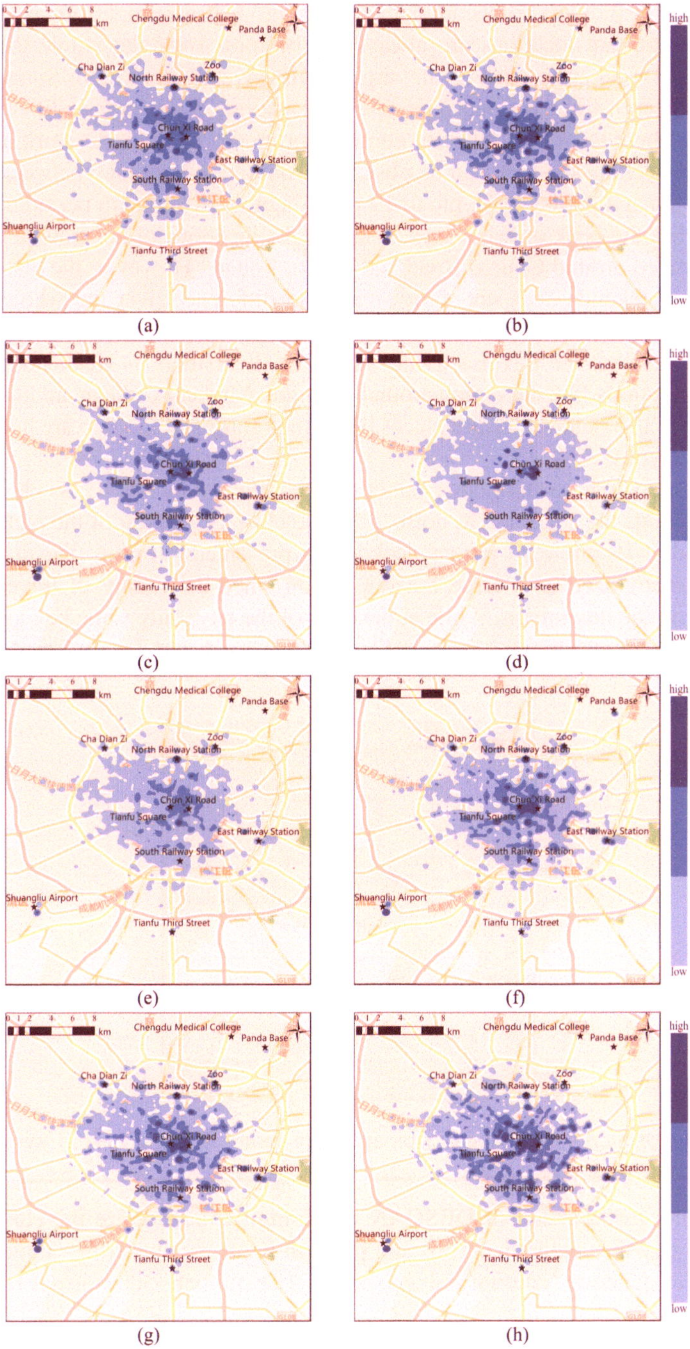

Fig. 3. Kernel density estimation of pick-up points (weekdays: (a) 7 am–11 am, (b) 11 am–15 pm, (c) 15 pm–19 pm, (d) 19 pm–23 pm, weekends: (e) 7 am–11 am, (f) 11 am–15 pm, (g) 15 pm–19 pm, (h) 19 pm–23 pm)

The KDE of pick-up points are depicted in Fig. 3. It is straightforward to observe that some important locations are always the hotspots over the whole day on both weekdays and weekends, such as the airport, railway stations, Tianfu Square, and the pedestrian street of Chunxi Road. The different locations of hotspots at different periods indicate that there exist two different commuting modes of going to work and getting off work on weekdays, which can be observed from the discrepancy between period of 7 am to 11 am and period of 15 pm to 19 pm. The travel volume in period of 19 pm to 23 pm reaches the peak on weekdays. Whereas the locations of hotspots in this period are concentrated around Chunxi Road and Tianfu Square. However, the distribution of hotspots on weekends does not show significant difference over the whole day. Different from weekdays, the hotspots in the period of 19 pm to 23 pm are scattered, which indicates that the travel activities are more diversified on weekends. Additionally, some locations like the region around Chengdu Medical College are hotspots only on weekdays. The outdoor leisure places like Panda Base and the Zoo are hotspots only during the daytime.

The locations where OD points gather can be obtained by DBSCAN clustering and sparse OD points are excluded as noise. E.g., the clustering result of the pick-up points in the period of 19 pm to 23 pm on weekdays is presented in Fig. 4. The downtown of the city is recognized as a large hotspot region since the plenty of travel activities concentrate in the downtown. While at the edge of the city, only some key locations are recognized as hotspot regions such as Shuangliu Airport and Tianfu Third Street.

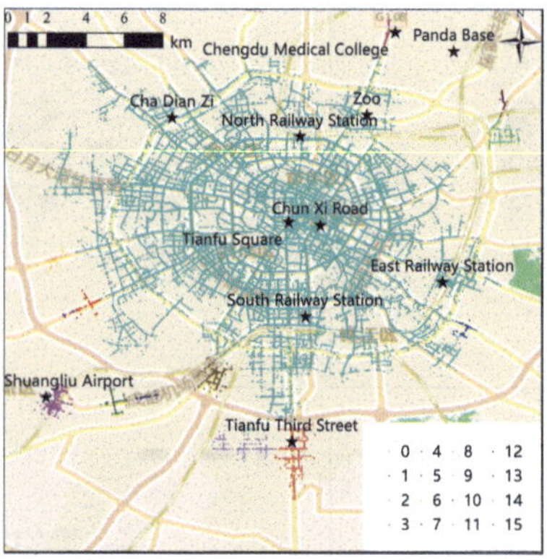

Fig. 4. DBSCAN clustering of pick-up points during 19 pm–23 pm of weekdays

According to the result of DBSCAN clustering, we categorize the clusters at the edge of the city into individual grids and perform K-means clustering on the cluster of downtown to generate Voronoi polygon grids. The ultimate 300 hotspot region girds are shown in Fig. 5.

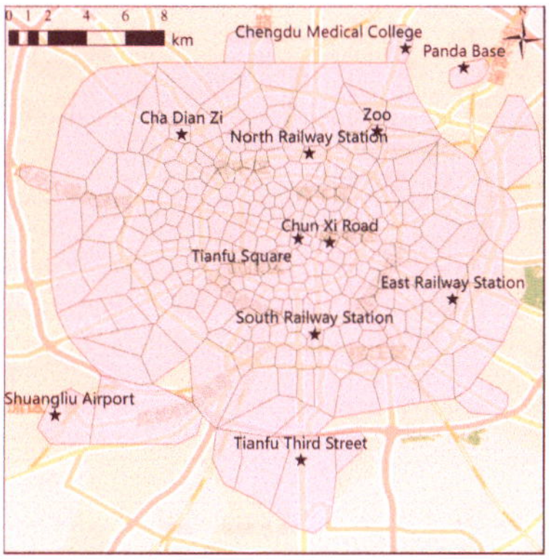

Fig. 5. Discretization of hotspot regions with 300 Voronoi polygon grids

3.2 Community Structure of Hotspot Regions

Hotspot region grids introduced above are regarded as the nodes of complex network. Node degree is a key factor for analyzing network, which quantifies the interaction of a node with other nodes. The degree can be categorized into the degree of entry and the degree of exit. In this paper, the degree of entry $d_{i,entry}$ and the degree of exit $d_{i,exit}$ of node v_i are presented by the number of pick-up points and drop-off points:

$$d_{i,entry} = \sum\nolimits_j N\{O \in v_j \cap D \in v_i\}, \tag{7}$$

$$d_{i,exit} = \sum\nolimits_j N\{O \in v_i \cap D \in v_j\}. \tag{8}$$

The source-sink feature of hotspot region nodes is indicated by $d_{i,entry}$ and $d_{i,exit}$:

$$NFR = \frac{d_{i,entry} - d_{i,exit}}{d_{i,entry} + d_{i,exit}}, \tag{9}$$

where NFR denotes the net flow ratio of node v_i. $NFR > 0$ indicates the sink feature of node v_i, on the contrary, $NFR < 0$ indicates the source feature.

As shown in Fig. 6, the hotspot regions where $NFR > 0$ and the number of inflow trips is larger than outflow trips are in warm red color. While the green color region is where $NFR < 0$ and the number of inflow trips is smaller than outflow trips. In the period of 7 am to 11 am on weekdays, the spatial distribution of source regions is dispersive and only a few hotspot regions show sink feature. The overall NFR of all the hotspot regions in the periods of 7 am to 11 am and 15 pm to 19 pm on weekdays form a temporal coherent state. The source-sink feature of plenty of hotspot regions are reversed between the two periods. We can observe obvious commuting pattern by comparing the overall NFR of these two periods on weekdays. However, it is also evident that the commuting effect is weaker on weekends. The NFR of regions, where transportation hubs are located, vary from positive to negative during a day. We can

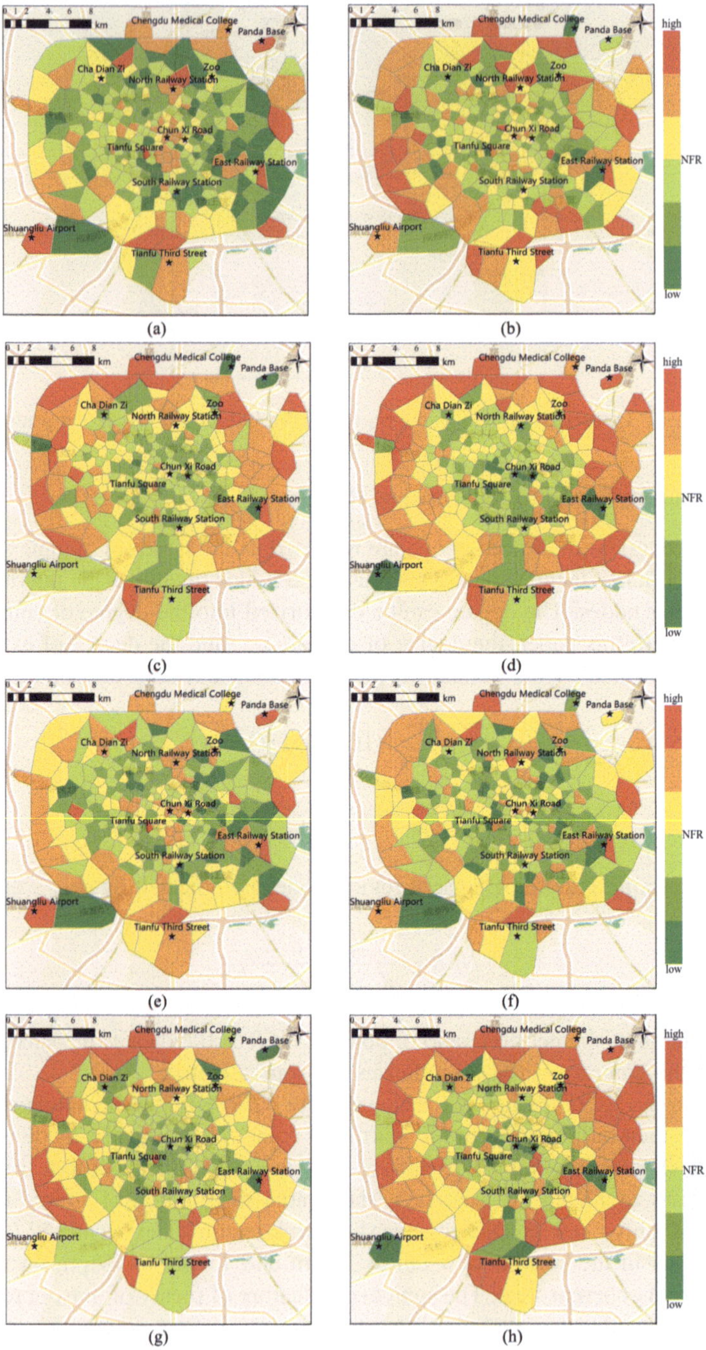

Fig. 6. Analysis of NFR of hotspot regions (weekdays: (a) 7 am–11 am, (b) 11 am–15 pm, (c) 15 pm–19 pm, (d) 19 pm–23 pm, weekends: (e) 7 am–11 am, (f) 11 am–15 pm, (g) 15 pm–19 pm, (h) 19 pm–23 pm)

deduce that residents tend to arrive transportation hubs and set out intercity travel at early time. In the period of 19 pm to 23 pm, the volume of intercity arrive is usually larger than that of depart. Additionally, the temporal evolution of *NFR* in some hotspot regions forms spatial coherent state. E.g., The *NFR* of the region where the Shuangliu Airport is located and its adjacent region show a negatively correlated regularity, as depicted in Fig. 7. This regularity indicates that there are plenty of trips back and forth between the spatial coherent regions all the time, leading to strong association between them.

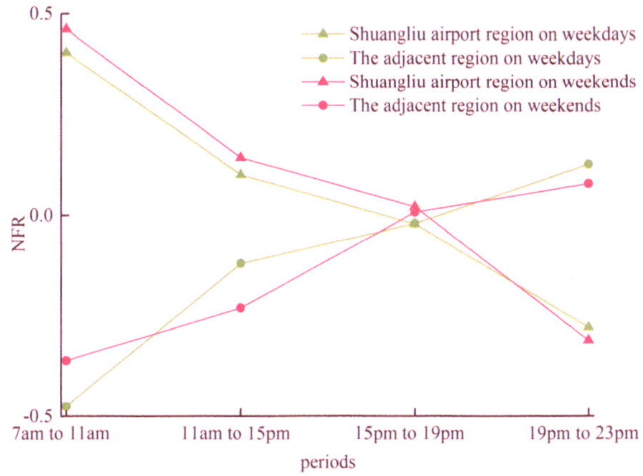

Fig. 7. Comparison of NFR of Shuangliu Airport region and the adjacent region

The Louvain algorithm is used to divide the graph constructed by hotspot region grids into sub-graphs. The detected hotspot region communities are depicted in Fig. 8. Plenty of trips start and terminate in the hotspot regions belonging to the same community. Thus, there are strong association among hotspot regions within the same community. Whereas the relations are weak for the hotspot regions of different communities since there are few trips connecting them.

The road network layout and the administrative division of Chengdu have some typical characteristics. The main roads are designed in ringed and radial layout. The 5 main urban districts are also arranged in radial layout. Influenced by the spatial layout of city, the overall community structure has some inherent spatial distribution feature in all the periods on both weekdays and weekends. The geographical boundaries are evident and these communities also form a ringed layout, as demonstrated in Fig. 8(a). The communities at the edge of the city are composed of adjacent hotspot regions since people tend to choose taxi for short trips. However, there are nonadjacent enclaves within the communities in the downtown. This phenomenon indicates that various urban function of hotspot regions can lead to strong connections among nonadjacent regions and break the limitation of geographical location and distance.

The number of detected communities during different periods on weekdays and weekends is presented in Table 1. Residents' travel behavior shows evident commuting

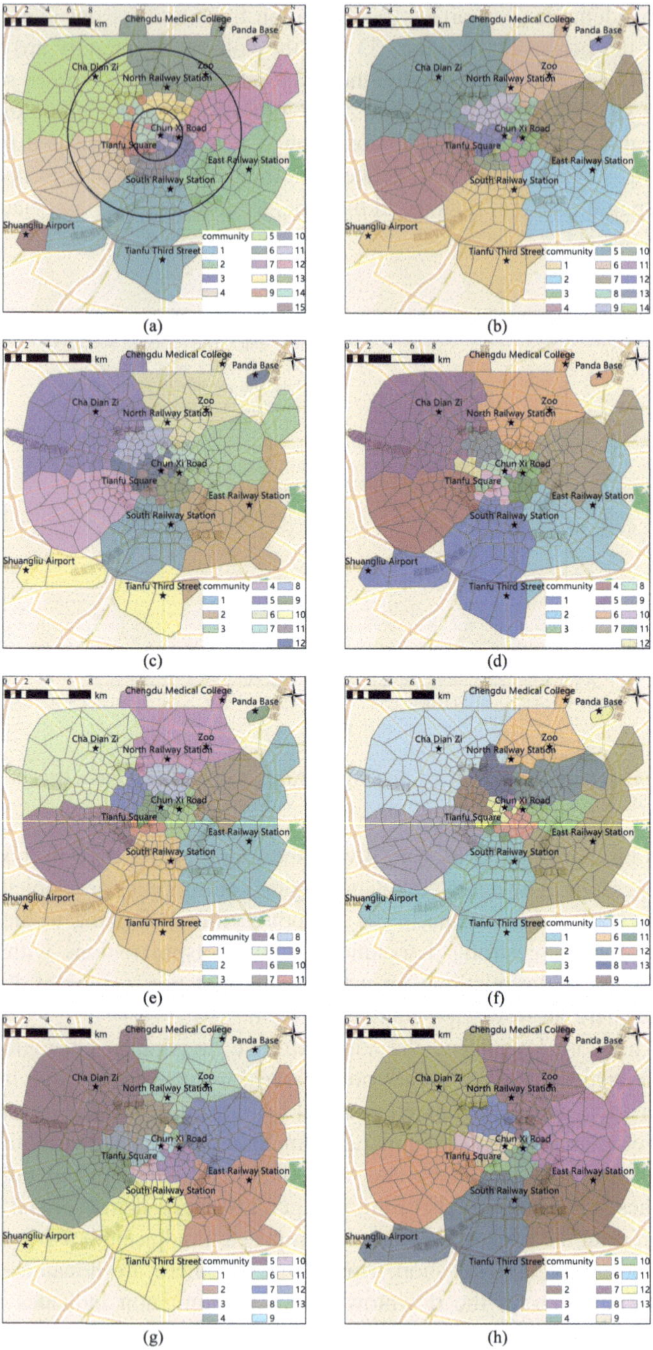

Fig. 8. Community structure of hotspot regions (weekdays: (a) 7 am–11 am, (b) 11 am–15 pm, (c) 15 pm–19 pm, (d) 19 pm–23 pm, weekends: (e) 7 am–11 am, (f) 11 am–15 pm, (g) 15 pm–19 pm, (h) 19 pm–23 pm)

pattern over the whole day on weekdays but shows steady pattern on weekends. Thus, the spatial distribution characteristics of hotspot region community structure varies greatly in different periods on weekdays but maintains similar feature on weekends. The spatial distribution of community structure is more compact at night. E.g., the region around Panda Base is tightly connected to the farther downtown hotspots during the daytime. However, it belongs to the same community with the nearest hotspots at night.

Table 1. Number of hotspot region communities at different periods of weekdays and weekends

Modes	Periods	Number of communities
Weekdays	7 am to 11 am	15
	11 am to 15 pm	11
	15 pm to 19 pm	12
	19 pm to 23 pm	12
Weekends	7 am to 11 am	11
	11 am to 15 pm	13
	15 pm to 19 pm	13
	19 pm to 23 pm	13

4 Discussion and Conclusion

The travel behavior of residents is influenced by the spatial layout of the city including the road network, infrastructure, and administrative division. Meanwhile, human activities interact with the objective environment resulting in the unique urban spatio-temporal characteristics. Thus, it is feasible to reveal the underlying spatio-temporal characteristics of environment by analyzing navigation big data generated in daily travel activities.

In this work, we propose to detect the spatio-temporal distribution and community structure of urban hotspot regions from navigation big data. We design a framework including parts of data preprocessing, hotspot region detection, spatial discretization, and community structure detection. We transform massive data into valuable knowledge by applying this framework to the taxi navigation data in Chengdu. The spatio-temporal distribution and community structure of hotspot regions show distinctive modes on weekdays and weekends and evolve with time during a day. The findings of this work could provide decision-making support for intelligent transportation system, smart city, and the city layout plan etc.

References

1. Zheng, Y., Liu, Y., Yuan, J., Xie, X.: Urban computing with taxicabs. In: International Conference on Ubiquitous Computing (2011)
2. Liu, J., Fang, Y., Guo, C., Gao, K.: Research progress in location big data analysis and processing. Geo Info Sci WHU **39**(4), 379–385 (2014)

3. Liu, S., Liu, Y., Ni, L.M., Fan, J., Li, M.: Towards mobility-based clustering. In: ACM SIGKDD international conference on KDD (2010)
4. Pan, G., Qi, G., Wu, Z., Zhang, D., Li, S.: Land-use classification using taxi GPS traces. IEEE Trans. Intell. Transp. Syst. **14**(1), 113–123 (2013)
5. Chen, R., Chen, M., Yao, X., Wang, J.: Detecting urban hotspot region association by navigation big data mining. J. Geo. Info. Sci. **21**(6), 826–835 (2019)
6. Xiao, L., Xu, W., Liu, J.: Detecting urban dynamics with taxi trip data for evaluation and optimizing of spatial planning: the example of Xiamen city, China. Int. Rev. Spat. Plan. Sustain. Dev. **4**(3), 14–26 (2016)
7. Qin, K., Zhou, Q., Xu, Y., Xu, W., Luo, P.: Spatial interaction network analysis of urban traffic hotspots. Prog. Geogr. **36**(9), 1149–1157 (2017)
8. Newman, M., Girvan, M.: Finding and evaluating community structure in networks. Phys. Rev. E Stat. Nonlinear Soft Matter Phys. **69**(2), 026113 (2004)
9. Green, P., Allan, H., Bernard, W.: Density estimation for statistics and data analysis. Appl. Stat. **37**(1), 120 (1988)
10. Martin, E., Hans-Peter, K., Jorg, S., Xu, X.: A density-based algorithm for discovering clusters in large spatial databases with noise. In: International Conference on KDD (1996)
11. Wang, Q., Wang, C., Feng, Z., Ye, J.: Review of K-means clustering algorithm. Electron. Des. Eng. **20**(7), 21–24 (2012)
12. Lambiotte, R., Delvenne, J., Barahona, M.: Laplacian dynamics and multiscale modular structure in networks. arXiv: arXiv:0812.1770v2 (2008)

FM and DTMB Signal Fingerprinting Positioning System Based on Multi-peak Gaussian Distribution Model

Hongyu Qiao[1,2], Hong Wu[1,2](✉), Menghuan Yang[1,2], Hongzhao Peng[1,2], Haixiao Yang[1,2], and Bin Zhao[1,2]

[1] College of Electronic Information and Optical Engineering, Nankai University, Tianjin, China
wuhong@nankai.edu.cn
[2] Tianjin Key Laboratory of Optoelectronic Sensor and Sensing Network Technology, Tianjin, China

Abstract. With the rapid development of wireless communication technology and the popularization of smart devices, all walks of life are constantly exploring location-based services in recent years. This paper proposes a combination of Frequency Modulation (FM) signal and Digital Television Terrestrial Multimedia Broadcasting (DTMB) signal to improve the indoor positioning accuracy of the new method for the disadvantages of partial positioning technology. The signal intensity of FM and DTMB is taken as the position fingerprint. The multi-peak Gaussian distribution model is used to fit the signal intensity, and then the probabilistic positioning matching algorithm is used to achieve high-precision positioning in the room. We verified the feasibility of the method through actual measurements in the indoor environment. Compared with the existing positioning technology, this method has the advantages of large coverage, high stability, low cost and high precision. The accuracy of the joint positioning of two signals is higher than that of the FM signal alone. It is verified that the positioning error of this system can reach 1.20 m when the multi-peak Gaussian distribution model is used. Compared with the single-peak Gaussian distribution model, the positioning accuracy is improved by about 0.80 m, which has certain reference significance in the field of indoor positioning.

Keywords: Indoor positioning · Location fingerprint · Multi-peak Gaussian distribution · Probability matching

1 Introduction

With the development of positioning and navigation technology, positioning technology based on GPS, Beidou and other systems has been able to achieve high-precision positioning in outdoor environments and meet people's needs for location services. However, in the indoor scene with much complexity, it is difficult for the receiver to obtain accurate satellite signals. At the same time, non-line-of-sight errors also have an impact on it. And it is difficult to achieve the ideal positioning effect by relying only on satellites. Compared with outdoor positioning, indoor positioning is more complicated

J. Sun et al. (Eds.): CSNC 2020, LNEE 650, pp. 279–288, 2020.
https://doi.org/10.1007/978-981-15-3707-3_27

and changeable in environmental conditions, and requires higher accuracy. At present, common indoor positioning technology methods include radio frequency tag positioning, ultrasonic positioning, Bluetooth positioning, ultra-wideband positioning, Wi-Fi positioning, etc. There are also some technologies for indoor positioning using light, geomagnetism, and images [1–3]. Among them, the cost of ultrasonic positioning and ultra-wideband positioning equipment is high, so the popularity of this system is low. Bluetooth positioning and radio frequency tag positioning related devices are relatively inexpensive to manufacture, but have short operating distances, so their applications are limited. Wi-Fi positioning technology is widely used, but in harsh environments, the positioning effect is not ideal. Wi-Fi signals' frequency is 2.4 GHz, susceptible to interference from other signals of the same frequency, and susceptible to the surrounding environment. The time-varying signal is strong and not stable enough, which has a greater impact on the positioning accuracy of the system [4]. On the other hand, every time when positioning is activated, Wi-Fi must be turned on to be in a networked state, which makes the positioning system dependent on Wi-Fi signals, and wireless access points need to be deployed densely, virtually increasing the cost of positioning. Therefore, the positioning technology is not fully mature, and there are still many problems to be solved.

This article proposes an indoor positioning method based on FM and DTMB signals. The strength of FM and DTMB signals is stable, the coverage area is wide, and the interference resistance is strong. It is an ideal positioning signal. At the same time, the experimental demonstration shows that the multi-peak Gaussian distribution fitting model is more accurate than the single-peak Gaussian distribution model, and the adjunction of DTMB signal can also improve the positioning accuracy. This method is of significance for indoor high-precision positioning. With the development of indoor positioning technology, its application can be extended to places with complex channel environments such as large shopping malls and parking lots. Finally, it will achieve precise positioning of targets in a wide area environment.

2 Location Fingerprint Positioning Based on FM and DTMB Signals

Before constructing the system, we first select the frequency of the FM signals and the DTMB signal. The system construction is divided into two parts, including the construction of offline database and online location fingerprint matching.

2.1 Signal Selection

FM signals and DTMB signals have the following characteristics: first, the positioning cost is low and it does not require additional deployment facilities; second, the coverage area is large. Taking Tianjin as an example, the FM signal frequency is between

87–108 MHz, and the DTMB signal frequency is 474–858 MHz. The lower the frequency is, the longer the wavelength is. So it can propagate longer distances; third, the signal source is stable, and the signal strength does not change much when it encounters obstacles, which makes it less susceptible to interference. In addition, when the FM signal and the DTMB signal are further compared, the wavelength of the FM signal is about 3 m. In an indoor environment, the size of the obstacle is usually between 1 m and 3 m. There is no refraction when obstacles occur, so the signal loss is small and the signal intensity changes are not obvious enough. That is, the FM signal intensity values collected at the points before and after the obstacles have basically not changed. But the system may generate additional positioning error of the obstacle size, which has a certain impact on positioning accuracy [5]. The DTMB signal has a wavelength of about 0.4 m and is more sensitive to obstacles of the above size. After the signal passes through the obstacle, its intensity value changes greatly, which helps the system to distinguish the points before and after the obstacle. The DTMB signal provides a difference compared to the FM signal. Therefore, this paper divides into two signal source positioning situations, one using only FM signals and the other using FM signals plus DTMB signals for joint positioning based on the above characteristics.

2.2 System Construction

There are two main parts of the entire positioning process, including the construction of an offline database and online location fingerprint matching. Specifically, the offline section has the following steps: First, the signal sampling point is selected according to the indoor area, and different frequencies of the FM signal and the DTMB signal are selected as the reference frequency. The frequencies at each sampling point are independent of each other, so all frequencies do not affect each other, and the probability information at the same sampling point can be multiplied. Second, we use the antenna to receive signals at various sampling points, record the signal strength value of each frequency, and build a position fingerprint database. The database contains the sampling frequency, signal strength, signal fitting model of FM and DTMB signals, and each sampling point's location information. During online matching, the FM and DTMB signals are received at the points to be located, and their intensity values are matched with the fitting model in the database to obtain the probability of their occurrence at the sampling points. The positioning results of the points to be located are obtained from the true positions of the sampling points. The system schematic diagram is shown in Fig. 1 below.

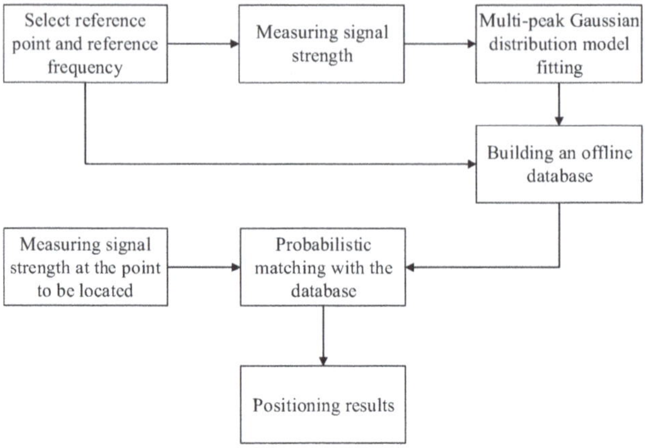

Fig. 1. System schematic diagram

In terms of signal reception, as shown in Fig. 2. The receiving frequency of the FM signal is 87–106.9 MHz, and the receiving frequency resolution is 0.1 MHz. At the same time, the receiving frequency of the DTMB signal is 582–590 MHz, and the frequency resolution is 0.1 MHz [6]. After receiving the FM and DTMB signals, analyzing signals and selecting a representative signal frequency for simulation.

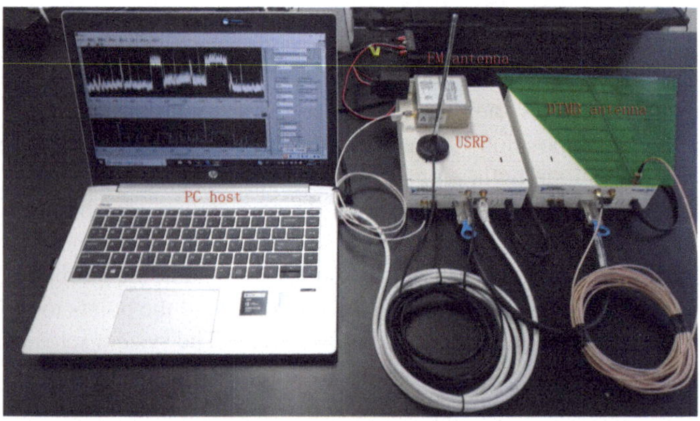

Fig. 2. FM signal and DTMB signal receiving system

3 Probabilistic Location Matching Algorithm

The probabilistic matching algorithm is a position estimation algorithm, which can provide a fitting model that is very close to the distribution of the actual signal strength, and provides probabilistic prediction estimates. In the online positioning, the probability characteristic of the multi-peak Gaussian distribution is used, and the experimental result position of the point to be located is obtained from the correspondence between the probability and the position. This method can directly obtain statistical rules from a large amount of data and is easy to carry out.

3.1 Construction of Offline Location Fingerprint Database

In the probabilistic location matching algorithm, the most critical step is the construction of an offline location fingerprint database. Due to the time-varying signal strength, for example, the measurement results during the day and night are different, and the results are also different in the morning and afternoon respectively. Therefore, when building a data model during the offline training section, a large number of repeated measurements of the signal strength of each frequency at the sampling point and repeated measurements during the day and night are performed to ensure the non-specificity of the data and better reflect the true signal strength distribution. When storing the data, it is necessary to include the measured signal strength, the type of the fitted distribution and the location coordinates of the sampling points. But this also undoubtedly increases the workload of the offline training section and increases the necessary storage space of the system. In the actual positioning environment, the probability distribution of most signal intensities approximates a Gaussian distribution. Therefore, we choose to use a Gaussian fitting model to fit the signal intensity distribution corresponding at each sampling point. At this time,

$$P(rss_p|LOC_r) = \frac{1}{Q}\sum_{q=1}^{Q}\left[\frac{a}{\sqrt{2\pi\sigma^2}}\exp(\frac{-(RSS_{pq} - rss_{pr})^2}{2\sigma^2})\right] \tag{3.1}$$

Here $P(rss_p|LOC_r)$ is the probability of the signal strength rss_p appearing at the LOC_r point. Q is the actual number of samples, and RSS_{pq} at the LOC_r position follows a normal distribution with a peak a and $N(rss_{pr}, \sigma)$. When the single-peak Gaussian distribution model is used, the fitting effect is shown in Fig. 3. The abscissa of the example is the normalized value of the signal strength collected at a frequency of 104.6 MHz. The signals collected at each frequency are regarded as obeying a single-peak Gaussian distribution, and the corresponding model is fitted. In this case, parameters such as the positioning accuracy and positioning error of the system are examined.

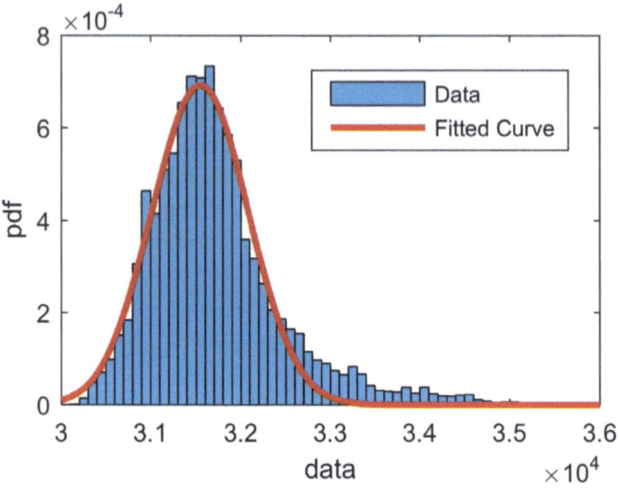

Fig. 3. Single-peak Gaussian fitting distribution curve

3.2 Multi-peak Gaussian Distribution Model

Probabilistic matching positioning algorithm based on Gaussian processes (GP) [7] can effectively improve the system's working efficiency and reduce the required storage space. However, during many experiments, it was found that a small part of the signal intensity appeared multiple peaks. If all the peaks are fitted with a single-peak Gaussian curve, it will affect the authenticity of the fingerprint database and eventually lead to poor positioning results. Therefore, we consider using a multi-peak Gaussian distribution model to fit the signal intensity. $P(rss_p|LOC_r)$ could be expressed,

$$P\left(rss_p|LOC_r\right) = \frac{1}{Q}\sum_{q=1}^{Q}\sum_{m=1}^{n}\frac{a_m}{\sqrt{2\pi\sigma_m^2}}\exp\left(\frac{-\left(RSS_{pq} - rss_{prm}\right)^2}{2\sigma_m^2}\right) \quad (3.2)$$

Here Q is the actual number of samples, and RSS_{pq} at the LOC_r position follows a normal distribution with a peak a_m and $N(rss_{prm}, \sigma_m)$. As shown in Fig. 4, a fitting curve of a bimodal Gaussian distribution model to the signal intensity is selected. The example abscissa is a normalized value of the signal strength collected at a frequency of 87.8 MHz.

The determination of the number of peaks is determined by fitting parameters. The more peaks, the closer the fitted Gaussian model is to the distribution of the true signal, and the more accurate the positioning result is. However, the workload of the offline training phase will be increased to a certain extent with higher complexity of the fitted model, and the complexity of the system will also increase accordingly. Therefore, the fitting method should be balanced between the positioning accuracy and the workload to select the method that is most needed. In the next section, the measured results of the single-peak and multi-peak Gaussian fitting systems will be measured and analyzed.

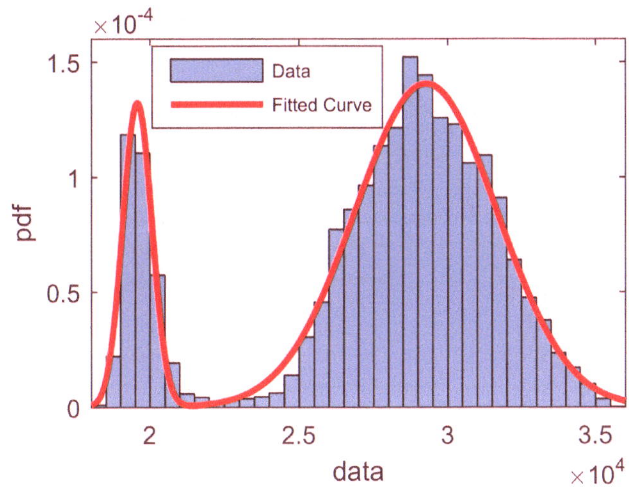

Fig. 4. Bimodal Gaussian fitting distribution curve

4 Analysis of Experimental Results Under Gaussian Distribution Fitting

After measuring data and fully understanding the principle, we built the system for simulation and got experimental results.

4.1 Only FM Signal as Signal Source

When FM signal is used as the only signal source for positioning, taking Tianjin as an example, 24 iconic frequencies of FM signal are selected, including 87.8 MHz, 91.1 MHz, 92.1 MHz, 98 MHz, 104.6 MHz and so on. We selected an area of about 200 m^2 in three laboratories with a total of 151 sampling points. The location information of the sampling points is known. After measuring the signal intensity information received at each frequency of each sampling point, single-peak and multi-peak Gaussian distribution fitting is performed to construct an offline database. Considering the instability of the signal at each moment, the signal intensity of each sampling point is measured multiple times during the actual measurement and then its distribution is estimated by fitted models [8–10].

Next, we use the points to be located for verification. We select 39 points to be located in the area, receive FM signals at the points to be located, and perform probability matching with the offline database to obtain the probability distribution model of the points to be located at each sampling point. The corresponding relationship with the position is used to obtain the experimental result position of the point to be located, and compared with the actual position, parameters such as positioning accuracy and positioning errors are obtained. Figure 5 below shows the cumulative probability distribution of positioning errors under the single-peak Gaussian fitting and multi-peak Gaussian fitting models when using only FM signals for positioning.

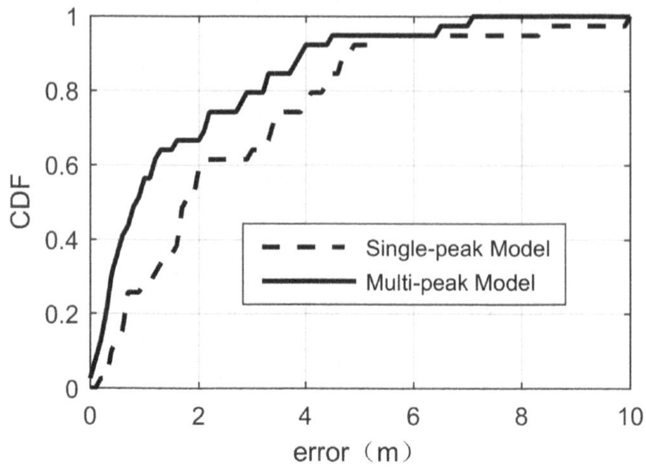

Fig. 5. Positioning error cumulative probability curve

With single-peak and multi-peak Gaussian models, the system's positioning accuracy can reach 4.80 m and 3.90 m with a 90% probability, and the average positioning errors are 2.57 m and 1.66 m respectively. It can be seen that the multi-peak Gaussian model can improve the positioning accuracy.

4.2 FM Signal Plus DTMB Signal Joint Positioning

The frequency of the DTMB signal is higher than other signals. We selected the 675 MHz signal and performed joint positioning with the FM signal to obtain the system positioning accuracy under the fitting of single-peak and multi-peak Gaussian distributions, which are 3.90 m and 3.30 m respectively. The cumulative probability distribution is as follows shown in Fig. 6.

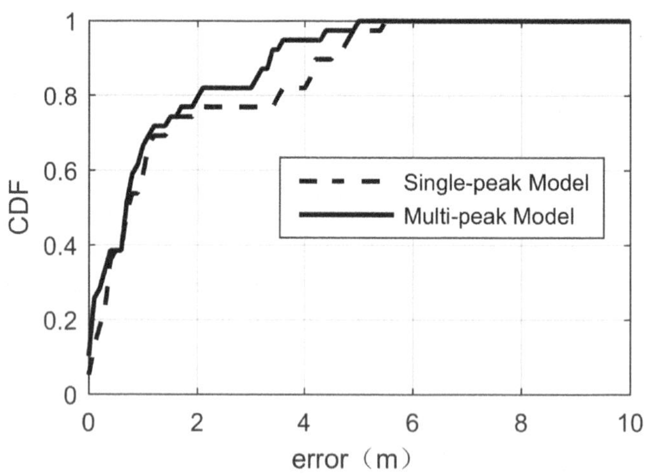

Fig. 6. Positioning error cumulative probability curve

When the signal sources are the same, the positioning errors of the single-peak Gaussian and multi-peak Gaussian distribution fitting models are 1.52 m and 1.20 m respectively. Under the same distribution model, the addition of DTMB signal can also improve the positioning accuracy of the system. However, at the same time, the complexity of building the system and the algorithm have also increased. Therefore, when high-precision positioning is required, a DTMB signal can be added to improve the corresponding system's positioning accuracy.

5 Conclusion

In this paper, a Gaussian distribution model and a probabilistic matching method are used for indoor positioning. The FM and DTMB signal strength are used as position fingerprints, and a positioning system is built for actual measurement and analysis. In the end, when the distribution model uses a multi-peak Gaussian distribution model, the average positioning error reaches 1.20 m and the cumulative probability error reaches 3.3 m, which is more accurate than the positioning accuracy when using a single-peak Gaussian model. This shows that the new fitting method is closer to the real signal distribution, and the cost is low. It also has a good resistance to the fluctuation of the signal in the indoor environment and can be used for accurate indoor positioning. The indoor environment is complex and changeable, and the outdoor signals are also complex. Besides, FM signals and DTMB signals can also be received outdoors. Therefore, in the future, we can consider combining the method of fitting the signal strength of multi-peak Gaussian distribution with outdoor positioning technology to achieve high-precision positioning in a wide area environment.

Acknowledgments. This study was supported in part by the National Natural Science Foundation of China under grant nos. 61571244 and 61871239, in part by Tianjin Science and Technology Plan Project under grant no. 18YFZCGX00480.

References

1. Liu, X., Liu, R., Wen, M.: Research on RFID-based positioning technology. Commun. Technol. **49**(7), 880–884 (2016)
2. Zhang, C., Wang, X., Dong, Y.: Indoor positioning and map construction based on geomagnetic field. J. Instrum. **2015**, 181–186 (2015)
3. Luo, L., Qin, Y.: Research on moving target tracking and positioning technology based on video image sequences. Command Inf. Syst. Technol. **1**(3), 70–73 (2010)
4. Luo, J.: WiFi indoor positioning technology and implementation based on location fingerprint. Shanghai Jiaotong University, Shanghai (2014)
5. Zhang, Y., Deng, L., Yang, Z.: Indoor positioning based on FM radio signals strength. In: 2017 First International Conference on Electronics Instrumentation & Information Systems (EIIS), pp. 1–5. IEEE (2017)
6. Yang, M., Wu, H., Liu, Z., et al.: Indoor positioning using public FM and DTMB signals based on compressive sensing. China Commun. **16**(5), 171–180 (2019)

7. Williams, C.K.I., Rasmussen, C.E.: Gaussian Processes for Machine Learning. MIT press, Cambridge (2006)
8. Yen, H.C., Wang, C.C.: Adapting Gaussian processes for cross-device Wi-Fi localization. In: International Conference on Indoor Positioning and Indoor Navigation, pp. 1–8. IEEE (2013)
9. Jørgensen, P.E.T.: A universal envelope for Gaussian processes and their kernels. J. Appl. Math. Comput. **44**(1–2), 1–38 (2014)
10. Kim, M., De la Torre, F.: Multiple instance learning via Gaussian processes. Data Min. Knowl. Discov. **28**(4), 1078–1106 (2014)

Kinematic Positioning Algorithm Based on the Grey Prediction Model for Urban Navigation

Tianhang Gao, Xianqiang Cui(✉), and Xun Wang

Department of Surveying and Geo-Informatics, Central South University, Changsha 410083, China
cuixianqiang@csu.edu.cn

Abstract. Lack of visible satellites in urban environment will have an adverse impact on positioning and navigation. The kinematic positioning will have a degrading accuracy and even fail to gain the consequences of positioning when in serious conditions. In view of the shortage of observable satellites, this paper, based on the grey prediction model, respectively uses the pseudo-range observation values of the latest epochs and the estimation vector of state parameters to forecast the pseudo-range observation values and state parameters of the current epoch. The experimental results show that it is feasible to use the grey prediction model to predict the state value as the navigation and positioning result or pseudo-range observation value for positioning in the epochs with insufficient observable satellites. The actual calculation example shows that the accuracy of predicted observation values in the absence of available satellites in the field is improved by 61.9%, 53.6% and 18.9% respectively in the ENU direction, and the predicted state parameters are improved by 38.1%, 18.1% and 11.3% respectively in the ENU direction. Obviously, the algorithm proposed in this paper can effectively solve the problem that the kinematic positioning accuracy is reduced or even cannot be located when the signal is blocked for a short time in urban environment, and improve the accuracy and reliability of the kinematic positioning.

Keywords: Urban environment · Missing data · Kinematic positioning · Kalman filter · Grey forecast

1 Introduction

GNSS measurements in urban environment, especially during kinematic data acquisition, data loss often occurs due to the blockage of tall buildings, trees and other obstacles, especially in the case of tunnel culverts and other circumstances where satellite signals cannot be tracked at all, resulting in positioning failure [1]. There are many methods currently available for the processing of missing visible satellites. The deletion method is the simplest. When the observed satellite data is not enough for positioning, the data is deleted and jumped to the next epoch for positioning, however, the use of deletion in kinematic positioning in urban environments can result in instable positioning results and waste some of the observed satellite data. The filling method is

J. Sun et al. (Eds.): CSNC 2020, LNEE 650, pp. 289–300, 2020.
https://doi.org/10.1007/978-981-15-3707-3_28

divided into a single filling method and multiple filling method [2]. The single filling method has a large error. The commonly used multi-fill method is used to estimate the missing values by different single interpolation methods, and then the optimization values are combined to obtain the best estimation. However, the multi-fill method has a large calculation quantity and a long calculation time. The EM algorithm is relatively simple to apply, but the convergence speed is so slow that it is not suitable for kinematic positioning [3]. Kalman filtering is a commonly used algorithm for kinematic navigation and positioning data processing, and when satellite data cannot be received, the dynamic model can also be used to give a predicted value, but the accuracy of the predicted value is not high [4–8]. In addition, multi-system combination and multi-sensor fusion method methods are also widely used. Compared with single system positioning, multi-system combination can effectively improve the visibility, availability, reliability and accuracy in urban environment, but the data quality will still be worse in the case of severe occlusion. In the tunnel environments, even the four-system combination cannot receive the satellite signals [9–11]. Multi-sensor fusion methods, such as the combination of satellite navigation system and inertial navigation system, can be used for positioning even when no satellite is tracked, such as in a tunnel. However, the positioning error of inertial navigation accumulates too fast and the cost will increase [12, 13].

In order to solve the problem of poor data acquisition, a gray prediction method based on a small amount of available information, USES the gray operator to improve the smoothness and quasi-index of the sequence and generate a new sequence, so as to realize the prediction and improve the accuracy of the prediction. In the 1960s, Professor Julong first proposed the theory of *remaining control*, then he proposed grey control theory in 1982 [14]. After thirty years of development, the grey system theory has been continuously improved, the branch structure has become clearer, and the practical application has become more and more broad. The application fields have been involved on many aspects such as hydrology, energy, economy, management, engineering, transportation and so on [15]. In 2005, Cui et al. analyzed the shortcomings of the quadratic polynomial model in the long-term prediction of satellite clock errors, and proposed to use the grey system to predict the satellite clock errors [16]. As one of the important contents of the grey system theory, the grey prediction theory is also a new modern prediction method [17]. However, there are few researches on the combination of grey prediction algorithm and kinematic positioning in urban environment.

Based on this, this paper tested the accuracy of grey prediction algorithm and Kalman filtering algorithm, combined the grey prediction algorithm and Kalman filtering algorithm to improve the urban navigation and positioning results in the case of serious data loss.

2 Grey Prediction Model

Grey prediction models are widely used to deal with systems with uncertain and fuzzy information. This advantage comes from its ability to predict future unknowns using only a limited amount of data (at least four observations). Grey prediction model has three basic operations: cumulative generation operation (AGO), first-order differential equations, and inverse cumulative generation operations (IAGO) [15]. The general process of the grey prediction model is summarized as follows.

Set the original number of samples be $X^{(0)} = \left(X_1^{(0)}, X_2^{(0)}, \ldots, X_n^{(0)}\right)$, the generated sequence of the original series can be obtained by accumulating one time $X^{(1)} = \left(X_1^{(1)}, X_2^{(1)}, \ldots, X_n^{(1)}\right)$, where $X_k^{(1)} = \sum_{i=1}^{k} X_i^{(0)}$. Use this to generate a series of discrete prediction models.

The differential equation of the grey system model is:

$$\frac{dX^{(1)}}{dt} + aX^{(1)} = b \tag{1}$$

This can be simplified as follows:

$$X_k^{(0)} + az_k^{(1)} = b \tag{2}$$

Where a is the parameter to be identified, b is the amount of grey action, and z is the immediate mean of $X_k^{(1)}$:$z_k^{(1)} = 0.5X_{k-1}^{(1)} + 0.5X_k^{(0)}$, 0.5 is the empirical value.

The values of a, b can be solved using the least squares method: $\begin{bmatrix} a \\ b \end{bmatrix} = (B^T B)^{-1} B^T Y$, where, $B = \begin{bmatrix} -z_2^{(1)} & 1 \\ -z_3^{(1)} & 1 \\ \vdots & \vdots \\ -z_n^{(1)} & 1 \end{bmatrix}$, $Y = \begin{bmatrix} X_2^{(0)} \\ X_3^{(0)} \\ \vdots \\ X_n^{(0)} \end{bmatrix}$.

The grey prediction model established by using the accumulated series of samples is directly provided below:

$$\hat{X}_{k+1}^{(1)} = \left(X_k^{(0)} - \frac{b}{a}\right)e^{-ak} + \frac{b}{a} \tag{3}$$

The model can be used to obtain the parameter predicted value $\hat{X}_k^{(1)}$ at any later time. At the same time, the predicted value of the original observation sample $X_k^{(0)}$ can be obtained through reduction and reduction:

$$\hat{X}_{k+1}^{(0)} = \left(X_k^{(0)} - \frac{b}{a}\right)e^{-ak(1-e^{a})} \tag{4}$$

When the grey prediction model is used to deal with the missing observation data, the number of observed data should not be less than 4. According to the characteristics of time series data, only a few or a dozen data close to the missing data epoch are used for estimation, which can improve the accuracy. In urban environments, the quality of satellite data will deteriorate due to signal block, and even satellite data cannot be received at all. The predicted accuracy of the Kalman filter kinematic model is low when there is no visible satellite. So it is feasible to use the grey prediction model to replace the Kalman filter kinematic model or use the predicted state parameters as the pseudo-observations to correct the predicted value of Kalman filter kinematic model.

3 Kinematic Positioning Filtering Algorithm Based on Prediction

3.1 Kinematic Positioning Filtering Based on Forecast State Parameters

Firstly, the pseudo-range single point positioning is applied to the original observation data, and the filtering algorithm is standard constant velocity Kalman filter. The Kalman filter kinematic model and observation model of the linear system are expressed as follows:

$$x_i = \Phi_{i,i-1}x_{i-1} + w_i \tag{5}$$

$$L_i = H_i x_i + v_i \tag{6}$$

Where $x_k \in R^n$ is the state parameter vector, $L_k \in R^m$ is the observation vector, $\Phi_{k,k-1}$ is an n × n-dimensional state transition matrix, H_k is an m × n-dimensional design matrix, $w_k \in R^n$ and $v_k \in R^n$ is the kinematic model noise vector and the observed noise vector, respectively, and satisfies the following equations

$$\begin{cases} Ew_i = 0,\ Ev_i = 0 \\ Ew_i w_i^T = \delta_{i,j} q_i,\ Ev_i v_j^T = \delta_{i,j} r_i \end{cases}$$

Where $\delta_{i,j} = \begin{cases} 1,\ i = j \\ 0,\ i \neq j \end{cases}$, q_i is the covariance matrix of the process noise, and r_i is the covariance matrix of the observed noise [18].

In the case of complete absence of observation data, the predicted value obtained by Kalman filter kinematic model is at low accuracy, and the error will increase significantly over time. At this time, the grey prediction model can be introduced, and the state parameters of the previous several epochs are used for modelling. As a result, a more accurate predicted value can be obtained. Then, the predicted value of the kinematic model of Kalman filter can be corrected by filter algorithm, and the more reliable estimated state parameters can be obtained when there is no observation data.

The predicted value of filtering in the absence of observation data was obtained by standard Kalman filter, and then the observation model of Kalman filter was reconstructed:

$$X_i^g = X_i + V_i \tag{7}$$

Where, the observation vector $X_i^g \in R^n$ is the grey prediction state parameter calculated by formula (5), and $V_i \in R^n$ is the grey prediction pseudo-observation noise vector. According to the following formula, further filter processing can obtain more accurate position parameters.

$$\bar{X}_i = \Phi_{i,i-1} x_{i-1} \tag{8}$$

$$\bar{P}_i = \Phi_{i,i-1} \mathrm{P}_{i-1} \Phi_{i,i-1}^T + Q_i \tag{9}$$

$$K_i = \bar{P}_i (\bar{P}_i + R_i)^{-1} \tag{10}$$

$$\hat{X}_i = \bar{X}_i + K_i (X_i^g - \bar{X}_i) \tag{11}$$

$$P_i = (I - K_i)\bar{P}_i \tag{12}$$

Where, $\bar{X}_i$ is the best one-step prediction value of the state, and K_i is the gain matrix. Q_i is the covariance matrix of the process noise, and R_i is the covariance matrix of the observed noise. $\bar{P}_i$ is predicted state covariance matrix. P_i is the state estimation covariance matrix. I is unit matrix.

Through the above formulas, the state values obtained by the grey prediction model are processed by filter algorithm, and the results are taken as the final state values. The difference between the filter algorithm based on grey prediction and the standard Kalman filter is that uses the predicted state parameters as the pseudo-observations to correct the predicted value of Kalman filter kinematic model, and then can improve the positioning accuracy when there is no observation data.

3.2 Dynamic Positioning Filtering Based on Forecast Pseudo-range Observations

In the case that the satellite signal is blocked, the corresponding pseudo-range observation value cannot be obtained. At this point, the pseudo-distance can be predicted by the grey prediction model and used in the pseudo-distance single-point positioning equation. The pseudo-distance single-point positioning equation is as follows:

$$\rho'(t) = \rho_i^s(t) + c \cdot \delta t_i(t) - c \cdot \delta t^s(t) + \delta I_i^s(t) + \delta T_i^s(t)$$

Where $\rho'(t)$ is the observation of satellite pseudorange observation value in epoch t, $\delta t_i(t)$ is the receiver clock difference of the satellite receiver in t epoch, $\delta t^s(t)$ is the satellite clock difference of the S satellite in t epoch, $\delta I_i^s(t)$ is the ionospheric delay error, $\delta T_i^s(t)$ is the tropospheric delay error. $\rho_i^s(t)$ is the ideal distance between the station and the satellite, can be represented as:

$$\rho_i^s(t) = c \cdot (t_i(t) - t^s(t)) = \left[(X^s(t) - X_i)^2 + (Y^s(t) - Y_i)^2 + (Z^s(t) - Z_i)^2\right]^{1/2}$$

In case of missing epoch of observation satellite, pseudo-distance observation value of S satellite can be obtained by grey prediction model according to pseudo-distance observation value of previous epochs that without missing satellite signal. The clock difference, ionospheric delay error and tropospheric delay error do not change much, so the data of previous epoch can be used.

Through the above formula, when the observation data is insufficient, the missing observation data can be predicted according to the grey prediction model and then be positioned, so that the coordinate results with high positioning accuracy can be obtained continuously.

4 Experiment and Analysis

Kinematic data was collected during a road test conducted on March 28, 2018 in Changsha. The measurement route is shown in Fig. 1. After a period of silence in front of the library in the main campus of Central South University, we started to drive on the campus. After a period of time, we drove along the south second ring road in Changsha city, which took about 35 min. The Central South University campus is heavily wooded, so there will be serious signal blocking. There is no obstructions basically in the second ring road, but there exists signal loss when passing some overpasses.

The test car is shown in Fig. 2. The used receiver is the Trimble net R9, which is placed inside the measuring car. The receiver can simultaneously receive data from four systems, GPS, GLONASS, Galileo and BDS, with a sampling interval of 1 s. The receiver antenna type is TRM55971.00, which is connected to the receiver through the sunroof and placed on the car roof.

The reference value of the car route is the result calculated using GPS RTK [19], and the coordinate system used is Earth-Fixed Coordinate System. The base station is installed on the roof of the Wenfa Building of Central South University. The receiver and antenna are the same type of receiver and antenna used by the rover. The sampling interval is also set to 1 s. Broadcast ephemeris, precise ephemeris, and precision clock files are downloaded from the IGS Data Center of Wuhan University.

Fig. 1. Measurement route

Fig. 2. Measuring car

4.1 Grey Prediction Accuracy Analysis

The accuracy of pseudo-distance observation value prediction by grey prediction algorithm was analyzed. A section of data with good observation quality was selected, and a total of 7 satellites of 4 systems were selected for testing. The pseudo-distance observed values of 6 epochs were selected for grey prediction modeling, and the predicted pseudo-distance values were compared with the true values. The resulting errors are shown in Fig. 3.

The grey prediction algorithm is compared with Kalman filtering algorithm. The coordinate data in the X direction of the measured data with five segments of different road shapes were used for testing, including the data of the five segments of the ephemeris of 400–410, 800–810, 1200–1210, 1600–1610 and 2000–2010 respectively. Among them, the ephemeris of 400–410, 1600–1610 and 2000–2010 are curved roads, and the ephemeris of 800–810 and 1200–1210 are straight lines. Both grey prediction and Kalman filtering approaches use the most recent 5 s data for modeling and then predict the data of 10 s. Then, the difference between the predicted value and the measured value is made to calculate the RMS value of the prediction error of 10 s, and the results of grey prediction and Kalman filtering prediction are shown as follows.

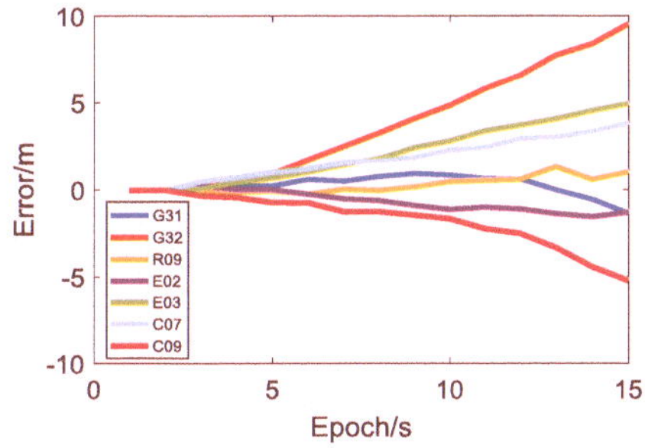

Fig. 3. Predicted pseudo-distance accuracy

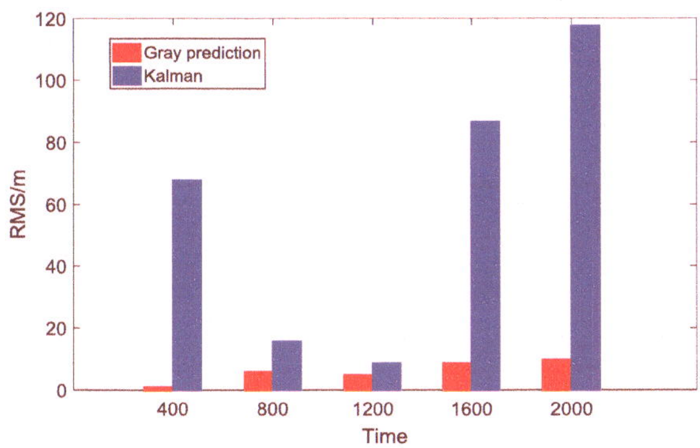

Fig. 4. RMS value of Grey prediction and Kalman filter prediction error

It can be seen from Fig. 3 that the pseudo-distance of each satellite predicted by the grey prediction algorithm is within 10 m within 15 s, and the error of the predicted pseudo-distance of C09 and G32 is the largest, and the error of G32 reaches 10 m quickly in 15 epochs. However, the prediction accuracy of other satellites is still good, with an error of about 5 m, among which the pseudo-distance prediction error curves of R09, E02 and G31 are all around 0 m. The reason for this is related to the data quality of the satellite. The satellite has high altitude angle, high signal-to-noise ratio, good data quality and higher pseudo-range accuracy predicted.

It can be seen from Fig. 4 that the error RMS values of the grey prediction in the five segments are all significantly smaller than the error RMS values predicted by Kalman filtering. The RMS values predicted by Kalman filtering in both straight paths are less than 20 m, while the prediction errors in the three curve segments are large relatively. The largest one exceeds 100 m, which is significantly larger than the error RMS value in the straight path. Grey prediction can achieve good prediction results in both straight and curved paths, and the maximum error RMS value is 10 m. The error of grey prediction of straight section is less than that of the last two curve sections, and the error of the first curve is less than that of other sections, which is probably because the modeling effect is in good agreement with the curve. Although the grey prediction algorithm is also affected by the terrain, it is less affected by the terrain when compared with the Kalman filtering prediction Thus, the prediction result is better than the Kalman filtering prediction result.

4.2 Positioning Test Under Ideal Conditions

For BDS data, pseudo-distance single point positioning was used, the height of satellite cut-off angle was selected as 10°, the ionosphere model was selected as Klobuchar model, and the troposphere model as Saastamoinen model.

Three processing schemes are designed for the insufficient epoch of the observation satellite. Scheme one is that the observation value is not predicted and only kalman filter is used for processing. Scheme 2: use the grey prediction algorithm in chapter 2.2 to predict the state value; Scheme 3: the grey prediction model in chapter 2.3 is used to predict the observed values, and the predicted observed values are used for kalman filtering.

Continuous lack the same in the same number of satellites, satellite and satellite on positioning accuracy of the prediction values may also be affected, choose a good quality data, excluding 15 consecutive epoch data to the number of available satellites only three, and then design the two kinds of test, the first 15 consecutive epoch for three of the same satellite, the second test each epoch keep different satellites. Three processing schemes were used for positioning respectively, and RTK results of the four systems when the observation data were complete were compared and converted into ENU coordinates. The positioning errors obtained were shown in the figures and chart (Figs. 5 and 6).

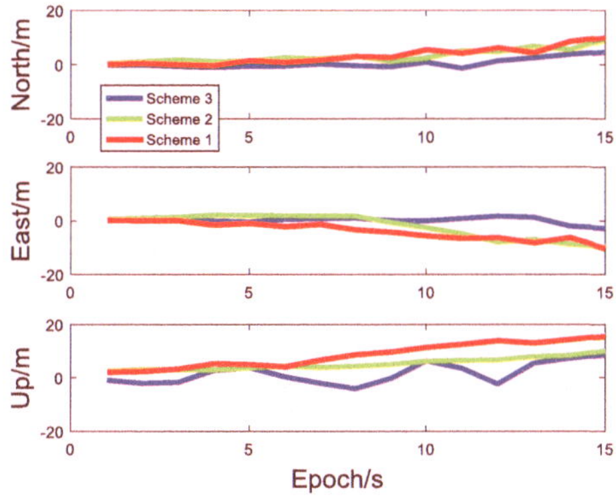

Fig. 5. Continuous loss of the same satellite at the same number of satellites

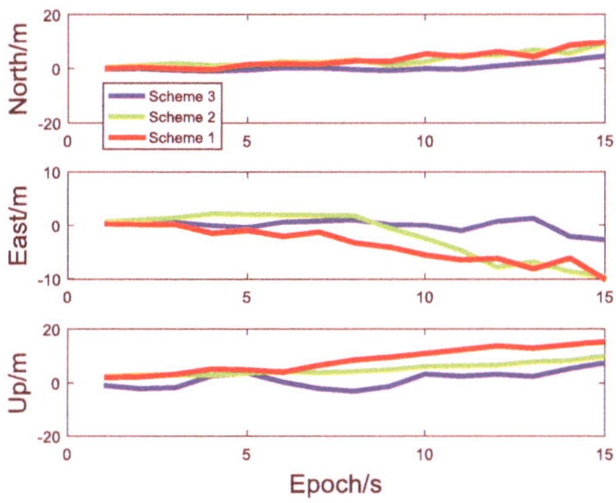

Fig. 6. Missing different satellites at the same number of satellites

When the number of visible satellites is the same, the positioning accuracy of the predicted observation values of different satellites is higher than that of the same continuous missing satellites. The reason is that the error accumulation of pseudo-range prediction in the case of continuous missing of the same satellite is larger than that in the case of missing of different satellites, so the accuracy of three-dimensional point position in the case of continuous missing of different satellites is 18.8% higher than that in the case of missing of the same satellite. The point position accuracy of the predicted state value remains unchanged, while the point position accuracy of the scheme without predicted pseudo-distance observation value and state value does not change much (Table 1).

Table 1. Positioning error RMS (/m)

Scheme		E/m	N/m	U/m	P/m
Scheme 3	Same satellites	1.85	1.17	4.74	4.21
	Different satellites	1.61	1.12	3.85	3.31
Scheme 2	Same satellites	4.95	3.73	8.83	6.30
	Different satellites	4.95	3.73	8.83	6.30
Scheme 1	Same satellites	4.62	4.81	11.78	9.72
	Different satellites	4.60	4.79	11.75	9.69

4.3 Data Processing and Analysis Under Real Environment

The positioning method and model used above are used for positioning the BDS data of the whole route. When the number of observable satellites is less than 4, the three processing schemes proposed above are used for processing. Compared with RTK results of four systems when the observation data are sufficient, and converted the coordinates into ENU coordinates, the positioning errors obtained are shown in the chart (Fig. 7).

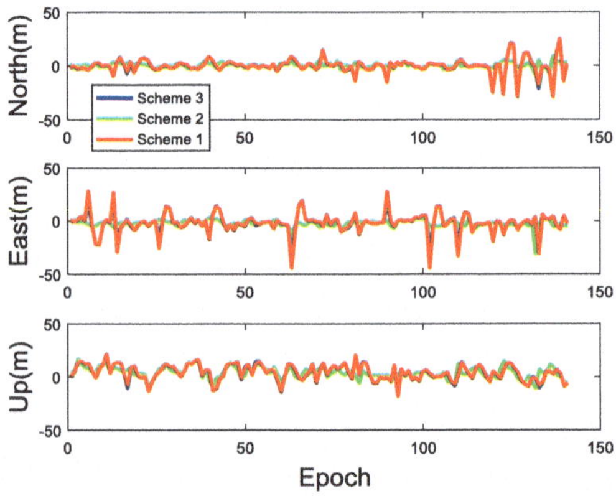

Fig. 7. Error curve of no visible satellite positioning

Table 2 Error RMS of no visible satellite positioning (/m)

Scheme	E/m	N/m	U/m	P/m
Scheme 1	7.12	8.17	9.01	14.09
Scheme 2	4.41	6.69	7.99	11.32
Scheme 3	2.73	3.77	7.31	8.66

As can be seen from the positioning error diagram, when there are not enough observable satellites, the positioning result solved by Kalman filtering has a large error, while the state value predicted by the grey prediction model and the predicted observation value can effectively reduce the positioning error. In some epoch, the predicted observation value still fluctuates, because the signal is blocked by trees, the multi-path effect is serious, and the data quality is low, so the predicted pseudo-distance observation value quality is not high, leading to a large positioning error in the end, but the error is still smaller than the unpredicted positioning error.

As can be seen from Table 2, compared with the unpredicted accuracy, the predicted observation value in the case of the absence of available satellites in the measured test has increased by 61.9%, 53.6% and 18.9% respectively in the ENU direction, and the three-dimensional position accuracy has increased by 38.5%. Compared with the unpredicted accuracy, the predicted status value has increased by 38.1%, 18.1% and 11.3% respectively in the ENU direction, and the three-dimensional position accuracy has increased by 19.7%. When the grey prediction model directly predicts the state parameters, the errors are directly accumulated on the state parameters. In the case of the absence of available satellites in the field test, the accuracy of predicted pseudo-range observation values has been improved by 38.1%, 43.7% and 8.5% respectively in the ENU direction, and the accuracy of three-dimensional position has been improved by 23.5%.

5 Conclusion

In urban environments, the quality of satellite data will deteriorate due to signal block, and even satellite data cannot be received at all. In case of insufficient observable satellites, it is feasible to use the grey prediction model to predict the state value of the epoch with insufficient observable satellites as the navigation and positioning result or the predicted observation value for positioning. When the grey prediction model directly predicts the state parameters, the errors are directly accumulated on the state parameters, and the errors can be effectively reduced by predicting the observed values such as pseudo-range and satellite position. Therefore, the positioning accuracy of the predicted pseudo-range observation values is better than that of the directly predicted state values. However, when there are no continuous observation satellites, state parameters can be predicted to obtain fast and continuous positioning.

However, the grey prediction error will gradually become larger due to the accumulation of errors in the case of a long time lack of visible satellites, especially in the road sections with complex topographic changes such as turning, uphill, downhill. Therefore, the continuous prediction using the grey model should be limited within ten epochs in the practical application.

Acknowledgements. This research was funded by the National Natural Science Foundation of China (No. 41674012).

References

1. Isaksson, A.J.: Identification of ARX-models subject to missing data. IEEE Trans. Autom. Control **38**(5), 813–819 (2002)
2. Allison, P.D.: Multiple imputation for missing data a cautionary tale. Sociol. Methods Res. **28**(3), 301–309 (2000)
3. Gibson, S., Ninness, B.: Robust maximum-likelihood estimation of multivariable kinematic systems. Automatica **41**(10), 1667–1682 (2005)
4. Yang, Y., Xu, T.: An adaptive Kalman filter based on sage windowing weights and variance components. J. Navig. **56**(2), 231–240 (2003)
5. Yang, Y., Weiguang, G.: A new learning statistic for adaptive filter based on predicted residuals. Prog. Nat. Sci.: Mater. Int. **16**(8), 833–837 (2006)
6. Xianqiang, C., Yuanxi, Y.: Adaptively robust filtering with classified adaptive factors. Prog. Nat. Sci.: Mater. Int. **16**(8), 846–851 (2006)
7. Yang, Y., Cui, X., Gao, W.: Adaptive integrated navigation for multi-sensor adjustment outputs. J. Navig. **57**(2), 287–295 (2004)
8. Chang, T.H., Wang, L.S., Chang, F.R.: A solution to the ill-conditioned GPS positioning problem in an urban environment. IEEE Trans. Intell. Transp. Syst. **10**(1), 135–145 (2009)
9. Liu, H., Xu, L., Shu, B., et al.: A new method to improve the performance of multi-GNSS pseudorange positioning in signal-degraded environment. Adv. Space Res. **58**(4), 577–586 (2016)
10. Cai, C., Gao, Y.: A combined GPS/GLONASS navigation algorithm for use with limited satellite visibility. J. Navig. **62**(4), 671–685 (2009)
11. Angrisano, A., Gaglione, S., Gioia, C.: Performance assessment of GPS/GLONASS single point positioning in an urban environment. Acta Geodaet. Geophys. **48**(2), 149–161 (2013)
12. Li, T.: High-accuracy positioning in urban environments using single-frequency multi-GNSS RTK/MEMS-IMU integration. Remote Sens. **10**(2), 205 (2018)
13. Gao, Z., Li, T., Zhang, H., et al.: Evaluation on real-time kinematic performance of BDS in PPP, RTK, and INS tightly aided modes. Adv. Space Res. **61**(9), 2393–2405 (2018)
14. Julong, D.: Grey Control System. Huazhong Institute of Technology Press (1993)
15. Zuo-Ya, Z., Yong-Qi, C., Xiu-Shan, L.: An improved grey model and its application research on the prediction of real-time GPS satellite clock errors. Chin. Astron. Astrophys. **33**(1), 72–89 (2009)
16. Cui, X., Jiao, W.: Application of grey system model in satellite clock bias prediction. J. Wuhan Univ. (Inf. Sci. Ed.) **30**(5), 75–78 (2005)
17. Lu, X., Yang, Z., Jia, X., et al.: Optimization method of grey system theory and its application in satellite clock error prediction. J. Wuhan Univ. (Inf. Sci. Ed.) **33**(5), 492–495 (2008)
18. Shumway, R.H., Stoffer, D.S.: An approach to time series smoothing and forecasting using the EM algorithm. J. Time Ser. Anal. **3**(4), 12 (1982)
19. Deng, C., Tang, W., Liu, J., et al.: Reliable single-epoch ambiguity resolution for short baselines using combined GPS/BeiDou system. GPS Solutions **18**(3), 375–386 (2014)

A New Adaptive Estimation Algorithm Based on CT Model and Ellipsoid Constraint

Xun Wang, Xianqiang Cui(✉), and Tianhang Gao

Department of Surveying and Geo-Informatics, Central South University, Changsha, China
1106736647@qq.com, cuixianqiang@csu.edu.cn, 526822189@qq.com

Abstract. Kalman filtering algorithm is frequently used in the kinematic measurement of Global Navigation Satellite System (GNSS) for positioning calculation. The vehicle often appears maneuvering turns in GNSS navigation and positioning with randomness and unpredictability, which makes it impossible to be described with an accurate models. If the Kalman filtering algorithm based on Constant Velocity (CV) or Constant Acceleration (CA) model is still applied for navigation solution, the accuracy and reliability of the filtering results will be significantly reduced. Aiming at the problem that the current dynamic model has a large error under the maneuvering turns, this paper combines the Coordinated Turn (CT) model and the fixed elevation algorithm to estimate the horizontal turning rate of the vehicle in real time, and limits the deviation of Z direction through the ellipsoid constraint, which ameliorates the impact of the dynamic model deviation on navigation solutions. Experiments show that no matter in left turn, right turn or compound road section, the adaptive filtering algorithm proposed in this paper can better control the influence of dynamic model deviation, and effectively improve the positioning accuracy and reliability of the vehicle during maneuvering turns.

Keywords: Kinematic measurement of GNSS · Coordinated turn model · Kalman filter · Ellipsoid constraint · Reliability

1 Introduction

Kalman filter algorithm is frequently applied to solve the spatial position of carrier in the kinematic measurement of GNSS to provide users with continuous navigation and positioning services [1, 2]. Kalman filter is a linear filter whose dynamic model is generally a constant velocity or constant acceleration model. This algorithm has higher precision when the vehicle does not appear or small maneuver occurs. The accuracy will be decline if the target arises some larger maneuvers because the dynamic model will not exactly deliver the state of object [3, 4, 7, 18]. In navigation and positioning of GNSS, accurate mathematical model is difficult to establish due to the irregular movement of the vehicle, especially in case of maneuvering. How to obtain more reliable positioning solutions under these maneuvers has been a hot topic of research.

J. Sun et al. (Eds.): CSNC 2020, LNEE 650, pp. 301–312, 2020.
https://doi.org/10.1007/978-981-15-3707-3_29

At present, CT model is regularly utilized to track the ground objects because the model is simple and convenient for data processing, meanwhile, this method has high accuracy. But it is mainly used to solve the turning maneuvers of vehicle in 2D plane [3–5]. At the same time, errors accumulate with time growth since the model does not estimate the turn rate in real time which adopts a fixed turn rate. For tracking of the civilian targets in space, the movement is considered to be a uniform turning motion in the plane and a uniform motion in vertical. In other words, the motion state is updated through CT combined with CV model what is commonly applied to track the aircraft or other air targets [4]. Interacting Multiple Model (IMM) algorithm estimate the state of target by switching between different models. Due to the continuity and irregularity of the movement, if you want to make accurate and reliable estimate which requires a huge dynamic models to express the motion of the object, this will increase the difficulty of calculation and cut down the efficiency. In addition, even if you have a huge set of models, the filter often applies a group of dynamic models during the data processing of some carriers with changing turning rates. Since the turning rate of the target is unknown and random, the accuracy and reliability of the state estimation will be reduced if the difference between the turn rate in the model and the actual turn rate is large [3–7]. The integrated navigation technology utilizes attitude parameters received by sensors to update state of the target which can solve the maneuver problem of vehicle. However, the precision of this technology depends on the sensors. Low-cost sensors error accumulation quickly and the sensors with high accuracy are expensive [8–11]. The development of robust adaptive filtering not only resist the influence of the gross errors of observations, but also adjust the weight of the dynamic model forecast information and the observed information. This method could resist the influences of dynamic disturbances on the filtering solution simultaneously [12–18]. However, this approach is not reconstruct the dynamic model when the target is maneuvering that only adjusts the weight of predict information and the observed information by adaptive factor, then make an estimate to the state of the target.

Aiming at the problem that the accuracy and reliability of Kalman filter solutions decrease when the object is maneuvering in GNSS kinematic measurement, a dynamic model adaptive estimation algorithm combining CT model and ellipsoid constraint is proposed that can estimate the turn rate in real time, which overcomes some disadvantages of the fixed turn rate and improves the accuracy and reliability of solutions.

2 Kalman Filter Algorithm

Kalman filter is an optimized auto-regressive algorithm what updates the motion state by recursive operation. Because it does not need to store a large amount of historical information and has a high computational efficiency, which making Kalman filter is widely applied [12–18]. In GNSS kinematic measurement, Kalman filter positioning algorithm is frequently used to compute the position of carrier to provide users with continuous navigation and positioning services. In this method, the pseudo-range observation equation is linearized first, then combined with CV or CA model for positioning. CV model is utilized in this paper. The dynamic model and observation model can be expressed as follows:

$$\boldsymbol{X}_k = \boldsymbol{F}_{\mathrm{k,k-1}}\boldsymbol{X}_{\mathrm{k-1}} + \boldsymbol{w}_{\mathrm{k}} \tag{1}$$

$$\boldsymbol{Z}_{\mathrm{k}} = \boldsymbol{H}_{\mathrm{k}}\boldsymbol{X}_{\mathrm{k}} + \boldsymbol{e}_{\mathrm{k}} \tag{2}$$

In the formulas: $\boldsymbol{X}_{\mathrm{k}}$ and $\boldsymbol{X}_{\mathrm{k-1}}$ represent the n dimensional state vector at k and k − 1 time respectively; $\boldsymbol{F}_{\mathrm{k,k-1}}$ represents the state transition matrix; $\boldsymbol{w}_{\mathrm{k}}$ and $\boldsymbol{e}_{\mathrm{k}}$ represents the process noise and observation noise at time k respectively; $\boldsymbol{Z}_{\mathrm{k}}$ represents the n dimensional observation vector at the time k; $\boldsymbol{H}_{\mathrm{k}}$ represents the observation matrix.

Kalman filtering is a state optimal estimation algorithm what adopts recursive operation to find a balance between the predicted value and observed value to estimate the motion state accurately. In GNSS navigation and positioning, recursive formulas (3)–(8) of Kalman filtering algorithm is often applied for computing the positions, then make an estimate of the carrier [3, 17, 18].

$$\bar{\boldsymbol{X}}_{\mathrm{k}} = \boldsymbol{F}_{\mathrm{k,k-1}}\hat{\boldsymbol{X}}_{\mathrm{k-1}} \tag{3}$$

$$\boldsymbol{\Sigma}_{\bar{X}_{\mathrm{k}}} = \boldsymbol{F}_{\mathrm{k,k-1}}\boldsymbol{\Sigma}_{\hat{X}_{\mathrm{k-1}}}\boldsymbol{F}^{\mathrm{T}}_{\mathrm{k,k-1}} + \boldsymbol{\Sigma}_{\mathrm{w_k}} \tag{4}$$

$$\bar{\boldsymbol{V}}_{\mathrm{k}} = \boldsymbol{H}_{\mathrm{k}} \cdot \bar{\boldsymbol{X}}_{\mathrm{k}} - \boldsymbol{Z}_{\mathrm{k}} \tag{5}$$

$$\boldsymbol{K}_{\mathrm{k}} = \boldsymbol{F}_{\mathrm{k,k-1}}\boldsymbol{H}_{\mathrm{k}}\boldsymbol{\Sigma}^{-1}_{\bar{V}_{\mathrm{k}}} \tag{6}$$

$$\hat{\boldsymbol{X}}_{\mathrm{k}} = \bar{\boldsymbol{X}}_{\mathrm{k}} - \boldsymbol{K}_{\mathrm{k}}\bar{\boldsymbol{V}}_{\mathrm{k}} \tag{7}$$

$$\boldsymbol{\Sigma}_{\hat{X}_{\mathrm{k}}} = (\boldsymbol{I} - \boldsymbol{K}_{\mathrm{k}}\boldsymbol{H}_{\mathrm{k}})\boldsymbol{\Sigma}_{\bar{X}_{\mathrm{k}}}(\boldsymbol{I} - \boldsymbol{H}^{\mathrm{T}}_{\mathrm{k}}\boldsymbol{K}^{\mathrm{T}}_{\mathrm{k}}) + \boldsymbol{K}_{\mathrm{k}}\boldsymbol{\Sigma}_{\mathrm{k}}\boldsymbol{K}^{\mathrm{T}}_{\mathrm{k}} \tag{8}$$

Where: $\bar{\boldsymbol{X}}_{\mathrm{k}}$ and $\hat{\boldsymbol{X}}_{\mathrm{k}}$ represent the state prediction values and parameter filtering estimate respectively at time k, $\boldsymbol{\Sigma}$ is the covariance matrix, $\bar{\boldsymbol{V}}_{\mathrm{k}}$ represent the n dimensional prediction residual at time k, $\boldsymbol{\Sigma}_{\mathrm{k}}$ is the observation noise covariance matrix, $\boldsymbol{I}$ is a n dimensional unit matrix.

Kalman is a linear filter, the positioning accuracy is high if the target does not or relatively small maneuver occurs. otherwise, the algorithm precision will be decreased. This is because reliable Kalman algorithm requires not only accurate observation model and stochastic model, but also precise dynamic model and reasonable estimation algorithm. However, the motion of carrier is difficult to ensure regular movement in actual measurement. It is impossible to establish an accurate dynamic model, especially in the case of maneuvering. The precision and reliability of the solutions will be reduced if an appropriate estimation algorithm is not adopted at this time.

3 Filtering Algorithm Based on CT Combined with CV

If the target makes a turn in 2D plane, it can be argued that the velocity and the turn rate are fixed in a short time. The movement of the target is known as Coordinated Turn

(CT). If the target state vector under this model is: $Y = [x, y, v_x, v_y]$. The state transition matrix can be expressed as [3–5]:

$$\begin{pmatrix} x_k \\ y_k \\ \dot{x}_k \\ \dot{y}_k \end{pmatrix} = \begin{pmatrix} 1 & 0 & \frac{\sin \varphi T}{\varphi} & \frac{\cos \varphi T - 1}{\varphi} \\ 0 & 1 & \frac{1 - \cos \varphi T}{\varphi} & \frac{\sin \varphi T}{\varphi} \\ 0 & 0 & \cos \varphi T & -\sin \varphi T \\ 0 & 0 & \sin \varphi T & \cos \varphi T \end{pmatrix} \begin{pmatrix} x_{k-1} \\ y_{k-1} \\ \dot{x}_{k-1} \\ \dot{y}_{k-1} \end{pmatrix} + \begin{pmatrix} \frac{T^2}{2} & 0 \\ 0 & \frac{T^2}{2} \\ T & 0 \\ 0 & T \end{pmatrix} \cdot Q_{\mathrm{k}-1} \tag{9}$$

Where: T indicates the sampling interval, $Q_{\mathrm{k}-1}$ represents the process noise, φ is the turning rate. If $\varphi > 0$, it indicates that the target is making a left turn. If $\varphi < 0$, it illustrates a right turn. If φ infinitely approaches to 0, the motion model of the target approximates CV model according to the principles of limit.

In order to estimate the 3D state when the vehicle is turning, the movement is generally decomposed into the uniform turning motion in plane and the uniform velocity motion in vertical. The positions are computed by Kalman filter based on CT combined with CV model. Suppose the 3D state vector is $[x, y, z, v_x, v_y, v_z]$, the state transition matrix of CT combined with CV model can be expressed as:

$$\begin{pmatrix} 1 & 0 & \frac{\sin \varphi T}{\varphi} & \frac{\cos \varphi T - 1}{\varphi} & 0 & 0 \\ 0 & 1 & \frac{1 - \cos \varphi T}{\varphi} & \frac{\sin \varphi T}{\varphi} & 0 & 0 \\ 0 & 0 & 1 & 0 & 0 & 1 \\ 0 & 0 & 0 & \cos \varphi T & -\sin \varphi T & 0 \\ 0 & 0 & 0 & \sin \varphi T & \cos \varphi T & 0 \\ 0 & 0 & 0 & 0 & 0 & 1 \end{pmatrix} \tag{10}$$

4 Adaptive Estimation Algorithm of the Dynamic Model

CT model is generally utilized to solve the problem of maneuvering in 2D plane, besides, this model adopts a fixed turning rate model which will causes the error increase with the growth of maneuvering time. In order to realize the update of the 3D state in GNSS navigation and positioning. This paper applies CT model to update the plane coordinates and considered that the elevation of before and after epochs remain unchanged in a short time. Then the 3D state prediction can be updated by ellipsoid constraint equation. This method effectively attenuates the accumulation of errors through the calculated turn rate simultaneously. The coordinates (x, y, z) of a point on ellipsoid fulfill the following ellipsoid constraint equation [20, 21]:

$$\frac{x^2 + y^2}{(a+h)^2} + \frac{z^2}{(b+h)^2} = 1 \tag{11}$$

In the formula: h indicates that the elevation value of the point, a, b represent the long and short half axes of the earth respectively.

The state estimates of latest epochs are used to compute the elevation of these epochs according to ellipsoid constraint equation firstly. Then predict the plane coordinates of the current epoch through CT model and calculate the elevation forecast value under the condition that the elevation of current epoch is consistent with previous epoch. Computing the Z-direction prediction to update the states by the predicted coordinates and elevation according to ellipsoid constraint equation. This method could estimate the turning rate of target in real time which overcome the problem of errors accumulate caused by fixed turning rate and improve the prediction accuracy of dynamic model. The algorithm flow is shown in Fig. 1. The judgment standard of this paper is that if the turn rate of past three consecutive epochs is less than given threshold (set as 10° in this paper), The vehicle is not considered to make a turn, otherwise, the carrier is supposed to be in a turn maneuver.

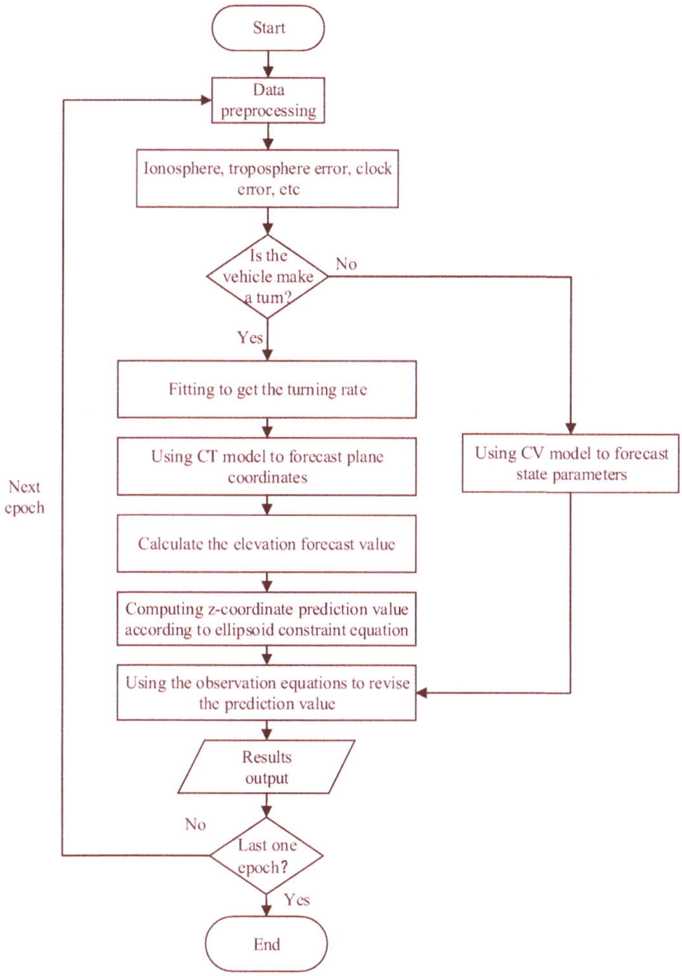

Fig. 1. Algorithm flow chart

5 Experiment and Analysis

The data used are from dynamic experiment conducted on March 28, 2018. Reference station is located at the top of WenFa Building in Central South University. The receiver can obtain observation data of multiple satellite navigation systems. Sampling interval was set to 1 s. Real Time Kinematic (RTK) results of GPS and BDS double systems were regarded as the reference value in this experiment. The pseudo-range single point positioning solution obtained by the combination of GPS and BDS is analyzed.

In order to verify the feasibility and superiority of the algorithm proposed in this paper under different maneuvering conditions, three sections will be intercepted for analysis in the experiment, which are left turn driving section (Fig. 2a), right turn driving section (Fig. 2b) and Compound driving section (Fig. 2c, including left and right turns and straight driving section). Three sections selected will be compared and analyzed using three methods as follows.

Method 1: standard Kalman filtering based on CV model;
Method 2: Kalman filtering based on CT and CV model;
Method 3: Kalman filtering based on CT model and ellipsoid constraint;

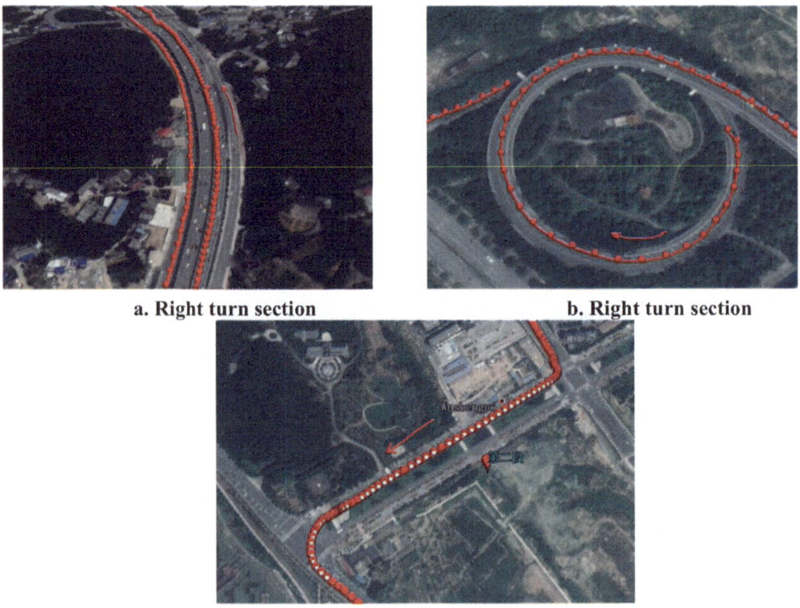

a. Right turn section **b. Right turn section**

c. Composite section

Fig. 2. Selection of experimental road section

In order to analyze the performance of the dynamic model adaptive estimation algorithm in left turn. Three schemes are used to process the observation data. The error curves and Root Mean Square Error (RMSE) statistics table are listed, refer to Fig. 3 and Table 1 for details.

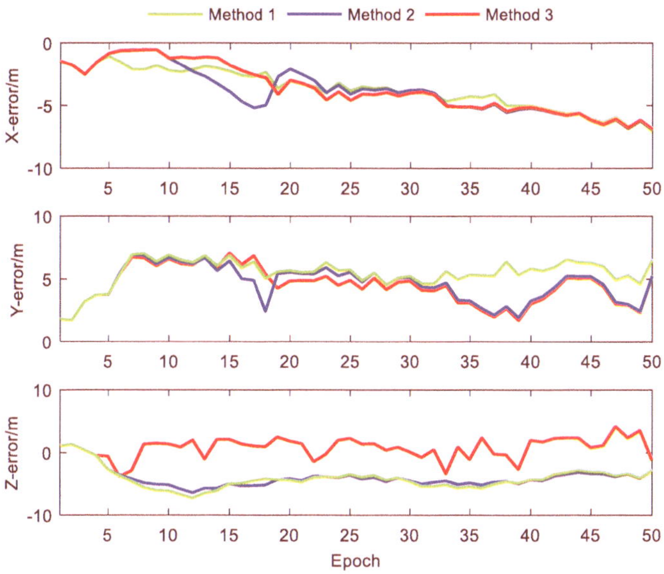

Fig. 3. Error curve of left turn section

Table 1. RMSE statistical table of left turn section

Methods	RMSE in different directions/m			
	RMSE-X	RMSE-X	RMSE-Z	RMSE-P
Method 1	3.96	5.65	4.40	8.18
Method 2	4.20	4.82	4.25	7.68
Method 3	4.09	4.75	1.88	6.55

It can be known from Fig. 3 and Table 1, the errors of CT model combined with ellipsoid constraint algorithm is smaller than other two methods in three directions whose change is more stable. It may be that the hysteresis of this scheme causes the error of this method is higher than the other two methods in some epochs. According to the statistical results in Table 1, CT model combined with ellipsoid constraint algorithm has the highest positioning accuracy. The precision of CT combined with CV model is next. The standard Kalman filter based on CV model is the worst. At the same time, method 2 and 3 have similar precision in X and Y because both of them adopt CT model in the plane. CT model combined with ellipsoid constraint algorithm has higher accuracy in Z direction because the road is flat in this section which makes the Z coordinate predicted value more accurate. The standard Kalman filter based on CV model has the lowest accuracy because the dynamic model in present can not accurately predict the state of the carrier, which resulting in a decrease in accuracy and reliability of positions.

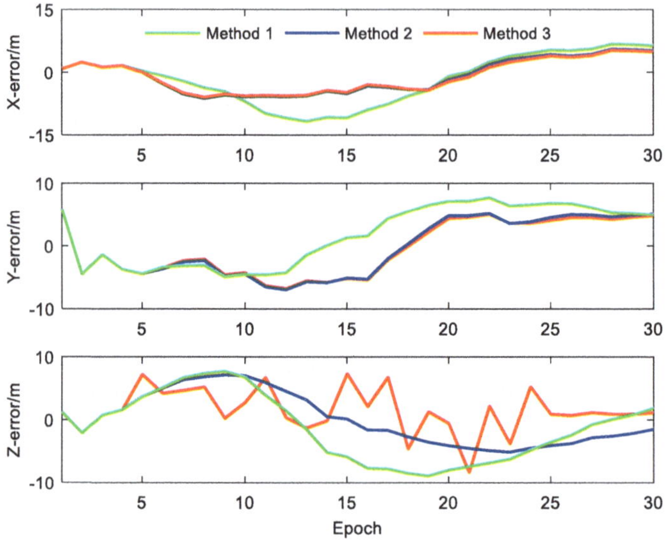

Fig. 4. Error curve of right turn section

Table 2. RMSE statistical table of right turn section

Methods	RMSE in different directions/m			
	RMSE-X	RMSE-Y	RMSE-Z	RMSE-P
Method 1	6.14	5.07	5.33	9.70
Method 2	4.18	4.55	4.03	7.39
Method 3	3.98	4.34	3.76	7.02

Figure 4 represents the error curves of three schemes in right turn section. Table 2 shows the statistical results. According to Fig. 4 and Table 2, the precision of the dynamic model adaptive estimation algorithm is higher than the other two schemes in three directions which has the highest positioning accuracy. The precision of this algorithm in Z direction is not significantly improved compared with that in left turn section. Because this section is a spiral rising segment of overpass, errors will be increase when fixed elevation algorithm adopted. Because this approach holds that the elevation of front and after epochs remain unchanged in a short time, so the elevation prediction value is more accurate in the place where the road is flat and the Z-direction prediction value solved by ellipsoid constraint equation has high precision simultaneously. But the accuracy will be decreased in the area with large fluctuations. Even so, the precision of this algorithm in Z direction is still better than the other two methods according to the experimental results of the right turn. The precision of CT combined with CV model algorithm is better than that of standard Kalman filtering based on the CV model in three directions, but the position accuracy of this algorithm is lower than the adaptive filtering algorithm proposed in this paper.

The method proposed in this paper is used to deal with the data under the compound road section and the error of the results is counted. Error curves is shown in Fig. 5. Statistical results are shown in Table 3. As can be seen from Fig. 5, the errors of three methods in X and Y directions are very similar in middle of compound section. Because this segment is a straight road, turn rate of many epochs is smaller than the threshold given in this paper which lead to the dynamic model consistent with Kalman filtering in X and Y directions when CT model is adopted, so the errors of three methods are very close in many epochs. The precision of CT model combined with fixed elevation algorithm is lower than the other two algorithms in Y direction according to the results in Table 3, but the accuracy in X and Z direction is higher. Method 3 has the highest positioning accuracy in three algorithms. The precision of CT combined with CV model algorithm is second. The standard Kalman filtering algorithm based on CV model has the lowest accuracy. At the same time, the precision of method 2 and 3 in X and Y directions is consistent since two methods adopt same dynamic model.

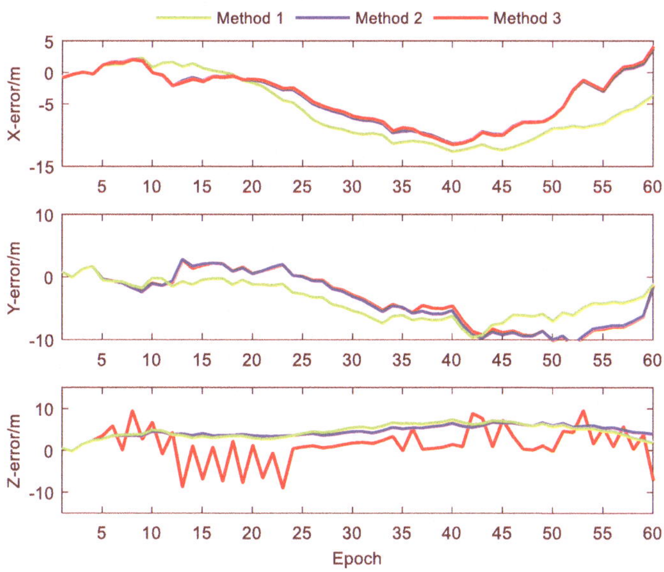

Fig. 5. Error curve of compound section

Table 3. RMSE statistical table of compound section

Methods	RMSE in different directions/m			
	RMSE-X	RMSE-Y	RMSE-Z	RMSE-P
Method 1	7.64	4.60	4.93	10.19
Method 2	5.93	5.20	5.16	9.29
Method 3	5.78	5.32	4.35	8.97

In order to verify the adaptability of the proposed algorithm, this method is utilized to solve the observed data of this experiment and the error of first 700 epochs is counted. Positioning results were compared with the other two methods. Error curves is expressed in Fig. 6 and statistical results are shown in Table 4. From the statistical results in Fig. 6 and Table 4. It can be seen that the CT model combined with ellipsoid constraint algorithm is better than method 1 in three directions whose positioning accuracy is the highest. Meanwhile, the error of three methods is large and the precision is low in first 300 epochs since the vehicle is traveling on the North Campus of Central South University in this time. Because there are many trees and serious occlusion and the data quality is poor, which causes the error of three methods to increase. It indicates that the signal occlusion will lead to the reduction of the accuracy of three algorithms. In the section after 300 epochs, the error of three schemes are consistent or close in X and Y directions in many epochs which illustrate that there are many straight sections in this test which causes three methods have the same mathematical model in X and Y directions, so the error of three algorithms are very similar in many epochs. At the same

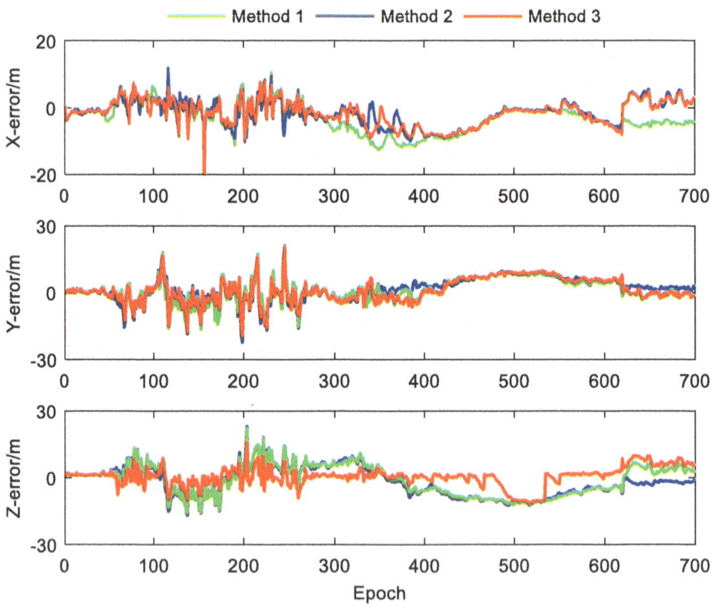

Fig. 6. Error curve

time, it can be seen that the CT model combined with the ellipsoid constraint algorithm improves significantly in the Z direction, indicating that the ellipsoid constraint equation effectively limits the influences of the deviation in Z direction on solutions.

Table 4. Statistical Table of RMS

Methods	RMSE in different directions/m			
	RMSE-X	RMSE-Y	RMSE-Z	RMSE-P
Method 1	5.31	5.57	6.84	10.29
Method 2	4.11	5.42	6.88	9.69
Method 3	4.03	5.51	4.42	8.14

6 Conclusion

Vehicles frequently emerge maneuvers in the kinematic measurement of GNSS. At the same time, the accuracy and reliability of solutions will be decreased if the conventional CA or CV model is adopted. In view of this, the paper combines CT model with ellipsoid constraint algorithm. CT model is applied to update the X and Y coordinates and the Z direction coordinate prediction is achieved according to ellipsoid constraint equation. Meanwhile, this method can estimate the turning rate in real time which overcome some shortcomings of the fixed turning rate model and improves the reliability of the solutions. The calculation results show that the positioning accuracy of the adaptive filtering algorithm proposed in this paper is better than Kalman filter in left turn, right turn, and composite road section, which effectively improves the positioning accuracy and reliability of the vehicle during maneuvering turns.

References

1. Yang, Y.: Concepts of comprehensive PNT and related key technologie. Acta Geodaetica et Cartographica Sinica **45**(5), 505–510 (2016)
2. Yang, Y., Lu, M., Han, C.: Some notes on interoperability of GNSS. Acta Geodaetica et Cartographica Sinica **45**(03), 253–259 (2016)
3. Sun, F., Wu, X., Wang, P.: Two-layer IMM tracking algorithm for turn maneuver. Control Theory Appl. **25**(2), 233–236 (2008)
4. Sánchez, E.E., et al.: A four-model based IMM algorithm for real-time visual tracking of high-speed maneuvering targets. J. Intell. Robot. Syst. **95**(2), 761–775 (2018)
5. Fan, X., Liu, F.: A new IMM method for tracking maneuvering target. J. Electron. Inf. Technol. **29**(03), 532–535 (2007)
6. Gong, S., Wang, B., Wu, H., et al.: Tracking of moving targets on airport surface based on IMM algorithm. Syst. Eng. Electron. **33**(10), 2322–2326 (2011)
7. Gao, L., Xing, J., Ma, Z., et al.: Improved IMM algorithm for nonlinear maneuvering target tracking. Procedia Eng. **29**(4), 4117–4123 (2012)
8. Zhou, K., Wu, X., Kong, J.: Modular characteristics and key technology of tightly-coupled GPS/INS. J. Chin. Inert. Technol. **17**(01), 42–45 (2009)

9. Ma, H., Cheng, P., Huang, H.: Research on the complete integrated GPS/INS navigation system of position, velocity and attitude. Bull. Surv. Mapp. (03), 10–14 (2016)
10. Mu, M., Long, Z.: A GNSS/INS integrated system for an arbitrarily mounted land vehicle navigation device. GPS Solutions **23**(4), 112 (2019)
11. Gao, W., Yang, Y., Cui, X., et al.: Application of adaptive Kalman filtering algorithm in IMU/GPS integrated navigation system. J. Wuhan Univ. (Inf. Sci. Ed.) **10**(05), 466–469 (2006)
12. Liu, M., Yin, X., Lv, Z., et al.: Research of robust adaptive filtering based on moving window. J. Geodesy Geodyn. **34**(06), 140–143 (2014)
13. Yang, Y., Cui, X.: Adaptively robust filter with multi adaptive factors. Surv. Rev. **40**(309), 260–270 (2008)
14. Su, T., Wen, Y., Zhu, J.: The algorithm of the dual adaptive factors filtering. Acta Geodaetica et Cartographica Sinica **44**(01), 26–31 (2015)
15. Cui, X., Yang, Y., Gao, W.: Comparison of adaptive filter arithmetics in controlling influence of colored noises. J. Wuhan Univ. (Inf. Sci. Ed.) **31**(08), 731–735 (2006)
16. Ren, X., Li, Z., Ding, Y.: Application of robust adaptive Kalman filtering on real-time orbit determination. J. Navig. Position. **4**(02), 26–28+35 (2016)
17. Cui, X., Yang, Y.: Adaptively robust filtering with classified adaptive factors. Prog. Nat. Sci.: Mater. Int. **16**(8), 846–851 (2006)
18. Yang, Y., He, H., Xu, G.: Adaptively robust filtering for kinematic geodetic positioning. J. Geodesy **75**(2), 109–116 (2001)
19. Yang, Y.: Adaptive Navigation and Kinematic Positioning. Surveying and Mapping Press (2017)
20. Xu, J., Wu, J.: Dilution of precision analytical method for satellite navigation system subject to height constraint. J. Time Freq. **41**(2), 120–129 (2018)
21. Liu, C., Du, X., Wang, M.: Positioning accuracy of CAPS subject to earth height constraint. GNSS World China **36**(01), 18–22 (2011)

Fault-Tolerant Navigation Method for Unmanned Aerial Vehicle Based on Heterogeneous Pseudorange Augmentation

Xin Chen[1], Rong Wang[1(✉)], Zhi Xiong[1], Jianye Liu[1], Weixing Qian[2], and Junnan Du[1]

[1] Navigation Research Center, College of Automation Engineering, Nanjing University of Aeronautics and Astronautics, Nanjing 211106, China
rongwang@nuaa.edu.cn
[2] College of Electrical and Automation Engineering, Nanjing Normal University, Nanjing 210023, China

Abstract. With the increasing application field of unmanned aerial vehicles, the integrity guarantee capability of satellite navigation system is constantly challenged by complex environments. The introduction of multi-source auxiliary information is an effective mean to enhance the availability of satellite navigation. As a method of land-based long-range radio navigation, Loran-C navigation system has good stability and can provide pseudorange auxiliary measurement information in specific areas to assist satellite receiver in fault detection. Aiming at the problems that the protection level provided by satellite navigation alone may not meet the requirements and the accuracy and reliability of fault detection result may reduce in complex environments, this paper proposed a fault-tolerant navigation method for unmanned aerial vehicle based on pseudorange augmentation. This method introduces the augmented pseudorange information of Loran-C system to assist the inertial/Beidou integrated navigation in integrity detection. Considering the difference of pseudorange measurement characteristics between Loran-C system and Beidou navigation system, this paper uses the weighted least square method to overcome heteroscedasticity and establishes the Loran-C/Beidou augmented pseudorange observation model. Based on this model this paper studies the augmented pseudorange system integrity protection level algorithm when introducing the Loran-C information, and constructs the fault detection and system fault-tolerance algorithm on this basis. The simulation results show that, compared with traditional satellite navigation, this method can effectively improve the integrity guarantee level of system and enhance the fault-tolerance ability of the system under the condition of partial limitation of satellite navigation.

Keywords: Fault-tolerant navigation · Integrity detection · Augmented pseudorange · Satellite navigation

J. Sun et al. (Eds.): CSNC 2020, LNEE 650, pp. 313–323, 2020.
https://doi.org/10.1007/978-981-15-3707-3_30

1 Introduction

With the increasing application of unmanned aerial vehicles, when the aircraft works in harsh conditions, the integrity guarantee capability of satellite navigation system may not meet the requirements. The introduction of multi-source auxiliary information is an effective means to enhance the availability of satellite navigation system. It can effectively reduce the risk in the use of satellite navigation system. The Loran-C system is a land-based long-range radio navigation system with large coverage, strong anti-interference ability and high reliability. With the ability to broadcast data, the Loran-C station can be used as a pseudo-satellite [1], and becomes a good complement to satellite navigation system. When the integrity protection level provided by satellite navigation receiver alone cannot meet the demand, the augmented pseudorange measurement information of Loran-C system can be introduced to assist the satellite receiver in fault detection. This mean is expected to effectively improve the geometry configuration of visible satellite and improve the fault-tolerant ability of the system. Tong [2] combined the Loran-C system with Beidou navigation system. It reduced the geometric dilution of precision and improved the effective positioning range of Beidou navigation system. Xu et al. [3] proposed an integrity monitoring algorithm based on pseudolite assisted Beidou positioning mode. It effectively improved the probability of fault detection and reduced the integrity protection level of system. Kim et al. [4] proposed the multi-radio integrated navigation system containing Loran-C system software design for GNSS backup and verified the accuracy and availability of multi-radio integrated system under the attack of jamming to GPS.

Considering the differences of pseudorange measurement characteristics between Loran-C system and satellite navigation system, this paper uses the weighted least squares method to overcome heteroscedasticity, and establishes the Loran-C/Beidou augmentation pseudorange observation model. Based on this model, the integrity protection level of the augmented pseudorange system is calculated, and Loran-C information is introduced to assist satellite receiver in fault detection to improve the accuracy and reliability of fault detection result.

2 Fault-Tolerant Navigation Scheme Based on Pseudorange Augmentation

The principle diagram of fault-tolerant navigation based on pseudorange augmentation is shown in Fig. 1.

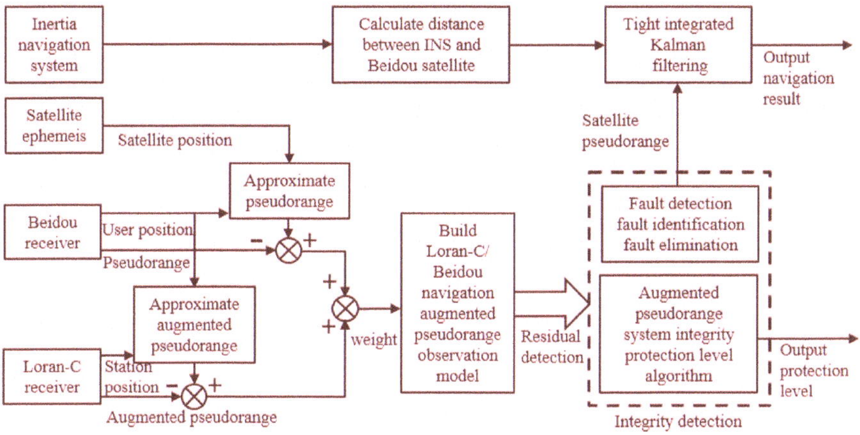

Fig. 1. The principle diagram of fault-tolerant navigation based on pseudorange augmentation

When the integrity protection level provided by Beidou receiver cannot meet the requirements, the augmented pseudorange measurement information of Loran-C system is introduced. By using the weighted least square method to overcome heteroscedasticity, the Loran-C/Beidou augmented pseudorange observation model is established. Because the measurement noise in this model obey the standard normal distribution, the sum of squared residuals can be used to construct detection statistics. Then, based on the chi-square distribution theory, the integrity protection level of augmented pseudorange system is obtained. When the integrity protection level are less than corresponding alarm limit, the Loran-C information is used to assist Beidou receivers in satellite fault detection by method of residual test. After eliminating the fault satellite, the Kalman filter is used to correct inertial navigation information. Finally, the integrated navigation result is output.

3 Loran-C/Beidou Augmented Pseudorange Observation Model

In order to combine with Beidou navigation system, the Loran-C system works in direct ranging mode. The Loran-C assisted Beidou satellite positioning model is extended based on Beidou satellite independent positioning model. Suppose that the Loran-C station clock is synchronized with Beidou satellite clock. The augmented pseudorange observation model can be expressed as

$$\boldsymbol{y} = \boldsymbol{G}\boldsymbol{X} + \boldsymbol{\varepsilon} \tag{1}$$

Suppose that m is the number of visible satellite, u is the number of Loran-C station, then the total number of satellites and Loran-C stations is $n = m + u$.

Where, $\boldsymbol{y} = \boldsymbol{R} - \boldsymbol{P}$, $\boldsymbol{R} = \begin{pmatrix} r_1 & r_2 & \cdots & r_n \end{pmatrix}^T$, r_i is the geometric distance from the user to visible satellite $i(1 \le i \le m)$ or Loran-C station $i(m+1 \le i \le n)$;

$P = (\rho_1 \quad \rho_2 \quad \cdots \quad \rho_n)^T$, ρ_i is the measured pseudorange from the user to visible satellite $i(1 \leq i \leq m)$ or Loran-C station $i(m+1 \leq i \leq n)$.

$G = \begin{bmatrix} -h_x^1 & -h_y^1 & -h_z^1 & 1 \\ -h_x^2 & -h_y^2 & -h_z^2 & 1 \\ \vdots & \vdots & \vdots & \vdots \\ -h_x^n & -h_y^n & -h_z^n & 1 \end{bmatrix}$, $(h_x^i \quad h_y^i \quad h_z^i)$ is the direction cosine from the user to visible satellite $i(1 \leq i \leq m)$ or Loran-C station $i(m+1 \leq i \leq n)$.

$X = [\delta x \quad \delta y \quad \delta z \quad \delta u_t]^T$, $(\delta x \quad \delta y \quad \delta z)$ is the position correction parameter of user receiver, δu_t is the distance error caused by clock difference; $\boldsymbol{\varepsilon}$ is the n dimensional noise vector.

Considering the difference of satellite pseudorange standard deviation and Loran-C augmented pseudorange standard deviation, the weighted least squares method is used to overcome heteroscedasticity. Multiply each term of Eq. (1) by a weight transformation matrix D to obtain the Loran-C/Beidou navigation augmentation pseudorange observation model as Eq. (2).

$$Dy = DG \cdot X + D\varepsilon \tag{2}$$

Where, $D = diag(1/\sigma_s, \cdots 1/\sigma_s, 1/\sigma_c, \cdots 1/\sigma_c)$, then weight matrix $W = D^T D$, σ_s, σ_c are respectively ranging standard deviations of the satellite and Loran-C.

In Eq. (2), the covariance matrix of noise vector $D\varepsilon$ is

$$\text{cov}(D\varepsilon) = \sigma^2 I_n = 1 \cdot I_n \tag{3}$$

where, I_n is the n order identity matrix.

Equation (2) transforms Eq. (1) into a model that satisfies the ordinary least squares assumption. The components of noise vector $D\varepsilon$ obey the standard normal distribution, that is, $D\varepsilon \sim N(0, 1)$. Therefore, the integrity protection level of augmented pseudorange system can be calculated based on this model.

4 Loran-C Assisted Integrity Detection Algorithm

4.1 Augmented Pseudorange System Integrity Protection Level Algorithm

In Eq. (2), the weighted least squares solution of estimated parameters X is

$$\hat{X} = (G^T WG)^{-1} G^T Wy \tag{4}$$

Where, $W = diag(1/\sigma_s^2, \cdots 1/\sigma_s^2, 1/\sigma_c^2, \cdots 1/\sigma_c^2)$.

The sum of squared residuals is

$$SSE = (y - G\hat{X})^T W (y - G\hat{X}) \tag{5}$$

Substitute Eq. (4) into Eq. (5), then

$$SSE = \boldsymbol{\varepsilon}^T(\boldsymbol{I} - \boldsymbol{G}(\boldsymbol{G}^T\boldsymbol{W}\boldsymbol{G})^{-1}\boldsymbol{G}^T\boldsymbol{W})^T\boldsymbol{W}(\boldsymbol{I} - \boldsymbol{G}(\boldsymbol{G}^T\boldsymbol{W}\boldsymbol{G})^{-1}\boldsymbol{G}^T\boldsymbol{W})\boldsymbol{\varepsilon} \quad (6)$$

Let $\boldsymbol{S} = \boldsymbol{I} - \boldsymbol{G}(\boldsymbol{G}^T\boldsymbol{W}\boldsymbol{G})^{-1}\boldsymbol{G}^T\boldsymbol{W}$, then

$$SSE = \boldsymbol{\varepsilon}^T\boldsymbol{W}\boldsymbol{S}\boldsymbol{\varepsilon} \quad (7)$$

According to the chi-square distribution theory, the sum of squared residuals can be as the detection statistic, then

In case of no satellite fault, E($\boldsymbol{D}\boldsymbol{\varepsilon}$) = 0, then $SSE \sim \chi^2(n-4)$;

In case of existing satellite fault, E($\boldsymbol{D}\boldsymbol{\varepsilon}$) $\neq$ 0, then $SSE \sim \chi^2(n-4, \lambda)$.

Where, λ is the decentralized parameter, $\lambda = \mathrm{E}(SSE)$.

When there is no satellite fault, the system is in a normal state. If an alarm occurs, it is a false alarm. Given the false alarm probability P_{FA}, there is the probability equation as follows

$$P_r(SSE < T) = \int_0^T f_{\chi^2(n-4)}(x)dx = 1 - P_{FA} \quad (8)$$

The detection threshold T of SSE can be obtained from the above equation.

When there is satellite fault, SSE should be greater than T. If SSE is less than T, it is missed. Given the missed detection probability P_{MD}, there is the probability equation as follows

$$P_r(SSE < T) = \int_0^T f_{\chi^2(n-4,\lambda)}(x)dx = P_{MD} \quad (9)$$

The decentralized parameters λ can be calculated from the above equation.

Assume that the i-th visible satellite exists fault, and its pseudorange deviation is $\boldsymbol{b}_i$, then the decentralization parameter λ is

$$\lambda = E(SSE) = E(\boldsymbol{\varepsilon}^T\boldsymbol{W}\boldsymbol{S}\boldsymbol{\varepsilon}) = \boldsymbol{W}\boldsymbol{S}_{ii}\boldsymbol{b}_i^2 \quad (10)$$

The horizontal positioning error caused by the fault is

$$\boldsymbol{RPE}_i = (\boldsymbol{A}_{1i}^2 + \boldsymbol{A}_{2i}^2)\boldsymbol{b}_i^2 \quad (11)$$

Where, $\boldsymbol{A} = (\boldsymbol{G}^T\boldsymbol{W}\boldsymbol{G})^{-1}\boldsymbol{G}^T\boldsymbol{W}$.

Multiply the numerator and denominator of Eq. (10) by $\boldsymbol{A}_{1i}^2 + \boldsymbol{A}_{2i}^2$ at the same time and get

$$\lambda = \boldsymbol{RPE}_i \Big/ \left(\frac{\boldsymbol{A}_{1i}^2 + \boldsymbol{A}_{2i}^2}{\boldsymbol{W}_{ii}\boldsymbol{S}_{ii}}\right) \quad (12)$$

Define $\boldsymbol{\delta HDOP}_i$ as the variable quantity of HDOP (Horizontal Dilution Of Precision), it can be expressed as

$$\boldsymbol{\delta HDOP}_i = \sqrt{\frac{\boldsymbol{A}_{1i}^2 + \boldsymbol{A}_{2i}^2}{\boldsymbol{W}_{ii}\boldsymbol{S}_{ii}}} \tag{13}$$

Replace the $\boldsymbol{RPE}_i$ in Eq. (12) with HAL (Horizontal Alarm Limit), then the variation limit of $\boldsymbol{\delta HDOP}_{\rm i}$ can be expressed as

$$\delta HDOP_T = \mathrm{HAL}\Big/\sqrt{\lambda} \tag{14}$$

When $\delta HDOP_{\max} < \delta HDOP_T$, in case that a satellite which is most difficult to detect fault is faulty, the missed detection probability P_{MD} can still be guaranteed. Therefore, $\delta HDOP_{\max}$ can be used as the availability guarantee in fault detection.

The HPL (Horizontal Protection Level) and VPL (Vertical Protection Level) of augmented pseudorange system is

$$\mathrm{HPL} = \max(\boldsymbol{\delta HDOP}_i) \times \sqrt{\lambda} \tag{15}$$

$$\mathrm{VPL} = \max(\sqrt{\frac{\boldsymbol{A}_{3i}^2}{\boldsymbol{W}_{ii}\boldsymbol{S}_{ii}}}) \times \sqrt{\lambda} \tag{16}$$

Compare HPL, VPL with corresponding alarm limits HAL and VAL (Vertical Alarm Limit), and get the integrity guarantee of system.

4.2 Loran-C Assisted Fault Detection and System Fault-Tolerance Algorithm

In Eq. (2), the post-test weight error of residual vector is

$$\hat{\sigma} = \sqrt{SSE/n - 4} \tag{17}$$

According to the detection threshold T obtained by Eq. (8), the detection limit of $\hat{\sigma}$ is

$$\sigma_T = \sqrt{T/n - 4} \tag{18}$$

During navigation, compare $\hat{\sigma}$ with σ_T in real time. If $\hat{\sigma} > \sigma_T$, then there is pseudorange fault in system.

The fault identification adopts the Badar data detection method. The residual are used to construct detection statistics to determine whether there is a gross error in a residual. Let the statistic be

$$\boldsymbol{d}_i = \frac{|\boldsymbol{v}_i|}{\sqrt{\boldsymbol{W}_{ii}\boldsymbol{S}_{ii}}} \tag{19}$$

There are a total of n statistics in the observation model. Given the total false alarm probability P_{FA}, the false alarm probability of a single statistic is P_{FA}/n, then there is the probability equation as follows

$$P_r(d > T_d) = \frac{2}{2\pi}\int_{T_d}^{\infty} e^{-\frac{x^2}{2}}dx = P_{FA}/n \tag{20}$$

The detection limit T_d can be obtained by the above equation.

Compare the statistics of the i-th satellite with the detection limit T_d. If $\boldsymbol{d}_i > T_d$, it indicates that the i-th satellite exists fault, and this satellite should be isolated.

After eliminating the fault satellite, the Kalman filter is used to correct inertial navigation information. Finally, the integrated navigation result is output.

5 Simulation Results and Analysis

To verify the effectiveness of the Loran-C assisted integrity detection algorithm, in the condition of inertia/Beidou and Loran-C enhanced inertia/Beidou integrated navigation, the algorithm was simulated. The simulation of Beidou navigation system is to simulate satellite constellation by setting satellite orbit parameters. The observation frequency is 1 Hz, the standard deviation of equivalent pseudorange error is 3 m [5], and the lowest elevation angle of visible satellite is 30°. In simulation, the gyroscope accuracy is 0.1°/h, the accelerometer bias is 300 μg, and the sampling frequency is 50 Hz. The location of Loran-C station is Xuancheng station in China (118.886° east longitude and 31.069° north latitude) [6], and the standard deviation of equivalent augmented pseudorange error is 10 m. The flight trace of the aircraft is dynamic flight trace, the initial position is 118.654° east longitude, 31.928° north latitude, and the altitude is 300 m. The simulation time is 2500 s. Figure 2 shows the flight trace of aircraft and the position of Loran-C station.

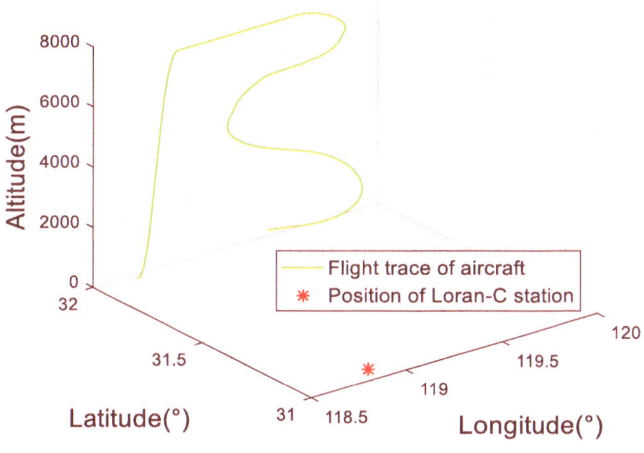

Fig. 2. The flight trace of aircraft and the position of Loran-C station

To verify the effectiveness of introduction of Loran-C information, during the flight, the integrity protection level HPL and VPL of Loran-C/Beidou augmented pseudorange system and Beidou system were calculated respectively, and were shown in Figs. 3 and 4. According to the performance requirements of civil aviation navigation, the HAL of NPA (non-precision approach) phase is 556 m, and the VPL of this phase is not explicitly required [7]. And the main indicator for judging the integrity of navigation system is HPL. Therefore, this paper mainly analyzes the HPL of system.

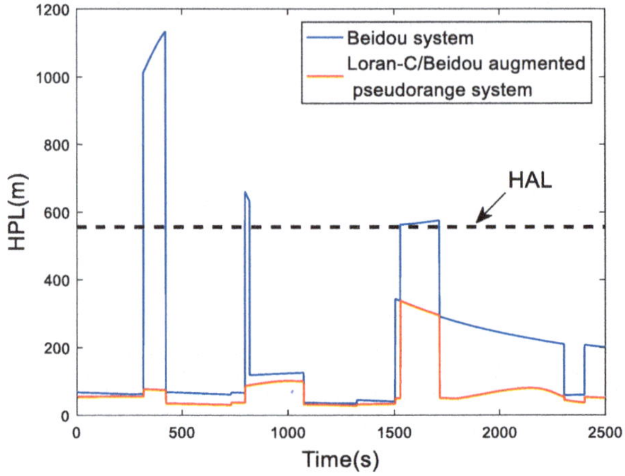

Fig. 3. HPL of Beidou system and Loran-C/Beidou augmented pseudorange system

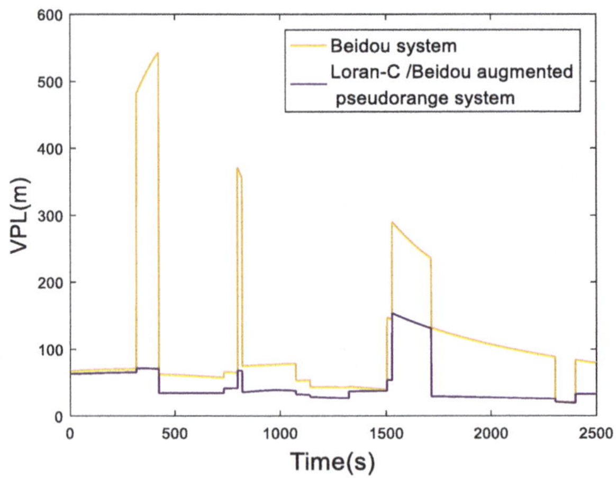

Fig. 4. VPL of Beidou system and Loran-C/Beidou augmented pseudorange system

From Figs. 3 and 4 we can see that, compared with adopting Beidou navigation alone, the HPL and VPL of Loran-C/Beidou augmented pseudorange system are significantly reduced. In some time, the HPL of Beidou system has exceeded HAL, while the HPL of Loran-C/Beidou augmented pseudorange system is less than HAL. The mean value of HPL in two periods is shown in Table 1. The result shows that the introduction of Loran-C information can effectively improve the integrity of Beidou navigation system.

Table 1. Average HPL in two periods during simulation

Number	Period	Average HPL in the period	
		Beidou navigation system	Loran-C/Beidou augmented pseudorange system
1	320–420 s	1078 m	77 m
2	1800–1900 s	268 m	57 m

To compare the fault detection ability of traditional Beidou receiver fault detection algorithm and Loran-C assisted fault detection algorithm at different HPL, the same visible satellite was injected with 40 m pseudorange fault at the periods in Table 1. The position error before fault detection and after adopting two algorithms are respectively shown in Fig. 5.

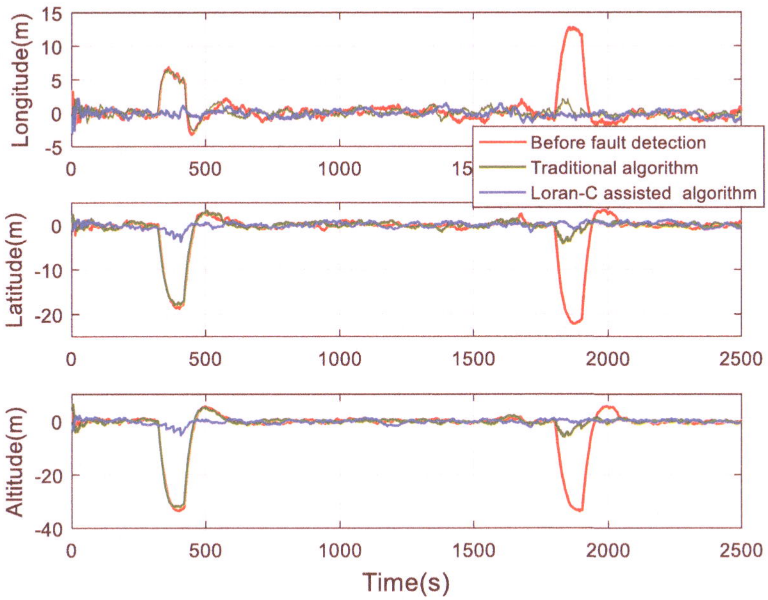

Fig. 5. Position error of before fault detection and after adopting traditional Beidou receiver fault detection algorithm and Loran-C assisted fault detection algorithm

From Fig. 5 we can see that, in 320–420 s, both algorithms can eliminate faulty satellite, and the position error remain unchanged. However, in 1800–1900 s, the traditional Beidou receiver fault detection algorithm could not identify faulty satellite, so the position error became significantly larger. While the Loran-C assisted fault detection algorithm can still eliminate faulty satellite. Therefore, when the HPL of Beidou Navigation far exceeds the alarm limit, the faulty satellite may not be detected and identified. In this case, the introduction of Loran-C information greatly reduced HPL. The HPL did not exceed the HAL. It can meet the missed detection probability of system. Therefore, this method can accurately identify faulty satellite, and effectively improve the fault-tolerance ability of system.

6 Conclusion

In this paper, when the integrity protection level provided by traditional satellite navigation can not meet the requirement of system, the Loran-C augmented pseudorange information is introduced. And by using the weighted least square method, the Loran-C/Beidou augmented pseudorange observation model was established. Based on this model, the integrity protection level of augmented pseudorange system was calculated, and the Loran-C information was used to assist the satellite receiver in fault detection. Simulation results show that compared with the method of traditionally adopting satellite navigation alone, this method can reduce the horizontal protection level of system and improve the integrity and fault-tolerance capability of navigation system.

Acknowlegements. This work was partially supported by the National Natural Science Foundation of China (Grant No. 61703208, 61673208, 61873125, 61533008, 61533009), the Foundation Research Project of Jiangsu Province (The Natural Science Foundation of Jiangsu Province, Grant No. BK20170815, BK20170767, BK20181291), the Aeronautic Science Foundation of China (Grant No. 20165552043, 20165852052), the Science and Technology Innovation Project for the Selected Returned Overseas Chinese Scholars in Nanjing, the Fundamental Research Funds for the Central Universities (Grant No. NS2017016, NP2018108, NJ20170005, NP2017209), the Advanced Research Project of the Equipment Development (30102080101), the "333 project" in Jiangsu Province (Grant No. BRA2016405), the Scientific Research Foundation for the Selected Returned Overseas Chinese Scholars (Grant No. 2016).

References

1. Hu, A.: Combined application of Loran-C and satellite navigation. Navigation of China (2005)
2. Tong, H., Lu, H., Liu, Z.: GDOP simulation analysis of Loran C/Beidou integrated navigation system in complex environment. Ship Sci. Technol. **34**(05), 108–112 (2012)
3. Xu, B., Xiao, W., Liu, W., Ou, G.: Pseudosat assisted Beidou positioning integrity monitoring algorithm based on weighted odd and even vector. Bull. Surv. Mapp. **11**, 9–11 (2016)
4. Kim, H., Lee, J., Oh, S.H., et al.: Multi-radio integrated navigation system M&S software design for GNSS backup under navigation warfare. Electronics **8**(2), 188 (2019)

5. Hongyu, Z., Li, H., Jing, L.: An optimal weighted least squares RAIM algorithm. In: 2017 Forum on Cooperative Positioning and Service (CPGPS), pp. 122–127 (2017)
6. Yin, M., Wang, S.: Availability analysis of GNSS RAIM with multiple outliers based on MUPB. In: 2017 International Conference on Computer Technology, Electronics and Communication (ICCTEC), pp. 534–539 (2017)
7. Wang, Y.: Design and implementation of Roland C navigation system and satellite navigation combined application software. University of Electronic Science and Technology of China (2014)

A Method for PPP Ambiguity Resolution Based on Bayesian Posterior Probability

Zhenqiang Du, Hongzhou Chai(✉), Xiao Yin, Chunhe Liu, and Mingchen Shi

Information Engineering University, Zhengzhou, China
gnsser1996@163.com, chaihz1969@163.com

Abstract. After the corrected by the uncalibrated hardware delay fractional part (FCB) product, the integer characteristic of precise point positioning (PPP) ambiguity can be recovered, which can significantly shorten the convergence time. However, the incorrectly fixed ambiguity seriously affects the accuracy of the results. In the convergence stage of PPP, the traditional method based on ratio test is prone to miss detection and fixed ambiguity incorrectly, which brings an extremely deviation to the results at the user end. Therefore, we propose a step-by-step PPP ambiguity resolution method based on Bayesian posterior probability. The method is deduced from the rigorous mathematical theory, which is guaranteed by the confidence level and can be adjusted according different scenarios. 116 global MGEX stations are selected as the server end to generate two kinds of FCB products, i.e., ionosphere-free combined FCB and undifferenced and uncombined FCB products, and another 50 stations are selected as the user end to test the proposed method. Compared with the traditional method, the success rate of the new method increases from 95.0% to 99.5% in static condition, and the missing detection rate detection reduces from 6.5% to 0.2%; the success rate of the new method increases from 82.9% to 95.9% in kinematic condition, and the missing detection rate detection reduces from 13.1% to 1.9%.

Keywords: PPP · FCB · Ambiguity resolution · Bayesian posterior probability

1 Introduction

Precise point positioning ambiguity resolution (PPP-AR) can significantly shorten the convergence time and improve the accuracy of position after the ambiguity is corrected by FCB product [1, 2, 4]. At present, the PPP models of FCB products at server end are mainly divided into two types, i.e., the ionosphere-free combined model and the undifferenced and uncombined model [5, 8]. The user end usually adopts the method of wide lane (WL) and narrow lane (NL) ambiguity resolution based on the ratio test [9].

The key of PPP-AR is the accurate separation of fractional cycle bias (FCB) and ambiguity. There are two kinds of separation methods of FCB, one is the single-difference between satellites method [10]. The FCB products are estimated by global MGEX observation data, and the receiver FCB is eliminated by the method of single-difference between satellites. The other is the undifferenced method [11]. By

J. Sun et al. (Eds.): CSNC 2020, LNEE 650, pp. 324–337, 2020.
https://doi.org/10.1007/978-981-15-3707-3_31

introducing the satellite end or receiver end FCB datum, the whole network adjustment can calculate the satellite end FCB and receiver end FCB at the same time. Due to the time-consuming adjustment of hundreds of MGEX stations around the world, Kalman filter is used to estimate FCB [12]. In PPP initialization stage, due to the influence of pseudo-range observation noise and multipath, the ambiguity accuracy of different satellites is different [13]. PPP ambiguity quality control method is used to fix ambiguity subsets to ensure the reliability of the fixed ambiguity [14]. However, the above research mainly focuses on FCB estimation and the user end adopts the ratio test to verify. The ratio test threshold is given by experience, it does not have a rigorous mathematical theoretical basis. In the initial stage of PPP ambiguity resolution, the method of PPP-AR based on ratio test is prone to miss detection and fixed ambiguity incorrectly, which brings a serious effect to the accuracy of results.

In view of the above problem, a step-by-step PPP ambiguity resolution method based on Bayesian posterior probability is proposed. The assumption is that the posterior probability of parameter estimation is higher than the given confidence level. The Bayesian posterior probability of ambiguity is deduced under the rigorous Bayesian theoretical framework. According to different scenarios, combining ratio value and Bayesian posterior probability to set the threshold, and the partial ambiguity is fixed based on the size of ambiguity covariance. Two kinds of FCB products, i.e., ionosphere-free FCB products and undifferenced and uncombined FCB products, are generated from 116 MGEX stations around the world, and another 50 MGEX stations as the user end to test the validity.

2 Method

The raw pseudo-range and carrier phase observation equation of GNSS can be expressed as [15]:

$$\begin{cases} P^i_{m,k} = \boldsymbol{\mu}^i_k \boldsymbol{r} + c\delta\hat{t}_k - c\delta t^i + \gamma_m I^i_1 + M^i_k \cdot d_{trop,k} + d_{m,k} - d^i_m + \varepsilon\left(P^i_{m,k}\right) \\ L^i_{m,k} = \boldsymbol{\mu}^i_k \boldsymbol{r} + c\delta\hat{t}_k - c\delta t^i + \gamma_m I^i_1 + M^i_k \cdot d_{trop,k} + b_{m,k} - b^i_m + \lambda_m N^i_{m,k} + \varepsilon\left(L^i_{m,k}\right) \end{cases} \tag{1}$$

where, P and L respectively represents the pseudo-range and carrier phase observation. subscript m, k, respectively represents frequency and station number, superscript i represents satellite number, $\boldsymbol{\mu}^i_k = (\alpha^i, \beta^i, \gamma^i)$ represents unit direction vector between stations and satellites, and $\boldsymbol{r} = (\delta x_r, \delta y_r, \delta z_r)^T$ represents coordinate correction. $\gamma_m = f_1^2/f_m^2$, I^i_1 represents ionosphere delay of L_1 frequency, M^i_k represents mapping function of tropospheric delay, d represents pseudo-range hardware delay deviation, b represents carrier phase hardware delay deviation, $\varepsilon(*)$ represents observation noise and other unmodeled errors.

The ionosphere-free combination eliminates the influence of the first-order ionosphere delay. The precise clock offset products released by IGS (International GNSS service) can be directly used to eliminate the influence of satellite clock offset [16]. Due to the ambiguity parameter absorb the hardware delay of pseudo-range and carrier phase, it loses the integer characterise.

$$\lambda_{IF} B^i_{IF,k} = b_{IF,k} - d_{IF,k} + d^i_{IF} - b^i_{IF} + \lambda_{IF} N^i_{IF,k} \tag{2}$$

where, the subscript *IF* represents the ionosphere-free combination, $B^i_{IF,k}$ represents the float solution of the ambiguity after absorbing the hardware delay. Because the ambiguity of the ionosphere-free combination loses the integer characteristic, it is usually divided into wide lane and narrow lane.

$$B^i_{IF,k} = \frac{f_1}{f_1 + f_2} B^i_{NL,k} + \frac{f_1 f_2}{f_1^2 - f_2^2} B^i_{WL,k} \tag{3}$$

where, the subscript *WL* and *NL* respectively represent the ambiguity of wide lane and narrow lane, $B^i_{WL,k} = B^i_{1,k} - B^i_{2,k}$, $B^i_{NL,k} = B^i_{1,k}$. Compared with the ionosphere-free combined model, the undifferenced and uncombined model can avoid the amplification of observation noise, consider the time-varying characteristics of the ionosphere, and shorten the convergence time of PPP [17, 18]. The global ionosphere grid model GIM released by code (Center for orbit determination in Europe, code) is selected for correction. Due to the satellite DCB is relatively stable one day, it can be corrected by the product. The receiver DCB is be estimate as parameters.

$$\bar{I}^i_{GIM} = \bar{I}^i_1 \tag{4}$$

$$\lambda_m B^i_{m,k} = (d^i_{IF} - d_{IF,k}) + (b_{m,k} - b^i_m) + \lambda_m N^i_{m,k} \tag{5}$$

where, $\bar{I}^i_{GIM}$ represents GIM grid model. Due to the limitation of decimeter precision of global ionosphere model GIM, the PPP-AR method of WL-NL ambiguity resolution is usually adopted.

2.1 The Traditional Ambiguity Resolution of the Client

The client usually fixes the PPP ambiguity of wide lane and narrow lane step by step. The influence of pseudo-range noise and multipath has been weakened after smoothing the MW combined, so the wide lane ambiguity of 0.86 m wavelength can be directly rounded and fixed after corrected by the FCB product. The round success rate is calculated according to the recommended formula of Dong and block [19]. If the success rate is higher than 0.999 and the dimension is higher than 4, it is considered that the ambiguity of wide lane is fixed successfully.

$$P_0 = 1 - \sum_{i=1}^{\infty} [erfc(\frac{i - |B - N|}{\sqrt{2}\sigma}) - erfc(\frac{i - |B - N|}{\sqrt{2}\sigma})] \tag{6}$$

$$erfc(x) = \frac{2}{\sqrt{\pi}} \int_x^{\infty} e^{-t^2} dt \tag{7}$$

where, B is the float solution of integer ambiguity, σ is the mean square error, and N is its nearest integer. The closer between the ambiguity of float solution and the fixed solution, its covariance is smaller, and the higher of round success rate of wide lane ambiguity. Due to the short wavelength of narrow lane ambiguity and the strong correlation between the ambiguity parameters, Lambda algorithm is usually used to fix the ambiguity of narrow lane. The search criteria are as follows:

$$\tilde{N} = \arg\min_{N \in Z}\left(\left(N - \hat{N}\right)^T Q_{\hat{N}}^{-1}\left(N - \hat{N}\right)\right) \tag{8}$$

where, $\hat{N}$ is the float solution vector of the ambiguity and N is the integer solution vector of ambiguity. It searches in the ambiguity domain that meets the search criteria. It is considered that the integer combination that makes the function value reach the minimum is the correct integer ambiguity. The optimal solution and the suboptimal solution are selected for the ratio test.

$$ratio = \frac{(\tilde{N}_{\text{sec}} - \hat{N})^T Q_{\hat{N}}^{-1}(\tilde{N}_{\text{sec}} - \hat{N})}{(\tilde{N}_{\min} - \hat{N})^T Q_{\hat{N}}^{-1}(\tilde{N}_{\min} - \hat{N})} > k \tag{9}$$

where, k is the ratio threshold value. Although the ratio test is widely used, it lacks a rigorous mathematical theoretical basis. Its threshold value is usually given by experience, usually 1.5, 2 and 3. The adaptability to different scenarios is insufficient, especially in the dynamic situation and PPP ambiguity resolution.

2.2 The Ambiguity Resolution Method Based on Bayesian

As an important branch of mathematical statistics, Bayesian theory assumes that the unknown parameter is a random variable with prior distribution information, and its prior probability is $h(\theta)$. After sampling $x_1, x_2, \ldots, x_n$, the prior distribution of θ is adjusted to obtain the Bayesian posterior probability density $h(\theta|(x_1, x_2, \ldots, x_n))$. When the Bayesian posteriori probability of the estimated parameters is higher than the given confidence level, the original assumption is accepted. The problem of correctly fixed PPP ambiguity is a hypothesis test in the integer domain, which assumes that the original hypothesis H_0 is that the ambiguity N is equal to the integer set $\tilde{N}$.

$$H_0 : N = \tilde{N} \cdots H_1 : N \neq \hat{N} \tag{10}$$

$$P(N = \tilde{N}|\beta) \geq c_p \tag{11}$$

where, H_1 is an alternative hypothesis. $P(N = \tilde{N}|\beta)$ represents the Bayesian posterior probability under the condition β, and c_p represents the confidence level. When the posterior probability P of Bayesian is higher than the given confidence level, accepting the original hypothesis H_0, otherwise accepting the alternative hypothesis. For the GNSS observation equation, Eq. (1) is simplified as:

$$L = AX + BN + \varepsilon \tag{12}$$

where, L is pseudo-range and carrier observation. N is integer ambiguity parameter. X is other system parameters except ambiguity, ε is noise of observation. A and B is corresponding coefficient matrix respectively. It is assumed that the probability of sample L distribution is $P((L, Q_L)|(N, X))$ and follows the normal distribution [20].

$$P((L, Q_L|(N,X)) = (2\pi)^{-\frac{m}{2}}|Q_L|^{-\frac{1}{2}} \times \exp\{-\frac{1}{2}||L - AX - BN||^2_{QL}\} \tag{13}$$

where, $||\bullet||^2_{Q_L}$ is $(\bullet)^T Q_y^{-1}(\bullet)$, $N \in Z^m$, $X \in R^m$. The least-squares orthogonal decomposition [21] is carried out for $||L - AX - BN||^2_{Q_L}$.

$$||L - AX - BN||^2_{Q_L} = ||\hat{N} - N||^2_{Q_N} + ||\hat{X}(N) - X||^2_{Q_{\hat{X}(N)}} + ||\varepsilon||^2_{Q_L} \tag{14}$$

where, $\hat{X}(N) = \hat{X} - Q_{\hat{X}\hat{N}} Q^{-1}_{\hat{N}}(\hat{N} - N)$, $Q_{\hat{X}(N)} = Q_{\hat{X}} - Q_{\hat{X}\hat{N}} Q^{-1}_{\hat{N}} Q_{\hat{N}\hat{X}}$. The Bayesian posterior probability $P((N, X)|(L, Q_L))$ of the parameter (N, X) is obtained when the observation value is known by combining Eqs. (11) and (14):

$$P((N,X)|(L,Q_L)) = \frac{P(L,Q_L)|(N,X)P(N,X)}{\sum\limits_{N \in Z^m} \int\limits_{X \in R^m} P(L,Q_L)|(N,X)P(N,X)d(X)} \tag{15}$$

Before observation, the ambiguity parameters are independent of other parameters. There is no prior information for the parameters, which means the ambiguity parameters can be selected arbitrarily in the whole integer space before observation. Similarly, system parameters can select any value in the whole real number before observation.

$$P(N,X) = P(N)P(X), \quad P(N) \propto 1, \quad N \in Z^m, \quad P(X) \propto 1 \tag{16}$$

Combining Eqs. (13), (15) and (16):

$$P((N,X)|(L,Q_L)) = \frac{\exp(-0.5||\hat{N} - N||^2_{Q_{\hat{N}}}) \exp(-0.5||\hat{X}(N) - X||^2_{Q_{\hat{N}}}) P(X)}{\sum\limits_{N \in Z^m} \exp(||\hat{N} - N||^2_{Q_{\hat{N}}}) \int\limits_{X \in R^m} \exp(-0.5||\hat{X}(N) - X||^2_{Q_{\hat{N}}}) P(X) d(X)} \tag{17}$$

The Bayesian posteriori probability of the ambiguity parameter when the float solution is known, which is consistent with that when the observation value is known.

$$P((N,X)|(L,Q_L)) = P((N,X)|(\hat{X}, Q_{\hat{X}(N)})) \tag{18}$$

In Eq. (17), when the float solution of the ambiguity is known, the edge distribution of the ambiguity parameter can be obtained. Thus, the Bayesian posteriori probability of the ambiguity parameter is obtained.

$$P(N = N_{fixed}|(\hat{X}, Q_{\hat{X}(N)})) = \int_{x \in R^m} P(N, X)P((N, X)|(\hat{X}, Q_{\hat{X}(N)}))d(X) = \frac{\exp(-0.5||\hat{N} - N||^2_{Q_{\hat{N}}})}{\sum_{N \in Z^m} \exp(||\hat{N} - N||^2_{Q_{\hat{N}}})} \tag{19}$$

Because $\sum_{N \in Z^m} *$ represents the whole ambiguity integer space, it is impossible to calculate all the candidate values of the ambiguity parameters. Considering the correct ambiguity should be near the float solution and the decreasing characteristic of exponential function, the integer space $S(N)$ of partial ambiguity should be selected instead [22]. The posterior probability of Bayesian ambiguity is the ratio of the maximum likelihood function of the optimal candidate value to the sum of all the candidates in the integer space of the whole ambiguity.

$$S(N) = \{\tilde{N}_{(i)}| \exp(-0.5||\hat{N} - \tilde{N}_{(i)}||^2_{Q_{\hat{N}}})\} \geq 10^{-8} \sum_{j=1}^{i-1} \exp(-0.5||\hat{N} - \tilde{N}_{(j)}||^2_{Q_{\hat{N}}}) \tag{20}$$

Compared with the traditional ratio test, Bayesian posterior probability has strict mathematical theory support, and can adjust different confidence levels according to different scenarios. For the PPP ambiguity resolution using FCB products, the miss detection error is prone to occur in the initial stage. It has passed the ratio test but has not converged. In order to improve its reliability and success rate, the ambiguity resolution can be carried out as follows:

Step 1: fix ambiguity of wide lane and sort ambiguity of narrow lane

According to formula (6) and (7), the ambiguity of wide lane is directly fixed and tested, and the ambiguity and covariance matrix of narrow lane are calculated by using the fixed ambiguity of wide lane. The ambiguity of narrow lane is sorted according $\sigma^2_{\hat{N}_1} \leq \sigma^2_{\hat{N}_2} \leq \cdots \leq \sigma^2_{\hat{N}_i} \leq \cdots \leq \sigma^2_{\hat{N}_m}$.

$$\hat{N} = \begin{bmatrix} \hat{N}_1 \\ \hat{N}_2 \\ \vdots \\ \hat{N}_i \\ \vdots \\ \hat{N}_m \end{bmatrix} \quad \cdots \quad \sigma^2_{\hat{N}} = \begin{bmatrix} \sigma^2_{\hat{N}_1} & \sigma^2_{\hat{N}_1\hat{N}_2} & \cdots & \sigma^2_{\hat{N}_1\hat{N}_i} & \cdots & \sigma^2_{\hat{N}_1\hat{N}_m} \\ & \sigma^2_{\hat{N}_2} & \cdots & \sigma^2_{\hat{N}_2\hat{N}_i} & \cdots & \sigma^2_{\hat{N}_2\hat{N}_m} \\ & & \ddots & & & \vdots \\ & & & \sigma^2_{\hat{N}_i} & \cdots & \sigma^2_{\hat{N}_i\hat{N}_m} \\ & & & & \ddots & \vdots \\ & & & & & \sigma^2_{\hat{N}_m} \end{bmatrix} \tag{21}$$

Step 2: calculate Bayesian posterior probability and ratio value
Lambda method is used to search and fix the ambiguity of narrow lane, ratio test is carried out according to formula (9), Bayesian posterior probability is calculated according to formula (19) and (22), different test thresholds are selected according to different scenarios.

$$P(N = N_{fixed}|(\hat{X}, Q_{\hat{X}(N)})) = \frac{\exp(-0.5||\hat{N} - N||^2_{Q_{\hat{N}}})}{\sum\limits_{N \in S(N)} \exp(||\hat{N} - N||^2_{Q_{\hat{N}}})} \tag{22}$$

Step 3: ambiguity subset resolution
The calculated Bayesian posterior probability is tested by hypothesis. If the test shows that the ambiguity is fixed successfully and the fixed solution is output, the confidence probability is c_p. If it fails to pass the test, reduce the dimension of the ambiguity vector and the corresponding covariance matrix. Delete the ambiguity with the largest covariance and its covariance matrix, and repeat step1, Step2 and Step3 until it is output as the fixed solution of PPP, or output as the float solution of PPP if the ambiguity dimension is less than 4.

3 Experiment and Analysis

In order to verify the effective of the method, the server end consists of 116 MGEX stations around the world, as shown the red pentagram in Fig. 1. Two kinds of FCB products, i.e., ionosphere-free combined FCB products and uncombined FCB products are generated. The user end is composed of 50 MGEX stations that are not involved in the server end, as shown the yellow triangle in Fig. 1. The time of observation data is

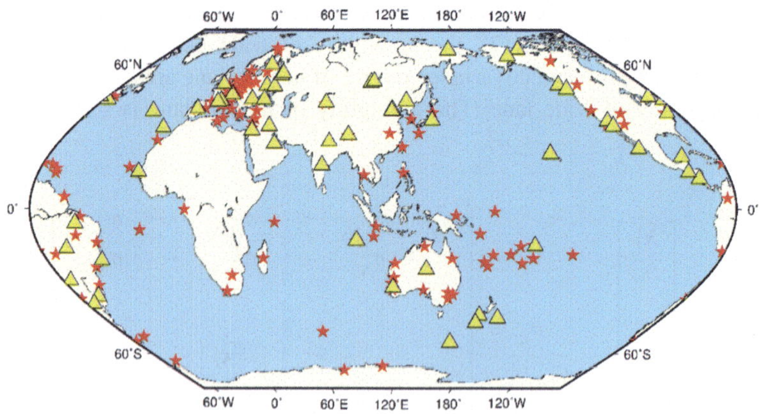

Fig. 1. Distribution of the server end network and user end stations. The red pentagrams denote the 116 MGEX stations with dual-frequency observation, while the yellow triangles denote the 50 MGEX stations for investigating the performance of PPP-AR.

2019, DOY 81-87, and the sampling interval is 30 s. The adjustment weekly result of GFZ network is selected as the true coordinate of MGEX station.

The main difference between the two kinds of PPP models at the server end is that the ionosphere-free combination PPP combines the observation to eliminate the influence of the ionospheric delay. It does not consider the pseudo-range hardware delay deviation of the satellite and the receiver. Based on the raw observation, the undifferenced and uncombined PPP estimates the ionospheric delay deviation, corrects the satellite DCB, and estimates the receiver DCB with $1.0\,e^{-2}m^2/s \times \Delta t$ processing noise. In this paper, the global ionospheric grid model GIM is selected as a prior constraint, the initial variance is $150\,\mathrm{m}^2$, and the process noise is $0.04\,\mathrm{m}^2/30\ \mathrm{s} \times \Delta t$.

3.1 Two Kinds of FCB Product

The server end uses the undifferenced method to calculate two sets of products, i.e., the ionospheric free combination FCB and uncombined FCB. Wide lane FCB products is estimated one day and narrow lane FCB products is estimated in 10 min. In order to test the accuracy of FCB products, two sets of FCB products are applied to 116 MGEX stations at the server end. After corrected by the FCB products, the distribution of ambiguity residuals in wide lane and narrow lane is shown in Fig. 2.

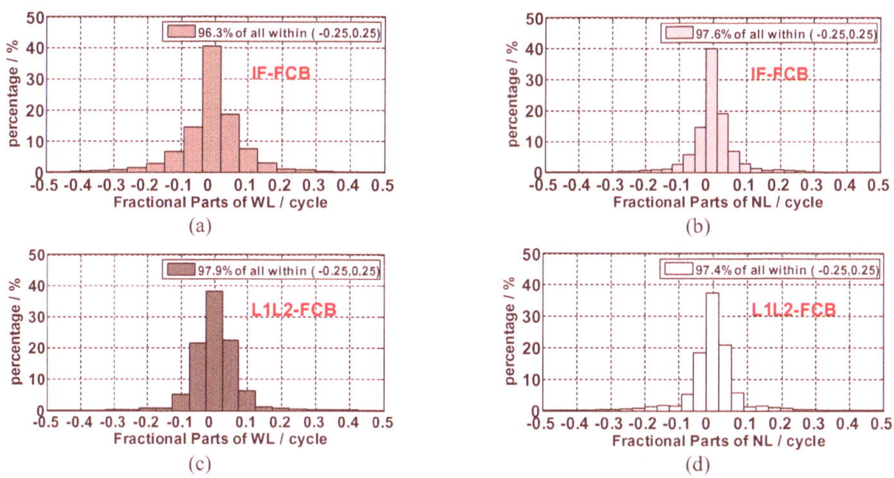

Fig. 2. Distributions of residuals of the wide lane (left panel) and narrow lane (right panel) ambiguity bias(cycle). The residual is the difference between the input float ambiguity used for FCB estimation and the calculates ambiguity with the final FCB estimates and the integer part of the ambiguity.

Figure 2(a) and (b) are the residual values of wide lane and narrow lane respectively using the ionosphere-free combination FCB. It can be seen that for 116 MGEX stations at the srever end, after FCB correction, 96.3% of the residual values of wide lane ambiguity are distributed between ±0.25, and 97.6% of the residual values of narrow lane ambiguity are distributed between ±0.25. (c) and (d) are the residuals of wide lane and narrow lane respectively using undifferenced and uncombined FCB. After being corrected by FCB products, 97.9% of the residuals of wide lane ambiguity are distributed between ±0.25, and 97.4% of the residuals of narrow lane ambiguity are distributed between ±0.25. It shows that the two FCB products generated by the server end have well accuracy.

3.2 Performance of Single Station PPP AR

In order to compare the difference between the ratio test and Bayesian posterior probability test, the output results are PPP float solution, PPP fixed solution, corresponding ratio value and Bayesian posterior probability. Figures 3 and 4 respectively show the results of ambiguity resolution with undifferenced and uncombined FCB products under static and dynamic conditions. The black line is the float solution of PPP and the red line is the fixed solution of PPP. The green dot on the left represents the ratio value of the corresponding epoch, and the green line on the right represents the Bayesian posterior probability.

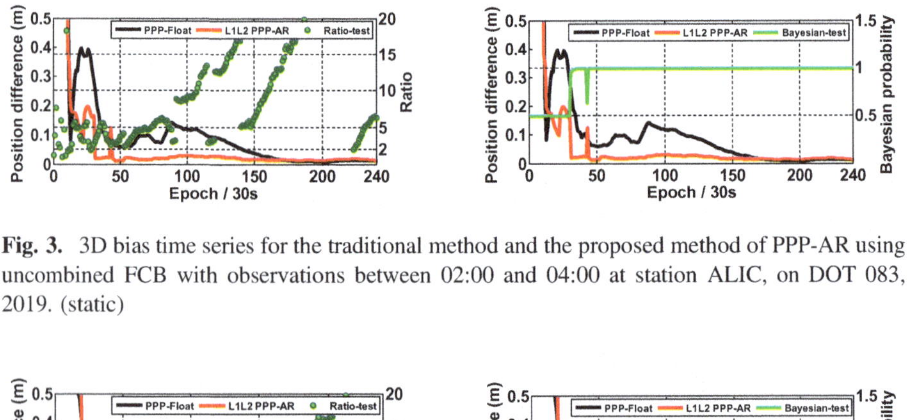

Fig. 3. 3D bias time series for the traditional method and the proposed method of PPP-AR using uncombined FCB with observations between 02:00 and 04:00 at station ALIC, on DOT 083, 2019. (static)

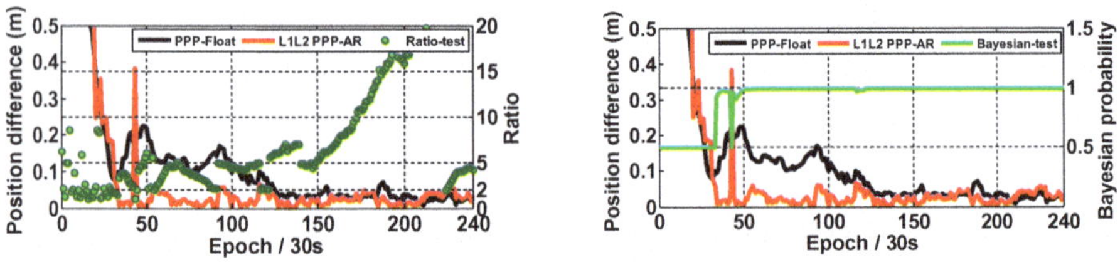

Fig. 4. 3D bias time series for the traditional method and the proposed method of PPP-AR using ionosphere-free combination FCB with observations between 02:00 and 04:00 at station ALIC, on DOT 083, 2019. (dynamic)

Whether static or dynamic, compared with the float solution of PPP, the fixed ambiguity can significantly improve the accuracy of position and shorten the convergence time. In the convergence stage of PPP, there often occurs that when the ratio test value reaches 3 or even more than 5, which brings a extremely interfere to the user end. Compared with the static situation, the dynamic situation is more serious, which makes it difficult to fix the ambiguity at user end. For the Bayesian posteriori probability test, when the ambiguity is not fixed correctly in the convergence stage of PPP, the Bayesian posteriori probability is always below the threshold value. When the ambiguity is fixed correctly, the posterior probability of Bayesian significantly increases and passes the threshold. It should be noted that when there is a fixed ambiguity error in the subsequent epoch, the corresponding Bayesian posterior probability will change significantly and fall below the threshold value.

Similarly, Figs. 5 and 6 show the results of ambiguity resolution under the static and dynamic conditions with the ionosphere-free FCB products in the ALIC station. The ionospheric FCB product correction can significantly shorten the convergence time of PPP and improve the accuracy of solution. However, whether under static or dynamic condition, there occurs frequently that the ratio value still passes the test when the ambiguity is not correctly fixed. In contrast, the Bayesian posterior probability test is still consistent with the correctly fixed ambiguity of PPP. When the ambiguity is not correctly fixed occurs in the subsequent epoch, the variation of the corresponding Bayesian posterior probability is small comparing with IF FCB, but still within the detection range. Therefore, compared with the traditional ratio test, Bayesian posterior probability has better detection effect for PPP ambiguity resolution, and can ensure the reliability of results.

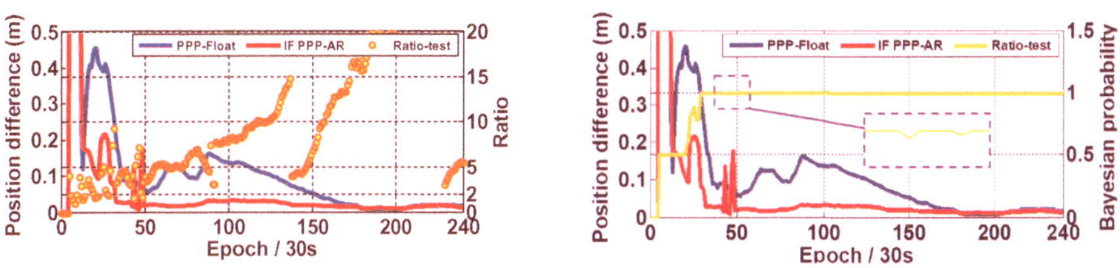

Fig. 5. 3D bias time series for the traditional method and the proposed method of PPP-AR using ionosphere-free combination FCB with observations between 02:00 and 04:00 at station ALIC, on DOT 083, 2019. (static)

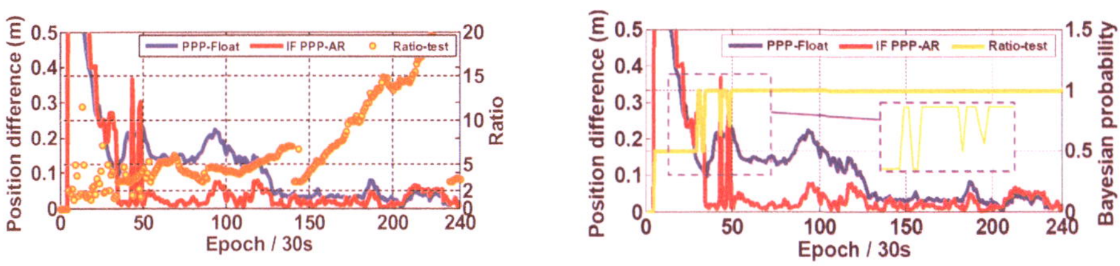

Fig. 6. 3D bias time series for the traditional method and the proposed method of PPP-AR using ionosphere-free combination FCB with observations between 02:00 and 04:00 at station ALIC, on DOT 083, 2019. (dynamic)

3.3 The Success Rate and Missing Detection Rate of Ambiguity Resolution

The success rate and missing rate are important indexes to the PPP-AR. The proportion of correctly fixed ambiguity in all the tested epochs is defined as the success rate of ambiguity resolution. The proportion of incorrectly fixed ambiguity but pass the test in all epochs is defined as the missing detection rate. According to the five strategies proposed in Table 1, the traditional ambiguity resolution method is compared with the proposed ambiguity resolution method based on Bayesian posterior probability.

Figure 7 shows the statistical results of the success rate of PPP-AR for the selected 50 client stations with five strategies. (a) and (b) is under the static and dynamic condition using the undifferenced and uncombined FCB products, respectively. (c) and (d) is under static and dynamic condition using the of ionosphere-free combination FCB products, respectively. It can be seen from (a) and (b) that the ambiguity resolution success rate of PPP-AR using undifferenced and uncombined FCB products increases as time goes on. In 30 min, the average success rate of traditional methods (test1 and test2) is 60.1%, and the average success rate of the proposed methods (test3, test4 and test5) is 95.5%. Similarly, it can be seen from (c) and (d) that compared with the traditional method, the success rate of ambiguity resolution using the proposed method in 30 min is increased from 64.1% to 84.1% by using ionosphere-free combined FCB products for PPP-AR. Therefore, the success rate of ambiguity resolution is significantly improved in a short time by using the proposed method, especially in the dynamic case.

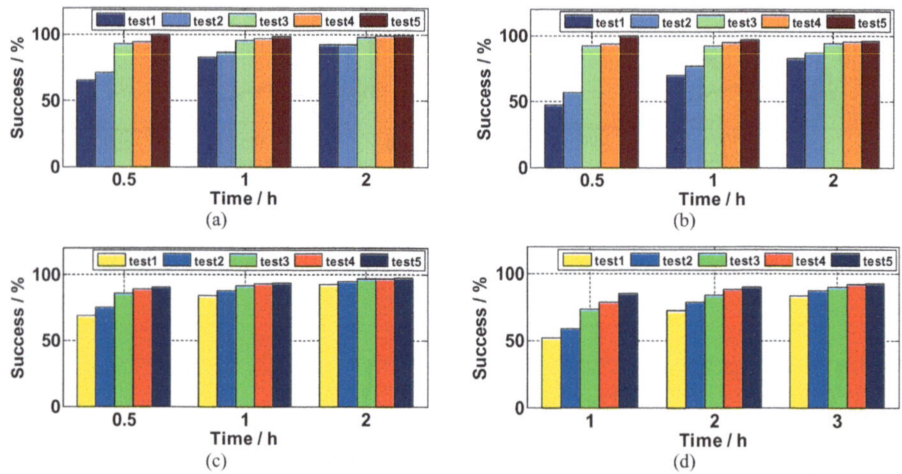

Fig. 7. Average success rate in five strategies of PPP-AR at 50 user end stations using the undifferenced and uncombined FCB products (up panels) and the ionosphere-free FCB products (bottom panels).

Table 1. Five test strategies of PPP ambiguity resolution. The Test1 and Test2 are the traditional threshold and the Test3, Test4, Test5 are the proposed threshold.

Scheme	Test conditions
Test1	Ratio >2
Test2	Ratio >3
Test3	Ratio >2 && Bayesian >0.9
Test4	Ratio >3 && Bayesian >0.90
Test5	Ratio >3 && Bayesian >0.99

In the convergence stage of PPP, the missing detection rate is even more important than the success rate for the client end. The incorrectly fixed ambiguity is regarded as a fixed solution, which will bring an extreme positioning error. Figure 8 shows the comparison of the missed detection rates in the five strategies at the 50 client end stations. (a) and (b) is under static and dynamic condition using the undifferenced and uncombined FCB products, respectively. (c) and (d) is under static and dynamic condition using the ionospheric-free combination FCB products, respectively.

Using the undifferenced and uncombined FCB products, the missed detection rate of traditional methods (test1, test2) are 22.1% and 16.1% respectively in 30 min under static conditions, while the missed detection rates of proposed methods (test3, test4, test5) are 17.5%, 10.0% and 0%. As shown in Fig. 8(b), compared with the traditional method, the proposed method reduces the missed detection rate from 28.8% and 18.3% to 5%, 2% and 0% under the dynamic condition, which has similar effect on using the ionosphere-free FCB product. The missing rate of the proposed method is significantly lower than the traditional method, especially in the dynamic condition.

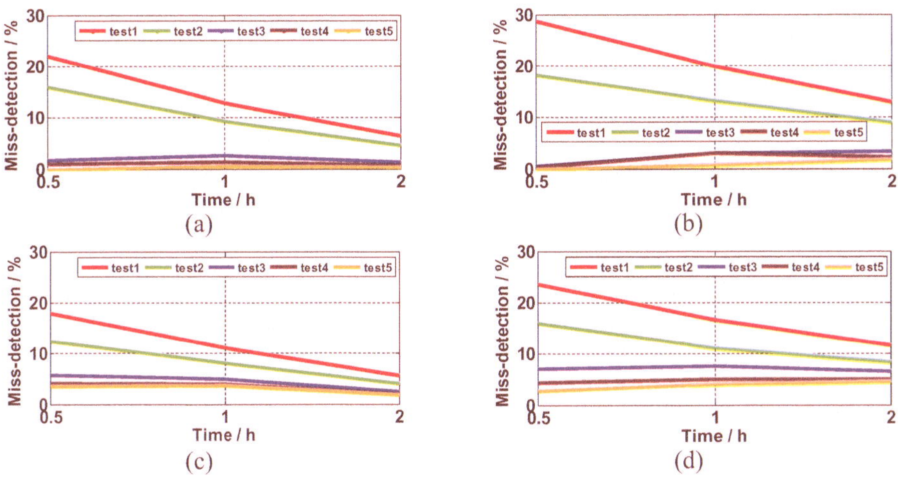

Fig. 8. Average missing detection rate in five strategies of PPP-AR at 50 user end stations for DOY 083, 2019, using the undifferenced and uncombined FCB products (up panels) and the ionosphere-free FCB products (bottom panels).

Table 2. The missing detection rate in five strategies for PPP-AR at user end in 30 min, 1 h and 2 h under static condition and dynamic condition.

Test	30 min (static)	30 min (kinematic)	1 h (static)	1 h (kinematic)	2 h (static)	2 h (kinematic)
Test 1	20.1%	26.3%	12.1%	18.4%	6.1%	12.4%
Test 2	14.3%	17.2%	8.8%	12.3%	4.4%	8.8%
Test 3	3.8%	3.9%	3.9%	5.4%	2.0%	5.0%
Test 4	2.6%	2.4%	2.7%	3.3%	1.4%	3.7%
Test 5	1.8%	1.4%	2.2%	2.5%	1.1%	3.2%

Table 2 shows the statistical results of the missed detection rate for PPP-AR at user end in 30 min, 1 h and 2 h. With the time goes on, the missing detection rate keeps decreasing. Because in the stage of PPP convergence, the geometry of the satellite changes slowly, and the noise of the ambiguity parameter is too large to converge. In the static condition, the miss detection rate of the traditional methods (test1 and test2) is 20.1% and 14.3% respectively, while the proposed methods (test3, test4 and test5) are reduced to 3.8%, 2.6% and 1.8%. In 30 min under the dynamic condition, the missed detection rate using the proposed method decreased from 26.3% and 17.2% to 3.9%, 2.4% and 1.4%. Therefore, compared with the traditional method, the proposed method can effectively reduce the missed detection rate of PPP-AR, and improve the reliability of the results.

4 Summary

Traditional ratio test for PPP-AR lacks the rigorous mathematical theory, usually fixing ambiguity incorrectly in the convergence stage of PPP, which brings an extreme deviation to the results. We proposed a step-by-step ambiguity resolution method based on Bayesian posterior probability. The method is deduced by rigorous mathematical theory, and combined with ration value to set the threshold value in five different scenarios. The ambiguity is fixed step by step according to the size of covariance matrix until output as the fixed or float solution of PPP. The experimental results show that the correctly fixed ambiguity can significantly shorten the convergence time of PPP and improve the accuracy of solution. Compared with the traditional method, the success rate of the new method increases from 95.0% to 99.5% in static condition, and the rate of missing detection reduces from 6.5% to 0.2%; the success rate of the new method increases from 82.9% to 95.9% in kinematic condition, and the rate of missing detection reduces from 13.1% to 1.9%.

Acknowledgement. This paper is funded by the National Natural Science Foundation of China (Fund No. 41574010, 41604013) and the National Key Laboratory of Geographic Information Engineering (ID: SKLGIE2015-Z-1-1).

References

1. Kouba, J., Héroux, P.: Precise point positioning using IGS orbit and clock products. GPS Solutions **5**(2), 12–28 (2001)
2. Navarro-Gallardo, M., Bernhardt, N., Kirchner, M., Musial, J.R., Sunkevic, M.: Assessing Galileo readiness in android devices using raw measurements. In: Proceeding of ION GNSS 2017, Portland, Oregon, September 2017, pp. 85–100 (2017)
3. Gao, Y., Wojciechowski, A., Chen, K.: Airborne kinematic positioning using precise point positioning methodology. Geomatica **59**(1), 29–36 (2005)
4. Geng, J.: Rapid integer ambiguity resolution in GPS precise point positioning. University of Nottingham (2011)
5. Geng, J., Teferle, F.N., Meng, X., et al.: Kinematic precise point positioning at remote marine platforms. GPS Solutions **14**(4), 343–350 (2010)
6. Chen, K., Ge, M., Babeyko, A., et al.: Retrieving real-time co-seismic displacements using GPS/GLONASS: a preliminary report from the September 2015 M w 8.3 Illapel earthquake in Chile. Geophys. J. Int. **206**(2), 941–953 (2016)
7. Li, P., Zhang, X., Ge, M., et al.: Three-frequency BDS precise point positioning ambiguity resolution based on raw observables. J. Geodesy **92**(12), 1357–1369 (2018)
8. Li, X., Li, X., Liu, G., et al.: Triple-frequency PPP ambiguity resolution with multi-constellation GNSS: BDS and Galileo. J. Geodesy **93**, 1–18 (2019)
9. Geng, J., Chen, X., Pan, Y., et al.: PRIDE PPP-AR: an open-source software for GPS PPP ambiguity resolution. GPS Solutions **23**(4), 91 (2019)
10. Ge, M., Gendt, G., Rothacher, M., et al.: Resolution of GPS carrier-phase ambiguities in precise point positioning (PPP) with daily observations. J. Geodesy **82**(7), 389–399 (2008)
11. Li, X., Zhang, X.: Improving the estimation of uncalibrated fractional phase offsets for PPP ambiguity resolution. J. Navig. **65**(3), 513–529 (2012)
12. Xiao, G., Sui, L., Heck, B., et al.: Estimating satellite phase fractional cycle biases based on Kalman filter. GPS Solutions **22**(3), 82 (2018)
13. Pan, Z., Chai, H., Kong, Y.: Integrating multi-GNSS to improve the performance of precise point positioning. Adv. Space Res. **60**(12), 2596–2606 (2017)
14. Li, P., Zhang, X.: Precise point positioning with partial ambiguity fixing. Sensors **15**(6), 13627–13643 (2015)
15. Leick, A., Rapoport, L., Tatarnikov, D.: GPS satellite surveying. Wiley (2015)
16. Bock, H., Dach, R., Yoon, Y., et al.: GPS clock correction estimation for near real-time orbit determination applications. Aerosp. Sci. Technol. **13**(7), 415–422 (2009)
17. Liu, T., Zhang, B., Yuan, Y., et al.: Multi-GNSS triple-frequency differential code bias (DCB) determination with precise point positioning (PPP). J. Geodesy **93**(5), 765–784 (2019)
18. Zhang, B.: Study on the theoretical methodology and applications of precise point positioning using undifferenced and uncombined GNSS data. Acta Geodaetica et Cartographica Sinica **43**(10), 1099 (2014)
19. Dong, D.N., Bock, Y.: Global positioning system network analysis with phase ambiguity resolution applied to crustal deformation studies in California. J. Geophys. Res. Solid Earth **94**(B4), 3949–3966 (1989)
20. Koch, K.R.: Bayesian Inference with Geodetic Applications. Springer (2006)
21. Teunissen, P.J.G.: Least-squares estimation of the integer GPS ambiguities. Invited lecture, section IV theory and methodology, IAG general meeting, Beijing, China (1993)
22. Wu, Z., Bian, S.: GNSS integer ambiguity validation based on posterior probability. J. Geodesy **89**(10), 961–977 (2015)

Convergence Analysis on Iterative Algorithm in Ultra-Wideband Positioning Under Ill-Conditioned Configuration

Chuanyang Wang[1,2], Jian Wang[3(✉)], Hang Yu[1,2], Yipeng Ning[4], and Feng Xu[1,2]

[1] NASG Key Laboratory for Land Environment and Disaster Monitoring, China University of Mining and Technology, Xuzhou, China
wangchuanyang@cumt.edu.cn, yhecit@163.com, FengXuWH@163.com

[2] School of Environment Science and Spatial Informatics, China University of Mining and Technology, Xuzhou, China

[3] Beijing University of Civil Engineering and Architecture, Beijing, China
wjiancumt@163.com

[4] School of Surveying and Geo-Informatics, Shandong Jianzhu University, Jinan, China
891197ning@163.com

Abstract. Since the functional model of UWB positioning is nonlinear, iterative algorithms are often considered for solving the localization problem. With a rough initial value, we can obtain the optimal solution by the way of continuous iteration. However, in the UWB indoor positioning, the positioning system is prone to become ill-posed. It resulting in iterative algorithm cannot easily converge to a global optimal solution. In this paper, the convergence on iterative algorithm is analyzed. First of all, the nonlinear least-squares solution of distance equations in UWB positioning is given, then, four optimization iterative methods are presented. Finally, the four methods are applied to UWB static and dynamic positioning under ill-conditioned positioning configuration, the convergence property of the four methods are compared. For the iteration, three types of initial values are selected, and the three cases represent bad, general and good initial value respectively. Experimental results are given to demonstrate that although the barycenter method can converge correctly, it is inefficient with too more iterations. In addition, with a good initial value, the Gauss-Newton method can converge effectively, its iterations increase a lot with a general initial value, and this method rarely converges successfully, and sometimes converges to a false local optimization solution when selecting a bad initial value. Moreover, both the regularized Gauss-Newton method and closed-form Newton method work and converge to the global optimum effectively under the three types of initial values, and the closed-form Newton method has fewer iterations. The study shows that the closed-form of Newton method has higher efficiency of convergence than the other methods in Ultra-wideband positioning under ill-conditioned configuration.

Keywords: Ultra-wideband · Convergence · Iterative · Positioning · Ill-conditioned

J. Sun et al. (Eds.): CSNC 2020, LNEE 650, pp. 338–347, 2020.
https://doi.org/10.1007/978-981-15-3707-3_32

1 Introduction

Urban dwellers spend more than 80% of their time indoors, however, the Global Navigation Satellite System (GNSS) signal doesn't strong sufficiently to penetrate through different materials used in construction, also the phenomena of reflection and multipath fading limit the utility of GNSS in dense urban or in the indoor environments, which makes the indoor positioning demand growing [1]. In general, Indoor localization technologies include: Bluetooth, WIFI, RFID, Ultra-wideband (UWB) and so on. Compared with other localization technologies, UWB is capable of providing robust signal through-wall propagation and a large bandwidth that allows high-resolution ranging even in harsh environments [2]. It is well-suited for indoor positioning applications.

The positioning model based on UWB observations is nonlinear. In general, the nonlinear algorithms can be divided into two families, non-iterative and iterative. The non-iterative algorithm often employs clever algebraic techniques in order to produce a simple direct solution [3, 4]. The most straightforward way to estimate the position may be directly solving a set of simultaneous equations [5]. Although the algebraic techniques solve distance nonlinear system of equations without linearization and iteration, they do not solve all general types of nonlinear equations but only those nonlinear system of equations that can be converted into algebraic form, because most nonlinear least-squares problem cannot be solved by algebraic methods. Moreover, they cannot make full use of extra measurements available to improve the accuracy. The traditional positioning algorithms solve the nonlinear distance equations by linearizing them, and then, the solution can be found iteratively [6]. Iterative algorithms are the procedures that are from a rough initial guess to generate a sequence of improving approximate solutions for nonlinear equations [7]. The iterative process requires and relies heavily on the initial starting values to ensure global minimum and faster convergence to the correct solution [8].

A standard method for solving the nonlinear equations is the Gauss-Newton iteration which is efficient and has a linear rate of convergence for points close to the solution [9]. However, the Gauss-Newton method is based on linearization, and may thus perform poorly when the base stations are ill-conditioned and the measurement equations are strongly nonlinear [10]. Tikhonov regularization is the most commonly used method for ill-posed problems, the chosen Tikhonov matrix gives preference to solutions with smaller norms which is known as L2 regularization [11]. The regularization and Gauss-Newton method are combined into the iteratively regularized Gauss-Newton method to deal with the matrix rank deficient problems [12]. The Gauss–Newton method strengthened with a regularization term is found to be as accurate as any of the closed-form methods through simulation comparison, while being simpler to implement and avoiding most of the pitfalls [13]. As one kind of steepest decent method, the barycenter method is discussed. It can stabilize the iteration without the matrix inversion and simplify the procedure without calculating the Hessian matrix [14]. The closed-form of Newton's method is connected by the Newton method and Gauss–Newton method, and it has been obtained for solving the distance equations [15].

The main contribution of this work is to analyze the convergence on iterative algorithm in UWB positioning under ill-conditioned configuration. This paper is organized as follows. In Sect. 2, the orthogonality condition to the nonlinear least squares solution of distance equations in UWB positioning is given. In Sect. 3, four kinds of optimization iterative methods are presented. Section 4, the convergence of four kinds of methods applied to UWB Positioning is analyzed by the numerical examples. In Sect. 5, the conclusions are summarized.

2 The Orthogonality Condition to the Nonlinear Least Squares Solution of Distance Equations in UWB Positioning

We evaluate the UWB sensor network positioning system based on the DW1000 module. By determining the time of flight (TOF) of signals travelling between the anchor and tag, the ranging measurements in these transceivers is performed based on the two way ranging (TWR).

The nonlinear observation model of distance equation in UWB positioning is given as

$$L_i = d_i(\mathbf{X}) + \varepsilon_i \; (i = 1, 2, \ldots, n) \tag{1}$$

where $\mathbf{L} = \begin{bmatrix} L_1 & L_2 & \cdots & L_n \end{bmatrix}^{\mathrm{T}}$ denotes the observation vector; $\mathbf{X} = [x, y, z]$ is the coordinate vector of unknown point; ε_i represents the corresponding ranging random error; $d_i(\mathbf{X})=\sqrt{(x_i - x)^2 + (y_i - y)^2 + (z_i - z)^2}$ is the Euclidean distance, (x_i, y_i, z_i) is the known coordinate of anchor.

The tag coordinates can be estimated by the nonlinear equations. If the nonlinear equations are over-determined, then the nonlinear least-squares solution is to find $\hat{\mathbf{X}} = \arg\min_{\mathbf{X}} g(\mathbf{X})$, where $g(\mathbf{x}) = \mathbf{V}^{\mathbf{T}}\mathbf{PV}/2$ is the least squares objective function, in which $\mathbf{V}(\mathbf{X}) = d(\mathbf{X}) - \mathbf{L}$ represents the residual vector, $\mathbf{P} = diag(p_1, p_2, \ldots, p_n)$ is the weight matrix. The solution of the objective function can be then obtained by solving the orthogonal condition as [11]

$$\mathbf{h}(\mathbf{X}) = g'(\mathbf{X}) = \mathbf{J}^{\mathrm{T}}(\mathbf{X})\mathbf{PV}(\mathbf{X}) = 0 \tag{2}$$

where $\mathbf{J} = \begin{bmatrix} \mathbf{e}_1^{\mathrm{T}} & \mathbf{e}_2^{\mathrm{T}} & \cdots & \mathbf{e}_n^{\mathrm{T}} \end{bmatrix}^{\mathrm{T}}$ represents the Jacobian matrix, in which $\mathbf{e}_i = [x - x_i, y - y_i, z - z_i]/d_i(\mathbf{X})$ is the direction cosine vector from the unknown point to the known point.

The nonlinear Eq. (2) is a base to discuss the properties of the nonlinear least-squares solution and develop iterative methods and even algebraic approaches for solving the over-determined distance equations [15]. The nonlinearity measures may be obtained by the Hessian matrix of the form

$$\mathbf{H}(\hat{\mathbf{X}}) = \mathbf{h}'(\hat{\mathbf{X}}) = g''(\hat{\mathbf{X}}) = \mathbf{J}^{\mathrm{T}}(\hat{\mathbf{X}})\mathbf{P}\mathbf{J}(\hat{\mathbf{X}}) + \sum_{i=1}^{n} V_i(\hat{\mathbf{X}})p_i V_i''(\hat{\mathbf{X}}) \\ = \mathbf{N}(\hat{\mathbf{X}}) + \mathbf{S}(\hat{\mathbf{X}}) \tag{3}$$

where $V_i''(\hat{\mathbf{X}})$ contains the second derivatives of the distance equation [9].

3 The Optimization Iterative Methods for UWB Positioning

There exist many optimization methods to minimize the objective function to achieve the optimal position estimate. Next in the following discussion we will use the barycenter method, the Gauss–Newton method, the regularized Gauss-Newton method and the closed-form Newton method for investigation to solve the UWB positioning model. We restrict ourselves to gradient descent methods for which the required derivatives can be obtained analytically. Gradient descent methods take the general form [9]

$$\mathbf{X}_{k+1} = \mathbf{X}_k + t_k\mathbf{Q}(\mathbf{X}_k)\mathbf{h}(\mathbf{X}_k) \quad (k = 0, 1, 2, \ldots) \tag{4}$$

where t_k is the positive scalar, and $\mathbf{Q}(\mathbf{X}_k)$ is the positive defined matrix. Different choices for t_k and $\mathbf{Q}(\mathbf{X}_k)$ correspond with different gradient descent methods.

3.1 The Barycenter Method

The barycenter method is structured by the choice $t_k\mathbf{Q}(\mathbf{X}_k) = \mathbf{I}/tr(\mathbf{P})$, the iterative equation can be written as

$$\mathbf{X}_{k+1} = \mathbf{X}_k + \frac{1}{tr(\mathbf{P})}\mathbf{h}(\mathbf{X}_k) \quad (k = 0, 1, 2, \ldots) \tag{5}$$

The relation indicates that the barycenter method is one kind of steepest descent method, the advantage of this method is its simplicity, no higher order partial derivatives that need to be calculated and no matrix inverse operation [14].

3.2 The Gauss–Newton Method

For small residual and low nonlinearity problems in nonlinear adjustment models, a linearized model is a sufficient approximation $\sum_{i=1}^{n} V_i(\mathbf{X})p_i V_i''(\mathbf{X}) \to 0$, the Hessian matrix $\mathbf{H}(\mathbf{X})$ can be approximated to its linear term $\mathbf{N}(\mathbf{X})$, with the choice $t_k\mathbf{Q}(\mathbf{X}_k) = \mathbf{N}^{-1}(\mathbf{X}_k)$, we have the Gauss-Newton method as

$$\mathbf{X}_{k+1} = \mathbf{X}_k + \mathbf{N}^{-1}(\mathbf{X}_k)\mathbf{h}(\mathbf{X}_k) \tag{6}$$

However, the issues come in when the approximation is not adequate, if $\mathbf{H}(\mathbf{X})$ is poor conditioned, $\mathbf{N}(\mathbf{X})$ will not be a sufficient approximation to the Hessian matrix, the solution of the least squares will be unstable.

3.3 The Regularized Gauss–Newton Method

To stabilize the solution of the ill-conditioned least squares, the Gauss–Newton method can be corrected by using prior information, this kind of prior assumption can easily be incorporated in the Gauss–Newton method as a regularization term, the regularized Gauss-Newton method is [13, 15]

$$\mathbf{X}_{k+1} = \mathbf{X}_k + (\mathbf{N}(\mathbf{X}) + \beta\mathbf{I})^{-1}(\mathbf{h}(\mathbf{X}_k) + \beta(\mathbf{X}_k - \mathbf{X}_r)) \tag{7}$$

where $\mathbf{X}_r$ is the prior location, β is regularization parameter, which corresponds to the prior probability distribution centered at $\mathbf{X}_r$.

For indoor applications, the preliminary information about the user position is limited, the ill-posed problems will occur when the anchors are configured close to coplanar and the measurement errors are large.

3.4 The Closed-Form Newton Method

To solve the ill-conditioned configurations problems, a new iterative descent method called the closed-form Newton method is proposed [15]. The Hessian matrix can be derived from $\mathbf{h}(\mathbf{X})$, it is expressed as

$$\mathbf{H}(\mathbf{X}) = \mathbf{J}^{\mathrm{T}}(\hat{\mathbf{X}})\mathbf{PRJ}(\mathbf{X}) + \alpha\mathbf{I} \tag{8}$$

where $tr(\cdot)$ is the trace of matrix, $\alpha = tr(\mathbf{P}) - tr(\mathbf{PR})$, $\mathbf{R} = diag(L_1/d_1(\mathbf{X}), L_2/d_2(\mathbf{X}), \ldots, L_n/d_n(\mathbf{X}))$.

The closed-form Newton iterative formula can be given

$$\mathbf{X}_{k+1} = \mathbf{X}_k + \mathbf{H}(\mathbf{X}_k)^{-1}\mathbf{h}(\mathbf{X}_k) \tag{9}$$

If the residual are small enough, $\mathbf{R} \to \mathbf{I}$, the closed-form Newton will be reduced to the Gauss–Newton method, it can be treated as a damped Gauss–Newton method.

4 Numerical Examples

In this section, we will compare the convergence of four kinds of methods described above by taking the static and dynamic tests. The hardware devices used are the DecaWave Mini2016 suite, including the anchor and tag nodes. The anchor nodes are placed on a tripod at the same antenna height. The experimental scene is an empty hall with 10 m × 18 m space area. During the tests, no relevant personnel entered. The data sampling interval of the positioning system is about 0.28 s.

For the initial values for iteration, three sets of coordinates are conducted: [360, 360, 86.75] (Case 1), [360, 360, 96] (Case 2) and [360, 360, 130] (Case 3). The three cases 1, 2 and 3 represent bad, general and good initial value respectively. In the following figures, the barycenter method is represented by the abbreviation BCT, the Gauss-Newton method is represented by the abbreviation GN, the regularized Gauss-Newton method is represented by the abbreviation RGN, the closed-form Newton method is represented by the abbreviation CFN.

4.1 Static Test

The main goal of this test is to evaluate the convergence of each method with four test points in a static situation. The experimental scene and the distribution of the anchors and test points are shown in Fig. 1. The positions of test points were calibrated ahead of time. In the static test, 300 epoch data was sampled at each test point. The tag node was on the top of the helmet, which was placed on the floor.

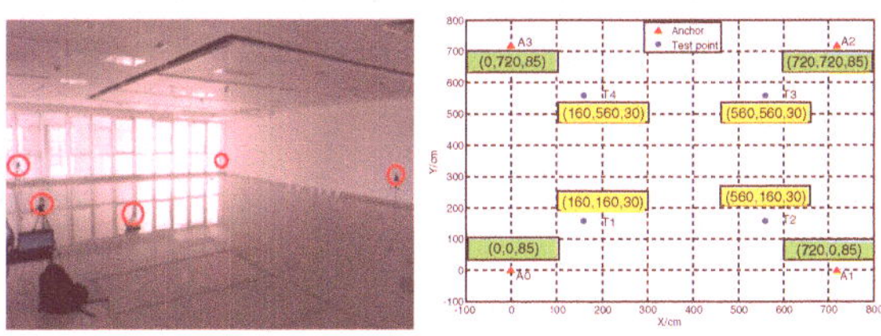

(a) (b)

Fig. 1. The experimental scene (a), the coordinates of the anchors and test points (b).

As is shown in the Fig. 2, the barycenter method converges to the global optimal solution in the three cases, but with too many iterations or too much computation load than other methods even with a good initial value (Case 3). The iteration success rate of the Gauss-Newton method is lowest with a bad initial value (Case 1), and all of the successful iterations are false convergence with the local optimal solution. Due to the deficiency of the Jacobian matrix, this method converges to a false solution in the end. The iteration success rate of the Gauss-Newton method is improved obviously with a general initial value (Case 2), and most epochs converge to the true solution, but converge with many iterations or diverge sometimes. All the iterations of the Gauss-Newton method converge to the true solutions with a good initial value (Case 3), and with few iterations. All the iterations of the regularized Gauss-Newton method and closed-form Newton method converge to the true solutions in the three cases, both of them work and converge to the global optimum effectively under the three types of initial values. The closed-form Newton method has the approximately equal iterations in the three cases and can converge to the global optimal solution with better convergence and higher efficiency. Both the regularized Gauss-Newton method and closed-form Newton method can be applied in Ultra-wideband positioning under ill-conditioned configuration.

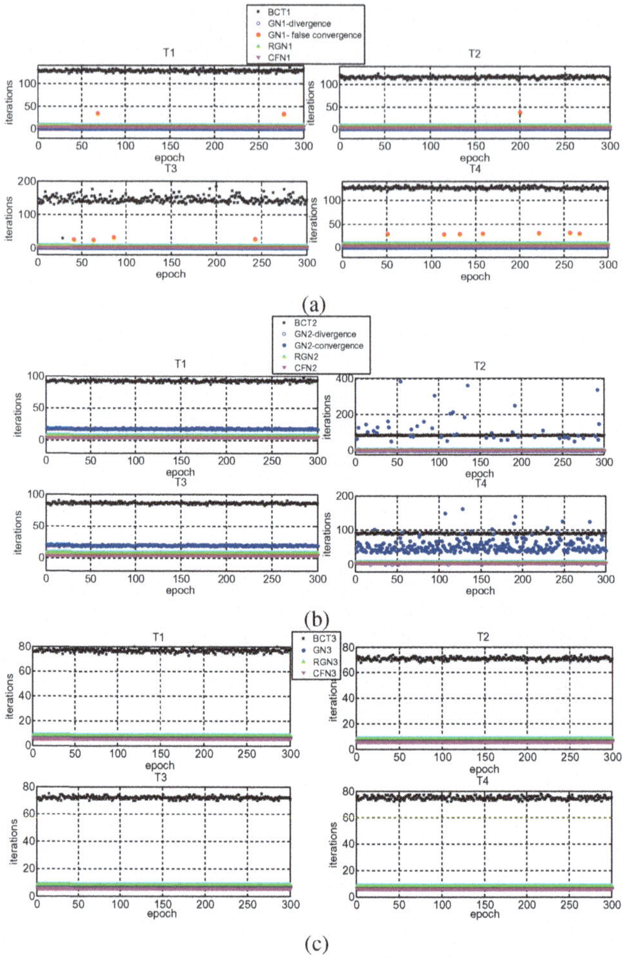

Fig. 2. The comparison of the four methods for the iterations in Case 1(a), Case 2(b), Case 3(c).

4.2 Dynamic Test

The evaluation of this test is in a dynamic situation. The test trajectory and the experimental scene are shown in Fig. 3. The tag node was mounted on the head of the human body at an antenna height of 1.95 m, the tester walked through the predetermined trajectory. The convergence property and efficiency of the four methods are analyzed.

As is shown in Fig. 4, when working with the with a bad initial value (Case 1), the Gauss-Newton method rarely converges successfully, and sometimes even converges to a false local optimization solution. With a general initial value (Case 2), although there is still an epoch with iterative divergence, the Gauss-Newton method has no false convergence problem. When selecting a good initial value (Case 3), all the four

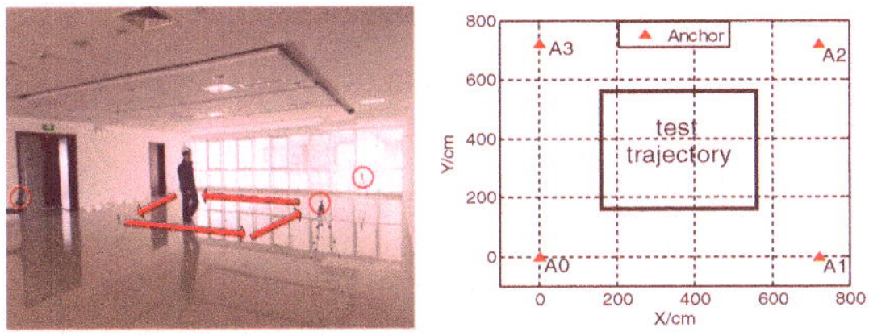

(a) (b)

Fig. 3. The experimental scene (a) and the test trajectory (b).

methods work and converge to the global optimum. The Gauss-Newton method can also fast coverage to the global optimization solution, and it even has fewer iteration times compared with the regularized Gauss-Newton method. The barycenter method converges very slowly in all the cases, but still has the global optimal solution in the end. The regularized Gauss-Newton method can work well in all the cases, with more iterations than the closed-form Newton method. The closed-form Newton method has the highest success rate and the minimum average number of iteration. It confirms that the closed-form Newton method performs better than the other three methods for all the cases in Ultra-wideband positioning under ill-conditioned configuration.

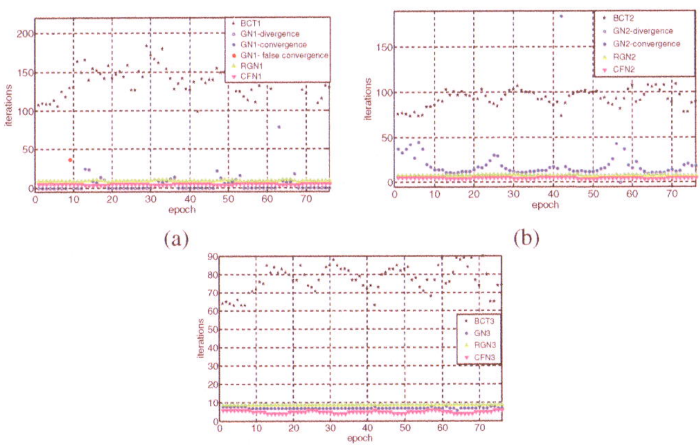

(a) (b)

(c)

Fig. 4. The comparison of the four methods for the iterations in Case 1(a), Case 2(b), Case 3(c).

5 References

In this paper, the convergence on four optimization iterative descent methods has been compared in the UWB positioning experiment under ill-conditioned positioning configuration. In the UWB indoor positioning, since the information on the user position is rather limited, it is not easy to get a good initial value. Moreover, the signals may be blocked, resulting in an increase in relative ranging error. The positioning system is prone to become ill-posed. The barycenter method can stabilize the iteration, but its iteration is inefficient. The Gauss-Newton method ignores the terms with high order so that it has higher computational efficiency than the other methods. But when selecting a bad initial value, the Gauss-Newton method seldom converges or converges to a false optimum. The regularized Gauss-Newton method constrains the ill-conditioned equations and takes advantage of the prior information which is not easy to obtain sometimes. The closed-form Newton method which considers the high order partial derivatives is more stable and effective. Both of them can converge to a true optimum with increasing the computational cost a little. Therefore, when the positioning configuration is ill-conditioned and the initial value cannot be obtained, we can choose to use the closed-form Newton method. Meanwhile, we generally try to avoid the ill-posed problems on the deployment of positioning system.

Acknowledgments. This study is funded by the National Key Research and Development Program of China (2016YFC0803103).

References

1. Li, X., Yang, S.: The indoor real-time 3D localization algorithm using UWB. In: International Conference on Advanced Mechatronic Systems, pp. 337–342 (2015). https://doi.org/10.1109/icamechs.2015.7287085
2. Ferreira, A.G., Fernandes, D., Catarino, A.P., et al.: Performance analysis of ToA-based positioning algorithms for static and dynamic targets with low ranging measurements. Sensors **17**(8), 1915 (2017). https://doi.org/10.3390/s17081915
3. Awange, J.L., Fukuda, Y.: Resultant optimization of the three-dimensional intersection problem. Surv. Rev. **39**(304), 100–108 (2007). https://doi.org/10.1179/003962607x165050
4. Sirola, N.A.: A versatile algorithm for local positioning in closed form. In: Proceedings of the 8th European Navigation Conference GNSS, pp. 16–19 (2004)
5. Fang, B.T.: Simple solutions for hyperbolic and related position fixes. IEEE Trans. Aerosp. Electron. Syst. **26**(5), 748–753 (1990). https://doi.org/10.1109/7.102710
6. Foy, W.H.: Position-location solutions by Taylor-series estimation. IEEE Trans. Aerosp. Electron. Syst. **2**, 187–194 (1976). https://doi.org/10.1109/TAES.1976.308294
7. Cheung, K.W., So, H.C., Ma, W.K., et al.: Least squares algorithms for time-of-arrival-based mobile location. IEEE Trans. Signal Process. **52**(4), 1121–1130 (2004). https://doi.org/10.1109/TSP.2004.823465
8. Xu, P.: A hybrid global optimization method: the multi-dimensional case. J. Comput. Appl. Math. **155**(2), 423–446 (2003). https://doi.org/10.1016/s0377-0427(02)00878-6
9. Teunissen, P.J.G.: Nonlinear least squares. Manuscripta Geod. **15**, 137–150 (1990)
10. Yan, J., Tiberius, C.C.J.M., Bellusci, G., et al.: Feasibility of Gauss-Newton method for indoor positioning. IEEE Position Location and Navigation Symposium (2008). https://doi.org/10.1109/PLANS.2008.4569986
11. Ng, A.Y.: Feature selection, L1 vs. L2 regularization, and rotational invariance. In: Proceedings of the Twenty-First International Conference on Machine Learning, p. 78 (2004). https://doi.org/10.1145/1015330.1015435

12. Madsen, K., Nielsen, H.B., Tingleff, O.: Methods for Non-linear Least Squares Problems, 2nd edn. Informatics and Mathematical Modelling (IMM), Technical University of Denmark (DTU), Lyngby (2004)
13. Sirola, N.: Closed-form algorithms in mobile positioning: myths and misconceptions. In: 2010 7th Workshop on Positioning Navigation and Communication (WPNC). IEEE, pp. 38–44 (2010). https://doi.org/10.1109/wpnc.2010.5653789
14. Xue, S., Yang, Y., Dang, Y.: Barycenter method for solving distance equations. Surv. Rev. **48**(348), 188–194 (2016). https://doi.org/10.1179/1752270615Y.0000000020
15. Xue, S., Yang, Y., Dang, Y.: A closed-form of Newton method for solving over-determined pseudo-distance equations. J. Geodesy **88**(5), 441–448 (2014). https://doi.org/10.1007/s00190-014-0695-y

Comparison of RDSS Timing for BD-2 and BD-3 System

Dongxia Wang[1,2](✉), Rui Guo[1](✉), Tianqiao Zhang[1], Zhijun Liu[1], and Jie Xin[1]

[1] 32021 Troops, Beijing 100094, China
wdx2008abc@163.com, shimbarsalon@163.com
[2] Shanghai Astronomical Observatory, Chinese Academy of Sciences, Shanghai 200030, China

Abstract. With the launching of BD-3 satellites, Beidou navigation satellite system is going to the global service operation. Therefore, this paper study the RDSS timing principle of BD-2 and BD-3 systematically. By analysing the one-way and two-way timing precision, we have the conclusions: for RDSS one-way timing, the precision of BD-2 is less than 30 ns, and its RMS is less than 6.81 ns, the precision of BD-3 is less than 20 ns, and its RMS is less than 4.26 ns; for RDSS two-way timing, the precision of BD-2 is less than 20 ns, and its RMS is less than 3.60 ns, the precision of BD-3 is less than 10 ns, and its RMS is less than 2.09 ns. The results reflect timing performance for Beidou regional and global system.

Keywords: RDSS timing · Performance evaluation · Beidou system

1 Introduction

Beidou navigation satellite system is developed independently by China, which has two modes as radio determination satellite service (RDSS) and radio navigation satellite service (RNSS). RDSS service is not only the important component and distinctive feature of Beidou system, but also the advantage and competitiveness which are the differences from GPS, GLONASS, Galileo and other systems [1–4].

RDSS time service adopts the four ranges mode, which is calculated by the ground master station to obtain the clock difference information required by the user. At present, Beidou global satellite navigation system has officially opened basic navigation services on December 27, 2018. Like Beidou regional system, its RDSS timing methods include two modes: one-way RDSS timing and two-way RDSS timing [5, 6]. Based on the measured data of Beidou system, this paper studies the one-way and two-way timing service performance of RDSS. On the basis of literature [7–9], the paper systematically compares the RDSS timing performance BD-2 and BD-3, which provides a difference reference of Beidou RDSS Consistency.

J. Sun et al. (Eds.): CSNC 2020, LNEE 650, pp. 348–357, 2020.
https://doi.org/10.1007/978-981-15-3707-3_33

2 RDSS Timing Principle and Data

2.1 One-Way Timing Principle

RDSS one-way timing of Beidou system is that the main control center (MCC) sends the navigation message information to the geosynchronous orbit satellite (GEO), and then satellite receives and transmits it to the user terminal. After receiving the navigation message parameters, the user terminal independently calculates the clock difference information to ensure the synchronization of local time and satellite time [3, 5]. The principle of one-way timing is shown in Fig. 1.

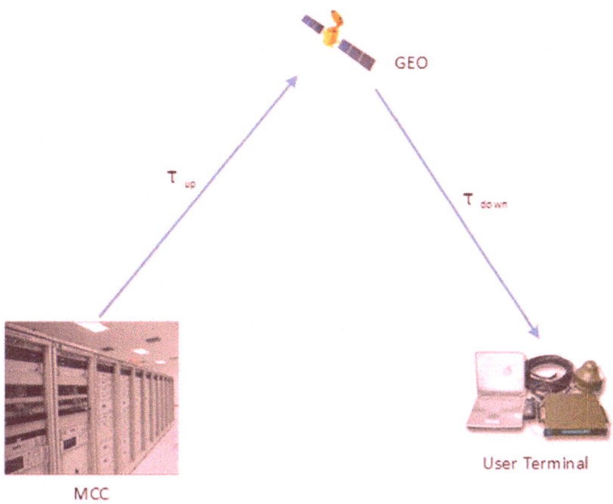

Fig. 1. RDSS one-way timing principle

RDSS one-way timing can be obtained as Eq. (1) [9].

$$\begin{aligned}\Delta t_{\mathrm{u}} &= \tau_{\mathrm{sys}} - \tau_{\mathrm{rev0}} \\ &= \left(\tau_{\mathrm{rev1}} - \tau_{\mathrm{up}} - \tau_{\mathrm{down}} - \tau_{\mathrm{other}}\right) - \left(\tau_{\mathrm{rev1}} - \tau_{\mathrm{delay}}\right) \\ &= \tau_{\mathrm{delay}} - \tau_{\mathrm{up}} - \tau_{\mathrm{down}} - \tau_{\mathrm{other}}\end{aligned} \tag{1}$$

where τ_{delay} is the total delay of the receiver signal, τ_{up} is the time delay from ground station to satellite, τ_{down} is the time delay from satellite to user, and τ_{other} is the other time delays such as ionosphere and troposphere on the link.

2.2 Two-Way Timing Principle

RDSS precision can be affected by many uncertain factors such as atmospheric time delay, it is difficult to calculate accurately. In order to achieve the higher precision, the unidirectional propagation delay between transmitters and receivers is determined by the bidirectional comparison method, and the Beidou high-precision RDSS bidirectional time service method based on the RDSS response ranging is established. This method requires that the user terminal has the receiving and response transmitting capabilities at the same time [3, 5]. The principle of two-way time service is shown in Fig. 2.

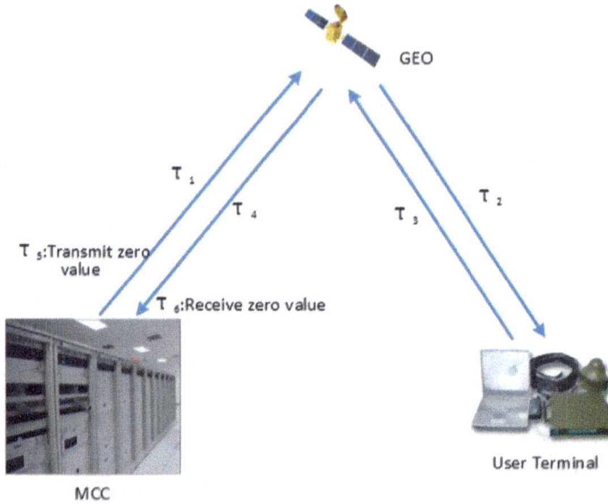

Fig. 2. RDSS two-way timing principle

Assuming τ_8 is the delay of the user terminal equipment (including the one-way transmission time delay and clock difference), RDSS two-way timing can be obtained as [9].

$$\Delta t = \rho - \tau_+ - \tau_8 = \rho - \frac{\tau + \tau_7 + \tau_5 - \tau_6}{2} - \tau_8 \tag{2}$$

where τ is the total two-way transmission delay, τ_+ is the forward transmission delay and τ_- is the reverse transmission delay, which can be expressed as

$$\left.\begin{aligned} \tau &= \tau_1 + \tau_2 + \tau_3 + \tau_4 + \tau_5 + \tau_6 \\ \tau_+ &= \tau_1 + \tau_2 + \tau_5 \\ \tau_- &= \tau_3 + \tau_4 + \tau_6 \end{aligned}\right\} \tag{3}$$

And the inconsistency between the forward and the reverse transmission delay can be expressed as

$$\tau_7 = \tau_1 + \tau_2 - \tau_3 - \tau_4 \tag{4}$$

2.3 RDSS Measured Data

In order to obtain comprehensive and sufficient calculation results, this paper selects RDSS data of Beidou regional system from October 2 to October 7, 2019. All five in-orbit GEO satellites are connected to the RDSS service, and each satellite in the Beidou regional system broadcasts two beams. Only the seven beams are considered here, because other three beams cannot receive data from Beijing station.

At present, Beidou global system have launched only one GEO satellite which is still at the test. Therefore, we just consider one beam of RDSS clock difference data for Beidou global system at Beijing station from October 2 to October 7, 2019.

3 One-Way Timing Analysis of BD-2 and BD-3

3.1 One-Way Timing Results of Regional System

For the analysis of the regional system, the one-way clock difference curve of Beijing station can be obtained as shown in Fig. 3.

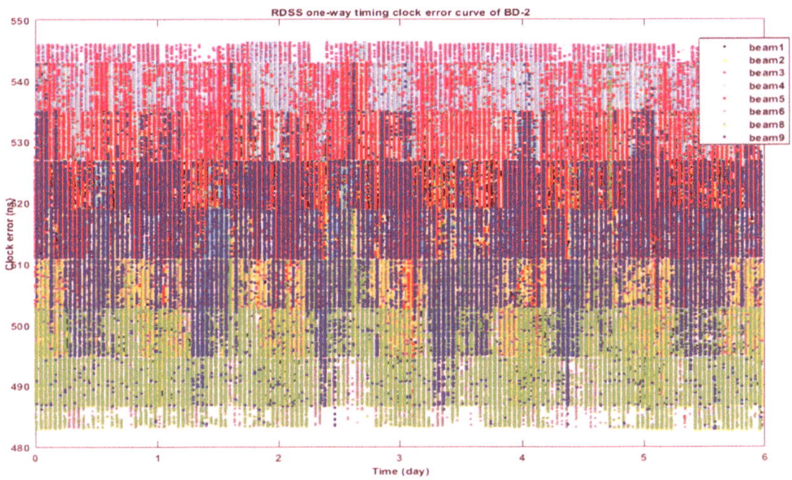

Fig. 3. RDSS one-way timing clock error curve of BD-2

It can be seen that there is a one day periodicity for different beams, and there is jump between beams. According to the statistical analysis, there are 1230 beam segments and 1230 beam switching in 6 days. Average the clock difference of each beam, we can have the mean value curve in Fig. 4.

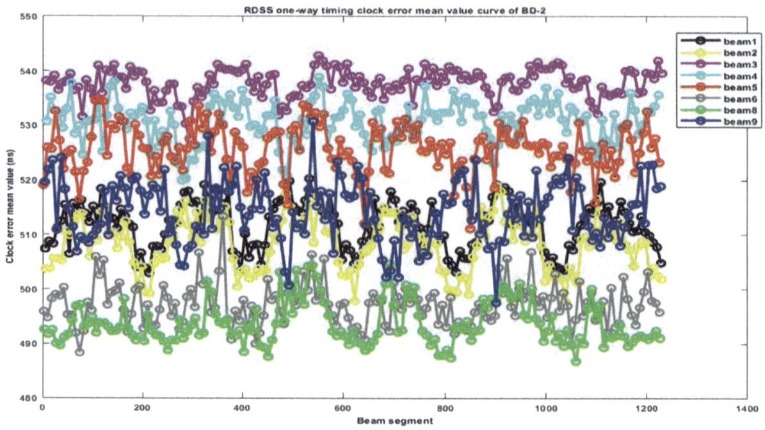

Fig. 4. RDSS one-way clock difference zero value curve of BD-2

It can be seen from Fig. 4 that the clock difference zero value of each beam is stratified. Remove the zero value for each beam, and the one-way timing precision of each beam in Beijing station can be obtained as shown in Fig. 5. It can be seen that the one-way timing precision is less than 30 ns, and RMS average value of all beams is 6.81 ns.

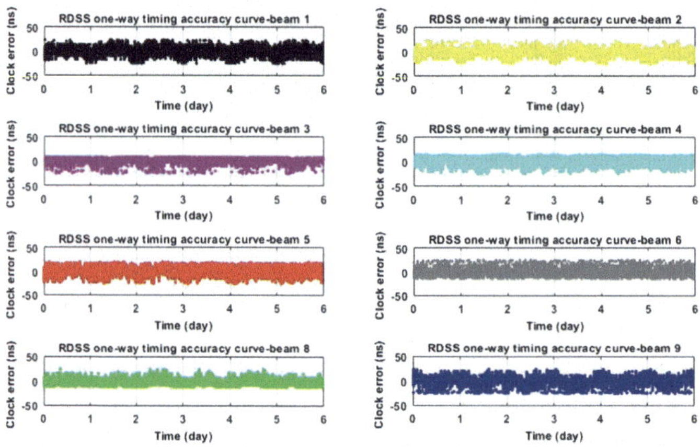

Fig. 5. RDSS one-way timing precision curve of BD-2

3.2 One-Way Timing Results of Global System

For the analysis of the global system, the one-way clock difference curve of Beijing station can be obtained as shown in Fig. 6.

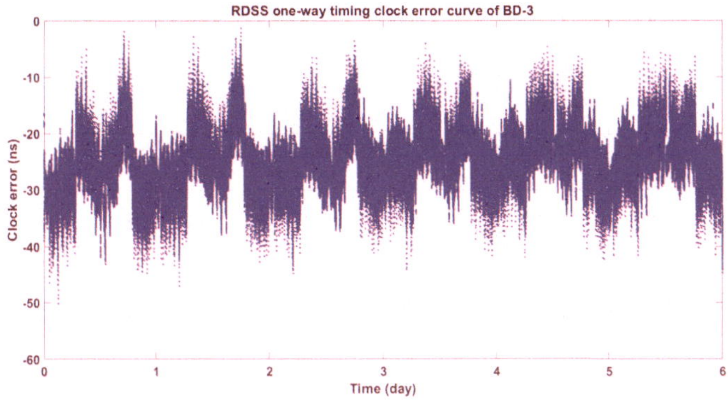

Fig. 6. RDSS one-way timing clock error curve of BD-3

It can be seen that there is a one day periodicity. Removing the zero value of the beam, the one-way timing precision of Beijing station can be obtained as shown in Fig. 7. It can be seen that the one-way timing precision is less than 20 ns, and RMS is 4.26 ns.

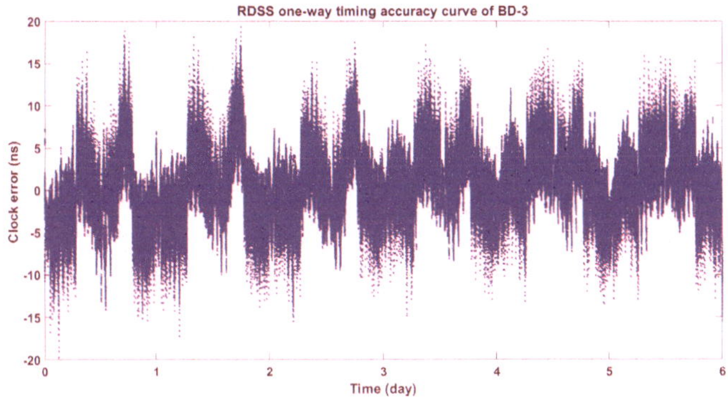

Fig. 7. RDSS one-way timing precision curve of BD-3

3.3 Summary

By studying the results of one-way timing of Beidou global system and regional system, this paper draws the following conclusions:

(1) the one-way timing precision of each beam in the regional system is less than 30 ns, RMS is 6.81 ns; the one-way timing precision of RDSS in the Beidou global system is less than 20 ns, RMS is 4.26 ns. As a result, the precision of one-way timing of global system has been improved by 40%.
(2) the time service data of regional system and global system have a period of one day. The reasons are as follows: satellite motion and ionosphere are all periodic changes of one day, so the precision of one-way time service should be improved by optimizing ionosphere model and one-way time service model.

4 Two-Way Timing Analysis of BD-2 and BD-3

4.1 Two-Way Timing Results of Regional System

For the analysis of the regional system, the two-way clock difference curve of Beijing station can be obtained as shown in Fig. 8.

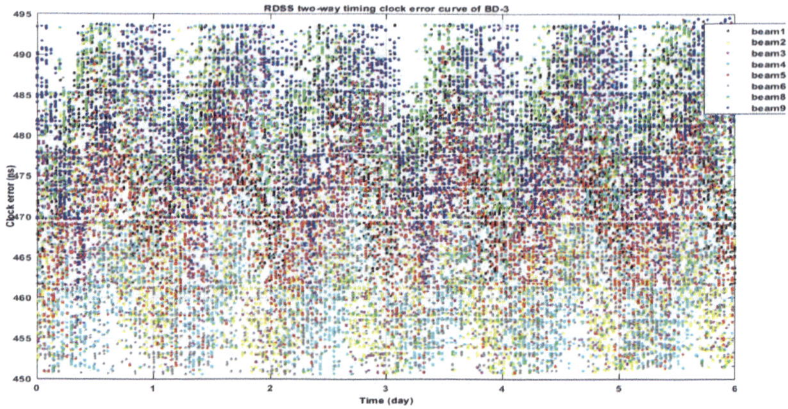

Fig. 8. RDSS two-way timing clock error curve of BD-2

It can be seen that there is a one day periodicity for different beams, and there is jump between beams. According to the statistical analysis, there are 1235 beam segments and 1234 beam switching in 6 days. Average the clock difference of each beam, we can have the mean value curve in Fig. 9.

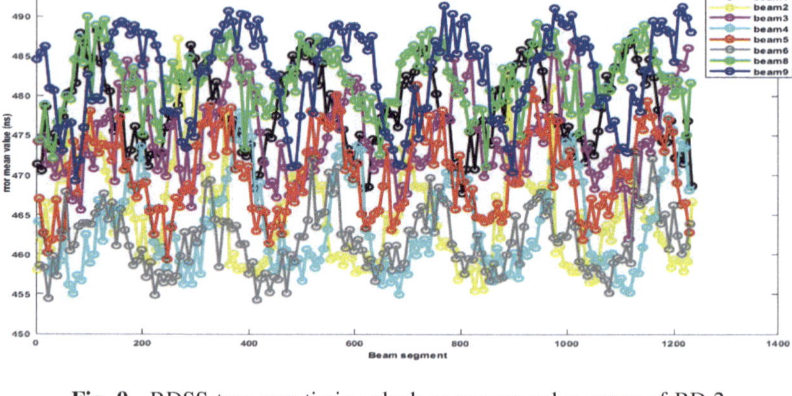

Fig. 9. RDSS two-way timing clock error zero value curve of BD-2

It can be seen from Fig. 10 that the clock difference zero value of each beam is stratified. Remove the zero value for each beam, and the two-way timing precision of each beam in Beijing station can be obtained as shown in Fig. 10. It can be seen that the two-way timing precision is less than 20 ns, and all beam RMS are 3.60 ns.

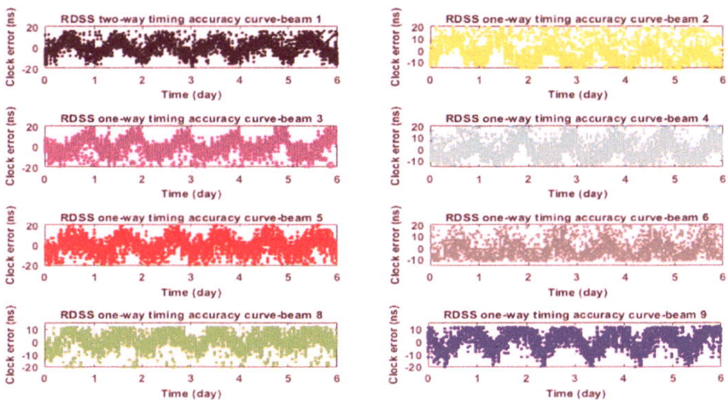

Fig. 10. RDSS two-way timing precision curve of BD-2

4.2 Two-Way Timing Results of Global System

For the analysis of the global system, the two-way clock difference curve of Beijing station can be obtained as shown in Fig. 11.

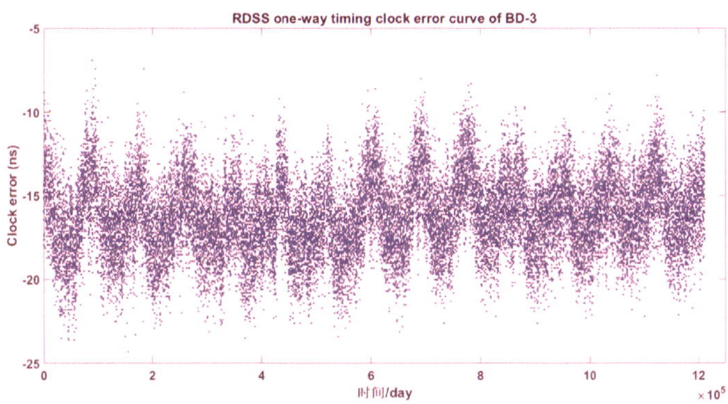

Fig. 11. RDSS two-way timing clock error curve of BD-3

It can be seen that there is a one day periodicity. By removing the zero value of beam, the two-way timing precision of Beijing station is less than 10 ns, RMS is 2.09 ns (Fig. 12).

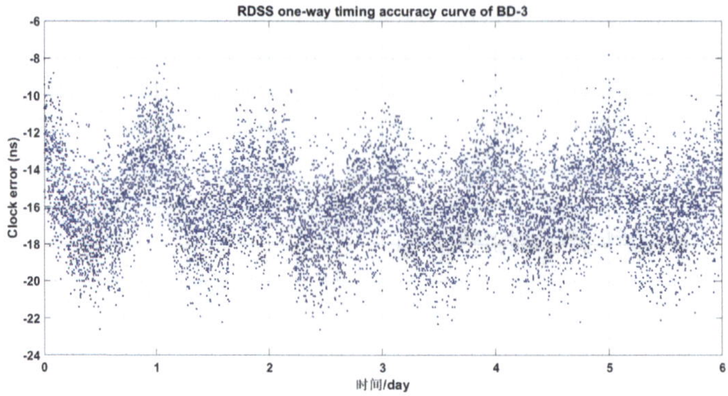

Fig. 12. RDSS two-way timing precision curve of BD-3

4.3 Summary

By studying the results of two-way timing of Beidou global system and regional system, this paper draws the following conclusions:

(1) the two-way timing precision of each beam is less than 20 ns, RMS is 3.60 ns; RDSS of Beidou global system is less than 10 ns, RMS is 2.09 ns. Therefore, the two-way timing precision of the global system has been improved by 42%.
(2) there are periodic phenomena in the two-way time service data of regional system and global system. The reasons are as follows: satellite motion and ionosphere are all periodic changes of one day, so the precision of two-way time service should be improved by optimizing ionosphere model and two-way time service model.

By comparing the results of two-way and one-way timing precision evaluation, the following rules can be obtained:

(1) Compared with the one-way clock difference curve, the difference of each beam of the two-way clock difference is small, the difference of each beam switching is small, that means the consistency is better. The results show that the two-way time service method can eliminate the link delay error and other factors in principle, and has higher precision than the one-way time service method.
(2) There is periodic phenomenon in one-way and two-way time service, and the cycle is 1 day. The results show that the two-way timing method cannot eliminate the errors caused by satellite motion, ionosphere and other periodic changes. RDSS timing precision needs to start from the perspective of error model refinement to further improve the timing precision of the system.

5 Conclusions

With the development and application of Beidou globalization, the RDSS service has attracted more and more attention, while the fast and efficient performance is the key to improve RDSS service. Based on the actual RDSS data of five GEO satellites of regional system and one GEO satellite of global system, this paper systematically analyzes the service performance of one-way and two-way timing performance, which provides a theoretical basis for improving the RDSS timing precision. The research results of this paper enrich the precision evaluation method system of RDSS.

Acknowledgments. Research supported by the National Nature Science Foundation of China (Nos. 61603397, 41874043, 41204022, 41374038) and the State Key Laboratory of Geodesy and Earth's Dynamics Foundation (Institute of Geodesy and Geophysics, CAS) (No. SKLGED2017-3-3-E).

References

1. Tan, S.: Innovative development and forecast of BeiDou system. Acta Geodaetica Cartogr. Sin. **46**(10), 1284–1289 (2017)
2. Tan, S.: Theory and application of comprehensive RDSS position and report. Acta Geodaetica Cartogr. Sin. **38**(1), 1–5 (2009)
3. Wang, T.: Study on the Timing Performance Evaluation of Beidou Satellite Navigation System. Chang'an University, Xi'an (2014)
4. Zhao, J., Qu, J., Yuan, H.: A new ambiguity resolution method using combined RNSS-RDSS of BeiDou. Acta Geodaetica Cartogr. Sin. **45**(4), 404–410 (2016)
5. Sha, H., Zhan, J., Wang, J., et al.: Research and compare of timing method application for compass satellite navigation system. In: Proceedings of the fourth CSNC. [S.l.]: [s.n.], pp. 112–116 (2013)
6. Zhan, J., Pang, J., Zhang, G., et al.: Modeling and simulation testing of RDSS timing. Sci. Sin. (Physica, Mechanica & Astronomica) **41**(5), 620–628 (2011)
7. Wang, D., Guo, R., Zhang, T., Hu, X.: Timing performance evaluation of Radio Determination Satellite Service (RDSS) for Beidou system. Acta Astronaut. **156**, 125–133 (2019)
8. Tan, S.: The Engineering of Satellite Navigation and Positioning, 2nd edn. National Defend Industry Press, Beijing (2010)
9. Wang, D., Guo, R., Xin, J., Yuan, Y.: Performance evaluation of RDSS timing service for Beidou system. Sci. Surveying Mapp. **44**(8), 132–141+148 (2019)

A Multi-constellation Positioning Method Based on Optimal Stochastic Modelling

Jian Wang[1], Zijian Zhou[2], Wei Jiang[1(✉)], Baigen Cai[1], and Wei Shangguan[1]

[1] Beijing Engineering Research Center of EMC and GNSS Technology for Rail Transportation, State Key Laboratory of Rail Traffic Control and Safety, School of Electronic and Engineering, Beijing Jiaotong University, 100044 Beijing, China
weijiang@bjtu.edu.cn

[2] School of Electronic and Engineering, Beijing Jiaotong University, 100044 Beijing, China

Abstract. Recent years, with the continuous improvement of positioning requirements, the inherent deficiency of single constellation positioning is increasingly magnified, the number and distribution of satellites seriously restrict the positioning effect, which impair the availability of the navigation system. In view of this phenomenon, the space-time unification of multi-constellation is carried out to realize the orbit determination error analysis. The weighted least square algorithm is combined with Helmert variance component estimation method to optimize stochastic model, single-constellation dynamic positioning and multi-constellation dynamic positioning are realized respectively in the scene of opening and sheltering. The error and accuracy of the two positioning methods are compared and analyzed to verify the improvement of multi-constellation. The experimental results demonstrate that the orbit errors of the broadcast ephemeris within 10 m. Compared with the traditional single constellation positioning method, the multi-constellation positioning method based on optimal modelling can effectively increase the number of visible satellites, ameliorate the geometric distribution of satellites and improve the positioning accuracy.

Keywords: Multi-constellation positioning · Orbit determination · Helmert variance component estimation method · Stochastic model

1 Introduction

With the continuous improvement of Global Navigation Satellite System (GNSS), multi-constellation based on GNSS interoperability has become a research focus. On account of multi-observations, joint positioning can effectively raise the number of visible satellites and optimize the geometric distribution, so that the system can get rid of dependence on a single constellation.

According to the parameters of navigation, there are inherent physical differences between different systems. Settling differences and improve the interoperability are the premise of multi-constellation positioning [1]. As early as 1990, Ashtech of the United

J. Sun et al. (Eds.): CSNC 2020, LNEE 650, pp. 358–367, 2020.
https://doi.org/10.1007/978-981-15-3707-3_34

States signed the "GPS/GLONASS joint development agreement" with the Soviet Aerospace Equipment Engineering Institute, laying the foundation for multi-system [2]. In 1992, Daly described the method of time transfer using GPS/GLONASS code phase in his paper, beginning the attempt of multi-system interoperability [3]. Since Chinese BeiDou navigation satellite system (BDS) was successfully established in 2012, more and more researchers begun to study multi-system positioning. Li proposed model and parameters of time and space among BDS, GPS and GLONASS, which advanced the unified method of GNSS integrated navigation and positioning [4].

What's more, the traditional stochastic model cannot describe the attributes of multi-mode system strictly, so it is essential to focus on the optimal stochastic modelling [5]. Shang put forward the weight matrix based on the elevation angle, finally obtained better positioning results [6]. Given the fixed prior weight for BDS/GPS/GLONASS multi-constellation system, Zhang got better positioning results than single GPS system and GPS/BDS system [7]. However, according to the changes of time and space, the fixed weight obviously does not conform to the characteristics of the stochastic model. Therefore, this study proposes a stochastic model optimization method with real-time weight adjustment.

In this paper, the GPS/BDS pseudo range single point positioning based on broadcast ephemeris is realized by using the dynamic data of vehicle driving. The satellite elevation angle model is taken as the factor of weight in the same system, the weight between different systems are determined by Helmert variance estimation. The stochastic model is adjusted in real time by using the satellite elevation angle and pseudo range residual. The results show that the optimal stochastic model can achieve better positioning effect.

2 Orbit Determination for GPS and BDS

2.1 Spatio Tempora Unification

The reference time of GPS follows UTC time of U.S. Naval Observatory, which is starting from 0:00 on January 6, 1980. The BDS time system is starting from 0:00 on January 1, 2006. The time system is converted according to formula 1.1.

$$\begin{cases} GPSweek = BDSweek + 1356 \\ GPSsec = BDSsec + 14 \end{cases} \tag{1.1}$$

GPS takes the world geodetic coordinate system in 1984 (WGS-84) as the spatial datum, and BDS takes the 2000 national geodetic coordinate system (CGCS2000) as the spatial datum. Except for the difference of some parameters' value, the basic definitions of the two protocols are similar, and the errors between protocols can be ignored for pseudo range level positioning. The space system parameters are shown in Table 1.

Table 1. Basic parameters of WGS-84 and CGCS2000

Basic parameters	WGS-84	CGCS2000
a	6378137 m	6378137 m
f	1/298.257223563	1/298.257222101
Ω	7.2921151467e-5 rad/s	7.2921150e-5 rad/s
μ	3.986005e14 m^3/s^2	3.986004418e14 m^3/s^2

a is semi-major axis, f is flattening of ellipsoid, Ω is rotational angular velocity of the earth and μ is product of gravity and mass

2.2 Orbit Determination Error Analysis

The accurate determination of satellite orbit is the basis of navigation and positioning. In this paper, the broadcast ephemeris data recorded on June 21, 2018 by UR380 multi-mode receiver. Fetching observation and navigation files in receiver independent exchange format (RINEX) version 3.02, calculating the position of GPS and BDS satellites, comparing the results with the precise ephemeris given by the German Helmholtz Potsdam Center (GFZ).

In WGS-84 coordinate system, the errors of GPS and BDS satellites in X, Y and Z directions are shown in Fig. 1.

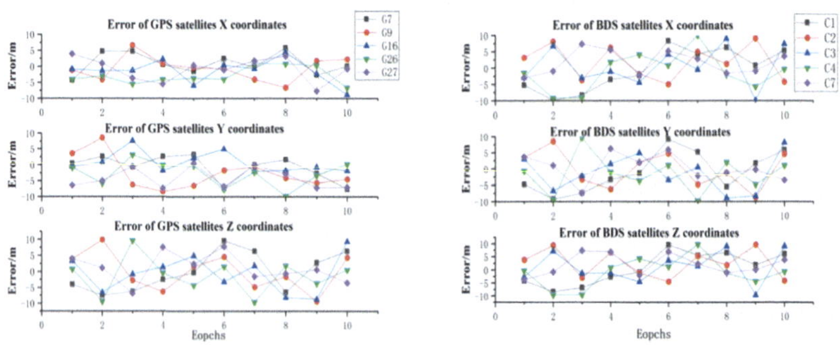

Fig. 1. Orbit error of GPS and BDS broadcast ephemeris satellites

3 Stochastic Model Analysis and Optimization

The stochastic model of multi-constellation system is directly related to the accuracy of location, more reliable positioning results can be obtained by optimizing the stochastic model. Two aspects need to be considered in optimization: First, the determination of satellite weight within the system. second, the determination of weight between different systems. The weight matrix of stochastic model W is generally determined by the unit weight variance $\hat{\sigma}_0^2$ and observation variance $\hat{\sigma}_O^2$.

$$W = \frac{\hat{\sigma}_0^2}{\hat{\sigma}_O^2} \tag{1.2}$$

3.1 Equal Weight Model

It is assumed that different satellites in different systems have equal observation quality, and the weight matrix W is expressed as:

$$W = I \tag{1.3}$$

Where, I is the unit matrix. Without considering the difference of observation in the system, the positioning error will increase, ignoring the difference among systems may lead to the failure of multi-constellation combination [8].

3.2 Prior Models

In a single system, the factors related to the quality of satellite signal are used as evaluation indexes to determine the weights of different observations. At present, the commonly factors include: Elevation angle, Signal-to-noise ratio, Pseudo range residual and User Ranging Accuracy (URA) [9, 10].

Due to the ionosphere and troposphere in satellite signal propagation process, signal quality is related to the elevation angle, which means higher elevation angle corresponds to smaller atmospheric delay. On the other hand, the satellite with low elevation angle is more vulnerable to the influence of multipath. The observation variance under the elevation angle model can be expressed as:

$$\hat{\sigma}_O^2 = \frac{1}{\sin^2(E)} \tag{1.4}$$

According to the literature, the optimal elevation angle model can be written as:

$$\hat{\sigma}_O^2 = a_0 + a_1 \exp(-E/E_0) \tag{1.5}$$

Where, E is the elevation angle of the corresponding satellite, E_0 is the preset elevation angle threshold, a_0 and a_1 can take 0.07 m and 0.6 m respectively [11].

Signal-to-noise ratio refers to the ratio of GNSS signal strength to noise strength, which means higher signal-to-noise ratio with better signal quality. The observation variance under the SNR model can be expressed as:

$$\sigma_{CN}^2 = \sigma_O^2 = 10 + 22500 \times 10^{-\frac{CN_0}{10}} \tag{1.6}$$

Where, CN_0 is the signal-to-noise ratio value.

Pseudo range residual model can be divided into pseudo range residual mean value method and pseudo range residual median method.

The URA is the statistical result of the error caused by the ground monitoring stations and satellites, reflecting the ranging accuracy of the signal, which is provided by 4-bit data N in the navigation message, and the value of URA can be determined according to N:

$$URA = \begin{cases} 2^{1+\frac{N}{2}}, & 0 \leq N \leq 6 \\ 2^{N-2}, & 6 < N \leq 15 \end{cases} \tag{1.7}$$

Therefore, the URA based model can be written as follows:

$$\hat{\sigma}_O^2 = URA^2 \tag{1.8}$$

3.3 Posterior Model

For multi-constellation navigation, the difference of unit variance must be considered: $\hat{\sigma}_{0GPS}^2 \neq \hat{\sigma}_{0BDS}^2$ [12]. Helmert variance estimation method is used to determine the best weight matrix.

Based on the GNSS adjustment model of weighted least squares, it is assumed that there are two kinds of uncorrelated factors. The adjustment function between factors can be written as follows:

$$Z_i = H_i\hat{x} - \Delta\rho_i \tag{1.9}$$

Where Z_i is the ith observation, H_i is the corresponding observation matrix, $\hat{x}$ is the state estimation, $\Delta\rho_i$ is the corresponding measurement error.

The stochastic model is:

$$D_i(Z_i) = D_i(\Delta\rho_i) = \sigma_0^2 W_i^{-1} \tag{1.10}$$

Therefore, the error variance is written as:

$$V_i = H_i\hat{x} - Z_i \tag{1.11}$$

The strict formula of Helmert variance estimation is written as follows:

$$S\hat{\theta} = R_\theta \tag{1.12}$$

Where: $\hat{\theta} = [\hat{\sigma}_{0GPS}^2\ \hat{\sigma}_{0BDS}^2]$, $R_\theta = [V_G^T P_G V_G\ V_B^T P_B V_B]$, S can be expressed as a matrix: $S = \begin{bmatrix} n_1 - 2tr(N_1) + tr(N_1)^2 & tr(N_1N_2) \\ tr(N_1N_2) & n_2 - 2tr(N_2) + tr(N_2)^2 \end{bmatrix}$.

Where: n_1 and n_2 refer to the number of GPS and BDS satellites, $N_1 = N_G N^{-1}$, $N_2 = N_B N^{-1}$, $N = N_G + N_B$, $N_G = H_G^T W_G H_G$, $N_B = H_B^T W_B H_B$.

Strict formula has a large amount of calculation. In practical application, it is generally calculated by simplified formula:

$$\sigma_{0G/B}^2 = \frac{V_{G/B}^T W_{G/B} V_{G/B}}{n_{1/2} - tr(N_{1/2})} \tag{1.13}$$

3.4 Stochastic Model Optimization Steps

In this paper, elevation angle model is used to determine weights in a single system, the initial proportion between systems are given by experience. Finally, the Helmert model is used to obtain the optimal weight matrix. The algorithm steps are as follows:

Step 1. According to the empirical value, take $\hat{\sigma}_{0G}^2 = 1$, $\hat{\sigma}_{0B}^2 = 2$, calculate the weight matrix of GPS and BDS at the initial time in combination with the elevation angle model.
Step 2. Calculate the residual vector of GPS and BDS satellites V_G, V_B.
Step 3. Calculate the updated unit weight variance using Eq. (1.13).
Step 4. Update the weight matrix W_G and W_B:

$$W_B = \frac{\sigma_{0G}^2}{\sigma_{0B}^2} W_{0B}, \; W_G = \frac{\sigma_{0G}^2}{\sigma_{0G}^2} W_{0G} \tag{1.14}$$

Step 5. Stop the cycle and output the optimal weight matrix while $\sigma_{0G}^2/\sigma_{0B}^2$ approximately equal to 1. If the cycle stop condition is not met, use the updated weight matrix to Step 2 to continue iteration.
Step 6. After the iteration, the optimal weight matrix is constructed:

$$W = \begin{bmatrix} W_G & 0 \\ 0 & W_B \end{bmatrix} \tag{1.15}$$

Using observation b, combined with the weighted least square model, the state matrix can be expressed as:

$$\Delta x_i = (H_i^T W_i H_i) H_i^T b \tag{1.16}$$

4 Experimental Results and Analysis

The West Campus of Beijing Jiaotong University is located in the core area of Beijing. The surrounding is mainly high-rise buildings. The playground is an open area, and the south gate is a sheltered area. In the experiment, Huace I90 RTK is taken as base station and UR380 is taken as mobile station, the frequency of both receivers are set to 1 Hz. The starting point is the open area near the playground, through the high-rise sheltered area, and finally back to the starting point.

The CGO2 software is used to import the observation files of mobile station and base station, which can carry out high-precision motion trajectory. The experimental environment and software output results are shown in Fig. 2.

Fig. 2. Experimental environment and software output results

4.1 Number of Visible Satellites

In this paper, the cut-off angle is set to 15°. The results show that the average number of visible satellites for GPS is 3.96, for BDS is 5.77, for GPS/BDS joint is 9.73. Since navigation and positioning depends on the number of satellites, It can maintain that the availability of multi-constellation system is enhanced.

The number of visible satellites during the experimental period is shown in Fig. 3.

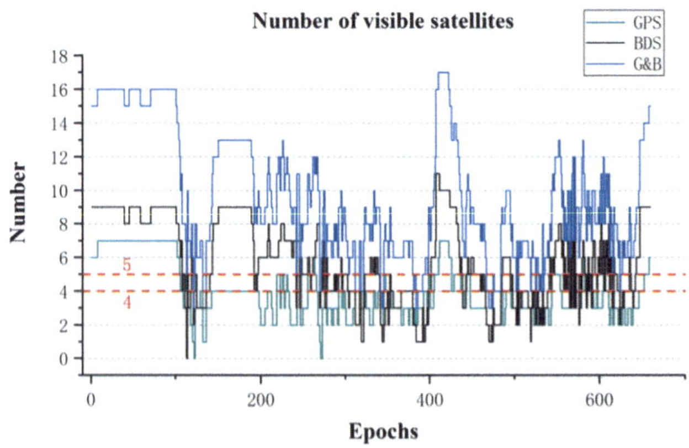

Fig. 3. Number of visible satellites

4.2 Geometric Distribution

The distribution of satellite signal is important to the accuracy of navigation, which is usually expressed by the division of precision (DOP). In the sheltered area, the number of satellites is not enough to form the observation matrix, therefore, the position division of precision (PDOP), horizontal division of precision (HDOP) and vertical division of precision (VDOP) of the three groups are calculated in the first 120 s in open area (Fig. 4).

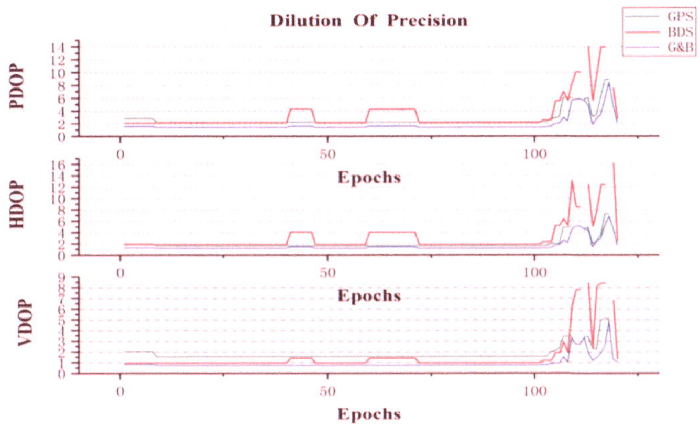

Fig. 4. Dilution of precision

Averaging the DOP values of three models, which reflecting the level of satellites distribution in three-dimensional, horizontal and elevation direction (Table 2).

Table 2. Mean value of DOP

Model	PDOP	HDOP	VDOP
GPS	2.65	1.94	1.79
BDS	3.28	3.08	1.58
GPS/BDS	1.82	1.56	0.93

4.3 Positioning Results and Availability Analysis

This paper realized GPS single point positioning, BDS single point positioning, GPS/BDS joint positioning based on elevation angle and GPS/BDS positioning based on posteriori Helmert model respectively. Using RTK results as reference, obtaining positioning errors in the North-East-Down coordinate.

At the beginning, the satellite signal is good, single constellation positioning can achieve the results within 5 m in the horizontal direction and 10 m in the elevation direction; meanwhile, the multi-constellation positioning can achieve the results within 3 m in the horizontal direction and 5 m in the elevation direction. During sheltered area, due to the interference of the signal, the multi-path effect is enhanced. However, the joint positioning results are better than the single point positioning, while the Helmert multi-constellation positioning model is better than the other three models, and the positioning results are as follows.

The corresponding root mean square (RMS) of the errors were calculated, which reflecting the positioning accuracy of different models (Table 3).

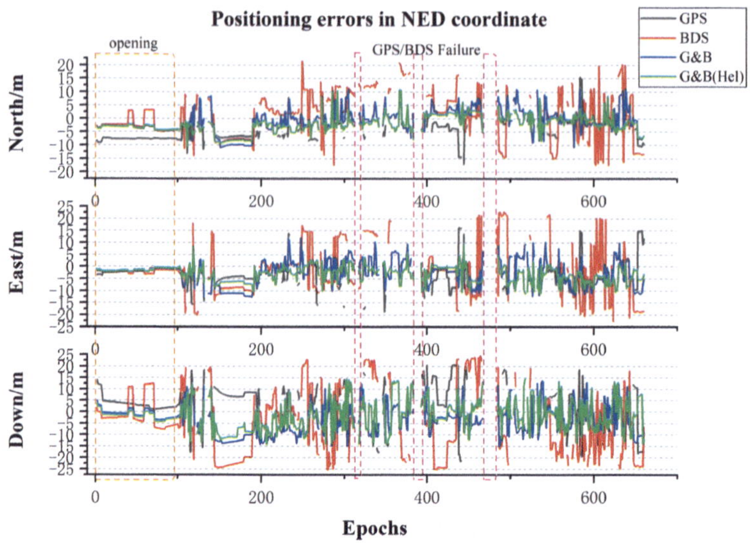

Fig. 5. Positioning errors in the NED coordinate

Table 3. RMS of positioning errors

Models	North error/m	East error/m	Down error/m
GPS(Ele)	6.7253	6.8704	9.7756
BDS(Ele)	8.4458	9.5938	14.8981
G&B(Ele)	4.4857	5.6563	7.0700
G&B(Hel)	3.7518	4.2050	6.5260

For single system positioning, the three-dimensional position and receiver clock difference constitute the state vector together, hence at least four visible satellites are needed for the basic positioning. Meanwhile, two clock differences exist in GPS/BDS positioning, so at least five satellites are needed for basic positioning. The red boxes in Fig. 5 are the failure time of GPS/BDS positioning, compared with single system positioning, the availability of multi-constellation model is greatly improved (Table 4).

Table 4. Positioning failure results

Models	GPS	BDS	GPS/BDS
Failure epochs	305	154	35
Failure rate	46.21%	23.33%	5.30%

5 Conclusion

In this paper, the scene is divided into open area and high-rise sheltered area by using the data of car around the campus. The experiment analyzed the single point positioning based on broadcast ephemeris pseudo range and multi constellation single point positioning. The results show that the orbit determination error of broadcast ephemeris satellite is within 10 m; the multi-constellation positioning optimized by Helmert model can improve the navigation system level in all directions. Compared with GPS single point positioning, BDS single point positioning and GPS/BDS joint positioning based on altitude angle, the optimal positioning model can improve the positioning accuracy, enhance the system availability, increase the number of visible satellites, and optimize the geometric distribution of signals.

Acknowledgements. This research was supported by National Key R&D Program of China (2018YFB1201500), The National Natural Science Foundation of China (61703034).

References

1. Xu, B., Guo, C.: Optimal positioning for GPS and Galileo satellite integrated navigation. Comput. Simul. **36**(3), 110–114 (2019)
2. Wang, H.: The Soviet Union and the United States signed the GPS/GLONASS joint development agreement. Sci. Surveying Mapp. (4), 50 (1990)
3. Daly, P.: Time transfer using GPS/GLONASS code phase. In: IEEE Colloquium on Precise Time & Frequency-the Beat of a Single Drum. London, UK (1992)
4. Li, H., Dang, Y., Mi, J., et al.: Research on spatio tempora unification of BDS/GPS/ GLONASS multi-mode fusion navigation and positioning. J. Geodesy Geodyn. **33**(4), 73–78 (2013)
5. Tabatabaei, A., Mosavi, M., Khavari, A., et al.: Reliable urban canyon navigation solution in GPS and GLONASS integrated receiver using improved fuzzy weighted least-square method. Wireless Pers. Commun. **94**(4), 3181–3196 (2017)
6. Shang, M., Gao, H., Chang, Q.: Weighting methods in GPS/GLONASS/BDS integrated positioning. J. Terahertz Sci. Electron. Inf. Technol. **12**(03), 374–379 (2014)
7. Zhang, C., Ten, J.: Study on priori weight ratio of combined GPS/BDS/GLONASS system. Electric Power Surv. Des. **05**, 52–55 (2019)
8. Yan, L., Huang, D., Feng, W., et al.: Performance comparison of pseudorange point positioning between COMPASS and GPS. In: Electronic Collection of the Third CSNC - S06 Beidou/GNSS Test and Evaluation Technology. Guangzhou, Guangdong, China (2012)
9. Yang, K., Cai, C., Zhang, S.: A high-precision weighted least squares algorithm for GNSS pseudo-range single point positioning. Comput. Simul. **36**(4), 229–233+239 (2019)
10. Zhu, H.: Practical weight estimation model for satellite navigation. Mod. Navig. **7**(6), 396–400 (2016)
11. Zhao, Y.: Research on Stochastical Model for Single Point Positioning of GPS&BDS Combined System. Southwest Jiaotong University, Chengdu (2017)
12. Zhang, Q., Zhao, L., Zhou, J.: A novel weighting approach for variance component estimation in GPS/BDS PPP. IEEE Semsor J. **19**(10), 3763–3771 (2019)

BDS-3/GNSS Data Quality and Positioning Performance Analysis

Renhai Mu(✉), Yamin Dang, and Changhui Xu

Chinese Academy of Survey and Mapping, Beijing, China
1017950536@qq.com

Abstract. Since November 2017, China has launched 24 Beidou-3 (BDS-3) navigation satellites. So far, China's independently developed Beidou satellite navigation system has officially entered the global service stage. Compared with the Beidou-2 satellite navigation system (BDS-2), the MEO satellite of the Beidou-3 satellite navigation system (BDS-3) adds two new signals, B1C and B2a. The quality of the original observation data can directly affect the navigation and positioning system. Positioning accuracy and service performance. This paper selects the original 182-day raw observation data of 23 tracking stations of the International GNSS Monitoring and Assessment System (iGMAS), and analyzes the data quality from four aspects: data integrity rate, signal-noise ratio, cycle slip ratio and multipath effect. The quality of the observed data of GLONASS and GALILEO is compared, and finally combined with the point positioning accuracy of each system for comprehensive evaluation. The results show that in terms of data integrity rate, the data integrity rate of BDS-3 is comparable to that of GPS, GLONASS and GALILEO, and can reach more than 80%. In terms of signal-to-noise ratio, the new BDS-3 satellite The signal strengths of the signals B1C and B2a are not much different from those of the old signals. Compared with other systems, the signal strength of the new BDS signals is better. In terms of the cycle-to-hop ratio, the cycle-bounce ratio of the new signal B2a of the BDS is higher than that of other signals. High, and superior to GPS and GLONASS, it is equivalent to GALILEO's E5b signal; in terms of multipath effect, the multipath RMS value of each signal of the four systems is within 0.5 m, and the multipath error of BDS-3 satellite B2a signal is slightly the multipath error below B1I and B3I is about 0.1 m. In terms of pseudo-single point positioning and precise single-point positioning, the BDS-3 has a slightly lower positioning accuracy than GPS and GALILEO, but is superior to GLONASS. The results of this paper have certain reference value for the location and maintenance of iGMAS observatory, GNSS data processing and product generation, the development of Beidou GNSS receiver and the positioning performance of Beidou satellite navigation system.

Keywords: BDS-3 · B1C and B2a · Data quality · Positioning performance

1 Introduction

As a global satellite navigation system independently developed by China, BeiDou navigation satellite system (BDS) has now developed to the third stage [1, 2]. As of November 2019, China has successfully launched 26 Beidou-3 navigation satellites,

J. Sun et al. (Eds.): CSNC 2020, LNEE 650, pp. 368–379, 2020.
https://doi.org/10.1007/978-981-15-3707-3_35

including 22 Beidou-3 Medium Earth Orbit (MEO) satellites. Table 1 shows the basic situation of the BDS-3 satellites. In order to improve compatibility and interoperability with other satellite navigation systems, BDS-3 adds new system signals B1C and B2a to the B1I and B3I signals broadcast by BDS-2 [3]. Among them, the B1C signal of BDS-3 is compatible and interoperable with the L1 signal of GPS and E1 signal of Galileo, and the B2a signal of BDS-3 is compatible and interoperable with L5 signal of GPS and Galileo E5a signal [4]. Because the frequency of the GLONASS signal is frequency division multiple access (FDMA), the signal frequency of G1 is 1602 + k*0.5625, the signal frequency of G2 is 1246 + k*0.4375, and k is the frequency number of 1 to 24 satellites of GLONASS. Due to limited space, this study will take one of the values of k to deal with the signal frequencies of two frequency bands. See Table 2 for the frequency of the frequency broadcast by the four major navigation system satellites.

Table 1. Basic situation of BDS-3 satellite launch

Satellites	Launch time	Orbits	PRN
1st and 2nd BDS-3 satellites	2017-11-5	MEO	C19, C20
3rd and 4th BDS-3 satellites	2018-01-11	MEO	C27, C28
5th and 6th BDS-3 satellites	2018-02-12	MEO	C22, C21
7th and 8th BDS-3 satellites	2018-03-29	MEO	C29, C30
9th and 10th BDS-3 satellites	2018-07-29	MEO	C23, C24
11th and 12th BDS-3 satellites	2018-08-24	MEO	C26, C25
13th and 14th BDS-3 satellites	2018-09-19	MEO	C32, C33
15th and 16th BDS-3 satellites	2018-10-15	MEO	C35, C34
17th BDS-3 satellites	2018-11-01	GEO	C59
18th and 19th BDS-3 satellites	2018-11-19	MEO	C36, C37
20th BDS-3 satellites	2019-04-20	IGSO	C38
21th BDS-3 satellites	2019-06-25	IGSO	Undisclosed
22th and 23th BDS-3 satellites	2019-09-23	MEO	Undisclosed
24th BDS-3 satellites	2109-11-05	IGSO	Undisclosed
25th and 26th BDS-3 satellites	2019-11-23	MEO	Undisclosed

Table 2. Frequency and wavelength of the four systems

Navigation system	Channel	Frequency/MHz	Wavelength/cm
GPS	L1	1575.420	19.03
	L2	1227.600	24.42
	L5	1176.450	25.48
BDS	B1I	1561.098	19.20
	B1C	1575.420	19.03
	B2a	1176.450	25.48
	B3I	1268.520	23.63
GAL	E1	1575.420	19.03
	E5a	1176.450	25.48
	E5b	1207.140	24.83
GLO	G1	1602.5625	18.70
	G2	1246.4375	24.05

In the process of GNSS positioning, measurement instruments, surrounding environment, weather and other reasons will directly affect the quality of GNSS data, and the availability, reliability and accuracy of GNSS positioning largely depend on the data quality of GNSS observation [5]. Therefore, GNSS data quality analysis is particularly important in data processing. Now internationally recognized better data quality analysis software is developed by UNAVCO TEQC, its advantage lies in the powerful and the operation is simple, but currently only supports processing GPS and GLONASS satellite data, two systems does not support the BDS-3 analysis of the data quality. Although the BDS-3 satellite has been put into use, both at home and abroad of BDS-3 new satellite signal data quality analysis is less, Cheng jun, dragons, etc. [6] through the carrier to noise ratio, the multipath effect, etc., and combined with single point positioning and relative positioning of the positioning results of BDS with BDS-3-2, comparing the data quality analysis of the BDS than BDS-2-3 data quality data quality is high, at the same time after joining BDS-3 satellite data location accuracy is improved; Guo et al. [7] analyzed part of the data quality of the four systems and concluded that the data quality of BDS is equal to that of GPS, and better than GLONASS in some cases. In this paper, combined with the actual observation data of iGMAS station, the quality comparison was made on the measured data of bds-3 and four global navigation satellite systems, and the comprehensive performance comparison and analysis were made on the pseudo-distance single point positioning and precise single point positioning results of the four systems.

2 GNSS Data Quality Analysis Indicators

2.1 Data Integrity Rate

The data integrity rate refers to the ratio between the actual epoch measurement and the theoretical epoch measurement received by the receiver [8]. The data integrity rate can directly reflect the integrity of the data received by the receiver during the observation period, and is an important indicator for evaluating the availability of data. The data integrity rate is mainly affected by the software and hardware performance of the receiver and the surrounding environment of the observation station. The calculation formula for the data integrity rate is:

$$DI_f = \sum_{i=1}^{n} B_i / \sum_{i=1}^{n} A_i \tag{1}$$

In which: DI_f is the completeness rate of the single system satellite observation data (%); n is the total number of single system satellites observed during the observation period; A_i is the total number of epoch observations theoretically received by the receiver during the observation period; B_i is total number of epoch observations received [9].

2.2 Signal-Noise Ratio

The signal-noise ratio (SNR) refers to the ratio of the power of the carrier signal to the sound power received by the receiver from the satellite [7]. The signal-noise ratio is one of the important indicators to measure the signal received by the GNSS receiver. Its numerical value also indirectly reflects the ranging accuracy of the carrier phase. The signal-noise ratio calculation formula is:

$$SNR = P_R/(kTB_n) \tag{2}$$

In which: is the noise bandwidth (Hz); k is the Bolman constant, and is $1.38 \times 10^{-23} J/K$. Because the noise power and the signal-noise ratio are related to the noise bandwidth B_n, the signal-noise ratio is generally used to C/N_0 describe the quantity independent of the noise bandwidth B_n [8]. The signal-noise ratio is usually measured in dBHz. Generally speaking, the higher the signal-noise ratio, the better the quality of the observed data.

2.3 Cycle Slip Ratio

Cycle slip refers to the whole cycle jump that occurs in the whole week counter in the receiver when the satellite loses lock or the signal is interrupted [11]. Cycle slip ratio refers to the ratio of the theoretical observations of the receiver to the number of observed cycle slips within a certain observation time [8]. Cycle slip ratio is one of the standards for measuring the carrier phase observation quality of receivers. The more severe the cycle slip, the smaller the ratio of the cycle slip ratio, and the worse the receiver's receiving quality is. There are many methods for cycle slip detection. The method used this time is the LG combination detection method. The LG combination calculation formula is:

$$\varphi_{LG} = \lambda_1\varphi_1 - \lambda_2\varphi_2 = (D_{iono}/f_2^2 - D_{iono}/f_1^2) + (\lambda_1 N_1 - \lambda_2 N_2) \tag{3}$$

$$\Delta\varphi_{LG}^i = (\varphi_{LG}^i - \varphi_{LG}^{i-1}) \bullet \sin(Ele)/\Delta t = \left(\frac{1}{f_2} - \frac{1}{f_1}\right) \bullet \frac{\Delta D_{iono} \bullet \sin(Ele)}{\Delta t} - \frac{\lambda_1 \Delta N_1^i - \lambda_2 \Delta N_2^i}{\Delta t} \bullet \sin(Ele) \tag{4}$$

In which: φ_{LG} is the phase observation value of LG combination; λ_i is the wavelength; φ_i is the phase observation; f_i is the frequency; D_{iono} is the frequency-independent ionospheric delay effect; N_i is the Integer ambiguity; $\Delta\varphi_{LG}^i$ represents the difference in phase observations of the LG combination between epochs [12]. Considering that the ionospheric change between epochs is not strong, the value of $\Delta\varphi_{LG}$ changes is small under normal circumstances, and $\Delta\varphi_{LG}^i$ will occur when the cycle slip occurs. The basis of this cycle slip detection is that when $\Delta\varphi_{LG}^i > 2m$, a cycle slip is considered to have occurred.

2.4 Multipath Error

Multipath error means that the receiver receives signals from the satellite broadcast, and also receives one or more reflected signals from the surrounding environment of the station at the same time. All signals are superimposed to produce a delay effect. The error caused by this phenomenon is called Is multipath error. At present, it is generally believed that the multipath error of pseudorange is much larger than the multipath error of carrier phase, so the focus of this research is the multipath error of pseudorange. For multi-frequency signals, the calculation formula of multipath error is:

$$MP_j = p_j - \frac{f_j^2 + f_i^2}{f_j^2 - f_i^2}\lambda_j\varphi_j + \frac{2f_i^2}{f_j^2 - f_i^2}\lambda_i\varphi_i + C_{ij} \tag{5}$$

In which: i and $j(i \neq j)$ represent different frequency points; MP_j is the sum of the pseudorange multipath error and noise at the frequency j, the unit is m; P_j is the pseudorange observation; λ_i and λ_j are the carrier wavelengths of different frequencies; f_i and f_j represent different frequencies; φ_i and φ_j are the observations of different frequencies and phases; C_{ij} is the combination of the ambiguity term and the hardware delay term, which is considered constant without cycle slip [5]. MP combined observations can eliminate geometrically related errors such as tropospheric delay errors and ionospheric delay errors, but they cannot eliminate errors such as phase ambiguity and hardware delays [6]. Therefore, by calculating the average of multiple epochs and subtracting the mean from a time series containing this parameter, a multipath error can be obtained. The multipath error is usually in m. The smaller the value, the smaller the multipath error.

3 Data Quality Analysis

3.1 Data Source

In order to verify the availability and reliability of the new signal data of BDS-3, to evaluate the quality of the observation data of the four major navigation systems, and to avoid the accidentalness of the experimental data, the data of 23 GNSS tracking stations distributed globally by iGMAS was used to select the observation time From May 13,2019 to November 8,2019 (doy132-313), a total of 182 consecutive days of observation data for 182 days, with a sampling interval of 30 s. The receiver manufacturers of the 23 GNSS tracking stations of iGMAS are shown in Table 3, and the global distribution is shown in Fig. 1.

Table 3. iGMAS global tracking station information

Receiver producer	Station list
CETC-20	abja, chu1, hmns, taht, xia1
CETC-54	bjf1, brch, cabn, clgy, dwin, icuk, kndy, lha1, peth, wuh1, zhon
UNICORE	byns, cnyr, gua1, krch, kun1, rdjn, sha1

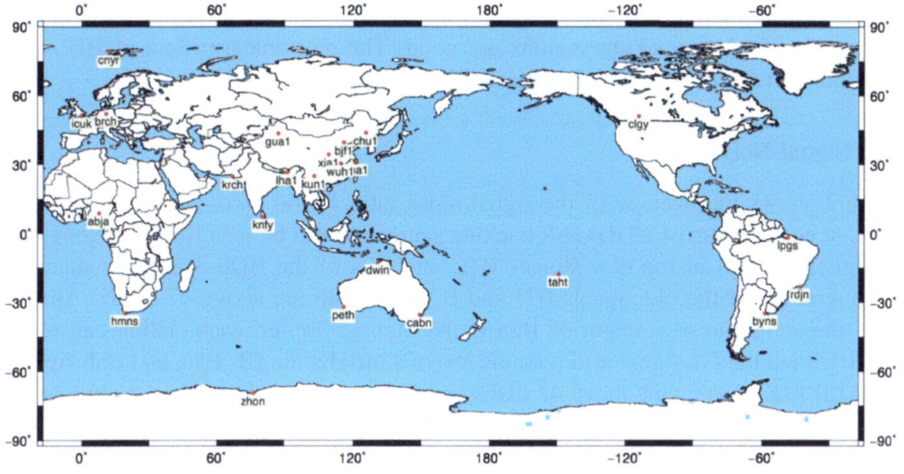

Fig. 1. iGMAS tracking station distribution map

3.2 Data Integrity Rate

Calculate the average value of the data integrity rate of each system in the 23 iGMAS tracking stations on days 132-313 of 2019, as shown in Fig. 2. The data integrity rate of each station and system can reach 80%. The GALILEO system performs best, and each station can achieve an integrity rate of more than 90%; GPS and BDS-3 can be maintained at more than 85%, and GLONASS's relative performance is average. From the tracking point of view, the integrity rate of the GNSS data received by the abja station is relatively low, but the GPS data signal reception is relatively stable. From the perspective of receiver type, the receiver of CECT-54 has better performance in

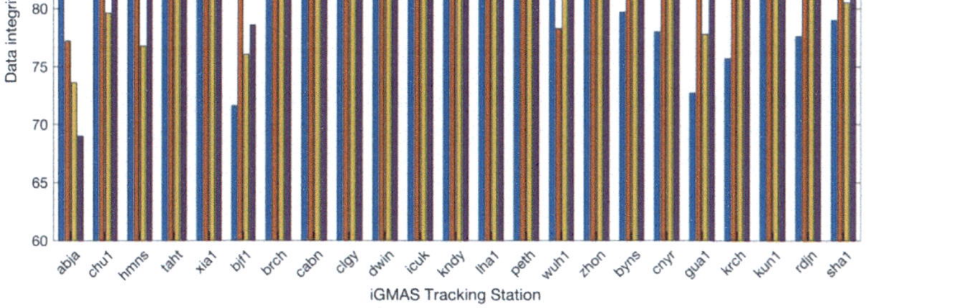

Fig. 2. Mean value of data integrity rate of each system in each station

receiving data. In general, the working status and observation conditions of the receivers of iGMAS tracking stations are good. The data integrity rate of BDS-3 is better than GLONASS, which is comparable to GPS and GALILEO.

3.3 Signal-Noise Ratio

Figure 3 shows the average of the signal-noise ratio of the corresponding frequency points of each system of 23 iGMAS tracking stations. It can be seen from the figure that the signal strength of the new signals B1C and B2a of the BDS-3 is better than the signal strength of the old signals B1I and B3I, but both are above 42 dBHz. Among them, the signal-noise strength of B2a is the highest; Better, each station can reach 44 dBHz, and the G2 signal also remains above 42dBHz; the E1, E5a, and E5b signals of GALILEO also remain above 42 dBHz; and the L2 signal on GPS is relative to L1 and L5 and other system signals. The signal-noise ratio performs poorly, but also meets a good threshold of 34 dBHz. Overall, the two newly broadcast signals of BDS-3 performed better in terms of signal-noise ratio.

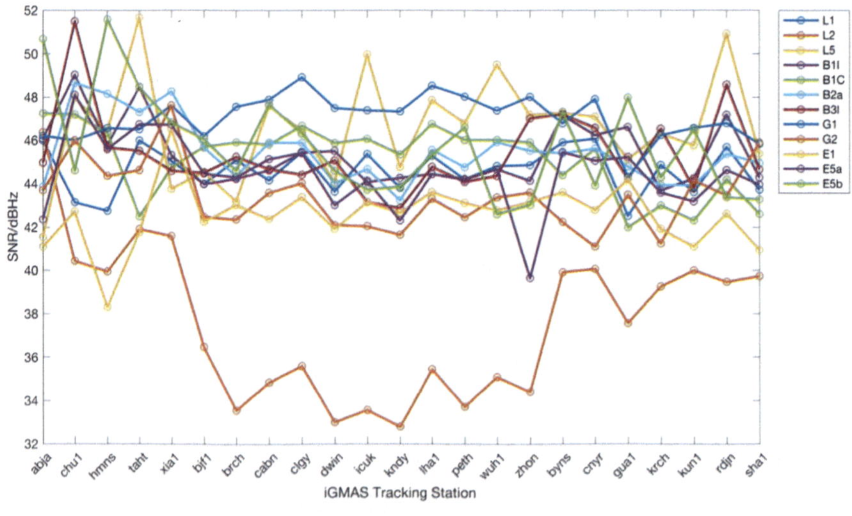

Fig. 3. Average signal-to-noise ratio of each signal in each station

3.4 Cycle Slip Ratio

Figure 4 shows the mean value of the cycle slip ratios of the corresponding frequency points of each system of the 23 iGMAS tracking stations. The reference value of the cycle slip ratio (o/slps) is 200. The smaller the value, the more serious the cycle slip [13]. It can be obtained from Fig. 4 that, except for the respective tracking stations, the cycle slip ratios of the corresponding frequency points of the four systems observed by most of the tracking stations are all greater than 200. All BDS-3 signals are above 500. Among them, the new signal B2a performs best in cycle slip. The cycle slip of E5a and E5b signals of GALILEO is more serious than that of E1. The cycle slip of L5 signal of GPS is better than L1 and L2 The G1 and G2 signals of GLONASS are more severe than the signal cycle slippage of other systems. Generally speaking, the new BDS-3 signal performs better than other system cycle slips.

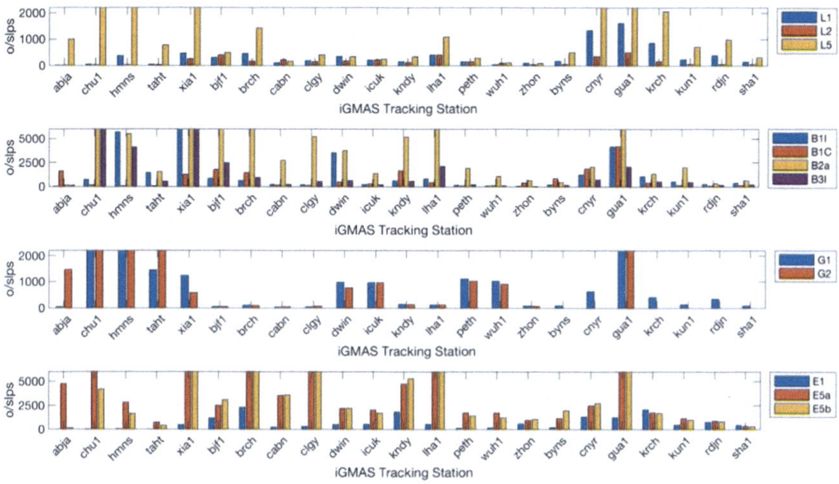

Fig. 4. Average cycle slip of each signal in each station

3.5 Multipath Error

Figure 5 shows the average multipath error of the corresponding frequency points of each system of 23 iGMAS tracking stations. From the perspective of the system alone, the corresponding multi-path error of each system is within 0.5 m. The multipath error of the B2a signal of BDS-3 is smaller than that of B1C and B1I, which is because there is no longer systematic deviation between BDS-2 and height Angle in BDS-3; the multipath error of L1 and L5 signals of GPS is smaller than that of L2 signal; G1 and G2 of GLONASS perform equally GALILEO's E5a signal is smaller than the multi-path error of E1 and E5b signals. Generally speaking, no matter which manufacturer's receiver, the multipath effect of the new BDS-3 signal is better than the other three systems.

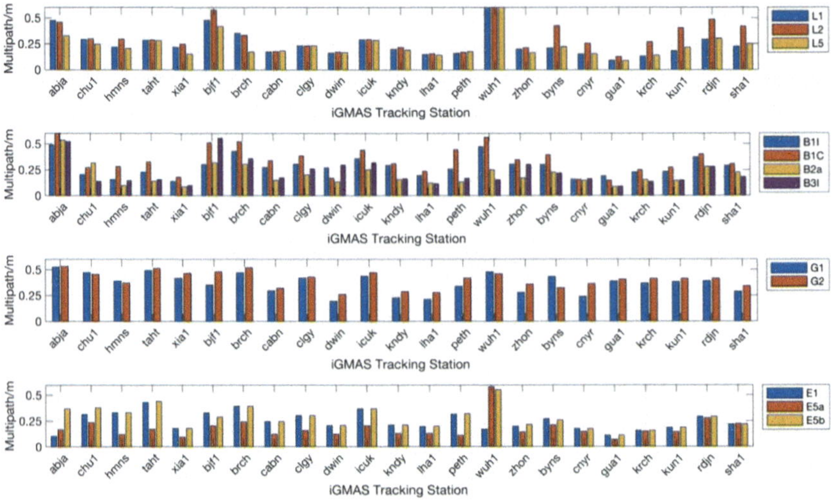

Fig. 5. Mean value of multipath error of each signal in each station

4 Positioning Accuracy Analysis

4.1 Single-Point Positioning Analysis

The single-point-positioning (SPP) calculation accuracy of the measured data of the four systems is compared. Due to the limited space, this time only the positioning of the data collected at the Antarctic Zhongshan Station (zhon) on the 294th day of 2019 is listed. As a result, the data integrity rate, signal-to-noise ratio, cycle slip ratio, and

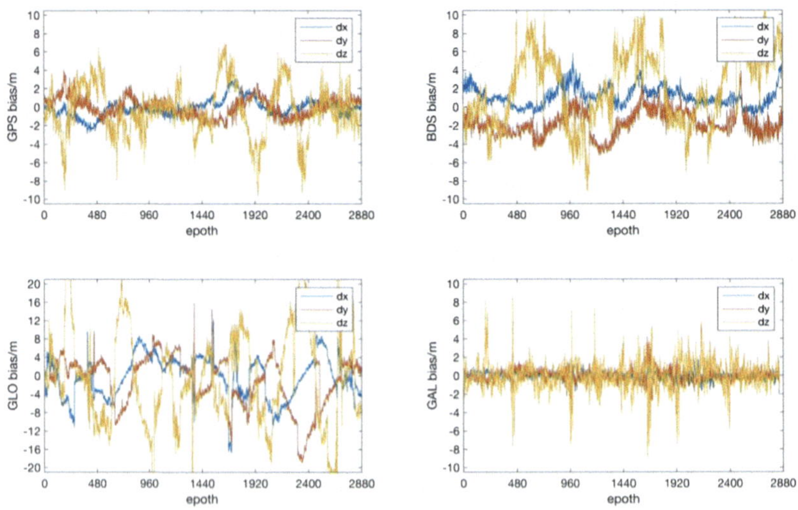

Fig. 6. SPP positioning deviation sequence of each system

Table 4. SPP positioning accuracy RMS of each system

System	dX/m	dY/m	dZ/m
GPS	0.9328	1.0200	2.8048
BDS	1.5653	2.2173	4.5986
GLO	4.5944	5.6416	11.3304
GAL	0.4185	0.5969	1.6242

multipath effects of the day performed better. The RTKLIB software of Ocean University of Tokyo, Japan is used to calculate the four system X, Y, and Z errors. See Table 4 for the calculation accuracy of the four systems in all directions (Fig. 6).

4.2 Precise-Point-Positioning Analysis

Compare precise-point-positioning (PPP) solution accuracy of measured data of the four systems, and calculate the first 300 epochs of the four systems in X, Y, and Z directions by RTKLIB software of Tokyo Ocean University of Japan. The error is shown in Fig. 7. The convergence time in each direction of each system is counted. The convergence is based on the coordinate deviation of less than 10 cm as the condition for convergence. At the same time, to ensure the reliability of the results, check the 20 epochs after convergence. Only when the coordinate deviation of 20 consecutive epochs is less than 10 cm, the current epoch is considered to converge [14]. See Table 5 for the convergence time of each system in each direction, and see Table 6 for the calculation accuracy RMS values of the four systems in all directions.

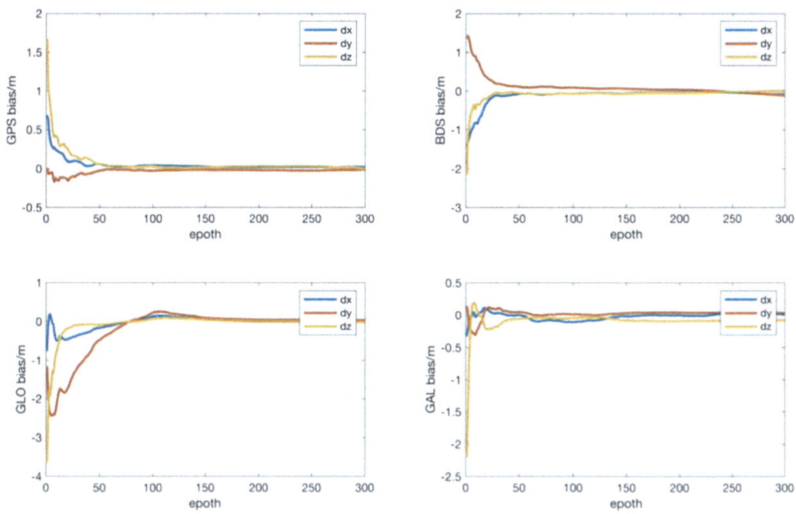

Fig. 7. PPP positioning deviation sequence of each system

Table 5. PPP positioning convergence time of each system

System	X/min	Y/min	Z/min
GPS	15	13	22
BDS	21	29	13
GLO	31	36	13
GAL	10	13	17

Table 6. PPP positioning accuracy RMS of each system

System	X/m	Y/m	Z/m
GPS	0.0294	0.0197	0.0767
BDS	0.0817	0.0812	0.0904
GLO	0.0522	0.2145	0.4157
GAL	0.0206	0.0313	0.0906

It can be seen from Fig. 7, Tables 5 and 6 that the four systems can converge within 100 epochs. In terms of positioning accuracy, the accuracy of the four systems in the horizontal direction is higher than the accuracy in the vertical direction. GPS and GALILEO have shorter convergence time and higher convergence accuracy. Compared with GPS, BDS-3 still has a certain gap. The convergence time in the Y direction is slightly longer, and the positioning accuracy is slightly lower, but it remains within 0.1 m. GLONASS has a long convergence time in the horizontal direction, and the accuracy is relatively low.

5 Conclusion

The quality of GNSS data determines the service performance of GNSS. This paper uses the measured data of 23 global tracking stations of iGMAS to evaluate the data quality of BDS-3 and GNSS from the aspects of data integrity, signal-to-noise ratio, cycle slip ratio, and multipath error. The two solutions of single point positioning and precise point positioning are used to compare the positioning performance of the four systems. Experimental research shows that the comprehensiveness of BDS-3 is equivalent to that of GPS and GALILEO. The specific results are as follows:

(1) In terms of data integrity rate, the data integrity rate of BDS-3 remains above 80%, which is comparable to the GPS and GLONASS integrity rates. GALILEO has the highest data integrity rate. In terms of signal-noise ratio, BDS-3's new signal B1C and B2a signal-noise ratio can reach more than 42 dBHz, which is better than the signal-noise ratio of the old signals B1I and B3I and GPS signals and GLONASS signals. In terms of cycle slip ratio, each BDS-3 signal can reach more than 500. In other systems, the BDS-3 new signal B2a cycle slip phenomenon is the least obvious; in terms of multipath effects, each signal of the four systems can reach less than 0.5 m, and the BDS-3 new signal B2a multipath error

is more than the other three signals of the BDS-3. The path error is small, and the multipath effect of BDS-3 is equivalent to that of GPS and GALILEO.

(2) Judging from the results of single point positioning, the positioning accuracy of BDS-3 is slightly lower than that of GPS and GALILEO, but better than GLONASS, and it can meet the requirements of normal use; judging from the precise point positioning results, the current convergence time and accuracy performance of BDS-3 are also worse than those of GPS and GALILEO.

Acknowledgements. Thanks to the iGMAS Beidou Analysis Centre of Chinese Academy of Survey and Mapping for the help and support.

References

1. Yang, Y., Xu, Y., Li, J., Yang, C.: Progress and performance evaluation of BeiDou global navigation satellite system: data analysis based on BDS-3 demonstration system. Sci China Earth Sci. **61**, 614–624 (2018). https://doi.org/10.1007/s11430-017-9186-9
2. Yang, Y., Mao, Y., Sun, B.: Basic performance and future developments of BeiDou global navigation satellite system. Satell. Navig. **1**, 1–8 (2020). https://doi.org/10.1186/s43020-019-0006-0
3. Comparison and Advantage Analysis of Beidou Navigation System and Global Positioning System (GPS). China CIO News (6), 141+147 (2014)
4. China Satellite Navigation Office (CSNO). BeiDou Navigation Satellite System Signal in Space Interface Control Document Open Service Signal B2a(Version 1.0). December, 2017
5. Liu, Q., Guo B., Hua X.: Quality analysis of the second generation BDS double—frequency observations. Geomatics Spat. Inf. Technol. **41**(7), 122–125 (2018)
6. Cheng, J., Wang, W., Ma, L., Liu, W.: Preliminary analysis of observation quality and positioning precision for BDS-3 satellites. Bull. Surveying Mapp. (8), 1–7 (2019)
7. Guo, L., Li, J., Liu, R.: Multi-system GNSS observation data quality detection and analysis. Geospatial Inf. **16**(3), 39–42+9 (2018)
8. Guo, L.: Development and Applications of GNSS Data Quality Assessment Software. Information Engineering University (2017)
9. Li, Y., Li, J., Pan, L., et al.: Quality assessment of the Beidou-3 new signal B1C and B2a observation data. Adv. Earth Sci. **33**(11), 1161–1168 (2018)
10. Zhang, X., Ding, L.: Quality analysis of the second generation compass observables and stochastic model refining. Geomatics and Inf. Sci. Wuhan Univ. **38**(7), 832–836 (2013)
11. Dang, Y., Bei, J., Cheng, Y.: Principles and Applications of Global Navigation Satellite System. Surveying and Mapping Press, Beijing (2007)
12. Cai, H., Zhao, Q., Sun, H., Hu, Z.: GNSS real-time data quality control. Geomatics Inf. Sci. Wuhan Univ. **36**(7), 820–824 (2011)
13. Bu, J., Li, X., Zuo, X., et al.: Quality comparison and analysis of Beidou/GPS/GLONASS multi system satellite positioning data. Prog. Geophys. **33**(1), 0001–0009 (2018) (in Chinese)
14. Li, L., Lv, Z., Di, S., Huang, X., Zhang, X.: Comparison and analysis of convergence time and positioning accuracy between Galileo and GPS precise point positioning. J. Geomatics Sci. Technol. **35**(2), 159–164 (2018)

A BEIDOU Short Message Based Method for Position Information Distribution of Reentry Vehicles

Yuanqing Zhao[1(✉)], Luyuan Wang[1], Xiangyu Li[2], Haogong Wei[1], Gang Chen[1], and JingShuang Cheng[1]

[1] Beijing Institute of Spacecraft System Engineering, CAST, Beijing, China
tracymy_29@126.com
[2] Space Star Technology Co., Ltd., CAST, Tianjin, China

Abstract. The traditional reentry vehicles, including recoverable satellites, vehicles for deep space mission and manned spacecrafts, utilize the way of transmitting typical beacon signals to the ground part to realize positioning. However, it usually takes long time for seeking as only direction information is available, which may push the risk to a high level. In this paper, a novel method depending on BEIDOU short message communication is proposed. Firstly, the reentry vehicles are capable of acquiring self-position information by receiving and analyzing signals from GPS or BEIDOU system. After that, BEIDOU short message communication chain is utilized to broadcast the accurate position information to the ground. Compared with the traditional method, the efficiency of seeking procedure will be significantly improved as more accurate position information is now available. Meanwhile, the risk of seeking will be reduced as well. The paper consists of three parts. In the first part, the whole information flow including the space part and the ground part is introduced. Then, the design of the system and the key instruments onboard is illustrated, as well as the feasibility in a typical application. At last, the verification results of the method proposed in an experiment are displayed. By comparing with the traditional method, the prospects and potential applications are concluded.

Keywords: BEIDOU short message · Reentry vehicles · Position information · Distribution

1 Introduction

The method based on radio direction finding has already been applied in many reentry vehicles for positioning. A radio beacon equipped on SHENZHOU 2 operating at 243 MHz was capable of being switched on after landing for the sake of assisting the search and rescue. Similar method was utilized in SHENZHOU 3 and SJ-10, the radio beacon onboard was switched on before landing, which made the search and rescue possible after the reentry vehicle entering the blackout. However, in the subsequent mission, the radio beacon operating at another frequency of 406 MHz had already been equipped onboard to control the risk of the search and rescue [1, 2].

J. Sun et al. (Eds.): CSNC 2020, LNEE 650, pp. 380–389, 2020.
https://doi.org/10.1007/978-981-15-3707-3_36

Given to the fact that traditional radio direction finding provides few accurate position information, while only direction information is available, which results in long time taken for the ground staff to search. On the other hand, the radio beacon normally is heavy and large in size as the operating frequency is in UHF, which increases the difficulty and cost for the search as the helicopters are often used.

In this paper, a novel method for position information distribution of reentry vehicles based on BEIDOU short message communication is proposed. The accurate position information is capable of being transmitted back to the user terminal on the ground through BEIDOU inter-satellite link, assisting to determine the landing position of the vehicles. Compared to the traditional method, the equipment onboard is much lighter and smaller, which is convenient to carry. Therefore, the method proposed in this paper is capable of reducing the difficulty and cost of the search and rescue on the ground, which has a significant impact on the design of the reentry vehicles in the future.

2 System Design

2.1 Satellite-Ground Link Design

The operating procedure of the method proposed consists of three steps. Firstly, the system onboard acquires the accurate position information of the reentry vehicles by receiving and analyzing the navigation signals from GPS or BEIDOU system. Then the position information framed according to the requirement of the formal protocol and modulated on carrier is transmitted through BEIDOU inter-satellite link. Finally, the user terminal is capable of extracting the position information of reentry vehicles by receiving and demodulating the carrier. The accurate position information will support the search and rescue for staffs on the ground.

Three different satellite-ground links are utilized in the design including navigation signal receiving link, beacon signal forward link and beacon signal return link, which are illustrated in Fig. 1.

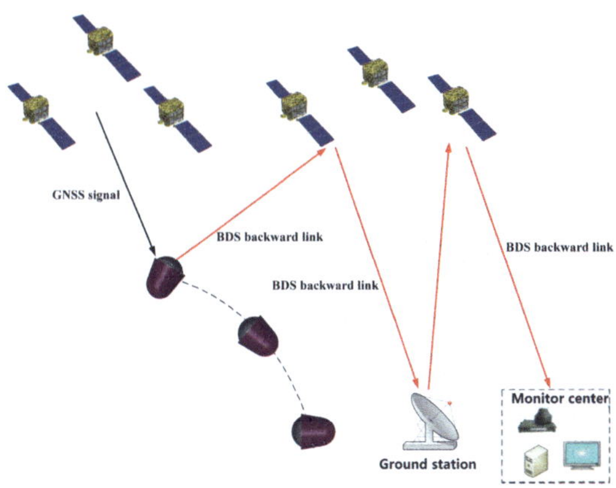

Fig. 1. Chains equipped onboard

2.2 System Onboard Design

The system onboard is constituted of three parts including a receiver, a GNSS antenna and three BDS (BEIDOU short message) antennas. The receiver equipped is utilized to demodulate and analyze the navigation signals transmitted from the GNSS antenna. After extracting the position information, the receiver is capable of framing according to the protocol and transmitting the position information from BDS antennas. Three antennas operate together to cover the whole space area. The schematic diagram is displayed in Fig. 2.

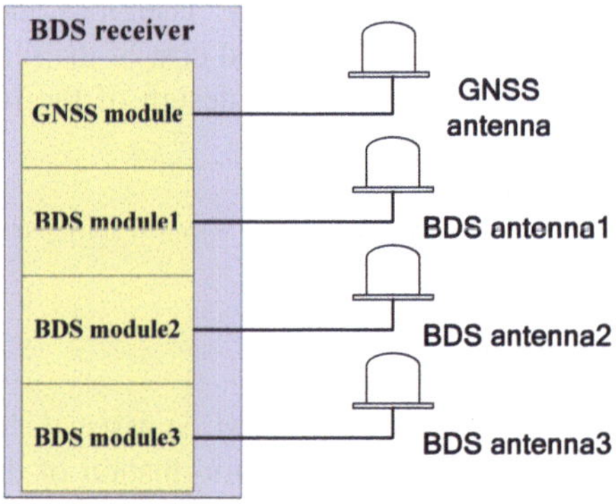

Fig. 2. Design of the system onboard

As the time interval of BEIDOU short message communication is strictly constrained [3, 4], more ID cards are essential for the sake of acquiring the position information as soon as possible. In the proposed design, 9 cards are equipped in the receiver, in which 3 cards correspond to 1 module. The time interval is consequently improved from 1 min to 7 s, enhancing the timeliness of the position information extracted on the ground. Thus the staffs on the ground are capable of determining the landing position of the reentry vehicles more conveniently.

2.3 Key Devices Design

2.3.1 BDS Receiver Design

The BDS receiver is constituted of three different parts including a power module, three BDS modules and a GNSS module. The GNSS module in the BDS receiver is mainly responsible for capturing, tracking and analyzing the GNSS signal to acquire self-position information, while the BDS modules then frame and modulate for the self-position information, and broadcast in the end through BDS antennas. The schematic diagram of the BDS receiver is displayed in Fig. 3.

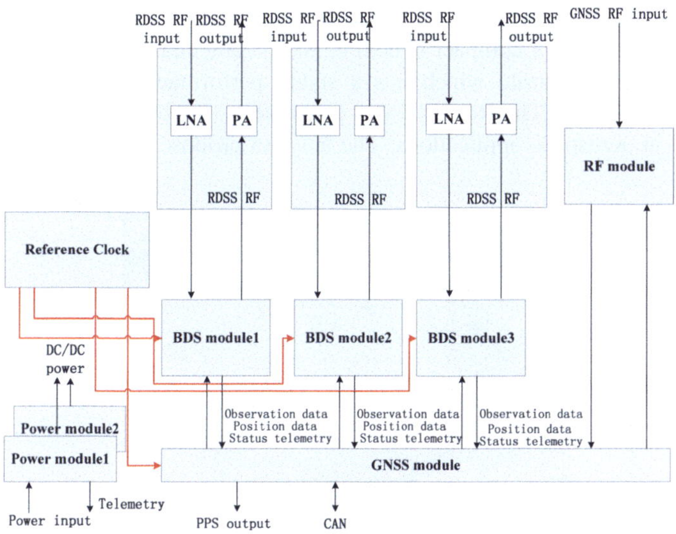

Fig. 3. Design scheme of the BEIDOU short message based receiver

The RF module displayed in Fig. 3 achieves low noise amplification and down-conversion. After AD conversion for IF signal, the receiver is capable of capturing, tracking and demodulating by using two-dimensional searching algorithm. The accurate position information of reentry vehicles can finally be determined according to the demodulated navigation messages.

The BDS modules in the receiver are capable of capturing and tracking the available BEIDOU satellites after the RDSS RF modules magnifying the navigation signal. Then the BDS modules put the accurate position information into a frame according to the protocol [5] displayed in Table 1. In the end, the frame is modulated and broadcasted to the ground through the corresponding BDS antenna. An unique clock is utilized by all modules in the receiver for achieving synchronization.

Table 1. Data protocol of BEIDOU short message

Command	Length	User address 1	Content					Check sum
$TXSQ	16 bit	24 bit	Type 8 bit	User address 2 24 bit	Message length 16 bit	Respond control 8 bit	Content	8 bit

2.3.2 BDS Antenna Design

The BDS antenna onboard should complete receiving beacon signals and transmitting BDS signals.

Considering the special configuration of the reentry vehicles, the antenna should have a low profile and conformal structure, which also contributes to decrease the impact of aerodynamic force.

The proposed antenna operating at two frequencies is in the form of double layers microstrip for realizing a compact structure. As the antenna has a narrow bandwidth, the Rogers TMM substrate which has a stable performance with the variation of temperature is adopted. The Rogers TMM substrate is also approved by NASA for its high quality in aerospace applications. The antenna proposed is displayed in Fig. 4.

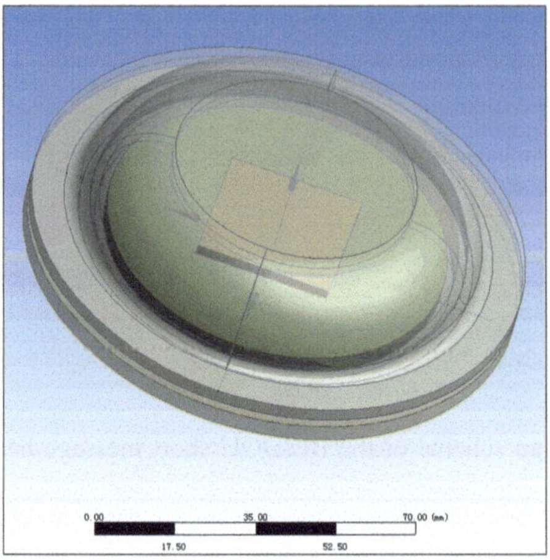

Fig. 4. Configuration of the BEIDOU short message antenna

The simulated results of radiation pattern and axial ratio are displayed in Figs. 5, 6, 7 and 8. The results shown indicate that the gain is higher than −4.5 dBi in the area of $0^\circ \leq \theta \leq 60^\circ$ and $0^\circ \leq \phi < 360^\circ$, while the axial ratio is lower than 5 dBi. The antenna can be applied to most missions.

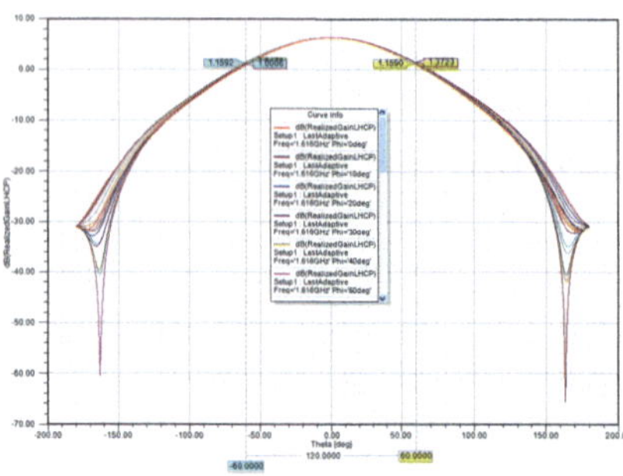

Fig. 5. Simulated radiation patterns of the antenna at the transmitting frequency

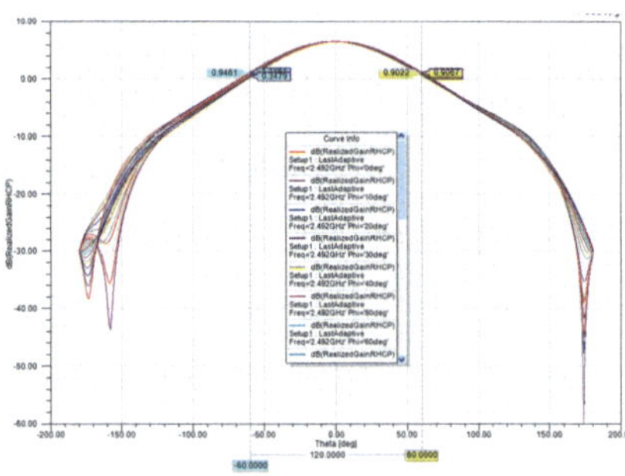

Fig. 6. Simulated radiation patterns of the antenna at the receiving frequency

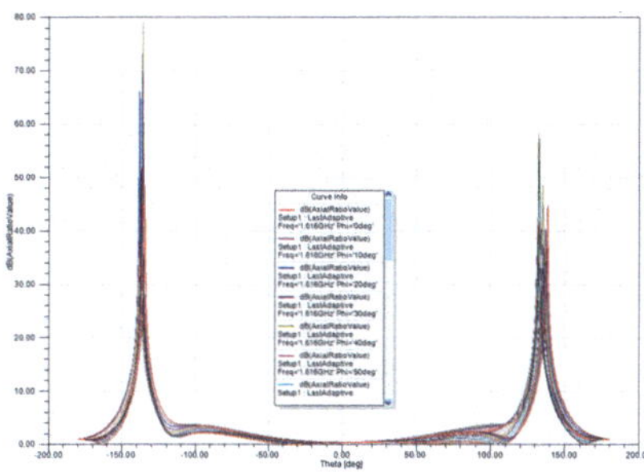

Fig. 7. Simulated axial ratio results of the antenna at the transmitting frequency

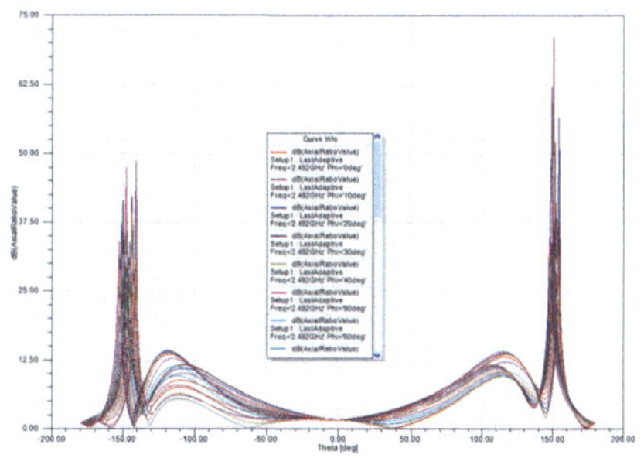

Fig. 8. Simulated axial ratio results of the antenna at the receiving frequency

2.3.3 GNSS Antenna Design

The GNSS antenna with a conformal structure is set at the bottom of the vehicle for weakening the air scour and realizing ablation resistance.

The antenna is in the form of helix. The number of turns is about 2 to 4, and the radius is about $\lambda/8$ to $\lambda/6$. The distance between two adjacent turns is about $\lambda/8$ to $\lambda/6$, while the aperture of the cavity is around $\lambda/4$ to $\lambda/3$. This design makes the antenna have a compact structure along with a wide bandwidth [6].

The electromagnetic window of the antenna is designed by SiO_2 based composite material for achieving thermal insulation, wide bandwidth as well as compact structure.

The simulated results of gain performance are displayed in Fig. 9, which indicate that the gain of the antenna is higher than −3.5 dBi in the area of $0° \leq \theta \leq 75°$ and $0° \leq \Phi \leq 360°$ in the whole band.

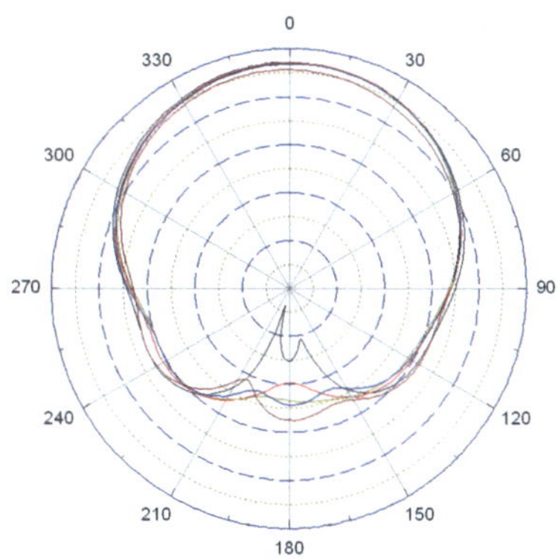

Fig. 9. Simulated radiation patterns of the GNSS antenna

3 Availability in a Typical Application

3.1 Doppler Adaptation

The doppler effect is analyzed for this case and the result is shown in Fig. 10, which indicates that the maximum frequency shift is about 40 kHz in the reentry process.

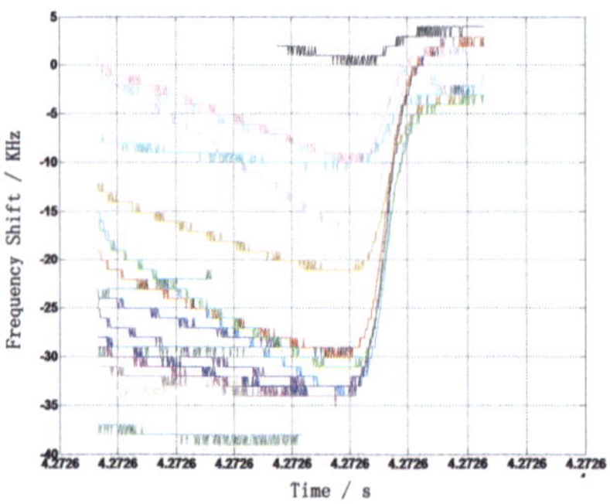

Fig. 10. Doppler effect analysis of the navigation signal

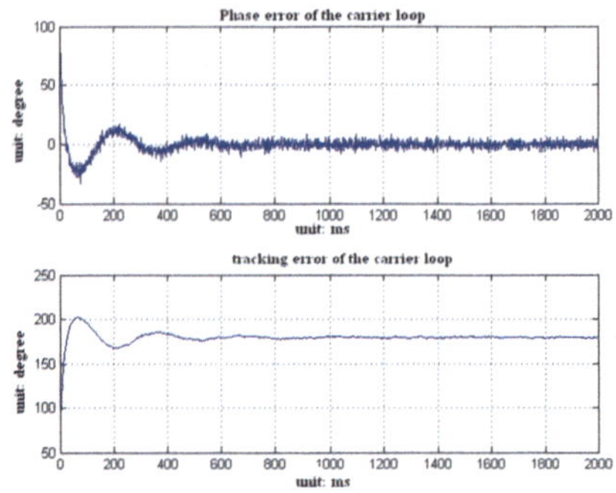

Fig. 11. Performance analysis of PLL

The design of a second-order FLL combined with a third-order PLL is utilized for fast signal capturing and steady signal tracking. In the design, the third-order PLL is utilized to compensate the impact of differential of acceleration, while the second-order FLL is adopted to realize fast signal capturing. The performance of the designed loop is displayed in Fig. 11.

As the frequency deviation of the received signal for the ground stations of BEIDOU system should be less than 800 Hz, the frequency shift acquired from the carrier loop is added to the transmitting frequency to compensate the frequency deviation. After the compensation, the frequency deviation is less than 800 Hz. The signal then can be used by the ground stations.

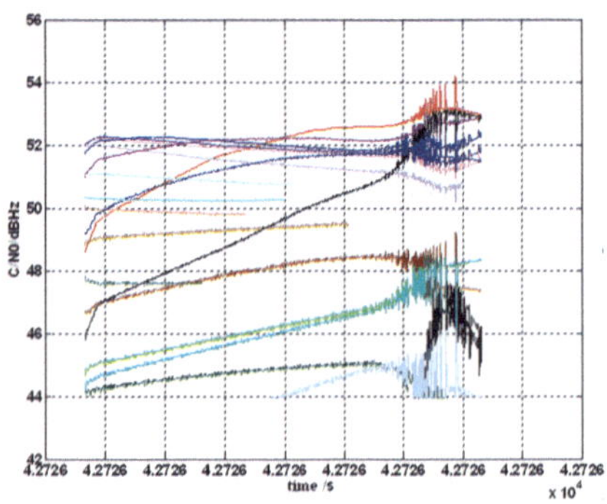

Fig. 12. Variation of the input signal power

3.2 Adaptation for the Power Level Fluctuation

The variation of C/N0 at the input RF port of the receiver during the whole reentry procedure is simulated and displayed in Fig. 12. The result indicates that the maximum variation of C/N0 is about 20 dB during the whole procedure.

The receiving channel of the receiver is determined as superheterodyne. The first IF is determined as 178.5 ± 4.08 MHz, while the second IF is determined as 12.24 MHz. The LNA of the receiving channel has a gain of 18 dB with a noise factor of 2.5 dB. The range of the gain adjustment of the AGC strategy is 45 dB, which is appropriate for this case.

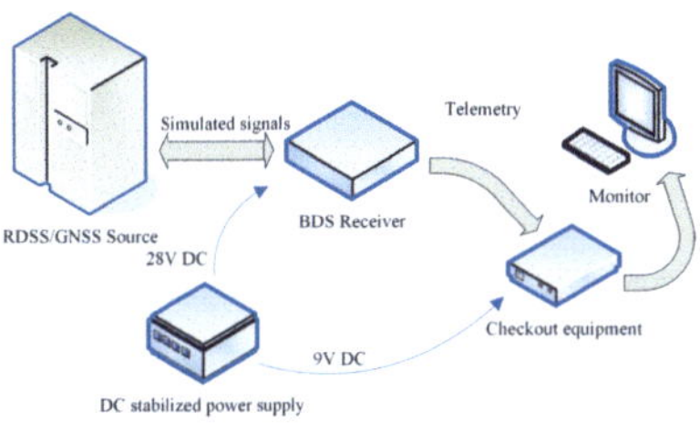

Fig. 13. Indoor test environment

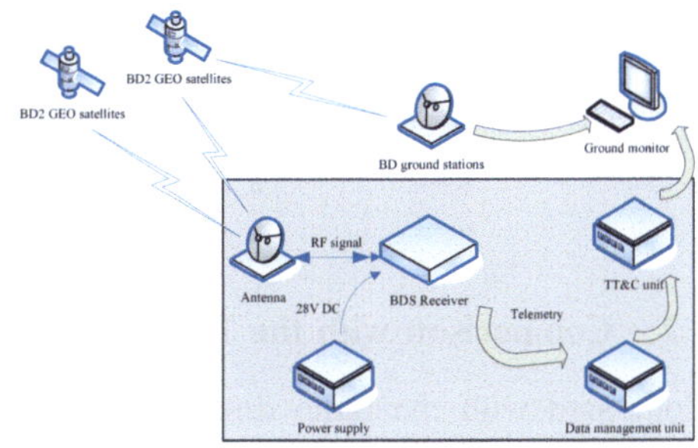

Fig. 14. Outdoor test environment

4 Experiment Verifications

4.1 Introduction

The experiment verifications are conducted indoor and outdoor respectively. Tests for the sensitivity, time of accomplishing positioning, dynamic performance and BEIDOU short message communication were conducted indoor. The dynamic performance test was conducted by utilizing the GNSS signal source, in which the source was driven by a specifical script to produce dynamic GNSS signals. The dynamic GNSS signals produced by the source were received and processed by the receiver for verifying the continuity of the signal tracking and analyzing ability. In contrast to the theoretical file, the accuracy of position can be verified. In BEIDOU short message communication test, the RF performance and BER of the link were verified respectively by utilizing a RDSS signal source which is driven by a data file created according to the STK simulation results. As the trace and attitude of the reentry vehicle can be imitated, the verification results are of great significance for the design. The indoor test environment is displayed in Fig. 13.

Tests including the time of accomplishing positioning, accuracy of position and BEIDOU short message communication were also conducted outdoor in wireless environment. A specialized user terminal was employed to receive short messages from the BDS receiver for verifying the performance. The outdoor test environment was displayed in Fig. 14.

The test results are displayed in Table 2. It indicates that the design proposed is appropriate for the case.

Table 2. Test results

Test scenario	Time interval of capturing	Sensitivity	Package loss
Indoor (high dynamic scenario)	1:4 s	Better than −126 dBm	No
	2:5 s	Better than −126 dBm	No
	3:4 s	Better than −126 dBm	No
	4:5 s	Better than −126 dBm	No
Outdoor (static scenario)	1:1 s	/	No
	2:1 s	/	No
	3:1 s	/	No
	4:1 s	/	No

4.2 Comparison with the Traditional Design

Compared with the radio direction finding based method, the proposed design has advantages as listed below.

1. The time cost of the search and rescue is significantly reduced for the higher efficiency.

The proposed design in this paper is capable of providing accurate position information in WGS84 coordinates rather than just direction information, which contributes to complete the search and rescue in time.

2. The equipments for the searching group on the ground are portable.

The equipments required by the traditional method are usually carried by helicopters for mobile search, while in the proposed design the essential equipments can be carried by just moving vehicles as the equipments are light and small in size.

5 Conclusion

In this paper, a novel method for position information distribution suitable for reentry vehicles based on BEIDOU short message communication is presented. In this design, the BEIDOU short message communication is utilized for transmitting position information of reentry vehicles to the specified user terminal. Compared to the traditional method, the efficiency is significantly improved. Firstly, the design of the proposed system and key devices is illustrated. Then, the adaptation analysis and verification results of the proposed design for a typical application are introduced. In the end, a comparison is made to demonstrate the advantages of the proposed design.

References

1. Shi, Q.: The Research and Realization of Beacon Based on Short Message Communication and Landing Prediction. North University of China, Taiyuan (2018)
2. Gao, S., Huang, W.: The development and prospect of Chinese spacecraft recovery and landing technology in the last sixty years. Spacecraft Recovery Remote Sens. **39**(4), 70–78 (2018)
3. Yu, L., Wang, X., Li, S.: Positioning data compression and reliable transmission based on Beidou short message. Appl. Electron. Tech. **38**(11), 108–111 (2012)
4. Chen, H., Guo, X., Wang, F., Lu, H.: Fishery harvesting information compressing and transmitting method based on Beidou short message. Trans. CSAE **31**(22), 155–160 (2015)
5. Xu, Y., Zhang, C., Li, Y., Zhang, Y., Ge, B., Song, H.: Research on Beidou voice communication system based on Beidou RDSS communication technology. In: The 10th China Satellite Navigation Conference (2019)
6. Chen, G., Yang, C., Sun, D., Zhi, G.: The ablation antenna design and verification of circumlunar free return and reentry flight vehicle. Sci. Sin. Tech. **45**(3), 268–275 (2015)

A Two-Dimensional Point Cloud Matching Method Based on ICP Improvement

Huchao Xu, Letao Zhou(✉), Yinghao Zhao, and Zheng Yuan

Faculty of Geosciences and Environmental Engineering,
Southwest Jiaotong University, Chengdu, China
ltzhou@home.swjtu.edu.cn

Abstract. The indoor map is the basis of indoor navigation and positioning. The core problem of indoor map construction by using two-dimensional lidar is to solve the matching problem between different point clouds. The standard ICP algorithm is easily affected by the initial position and interference points, so this paper proposes an improved algorithm based on ICP for 2D point cloud matching. This paper proposes a dynamic initial value matching method based on the highly correlated characteristics of the transformation parameters of adjacent point clouds. Then set the radius threshold to filter out the corresponding points to be matched to provide more ideal data input for the ICP algorithm. In this paper, the indoor parking lot with an area of about 2500 m^2 can be collected by using the vehicle 2D lidar equipment, which takes about 85 s to collect. And the global map can be obtained by using the improved ICP algorithm, which takes about 100 s to process the data. Experimental results show that this method has the remarkable characteristics of low cost and high efficiency. Compared with the total station measurement, the method can meet the requirements of the plane map with decimeter accuracy required by the navigation and positioning of the indoor parking lot, and it has a better matching effect in the practical application.

Keywords: Point cloud matching · ICP algorithm · Lidar · Indoor map construction

1 Introduction

With the rapid development of science and technology, lidar sensors are widely used in mapping, unmanned driving, robotics, reverse engineering [1] and other fields due to their high efficiency, high precision, portability and strong environmental adaptability. The collected point cloud data can effectively reconstruct the space objects. However, due to the complexity of the real space structure and the existence of occlusion phenomenon, we cannot measure the global point cloud data at one time [2]. In order to completely reconstruct the surface morphology of the space object, we need to set up multiple stations to scan the object from different angles. The process of transforming point cloud data in different coordinate systems into the same coordinate system is defined as point cloud matching or registration [3].

J. Sun et al. (Eds.): CSNC 2020, LNEE 650, pp. 390–398, 2020.
https://doi.org/10.1007/978-981-15-3707-3_37

Point cloud matching is a key technology in the field of digital detection. Among them, the Iterative Closest Point (ICP) is made up by computer vision researchers Besl and McKay in 1992 [4, 5]. However, the standard ICP algorithm has disadvantages such as susceptibility to some isolated values and new points of interference in actual use. In order to solve these problems, many scholars at home and abroad have studied from many aspects and formed variations of ICP algorithms [6, 7], such as Chetverikov [8] proposed an improved Trimmed ICP method based on the correlation function, and Bergevin [9] proposed a point-to-face nearest search method.

2 Improved ICP Algorithm

The purpose of point cloud matching is to solve the optimal rotation Angle and translation parameters between two groups of point cloud data, so that multiple groups of point cloud data in different coordinate systems can be transformed into the same global coordinate system to generate the global point cloud data we need.

The ICP algorithm is a point-to-point matching algorithm. So this paper proposes a dynamic initial value matching method based on the highly correlated characteristics of the transformation parameters of adjacent point clouds. Then set the radius threshold to filter out the corresponding points to be matched to provide more ideal data input for the ICP algorithm.

2.1 Dynamic Initial Value Matching Method

As shown in Fig. 1 below, we randomly select 100 frames of adjacent point cloud data and display the original observation data of these 100 frames in the same figure. Then, we randomly selected the adjacent point cloud data of 901 frames and plotted the first derivative graph of the difference of transformation parameters of their adjacent frames. Figure 2 is the change rate graph of X-axis translation parameter difference between two adjacent frames after matching the adjacent point cloud data of frame 901; Fig. 3 is the change rate diagram of the translation parameter difference on the Y-axis; Fig. 4 is the change rate diagram of the difference of Angle rotation parameters. From Figs. 2, 3 and 4, we can intuitively see that the transformation parameters of adjacent frames have a relatively centralized trend. Through the analysis of the statistical Tables 1, 2 and 3, the X axis translation parameters with 68.33% of the adjacent point cloud rate between 0 and 2, Y axis translation parameters with 74.22% of the adjacent point cloud rate between 0 and 4, 79.89% of Angle rotation parameters between adjacent point cloud rate at 0–0.4, so we can think of adjacent frames, Angle rotation parameters and Y direction translation transformation parameters has strong stability, the X axis translation transformation parameters have certain relevance.

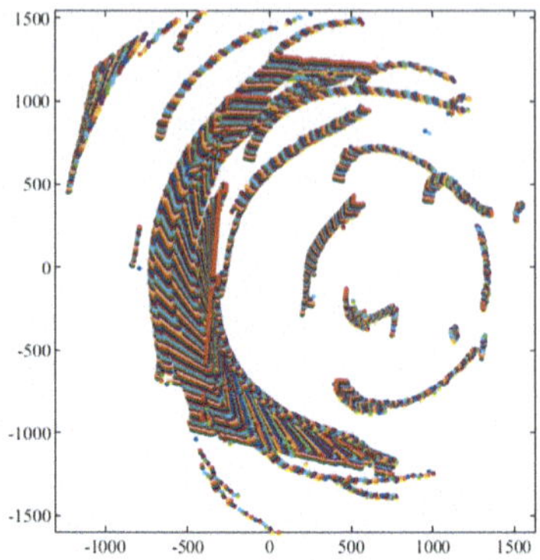

Fig. 1. 100 frames of original point cloud data

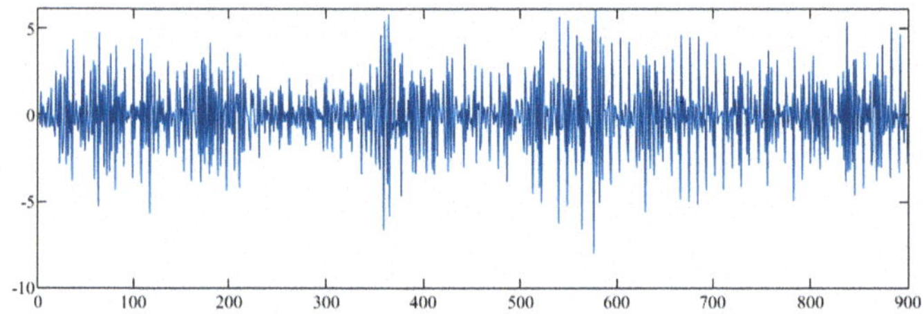

Fig. 2. Change rate chart of x-axis translation parameter difference of adjacent point cloud data in 900 frames

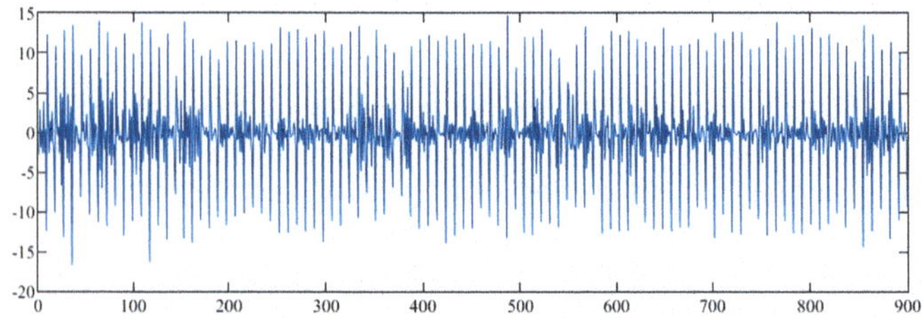

Fig. 3. Change rate chart of y-axis translation parameter difference of adjacent point cloud data in 900 frames

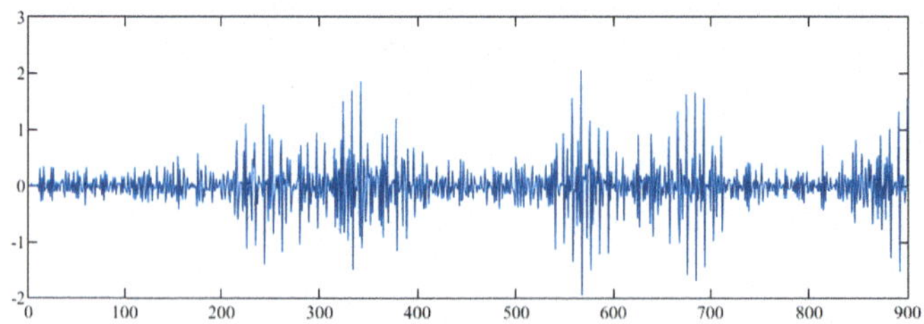

Fig. 4. Change rate chart of angle change difference of adjacent point cloud data in 900 frames

Table 1. Statistical table of change rate of x-axis translation parameter difference of adjacent point cloud data in 900 frames

|Range|	0–2	2–4	4–6	6–8
Number	615	227	53	5
Percentage	68.33	25.22	5.89	0.56

Table 2. Statistical table of change rate of y-axis translation parameter difference of adjacent point cloud data in 900 frames

|Range|	0–4	4–8	8–14	14–18
Number	668	42	184	6
Percentage	74.22	4.67	20.44	0.67

Table 3. Statistical table of change rate of difference value of angle change of adjacent point cloud data in 900 frames

|Range|	0–0.4	0.4–0.8	0.8–1.4	1.4–2
Number	719	114	49	17
Percentage	79.89	12.67	5.44	1.89

Therefore, this paper proposes a dynamic initial value method as follows: the optimal transformation parameter obtained in the previous frame is taken as the initial value of the cloud transformation parameter in the next frame to be matched, and the initial transformation is carried out. This process is also referred to as the process of rough matching in this paper. The point cloud after rough matching transformation is allowed to participate in the subsequent ICP matching as the new point cloud to be matched, thus reducing the number of iterations of ICP matching.

$$R'_{t+1} = R_t$$

$$Date'_{(t+1)} = R'_{t+1} \cdot Date_{(t+1)}$$

As can be seen from Figs. 5 and 6, this method can better match the subsequent two adjacent frame point cloud data together. To provide a more ideal initial matching state for the subsequent ICP algorithm, thus reducing the number of iterations and improving the matching efficiency.

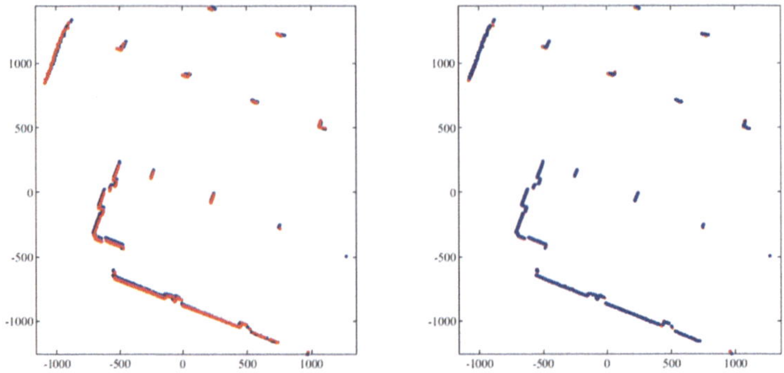

Fig. 5. Comparison between original data (left) and rough matching results (right)

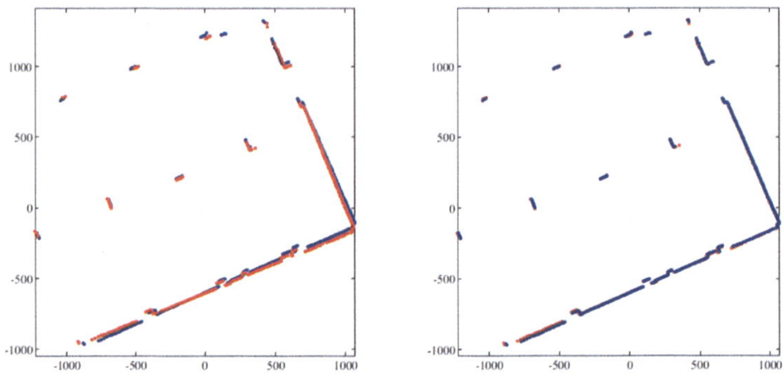

Fig. 6. Comparison between original data (left) and rough matching results (right)

2.2 Radius Threshold Screening Preprocessing Method

The standard ICP algorithm is derived under the ideal condition that two point cloud data can overlap completely. In the actual situation, due to the systematic nature of the lidar and the occlusion of the real space structure, it is impossible to guarantee that the number of measured frame point clouds is the same. In this paper, a preprocessing method is proposed to treat the matched two frames of point cloud data by setting the radius threshold, and filter them point by point, eliminate interference points, and only retain the points to be matched that are closer to the actual situation. The flow of this method is shown in Fig. 7:

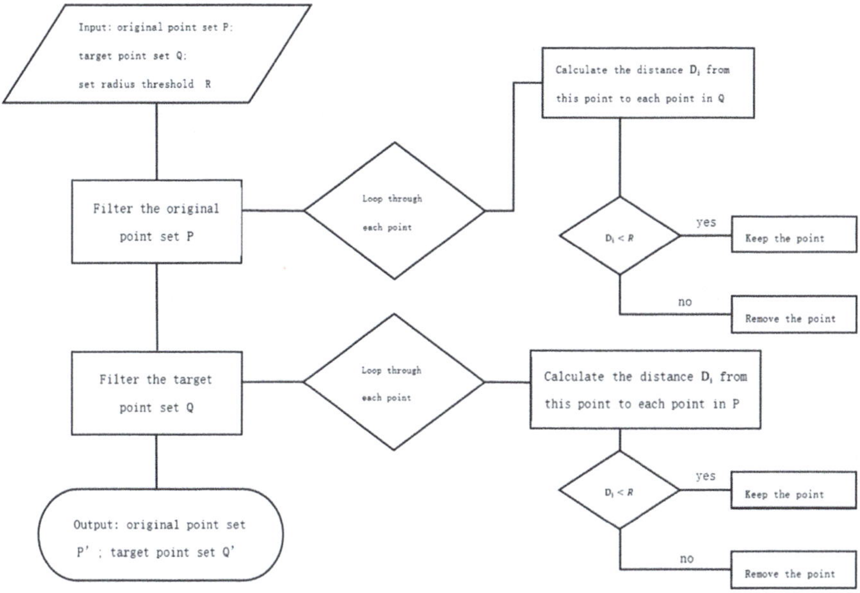

Fig. 7. Flow chart of radius threshold filtering

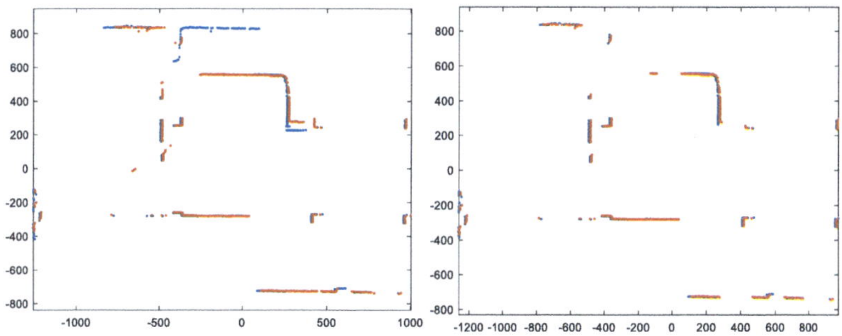

Fig. 8. Comparison between the original point cloud (left) and the preprocessing result (right)

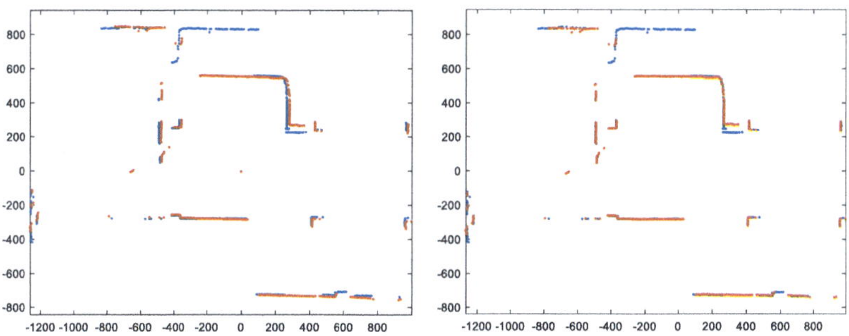

Fig. 9. Comparison of original point cloud matching (left) and pre-processing matching (right)

According to the Figs. 8 and 9, we can find that radius threshold filtering pre-processing method can better eliminate the interference between two frames point cloud point (including due to the spatial structure of new points and error caused by systemic and posture sensor point), retain more accord with the actual corresponding point cloud data, and its matching effect more ideal than without pretreatment method matching results.

Table 4. Comparison of matching performance between standard ICP and the improved algorithm

Algorithm	Average matching time of single frame (s)	Average matching error of single frame (cm)	Average number of iterations per frame
Standard ICP algorithm	0.091	107.715	13
Improved algorithm in this paper	0.095	14.592	8

2.3 Comparative Analysis of Algorithm Performance

A total of 1000 frame point cloud data were selected for the performance comparison analysis of this algorithm, and the average effective number of points per frame was about 430, which was matched and analyzed by Matlab software. In this paper, the average euclidean distance between the corresponding point pairs determined in ICP matching is taken as the error index of this algorithm. According to the experimental data, it can be obtained from Table 4: In terms of matching time, the algorithm in this paper is basically consistent with ICP; In terms of the number of iterative matches, since the dynamic initial value matching operation requires only one transformation to match two point clouds together roughly, the algorithm in this paper is approximately twice as fast as the standard ICP algorithm. On the precision of matching error, because the radius threshold filtering operations would obviously noise points (including due to the spatial structure of new points and error caused by systemic and posture sensor point), retain more accord with the actual corresponding point cloud data, the error of the proposed algorithm is about one seventh of that of the standard ICP algorithm. Finally, this paper provides a more ideal input data for ICP algorithm through dynamic initial value matching and radius threshold filtering, thus reducing the number of iterations and improving the matching effect.

3 Experimental Results and Analysis

3.1 Experimental Instruments and Scenes

The experimental instrument consists of a trolley and a 2D lidar (EAI G4). The maximum range of the lidar is 16 m, the scanning angle is 360°, the scanning frequency is 12 circles/s, and the sampling frequency is 9000 points/s. For the experimental scene, we chose a part of the parking lot at the bottom of the fourth teaching building of the school, with an area of about 2500 m^2. The experimental equipment and scene are shown in Fig. 10.

(a) EAI G4 Lidar **(b-d) Experimental scenario 1-3**

Fig. 10. Experimental instruments (**a**) and scenes (**b–d**)

3.2 Experimental Process and Results

In this experiment, the 2D lidar was fixed on the roof of the car, and the indoor parking lot with an area of about 2500 m^2 which at the bottom of the fourth education building of the school can be collected by using the vehicle 2D lidar equipment, which takes about 85 s to collect. And the global map can be obtained by using the improved ICP algorithm, which takes about 100 s to process the data. Figure 11 respectively represents the original point cloud data graph and the matching result graph. The figure on the left shows the adjacent original point cloud data of 1000 frames collected in this experiment in the same figure, and the figure on the right shows the global matching map generated with the last frame point cloud data as the global coordinate reference frame after the improved ICP matching algorithm in this paper. It can be seen that the global matching result diagram can better reflect the planar structure of the indoor parking lot and has better matching effect in practical application.

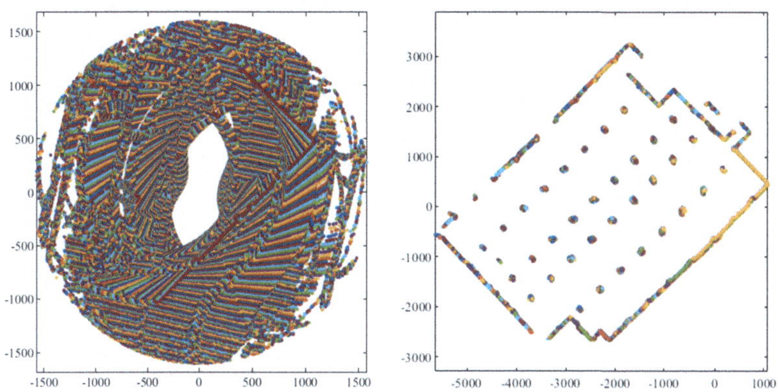

Fig. 11. Original data graph (left) and global matching result graph (right)

4 Conclusion

In order to solve the problem that the standard ICP algorithm has poor matching effect due to the influence of interference points, this paper provides a more ideal input data for the ICP algorithm through dynamic initial value matching and radius threshold filtering method, so as to reduce the number of iterations and improve the matching effect. Through the method of obtaining the plane map of indoor parking lot proposed in this paper, it takes about 85 s to collect the data of the indoor parking lot with an area of about 2500 m^2. Using the improved ICP algorithm, it takes about 100 s to process the data to obtain the global map of the experimental scene. Experimental results show that this method has the remarkable characteristics of low cost and high efficiency. Compared with the total station measurement, the method can meet the requirements of the plane map with decimeter accuracy required by the navigation and positioning of the indoor parking lot, and has a better matching effect in the practical application.

References

1. Imbert, M., Li, X.: Automatic registration of 3D point clouds for reverse engineering. J. Comput. Theor. Nanosci. **4**(6), 2431–2432 (2011)
2. Xu, W., Xu, H.: Application of improved ICP algorithm in point cloud registration. Syst. Simul. Technol. Appl. **14**, 4 (2012)
3. Akca, D., Armin, G.: Recent advances in least squares 3D surface matching. Opt. 3-D Measur. Tech. VII **2**, 197–206 (2005)
4. Besl, P.J., Mckay, N.D.: A method for registration of 3-D shapes. IEEE Trans. Pattern Anal. Mach. Intell. **14**(2), 239–256 (1992)
5. Schenker, P.S.: Method for registration of 3-D shapes. IEEE Trans. Pattern Anal. Mach. Intell. **14**(2), 239–256 (1992)
6. Minguez, J., Montesano, L., Lamiraux, F.: Metric-based iterative closest point scan matching for sensor displacement estimation. IEEE Trans. Rob. **22**(5), 1048–1054 (2006)
7. Pomerleau, F., Colas, F., Siegwart, R., et al.: Comparing ICP variants on real-world data sets: open source library and experimental protocol. Auton. Robot. **34**, 133–148 (2013)
8. Chetverikov, D., Svirko, D., Stepanov, D., et al.: The trimmed iterative closest point algorithm. In: International Conference on Pattern Recognition. IEEE, Quebec City, pp. 545–548 (2002)
9. Bergevin, R., Soucy, M., Gagnon, H., et al.: Toward a general multi-view registration technique. IEEE Trans. Pattern Anal. Mach. Intell. **18**(5), 540–547 (1996)

DTMB and FM Signals Indoor Fingerprint Positioning System Based on Compressive Sensing

Menghuan Yang[1,2], Hong Wu[1,2(✉)], Hongzhao Peng[1,2], and Zhuo Chen[1,2]

[1] College of Electronic Information and Optical Engineering, Nankai University, Tianjin, China
wuhong@nankai.edu.cn
[2] Tianjin Key Laboratory of Optoelectronic Sensor and Sensing Network Technology, Tianjin, China

Abstract. Location-based services have become an indispensable part of people's daily lives. GNSS can provide favorable outdoor location services, but it cannot meet the needs of indoor positioning. This paper proposes an indoor localization fingerprint system using DTMB and FM signals, which consists of offline and online stages. In the offline stage, the Received Signal Strength of the Reference Points is measured in the experimental area, and the data acquisition amount is reduced by Compressive Sensing theory. In the online stage, the coordinates of Test Points are reconstructed by using Compressive Sensing algorithm. The experiments show that the average positioning accuracy of the DTMB and FM signal indoor positioning system based on Compressive Sensing is about 2.14 m. The proposed system has larger coverage and lower cost, which can meet the actual indoor positioning requirements.

Keywords: Indoor positioning · Fingerprint match · Compressive Sensing · FM signal · DTMB signal

1 Introduction

With the rapid development of the social economy and the continuous deepening of the informatization process, people are increasingly eager to obtain location information. Global Navigation Satellite System can provide favorable outdoor location services, while indoor positioning has become an urgent problem to be solved. Common indoor positioning solutions include Wi-Fi positioning [1], radio frequency identification positioning [2], ultrasonic positioning [3], visible light positioning [4], Bluetooth positioning [5] and ultra-wideband positioning [6].

Wi-Fi positioning technology is a research hotspot in the field of indoor positioning. In good cases, the average error of Wi-Fi positioning can reach 3 m [1]. However, Wi-Fi signals are susceptible to environmental interference. Wi-Fi signals use the freely licensed ISM 2.4 GHz band and are susceptible to interference from other co-channel signals. At the same time, water molecules have absorption peaks in the 2.4 GHz band, which means that Wi-Fi signal strength is affected by air humidity.

J. Sun et al. (Eds.): CSNC 2020, LNEE 650, pp. 399–408, 2020.
https://doi.org/10.1007/978-981-15-3707-3_38

More importantly, water contributes up to 60% of an adult body [7]. Thus Wi-Fi signals are highly susceptible to indoor pedestrians. All in all, the Wi-Fi signal is unstable and time-varying. When Wi-Fi is used for indoor positioning, there are problems such as low positioning accuracy and weak resistance to external interference.

In 2011, Moghtadaiee et al. proposed using public FM broadcast signals for indoor positioning [8]. Compared with Wi-Fi signals, FM signals have the characteristics of large transmission power, wide coverage, and stable signals. However, the FM signal has a wavelength of about 3 m, which is equivalent to the size of indoor obstacles such as doors, windows, and walls. This means that the FM signal will produce significant diffraction in the room, weakening the ability to depict different positions. In 2015, Wu Hong et al. introduced Digital Television Terrestrial Multimedia Broadcasting (DTMB) signals to improve the accuracy of FM signal positioning [9]. The frequency of the DTMB signal is 470–860 MHz, with a wavelength range of 0.34–0.64 m. Combined with the FM signal with a wavelength of 3 m, the positioning accuracy can be improved.

Compressive sensing (CS) theory states that when the signal is sparse or compressible, we can obtain a compressed representation of the signal in a linear projection manner, and the obtained data can reconstruct the original digital signal with a high probability. Considering the natural sparsity of the location information, the CS can be applied to the field of indoor positioning. This paper proposes a fingerprint matching and positioning scheme based on CS for FM and DTMB signals. In addition, the huge workload of the offline stage data acquisition is also a big problem in the fingerprint positioning system. In this paper, the CS theory is used for reconstructing the RSS values of all RPs from that of partial RPs, thus reducing the workload in the offline stage.

The reminders of the paper are as follows: Sect. 2 introduces the principle of fingerprint localization algorithm; Sect. 3 proposes a fingerprint localization system based on CS; Sect. 4 conducts field experiments in indoor environment, and analyzes the positioning results; Sect. 5 summarizes the full text.

2 Principle of Fingerprint Matching Localization Algorithm

This section will introduce the principle of fingerprint matching localization algorithm. The basic idea is to compare the Received Signal Strength (RSS) with the pre-measured RSS of different locations and find the closest location. This means that prior knowledge, including coordinates and RSS values of all RPs, is required before location.

Fingerprint matching positioning system is consisted of an offline stage and an online stage. In the offline stage, the FM and DTMB signals are measured at the RPs in the location area. And the RSS of FM and DTMB at each RP is stored. The RSS value and coordinates of all RPs constitute a fingerprint database, also known as radio map. In the online stage, FM and DTMB signals are collected at a TP, and then matched with the database obtained in the offline stage to complete the position estimation of a TP.

In the offline stage, N RPs are set in the indoor area, and the position coordinates of each RP are given. At RP i, the position coordinate vector is P_i, and the RSS mean vector is Φ_i. The location fingerprint database can be represented as the following mapping relationship:

$$f : \Phi_i \rightarrow P_i \quad \Phi_i \in R^M, i = 1, 2, \ldots, N, \tag{1}$$

where M represents the signal dimension, which is the number of frequency points.

In the online stage, the RSS of a TP is matched with the RP fingerprint database. Common matching algorithms include Nearest Neighborhood (NN), K Nearest Neighborhood (KNN) and K Weighted Nearest Neighborhood (KWNN). These algorithms are based on the same scheme. Firstly, calculate the Euclidean distance between the RSS of TP and that of each RP. Then select one or more RPs with the smallest distance. Finally use the mean or weighted mean of the coordinates of selected RPs to obtain a position estimate of the TP.

Take the KWNN algorithm as an example. First, the Euclidean distance of the RSS in a TP and that of each RP is calculated. The Euclidean distance can be expressed as

$$d_i = \|y_0 - \Phi_i\|_2, i = 1, 2, \ldots, N, \tag{2}$$

where y_0 is the RSS measurements at a TP, and d_i represents the Euclidean distance between the RP i and the TP. And $\|\cdot\|_2$ is the l_2 norm of the vector. Take the K RPs with the smallest distance to form the set A, and A can be expressed as

$$A = \left\{ i | \min_i k \, d_i, i = 1, 2, \ldots N \right\}. \tag{3}$$

Then the positioning estimation is calculated as the weighted mean of the RPs' coordinates in the set A. The weight of each RP coordinate is

$$\omega_i = \frac{d_i^{-p}}{\sum_{j \in A} d_j^{-p}}, \tag{4}$$

where p is the weight adjustment factor. In our experiment, we set $p = 0.5$ as described in [10]. The positioning result can be expressed as

$$\hat{P} = \sum\nolimits_{i \in A} \omega_i P_i. \tag{5}$$

Nearest neighboring algorithms, such as KNN and KWNN, has the advantage of less computation. Thus they have been widely used in the field of indoor positioning. However, it is susceptible to interference from similar fingerprints, which increases positioning errors. The average accuracy of KWNN indoor positioning is generally about 3–4 m [1, 11], which cannot meet the requirements of high-precision indoor positioning.

3 CS-Based Fingerprint Positioning System

The applications of CS theory in fingerprint matching positioning system include two aspects. In the offline stage, the CS theory is applied to reducing the number of RPs to be measured, and the RSS of all RPs is reconstructed from that of partial RPs by CS reconstruction algorithms. In the online stage, the CS algorithms is used to accurately reconstruct the coordinates of the TPs from the RSS measurements.

3.1 Offline Stage

In a real experimental environment, the labor cost of offline phase is positively correlated with the number of RPs. Considering that the RSS values vary smoothly over the experimental area, the Wavelet coefficients of the RSS value have a sparse nature. That is, CS algorithms can be used to reconstruct the RSS of all RPs from that of partial RPs, thus speeding up the construction of radio map in offline stage.

Without loss of generality, for a certain frequency, the RSS of all the RPs are arranged in a particular order to form vector r. It can be expressed as

$$r = \left[\varphi_{1,j}, \varphi_{2,j}, \ldots, \varphi_{M,j}\right]^T, \tag{6}$$

In fact, r is not sparse and needs to be transformed into a sparse vector through a sparse transformation matrix. Common sparse transformations include Fourier Transform, Discrete Cosine Transform, and Wavelet Transform [12]. Given that the distribution of RSS value in space can be regarded as an image, and that the Discrete Wavelet Transform (DWT) gains great success in the image processing field [13], we use DWT to sparse r. As shown in the Eq. (7),

$$R = W \bullet r, \tag{7}$$

where W is the DWT matrix and R is a sparse vector. In order to select K RPs from the all N RPs ($K < N$), define the random selection matrix G. G is a $K \times N$ matrix. In each row of G, only the value of the selected RP is set to 1, and the other values are 0. If y represents the RSS value of the randomly selected K RPs, then,

$$y = Gr = GW^{-1}R = \Psi R, \tag{8}$$

where $\Psi = GW^{-1}$, Ψ is the sensing matrix in the theory of CS. As long as Ψ is approximately orthogonal, R can be reconstructed from y with a high probability. In order to convert the sensing matrix Ψ into an orthogonal matrix, the orthogonalization method described in [14] is applied. Define the orthogonalization factor T as

$$T = Q\Psi^+, \tag{9}$$

$$Q = orth\left(\Psi^T\right)^T, \tag{10}$$

where Ψ^{+} is the pseudo-inverse of Ψ, and orth(•) is the standard orthogonal basis of the linear space formed by the matrix column vector. So the formula can be converted to

$$z = Ty = Q\Psi^{+}\Psi R \approx QR, \tag{11}$$

where Q is approximately orthogonal. By solving the following l_1 optimization problems, the sparse vector R can be solved from the observation z.

$$R = \arg\min\|R\|_1, s.t.\ z = QR. \tag{12}$$

The above problem can be solved by the CS reconstruction algorithm. By performing inverse DWT on the reconstructed R, r can be solved.

Through adopting the above-described CS reconstruction method, the RSS of all N RPs can be reconstructed from that of the randomly selected K RPs, thereby reducing the labor cost of the offline stage and improving the practicality of the fingerprint matching positioning scheme.

3.2 Online Stage

The position of a TP in the space is unique, which means that the projection of its coordinates in a certain linear space is sparse. Assuming that there are N RPs in the area and the TP is located at the RP i, the position of TP can be represented as a sparse vector θ. θ is a sparse vector with a sparsity of 1. Except for $\theta\ (i) = 1$, the other values of θ are 0.

$$\theta = [0, 0, \ldots, 0, 1, 0, \ldots, 0]^T. \tag{13}$$

As described in [14], the observation equation of the system can be expressed as

$$y_0 = \Phi\theta + \varepsilon, \tag{14}$$

where y_0 is the RSS measurements collected at a TP, the column vectors of Φ are the mean RSS of the RPs, and ε is the observed noise. Since the restricted isometry property (RIP) of the CS reconstruction is not satisfied, it needs to be pre-orthogonalized. The pre-orthogonalization factor can be expressed as

$$T_0 = H\Phi^{+}, \tag{15}$$

$$H = orth(\Phi^T)^T. \tag{16}$$

The Eq. (14) is multiplied by (15) and becomes

$$z_0 = T_0 y_0 = T_0\Phi\theta + T_0\varepsilon \approx H\theta + \varepsilon'. \tag{17}$$

Then, H is an approximate orthogonal matrix and satisfies the RIP property. θ can be reconstructed by solving the following l_1 optimization problem.

$$\hat{\theta} = \arg\min\|\theta\|_1, s.t.\ z_0 = H\theta + \varepsilon'. \tag{18}$$

According to the theory of CS, θ can be reconstructed from observation z_0 with a high probability. Commonly used CS reconstruction algorithms include iterative greedy algorithm and convex optimization algorithm, such as Orthogonal Matching Pursuit (OMP) [15], Compressive Sampling Matching Pursuit (CoSaMP) [16], Base Pursuit (BP) [16], Iterative Hard Threshold (IHT) [16] and so on.

In most cases, the TP is not completely located at a certain RP, in which case the sparsity of θ is greater than one. θ has a plurality of non-zero coefficients, and these non-zero coefficients represent the probability that the corresponding RP is used as the position estimation. Therefore, in practical applications, we set a threshold τ. And only the coefficient exceeding the threshold is used for position estimation. That is, the RP set B for position estimation is

$$B = \left\{ i \,|\, \hat{\theta}(i) > \tau \right\}. \tag{19}$$

These selected coefficients are then normalized to obtain the weights of the corresponding RPs. Finally, the position estimation $\hat{P}$ is the weighted mean from the RPs in the set B, as shown in Eq. (20).

$$\hat{P} = \sum_{i \in B} \frac{\hat{\theta}(i)}{\sum_{j \in B} \hat{\theta}(j)} P_i. \tag{20}$$

4 Experimental Results and Analysis

In order to verify the proposed indoor positioning solution, we conducted field tests in a typical indoor environment. Figure 1 shows the actual indoor experimental environment consisting of three rooms and a corridor with an area of about 200 m^2, including 164 RPs and 41 TPs. The average interval between RPs is about 1 m. There are many obstacles and pedestrians in the experimental area. In this case, the fingerprint matching scheme is more practical since it is very difficult to establish a practical signal propagation model. In terms of signal reception, as shown in Fig. 2, two Universal Software Radio Peripherals (USRP) are used to simultaneously receive FM broadcast signals and DTMB signals. A simple whip antenna is used for FM signal reception, and a PCB Logarithmic period antenna is used for DTMB signal reception.

In the experiment of this paper, the FM signals are received at 87–106.9 MHz with a frequency resolution of 0.1 MHz. At the same time, the DTMB signals are received at 582–590 MHz and the frequency resolution is 0.1 MHz.

In the offline stage, the signal is acquired at all RPs for 2 min and 20 times per second. Figure 3 is the average of the RSS signal for RP 1 within 2 min.

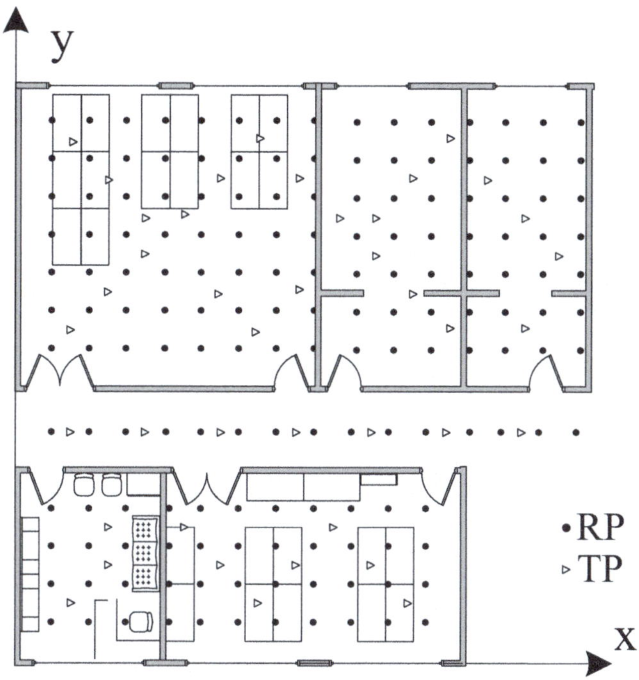

Fig. 1. Distribution map of RPs and points to be located in the experimental environment

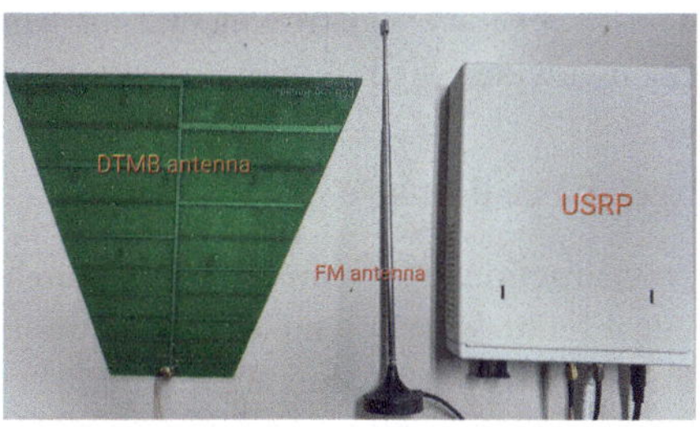

Fig. 2. FM signal and DTMB signal receiving system

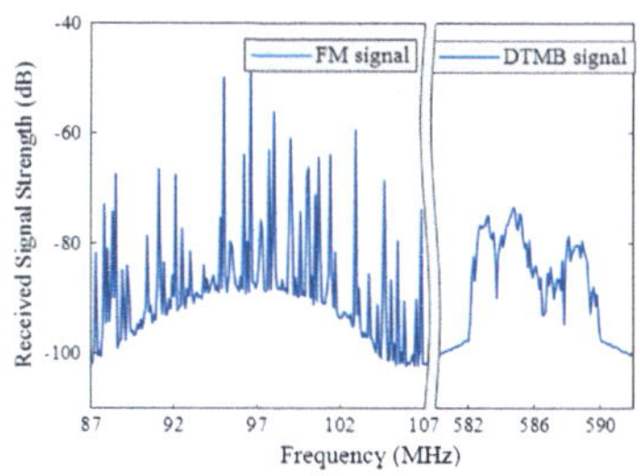

Fig. 3. RSS values of FM and DTMB signal at RP 1

4.1 Offline Stage CS Reconstruction Accuracy

As described in Sect. 3.1, in order to reduce the labor cost in offline stage, the CS algorithm is applied to reconstruct the radio map of all 164 RPs from that of the K RPs. Obviously, the smaller the K value, the smaller the amount of data collected during the offline stage. However, the accuracy of RSS reconstruction decreases as the value of K decreases. In order to evaluate the reconstruction accuracy, this paper measures the RSS values of all RPs, and randomly selects K points for reconstruction, and calculates the average reconstruction error.

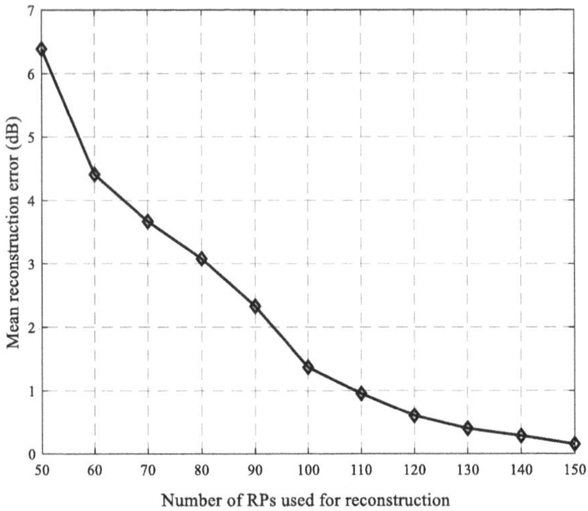

Fig. 4. Relationship between the number of RPs used for reconstruction and the average reconstruction error

As shown in Fig. 4, experiments have verified that it is feasible to reconstruct the RSS values of all RPs by CS using partial RP samples. When 100 RPs measurements are used for reconstruction, the average reconstruction error is 1.41 dB. Compared with the total of 164 RPs, the data collection is reduced by 39%. When K further decreases, the RSS reconstruction error increases rapidly. Therefore, we choose K = 100 at the cost of 1.41 dB reconstruction error in exchange for 39% workload reduction in radio map construction, which has obvious advantages in a wide range of indoor positioning.

4.2 Online Stage Compression Sensing Precise Positioning

As described in Sect. 3.2, this paper uses CS reconstruction algorithms to achieve high-precision positioning of the indoor environment. Figure 5 is a cumulative probability distribution curve of positioning errors of different CS reconstruction algorithms when using the actual measured radio map. And the reconstruction algorithms include OMP [15], CoSaMP [16], IHT [16] and BP [16] algorithms. The results show that the BP reconstruction algorithm has better positioning accuracy than others. The average positioning error is 1.83 m and the maximum positioning error is about 5 m. In contrast, the average positioning error of the KWNN algorithm is 2.98 m and the maximum positioning error is about 9 m.

In addition, when using the radio map which was reconstructed as describe in Sect. 4.1, the positioning accuracy would decrease. Table 1 shows the positioning error statistics of with BP algorithm when the radio map is reconstructed with different K values in the offline stage. As the value of K decreases, the positioning accuracy will decrease. When $K = 100$, the mean positioning error in online stage is 2.14 m. Positioning accuracy is reduced by 16% compared to using 164 actual measured RP data. Considering 39% workload reduction in the offline stage, a 16% reduction in accuracy is acceptable. When K is less than 100, the mean positioning error increases greatly. What's more, the standard deviation of the error increases greatly, which means that the accuracy and stability of the positioning system are drastically deteriorated. In summary, when using the radio map reconstructed from 100 RPs RSS values, a good balance between positioning accuracy and data collection is achieved.

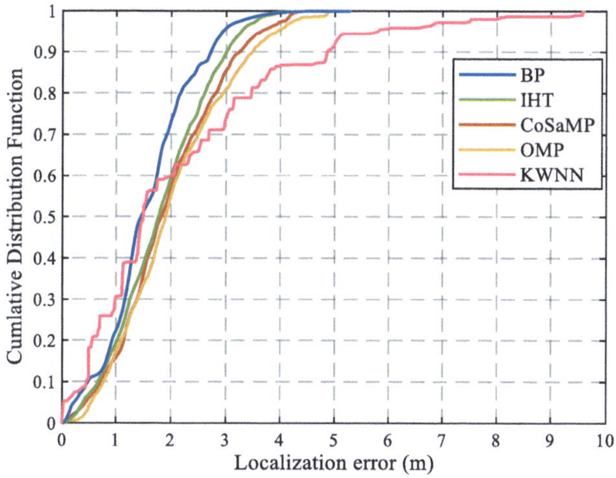

Fig. 5. Comparison of CDF curves using positioning errors of CS and KWNN algorithm

Table 1. The positioning error statistics of different recovered radio map with BP algorithm

K	60	80	100	120	140	150
Mean [m]	4.97	3.05	2.14	2.08	1.95	1.88
Median [m]	3.63	2.67	2.01	1.92	1.61	1.52
Standard deviation [m]	8.20	6.65	3.81	2.88	1.93	1.41

5 Conclusion

This paper presents a practical indoor positioning system. The system uses public FM and DTMB signals for positioning estimation. Considering the extensive coverage of public FM and DTMB signals, no additional signal transmitters are needed, which is a low cost, wide coverage solution.

The positioning scheme is divided into two parts, the offline stage and the online stage. In order to reduce the workload in the offline stage, we use CS algorithms to reconstruct the RSS values of all RPs from that of partial RPs. Compared with the actual measured radio map, the offline stage workload is reduced by 39%, and the average positioning error increases from 1.83 m to 2.14 m. As is known to all, the main shortcoming of fingerprint matching positioning scheme is the huge labor cost in offline stage. In the view of the fact that the proposed scheme can effectively reduce the workload of data acquisition, it is a suitable positioning solution in a wide range of indoor positioning environments.

Acknowledgements. This work was supported in part by the National Natural Science Foundation of China under grant No. 61571244, in part by Tianjin Research Program of Application Foundation and Advanced Technology under grant No. 16YFZCSF00540 and No. 18YFZCGX00480.

References

1. Xue, W., et al.: APs' virtual positions-based reference point clustering and physical distance-based weighting for indoor Wi-Fi positioning. IEEE Internet of Things J. **5**(4), 3031–3042 (2018)
2. Ma, H., et al.: The optimization for hyperbolic positioning of UHF passive RFID tags. IEEE Trans. Autom. Sci. Eng. **14**(4), 1590–1600 (2017)
3. Khyam, M.O., et al.: Multiple access chirp-based ultrasonic positioning. IEEE Trans. Instrum. Meas. **66**(12), 3126–3137 (2017)
4. Zhuang, Y., et al.: A survey of positioning systems using visible LED lights. IEEE Commun. Surv. Tutorials **20**(3), 1963–1988 (2018)
5. Yu, N., et al. A precise dead reckoning algorithm based on Bluetooth and multiple sensors. IEEE Internet of Things J. **5**(1), 336–351 (2018)
6. Xu, Y., Tian, G., Chen, X.: Enhancing INS/UWB integrated position estimation using federated EFIR filtering. IEEE Access **6**, 64461–64469 (2018)
7. Jéquier, E., Constant, F.: Water as an essential nutrient: the physiological basis of hydration. Eur. J. Clin. Nutr. **64**(2), 115 (2010)
8. Moghtadaiee, V., Dempster, A.G., Lim, S.: Indoor localization using FM radio signals: a fingerprinting approach. In: 2011 International Conference on Indoor Positioning and Indoor Navigation. IEEE (2011)
9. Wu, H., et al.: Indoor localization using FM radio and DTMB signals. Radio Sci. **51**(7), 1030–1037 (2016)
10. Chen, W., et al.: A novel clustering and KWNN-based strategy for Wi-Fi fingerprint indoor localization. In: 2015 4th International Conference on Computer Science and Network Technology (ICCSNT), vol. 1. IEEE (2015)
11. Hu, J., et al.: Experimental analysis on weight K-nearest neighbor indoor fingerprint positioning. IEEE Internet of Things J. **6**(1), 891–897 (2019)
12. Duarte, M.F., Eldar, Y.C.: Structured compressed sensing: from theory to applications. IEEE Trans. Sig. Process. **59**(9), 4053–4085 (2011)
13. Demirel, H., Anbarjafari, G.: Image resolution enhancement by using discrete and stationary wavelet decomposition. IEEE Trans. Image Process. **20**(5), 1458–1460 (2011)
14. Feng, C., et al.: Received-signal-strength-based indoor positioning using compressive sensing. IEEE Trans. Mob. Comput. **11**(12), 1983–1993 (2012)
15. Wang, J., Shim, B.: On the recovery limit of sparse signals using orthogonal matching pursuit. IEEE Trans. Signal Process. **60**(9), 4973–4976 (2012)
16. Giryes, R., Elad, M.: RIP-based near-oracle performance guarantees for SP, CoSaMP, and IHT. IEEE Trans. Signal Process. **60**(3), 1465–1468 (2012)

Daily Climatological Fields Based on GNSS Radio Occultation Measurements: A Feasibility Study

Zhen Shen, Qimin He, Longjiang Li, Kefei Zhang, and Suqin Wu(✉)

School of Environment Science and Spatial Information,
China University of Mining and Technology, Xuzhou, China
suqin_wu@cumt.edu.cn

Abstract. GNSS Radio occultation (RO) technique can provide high-quality measurements of the Earth's atmosphere. Monthly (or seasonly) RO climatological fields have been widely used in climate and weather studies. However, the low temporal resolution makes it difficult in resolving the rapid variations in the atmosphere. Here we established the daily climatological fields by GNSS RO measurements from the Constellation Observing System for Meteorology, Ionosphere, and Climate. By applying a "point" strategy, RO profiles are assimilated into a $3° \times 3°$ longitude-latitude global grid through the Gaussian weighting function. By comparison between the RO climatological fields and European Centre for Medium Range Weather Forecasts reanalysis, the differences display distinctive features in different layers of the Earth's atmosphere. The overall differences are about 2 K, 1.5% and 40 m (temperature, density, and geopotential height) in the lower troposphere; 1 K, 0.5% and 40 m in the upper troposphere and lower stratosphere region; and 2 K, 1% and 100 m in the upper stratosphere. The results have demonstrated that the observation-based RO climatological fields are reliable in both northern and southern hemispheres.

Keywords: Radio occultation · Atmospheric parameters · Daily climatological fields · "Point" strategy

1 Introduction

The atmosphere has been an interesting research topic for decades, and the changes in weather and climate influence every kind of creature on the planet. Radio occultation (RO), which is a relatively new technique, was first used to detect the structure of the atmosphere in Mars and Venus since the mid-l960s (Eshleman 1973; Fjeldbo and Eshleman 1968). The first GNSS RO mission to sound the Earth's atmosphere is the Global Positioning System/Meteorology (GPS/MET) experiment, which was launched in April 1995 and kept working until February 1997. The GPS/MET experiment proved the feasibility of the RO technique in the detection of the Earth's atmosphere and provided about 150 atmospheric profiles per day during its lifetime (Rocken et al. 1997). The next milestone project of RO was the Challenging Minisatellite Payload (CHAMP) program, which had been launched in July 2000 and completed in September 2010. The CHAMP mission provided more than four hundred thousand RO

J. Sun et al. (Eds.): CSNC 2020, LNEE 650, pp. 409–418, 2020.
https://doi.org/10.1007/978-981-15-3707-3_39

events in total (Reigber et al. 2002). In June 2006, the number of RO measurements had significant growth with the implementation of the Constellation Observing System for Meteorology, Ionosphere, and Climate (COSMIC) project. The COSMIC constellation, which consists of six low Earth orbit (LEO) satellites, provided more than 2000 profiles per day during its ideal working time (about 2007 to 2010) (Anthes et al. 2008; Wu and Shieh 2002).

RO measurements provide vertical profiles of refractivity, bending angles and dry temperature/pressure as a function of mean sea level (MSL) altitude. Wet/physical parameters (e.g., wet temperature/pressure and vapour pressure) could then derived by combining a priori or background information, using the one-dimensional variational method (Das and Pan 2014; Healy and Eyre 2000). The applications of RO atmospheric data have been widely investigated in recent years due to the advantages of the RO technique, such as global coverage, self-calibration, high quality, and high vertical resolution. Brunner et al. (2016) found that RO measurements robustly capture atmospheric blocking events, by establishing a 2.5° × 2.5° longitude-latitude climatological field using RO profiles from CHAMP, GRACE and COSMIC. Climate changes in the upper troposphere and lower stratosphere (UTLS) were detected by RO atmospheric profiles, and the results proved that RO atmospheric parameters are sensitive and complementary trend indicators at different UTLS height regions (Foelsche et al. 2008; Lackner 2010). The variability and trends of global tropopause height were discussed based on monthly mean RO climatological fields (Schmidt et al. 2008). Furthermore, RO data had been used in studies of El Nino-Southern Oscillation (i.e. ENSO) (Scherllin-Pirscher et al. 2012), atmospheric tides (Pirscher et al. 2010), and tropospheric dry layers (Rieckh et al. 2017). All these studies underlined the feasibility and potential of the GNSS RO technique in weather and climate studies.

Previous studies mostly based on large-area or long-time average RO climatological fields, which were difficult to detect the rapid or small-scale variations in the atmosphere. With the implementation of the COSMIC mission, the number of RO measurements experiences a revolutionary growth, that enables us to seek the possibility of constructing a higher spatio-temporal resolution climatological field. For this purpose, some improvements were proposed in this study: (i) The wetPrf files were used instead of atmPrf files. WetPrf files, which provide the wet atmospheric parameter profiles, describe the real structure of the atmosphere. The using of wetPrf files avoids the dry-wet parameter differences in the lower troposphere. (ii) A "point" strategy is adopted instead of the "bin" strategy. The "bin" strategy is widely used in previous studies. By applying a grid system formed by lines of latitude and longitude, the Earth's surface was divided into cells with different size, e.g., 5° × 5° longitude/latitude (Scherllin-Pirscher et al. 2014), 360°× 10° (Foelsche et al. 2008). The profiles of atmospheric parameters in each cell were then obtained by averaging the RO measurements inside. In the "point" strategy, however, climatological fields are calculated based on grid points, and the value of atmospheric parameters at every grid point was derived from nearby RO profiles (see Sect. 3 for more detail).

The structure of this paper is outlined below. Section 2 introduces the data sets used in this study, including RO data from COSMIC and the reanalysis data from European Centre for Medium Range Weather Forecasts (ECMWF). Section 3 describes the data processing methods from RO atmospheric profiles to global climatological fields.

Section 4 evaluates the reliability of RO climatological fields by comparing it with ECMWF reanalysis data. Section 5 discusses the conclusions and perspectives of future development.

2 Data

Two types of data were used in this study, including RO data from the COSMIC Data Analysis and Archive Center (CDAAC) and EAR-Interim reanalysis data from ECMWF.

2.1 RO Data

CDAAC provides three types of COSMIC products, including COSMICRT, COSMIC, and COSMIC2013. COSMICRT is the near real-time data series, which is a few days delayed than the real-time. Since COSMICRT uses global forecast system (GFS) 12-hour forecast data as background information, the precision of the data series is relatively lower than the other two. COSMIC and COSMIC2013 are post-processed and re-processed data series, respectively. Considering the number of COSMIC atmospheric profiles achieved a significant growth at the end of 2006 and started to decline at about the beginning of 2011, the atmospheric profiles of COSMIC2013 data series between January 2007 and December 2011 were adopted in this study. There are 3,116,746 high-quality atmospheric profiles during this period and 1707/day on average.

The geographical distribution of RO observations depends mainly on altitude and inclination of the orbit of the LEO satellite. COSMIC satellites with an orbit inclination of 72° provide more observations in the middle latitudes. The maximum number is found between about $\pm 40° \sim \pm 60°$ latitudes, and the minimum is in the polar regions. The spatial density of COSMIC RO profiles, however, is the highest in the polar regions and lowest near the equator due to the changes of geographical area with varying latitudes (Pirscher 2010).

2.2 Re-Analyses Data

ERA-Interim reanalysis data was used to test the reliability of RO climatological fields. ERA-Interim, managed by ECMWF, is a global atmospheric reanalysis from 1979 and continuously updates in real-time. ERA-Interim provides four global atmosphere reanalysis data (e.g., temperature, specific humidity, component of the wind, et al.) per day at 00, 06, 12 and 18 coordinated universal time (UTC) (Dee et al. 2011). We have to note that ERA-Interim is not independent with RO data because COSMIC measurements had been assimilated into ECMWF since December 2006, and the reanalysis data is used as background information in RO data retrieval. However, it is still valuable for comparisons because (i) the number of GPS-RO measurements is quite small compared to the total assimilated observations (less than 1%), and the effect of RO measurement on reanalysis is very limited; (ii) There are still some differences between them, and the investigation of the differences can play an essential role in improving the performance of the two data set (Padullés et al. 2018).

3 Construction of RO Climatological Fields

RO climatological fields are established in a global grid with a horizontal resolution of 3° × 3° longitude-latitude and temporal resolution of 1 day. The vertical profiles of atmospheric parameters at each grid point are derived by taking a weighted average of all RO measurements if the geographical distance between the location of RO measurements and grid points is less than 300 km (the search radius based on the horizontal resolution of RO observations).

$$x_{gird}(\lambda, \varphi, h) = \frac{\sum w_i x_i(\lambda_i, \varphi_i, h)}{\sum w_i} \tag{1}$$

Where x_{grid} (λ, φ, h) is the value of atmospheric parameters on grid points at longitude λ, latitude φ and the altitude of h; x_i (λ_i, φ_i, h) is the RO atmospheric profiles in search radius, and λ_i and φ_i are the geographic locations of RO measurements; w_i, which is the weight for each RO profile, is derived by Gaussian weighting function

$$w_i = \exp\left[-\left(\frac{\Delta d}{D}\right)^2\right] \tag{2}$$

Where Δd is the geographical distance between RO measurements and grid points, in km; D = 300 km is the search radius.

For most grid points, the vertical profiles of atmospheric parameters can be derived from adjacent RO measurements. However, because of the quantity and uneven distribution of RO measurements, there are still some grid points around which no RO observation exists. Those points are defined as "blank points". From 2007 to 2011, the majority of blank points are found in low latitudes (67.2%, 16.8% and 16.0% in low, middle and high latitudes, respectively) since the spatial density of RO measurements is the lowest in low latitudes.

4 Validation of RO Climatological Fields

In this section, the reliability of RO climatological fields was evaluated by comparing with ECMWF reanalysis data, including temperature, density, and geopotential height, but only the temperature results are presented in this paper to keep the paper reasonably concise. RO climatological fields were interpolated to the same pressure level as ECMWF reanalysis by cubic spline interpolation method (Schwärz et al. 2016). Since year to year details cannot be wholly shown and the results are consistent during the study period (2007–2011), only the differences between RO climatological fields and ECMWF reanalysis in 2008 were shown.

The differences between RO climatological fields and ECMWF reanalysis vary gradually with pressure (not show here). The differences are the strongest in the lower troposphere and decrease as the increasing height (decreasing pressure). Between about 400 hPa and 50 hPa pressure surfaces, the differences are relatively stable, while increase rapidly above 50 hPa. Here we selected three pressure surfaces, which represent three different layers in Earth's atmosphere: 800 hPa (low troposphere), 200 hPa (UTLS region), and 3 hPa (upper stratosphere). The geographic maps of temperature differences on the three pressure levels are shown in Figs. 1, 2 and 3 in different seasons according to seasonal determinations of the northern hemisphere (NH): winter (December to February), spring (March to May), summer (June to August) and autumn (September to November). We shall point out that the differences are not necessarily because of the deficiency of RO measurements but result from the inaccuracy of ECMWF reanalysis, which is attributable primarily to incomplete model and imperfect radiosonde observations.

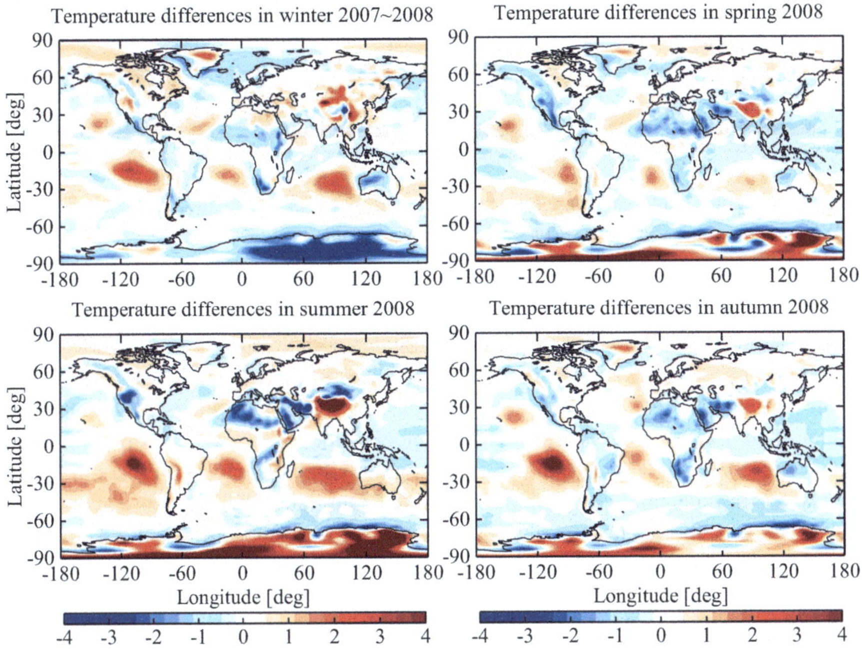

Fig. 1. Geographic maps of temperature differences between RO climatological fields and ECMWF reanalysis on 800 hPa surface in different seasons, unit: K.

On 800 hPa pressure surface, in most places, the mean temperature differences are less than 2 K. Notable differences (mainly between 2 K and 4 K) are found on the mainland of the Northern Hemisphere (NH), marine areas of the Southern Hemisphere (SH), and the Antarctic region.

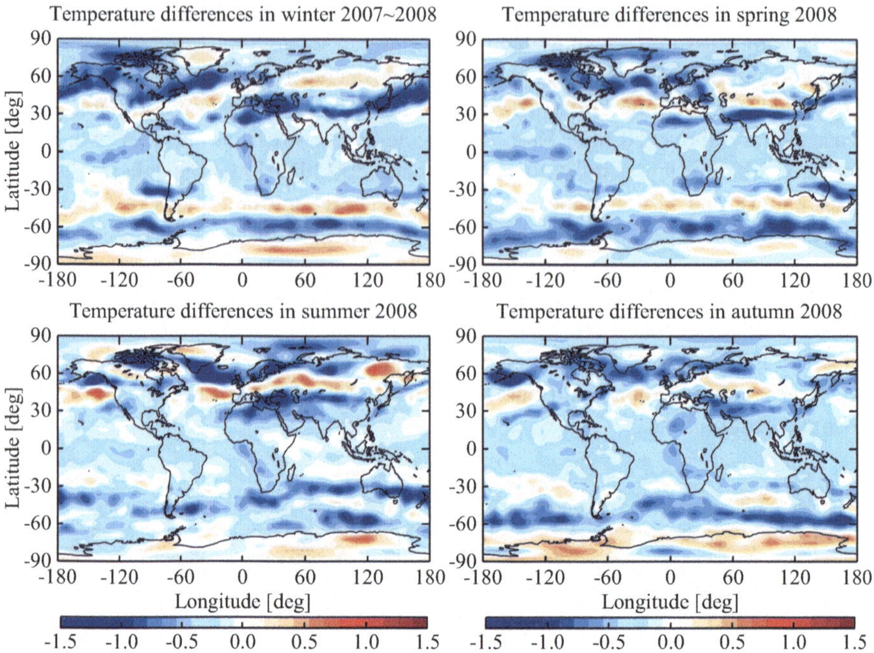

Fig. 2. Geographic maps of temperature differences between RO climatological fields and ECMWF reanalysis on 200 hPa surface in different seasons, unit: K.

In NH, small-scale negative differences (<1 K) are found in the low-middle latitudes mainland (e.g., west coast of the United States and North Africa) in winter, spring and autumn. The differences are much stronger in summer, e.g., in North Africa, the Arabian Peninsula, West Asia and the west coast of the United States, RO temperature is colder than ECMWF temperature with an extreme of 3 K. The most significant biased is more than 4 K and found in the Qinghai-Tibet Plateau, which is attributable primarily to the special topographic features of Qinghai-Tibet Plateau (average elevation exceeds 4,500 m and the surface pressure between 500 hPa and 700 hPa).

In SH, RO temperature is warmer than ECMWF in ocean areas, especially in the East Indian Ocean and the East Pacific Ocean with temperature differences exceeding 4 K in summer and autumn. In the Antarctic region, the evolution of temperature differences displays an annual cycle. RO temperature is warmer than ECMWF with an extreme of 5 K in spring, summer, and autumn, while it is colder with about 4 K in winter.

On a 200 hPa pressure surface, the distribution of the temperature differences is approximately symmetric with respect to the equator. RO temperature is colder than ECMWF in low latitude and the most of middle-high latitudes (<1 K). Three band-structure discontinuous positive biases are located at about 30°N, 40°S and 70°S latitude, varying with seasons. In winter and summer, the differences in NH is slightly stronger than in SH. Overall, temperature differences are less than 1 K on the 200 hPa surface in most places (notes the different color axis in Figs. 1.10 and 1.11). The small

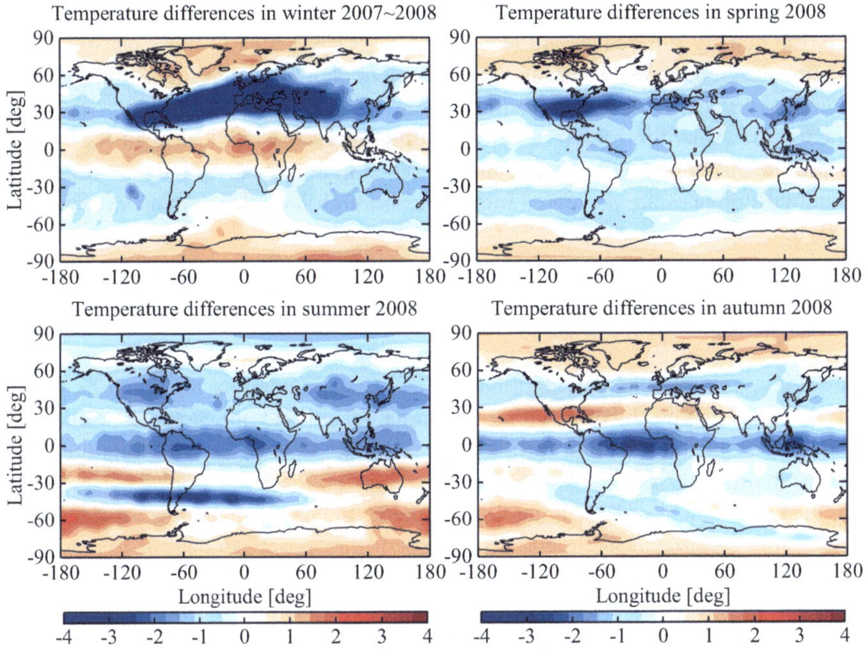

Fig. 3. Geographic maps of temperature differences between RO climatological fields and ECMWF reanalysis on 3 hPa surface in different seasons, unit: K.

differences indicate that RO measurements are most sensitive in the UTLS region, which generally agreed with early studies (Steiner and Kirchengast 2005).

On the 3 hPa surface, the temporal evolution displays an annual cycle. In NH, RO temperature is warmer than ECMWF at high latitudes and equatorial regions in winter (<2 K), while significant biases are found in middle latitudes with an extreme value of −5 K. In spring, the negative differences (<3 K) extend southward to the equator, and positive differences still cover the high latitudes. In summer, the positive differences in high latitude gradually disappear, and RO temperature is colder than ECMWF in most places with an extreme value of 2 K. In autumn, the region of negative differences decline as new positive differences emerge in the high latitudes.

In SH, during the winter season, the temperature differences are similar to that of NH. In spring and summer, the negative differences zone is compressed by the northern positive differences. In autumn, the range of positive differences is reduced, and negative biases reoccupy the middle latitudes. Overall speaking, on 3 hPa surface, the temperature differences are more evident in NH during winter and spring, while the opposite is true in summer and autumn.

The geographic distribution and seasonal variation of density differences and GH differences are roughly similar to that of temperature differences (not shown here). The overall results are summarized as follows: (i) Density. In the low troposphere, the density differences are about 1% in most places (RMS 2%), and the maximum differences are about 2% which is found in NH high latitudes around the surface. In the

UTLS region, density differences are less than 0.5% (RMS 1%). In the upper stratosphere, the density differences are less than 1% (RMS 1.5%) in most places, and the maximum differences reach 3% in NH; (ii) Geopotential height. The differences of geopotential height are less than 40 m between 1000 hPa and 50 hPa pressure surface except for the Antarctic region, where the differences exceed 120 m in the low troposphere. In the upper stratosphere, the overall differences are about 100 m, and the maximum differences are presented in middle and high latitudes with an extreme value of 200 m.

5 Summary and Conclusion

In this study, a number of global climatological fields, including temperature, density, and GH, have been established with a large quantity of RO atmospheric profiles from COSMIC. The spatial-temporal resolution is set to be 3° × 3° latitude-longitude and 1 day respectively.

In the construction of RO climatological fields, the "bin" strategy is replaced by the "points" strategy. One of the major advantages of the "point" strategy is that it enables us to establish a high spatial-temporal resolution field is established in places where both of the geographical area and the number of RO profiles are minimal (e.g., high latitudes). The profiles of atmospheric parameters on "blank points" are filled by an interpolation strategy of using nearby grid points. It works well for the most time from 2007 to 2009. But for 2010 and beyond, the blank grids rate is relatively higher, especially in low latitudes. The situation can be improved by assimilating measurements from other RO systems like CHAMP and GRACE.

By comparing with ECMWF reanalysis data, RO climatological fields are found to be very sensitive in revealing a detailed atmosphere structure. The differences between RO climatological fields and ECMWF reanalysis display distinctive features in different layers of the Earth's atmosphere. In the low troposphere, the RO climatological fields are well consistent with ECMWF in most places, while significant biases are presented in some independent regions (e.g., Qinghai-Tibet Plateau and Antarctica). In the UTLS region, the differences are the smallest, and approximately symmetric with respect to the equator. In the upper stratosphere, the characteristics display notable spatial and temporal differences in the two hemispheres. The overall differences are about 2 K, 1.5% and 40 m (temperature, density, and GH) in the low troposphere; 1 K, 0.5% and 40 m in the UTLS region; and 2 K, 1% and 100 m in the upper stratosphere.

Compared to ECMWF reanalysis, the observation-based RO climatological fields are reliable in both hemispheres. The daily climatological fields cannot represent an optimal sampling of all harmonics of the variations (e.g., the semidiurnal variations). However, it is efficient to sample the diurnal or seasonal cycle in atmospheric parameters and provide a strong contribution to weather and climate research. This work provides a first step towards the novel applications of GNSS RO technique for climatology.

Acknowledgments. We acknowledge CDAAC for the use of the COSMIC GPS-RO datasets. We would like to thank ECMWF for providing the ERA-Interim reanalysis data. This work was funded by the National Natural Science Foundation of China (Grant No. 41874040) and the State Key Program of National Natural Science Foundation of China (Grant No. 41730109).

References

Anthes, R.A., et al.: The COSMIC/FORMOSAT-3 mission: early results. Bull. Am. Meteorol. Soc. **89**(3), 313–334 (2008). https://doi.org/10.1175/BAMS-89-3-313

Brunner, L., Steiner, A.K., Scherllin-Pirscher, B., Jury, M.W.: Exploring atmospheric blocking with GPS radio occultation observations. Atmos. Chem. Phys. **16**(7), 4593–4604 (2016). https://doi.org/10.5194/acp-16-4593-2016

Das, U., Pan, C.J.: Validation of FORMOSAT-3/COSMIC level 2 "atmPrf" global temperature data in the stratosphere. Atmos. Meas. Tech. **7**(3), 341–365 (2014). https://doi.org/10.5194/amt-7-731-2014

Dee, D.P., Uppala, S.M., Simmons, A., Berrisford, P., Poli, P., Kobayashi, S., Andrae, U., Balmaseda, M., Balsamo, G., Bauer, D.P.: The ERA-Interim reanalysis: configuration and performance of the data assimilation system. Q. J. Roy. Meteorol. Soc. **137**(656), 553–597. https://doi.org/10.1002/qj.828 (2011)

Eshleman, V.R.: The radio occultation method for the study of planetary atmospheres. Planet. Space Sci. **21**(9), 1521–1531 (1973). https://doi.org/10.1016/0032-0633(73)90059-7

Fjeldbo, G., Eshleman, V.R.: The atmosphere of Mars analyzed by integral inversion of the Mariner IV occultation data. Planet. Space Sci. **16**(8), 1035–1059 (1968). https://doi.org/10.1016/0032-0633(68)90020-2

Foelsche, U., Kirchengast, G., Steiner, A.K., Kornblueh, L., Manzini, E., Bengtsson, L.: An observing system simulation experiment for climate monitoring with GNSS radio occultation data: setup and test bed study. J. Geophys. Res. Atmos. **113**(D11) (2008). https://doi.org/10.1029/2007jd009231

Healy, S., Eyre, J.R.: Retrieving temperature, water vapour and surface pressure information from refractive-index profiles derived by radio occultation: a simulation study. Q. J. Roy. Meteorol. Soc. **126**(566), 1661–1683 (2000). https://doi.org/10.1002/qj.49712656606

Lackner, B.C.: Exploring Trend Indicators of Climate Change from Radio Occultation and Optimal Trend Detection. University of Graz, 188 p. (2010)

Padullés, R., Cardellach, E., Wang, K.-N., Ao, C.O., Turk, F.J., de la Torre-Juárez, M.: Assessment of global navigation satellite system (GNSS) radio occultation refractivity under heavy precipitation. Atmos. Chem. Phys. **18**(16), 11697–11708 (2018). https://doi.org/10.5194/acp-18-11697-2018

Pirscher, B.: Multi-satellite climatologies of fundamental atmospheric variables from radio occulation and their validation. Ph.D. thesis, University of Graz, 238 p. (2010)

Pirscher, B., Foelsche, U., Borsche, M., Kirchengast, G., Kuo, Y.H.: Analysis of migrating diurnal tides detected in FORMOSAT-3/COSMIC temperature data. J. Geophys. Res. Atmos. **115**(D14). https://doi.org/10.1029/2009jd013008 (2010)

Reigber, C., Lühr, H., Schwintzer, P.: CHAMP mission status. Adv. Space Res. **30**(2), 129–134 (2002). https://doi.org/10.1016/S0273-1177(02)00276-4

Rieckh, T., Anthes, R., Randel, W., Ho, S.-P., Foelsche, U.: Tropospheric dry layers in the tropical western Pacific: comparisons of GPS radio occultation with multiple data sets. Atmos. Meas. Tech. **10**(3), 1093–1110 (2017). https://doi.org/10.5194/amt-10-1093-2017

Rocken, C., Anthes, R., Exner, M., Hunt, D., Sokolovskiy, S., Ware, R., Gorbunov, M., Schreiner, W., Feng, D., Herman, B.: Analysis and validation of GPS/MET data in the neutral atmosphere. J. Geophys. Res. **102**(D25), 29849–29866 (1997). https://doi.org/10.1029/97JD02400

Scherllin-Pirscher, B., Deser, C., Ho, S.P., Chou, C., Randel, W., Kuo, Y.H.: The vertical and spatial structure of ENSO in the upper troposphere and lower stratosphere from GPS radio occultation measurements. Geophys. Res. Lett. **39**(20). https://doi.org/10.1029/2012gl053071 (2012)

Scherllin-Pirscher, B., Steiner, A.K., Kirchengast, G.: Deriving dynamics from GPS radio occultation: three-dimensional wind fields for monitoring the climate. Geophys. Res. Lett. **41**(20), 7367–7374 (2014). https://doi.org/10.1002/2014gl061524

Schmidt, T., Wickert, J., Beyerle, G., Heise, S.: Global tropopause height trends estimated from GPS radio occultation data. Geophys. Res. Lett. **35**(11), 213–226 (2008). https://doi.org/10.1029/2008GL034012

Schwärz, M., Kirchengast, G., Scherllin-Pirscher, B., Schwarz, J., Ladstädter, F., Angerer, B.: Multi-mission Validation by Satellite Radio Occultation–Extension Project Report. University of Graz (2016)

Steiner, A.K., Kirchengast, G.: Error analysis for GNSS radio occultation data based on ensembles of profiles from end-to-end simulations. J. Geophys. Res. **110**(D15), D15307 (2005). https://doi.org/10.1029/2004JD005251

Wu, A., Shieh, C.: ROCSAT-3 constellation design and data simulation. Paper presented at IAF Abstracts, 34th COSPAR Scientific Assembly (2002)

Stochastic Modeling of BeiDou Double-Difference Observation and Impact Analysis

Zhongzhi Wang, Weikai Miao(✉), and Yunzhong Shen(✉)

College of Surveying and Geo-Informatics, Tongji University, Shanghai, China
{wkmiao,yzshen}@tongji.edu.cn

Abstract. Double-difference integer ambiguity resolution (IAR) is playing the vital part in the high precision GNSS positioning and navigation. The IAR success rate depends on three factors: the functional model, the stochastic model and the chosen method of the integer ambiguity estimation. Stochastic model plays an important role in parameter estimation of global navigation satellite system (GNSS). Only a correct stochastic model can be used to obtain the reliable integer ambiguity, the accurate positioning and the reliable baseline precisions. In this paper, a stochastic model with significantly sophisticated structure is designed, and the MINQUE method is utilized to estimate the cross correlations between different types of observations at arbitrary frequency and the time correlations for phase and code observations per frequency. In assessing the stochastic model, the short-length baseline and zero-length baseline with sampling frequency of 1 s are processed to analyze the impact of the realistic stochastic model considering the cross and time correlations on the IAR success rate and positioning. The results confirm that compared with the empirical stochastic model ignoring the cross and time correlations, the more realistic stochastic model can significantly improve the IAR theoretical and practical success rate, especially for single-frequency data, the practical success rate increases by 5%. The baseline precision that ignores cross and time correlation has a large difference from the theoretical ones. Namely, the baseline precision ignoring physical correlations is too optimistic and unrealistic. On the contrary, the baseline precision that considers physical correlation match the theoretical ones more well.

Keywords: BDS · Stochastic model · Cross correlation · Time correlation · Baseline precision · Integer ambiguity resolution success rate

1 Introduction

The key issue for high precision GNSS positioning and navigation with the precise GNSS carrier phase observations is the correct integer ambiguity resolution. The IAR success rate depends on three factors [1, 2]: (1) functional model; (2) stochastic model; (3) the chosen method of the integer ambiguity estimation. Realistic stochastic model plays an important role in parameter estimation which is effective only when the realistic stochastic model is used [3]. The stochastic models are even more important in term of the reliability because the statistical tests involve covariance matrices. Such statistical tests are sensitive to stochastic model [4]. The traditional stochastic model

J. Sun et al. (Eds.): CSNC 2020, LNEE 650, pp. 419–434, 2020.
https://doi.org/10.1007/978-981-15-3707-3_40

only considers the precision of the observations (diagonal elements), but the physical correlation between the observations is a priori or even be ignored [5]. The early studies on stochastic models of GNSS were based on the elevation dependence. With the further researches of GNSS stochastic models, the cross and time correlations between observations were took into account [6, 7].

Many works have been done to analyze the impact of stochastic model on the global navigation satellite system (GNSS) positioning and IAR success rate. Howind [8] and El-Rabbany [9, 10] considered time correlations of observations, and the results confirm that the time correlation has little effect on positioning and has significant effect on the evaluation of baseline precision. Amiri-Simkooei et al. constructed a more realistic GNSS stochastic model which was proved can significantly improve the IAR success rate on single frequency data. Li et al. constructed a very complex BeiDou triple-frequency observation stochastic model, taking into account the cross and time correlations between observations. The results confirm that a more realistic stochastic model can obtain more realistic baseline precision, but which without considering cross and time correlations is too optimistic and unrealistic.

China's BeiDou navigation satellite system (BDS) provides more services compared with other GNSS [11]. Many efforts have been conducted to understand and improve the capabilities of the integer ambiguity resolution and positioning of BeiDou [12–15]. However, the (co)variance matrix of a simple stochastic model studied usually only considered diagonal elements, ignoring the covariance factors, namely the cross and time correlations between observations. Even after the satellite elevation dependence is applied to the stochastic model, it still can't objectively reflect the actual situation of the correlations of the satellite observations. The contribution of this paper is to construct a more realistic stochastic model considering the physical correlations between observations. The paper is organized as follows. In Sect. 2, based on the existing research, a complex BeiDou DD (double-difference) observation stochastic model is designed, which considers the cross correlation between different types of observations and the time correlation between epochs. In Sect. 3, MINQUE, a variance component estimation method for estimating physical correlation, is introduced. In Sect. 4, single-frequency and double-frequency data of the zero-length and short-length baseline are processed to analyze the impacts of the realistic stochastic on positioning and IAR success rate.

2 Realistic Stochastic Model

The traditional stochastic model generally considers the satellite observations which are of equal precision, and ignoring the cross correlation between difference types of observations and the time correlation between observations of different epoch intervals which are obviously unreasonable. In this paper, we derive a complex stochastic model, which considering multiple correlations.

2.1 Elevation-Dependent Model

It is assumed that the two receivers simultaneously tracking four identical satellite s, k, l, m, in which satellite s has been considered as reference. The (co)variance matrix of the SD (single difference) observations is

$$Q_y^D = 2diag(\sigma_{\rho_s}^2 \ \sigma_{\rho_k}^2 \ \sigma_{\rho_l}^2 \ \sigma_{\rho_m}^2 \ \sigma_{\varphi_s}^2 \ \sigma_{\varphi_k}^2 \ \sigma_{\varphi_l}^2 \ \sigma_{\varphi_m}^2) \tag{1}$$

where, Q_y^D denotes the SD (co)variance matrix, σ_ρ and σ_φ denotes satellite undifferenced pseudo and phase observation precision, respectively. Applying the error propagation law, the DD covariance matrix is obtained as follows

$$Q_y^{DD} = JQ_y^D J^T = 2\begin{bmatrix} \sigma_{\rho_s}^2+\sigma_{\rho_k}^2 & \sigma_{\rho_s}^2 & \sigma_{\rho_s}^2 & & & \\ \sigma_{\rho_s}^2 & \sigma_{\rho_s}^2+\sigma_{\rho_l}^2 & \sigma_{\rho_s}^2 & & & \\ \sigma_{\rho_s}^2 & \sigma_{\rho_s}^2 & \sigma_{\rho_s}^2+\sigma_{\rho_m}^2 & & & \\ & & & \sigma_{\varphi_s}^2+\sigma_{\varphi_k}^2 & \sigma_{\varphi_s}^2 & \sigma_{\varphi_s}^2 \\ & & & \sigma_{\varphi_s}^2 & \sigma_{\varphi_s}^2+\sigma_{\varphi_l}^2 & \sigma_{\varphi_s}^2 \\ & & & \sigma_{\varphi_s}^2 & \sigma_{\varphi_s}^2 & \sigma_{\varphi_s}^2+\sigma_{\varphi_m}^2 \end{bmatrix} = \Sigma_E \tag{2}$$

where, Q_y^{DD} denotes the DD (co)variance matrix, the $J = DD$ operator

$$J = \begin{bmatrix} 1 & -1 & 0 & 0 & & & & \\ 1 & 0 & -1 & 0 & & & & \\ 1 & 0 & 0 & -1 & & & & \\ & & & & 1 & -1 & 0 & 0 \\ & & & & 1 & 0 & -1 & 0 \\ & & & & 1 & 0 & 0 & -1 \end{bmatrix} \tag{3}$$

expand the Q_y^{DD} as follows,

$$\begin{aligned} Q_y^{DD} = {} & \begin{bmatrix} 2&2&2&0&0&0\\ 2&2&2&0&0&0\\ 2&2&2&0&0&0\\ 0&0&0&0&0&0\\ 0&0&0&0&0&0\\ 0&0&0&0&0&0 \end{bmatrix}\sigma_{\rho_s}^2 + \begin{bmatrix} 2&0&0&0&0&0\\ 0&0&0&0&0&0\\ 0&0&0&0&0&0\\ 0&0&0&0&0&0\\ 0&0&0&0&0&0\\ 0&0&0&0&0&0 \end{bmatrix}\sigma_{\rho_k}^2 + \begin{bmatrix} 0&0&0&0&0&0\\ 0&2&0&0&0&0\\ 0&0&0&0&0&0\\ 0&0&0&0&0&0\\ 0&0&0&0&0&0\\ 0&0&0&0&0&0 \end{bmatrix}\sigma_{\rho_l}^2 + \begin{bmatrix} 0&0&0&0&0&0\\ 0&0&0&0&0&0\\ 0&0&2&0&0&0\\ 0&0&0&0&0&0\\ 0&0&0&0&0&0\\ 0&0&0&0&0&0 \end{bmatrix}\sigma_{\rho_m}^2 \\ & + \begin{bmatrix} 0&0&0&0&0&0\\ 0&0&0&0&0&0\\ 0&0&0&0&0&0\\ 0&0&0&2&2&2\\ 0&0&0&2&2&2\\ 0&0&0&2&2&2 \end{bmatrix}\sigma_{\varphi_s}^2 + \begin{bmatrix} 0&0&0&0&0&0\\ 0&0&0&0&0&0\\ 0&0&0&0&0&0\\ 0&0&0&2&0&0\\ 0&0&0&0&0&0\\ 0&0&0&0&0&0 \end{bmatrix}\sigma_{\varphi_k}^2 + \begin{bmatrix} 0&0&0&0&0&0\\ 0&0&0&0&0&0\\ 0&0&0&0&0&0\\ 0&0&0&0&0&0\\ 0&0&0&0&2&0\\ 0&0&0&0&0&0 \end{bmatrix}\sigma_{\varphi_l}^2 + \begin{bmatrix} 0&0&0&0&0&0\\ 0&0&0&0&0&0\\ 0&0&0&0&0&0\\ 0&0&0&0&0&0\\ 0&0&0&0&0&0\\ 0&0&0&0&0&2 \end{bmatrix}\sigma_{\varphi_m}^2 \end{aligned} \tag{4}$$

The empirical stochastic model set $\sigma^2_{\rho_s} = \sigma^2_{\rho_k} = \sigma^2_{\rho_l} = \sigma^2_{\rho_m} = \sigma^2_{\rho}$, $\sigma^2_{\varphi_s} = \sigma^2_{\varphi_k} = \sigma^2_{\varphi_l} = \sigma^2_{\varphi_m} = \sigma^2_{\varphi}$. This is obviously unreasonable. The parameters of the elevation-dependent model given by Eq. (5) are fitted by the estimated variance components.

$$Model: \sigma = \frac{b_1}{\sin\theta + b_2} \tag{5}$$

This model was proposed by Parkinson and Spilker in 1996, where σ is standard deviation of the observables and θ is elevation angles of the satellite.

Based on the variance data of two frequency phase and code observation of each BeiDou satellite, the above function model is used to fit the satellite precision.

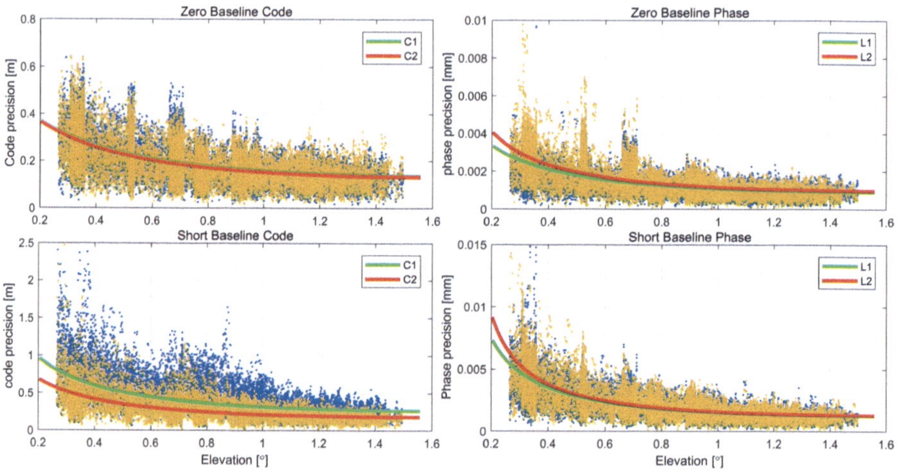

Fig. 1. BeiDou satellite elevation-dependent precision

In the Fig. 1, the green and red lines represent the frequency of B1(C2I, L2I) and B2(C7I, L7I) respectively. For the BeiDou code observation, the precision of B2 frequency is higher than that of B1 frequency. For the phase observation, the precision of B1 and B2 frequency is the same. Table 1 provides the fitting values of unknown parameters b1 and b2 of the elevation-dependent model in the short-length and zero-length baseline.

Table 1. Fitting values of unknown parameters

Data	Obs. types	B1		B2	
		b1	b2	b1	b2
Short baseline (8 m)	Code	0.2991	0.1065	0.2081	0.0988
	Phase	0.001268	−0.0308	0.001238	−0.06857
Zero baseline (0 m)	Code	0.1744	0.2707	0.1688	0.2569
	Phase	0.001069	0.1157	0.001061	0.05709

2.2 Time Correlation of Observables

In this section the time correlations between observations and the structure of the corresponding stochastic model is introduced. Only the three epoch of the two observation types (C2I, L2I) are considered (as previously assumed, the two receivers observe four identical satellite simultaneously).

$$Q_{y_T}^{D} = \begin{bmatrix} \sigma_{(0)}Q_y^D & \sigma_{(1)}Q_y^D & \sigma_{(2)}Q_y^D \\ \sigma_{(1)}Q_y^D & \sigma_{(0)}Q_y^D & \sigma_{(1)}Q_y^D \\ \sigma_{(2)}Q_y^D & \sigma_{(1)}Q_y^D & \sigma_{(0)}Q_y^D \end{bmatrix} = \Sigma_T \otimes Q_y^D \tag{6}$$

where, $Q_{y_T}^{D}$ denotes time-dependent (co)variance matrix of the SD observations, $\otimes$ is Kronecker product of two matrices, the unknown matrix is

$$\Sigma_T = \begin{bmatrix} \sigma_{(0)} & \sigma_{(1)} & \sigma_{(2)} \\ \sigma_{(1)} & \sigma_{(0)} & \sigma_{(1)} \\ \sigma_{(2)} & \sigma_{(1)} & \sigma_{(0)} \end{bmatrix} \tag{7}$$

where, $\sigma_{(i)}$ denotes the time correlation coefficient between observations with the interval of i epoch, Q_y^D is given by Eq. (2). Through the error propagation law, the DD covariance matrix is obtained as follows

$$Q_{y_T}^{DD} = \begin{bmatrix} J & 0 & 0 \\ 0 & J & 0 \\ 0 & 0 & J \end{bmatrix} \begin{bmatrix} \sigma_{(0)}Q_y^D & \sigma_{(1)}Q_y^D & \sigma_{(2)}Q_y^D \\ \sigma_{(1)}Q_y^D & \sigma_{(0)}Q_y^D & \sigma_{(1)}Q_y^D \\ \sigma_{(2)}Q_y^D & \sigma_{(1)}Q_y^D & \sigma_{(0)}Q_y^D \end{bmatrix} \begin{bmatrix} J^T & 0 & 0 \\ 0 & J^T & 0 \\ 0 & 0 & J^T \end{bmatrix} \tag{8}$$

which simplifies to

$$Q_{y_T}^{DD} = \begin{bmatrix} \sigma_{(0)}\Sigma_E & \sigma_{(1)}\Sigma_E & \sigma_{(2)}\Sigma_E \\ \sigma_{(1)}\Sigma_E & \sigma_{(0)}\Sigma_E & \sigma_{(1)}\Sigma_E \\ \sigma_{(2)}\Sigma_E & \sigma_{(1)}\Sigma_E & \sigma_{(0)}\Sigma_E \end{bmatrix} = \Sigma_T \otimes Q_y^{DD} = \Sigma_T \otimes \Sigma_E \tag{9}$$

where, $\Sigma_E = JQ_y^D J^T$ is given by Eq. (2). For the K epochs, the matrix Σ_T contains K unknown time correlation coefficients. Considering that when the time interval is 0, the time correlation coefficient of the observation is 1, therefore, $K-1$ unknown time correlation coefficients of K epochs need to be estimated. This (co)variance matrix structure is suitable for all types of observations.

2.3 Cross Correlation of Observables

In this section the correlation between different observation types is considered. Based on the previous assumptions, only two types of the observations (single frequency) C2I and L2I are considered.

$$Q_{y_C}^{D} = \begin{bmatrix} \sigma_\rho^2 Q_y^D & \sigma_{\rho\varphi}Q_y^D \\ \sigma_{\rho\varphi}Q_y^D & \sigma_\varphi^2 Q_y^D \end{bmatrix} = \Sigma_C \otimes Q_y^D \tag{10}$$

where, $Q_{y_C}^D$ denotes cross-dependent (co)variance of the SD observations, the unknown matrix is

$$\Sigma_C = \begin{bmatrix} \sigma_P^2 & \sigma_{P\varphi} \\ \sigma_{P\varphi} & \sigma_\varphi^2 \end{bmatrix} \tag{11}$$

Applying the error propagation law, the DD form covariance matrix is obtained as follows

$$Q_{y_C}^{DD} = \begin{bmatrix} J & 0 \\ 0 & J \end{bmatrix} \begin{bmatrix} \sigma_\rho^2 Q_y^D & \sigma_{\rho\varphi} Q_y^D \\ \sigma_{\rho\varphi} Q_y^D & \sigma_\varphi^2 Q_y^D \end{bmatrix} \begin{bmatrix} J^T & 0 \\ 0 & J^T \end{bmatrix} \tag{12}$$

which simplifies to

$$Q_{y_C}^{DD} = \begin{bmatrix} \sigma_\rho^2 \Sigma_E & \sigma_{\rho\varphi} \Sigma_E \\ \sigma_{\rho\varphi} \Sigma_E & \sigma_\varphi^2 \Sigma_E \end{bmatrix} = \Sigma_C \otimes \Sigma_E \tag{13}$$

where, $\Sigma_E = JQ_y^D J^T$. Ten (co)variance components, namely, four variances and six covariances need to be estimated for dealing four observation types (double frequency) data. Similarly, three (co)variance components, namely, two variances and one covariance need to be estimated for dealing two observation types (single frequency) data.

3 Estimation of Cross and Time Correlation Coefficient

MINQUE Variance Component Estimation

Many VCE methods such as MINQUE, BIQUE and LS-VCE based on different error distribution and estimation criteria have been widely used to estimate the correlations between satellite observations. Each has its own characteristics and advantages. In this paper, the classical MINQUE method is chosen to estimate the time and cross correlation coefficients.

The observation equation is $y = Ax + \varepsilon$, according to invariance, unbiasedness and minimum norm criterion, the estimation formula of *MINQUE* method is deduced

$$(W_0 \otimes W_0)vec\{\hat{\Sigma}_y\} = vec\{\Sigma_0^{-1} v_0 v_0^T \Sigma_0^{-1}\} \tag{14}$$

The expanded form is

$$\begin{bmatrix} tr(W_0U_1W_0U_1) & \cdots & tr(W_0U_1W_0U_1) \\ \vdots & \ddots & \vdots \\ tr(W_0U_1W_0U_1) & \cdots & tr(W_0U_1W_0U_1) \end{bmatrix} \begin{bmatrix} \hat{\theta}_1 \\ \vdots \\ \hat{\theta}_h \end{bmatrix} = S\theta = \begin{bmatrix} v_0^T W_0 U_1 W_0 v_0 \\ \vdots \\ v_0^T W_0 U_h W_0 v_0 \end{bmatrix} = G \tag{15}$$

where, $W_0 = \Sigma_0^{-1} R_0$. The cross and time correlation coefficients of observation are computed using a sliding window of K epochs. The (co)variance matrix of observation of K epochs is supposed to be set as $Q = \sum_{i=1}^{k} \sigma_i U_i$, where k denotes the number of (co)variance components σ_i,$U_{i(K \times K)}$ denotes the corresponding coefficient matrix of σ_i. Therefore,σ_i can be estimated by the following formula

$$\hat{\sigma} = S^{-1} G \tag{16}$$

4 Impact of Realistic Stochastic on Positioning and IAR

4.1 Impact of Cross Correlation on Positioning

Double-frequency (C2I, L2I, C7I, L7I) data of zero-length and short-length baseline are processed to analyze the impacts of the cross correlation on positioning. The elevation-dependent model constructed by Eq. (5) is employed while the fitting parameters are obtained through Table 1. The MINQUE method and estimation strategy proposed in Sect. 3 are utilized to estimate the cross correlation coefficients between observations (C2I-C7I, C2I-L2I, C2I-L7I, C7I-L2I, C7I-L7I, L2I-L7I) of the 9000 epochs (Fig. 2).

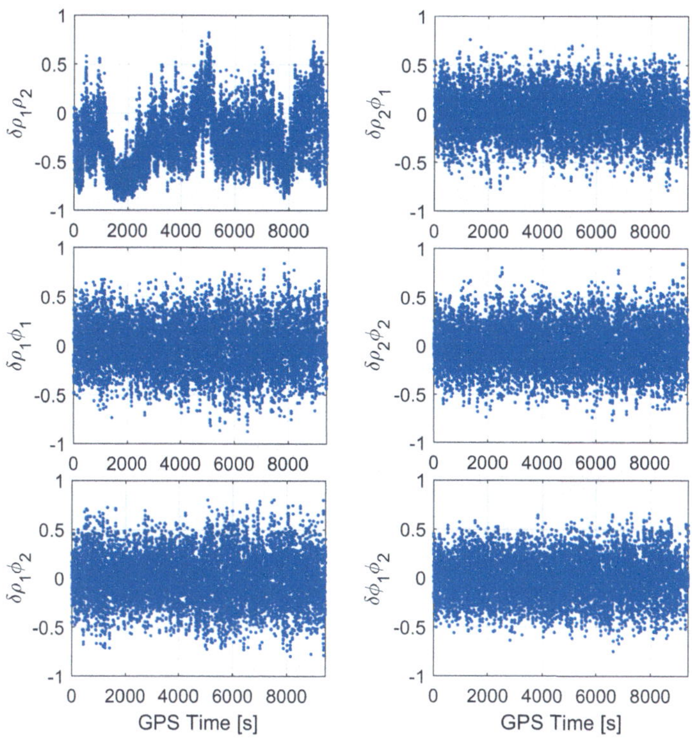

Fig. 2. Cross correlation coefficient of short baseline

Table 2. Mean cross correlation of short and zero baseline

Data	$\delta_{\rho_1\rho_2}$	$\delta_{\rho_1\varphi_1}$	$\delta_{\rho_1\varphi_2}$	$\delta_{\rho_2\varphi_1}$	$\delta_{\rho_2\varphi_2}$	$\delta_{\varphi_1\varphi_2}$
Short baseline (8 m)	−0.260	−0.006	0.0032	0.002	−0.011	−0.002
Zero baseline (0 m)	−0.025	−0.013	0.011	0.001	−0.011	0.009

The mean cross correlation coefficients (Table 2) are set as actual reference values to simplify the calculation. Then, the simple (ignoring the cross correlation) and realistic (considering cross correlation, following Eq. (17)) stochastic model are used to solve positioning and baseline precision of the 4 h data of short-length baseline (Fig. 3).

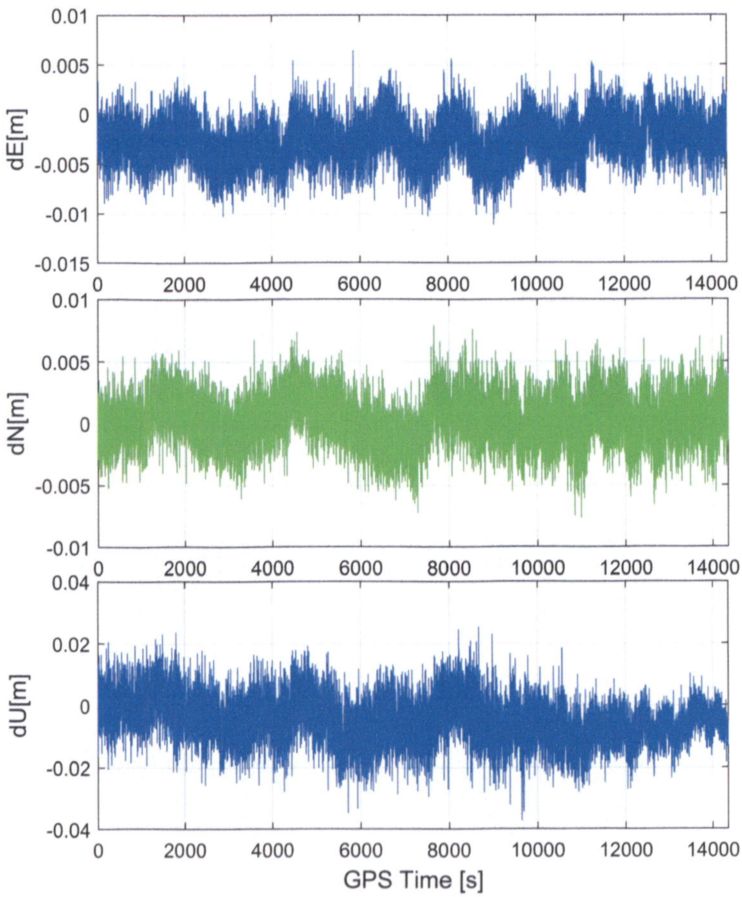

Fig. 3. Baseline solutions solved by stochastic model ignoring cross correlations for short-length baseline

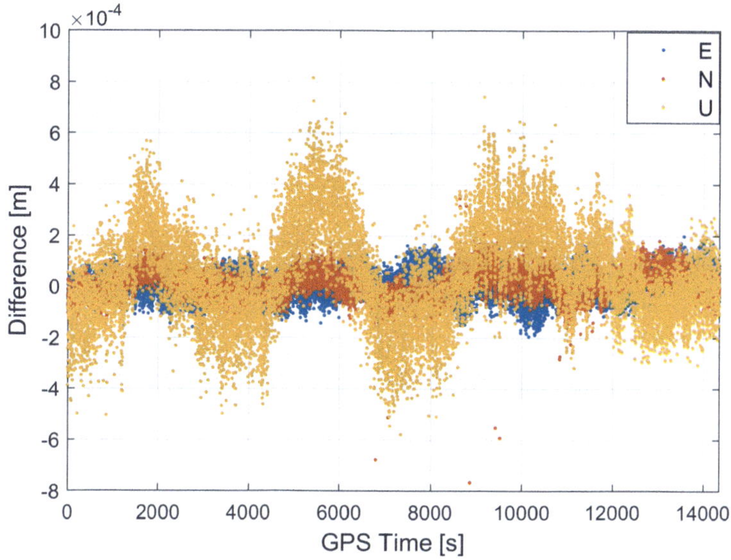

Fig. 4. The errors of baseline solutions solved by realistic and empirical stochastic model for zero-length baseline

$$Q_{yc}^{DD} = \begin{bmatrix} \sigma_{\rho_1}^2 \Sigma_E & \delta_{\rho_1\rho_2}\sigma_{\rho_1}\sigma_{\rho_2}\Sigma_E & \delta_{\rho_1\varphi_1}\sigma_{\rho_1}\sigma_{\varphi_1}\Sigma_E & \delta_{\rho_1\varphi_2}\sigma_{\rho_1}\sigma_{\varphi_2}\Sigma_E \\ \delta_{\rho_1\rho_2}\sigma_{\rho_1}\sigma_{\rho_2}\Sigma_E & \sigma_{\rho_2}^2\Sigma_E & \delta_{\rho_2\varphi_1}\sigma_{\rho_2}\sigma_{\varphi_1}\Sigma_E & \delta_{\rho_2\varphi_2}\sigma_{\rho_2}\sigma_{\varphi_2}\Sigma_E \\ \delta_{\rho_1\varphi_1}\sigma_{\rho_1}\sigma_{\varphi_1}\Sigma_E & \delta_{\rho_2\varphi_1}\sigma_{\rho_2}\sigma_{\varphi_1}\Sigma_E & \sigma_{\varphi_1}^2\Sigma_E & \delta_{\varphi_1\varphi_2}\sigma_{\varphi_1}\sigma_{\varphi_2}\Sigma_E \\ \delta_{\rho_1\varphi_2}\sigma_{\rho_1}\sigma_{\varphi_2}\Sigma_E & \delta_{\rho_2\varphi_2}\sigma_{\rho_2}\sigma_{\varphi_2}\Sigma_E & \delta_{\varphi_1\varphi_2}\sigma_{\varphi_1}\sigma_{\varphi_2}\Sigma_E & \sigma_{\varphi_2}^2\Sigma_E \end{bmatrix} \quad (17)$$

Obviously, the baseline solutions shown in Fig. 4 are equivalent for the realistic (considering cross correlation) and empirical (ignoring cross correlation) stochastic models. In order to analyze the impact of the cross correlations between observations on the baseline precision, a value that can better reflect the realistic baseline precision is required to calculate. Given the sliding window K = 20, the standard deviations of the baseline solutions are computed using the single epoch baseline solutions as the actual baseline precision. Then, we compare the baseline precision obtained from realistic and empirical stochastic models with the actual baseline precision.

The results in the Fig. 5 confirm that the baseline precision computed by the ignoring cross correlations' stochastic model has a large difference from the actual baseline precision. In other words, the baseline precision that ignoring cross correlation is too optimistic. On the contrary, the baseline precision that considering cross correlations is more in line with the actual situation.

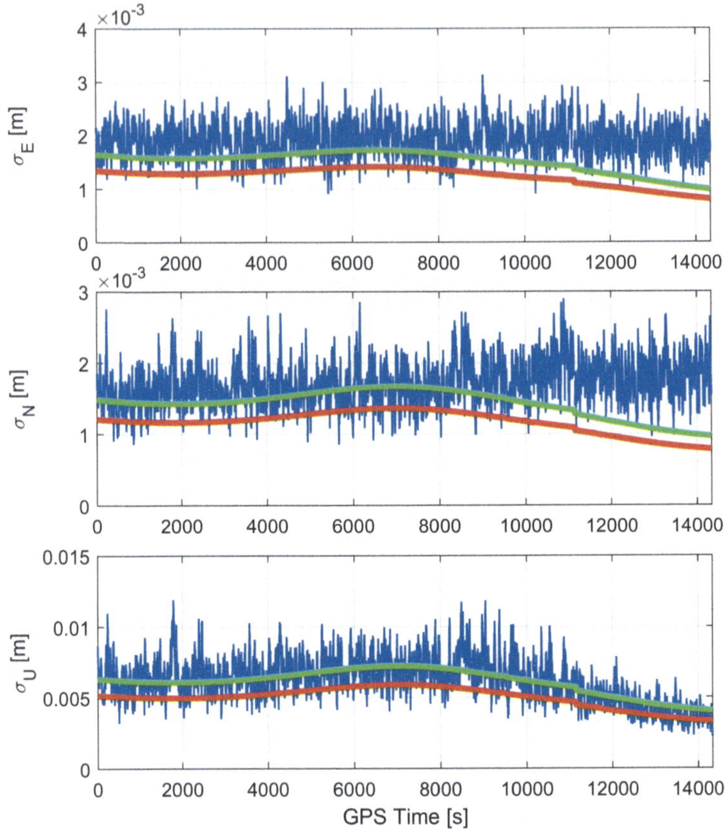

Fig. 5. Impact of cross correlations on baseline precision for short-length baseline. The green and red lines represent the theoretical precisions with considering and ignoring cross correlations, respectively, while the blue line represents the actual precisions.

4.2 Impact of Time Correlation on Positioning

The impact of the time correlations on positioning is similar to that of cross correlations. The focus of the research is to analyze the impact of the time correlations on baseline precision. Double frequency (C2I, L2I, C7I, L7I) data of the short-length baseline are processed to analyze the impact of time correlations on baseline precision. The elevation-dependent model is constructed by Eq. (5) while the fitting parameters are obtained by Table 1. The time window *K* is set to 20, and the MINQUE method is applied to estimate the time correlation coefficients of the time interval $\tau(\tau = 1, 2, \ldots, 19)$ of the 2000 consecutive epochs, as shown in the Fig. 6 below. And the mean time correlation coefficients of the 2000 epochs are shown in the Table 3.

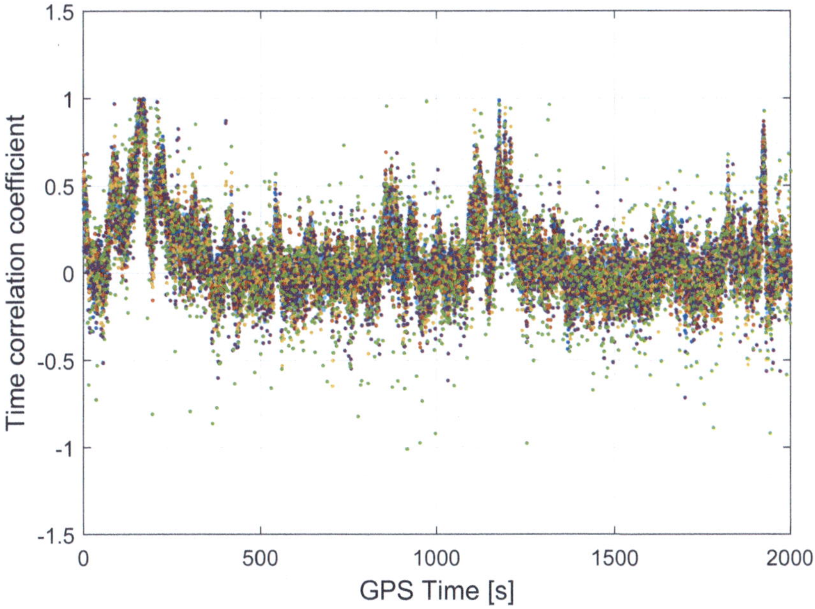

Fig. 6. Time correlation coefficient. Different colors denotes different time intervals

Table 3. Mean time correlation coefficient

Time interval	Time correlation coefficient	Time interval	Time correlation coefficient
1 s	0.200	11 s	0.084
2 s	0.120	12 s	0.080
3 s	0.110	13 s	0.087
4 s	0.100	14 s	0.070
5 s	0.100	15 s	0.072
6 s	0.090	16 s	0.068
7 s	0.090	17 s	0.060
8 s	0.080	18 s	0.060
9 s	0.090	19 s	0.055
10 s	0.086		

Based on the theory proposed by Li [4], considering that the time correlation is not available when solving a single epoch, so the theoretical baseline precision can be defined as

$$\sum\nolimits_{k=1}^{K} Q_{\hat{x}_k}/K \tag{18}$$

where, the $Q_{\hat{x}_k}(k = 1, \ldots, K)$ denotes variance matrices of $\hat{x}_k$. $\hat{x}_k$ is the single epoch baseline solutions over K epochs.

The baseline error is calculated from the time correlated observations, so it is also time correlated. The time correlated baseline error is applied to the empirical baseline precision, therefore if the time correlations of the observations in the stochastic model are ignored, the empirical and theoretical precision will not match well. The obtained empirical baseline precision will be less than the theoretical ones, which is too optimistic and does not fit the actual situation.

The least squares baseline solutions at epoch i and $i+\tau$ are $\hat{x}_i = Q_{\hat{x}_i} A_i^T Q_i^{-1} y_i$ and $\hat{x}_{i+\tau} = Q_{\hat{x}_{i+\tau}} A_{i+\tau}^T Q_{i+\tau}^{-1} y_{i+\tau}$. Since the observations y_i and $y_{i+\tau}$ are correlated with covariance matrix $Q_{i,i+\tau}$, the covariance matrix of $\hat{x}_i$ and $\hat{x}_{i+\tau}$ is derived as $Q_{\hat{x}_i\hat{x}_{i+\tau}} = Q_{\hat{x}_i} A_i^T Q_i^{-1} y_i Q_{i,i+\tau} Q_{\hat{x}_{i+\tau}} A_{i+\tau}^T Q_{i+\tau}^{-1} y_{i+\tau}$. In a short time span, it can be considered that the elevation angles of the same satellite being tracked is almost constant. It follows that $Q_i \approx Q_{i+\tau} \equiv Q$ and $A_i \approx A_{i+\tau} \equiv A$, and further $Q_{\hat{x}_i} \approx Q_{\hat{x}_{i+\tau}} \equiv Q_{\hat{x}}$ and $Q_{i,i+\tau} = \sigma_\tau Q$. Where,$\sigma_\tau$ is the time correlation coefficient for a time lag of τ.

$$D(X) = \begin{bmatrix} 1 & \sigma_1 & \cdots & \sigma_{K-1} \\ \sigma_1 & 1 & \ddots & \\ \vdots & \ddots & \ddots & \sigma_1 \\ \sigma_{K-1} & \cdots & \sigma_1 & 1 \end{bmatrix} \otimes Q_{\hat{x}} = R \otimes Q \tag{19}$$

where, $X = [\hat{x}_1, \ldots, \hat{x}_K]$, time correlation coefficients $\sigma_\tau, (\tau = 1, \ldots, K)$. Thus, the empirical baseline precision considering time correlations can be expressed as

$$\hat{Q}_{\hat{x}} = \frac{X\left[R^{-1} - R^{-1} e_K (e_K^T R^{-1} e_K)^{-1} e_K^T R^{-1}\right] X^T}{K - 1} \tag{20}$$

Obviously, if the time correlations are ignored

$$\hat{Q}_{\hat{x}} = X\left[I_K - e_K e_K^T / K\right] X^T / (K - 1) \tag{21}$$

Given the time window $K = 20$, the theoretical baseline precision can be calculated by Eq. (18), and the empirical baseline precision with considering and ignoring time correlations can be calculated by Eqs. (20) and (21), respectively. The results are shown in Fig. 7 confirm that compared with the empirical baseline precisions ignoring the time correlation, the ones with considering time correlation are closer to the theoretical value. The empirical precisions that ignores time correlations are smaller than the theoretical ones.

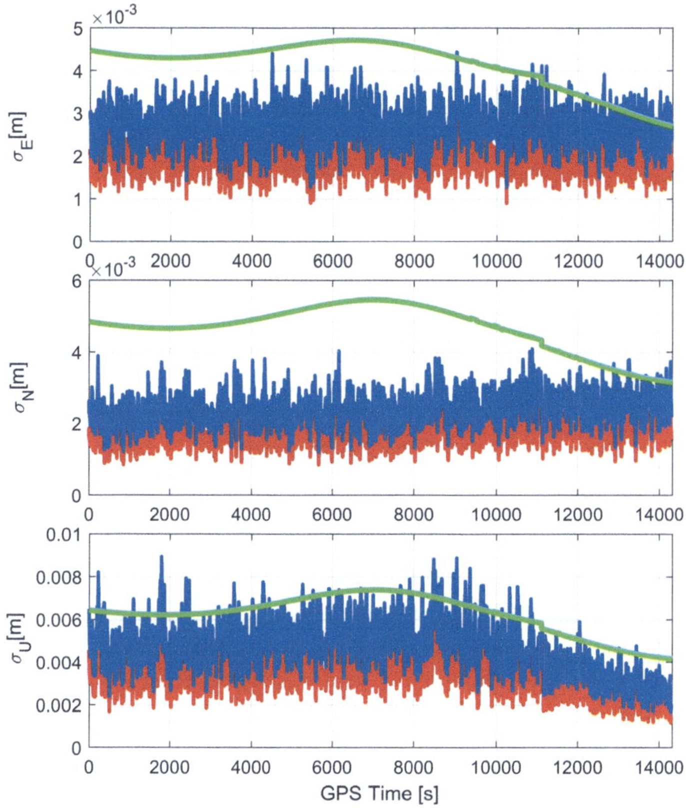

Fig. 7. Impact of time correlations on baseline precision for short-length baseline. The green line denotes the theoretical precision, the blue and red lines denote the practical precision with considering and ignoring the time correlations

4.3 Impact of Physical Correlation on IAR

Lambda algorithm proposed by Teunissen is used to solve integer ambiguity. Simultaneously consider cross and time correlations of observations, and the mean cross and time correlation coefficients are given by Tables 2 and 3. Three frequency patterns, four hours of observations are processed to analyze the impact of physical correlation on IAR. Considering the following three scenarios,

Case1: double frequency (C2I, L2I, C7I, L7I)
Case2: single frequency (C7I, L7I)
Case3: single frequency (C2I, L2I)

The realistic and empirical stochastic model (ignoring the physical correlation between observations) are used to calculate the theoretical and practical success rates of the IAR, respectively.

The IAR 'theoretical' success rate (SR^{theory}) is defined by Teunissen (1998).

$$SR^{theory} = \left[2\Phi(\frac{1}{2ADOP}) - 1\right]^n \tag{22}$$

where, $\Phi(x) = \int_{-\infty}^{x} \frac{1}{\sqrt{2\pi}} \exp\{-\frac{1}{2}z^2\}dz$, n denotes ambiguity vector's dimension. The SR^{theory} is a function of the ambiguity dilution of precision (ADOP), the ADOP in cycles is $ADOP = \sqrt{|Q_{\hat{a}}|^{\frac{1}{n}}}$, where, $|\cdot|$ represents a matrix determinant operator, and the $Q_{\hat{a}}$ is the covariance matrix of the float ambiguity estimate. Note that the SR^{theory} given in Eq. (22) is only an approximation (upper band) of the true success rate of the integer least squares.

The IAR 'practical' success rate is defined as

$$SR^{Realistic} = \frac{n_{same}}{n_{total}} \times 100\% \tag{23}$$

$$SR^{Nominal} = \frac{n_{same}}{n_{total}} \times 100\% \tag{24}$$

where, $SR^{Realistic}$ and $SR^{Nominal}$ represent the practical success rates of the realistic and empirical stochastic model, respectively, n_{same} denotes the number of epoch equal to the truth value of the integer ambiguity, and n_{total} is the total number of epoch. The kalman filter solution of DD integer ambiguity is taken as the truth value.

Table 4. The actual success rate of the realistic and empirical stochastic model

Type	Case1	Case2	Case3
$SR^{Realistic}$	98.67	93.33	64.04
$SR^{Nominal}$	97.14	89.0	57.91

The results are shown in Table 4 and Fig. 8. Clearly, a more realistic stochastic model can significantly improve the IAR success rate, especially single frequency data. In terms of theoretical success rate: the realistic stochastic is used. For single-frequency data, the success rate of the IAR is greatly increased. For dual-frequency data, the success rate is also improved, but the effect is not significant. In terms of practical success rate: the realistic stochastic model is used. For single-frequency data, the success rate of the IAR is increased by 5%.

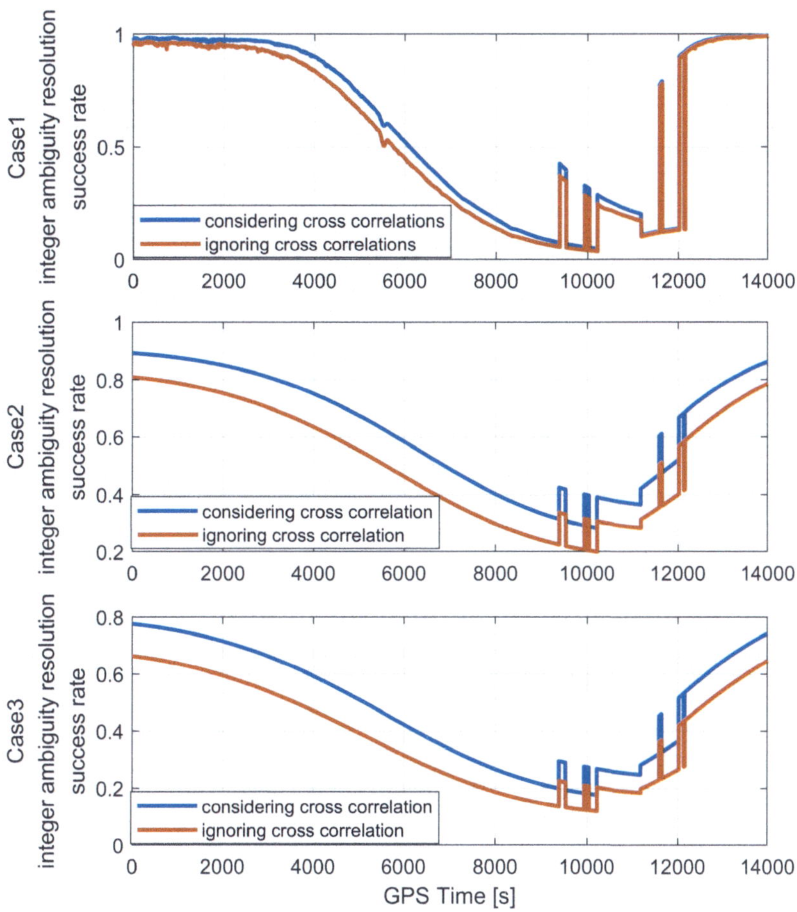

Fig. 8. The impact of physical correlation on integer ambiguity resolution success rate. The blue line denotes IAR success rate with considering physical correlation and the orange line denotes IAR success rate with ignoring physical correlation.

5 Concluding Remarks

This paper mainly studies the impact of cross and time correlation on BeiDou positioning and the IAR success rate. Construct a BeiDou DD observation stochastic model with complex structure, and the MINQUE method is selected to estimate cross and time correlation coefficients. Then, we compare the impacts of the stochastic considering and ignoring physical correlation on positioning and precision of the short baseline. The results confirm that compared with the empirical stochastic model ignoring the physical correlation, the realistic stochastic model has weak impact on positioning, but the more realistic stochastic can be applied to obtain more actual baseline precision. In terms of integer ambiguity resolution, the results are presented for two kinds of success rates, namely practical and theoretical success rates. The results, as expected, indicated that a more realistic stochastic model can significantly improve the IAR success rate on single-frequency with the success rate increasing by 5%.

References

1. Amiri-Simkooei, A.R., Jazaeri, S., Zangeneh-Nejad, F., et al.: Role of stochastic model on GPS integer ambiguity resolution success rate. GPS Solutions **20**(1), 51–61 (2016)
2. Hofmann-Wellenhof, B., Lichtenegger, H., Wasle, E.: GNSS Global Navigation Satellite Systems–GPS, GLONASS, Galileo & More. Springer, New York (2008)
3. Li, B., Shen, Y., Lou, L.: GNSS elevation-dependent stochastic modelling and its impacts on the statistic testing. J. Surv. Eng. (2015). https://doi.org/10.1061/(ASCE)SU.1943-5428.0000156
4. Li, B.: Stochastic modeling of triple-frequency BeiDou signals: estimation, assessment and impact analysis. J. Geodesy **90**(7), 593–610 (2016)
5. Teunissen, P.J.G.: Influence of ambiguity precision on the success rate of GNSS integer ambiguity bootstrapping. J. Geodesy **81**(5), 351–358 (2007)
6. Jin, X., de Jong, C.D.: Relationship between satellite elevation and precision of GPS code observations. J. Navig. **49**, 253–265 (1996)
7. Amiri-Simkooei, A.R., Teunissen, P.J.G., Tiberius, C.C.J.M.: Application of least-squares variance component estimation to GPS observables. J. Surv. Eng. **135**(4), 149–160 (2009)
8. Howind, J., Kutterer, H., Heck, B.: Impact of temporal correlation on GPS-derived relative point positions. J. Geodesy **73**, 246–258 (1999)
9. El-Rabbany, A., Kleusberg, A.: Effect of temporal physical correlation on accuracy estimation in GPS relative positioning. J. Surv. Eng. **129**(1), 28–32 (2003)
10. El-Rabbany, A.: The effect of physical correlations on the ambiguity resolution and accuracy estimation in GPS differential positioning. Technical Report No. 170, Department of Geodesy and Geomatics Engineering, University of New Brunswick, Fredericton, Canada (1994)
11. Yang, Y., Mao, Y., Sun, B.: Basic performance and future developments of BeiDou global navigation satellite system. Satell. Navig. **1**, 1–8 (2020). https://doi.org/10.1186/s43020-019-0006-0
12. Shi, C., Zhao, Q., Hu, Z., et al.: Precise relative positioning using real tracking data from COMPASS GEO and IGSO satellites. GPS Solutions **17**(1), 103–119 (2013)
13. Wang, K., Rothacher, M.: Ambiguity resolution for triple-frequency geometry-free and ionosphere-free combination tested with real data. J. Geodesy **87**(6), 539–553 (2013)
14. Li, B., Feng, Y., Gao, W., et al.: Real-time kinematic positioning over long baselines using triple-frequency BeiDou signals. IEEE Trans. Aerosp. Electron. Syst. **51**(4), 3254–3269 (2015)
15. Miao, W., Li, B., Zhang, Z., et al.: Combined BeiDou-2 and BeiDou-3 instantaneous RTK positioning: stochastic modeling and positioning performance assessment. J. Spat. Sci. (2019). https://doi.org/10.1080/14498596.2019.1642250

Research on Key Performance of BeiDou Global Short Message Communication Service

Xin Nie[1(✉)], Jun Xie[1], Tianxiong Liu[1], Chonghua Liu[1], and Kanglian Zhao[2]

[1] Beijing Institute of Spacecraft System Engineering, Youyi Road, Haidian District, Beijing 100094, China
nxbuaa@126.com
[2] Nanjing University, Nanjing, China

Abstract. By carrying global message communication payloads on some satellites of the BeiDou Global System, a two-way link between a user terminal and a ground control center can be constructed by using inter-satellite links and satellite-ground links. At the same time as providing PNT services, BeiDou system provides global short message communication service. On the basis of designing the system scheme, this paper analyzes the key performance of the global short message service, the coverage characteristics, and the system capacity through simulation.

Keywords: Global short message communication · BeiDou system

1 Introduction

According to the national development plan, the PNT service range of the BeiDou system will cover the world around 2020, with unique inter-satellite links to achieve inter-satellite interconnection, and it is the only constellation in China that can achieve global coverage [1, 2]. By carrying global message communication payloads on some satellites of the BeiDou global system, a two-way communication link between a user terminal and the ground control center can be constructed by using inter-satellite links and satellite-ground links to build a global short message communication service system. The system can provide global users with short message communication, location reporting, non-real-time voice and image data transmission services.

On the basis of sorting out related user needs and combining the characteristics of BeiDou global system, this paper proposes the BeiDou global short message communication system scheme, and conducts research in terms of demand analysis, system composition, constellation scheme, key indicators and service performance. Theoretical and simulation analysis shows that the system scheme is reasonable and feasible. When the BeiDou global satellite navigation system is completed around 2020, it will have global short message communication service capabilities [3, 4].

J. Sun et al. (Eds.): CSNC 2020, LNEE 650, pp. 435–442, 2020.
https://doi.org/10.1007/978-981-15-3707-3_41

2 System Solution Overview

The BeiDou global short message communication service system is an integrated information service system consisting of a space segment (BeiDou satellite), a ground segment (ground center station), and user segment (user terminal) [6].

Since the BDS satellite does not have continuous observation conditions at the ground control station, it is necessary to use the inter-satellite link for information relay. The overseas users transmit information to the visible satellite. The information is relayed to the domestic central station through one or more hops, which realizes the reverse link communication (overseas users to the domestic central station). At the same time, using the two-way communication function of the inter-satellite link, a forward link (domestic center station to overseas users) can be established.

(1) reverse link information flow: overseas user terminal-> BDS satellite …-> BDS satellite-> central station
(2) forward link information flow: central Station-> BDS Satellite…-> BDS Satellite-> Overseas User Terminal

The global data communication service system can provide services such as location reporting, search and rescue information, short message communication, non-real-time voice and image transmission, and can be interconnected with other information systems.

3 Constellation Design and Coverage Performance Analysis

The basic problem of the global short message communication service system is constellation design, which is related to many aspects such as system cost, service range, link characteristics, and communication quality. The BeiDou satellite has an orbital height of 21528 km, distributed on three orbital planes, and an orbital inclination angle of 55° [7, 8]. In order to ensure the reliability of the system, considering the fact that the communication performance is affected by rain attenuation and occlusion at different elevation angles, the constellation design is based on different elevation angles with global dual coverage as the goal.

3.1 Coverage Analysis

In order to make users have a better experience, the observation elevation angle is increased to 15° and 30° for analysis. The results show that the 15-satellite sub-constellations can achieve double continuous global coverage (100% time availability) at an elevation angle of 15°, as shown in Fig. 1. Global coverage (99% time availability) is achieved at 30° elevation.

If all satellites in the 24-satellite constellation are equipped with message communication loads, the coverage area is shown in Fig. 2, which shows that under the condition of a minimum observation elevation of 30°, continuous global double coverage can be achieved, and the time availability is 100%.

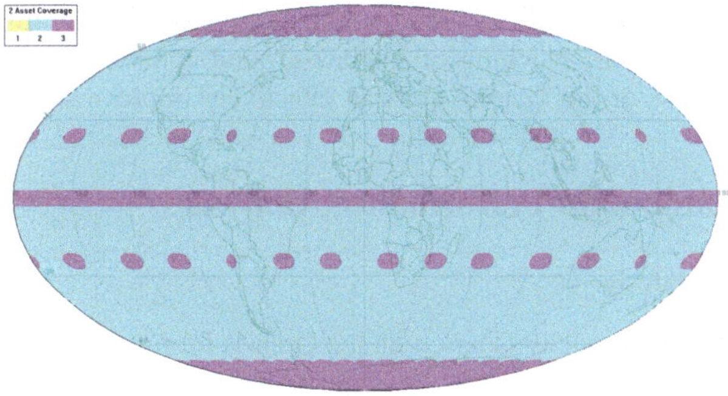

Fig. 1. Coverage performance of 15-satellite constellation (@15° elevation)

Fig. 2. Coverage performance of 24-satellite constellation (@30° elevation)

3.2 Analysis of Uninterrupted Visible Time

As an example, one of the constellation schemes in the 15-star constellation scheme obtained from the analysis above is used to further observe the satellite transit time (including the maximum value, minimum value, and average value) of the constellation at typical global position at an elevation angle of 15°. The simulation cycle is a regression cycle, as shown in Table 1.

Table 1. Statistics of transit times at typical locations around the world

Locations	Time (s)		
	Minimum	Maximum	Average
Canberra	2786	27136	17858
Damascus	1858	27599	18040
Jakarta	2269	39251	24056
Nairobi	2764	37732	26420
Paris	616	23824	15597
Washington	2072	26191	17485

Simulation results show that the shortest transit time of a single satellite is 616 s, which can avoid frequent switching of satellite-ground connection relationships in a single service process of the system, and can effectively improve service quality.

4 Key Indicators and Performance Analysis

4.1 Capacity Analysis

The analysis of system capacity is based on the Oran-B formula. It can be known from the traffic theory that for a system with N channels, the B blocking rate obeys the Oran distribution, where A is the traffic volume:

$$B(N,A) = \frac{A^N/N!}{\sum_{i=0}^{N} A/i!}$$

Since B is a function of N and A, it is written as B (N, A). The failure of the call means that all channels in the system have been occupied, so the new call cannot get the required channel.

4.1.1 Reverse Link Capacity

Assuming that the inter-satellite link allocates different number of frames for message communication every 3 s, the reverse link capacity calculated according to the Oran-B formula is shown in the table below (Table 2).

Table 2. Service capacity corresponding to different blocking rates

Frame (/3s)	Capacity		
	0.1%	1%	5%
1	0.24	0.72	1.92
2	2.16	4.32	7.68
3	5.76	9.6	14.88
4	10.32	15.6	22.8
5	15.48	22.44	31.2
6	18.24	25.92	35.52
12	21.24	29.52	39.84

4.1.2 Forward Link Capacity

The forward link capacity is mainly determined by the frequency of information injection and the number of users in a single message. According to the data injected into each target satellite, then the service capacity per hour per satellite is $9 * 2 * 2/10 * 3600 \approx 1.3E4$ times/hour. The system capacity is the product of the capacity of a single satellite and the number of satellites. The system capacity is $1.3 * 27 \approx 3.5E5$ times/hour.

4.2 Communication Delay and Hop Count

4.2.1 Reverse Link Capability Simulation

According to simulation of the regression period, the packet delivery delay and its cumulative probability density, frame forwarding hops, and distribution were obtained. The main results are as follows: 99% of reverse link frames arrive within 11 s and the longest delay is 64 s. Packets arriving within 2 hops account for 97.5%, and the maximum number of forwarding hops is 10 hops. The specific results are shown in Figs. 3, 4, 5 and 6. In the cumulative probability density graph, the horizontal axis is the frame arrival delay, and the vertical axis is the cumulative probability density. In the frame delivery delay graph, the horizontal axis is the simulation time, and the vertical axis is the frame arrival delay. In the frame forwarding hop count graph, the horizontal axis is the simulation time, and the vertical axis is the frame hop number.

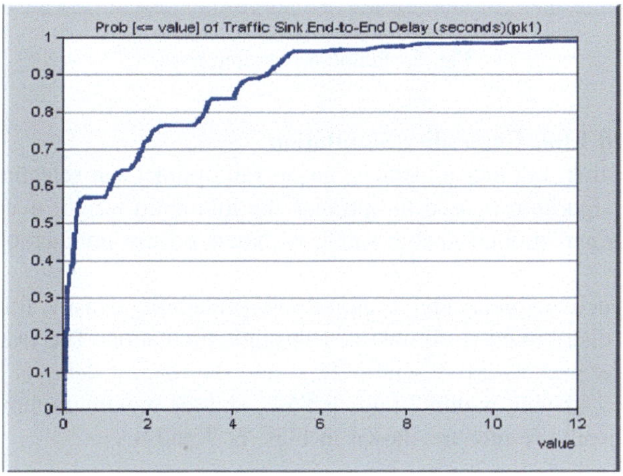

Fig. 3. Cumulative probability density (CDF) of frame delivery delay

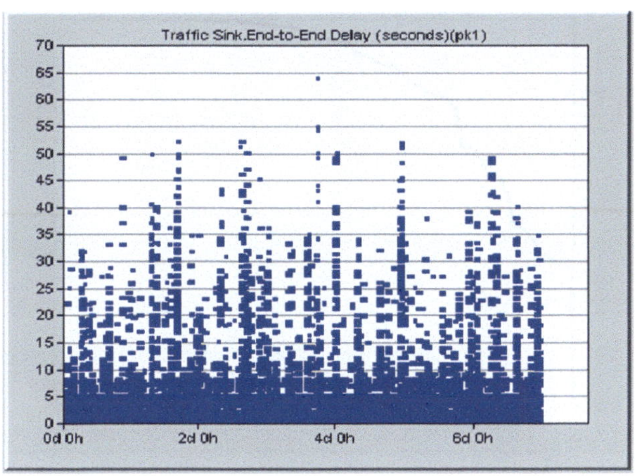

Fig. 4. Frame delivery end-to-end delay

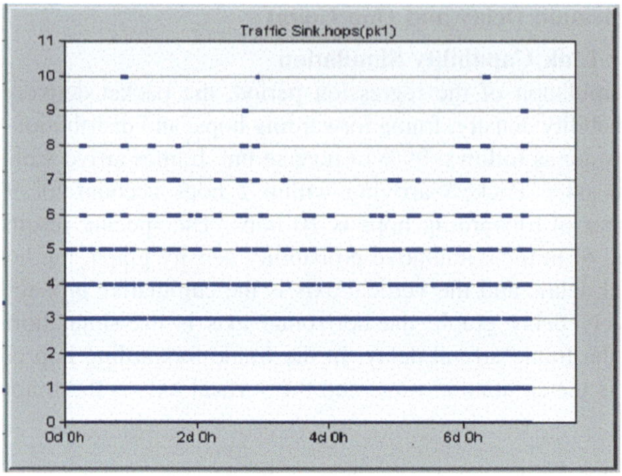

Fig. 5. Frame forwarding hops

4.2.2 Forward Link Capability Simulation

For the forward link, in order to achieve an overall optimal star selection strategy, the Kuhn-Munkras algorithm is used to optimize the minimum weight matching between forward packets and ground-visible satellites based on the number of waiting time slots.

The packets delivery delay and its cumulative probability density, frame forwarding hop count, and distribution were obtained through simulation. The results are as follows: 99% of the packets arrive within 18 s, and the longest delay is 53 s. The proportion of frames arriving within 2 hops is 85%, and the maximum number of hops is 10 hops. The specific results are shown in Figs. 6, 7 and 8.

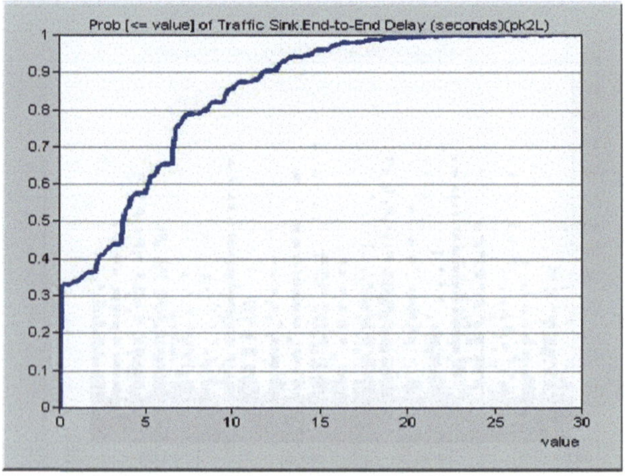

Fig. 6. Cumulative probability density (CDF) of delivery delay

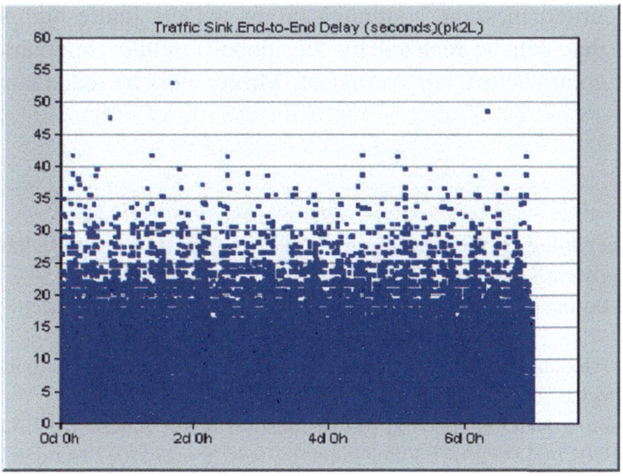

Fig. 7. Frame delivery end-to-end delay

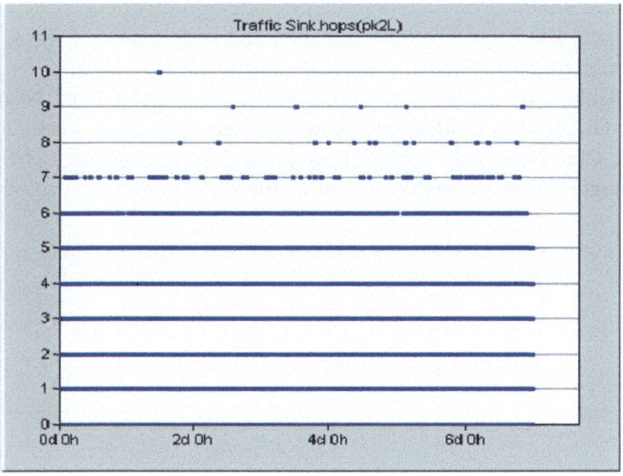

Fig. 8. Frame forwarding hop count

5 Conclusion

On the basis of designing the system scheme, this paper analyzes the key performance of the global short message service, the coverage characteristics, and the system capacity through simulation. As the BeiDou global short message communication service is a "value-added" service provided while ensuring the main task of the PNT service, the current bottleneck affecting the communication capability is mainly the rate constraint of the inter-satellite link. With the development and application of laser inter-satellite link technology, the communication capability will be greatly improved. In view of China's national conditions, the priority is given to the development of low-speed, high-concurrent message communications and the future high-speed communication system that will be realized by the global satellite communication system, which is neither contradictory nor redundant. Means must be redundantly backed up.

References

1. BeiDou: Key technologies of new generation BeiDou navigation satellites have been verified. www.BeiDou.gov.cn/NewsCenter/Text. Accessed 4 Feb 2016
2. Ran, C.: Operation and development of BeiDou satellite navigation system. Satell. Appl. **6**, 13–16 (2016)
3. Fan, B., et al.: Research on the development of global satellite navigation system data transmission service. Spacecr. Eng. **25**(3), 1–8 (2016)
4. BeiDou: Explaining BeiDou III: Add global location report function [EB/OL]. http://www.BeiDou.gov.cn/2017/03/08/20170308b21a480247fe4c289c27470c299e6e56.html. Accessed 1 Apr 2017/Accessed 8 Mar 2017
5. Fan, B., et al.: Application and suggestion of BeiDou satellite navigation system in Wenchuan earthquake. Spacecraft Eng. **17**(4), 6–13 (2008)
6. Tan, S.: Generalized satellite radio location reporting principles and application value. J. Surv. Mapp. (01), 1–5 (2009)
7. Ni, Y.: STK-based BDS constellation simulation and performance analysis. Comput. Meas. Control **24**(1) (2016)
8. Hofmann Wellenhof, B., Lichtenegger, H., Wasle, E.: GNSS-Global Navigation Satellite Systems GPS, GLONASS, Galileo, and More. SpringerWien, New York (2007)
9. Tencent News: China Research Satellite Link Phased Array Antenna, Let BeiDou Satellites Real-time Dialogue. news.qq.com. Accessed 6 Aug 2013

An Algorithm of Passive Location About Satellite Navigation Disturb Source Based on Combat Platforms Networking

Linxu Wu[1,2(✉)], Jiang Li[1,2], and Xiaoyu Wang[1]

[1] The 20th Research Institute of China Electronics Technology Corporation, Xi'an 710068, China
987362900@qq.com, lijiang0613@163.com, wangxiaoyu8866@163.com

[2] Shaanxi Key Laboratory of Integrated and Intelligent Navigation, Xi'an 710068, China

Abstract. Focusing on the problem that the satellite navigation and positioning of the combat members in the future electronic confrontation is disturbed, a solution for passive location of the interference source is proposed. Under the condition of members networking, the interference source is located accurately using least - square according to the measurement information of each member so that the weapon systems can be guided to implement over - the - horizon strikes. Finally, the simulation results show that the solution proposed in this paper can effectively locate the disturb source and reduce the number of false results. At the same time, this solution is small in calculation and suitable for engineering applications.

Keywords: Cross location · Satellite navigation disturb source · Networked battlefield service · Multiple-platform · Multiple-target

1 Introduction

Under modern war conditions, satellite navigation systems are widely used in military operations, which greatly improves the combat capabilities of combat units. Therefore, in the new situation of electronic warfare, interference to satellite navigation system has become an inevitable choice for current combat applications [1]. In order to effectively suppress the effects of satellite navigation interference sources on satellite navigation receiving equipment, researchers from various countries have conducted a lot of research in the anti-interference field of satellite navigation.

At present, the research on anti-influence of satellite navigation mainly focuses on that satellite navigation receivers form nulls in the direction and frequency of interfering signals to suppress the power of interference signal. Thus, the satellite navigation receiver can work effectively with this passive anti-interference method [2–6]. At the same time, there is a small amount of research work to detect satellite navigation interference signals by developing radio reconnaissance equipment to estimate the position of satellite navigation interference sources, and then use other methods to

J. Sun et al. (Eds.): CSNC 2020, LNEE 650, pp. 443–453, 2020.
https://doi.org/10.1007/978-981-15-3707-3_42

remove the interference sources. But this anti-jamming method for actively detecting satellite navigation interference sources easily exposes its own position, and also increases the equipment cost of combat application platforms. To this end, Wang et al. proposed a passive estimation method for satellite navigation interference sources based on anti-jamming antenna equipment [7]. This method can obtain the direction of the satellite navigation interference source but still cannot locate the interference source. According to the research in reference [8], this paper proposes a passive method for locating satellite navigation interference sources based on combat platform networking. This method draws on the idea of passive radar networking and positioning to realize the position estimation of satellite navigation interference source using a passive cross-location algorithm [9, 10] in combination with the least squares algorithm by using data communication links on the combat platform to share the satellite navigation interference source data acquired by multiple combat platforms [11, 12].

2 Design of Interference Source Location Method Based on Platform Networking

The satellite navigation interference source positioning method designed in this article is based on that the anti-jamming antenna of the combat platform equipment has a direction finding function and other sensors on the platform can locate the platform when satellite navigation is disturbed [13]. In the course of coordinated operations, multiple combat platforms complete the networking through communication systems to achieve data exchange between platforms within the network [14]. Then, the interference source location method proposed in this paper is used to accurately and real-time locate the enemy interference source in order to provide accurate warnings and guide other weapon systems to perform beyond-horizon attacks.

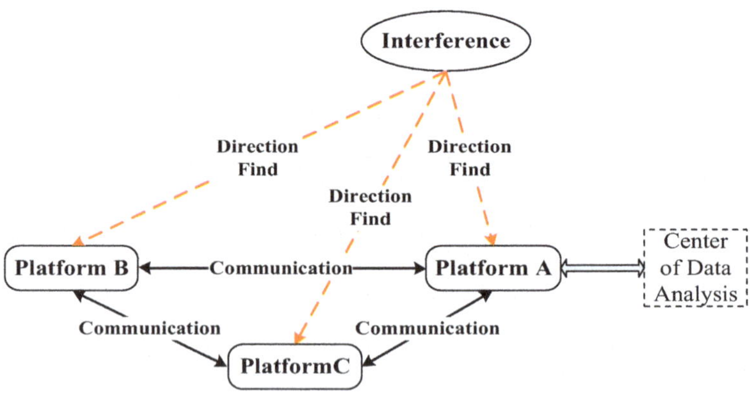

Fig. 1. Satellite navigation interference source positioning method based on multi-platform cooperation

The positioning diagram is shown in Fig. 1. When locating an enemy satellite navigation interference source, the interfered combat platforms A, B, and C will measure the information of the interference source signal. Then, use the data communication link on the platform to perform data interaction with other platforms and store the relevant data of all platforms on platform A equipped with a data processing center device for analysis and processing to achieve the positioning of satellite navigation interference sources.

Finally, the data processing center sends the position of the interference source to the combat decision unit for display. And the information is also sent to other platforms through the data communication link of platform A to guide the weapon system to conduct coordinated strike.

3 Multi-platform Passive Cooperative Positioning Algorithm

3.1 Basic Principles of Cooperative Positioning

In 3D space, suppose that there are N platforms in different positions locating an interference source in the method of cross location and the schematic is shown in Fig. 2.

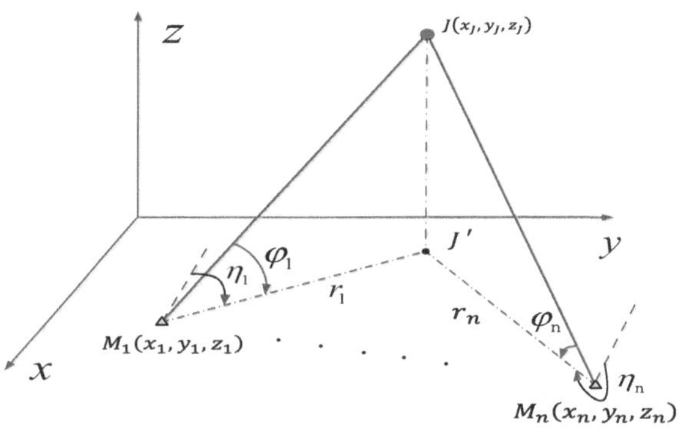

Fig. 2. Schematic diagram of cross positioning

The coordinates of the interference source T are (x_T, y_T, z_T) unknown. The coordinates of the two platforms are known as $A(x_1, y_1, z_1)$, $B(x_2, y_2, z_2)$ and the azimuths and pitch angles to the interference source are known as $(\eta_i, \varphi_i), i = 1, 2$. Under the condition of knowing the position of each platform and the angles to the interference source, the following equations can be obtained from Fig. 2.

$$\begin{cases} \tan\eta_1 = \frac{x_T - x_1}{y_T - y_1} \\ \tan\eta_2 = \frac{x_T - x_2}{y_T - y_2} \\ \tan\varphi_1 = \frac{z_T - z_1}{r_1} \\ \tan\varphi_2 = \frac{z_T - z_2}{r_2} \end{cases} \quad (1)$$

where

$$\begin{cases} r_1 = \sqrt{(x_T - x_1)^2 + (y_T - y_1)^2} \\ r_2 = \sqrt{(x_T - x_2)^2 + (y_T - y_2)^2} \end{cases} \tag{2}$$

Solve the above formulas to get the 3D position of the target [15]. For multi-platform positioning with N measurement terminals, there will be 2N relations and the position of the interference source can be calculated according to the same method [16].

3.2 Disadvantages of Two Platform Positioning

3.2.1 Special Point

There are some problems in two -platform positioning such as dead zone in position calculation.

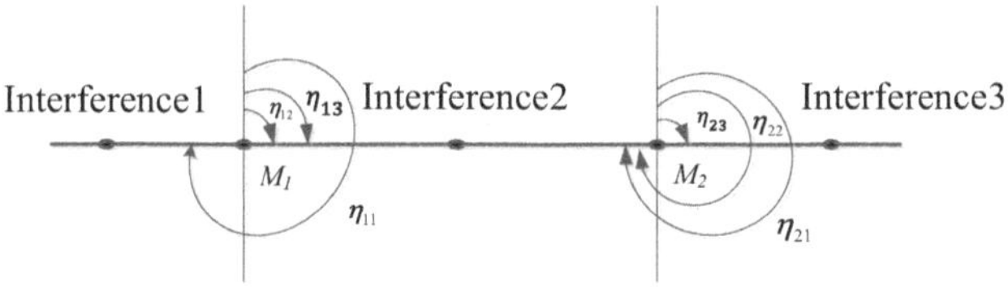

Fig. 3. Schematic diagram of two platform cross location

As shown in the Fig. 3, when the location of the interference source is in line with the two measurement terminals, that is, $\eta_{11} = \eta_{21}$, $\eta_{12} = 360 - \eta_{22}$, the angle information obtained is unable to calculate the position of the interference source.

3.2.2 False Points

In the case of multiple interference sources, there will be false points [17] which will affect the positioning accuracy of the interference sources. At the same time, with the increase of interference sources the number of false target points will increase rapidly so that the difficulty of accurate positioning will also increase greatly.

According to the principle of cross positioning, when there are two platforms and N targets there will be 2N links between the targets and the platforms and there will be at most N^2 intersections through cross positioning where $N(N - 1)$ results are false points [18]. As shown in the Fig. 4 below, there are 3 false points in this case

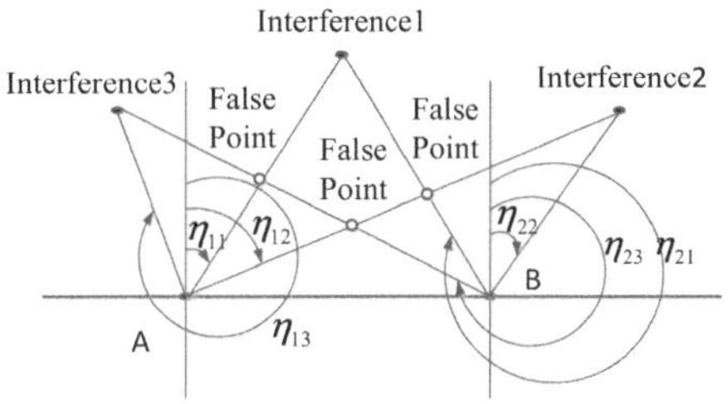

Fig. 4. Schematic diagram of two-platform with multi targets

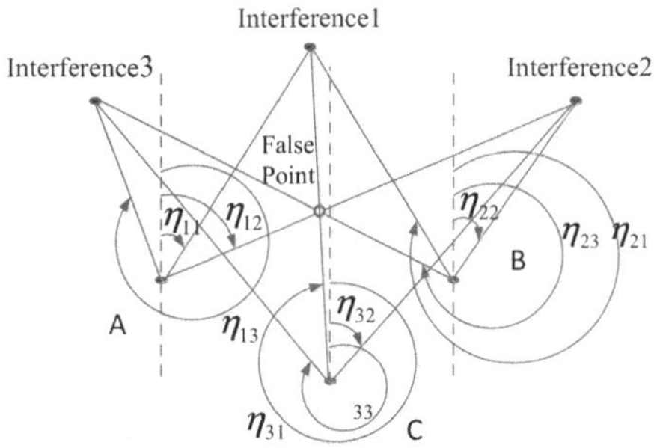

Fig. 5. Schematic diagram of multiple platform with multiple targets

To this end, the number of false points can be reduced to improve the accuracy [20] of the positioning of the interference source by increasing the number of measurement platforms [19] to get more measurement information and more constraints.

As shown in Fig. 5, the three measurement platforms A, B, and C are not on the same line. In this case, no matter where the interference source is, there will always be two measurement platforms that can give effective measurement information so that the interference sources can't be co-linear with all measurement platforms. Thus, there will not be dead zone and the number of false points is reduced to one which the accuracy of positioning of the interference source improves greatly by adding a measurement platform.

3.3 Multi-platform Collaborative Positioning Algorithm

As shown in Fig. 2, for multi-platform positioning, assume that the coordinates of the measurement platforms are (x_1, y_1, z_1), (x_2, y_2, z_2), $\cdots$, (x_n, y_n, z_n). And assume that the coordinate of the interference source is (x, y, z) and the measured value of the angle is Z so it derives

$$Z = H(x) + \Delta\theta. \quad (3)$$

where

$$H(x) = \begin{bmatrix} \arctan((x - x_1)/(z - z_1)) \\ \arctan((x - x_2)/(z - z_2)) \\ \vdots \\ \arctan((x - x_n)/(z - z_n)) \\ \arctan((y - y_1)/r_1) \\ \vdots \\ \arctan((y - y_2)/r_2) \\ \arctan((y - y_n)/n) \end{bmatrix} \quad (4)$$

$$Z = [\eta_1\, \eta_2 \cdots \eta_n\, \varphi_1\, \varphi_2 \cdots \varphi_n]^{\mathrm{T}}$$

$$\Delta\theta = [\Delta\eta_1\, \Delta\eta_2 \cdots \Delta\eta_n\, \Delta\varphi_1\, \Delta\varphi_2 \cdots \Delta\varphi_n]^{\mathrm{T}}$$

where

$$r_i = \sqrt{(x - x_i)^2 + (z - z_i)^2}, (i = 1, 2\ldots n)$$

Expand formula (3) at X_0, it derives

$$Z = H(X_0) + B(X - X_0) + \Delta\theta \quad (5)$$

where

$$\mathrm{B} = \begin{bmatrix} [(z_0 - z_1) - (x_0 - x_1)]/r_1^2 \\ [(z_0 - z_2) - (x_0 - x_2)]/r_2^2 \\ \vdots \\ [(z_0 - z_n) - (x_0 - x_n)]/r_n^2 \\ r_1/R_1^2 - [(z_0 - z_1) + (x_0 - x_1)]/\left(r_1 R_1^2\right) \\ r_2/R_2^2 - [(z_0 - z_2) + (x_0 - x_2)]/\left(r_1 R_2^2\right) \\ \vdots \\ r_n/R_n^2 - [(z_0 - z_n) + (x_0 - x_n)]/\left(r_n R_n^2\right) \end{bmatrix} \quad (6)$$

$$\begin{cases} r_i = \sqrt{(x - x_i)^2 + (z - z_i)^2} \\ R_i = \sqrt{(x - x_i)^2 + (y - y_i)^2 + (z - z_i)^2} \end{cases} \quad (7)$$

Let

$$\begin{cases} U = Z - H(X_0) \\ V = X - X_0 \end{cases} \quad (8)$$

so that

$$U = BV + \Delta\theta \quad (9)$$

For the time being the characteristics of the measurement error Δθ are not understood so they are not considered. According to the least squares estimate, we get

$$V^{g} = \left(B^T B\right)^{-1} B^T U \quad (10)$$

Take Eq. (8) in so the position of the interference source is shown as

$$X^g = \left(B^T B\right)^{-1} B^T (Z - H(X_0)) + X_0 \tag{11}$$

Where, X^g is the estimated value of the location of the interference source, and V_g is the deviation of the estimated value from the initial value.

Select a suitable initial value X_0 and make an iterative calculation. When the solution result stabilizes, it can be used as the final positioning result [21].

4 Experiment and Result Analysis

First, locate interference sources with passive location of double stations. The positions of the two measuring platforms are $(118.9 \quad 23.8 \quad 0)$ and $(119.1 \quad 23.8 \quad 0)$. The positions of the three interference sources of satellite are $(118.9 \quad 24 \quad 1000)$, $(119 \quad 24 \quad 1000)$, $(119.1 \quad 24 \quad 1000)$. The distribution of measurement platforms and satellite interference sources is shown in Fig. 6. Assume the accuracy of terminal's direction finding is 0.5°. Operating 1000 measurements and passive location the results are shown in Figs. 7, 8 and 9.

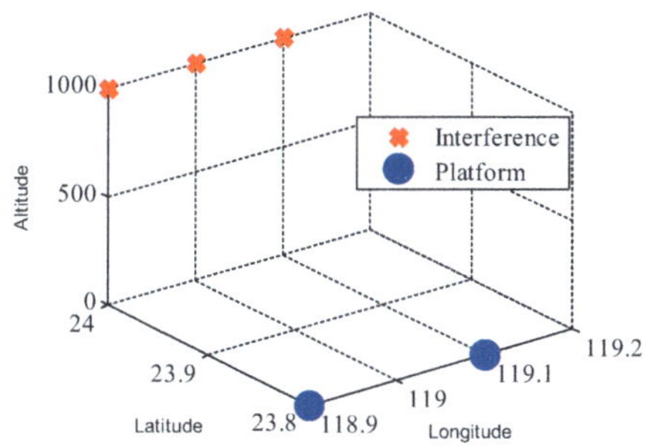

Fig. 6. Distribution of measurement platforms and satellite interference sources

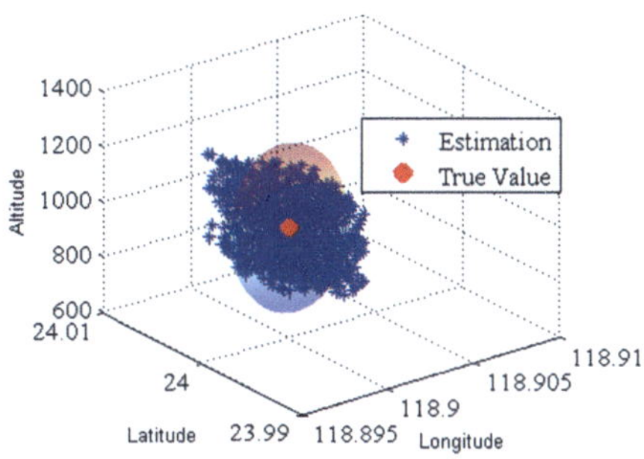

Fig. 7. Positioning results of No.1 interference source by double-platform location

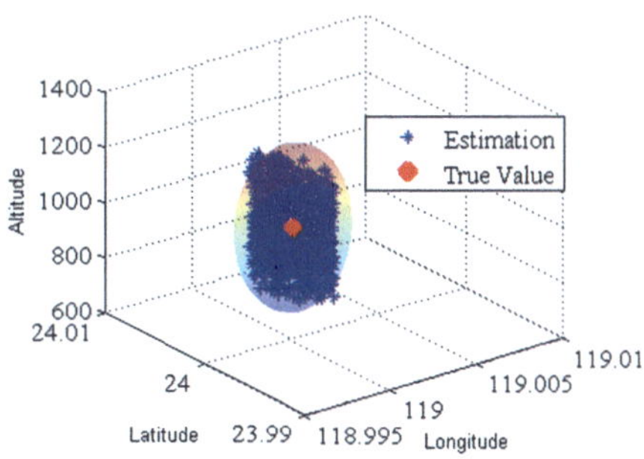

Fig. 8. Positioning results of No.2 interference source by double-platform location

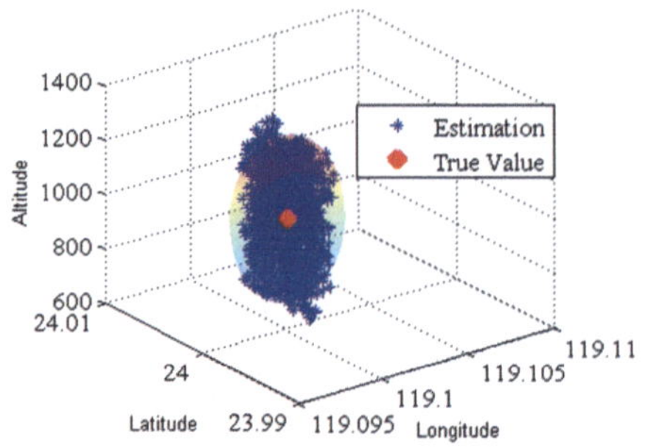

Fig. 9. Positioning results of No.3 interference source by double-platform location

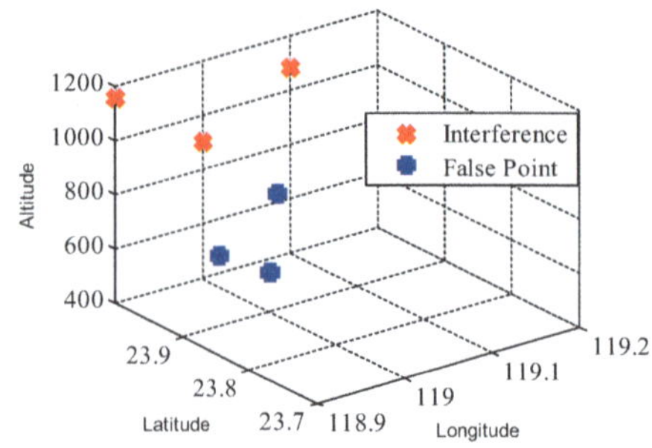

Fig. 10. Positioning results interference sources by double-platform location

Table 1. Double-station positioning solution result statistics table

	True value			Standard deviation			Error	2σ
	Longitude	Latitude	Altitude	Longitude	Latitude	Altitude		
Interference 1	118.9	24	1000	0.0011	0.0024	102.4	257.34	495 m
Interference 1	119	24	1000	8.7042e−04	0.0018	114.04	213.68	
Interference 1	119.1	24	1000	0.0011	0.0024	142.68	273.03	
False Point 1	118.9673	23.9327	598.0435	False value is the result of wrong solution without true value				
False Point 2	119.0004	23.8990	547.3503					
False Point 3	119.0337	23.9321	763.5439					

Remark: The error is the average of the distance from the calculated result to the true value with measuring 1000 times

As shown in the figures above, performing 1000 iteration calculations by passive location of double stations it can get relatively accurate location of interference source and the errors of latitude and longitude are less than 300 m, and the height error is within 200 m. And according to Table 1, the 2σ error is 495 m. But, known from Fig. 10, when several interference sources are at the same height and coplanar with the two measurement terminals, even if the pitch angle information of several satellite interference sources can be measured, the wrong calculation combination cannot be effectively excluded, which leads to the occurrence of false positioning points. As shown in Fig. 10 and Table 1, there are three false positioning results.

If multiple stations are used for positioning solution, the positions of the three satellite interference sources remain unchanged and an additional one whose coordinate is (119 23.7 0) is added in. The distribution of the measurement platforms and interference sources is shown in Fig. 11. At this time, The calculation results of the interference source position solved by multi-platform passive positioning are shown in Figs. 12, 13, 14, 15 and Table 2.

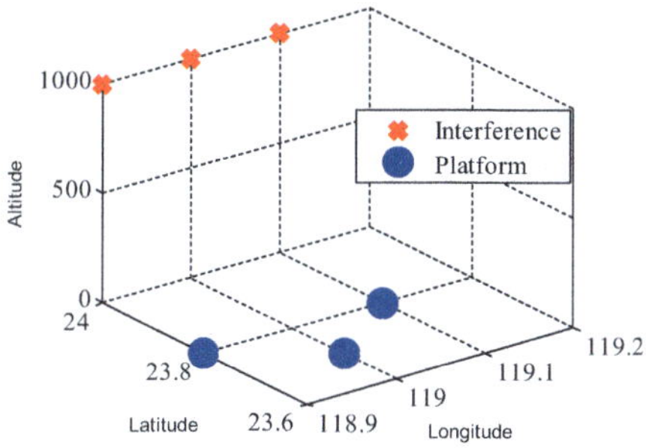

Fig. 11. Distribution of multi-station measurement platforms and satellite interference sources

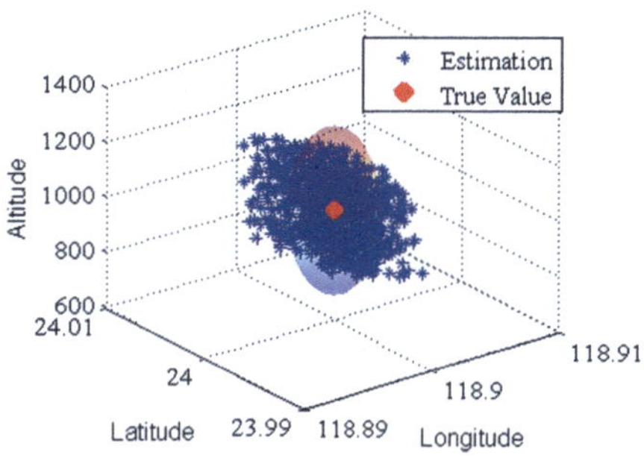

Fig. 12. Positioning results of No.1 interference source by multi-station location

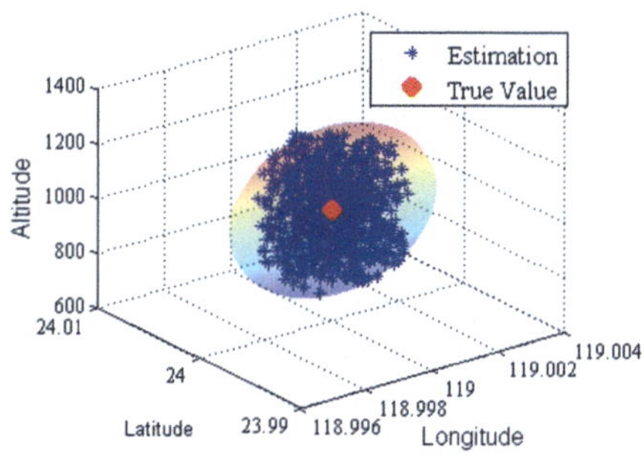

Fig. 13. Positioning results of No.2 interference source by multi-station location

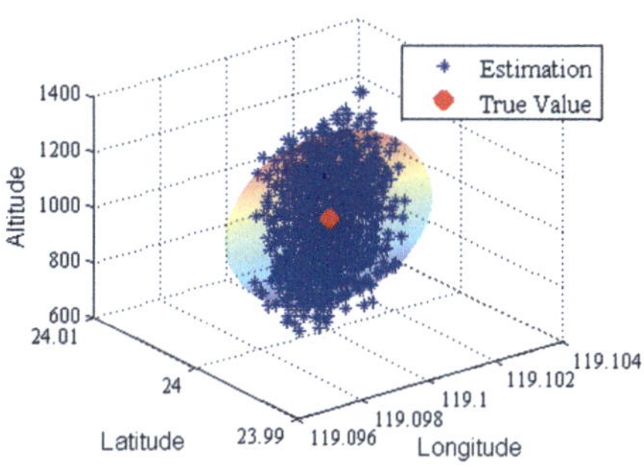

Fig. 14. Positioning results of No.3 interference source by multi-station location

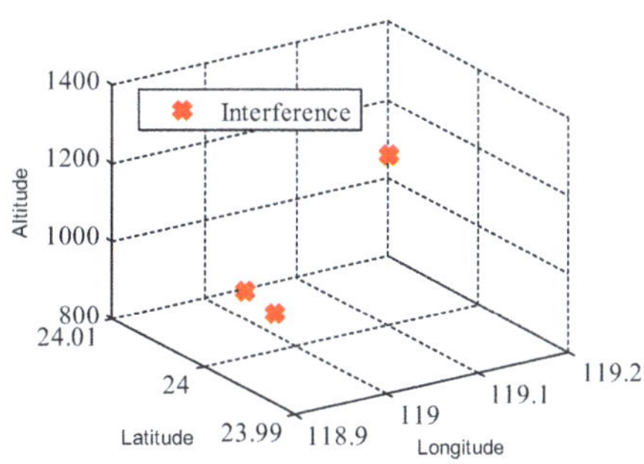

Fig. 15. Positioning results interference sources by multi-station location

Table 2. Multi-station positioning solution result statistics table

	True value			Standard deviation			Error	2σ
	Longitude	Latitude	Altitude	Longitude	Latitude	Altitude		
Interference 1	118.9	24	1000	0.001	0.0023	101.07	249.08	475 m
Interference 2	119	24	1000	7.7507e−04	0.0018	112.24	210.46	
Interference 3	119.1	24	1000	0.001	0.0023	142.09	268.02	

Remark: The error is the average of the distance from the calculated result to the true value with measuring 1000 times

As shown in Figs. 12, 13, 14 and 15, the position accuracy of the interference source solved by multi-station positioning is similar to that of the two-station positioning. The errors of longitude and latitude are within 300 m and the height error is within 200 m. There is not much difference between the two in terms of calculation accuracy. Meanwhile, it can be seen from Table 2 that the 2σ three-dimensional error of the positioning result of the interference source is 475 m.

Compared with double-station location, multi-station location can obtain more measurement information to provide more constraints. It can exclude false settlement combinations to improve the effectiveness of positioning results greatly. Comparing Figs. 10 and 15, it can be seen that in this case, the number of false positioning results is reduced effectively by multi-platform positioning method based on platforms networking. Besides, it can be seen from Table 2 that the results obtained in the article are all the positions of the actual interference sources and there will be no false positioning results. The positioning accuracy is better than two-station positioning.

5 Conclusion

This paper proposes a new cooperation scheme for combat members for the passive location of satellite interference sources. Compared with double-platform positioning, the multi-platform positioning method used in this scheme can obtain more measurement information and due to the existence of the battlefield operation decision center the battlefield data measured by each member can be comprehensively used to strengthen the effective utilization of measured data.

Experimental results show that the scheme can avoid dead zones during the positioning process. At the same time, due to the increase of constraints, the number of false positioning points is greatly reduced and the accuracy of the solution is improved. For example, there is no false point in the case in the article.

In addition, this scheme can also be used to locate other interference sources and it can also meet cooperative operations in terms of saturated attacks on interference sources, coordinated multi-weapon system strikes and relative positioning in battlefield members under the strong interference to satellite positioning system. Thus, the scheme has broad application prospects.

References

1. Cheng, L., Zhang, S., Fangling, Z.: A study on optimized deployment of satellite navigation jammers. Fire Control Command Control **40**(7), 43–46 (2015)
2. Kadlec, P., Stumpf, M., Raida, Z.: Adaptive beam forming in time domain. In: 2011 International Conference on Electromagnetics in Advanced Applications (ICEAA), pp. 299–302 (2011)
3. Fante, R.L., Vaccaro, J.J.: Wideband cancellation of interference in a GPS receive array. IEEE Trans. Aerosp. Electron. Syst. **36**(2), 549–564 (2000)
4. Dang, M.: Application of adaptive nulling antenna in integrated guidance anti-jamming, GNSS World China, pp. 26–32 (2008)
5. Zhuang, X., Cui, X., Lu, M.: Analysis of beamforming algorithms for GPS receivers in the jamming environment. Syst. Eng. Electron. **31**(3), 570–574 (2009)
6. Djigan, V.I.: Adaptive signal processing in multi-beam arrays. In: 2011 6th International Conference on Design & Technology of Integrated Systems in Nanoscale Era (DTIS), pp. 51–54 (2011)
7. Wang, X., Zhang, H., Xie, B.: A doa estimation method of satellite navigation jammers based on adaptive beamforming. Navig. Position. Timing **3**(2), 59–64 (2016)
8. Xing, Y.: Research on Network Positioning Technology of Airborne Passive Radar, pp. 8–29. National University of Defense Technology, Chang Sha (2005)
9. Gao, L., Si, X., Guo, L., Jiang, Y.: Algorithm research of passive location with two station to moving target. J. Proj. Rockets Missiles Guidance **25**(2), 648–651 (2005)
10. Li, H., Chen, P.: Algorithm and accuracy analysis for 3D direction-finding crossing location in multi-station. Command Control Simul. **29**(2), 54–59 (2007)
11. Ni, Z., Feng, G., Ma, B.: An algorithm of multiple-vessel localization based on weighting least squares. Trans. Beijing Inst. Technol. **25**(11), 971–974 (2005)
12. Shen, X., Xu, B., Zou, J., et al.: A positive restriction method used in crossing location. J. Univ. Electron. Sci. Technol. China **43**(6), 834–837 (2014)
13. Hou, Z., Xie, Z.: Application analysis of passive location technology in naval ships. Ship Electron. Eng. **39**(8), 63–65 (2018)
14. Jin, Z.: The network application of electronic warfare equipments based on data link. J. Sichuan Ordnance **38**(7), 89–93 (2017)
15. Wang, H., Mi, X., Huangfu, H.: precise geometrical analysis of ambiguous area in beam-crossing locating. Electron. Optics Control **19**(3), 17–20 (2012)
16. Xia, Y., Yu, H., Fang, M., Li, Z.: A new algorithm for eliminating false intersection points in jammer locating. Aerosp. Electron. Warf. **34**(4), 20–23 (2018)
17. Hu, H., Jing, Z.: Study on eliminating false intersection points in two platform cross location. Avionics Technol. **35**(4), 29–32, 38 (2004)
18. Chen, F., Ma, H., Wu, D.: Research on cross-localization technique in system of multiple-radar and multiple-target. Fire Control Command Control **36**(9), 85–87 (2011)
19. Mou, Z., Liu, B.: Research on multi-platform passive location for multi targets. J. Detect. Control **39**(1), 58–65 (2017)
20. Liu, D., Pang, H., Yin, B.: Study on layout optimization methods for multi-ship DOA location system. Comput. Simul. **32**(3), 27–32 (2015)
21. He, Q., Zhao, D.: Algorithm of direction-finding location of single airplane based on least square method. Ship Electron. Eng. **36**(1), 37–39 (2013)

Satellite Navigation Signal and Signal Processing

Real-Time Parallel Generation Method of Weil Code and Its Implementation in New GNSS Signal

Zhiqin Xue(✉) and Kun Liu

Space Star Technology Co., Ltd., Beijing 100194, China
xuezhiqin_celery@163.com

Abstract. Pseudo-codes with good autocorrelation properties and cross-correlation properties play a very important role in CDMA satellite navigation systems. For the navigation signal with the code length of 10230, the Weil code based on the Legendre sequence has better correlation characteristics than the traditional gold code. Therefore, in the new system signal, the GPS L1C and the BDS B1C frequency point both use the Weil code as the ranging code. The generation method of Weil code is complicated, and the calculation amount is large, which is difficult to generate in real time. Therefore, in the past, most of the receivers use a storage table to store the Weil code. This method is simple and easy to implement, but consumes a large amount of storage resources. Based on the generation method of Weil, this paper proposes a real-time parallel generation method of Weil code, which can not only generate Weil code in real time, but also can be used to generate Weil code in parallel for multiple channels. The method only needs to store the Legendre sequence with length N, which occupies less resources, and has high engineering application value in the receiver tracking channel.

Keywords: Weil code · Legendre code · Gold code · Pseudo code real-time generation · Multi-channel pseudo code generation · New system signal

1 Introduction

A code division multiple access system requires a binary sequence with good auto-correlation properties and cross-correlation properties as a ranging code. In the traditional satellite navigation system, the gold code is mostly used as the ranging code, and the gold code can be generated by using the shift register. The generation method is simple and the hardware implementation is easy. However, the period of the gold code generated by the n-level feedback shift register is $2^n - 1$, which is quite different from the code length of 10230. So it needs to be processed by interception, which will undoubtedly destroy the auto-correlation properties and cross-correlation properties of the gold code itself [1]. Unlike the gold code, the code length of the Weil code is a prime number. For a ranging code whose code length is far from $2^n - 1$ (for example, a ranging code with a code length of 10230), the Weil code only needs to be truncated or inserted into a dozen or even several chips can be used, and fewer chips are truncated or inserted, which maintains the good autocorrelation and non-correlation properties of

J. Sun et al. (Eds.): CSNC 2020, LNEE 650, pp. 457–465, 2020.
https://doi.org/10.1007/978-981-15-3707-3_43

the original Weil code [2]. Therefore, in the new system signal, the GPS L1C frequency point uses a Weil code of length 10223 to generate a ranging code of the L1C frequency point by inserting a fixed 7 chips. The BDS B1C frequency point obtains the ranging code with a length of 10230 by intercepting the Weil code with a length of 10243.

Weil code and gold code are generated in completely different ways. The gold code is generated by the linear feedback shift register, and the hardware implementation of gold code is rather simple. The generation of the Weil code is based on the Legendre sequence. The Legendre sequence is a finite-length random sequence constructed based on the congruence theory. The generation algorithm of Legendre sequence is complex, so it is rather difficult to directly generate Legendre sequence. The Weil code is formed by truncating a sequence index or inserting a fixed code sequence based on the Legendre sequence. The generation method is more complicated than the gold code and is difficult to generate in real time. In order to ensure the real-time use of the Weil code, most of the receivers in engineering applications do not directly generate the Weil code, but use the lookup table to store the Weil code of all satellites in the ROM. In the actual capture or tracking process, the Weil code sequence of the desired satellite is selected by means of a lookup table. The lookup table implements the Weil code, which is simple and easy to implement, and has high reliability. However, this method occupies a large amount of hardware resources [3], which is not conducive to the miniaturization design of the satellite navigation chip.

Based on the analysis of GPS L1C and BDS B1C Weil code generation methods, this paper proposes a real-time parallel generation method of Weil code. This method only needs to store the Legendre sequence corresponding to the Weil code, which occupies less resources. Through the ping-pong operation of two buffers, the method can realize the real-time generation of multi-channel Weil code at any sampling rate, and the algorithm is simple, easy to implement, and has strong engineering applicability.

2 Weil Code Generation

At present, the satellite navigation system uses the Weil code as the frequency of the ranging code, which has two frequency points, GPS L1C and BDS B1C. There are two main differences between the two frequency points ranging code generation. First, the GPS L1C frequency is based on the Legendre sequence with a length of 10223, and the BDS B1C is based on the Legendre sequence with a length of 10243; Second, the ranging code of GPS L1C frequency point is constructed on the basis of the Weil code with a length of 10223 bits by inserting an extended sequence with a length of 7 bits. The ranging code of the BDS B1C signal is obtained by cyclic interception based on the Weil code of length 10243 bits.

2.1 GPS L1C Signal Ranging Code

The GPS L1C signal has a ranging code length of 10,230 chips, which is generated based on a Legendre sequence with a specific length of 10223 and a fixed 7-bit extension sequence [4]. First, the fixed, N-length Legendre sequence $L(k)$ $(k = 0, 1, \ldots N-1)$ is generated as follows:

$$L(k) = \begin{cases} 0, & k = 0 \\ 1, & k \neq 0, \textit{ there is x such that } k = x^2 \bmod N \\ 0, & k \neq 0, \textit{ there is no x such that } k = x^2 \bmod N \end{cases} \quad (1)$$

It is known from the uniqueness of the fixed length Legendre code that the Legendre sequence is fixed and identical for different satellites at the same frequency.

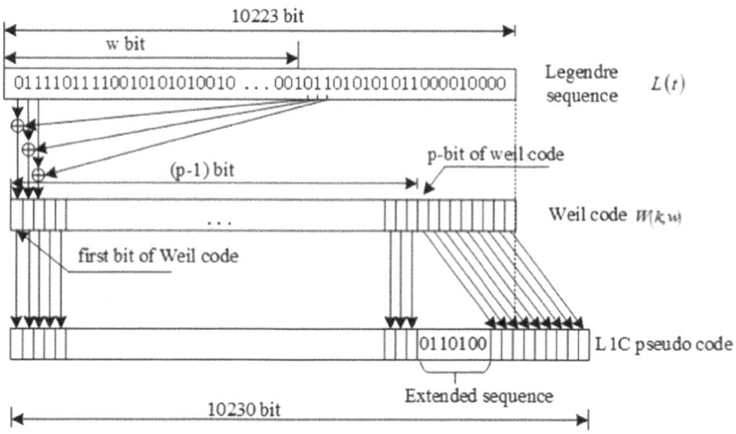

Fig. 1. Generation of GPS L1C PRN codes

Then, based on the Legendre sequence of length N (for GPS L1C signal $N = 10223$), a Weil code sequence of length 10223 bits $W(k, w)$ is generated as follows:

$$W(k, w) = L(k) \oplus L\big((k + w)_N\big) \quad k = 0, 1, \ldots, N-1 \quad (2)$$

Where w is the phase difference between the two Legendre sequences, also known as the Weil index, which actually quantifies the translation between two Legendre sequences.

Finally, a fixed extension sequence with a length of 7 bits is inserted into the Weil code sequence $W(k, w)$ with a length of 10223 bits, to obtain the ranging code of GPS L1C frequency point with a length of 10230.

$$c(k, w, p) = \{W(0, w), \cdots, W(p-2, w), 0, 1, 1, 0, 1, 0, 0, W(p-1, w), \cdots, W(10222, w)\} \quad (3)$$

Wherein, 0110100 is a 7-bit fixed extension sequence, and p is an insertion position of the extended sequence, that is, the extended sequence is inserted into the p-th chip position of the Weil code sequence $W(k, w)$. Figure 1 shows the block diagram of generating GPS L1C frequency point ranging code.

2.2 BDS B1C Signal Ranging Code

In the process of generating BDS B1C signal pseudo code, the generation of Legendre sequence and the generation of Weil code are the same as the GPS L1C signal, except that the length of the Legend sequence. In the GPS L1C signal pseudo code, the length of the Legendre sequence and the length of the Weil code sequence are $N = 10223$, and in the BDS B1C signal, $N = 10243$. The generation of Legendre sequence and Weil code sequence will not be provided here [5].

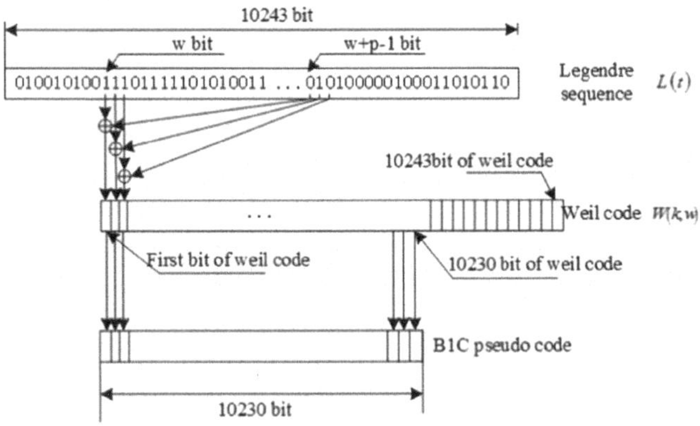

Fig. 2. Generation of BDS B1C PRN codes

In the process of generating the ranging code sequence by the Weil code sequence, the BDS B1C signal is different from the GPS L1C signal. The ranging code of the BDS B1C signal is obtained by truncating the Weil code of length 10243 instead of inserting the extended sequence. The method for generating the ranging code of the BDS B1C signal is as follows:

$$c(k; w; p) = W((k + p - 1) \bmod N; w) \; k = 0, 1, \ldots, 10229 \tag{4}$$

Where, p is the intercept point, indicating that the interception starts from the p-th bit of the Weil code sequence, and a total of 10,230 bits are intercepted. Figure 2 shows the block diagram of generating BDS B1C frequency point ranging code.

3 Real-Time Parallel Weil Code Generator

3.1 Current Receiver Processing Methods and Existing Problems

At present, most of the receivers used in engineering applications use the look-up table method to implement the Weil code, but the look-up table method needs to store the ranging codes of all satellites in the ROM. GPS L1C and BDS B1C each have 63 satellites. Each satellite has two sets of ranging codes for Data and Pilot. It takes a lot of storage space to store so many ranging codes in ROM, which consumes a lot of hardware resources. Therefore, it is necessary to study the real-time generation method of the Weil code. However, the generation definition of the Legendre code is complicated and cannot be generated in real time with a simple circuit structure. We know that the Legendre sequence of fixed length is unique, that is, the Legendre sequence of different satellites of the same frequency is fixed and identical. The generation of Weil code is based on this fixed Legendre sequence. Therefore, we can only store the Legendre sequence of length N, and use this sequence to generate the Weil code in real time in the channel, which can reduce the storage space to a certain extent.

In the process of receiver tracking, each channel accumulates the code NCO according to its own code frequency control word. The code NCO overflows to obtain the corresponding code clock. Under the action of the code clock, the corresponding code chip is read into the channel for pseudo-code stripping to obtain the accumulative amount. Therefore, in the process of pseudo-code generation, a way to configure a Legendre ROM in each tracking channel can be adopted. The Legendre ROM in each tracking channel is independent from each other, and the read operations of each ROM are independent of each other. In this way, it is only necessary to ensure that the generation time of each code chip is less than $1/f_c$ (f_c is the code clock frequency). That is, it is only necessary to ensure that a code chip is generated before the arrival of the next code clock pulse. However, with the rapid development of satellite navigation technology, the number of channels for parallel tracking is increasing. If each tracking channel is configured with an independent Legendre ROM, the storage space consumed is linearly related to the number of tracking channels. Therefore, the way in which all tracking channels store identical and independent Legendre sequences is no longer practical.

The Legendre sequence stored in the Legendre ROM in the tracking channel is completely identical, therefore, it can be considered to keep only one Legendre ROM. That is, all tracking channels reuse one Legendre ROM. When M channels are simultaneously tracked, in the limit case, M channels may need to generate pseudo code chips at the same time. The pseudo-code generation of each channel takes up at least 1 sampling clock, so M channel pseudo-code generation needs M/f_s time, where f_s is the sampling frequency. The time required for pseudo code generation must be less than the interval between two code clocks, that is, $M/f_s < 1/f_c$ is required. This condition is difficult to satisfy when the sampling rate f_s is low or the number of channels

M tracked at the same time is large. Based on this problem, this paper proposes a method of using ping-pong operation to realize real-time generation of multi-channel Weil code in the case of only one Legendre ROM.

3.2 Real-Time Parallel Weil Code Generator

The structure of the real-time parallel Weil code generator is shown in Fig. 3. When M channels are simultaneously tracked, the time required to generate the pseudo code of M channels is M/f_s. When the sampling rate f_s is small or the number of tracking channels M is large, this is obviously a long time. This time may exceed the clock interval $1/f_c$. Therefore, we use the method of generating N pieces of code at a time, and each channel is equipped with a N-bit buffer. The generated N chips are stored in the buffer for use by the tracking channel. In this way, after the N chips are used up, the time interval for re-generating the chips is extended to N/f_c, so we only need to guarantee $M/f_s < N/f_c$.

There is a special case where if the chips in the N-bit buffer of the M channels are used up at the same time, it is necessary to face the problem of simultaneously generating M channel pseudo-codes in one code interval $1/f_c$. In order to solve this special case, each tracking channel is equipped with two N-bit buffers, namely N-bit buffer 1 and N-bit buffer 2. The two N-bit buffers adopt ping-pong operation and read and write in turn. The specific process is as follows:

(1) Each channel is time-multiplexed to read two sets of N-bit Legendre sequences from memory, denoted as $L(k_1)$ and $L(k_2)$, respectively, which correspond to the Legendre sequence in the Weil structure and the Legendre sequence after translation. The two sets of Legendre sequences are XORed to obtain the Weil code sequence of the channel.
(2) Each channel is configured with two N-bit buffers, and each of the two buffers has a flag flag1 and flag2 indicating whether the data is valid. When the flag is 1, it indicates that the N-bit Weil code is valid, and the channel can be read. When flag is 0, it indicates that the Weil code is invalid, and the Weil code sequence can be written. This forms a ping-pong operation, that is, when the channel reads the Weil code in the buffer 1, the Weil code obtained in step (1) is written into the N-bit buffer 2. After the channel uses the N-bit Weil code in the buffer 1, the Weil code in the buffer 2 is read, and the new Weil generated in the step (1) at this time is written into buffer 1.
(3) Inside each channel, the read Weil is processed accordingly according to the frequency of processing. If the frequency of the tracking is the GPS L1C frequency point, it is necessary to suspend the use of the Weil code in the buffer at a specific position, and insert a fixed 7-bit extension sequence, and after the insertion is completed, continue to use the remaining Weil code in the buffer.

If the frequency of the tracking is BDS B1C frequency, it is necessary to reset the address of the read Legend memory after completing the generation of the ranging code (10230 chips) of one code week, and regenerate Pseudo code of the next code week.

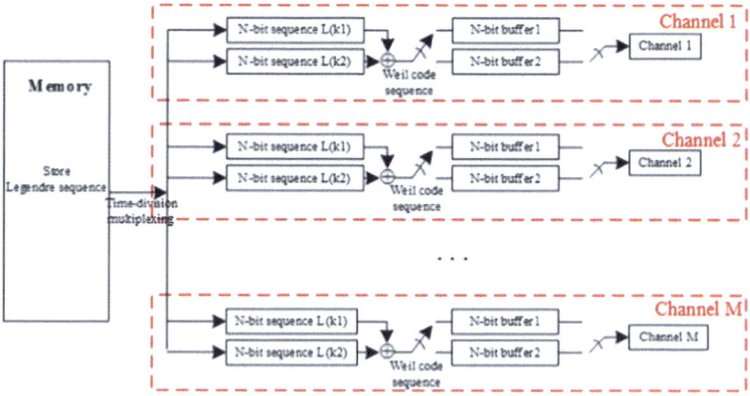

Fig. 3. The structure of real-time parallel Weil code generator

3.3 Weil Code Generator Implementation

In the baseband processing process, the satellite signal is captured first. After the capture is completed, the capture module transmits information such as the captured code phase and carrier Doppler to the tracking module. After receiving the information, the tracking module starts a tracking channel. This channel performs tracking of this satellite signal. Among M tracking channels, channel polling is adopted to generate their own ranging codes according to the real-time parallel Weil code generation method mentioned above.

This section takes the BDS B1C frequency point as an example to give the implementation process of the real-time parallel Weil code generator. In the implementation process of real-time parallel Weil code generation, it is necessary to preferentially respond to the satellite that has just been transmitted by the capture module, that is, the tracking channel that needs to respond to the new startup preferentially. This is mainly because the newly started tracking channel does not have any Weil code in the two N-bit buffers at this time. The available Weil code must be generated before its chip pulse arrives to ensure the normal tracking of the channel.

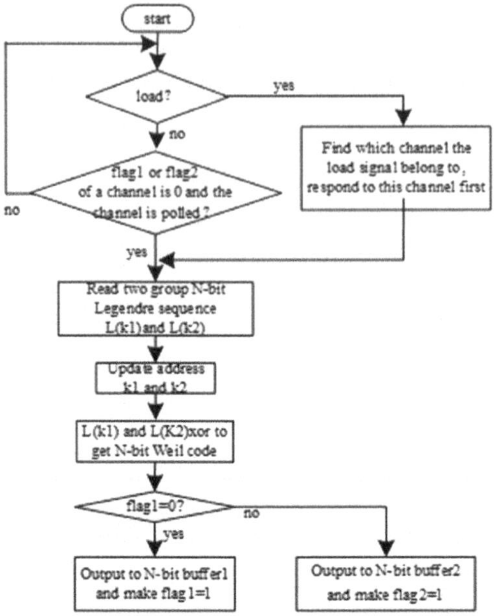

Fig. 4. Implementation flow chart of real-time parallel Weil code generator

The real-time parallel Weil code generation implementation flow chart is shown in Fig. 4. First, determine if there is a new capture tracking signal. Here, the start signal of the tracking is the load signal, that is, to judge whether there is a load signal. If there is a load signal, it is necessary to judge the channel number of the newly activated channel and give priority to this channel to generate the N-bit Weil code. Then, if there is no load signal at this time, all channels are polled in turn. When polling to a certain channel and detecting that its buffer valid flag flag1 or flag2 is 0 (flag1 and flag2 are initialized to 0), it indicates that the data in this buffer has been used up, and needs to generate and cache. According to the Legendre sequence addresses k_1 and k_2 of this channel, two sets of N-bit Legendre sequences $L(k_1)$ and $L(k_2)$ are read by time division multiplexing, and the addresses k_1 and k_2 are updated after reading, $k_1 = k_1 + N$, $k_2 = k_2 + N$. Finally, the two sets of Legendre sequences are XORed to form an N-bit Weil code. Then, it is judged whether flag1 is 0. If flag1 is 0, the generated N-bit Weil code is buffered into the buffer 1, otherwise it is buffered into the buffer 2, and the corresponding flag is set to 1, indicating that the data in the buffer is valid and available to the channel.

The Weil code generation formula of the BDS B1C frequency point shows that:

$$\begin{aligned} c(k;w;p) &= W((k+p-1) \bmod N; w) \\ &= L((k+p-1) \bmod 10243) \oplus L((k+p+w-1) \bmod 10243) \end{aligned} \tag{5}$$

Where $k = 0, 1, \ldots, 10229$. At the initial time of tracking, it is necessary to initialize the Legendre read addresses k_1 and k_2, the initial value of k_1 is $(p - 1)$ mod 10243, and the initial value of k_2 is $(p + w - 1)$ mod 10243.

In the tracking channel, under the action of the chip pulse, the chips in the buffer 1 are first read sequentially, and when the N-bit chips in the buffer 1 are used, the flag1 flag is set to 0. Then, the chips in the buffer 2 are sequentially read. Similarly, after the chips in the buffer 2 are read, the flag2 flag is set to 0, so that the buffer 1 and the buffer 2 are alternately read. This ensures real-time multi-channel parallel generation of the Weil code.

4 Conclusion

In modern satellite navigation signals, GPS L1C frequency point and BDS B1C frequency point no longer use gold code, but instead generate a more complicated Weil code. The Weil code is generated based on a fixed-length Legendre sequence. The generation method is complex, the calculation amount is large, and the calculation time is long. It is difficult to generate in real time in the receiver. Therefore, currently the receiver usually uses the method of directly storing the satellite ranging code. When the tracking channel is implemented, the required ranging code is sent into the tracking channel through a look-up table method. However, by storing the ranging code in the memory, a large amount of storage space is wasted, and more hardware resources are consumed. Based on the generation method of Weil code and the use characteristics of ranging code in the tracking channel, this paper proposes a real-time parallel generation method of Weil code. This method only needs to store the Legendre sequence with length N, which takes up little storage space. Besides, this Weil code generation method is not limited by sampling rate and the number of tracking channels. Finally, this paper gives the implementation of the universal real-time Weil code generator. The module has versatility in different receivers. It can generate the required pseudo-codes with simple configuration. The engineering implementation of this method is simple and has high engineering application values in the receiver tracking channel.

References

1. Deng, Z.-L., Ma, H.-T., Li, C.-M.: Research on characteristic analysis and optimization of Weil code in new GNSS signals. In: The 5th China Satellite Navigation Academic Annual Conference Electronic Collection
2. He, C.-L., Wang, Y.: Weil code correlation performance of GPS L1C signal. Radio Commun. Technol. **39**(1), 32–35 (2013)
3. Sun, D., Cui, X., Zhu, Y., Lu, M.: Design and implement of a real-time multi-channel weil-code generator. In: The 5th China Satellite Navigation Academic Annual Conference Electronic Collection
4. Global Positioning System Wing (GPSW) Systems Engineering & Integration, Interface Specification, IS-GPS-800, Revision A, Navstar GPS Space Segment/User Segment L1C Interface
5. BeiDou Navigation Satellite System Signal in Space, Interface Control Document, Open Service Signal B1C (Version 1.0)

Pseudorandom Code Error Monitoring Method for GNSS Signal

Yi Yang(✉), Lin Chen, Yuqi Liu, and Shaobin Guo

The 29th Research Institute of China Electronics Technology Group Corporation, Chengdu, China
yangyi355@163.com

Abstract. Pseudorandom code error of GNSS signal could degrade user receiver's data demodulation and ranging performance, it is usually caused by unexpected behaviour within the logic circuits of navigation signal generation in satellite payload. Due to the complexity of modernized GNSS signal structure, real-time pseudorandom code error monitoring has become an important and challenging issue. A novel method is proposed in this paper, by appending several calculation steps into conventional receiver's signal tracking loop and with the external aid of navigation message, rapid detection of GNSS single spreading code error can be realized. The simulation result shows that by taking this method, accidental pseudorandom code error can be instantly detected regardless of various code chip waveforms and multiplexing schemes, and it's applicable for both public and authorized signals. This method can be applied to the design of Satellite Autonomous Integrity Monitoring (SAIM) payload.

Keywords: Global Navigation Satellite System · Pseudorandom code · Signal quality monitoring

1 Introduction

Direct Sequence Spread Spectrum (DSSS) technique is widely adopted in various GNSS signal structure design, which makes GNSS signals own the following features: ranging based on code phase, Code Division Multiple Access (CDMA), and excellent anti-interference performance. Spreading sequences are pseudorandom noise (PRN) codes which can be repetitively generated by both transmitters and receivers, thus acquisition and tracking processing can be accomplished by users. Spreading sequences of open service signals are short codes with the cycle between 1 ms and 10 ms, user interface specification documents describe how they can be generated; while spreading sequences of authorized service signals are long codes which can be considered to be non-cyclic, and the generation method is not announced.

The non-ideal features and unexpected behaviours of navigation signal generation payload may deteriorate GNSS signal quality and thereby affect the performance of the whole navigation system. [1] analyses the effect of single event upset (SEU) on C/A code generators, and concludes that in the case SEU changes the status of registers, the output C/A code would become unused sequences or even sequences being used by other GPS satellites. Additionally, unexpected PRN code reversal can be resulted by

J. Sun et al. (Eds.): CSNC 2020, LNEE 650, pp. 466–473, 2020.
https://doi.org/10.1007/978-981-15-3707-3_44

imperfect timing design of baseband signal generator circuits within the on-board payload.

According to the current evaluation system of GNSS signal quality and corresponding testing method, the monitoring and evaluation of GNSS signal quality in time domain are limited to shape distortion of code chip waveforms [2–4], monitoring method of PRN code error has been rarely mentioned. [5] proposes a method to decide and extract PRN code, but it's only suitable for offline processing of QPSK signal. PRN code error monitoring method proposed by [6] has considered the impact of subcarrier introduced by BOC modulation, but it can't deal with the situation when multiple signals are multiplexed in a single frequency. Altogether, due to the complexity of modernized GNSS signal structure and the rapid variation of PRN code bit, we are still lack of efficiency method for real-time monitoring of pseudorandom code error.

Aiming at the requirement of the monitoring of PRN code error for modernized GNSS signals and based on the procedure of receiver signal processing, this paper proposes a method to monitor GNSS signal PRN code error. By generating incorrect multiplexed signals intentionally in receiver and correlating them with received signal, we can decide if each PRN code is correct in real-time. This method demands low calculation resources because it makes full use of information from the original receiver's tracking loop, and it can be applied to the design of SAIM payload.

2 Navigation Signal and Its Generation Procedure

Generation procedure of GNSS signal in a single frequency mainly includes spreading spectrum modulation of each signal, constant envelop multiplexing, and radio frequency modulation, as illustrated in Fig. 1.

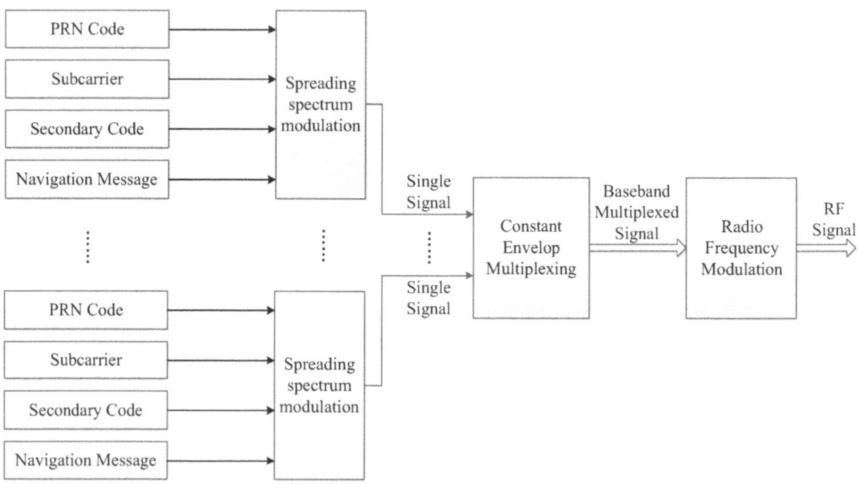

Fig. 1. Block diagram of GNSS signal generation

PRN codes are ±1 sequences with excellent auto-correlation and cross-correlation properties, they can be repetitively generated by both transmitters and receivers with the code rate between 1.023 Mcps and 10.23 Mcps. Subcarrier, as an optional element, defines the shape of PRN code waveform, it's a cyclic sequence and its frequency is usually equal to or several times of PRN code rate. Secondary code and navigation message act in a similar way for signal quality evaluation processing, they decide if PRN code should reverse, the data width is in the order of milliseconds.

Considering several signals are transmitted in one RF frequency, these separate signals have to be multiplexed into a complex signal during digital baseband signal processing according to certain power ratio and carrier phase relationship. In order to improve the efficiency of solar cells, the envelope of multiplexed signal remains constant by adding intermodulation items into them, so the working point of high power amplifier (HPA) can be set to saturation point.

We can conclude from above analysis that due to the complexity of GNSS signal structure, it's difficult to detect and precisely identify the fault when a PRN code error happens accidentally. Figure 2 visually shows an example of PRN code error by simulation. We assume that four signals are multiplexed together, their modulation schemes are TMBOC(6,1,4/33), BOC(10,5), BOC(1,1) and BPSK(1) respectively. If the second PRN code chip of BOC(1,1) signal is reversed as the red curve shows left, the envelop of multiplexed signal is still constant but the waveforms of real part and imaginary part are distorted irregularly, so we can't determine which signal is abnormal by simply judging the polarity of sampled equivalent baseband signal. In order to eliminate the effect of other signals in the same frequency, one resolution is to accumulate the target signal by PRN code cycle after compensating the modulation of navigation message and secondary code, based on the randomness of navigation message, signals except for the target one are counteracted so we can decide each PRN code chip by chip. The limitation of this method is that it's not suitable for non-cyclic authorized signal and it's not able to detect the error in real-time.

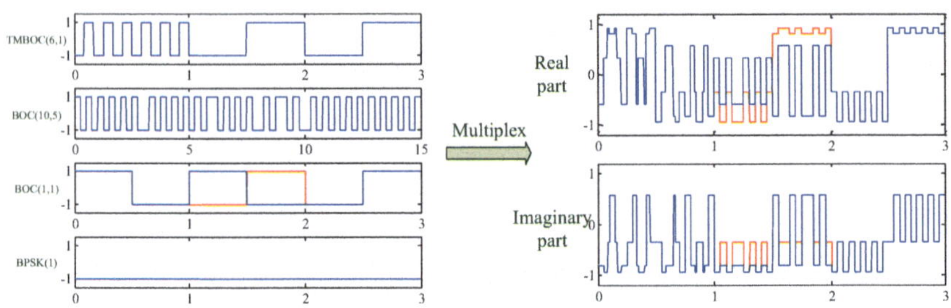

Fig. 2. Example of pseudorandom code error

3 Detection Method of Pseudorandom Code Error

In this section we propose an onboard real-time detection method of PRN code error for GNSS signal. SAIM receiver down-converts the received signal to zero-IF baseband signal in carrier tracking loop including real and imaginary parts, and the prompt branch in code tracking loop has generated each separate baseband signals and they are synchronized with the received signal. By making use of above existing signals and with the navigation message transmitted from another function unit of satellite payload, we can completely generate the constant envelop multiplexed baseband signal. Correlating the generated signal with received baseband signal in the range of one code chip and judging the result, we can determine if PRN code error has happened, the implementation diagram is shown in Fig. 3.

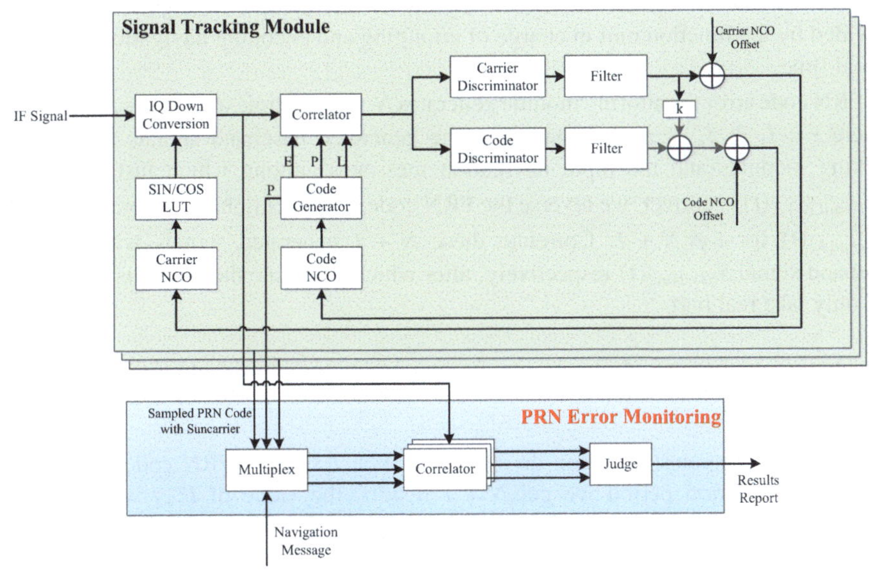

Fig. 3. Implementation of pseudorandom code error detection

The input IF signal of tracking modules can be expressed as

$$S_{IF}(t) = S_I(t) * \cos[2\pi(f_0 + f_d)t + \varphi_0] - S_Q(t) * \sin[2\pi(f_0 + f_d)t + \varphi_0] \quad (1)$$

where $S_I(t)$ and $S_Q(t)$ are real and imaginary parts of baseband signal respectively, f_0 is receiver's local intermediate frequency, f_d is carrier Doppler offset, φ_0 is a random carrier phase. Assume the number of signals multiplexed in this frequency is N, denoted as s_1, s_2 … s_N, for each of which can be either public or authorized signals. Choose one of these signal's tracking modules as a reference and export the received baseband signal $S_{baseband}(t)$ after down conversion to PRN error monitoring module, $S_{baseband}(t)$ can be expressed as

$$S_{baseband}(t) = \{S_I(t) + S_I(t) * \cos[2 * 2\pi(f_0 + f_d)t] - S_Q(t) * \sin[2 * 2\pi(f_0 + f_d)t] \\ + j * S_Q(t) - j * S_I(t) * \sin[2 * 2\pi(f_0 + f_d)t] - j * S_Q(t) * \cos[2 * 2\pi(f_0 + f_d)t]\} * e^{j\varphi} \tag{2}$$

where φ is the designed phase of the chosen reference signal. Considering high frequency components would be filtered in subsequent correlation processing, Eq. 2 can be simplified as below

$$S_{baseband}(t) = [S_I(t) + j * S_Q(t)] * e^{j\varphi} \tag{3}$$

We can see the transmitted baseband signal is recovered except for a constant phase bias φ which is known.

Each baseband signals s_1, s_2 … s_N including subcarrier are generated in their carrier tracking loop, they should also be exported to the PRN error monitoring module. At last, we still need navigation messages to generate the multiplexed signal, they can be provided by the function unit in charge of arranging and encoding navigation messages in real-time.

PRN code error monitoring module generates $N + 1$ multiplexed signals $\tilde{S}_{baseband,i}(t)$ (where $i = 1, 2, 3, \ldots N + 1$) based on the generated baseband signals from signal tracking modules and the input navigation messages, among which just one signal $\tilde{S}_{baseband,N+1}(t)$ is correct, we reverse the PRN code of s_i intentionally before generating $\tilde{S}_{baseband,i}(t)$ for $i \neq N + 1$. Correlate these $N + 1$ generated signals with received baseband signal $S_{baseband}(t)$ respectively, after which we rotate these results by angle φ and only take real part

$$I_i = real\left(\left(\int_{T_c} S_{baseband}(t) \bullet \tilde{S}^*_{baseband,i}(t)dt\right) \bullet e^{-j\varphi}\right) \tag{4}$$

where T_c is the greatest common divisor of all $N + 1$ signals' PRN code chip width. For each correlation period we get $N + 1$ results, the value of I_{N+1} would be the greatest if no PRN code error happens because $\tilde{S}_{baseband,N+1}(t)$ and $S_{baseband}(t)$ are best matched after phase φ compensation. Nevertheless, the value of I_i would be greatest if the PRN code of s_i is incorrect because $\tilde{S}_{baseband,i}(t)$ and $S_{baseband}(t)$ would be best matched instead of $\tilde{S}_{baseband,N+1}(t)$. By searching for the greatest value in $N + 1$ integration results, we can determine whether PRN code error happens and find out the incorrect signal if that error does happen.

4 Validation of the Method

In this chapter we present the validation experiment with a GNSS signal simulator which is able to generate GNSS signal with customized structure and simulate the PRN code error situation by programing. The signal structure we use is shown in Table 1, there're four signals multiplexed together with different modulation schemes and PRN code rate. The 287th PRN code of s_1, 4181st PRN code of s_2, 639th PRN code of s_3, and 572nd PRN code of s_4 are reversed as the test case.

Table 1. Signal structure of the validation test

Signal	Modulation	Power ratio	Phase	Code length	Error code
s_1	MBOC(6,1)	1/3	45°	5115	287
s_2	BOC(10,5)	1/3	−45°	1023	4181
s_3	BOC(1,1)	1/6	0°	1023	639
s_4	BPSK(1)	1/6	90°	1023	572

We sampled the signal transmitted by simulator with a high speed signal sampling and storage device, the sampling rate is 190 MHz. After the acquisition and tracking processing by a software receiver, we extract 1 ms baseband signal for PRN code error detection. Integration and dumping rate is set to 5.115 MHz according to the method described in last section, we can get five correlation values from I_1 to I_5 in each integration cycle which are plotted in Fig. 4. It's clearly shown that in most cases the value of I_5 is the greatest which means the PRN codes of received signal are correct during corresponding integration cycles. While four glitches can be observed in I_5 curve, where implies that I_5 is not the greatest among all correlation values and hence PRN code error is detected.

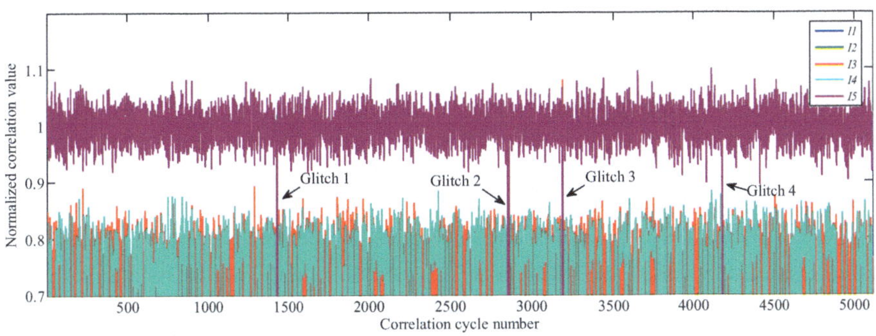

Fig. 4. Correlation values in each cycle

A detailed observation is preceded by zooming in for a closer view of these four glitches respectively as shown in Fig. 5. For glitch 1, the value of I_1 is the greatest and it lasts from 1431st to 1435th correlation cycle, which means the 287th PRN code of signal s_1 is incorrect; for glitch 2, the value of I_4 is the greatest and it lasts from 2856th to 2860th correlation cycle, which means the 572nd PRN code of signal s_4 is incorrect; for glitch 3, the value of I_3 is the greatest and it lasts from 3191st to 3195th correlation cycle, which means the 639th PRN code of signal s_3 is incorrect; for glitch 4, the value of I_2 is the greatest and it appears in 4181st correlation cycle, which means the 4181st PRN code of signal s_4 is incorrect.

The detection result we obtain from Fig. 5 has exactly matched with the settings given by Table 1, which proves that with this method we can not only detect the

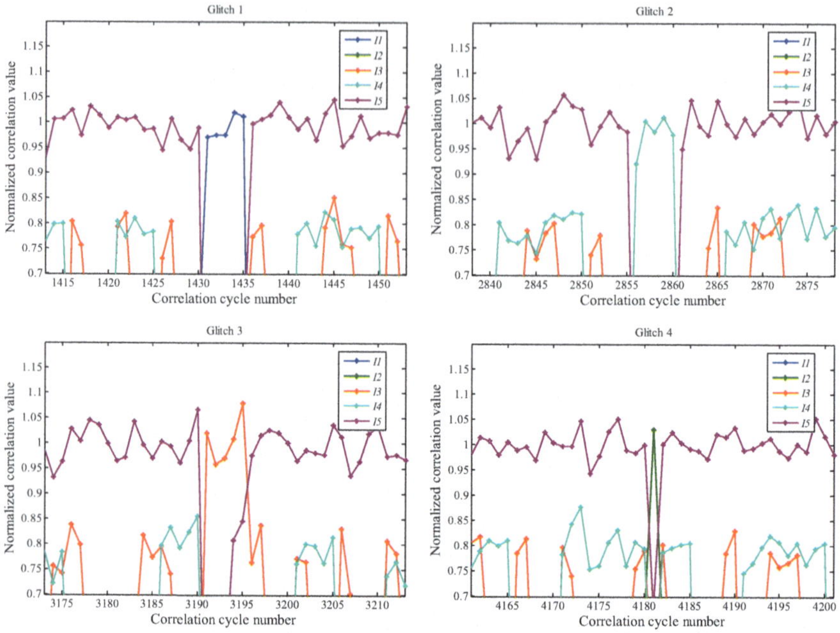

Fig. 5. Correlation values around abnormal areas

unexpected reversal of PRN code for GNSS signal, but also find out the corresponding incorrect signal and code chip.

5 Concluding Remarks

The unexpected reversal of PRN code in GNSS signal should be autonomously detected and reported because is an indicator of abnormal operation of satellite payload. This paper presents a novel method to detect PRN code error by generating multiplexed signals and correlating them with received baseband signal. This method is immune to the impact of multiplexing and subcarrier introduced by BOC modulation, and suitable for both public and authorized signals. This method asks for no extra device, automatic and continuous monitoring of PRN code error for GNSS signal can be achieved by just updating the software of SAIM payload.

References

1. Qu, B., Wei, J., Zhang, S., et al.: An assessment effect analysis of single event upset for C/A code generator. In: CSNC 2013 (2013)
2. Macabiau, C., Chatre, E.: Impact of evil waveform on GBAS performance. In: Proceedings of the IEEE PLANS 2000, SanDiego, California, pp. 22–29. IEEE (2000)

3. Eric, R., Phelts, R.E., Enge, P.K.: A simple method of signal quality monitoring for WAAS LNAV/VNAV. In: ION GPS 2002, Portland, pp. 800–801 (2002)
4. Soellner, M., Kurzhals, C.: GNSS offline signal quality assessment. In: 21st ION GNSS, 16–19 September 2008, pp. 909–920 (2008)
5. Zhao, H., Li, X.: Correlation loss assessment algorithm for long-code navigation signal. Commun. Technol. **51**(10), 23–27 (2018)
6. Xu, C., Wang, Y., Tang, X., et al.: Modernized GNSS signal abnormality model and Evil signal detection technique. In: CSNC 2014 (2014)

Research on MPSK Modulation Based GNSS Signals with High Data Rate

Tao Yan(✉), Ying Wang, Xiao Liu, Lang Bian, and Yansong Meng

China Academy of Space Technology (Xi'an), Xi'an 710000, China
iamyantao@126.com

Abstract. With the increase of the requirement on the fast time-to-fix and precise message, the high data rate becomes on the key feature of GNSS signals. The most existing GNSS signals are the DSSS signals with BPSK or BOC modulation. Thus, to increase the data rate would cause two results. One is that the spreading spectrum gain decreases and the equivalent PRN code length shortens. The other is that the PRN code rate increases while keeping code length unchanged, so the signal bandwidth increases. The CSK modulation can modulate data using the PRN code phase, and change flexible the data rate. Meanwhile, the PRN code length and code rate maintain constant. Due to the above advantages, CSK modulation has been applied in QZSS L6 signal. However, CSK is usually used to demodulate data information, and difficult to measure range. Aiming at this problem, this paper proposes the MPSK modulation based GNSS signals for broadcasting high rate data and ranging. Firstly, the MPSK modulation is introduced, and the scheme of MPSK modulation applied in GNSS signals is analyzed. Secondly, taking 8PSK as an example, the demodulation performance, code/carrier tracking schemes, and tracking performance are analyzed. On this basic, the advantages and disadvantages comparing MPSK modulation and CSK modulation are discussed. Then, the constant envelope multiplexing method combining the existing GNSS signals and MPSK modulation signal is studied. Finally, an application example of MPSK modulation based GNSS signals is presented, and the performance is analyzed with the help of simulation. The results verify the effectiveness of MPSK modulation as a candidate of GNSS signals with high data rate.

Keywords: MPSK modulation · High data rate · CSK modulation · GNSS signals

1 Introduction

With the continuous expansion of the satellite navigation applications field, more and more emerging applications and services based on GNSS signals are developed, and it is expected to improve the accuracy and availability. In this context, Both Galileo and QZSS broadcast wide area precise positioning signals on their basic navigation constellations, providing centimeter level augmentation services [1–3]. It can be seen that the development trend of centimeter level high-precision positioning service is more and more obvious.

J. Sun et al. (Eds.): CSNC 2020, LNEE 650, pp. 474–484, 2020.
https://doi.org/10.1007/978-981-15-3707-3_45

At present, high-precision service needs to broadcast additional data, such as orbit correction, clock correction, code bias correction, carrier phase bias correction, as well as integrity data. Taking QZSS L6 signal as an example, QZSS only provides augmentation services to the surrounding areas of Japan, and the data rate required is about 2 kbps, far greater than 50 bps of GPS C/A code signal.

In addition, in order to reduce the time-to-first-fix (TTFF), it is required to obtain the TTFF data (TTFFD) quickly. TTFFD is the required time to recovery the clock and ephemeris data (CED). For the current GNSS, the time is always higher than 15–20 s, which is no valuable in most emerging applications. Therefore, European researchers also believe that the next generation of navigation signals needs to meet the requirement of high data rate [4, 5].

In current BPSK/BOC modulation, to increase the data rate, either to increase the code rate, which will lead to a wider spectrum; or to reduce the code length, which will lead to a decrease in the isolation orthogonality of the PRN code. CSK modulation can provide flexible data rate, while ensuring the same code rate, and improving the data rate does not affect its demodulation performance and PRN code correlation performance [6, 7]. Therefore, it is applied in QZSS L6 signal. Because of the high data rate capability of CSK, the performance of CSK for satellite navigation has been widely studied [8, 9].

However, CSK uses the code phase to modulate information, and exploits the cyclic shifting of a fundamental code to get orthogonal code sequences. Therefore, CSK is not suitable for ranging. In each epoch, the receiver does not know which cyclic shifting version of fundamental PRN code is broadcasted. Moreover, the receiving complexity of CSK signal increases obviously.

In order to meet the demand of high data rate and high precision ranging for future navigation signals, this paper proposes a high data rate GNSS signal based on MPSK modulation, which can provide high data rate and the ranging capability, and the receiver complexity does not increase. The following paper is organized as follows: Sect. 2 describes the characteristics of GNSS signal based on MPSK modulation, Sect. 3 analyzes the performance of GNSS signal based on MPSK modulation, Sect. 4 studies the constant envelope multiplexing method of MPSK signal, Sect. 5 gives an application example, and finally gives the conclusion.

2 GNSS Signals Based on MPSK Modulation

In this paper, MPSK modulation signal is proposed to meet the needs of high data rate and high precision ranging ability. The goal of high data rate is to increase the data rate without increasing the code rate or bandwidth. The traditional BPSK signal cannot meet this requirement. CSK modulation meets the needs of high data rate, but it is difficult to range using CSK signal.

2.1 CSK Modulation

CSK is a direct sequence spread spectrum modulation signal, which overcomes the limitation of spread spectrum gain on data rate. It can be regarded as a M-ary orthogonal signal. In order to transmit $k = \log_2(M)$ bits, M orthogonal signal waveforms are required. Compared with the traditional M-ary orthogonal signal, the particularity of CSK modulation is that each waveform is obtained by different cyclic shifting of the same PRN code, and each cyclic shifting is an integer chip. Therefore, CSK modulates the data on the PRN code phase.

The spread spectrum code used by CSK is called the fundamental code, which is recorded as $c_d(t)$, and the code period is T_P. From the fundamental code $c_d(t)$, M cyclic shifts of the fundamental PRN code are generated, which are denoted as $c_d(t) \sim c_{d-M+1}(t)$, and the mathematical expression is [6]:

$$c_{d-x}(t) = c_d\left([t - m_x \cdot T_c]_{\mathrm{mod}(L \cdot T_c)}\right), \; x = 0, 1, \cdots, M-1 \tag{1}$$

where L is the code length of the fundamental code, T_c is the chip width, and m_x represents the chip number of cyclic shifting.

CSK modulation is generally denoted as CSK (k, n), which means that each PRN code phase represents k-bit information and lasts for n code periods, that is, the symbol width is $T_s = nT_p$. For a CSK modulation signal with a code length of L, a code phase can represent up to $\lfloor \log_2(L) \rfloor$ bits. In QZSS L6 signal, CSK (8, 1) is used.

2.2 MPSK Modulation

MPSK is a multi-ary phase-shifting keying modulation. However, it uses carrier phase to modulate data, and the carrier phase takes one of M possible values, i.e.

$$\theta_{MPSK}(t) \in \{\theta_m, \; m = 0, 1, 2, \cdots, M-1\} \tag{2}$$

Therefore, each phase symbol can modulate $k = \lfloor \log_2(M) \rfloor$ bit information.

For MPSK signals with uniform phase distribution, the phase value can be expressed as $\theta_m = \frac{2\pi \cdot m}{M}$, $m = 0, 1, 2, \cdots, M-1$.

Figure 1 shows the schematic diagram of 8PSK modulation constellation. Each modulation phase is mapped to 3-bit information. It is noted that in order to improve the demodulation performance, gray code is used in the figure, and the information of adjacent phase mapping is only 1 bit flipped [10].

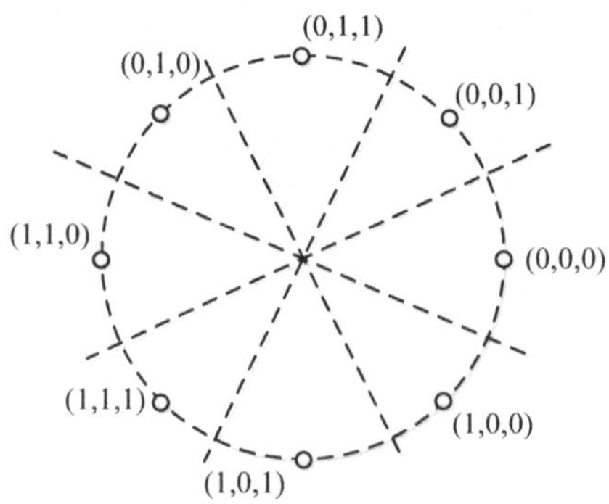

Fig. 1. 8PSK modulation constellation

2.3 GNSS Signals Based on MPSK

For the GNSS signals based on MPSK, the baseband signals are expressed as follows:

$$s_{MPSK}(t) = \sqrt{2P_{MPSK}}c_{MPSK}(t)e^{j\theta_{MPSK}(t)} \tag{3}$$

Where P_{MPSK} is the power of MPSK signal, $c_{MPSK}(t)$ is the PRN code waveform of MPSK signal, code period is T_P, $\theta_{MPSK}(t)$ is the modulation phase of MPSK signal, there are M possible phase values, the duration of each phase value is the symbol width T_s, T_s is an integral multiple of T_p, and the transformation of modulation phase is aligned with the beginning of the code waveform.

Compared with BPSK signal, the main advantage of MPSK modulation is that it can modulate more data rate while keeping the code rate and code period unchanged. For BPSK signal whose code period is T_p, the maximum information rate is $1/T_p$ when the message symbol does not cross the code period; for MPSK modulation, when $T_s = T_p$, the maximum information rate is k/T_p.

3 Performance Analysis of GNSS Signals Based on MPSK Modulation

The receiving of MPSK modulation signal is basically the same as that of BPSK signal. The baseband expression of IF signal received by the receiver is as follows:

$$s_{IF}(t) = \sqrt{2P_{MPSK}}c_{MPSK}(t-\tau)e^{j\theta_{MPSK}(t)}e^{j(2\pi f_{IF}t+\theta_0)} + N(t) \tag{4}$$

Where, f_{IF} is the carrier frequency, θ_0 is carrier initial phase, τ is propagation delay, $N(t)$ is the complex white noise. DLL is used for signal acquisition and tracking, and the receiver generates the early (E), prompt (P) and late (L) replicas of carrier and PRN code, respectively

$$\begin{aligned} &s_{ref,E}(t) = c_{MPSK}\left(t-\hat{\tau}+\frac{d}{2}\right)e^{j\left(2\pi\hat{f}_{IF}t+\hat{\theta}_0\right)};\ s_{ref,P}(t) = c_{MPSK}(t-\hat{\tau})e^{j\left(2\pi\hat{f}_{IF}t+\hat{\theta}_0\right)};\\ &s_{ref,L}(t) = c_{MPSK}\left(t-\hat{\tau}-\frac{d}{2}\right)e^{j\left(2\pi\hat{f}_{IF}t+\hat{\theta}_0\right)} \end{aligned} \tag{5}$$

Where $\hat{f}_{IF}$ is the IF frequency estimation, $\hat{\theta}_0$ is the carrier initial phase estimation, $\hat{\tau}$ is the propagation delay estimation. By correlating the early, prompt and late replicas with the received MPSK signal, the following results are obtained:

$$\begin{aligned} IE + jQE &= \sqrt{2P_{MPSK}}R\left(\Delta\tau-\frac{d}{2}\right)e^{j\theta_{MPSK}(t)}e^{j\Delta\theta} + N_E(t),\\ IP + jQP &= \sqrt{2P_{MPSK}}R(\Delta\tau)e^{j\theta_{MPSK}(t)}e^{j\Delta\theta} + N_P(t)\\ IL + jQL &= \sqrt{2P_{MPSK}}R\left(\Delta\tau+\frac{d}{2}\right)e^{j\theta_{MPSK}(t)}e^{j\Delta\theta} + N_L(t) \end{aligned} \tag{6}$$

Where, it is assumed that $\hat{f}_{IF} \approx f_{IF}$. $\Delta\tau = \hat{\tau} - \tau$ is the propagation delay estimation error, $\Delta\theta = \theta_0 - \hat{\theta}_0$ is the carrier phase estimation error, $N_E(t)$, $N_P(t)$ and $N_L(t)$ are the noise terms.

3.1 Tracking Performance Analysis

The code tracking process of MPSK signal is similar to that of BPSK signal, the main difference is the carrier tracking discrimination loop.

The noncoherent code discrimination function is

$$D(\Delta\tau) = \left(IE^2 + QE^2\right) - \left(IL^2 + QL^2\right) = 2P_{MPSK}\left(R^2\left(\Delta\tau - \frac{d}{2}\right) - R^2\left(\Delta\tau + \frac{d}{2}\right)\right) \quad (7)$$

It can be seen that MPSK signal has the same code tracking performance as BPSK signal.

Because the MPSK signal has M phase points, when tracking the carrier phase, the first step is to pre determine the phase points. Of course, the phase points are fuzzy and need to be corrected by the message preamble. At this time, the carrier tracking discrimination function can be written as:

$$D_{carrier} = \text{atan2}(QP, IP) - \hat{\theta}_{MPSK} \quad (8)$$

Where, $\text{atan2}(\cdot,\cdot)$ is the four quadrant inverse tangent function, and

$$\hat{\theta}_{MPSK} = \min|\theta_m - \text{atan2}(QP, IP)|, m = 0, 1, 2, \cdots, M-1 \quad (9)$$

According to the experience threshold, for BPSK signal with modulated data, the carrier tracking threshold is that the 1σ tracking error value is less than 15° [11]. For MPSK signal, in order to ensure the correctness of the pre decision symbol, the carrier tracking error cannot exceed $\pm\pi/M$. According to the thumb rule, the 1σ tracking error value is less than $\pi/(3M)$, obviously, leading to the decrease of the carrier tracking threshold. Figure 2 shows the code tracking error curve under thermal noise, with loop bandwidth of 18 Hz and integration time of 1 ms. It can be seen that for 8PSK, under this condition, the carrier tracking performance decreases by 4.4 dB.

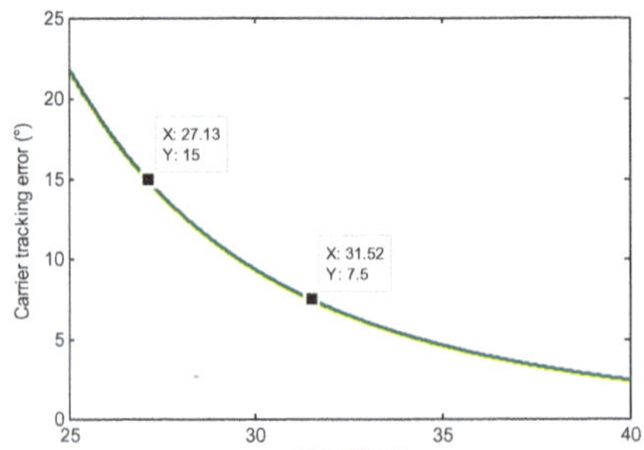

Fig. 2. Carrier tracking error curve

3.2 Demodulation Performance Analysis

There are some approximate formulas for evaluating the BER performance of MPSK signals. For MPSK signals with uniform phase distribution, the analytical expression can be analyzed [12].

Taking 8PSK shown in Fig. 1 as an example, it is assumed that for ideal carrier phase tracking, the probability of broadcasting eight phases is the same, and the bit error rate is the same. Without losing generality, the bit sequence (0, 0, 0) is broadcasted. The probability of correct decision is:

$$P_0 = P_{(0,0,0)\to(0,0,0)} = \frac{1}{\sqrt{2\pi}} \int_0^{\infty} e^{-\frac{\left(x - \sqrt{\frac{2CT_s}{N_0}} \cos\frac{\pi}{8}\right)^2}{2}} \left(1 - 2Q\left(x \cdot \tan\frac{\pi}{8}\right)\right) dx \tag{10}$$

If (0, 0, 0) is judged as (0, 0, 1) or (1, 0, 0), the probability of error 1 bit is:

$$P_1 = P_{(0,0,0)\to(0,0,1)} = P_{(0,0,0)\to(1,0,0)} = \frac{1}{\sqrt{2\pi}} \int_0^{\infty} e^{-\frac{\left(x - \sqrt{\frac{2CT_s}{N_0}}\right)^2}{2}} \begin{pmatrix} Q\left(x \cdot \tan\frac{\pi}{8}\right) \\ -Q\left(x \cdot \tan\frac{3\pi}{8}\right) \end{pmatrix} dx \tag{11}$$

The probability of judging (0, 0, 0) as (0, 1, 1) or (1, 0, 1) is as follows:

$$P_2 = P_{(0,0,0)\to(0,1,1)} = P_{(0,0,0)\to(1,0,1)} = \int_0^{\infty} \frac{1}{\sqrt{2\pi}} e^{-\left(x^2 + \frac{2CT_s}{N_0}\right)} Q\left(x \cdot \tan\frac{3\pi}{8}\right) dx \tag{12}$$

The probability of judging (0, 0, 0) as (1, 1, 1), error 3 bits and (0, 0, 0) as (0, 1, 0) error 1 bits are as follows:

$$P_3 = P_{(0,0,0)\to(0,1,0)} = P_{(0,0,0)\to(1,1,1)} = \int_{-\infty}^{0} \frac{1}{\sqrt{2\pi}} e^{-\frac{\left(x - \sqrt{\frac{2CT_s}{N_0}}\right)^2}{2}} \begin{pmatrix} Q\left(x \cdot \tan\frac{3\pi}{8}\right) \\ -Q\left(x \cdot \tan\frac{\pi}{8}\right) \end{pmatrix} dx \tag{13}$$

If (0, 0, 0) is determined as (1, 1, 0), the probability of error 2 bits is:

$$P_4 = P_{(0,0,0)\to(1,1,0)} = \int_0^{\infty} \frac{1}{\sqrt{2\pi}} e^{-\frac{\left(x + \sqrt{\frac{2CT_s}{N_0}}\right)^2}{2}} \left(1 - 2Q\left(x \cdot \tan\frac{\pi}{8}\right)\right) dx \tag{14}$$

Regardless of channel coding, BER of 8PSK signal is:

$$BER = \frac{1}{3}(2P_1 + 2 \cdot 2P_2 + P_3 + 3P_3 + 2P_4) \tag{15}$$

Figure 3 shows BER performance of 8PSK signal. As a comparison, BER performance of BPSK signal and CSK (3, 1) signal is given. For 8PSK and CSK (3, 1), $T_s = 1$ ms, and information rate is 3 kbps.

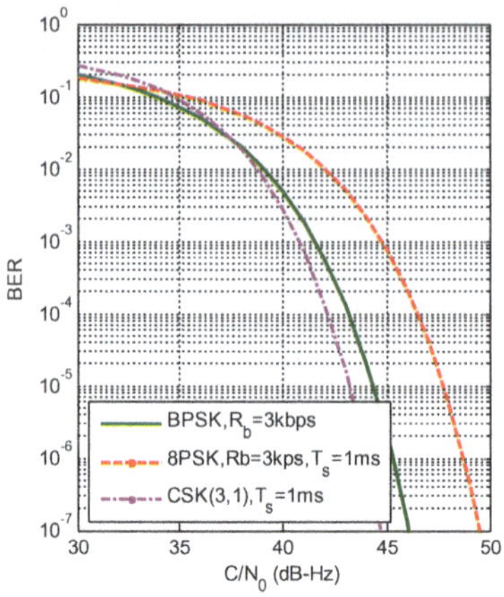

Fig. 3. BER performance

3.3 CSK vs. MPSK

Both MPSK and CSK have the ability of high data rate. Compared with CSK, the demodulation threshold of MPSK is increased by about 4 dB. However, MPSK signal has the ability of independent tracking and ranging. The receiving technology is mature, and the receiving complexity is similar to BPSK signal, significantly lower than CSK signal.

4 Constant Envelope Multiplexing Method of MPSK Signal

In GNSS, constant envelope multiplexing technology is usually used to combine multiple signal components into a composite signal.

The following case is analysed in this section. There are r signal components in total. When the first $r - 1$ signal component is BPSK and BOC modulation signal, the last one is MPSK modulation. The power of r signal components is P_r respectively, the power and phase of the first $r - 1$ signal components are fixed, while the phase of MPSK signal components is determined by the data bits, there are M possible combinations.

By modifying POCET method, constant envelope multiplexing method can be developed. The final constant envelope multiplexing signal can be generated based on a lookup table, as shown in Table 1. Table 1 shows the case of four signal components. The fourth signal component is MPSK modulation. The final phase lookup table (LUT) can be regarded as a two-dimensional lookup table.

Table 1. Phase look-up table

			$s_1(t)$	1	1	1	1	1	1	1	1	-1	-1	-1	-1	-1	-1	-1	-1
			$s_2(t)$	1	1	1	1	-1	-1	-1	-1	1	1	1	1	-1	-1	-1	-1
			$s_3(t)$	1	1	-1	-1	1	1	-1	-1	1	1	-1	-1	1	1	-1	-1
			$c_4(t)$	1	-1	1	-1	1	-1	1	-1	1	-1	1	-1	1	-1	1	-1
d$_4(t)$			Phase state $\theta_{CEM}(t)$, $s_{CEM}(t) = A\exp\left(j\theta_{CEM}(t)\right)$																
0	0	0		28.7757	90.4454	245.6444	69.2144	336.9673	29.7594	283.5916	320.3297	140.3297	103.5916	209.7594	156.9673	249.2144	65.6444	270.4454	208.7757
0	0	1		32.3135	88.2397	292.0053	116.0613	334.6867	28.6819	324.7723	281.2433	101.2433	144.7723	208.6819	154.6867	296.0613	112.0053	268.2397	212.3135
0	1	1		35.1184	91.0446	7.2968	191.3529	338.5858	22.1148	328.6714	274.6762	94.6762	148.6714	202.1148	158.5858	11.3529	187.2968	271.0446	215.1184
0	1	0		32.9127	94.5824	54.1437	237.7137	19.7665	343.0284	326.3909	273.5987	93.5987	146.3909	163.0284	199.7665	57.7137	234.1437	274.5824	212.9127
1	1	0		90.4453	28.7757	69.2145	245.6445	29.7595	336.9672	320.3297	283.5917	103.5917	140.3297	156.9672	209.7595	65.6445	249.2145	208.7757	270.4453
1	1	1		88.2397	32.3135	116.0613	292.0053	28.6819	334.6867	281.2433	324.7723	144.7723	101.2433	154.6867	208.6819	112.0053	296.0613	212.3135	268.2397
1	0	1		91.0446	35.1184	191.3529	7.2968	22.1148	338.5858	274.6762	328.6714	148.6714	94.6762	158.5858	202.1148	187.2968	11.3529	215.1184	271.0446
1	0	0		94.5824	32.9127	237.7137	54.1437	343.0284	19.7665	273.5987	326.3909	146.3909	93.5987	199.7665	163.0284	234.1437	57.7137	212.9127	274.5824

The phase value of each line is symmetrical, and the average correlation value between r signal components and the phase value of each line is:

$$corr_{r,m} = \frac{A}{2^N} \sum_{k=0}^{2^r-1} b_r(k) e^{j\theta_{m,k}} \tag{16}$$

where $b_r(k) = \pm 1$ is the value of the r-th signal component in the k-th column of the phase LUT.

Power constraints are expressed as:

$$P_r = \left|corr_{r,m}\right|^2 = \left|\frac{A}{2^N} \sum_{k=0}^{2^r-1} b_r(k) e^{j\theta_{m,k}}\right|^2, m = 0, 1, 2, \cdots, M-1 \tag{17}$$

Phase constraints is:

$$\begin{aligned} &\mathrm{Im}\left\{e^{-j\Delta\phi_{rm,qp}} corr_{r,m} \cdot corr_{q,p}^{*}\right\} = 0 \\ &\mathrm{Re}\left\{e^{-j\Delta\phi_{rm,qp}} corr_{r,\mathrm{m}} \cdot corr_{q,p}^{*}\right\} > 0 \end{aligned} \tag{18}$$

When $r = q$ and is not an MPSK signal component, $\Delta\phi_{rm,qp} = 0$. when $r = q$ and is an MPSK signal component, $\Delta\phi_{rm,qp}$ is determined by the mapping phase of the corresponding line. When $r \neq q$, $\Delta\phi_{rm,qp}$ can be obtained similarly.

Under the power and phase constraints, A^2 is minimized and the multiplexing efficiency is the highest.

5 Application Example

In this section, consider the following example: there are four binary signal components with equal power. Without considering phase constraints, the best result of constant envelope multiplexing of four signals is DualQPSK [13].

According to baseband expression of DualQPSK [13], it can be seen that the phases of the four signal components are 0°, 90°, 45° and 135° respectively, and the multiplexing efficiency is 85.36%.

Now, the fourth signal component is modulated with 8PSK, and denoted by $c_4(t)$, and the proposed method described above is used to obtain the phase lookup table, as shown in Table 1. At this time, the multiplexing efficiency of constant envelope multiplexing becomes 81.20%.

The fourth signal component adopts 8PSK modulation, and each phase symbol represents 3 bits. The phase of the first three signal components remains unchanged, which are 0°, 123.3581° and 61.6790°, respectively. The phase of the fourth signal component changes with the combination of data bits, uniformly distributed on the unit circle. For the eight different data combinations, the phases of the four signal components are −95.8210°, −50.8209°, −5.8209°, 39.1791°, 84.1791°, 129.1791°, 174.1791°, −140.8209°, respectively.

6 Conclusions

In this paper, a GNSS signal based on MPSK modulation is proposed to meet the demand of high data rate and high precision ranging for future GNSS signal. Compared with the traditional BPSK signal, MPSK signal can improve the information rate without changing the code rate and symbol width. The code tracking performance is the same as BPSK, but the carrier tracking performance has degradation. Compared with CSK modulation, although the demodulation performance will be reduced, MPSK can provide high information rate and high precision ranging capability, and the receiving complexity is similar to BPSK. Therefore, the GNSS signal based on MPSK modulation provides another choice for the future high data rate GNSS signal.

References

1. Hernandez, I.F., Rodriguex, I., Tobias, G., Calle, J.D., Carbonell, E., Seco-Granados, G., Simon, J., Blasi, R.: Galileo's commercial service testing GNSS high accuracy and authentication. InsideGNSS, pp. 38–48, January/February 2015
2. Aubault-Roudier, M., Laurichesse, D., Bitar, H.A., Raimondi, M., Lesage, P., Sfeir, A., Klein, M., Sihrener, M., Ramponi, N.: A demonstration of the Galileo E5b signal with real-time PPP information broadcast via the SES Astra 5B GEO satellite. InsideGNSS, pp. 58–65, January/February 2018
3. Quasi-Zenith Satellite System Interface Specification Centimeter Level Augmentation Service (IS-QZSS-L6-001). Cabinet Office (2018)
4. Garcia-Peña, A., Julien, O., Anghileri, M., Floch, J.J., Paonni, M.: Multi-purpose TDM component for GNSS. In: ION GNSS+ 2018, pp. 943–962 (2018)
5. Anghileri, M., Floch, J.J., Margaria, D., Motella, B., Garcia-Peña, A., Julien, O., Macabiau, C., Chauvat, R., Paonni, M.: FUNTIMES: future navigation and timing evolved signals. In: ION GNSS+ 2018 (2018)
6. Garcia-Peña, A., Salos, D., Julien, O., Ries, L., Grelier, T.: Analysis of the use of CSK for future GNSS Signals. In: ION GNSS+ 2013, pp. 1461–1479 (2013)
7. Garcia-Peña, A., Roudier, M. A., Ries, L., Boucheret, M.L., Poulliat, C., Julien, O.: Code shift keying prospects for improving GNSS signal designs. InsideGNSS, pp. 52–62, November/December 2015
8. Garcia-Peña, A., Paimblanc, P., Julien, O., Ries, L., Grelier, T.: Analysis of different CSK configurations in an urban environment when using non-coherent demodulation. In: NAVITEC 2014: 7th ESA Workshop on Satellite Navigation Technologies, Noordwijk, Netherlands (2014)
9. Andreotti, R., Emmanuele, A., Fontanella, D., Zanier, F., Luise, M.: Code-Shift-Keying (CSK) with advanced FEC coding for GNSS applications in satellite multipath channel. In: ION/IEEE PLANS 2014 (2014)
10. Lassing, J., Strom, E.G., Agrell, E., Ottosson, T.: Computation of the exact bit error rate of coherent M-ary PSK with Gray code bit mapping. IEEE Trans. Commun. **51**(11), 1758–1760 (2003)

11. Kaplan, E.D., Hegarty, C.J.: Understanding GPS: Principles and Applications, 2nd edn. Artech House Publishers, Norwood (2005)
12. Jeong, W., Lee, J., Yoon, D.: New BER Expression of MPSK. IEEE Tran. Aerosp. Electron. Syst. **60**(4), 1916–1923 (2011)
13. Zhang, K.: Generalised constant-envelope DualQPSK and AltBOC modulations for modern GNSS signals. Electron. Lett. **49**(21), 1335–1337 (2013)

The Base Stations' Networking Scheme and Spreading Code Optimization Strategy of TC-OFDM

Ke Han, Jiabing Yin(✉), Zhongliang Deng, Jieshu Dong, Shihao Tang, and Zhiyuan Ma

School of Electronic Engineering, Beijing University of Posts and Telecommunications, Beijing, China
1446479351@qq.com

Abstract. In order to achieve large-scale positioning by the ground positioning base station network, the mode of co-address and co-frequency broadcasting between the positioning base station and the existing communication base station can be considered. Under this circumstance, in order to meet the need of ultra-dense networking of integrated communication and navigation system, an optimized cellular positioning networking scheme is proposed in this paper. In the outdoor complex transmission environment, the boundary conditions that the proposed cellular network model needs to meet are analysed for the TC-OFDM system. Besides, based on the greedy algorithm of quadratic optimization, the spreading codes with better performance are selected for the reuse of cellular network modules. Theoretical analysis and simulation results show that the optimized cellular network model can well meet the large-scale positioning requirements of integrated communication and navigation system, and the selected spreading code has better performance.

Keywords: TC-OFDM · Cellular network model · Spreading code · The greedy algorithm of quadratic optimization

1 Introduction

With the continuous development of communication and Internet technology, location services based on various positioning technologies have been widely introduced into people's life. In the real environment, the Global Navigation Satellite System (GNSS) is subject to blockages and reflections from the building, which can easily cause the positioning accuracy to decrease or even fail to locate. However, the ground base station network has good indoor and outdoor signal coverage capability due to its good reliability, anti-interference and network flexibility [1]. The Time & Code Division-Orthogonal Frequency Division Multiplexing (TC-OFDM) system discussed in this paper is a positioning system that can realize communication and navigation integration, and can achieve high-precision indoor and outdoor seamless positioning [2]. Nevertheless, if TC-OFDM positioning system wants to achieve large-scale outdoor positioning, it is necessary to consider not only the interference of communication signals and outdoor complex environments, but also the reuse of spreading codes

J. Sun et al. (Eds.): CSNC 2020, LNEE 650, pp. 485–496, 2020.
https://doi.org/10.1007/978-981-15-3707-3_46

between base stations. Therefore, it is particularly important to study the network scheme of base stations and the optimization of spreading codes.

Liu et al. studied the effect of number and layout of pseudo satellites on system positioning performance [3]. Jian et al. analysed four positioning base station layout schemes, and studied the impact of factors such as the area scope, number and height of positioning base stations on geometric dilution of precision (GDOP) [4]. These algorithms have well analysed the relationship between the layout of base stations and the positioning performance in the local area, however, the influence of communication signal interference and inter-base station signal interference is not taken into account. Moreover, for the reuse and optimization of spreading codes, Wang et al. proposed a spreading code optimization algorithm based on the greedy algorithm [5], but the time complexity is large, and the result is hard to converge when the number of codes is large.

Aiming at the large-scale positioning networking in the outdoor scenario of TC-OFDM system, this paper applies the cellular node networking method, and analyses various conditions that the cellular network model needs to meet, in addition, the spreading codes with better performance can be obtained based on the greedy algorithm of quadratic optimization.

2 Theoretical Analysis

2.1 Characteristics of Spreading Codes

Spreading code is an important part of code division multiplexing system and is widely used in navigation and positioning system. $C_{TC-OFDM}$ is the spreading code of the positioning signal of the TC-OFDM system, and the code length is 10230. $C_{TC-OFDM}$ is generated by inserting a fixed-length information sequence $[\ 0 \quad 1 \quad 1 \quad 0 \quad 1 \quad 0 \quad 0]$ into the Weil code of length 10223. The generation structure diagram of $C_{TC-OFDM}$ is shown in Fig. 1.

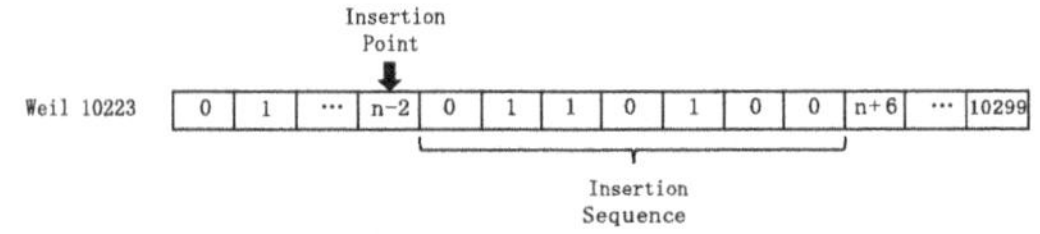

Fig. 1. Code generation structure of $C_{TC-OFDM}$

The balance of spreading code refers to the ratio of the absolute value of the difference between the number of '0' and '1' in the code to the total length of the spreading code sequence. The spreading code sequence with better balance has stronger carrier suppression ability, which can better prevent carrier leakage [6].

The correlation of the spreading code directly affects the accuracy of the navigation and positioning results, so the correlation is one of the most important indicators for the selection of the spreading code. The autocorrelation expression of the spreading code is defined as:

$$AC^{e,o}(l, f_{offs}) = \frac{1}{N}\sum_{k=0}^{N-1}(-1)^{\tau}a_{k,1}a_{k-l,1}e^{2\pi j\frac{f_{offs}}{f_s}(k-1)} \tag{1}$$

The cross-correlation expression is defined as:

$$CC^{e,o}(l, f_{offs}) = \frac{1}{N}\sum_{k=0}^{N-1}(-1)^{\tau}a_{k,1}a_{k-l,2}e^{2\pi j\frac{f_{offs}}{f_s}(k-1)} \tag{2}$$

Where N is the number of cycles, f_c denotes the code rate, and f_{offs} represents the doppler frequency shift. $a_{k,1}$ and $a_{k-l,1}$ are the original chip and the shifted chip respectively. The subscripts 1 and 2 represent code 1 and code 2. The superscript e and o represent even correlation and odd correlation respectively, the value of τ is 0 when even correlation, and the value of τ is $\frac{sign(l-k)+1}{2}$ when odd correlation.

We can measure the performance of the spreading code by the sidelobe energy, and the correlated sidelobe energy can be expressed as:

$$E_b = 20\log_{10}(\frac{E_{corr}}{length^2}) \tag{3}$$

Where $length$ represents the length of the spreading code sequence, and E_{corr} represents the maximum value of the autocorrelation sidelobe or the peak value of the cross-correlation. If the correlated sidelobe energy is larger, the ranging performance will be worse, and vice versa [7].

2.2 Okumura-Hata Model

Okumura-Hata model is a widely used channel transmission loss model. The applicable frequency range is 150 M–1500 MHz, the effective height of the base station antenna is 30 m–200 m, and the height of the receiver antenna is 0–1.5 m [8].

The empirical formula of Okumura-Hata model obtained through data statistics is as follows:

$$\begin{aligned} L_p(dB) = {} & 69.55 + 26.16\lg f_c - 13.82\lg h_{te} - \alpha(h_{re}) \\ & + (44.9 - 6.55\lg h_{re})\lg d + C_{cell} + C_{terrain} \end{aligned} \tag{4}$$

Where $f_c(MHz)$ represents the carrier frequency, $h_{te}(m)$ represents the effective height of the base station antenna, $h_{re}(m)$ represents the height of the mobile station antenna, $d(km)$ represents the distance between the transmitting and receiving antennas. And $\alpha(h_{re})$ represents the antenna height factor of the mobile station antenna, as defined below:

$$\alpha(h_{re}) = \begin{cases} (1.11 - \lg f_c - 0.7)h_{re} - (1.56\lg f_c - 0.8) \\ \quad (Small\ and\ medium - sized\ cities) \\ 8.29(\lg 1.54h_{re})^2 - 1.1 \\ \quad (big\ cities,\ suburbs,\ countryside \quad f_c \leq 300\,\text{MHz}) \\ 3.2(\lg 11.75h_{re})^2 - 4.97 \\ \quad (big\ cities,\ suburbs,\ countryside \quad f_c \geq 300\,\text{MHz}) \end{cases} \tag{5}$$

C_{cell} is the correction factor, specifically defined as follows:

$$C_{cell} = \begin{cases} 0 \quad (urban) \\ -2[\lg(f_c/28)]^2 - 5.4 \quad (suburb) \\ -4.78(\lg f_c)^2 + 18.33\lg f_c - 40.98 \quad (countryside) \end{cases} \tag{6}$$

$C_{terrain}$ is the terrain correction factor, which reflects the influence of some important terrain environment on transmission loss. The terrain correction factor can be artificially set by testing [9].

3 System Analysis

3.1 Cellular Networking Model

In the integrated communication and navigation positioning system, the base station layout needs to face the following four problems: I. The reuse of spreading code. II. Interference between communication signals and positioning signals. III. The basic conditions for positioning. IV. The autocorrelation interference between different base stations using the same spreading code. Therefore, we establish outdoor cellular network model and discuss these issues respectively.

I. It can be known from the generation method of the spreading codes that the number of spreading codes is limited, and the selected spreading code group with good comprehensive performance only accounts for a part of the total number of spreading codes. In order to solve the need of large-scale positioning networking, it is necessary to reuse the spreading code. Therefore, a modular design of the cellular network is required. In the design, different base stations use different spreading codes in each module, and the same spreading code group is reused between different modules. The schematic diagram of each module is shown in Fig. 2.

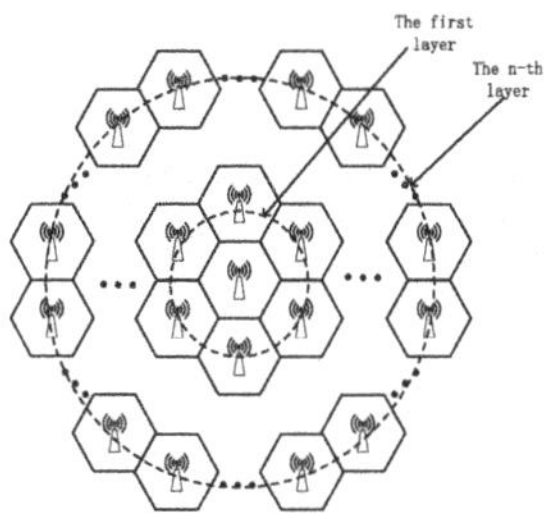

Fig. 2. Schematic diagram of cellular network module

As shown in the Fig. 2, n is the number of layers of the cellular network, and the side length of the hexagon in cellular network is set as r. From the geometric knowledge, the distance from the central base station to the first layer base station is $\sqrt{3}r$ and the distance from the central base station to the n-th layer base station is $\sqrt{3}nr$.

II. In the integrated communication and navigation system, the communication base station and the positioning base station are located at the same address, so it is necessary to consider the interference between the communication signal and the positioning signal. Assume that the absolute value of the communication signal power is P_c and the unit is dBm. Assume that the absolute value of the positioning signal power is P_s, and the unit is dBm. Because the spreading gain G_ρ of the spreading code is approximately equal to the ratio of the data code width T_D to the pseudo code width T_C, that is:

$$G_\rho = T_D/T_C \approx 43\,\text{dB} \tag{7}$$

Assume that the Signal to Noise Ratio (SNR) threshold that satisfies the positioning performance is SNR_s. If the positioning receiver wants to demodulate the positioning signal from the communication signal, it needs to meet:

$$P_s > P_c - G_\rho + SNR_s \tag{8}$$

Assume that the SNR threshold that satisfies the requirements of communication performance is SNR_c. To make the communication signal undisturbed, it needs to meet:

$$P_s < P_c - SNR_c \tag{9}$$

From Eq. (8), the lower bound of the power of the positioning base station is obtained, set as P_1; from Eq. (9), the upper bound of the power of the positioning base station is obtained, set as P_2. When $P_1 < P_s < P_2$, communication performance and positioning performance can be satisfied at the same time.

III. Since TC-OFDM system uses the principle of Time Difference of Arrival (TDOA) for positioning, it is necessary to ensure that the receiver can receive signals from at least four positioning base stations at any position in the area of cellular network. Assume that the receiver's reception threshold is Th. Because the distance between the adjacent base stations is $d_1 = \sqrt{3}r$, according to the empirical formula of the Okumura-Hata model Eq. (4), when the carrier frequency is f_C, the base station antenna is $h_b(m)$, the receiver antenna height is $h_m(m)$, and the transmission distance is $\sqrt{3}r$, The path transmission loss is expressed as $L(d = \sqrt{3}r)$ dB, which needs to meet:

$$P_s - L(d = \sqrt{3}r) \geq Th \tag{10}$$

According to Eq. (10), the upper bound of r can be obtained, which is set as r_1. When $r < r_1$ is satisfied, it can receive at least 4 base station signals at any position in the area of cellular network.

IV. When the number of cellular network layers is n, it is obtained from the geometric knowledge that the total number of base stations N in each module is:

$$N = 1 + 6 + 12 + \cdots + 6n = 1 + 3n(n+1) \tag{11}$$

The base stations are numbered $1, 2, \cdots, N-1$ in the same relative order in each module, and the same number of base stations in different modules reuse the same spreading code. From geometric knowledge, the maximum distance d_2 between adjacent modules of the same number satisfies $d_2 \leq 2\sqrt{3}nr + \sqrt{3}r$. Due to the strong autocorrelation characteristics of spreading code, it is necessary to ensure that the signals from the base stations with the same number of different modules cannot be received simultaneously at any position of the cellular network, so as to avoid the autocorrelation interference of spreading codes. According to the empirical formula of the Okumura-Hata model Eq. (4), it can be found that the path transmission loss is expressed as $L(d = (2\sqrt{3}nr + \sqrt{3}r))$ dB when the transmission distance is $\frac{2\sqrt{3}nr + \sqrt{3}r}{2}$, and it needs to meet:

$$P_s - L(d = \frac{(2n+1)\sqrt{3}r}{2}) < Th \tag{12}$$

According to Eq. (12), the lower bound of r can be obtained, which is set as r_2. When $r > r_2$ is satisfied, the receiver will not receive the same spreading code signal from different base stations at any position in the area of cellular network. In summary, we design cellular network using this modular method to reuse spreading codes. When the base station power satisfies Eqs. (8) and (9), the positioning performance and communication performance can be met simultaneously. When the side length of the hexagon in cellular network satisfies Eqs. (10) and (12), it can be guaranteed that at least four positioning base station signals are received and no autocorrelation interference exists at any position in the area.

3.2 Selection Strategy of Spreading Codes Based on Greedy Algorithm of Quadratic Optimization

As can be seen from the previous section, different modules of the cellular network reuse the same N spreading codes. Therefore, a number of N spreading code sets needs to be selected from the M candidate codes, and the upper bound of the algorithm complexity is C_M^N.

The balance, autocorrelation, and cross-correlation of the spreading code are taken as the preferred indexes of the spreading code. In the screening of spreading codes, since the complexity of the balance screening and the autocorrelation screening is linear, the most complex is the cross-correlation screening [10]. Thus, we focus on the algorithms of cross-correlation screening.

The cross-correlation matrix R is defined as Eq. (13), and the number of rows and columns correspond to code numbers. In addition, the cross-correlation peak of the two spreading codes are at the intersection of the rows and the columns.

$$R = \begin{Bmatrix} 0 & R_{1,2} & R_{1,3} & \cdots & R_{1,n} \\ R_{2,1} & 0 & R_{2,3} & \cdots & R_{2,n} \\ R_{3,1} & R_{3,2} & 0 & \cdots & R_{3,n} \\ \cdots & \cdots & \cdots & 0 & \cdots \\ R_{n,1} & R_{n,2} & R_{n,3} & \cdots & 0 \end{Bmatrix} \tag{13}$$

Use the maximum value in the cross-correlation matrix as the cross-correlation optimization criterion for the spreading code group:

$$Metric(a) = \max(R_{i,j}) \tag{14}$$

Where $i, j = 1, \ldots, n$ and $i \neq j$. Cross-correlation screening can be regarded as the largest cluster problem. We need to find a local optimal solution with a number of N and all cross-correlation peaks satisfying the conditions. The steps of screening the spreading code based on the greedy algorithm of quadratic optimization are as follows:

Step 1: Set the balance threshold, autocorrelation sidelobe peak threshold, and cross-correlation peak threshold, and put them into the new code set S that meet the conditions of the balance and autocorrelation threshold.
Step 2: Firstly, randomly select an initial code group Q with a number of N from S, and set the iteration factor to b_n.
Step 3: Randomly select one of the remaining codes from S to join the code group Q, and set the cross-correlation matrix to R. The greedy algorithm is used to optimize the objective function $Metric(a)$, find the maximum value in the current cross-correlation matrix R, and obtain its row number X and column number Y. Suppose that X rows and X columns are deleted from R, and the maximum value of the matrix after deletion is $R_{i1,j1}$. Suppose that Y rows and Y columns are deleted from R, and the maximum value of the matrix after deletion is $R_{i2,j2}$. Based on the greedy algorithm, if $R_{i1,j1} \geq R_{i2,j2}$, delete Y rows and Y columns from R as the

current matrix, otherwise delete X rows and X columns from R as the current matrix. Meanwhile, update the objective function $Metric(a)$.

Step 4: Check the values in the current matrix. If the cross-correlation peaks updated by step 3 are all less than the threshold, then go to step 6; otherwise, continue to iterate step 3.

Step 5: If step 3 iterates b_n times and it is not over yet, perform a quadratic optimization on the existing code group Q: select a qualified code set whose cross-correlation peaks with other spreading codes in Q are all less than the threshold. Remove some absolutely impossible codes, and continue to select the corresponding number of codes from S to join Q, and turn to step 3.

Step 6: Output optimized code group with a number of N and satisfying the conditions.

The spreading code selection process based on greedy algorithm of quadratic optimization is shown in Fig. 3. Compared with the traditional greedy algorithm, the current result set is quadratic optimized after a certain number of iterations, which makes the calculation result easier to converge.

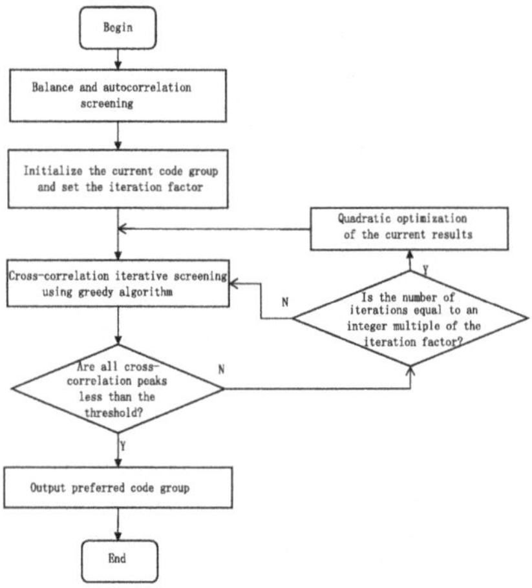

Fig. 3. Spreading code screening process based on greedy algorithm of quadratic optimization

4 Simulation and Analysis

4.1 Feasibility Analysis of Cellular Network Layout

According to the derivation in Sect. 3.1, we take the number of layers $n = 2$ of each module of the cellular network as an example for simulation analysis. The overall layout of the base station is shown in Fig. 4, where each color represents a cellular network module, and the center of each hexagon has a base station. It can be known from Fig. 4 that each module includes 19 base stations.

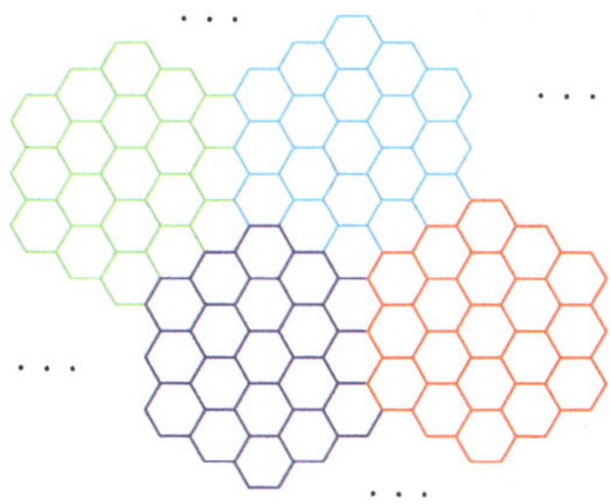

Fig. 4. Base station layout when n = 2

From Eq. (7), the spreading gain of spreading code is about 43 dB. When the spreading gain is 43 dB, the relationship between the symbol error rate (SER) and the SNR of the positioning signal is shown in Fig. 5a. Besides, the relationship between the SER and SNR of QPSK modulated communication signals under the condition of additive gaussian white noise channel is analyzed, as shown in Fig. 5b.

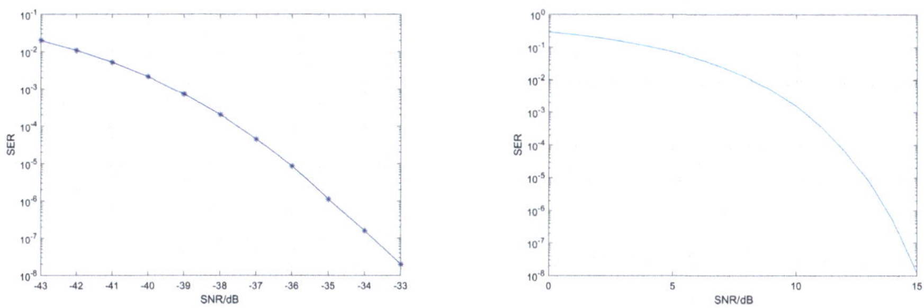

(a) The relationship between SER and SNR of positioning signal

(b) The relationship between SER and SNR of communication signal

Fig. 5. (a) The relationship between SER and SNR of positioning signal. (b) The relationship between SER and SNR of communication signal.

According to Figs. 5a and b, when the communication signal is 15 dB to 33 dB higher than the positioning signal, that is, $15\,\text{dB} < P_c - P_s < 33\,\text{dB}$, the symbol error rate of the communication signal and the positioning signal are both less than 10^{-7}. This can meet both communication and positioning needs.

Since TC-OFDM system uses the principle of TDOA, the receiver needs to receive at least 4 base station signals to achieve positioning. $\phi(r)$ is used to indicate whether the minimum number of base stations that meet the TDOA positioning conditions can be received at any location in the area, and r is the side length of hexagon in cellular network.

$$\phi(r) = \begin{cases} 1 & (All\ can\ meet\ the\ positioning\ conditions\ in\ the\ area) \\ 0 & (others) \end{cases} \tag{15}$$

$\psi(r)$ is used to indicate whether there is no autocorrelation interference at any position in the area, and r is the side length of hexagon in cellular network.

$$\psi(r) = \begin{cases} 0 & (There's\ no\ autocorrelation\ interference\ in\ the\ area) \\ 1 & (There\ exists\ autocorrelation\ interference) \end{cases} \tag{16}$$

Taking the 4G communication network in the urban environment as an example, the typical power of the communication base station is 30 W, which is 44.8 dBm. Set the base station power 30 dB lower than the communication base station, which is 14.8 dBm. In this case, the relationship between $\phi(r)$, $\psi(r)$ and r is shown in Fig. 6.

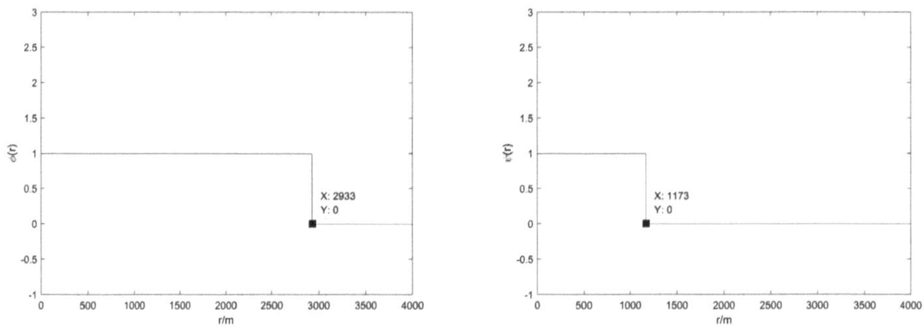

(a) Whether all can meet the positioning conditions in the area

(b) Whether there is autocorrelation interference in the area

Fig. 6. (a) Whether all can meet the positioning conditions in the area. (b) Whether there is autocorrelation interference in the area.

In this case, it can be known from Figs. 6a and b that when the side length of hexagon in cellular network meets $1173\,\text{m} < r < 2933\,\text{m}$, at least 4 base station signals can be received and there is no autocorrelation interference at any position in the area.

To sum up, the simulation results show that by reasonably adjusting the power of the positioning base station and the side length of hexagon in cellular network, the proposed model can well meet the requirements of large-scale positioning networking.

4.2 Analysis of Spreading Code Optimization Results

In order to verify the optimization algorithm of the spreading code proposed in this paper, under the 0 Hz Doppler frequency shift, 63 newly selected TC-OFDM spreading codes were compared with 63 existing spreading codes selected from GPS L1C and BDS B1C by Eq. (3). The statistical results of cross-correlation peaks of the three types are shown in the Table 1.

Table 1. Cross-correlation peaks of 63 spreading codes of three types

Type	Minimum cross-correlation peaks	Maximum cross-correlation peaks	Mean of cross-correlation peaks	Variance of cross-correlation peaks
TC-OFDM	−32.38	−30.72	−31.53	0.3863
GPS L1C	−31.99	−27.39	−30.12	0.8064
BDS B1C	−32.07	−26.72	−29.72	0.5629

From Table 1, we can see the cross-correlation peaks' range of different types of spreading codes. The comparison shows that the selected TC-OFDM spreading code has smaller cross-correlation peaks than GPS L1C and BDS B1C spreading codes, the mean value is 1.41 dB lower than GPS L1C codes, and 1.81 dB lower than BDS B1C codes, and has Smaller variance. Therefore, It can be seen that the performance of the spreading code selected by the greedy algorithm of quadratic optimization is better.

5 Summary

In this paper, the cellular networking scheme of TC-OFDM in the outdoor scenario is modeled and analyzed. In the integrated communication and navigation system, the various boundary conditions that the cellular network model needs to meet are calculated. In addition, the simulation results show that the cellular network model can well meet the requirements of large-scale positioning when the power of the base station and the side length of hexagon in cellular network are in an appropriate range. Moreover, for the problem of spreading code reuse faced by base station networking, this paper proposes greedy algorithm of quadratic optimization for cross-correlation screening of spreading codes. The performance of 63 selected TC-OFDM spreading codes is compared with GPS L1C and BDS B1C spreading codes, and the results show that the proposed screening algorithm is feasible.

Acknowledgements. This work is financially supported by National Key Research & Development Program under Grant No. 2016YFB0502003.

References

1. Liu, W., Bian, X., Deng, Z., Mo, J., Jia, B.: A novel carrier loop algorithm based on maximum likelihood estimation (MLE) and Kalman filter (KF) for weak TC-OFDM signals. Sensors **18**(7), 2256 (2018)
2. Hu, E., Deng, Z., Xu, Q., Yin, L., Liu, W.: Relative entropy-based Kalman filter for seamless indoor/outdoor multi-source fusion positioning with INS/TC-OFDM/GNSS. Cluster Comput. **22**(3), 8351–8361 (2018)
3. Teng, Y., Wang, J., Huang, Q.: Minimum of geometric dilution of precision (GDOP) for five satellites with dual-GNSS constellations. Adv. Space Res. **56**(2), 229–236 (2015)
4. Chuanyang, W., Jian, W.: Study of base station layout of ultra wideband emergency positioning. Sci. Surv. Mapp. (2019)

5. Wang, J., Wang, Y., Yang, Z.: A proposed selection and allocation of spread-spectrum code for compass B1C signal. In: 5th China Satellite Navigation Conference (CSNC) (2014)
6. Rushanan, J.J.: Weil sequences: a family of binary sequences with good correlation properties. In: 2006 IEEE International Symposium on Information Theory. IEEE (2006)
7. Jian-Feng, Z., Jian-Ping, A.N., Ai-Hua, W.: A flexible-length spreading code generator based on iterative interleaving. J. Navig. Position. (2015)
8. Singh, Y.: Comparison of Okumura, Hata and cost-231 models on the basis of path loss and signal strength. Int. J. Comput. Appl. **59**(11), 37–41 (2012)
9. Dube, R.R., Dhanashetti, A.G.: Analysis of wimax connectivity in rural and urban area using Okumura-Hata propagation model. Int. J. Innov. Res. Dev. **1**(1), 439–443 (2014)
10. Wang, J., Wang, Y., Yang, Z.: A proposed selection and allocation of spread-spectrum code for compass B1C signal (2014)

Frame Synchronization Method for BDS B2a Signal Under the Constraint of Non-binary LDPC Code

Bowen Jiang(✉), Lin Chen, Hongchen Pan, Yuqi Liu, Jiangyu Chen, and Shaobin Guo

The 29th Research Institute of China Electronics Technology Group Corporation, Chengdu, China
jawbone@buaa.edu.cn

Abstract. Low-Density Parity-Check (LDPC) Code has become one of the best coding technologies at present because of its low complexity iterative decoding algorithm and its performance with further close to the Shannon limit. Non-binary LDPC codes designed in high order Galois fields have shown better potential than binary LDPC codes with the same code length and code rate, therefore it was applied for the navigation message encoding in BeiDou satellite navigation system (BDS). The navigation message contains time information, ephemeris parameters, clock correction terms and so forth, which is necessary for navigation solution. As the synchronization head was not encoded, it is hard to deal with data synchronization and navigation data decoding when the navigation receiver works at a low Signal-to-Noise Ratio (SNR). Aiming at this problem, we present a synchronization scheme for BDS B2a signal by utilizing the information from the process of decoding. The synchronizer is based on syndrome checks using the decoding result of the Extended Min-Sum (EMS) algorithm. Ratios of satisfaction of all check restrictions are computed firstly, and then the boundary value is obtained through MonteCarlo method. Simulation results show that the proposed algorithm performs well at relatively low SNR compared with traditional method.

Keywords: Non-binary LDPC codes · Frame synchronization · Navigation message · Low SNR · Extended min-sum algorithm

1 Introduction

LDPC code was first proposed by Gallager in 1962, which is proved to have performance close to Shannon limits when the code length is long enough [1]. In the case of small or moderate codeword length, Non-binary LDPC code has better decoding performance than that for binary LDPC code [2], which has been applied to coding the B2a signal of BD navigation system. Frame synchronization is a necessary step for decoding navigation message correctly. However, in the case of low SNR, it is difficult to achieve frame synchronization quickly and reliably by only using the synchronous head information. In references [3, 4], a blind frame synchronization method using binary LDPC code is proposed. After the hard decision on the received symbol, the

J. Sun et al. (Eds.): CSNC 2020, LNEE 650, pp. 497–505, 2020.
https://doi.org/10.1007/978-981-15-3707-3_47

proportion that satisfies the parity-check equation of LDPC code was used to determine whether the frame synchronization is successful. This method does not take advantage of the coding gain of LDPC code and thus the performance is decreased obviously. In references [5–7], soft decoding information is used for frame synchronization, which shows a better performance than hard decision. However, it needs to traverse through all the possible starting positions and calculate the maximum value for synchronization, which undoubtedly increases the computational burden. Reference [8] comprehensively considered the performance and computational complexity of synchronization using synchronous head and LDPC code constraint for frame synchronization, but the research only focuses on binary LDPC codes. In reference [9], synchronization of Non-binary LDPC codes is studied respectively based on hard decision and soft decoding information. At the expense of increased complexity, the soft information synchronization algorithm performs better than hard syndrome based synchronization. A Fourier domain decoder called FFT-BP algorithm was implemented to reduce the complexity, but it is only convenient for messages expressed in the probability domain, which is not conducive to the hardware implementation. Among all existing decoding methods, the extended min-sum (EMS) algorithm reduces the decoding computation by truncating the length of the vector messages. By adjusting the truncated values, it improves the complexity-performance trade-off. Moreover, the EMS algorithm operated in the logarithm domain and multiplication is not involved, which is convenient for hardware implementation [10, 11].

In this paper, we propose a novel frame synchronization algorithm that uses the number of parity-check equations that are satisfied the LDPC code constraints to determine the frame starting point and the computational complexity of frame synchronization is reduced by using EMS algorithm. Moreover, the influence of the length of vector messages in EMS algorithm on the performance of frame synchronization is analyzed through simulation.

2 System Model

According to the Signal in Space Interface Control Document (ICD) of BDS, the detailed navigation message modulated on B2a signal is given as follows: the message length of each frame is 600 symbols, the symbol rate is 200 sps, and the transmission cycle is 3 s. The basic frame structure is given by (Fig. 1):

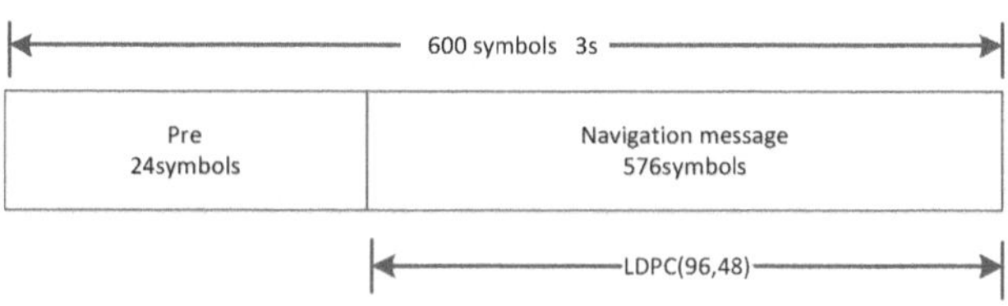

Fig. 1. Frame structure of B2a

The description of the Non-binary LDPC code used in B2a is defined as follows: the codeword generated after coding is $\boldsymbol{c} = (\boldsymbol{c_0}, \boldsymbol{c_1}, \cdots\cdots, \boldsymbol{c_{95}})$, the codeword transmitted over additive white Gaussian noise (AWGN) channel is $\boldsymbol{y} = (\boldsymbol{y_0}, \boldsymbol{y_1}, \cdots\cdots, \boldsymbol{y_{95}})$, each symbol over GF (64) is represented by six bits $\boldsymbol{y}_j = (y_{j,0}, y_{j,1}, \cdots\cdots, y_{j,5})$, and $\boldsymbol{H}(\boldsymbol{48} \times \boldsymbol{96})$ is the parity check matrix. In essence, the frame synchronization proposed in this paper is to find the threshold λ which meets the condition:

$$\hat{s} = \min_{s \in C} s \tag{1}$$

where $C = \{s | \Omega_s \geq \lambda, s \in [0, N-1]\}$, Ω_s represents the equation number that satisfies the parity-check equations calculated from the frame position s. In order to simplify the problem, the 180° phase ambiguity of data bits is not considered, but this does not affect the conclusion of our work.

3 Proposed Method

3.1 EMS Algorithm

In reference [9], the problem of frame synchronization in Non-binary system has been discussed, where the FFT-BP algorithm is used for frame synchronization.

To reduce the computational complexity, the vector messages are truncated by EMS algorithm, and the detailed steps are as follows:

1. Initialization

The initial log-density ratio (LDR) vector is $\boldsymbol{L}(z) = [L(0), L(1), \cdots\cdots, L(63)]$, and the components can be written as

$$L(x) = \frac{2}{\sigma^2} \sum_{b=0}^{5} |y_{j,b}| \Delta_{j,b} \tag{2}$$

where $x = (x_0, x_1, \cdots\cdots, x_5)$ is Galois field element, $\hat{x} = (\hat{x}_0, \hat{x}_1, \cdots\cdots, \hat{x}_5)$ is the value after the hard decision, $\Delta_{j,b} = \begin{cases} 0, & x_b = \hat{x}_b \\ 1, & x_b \neq \hat{x}_b \end{cases}$, and σ is the standard deviation of gaussian white noise.

The initial message $\boldsymbol{L}_j (0 \leq j < n)$ is sorted in ascending order and the smallest values of $n_m (n_m \leq 64)$ are stored as the initial value $V2C_{j \to i}$ of variable nodes VN_j.

2. Permutation step

In the permutation step, the coded symbol is multiplied by a non-binary parity check element. Using this transformation, the permutation of the messages due to non-zero values of $\boldsymbol{H}(\boldsymbol{48} \times \boldsymbol{96})$ has been done.

3. Check node update

In order to reduce the computational complexity, serial decoding is adopted. Furthermore, the forward-backward structure is used to update the check node. The recursive structure for B2a is as follows (Fig. 2):

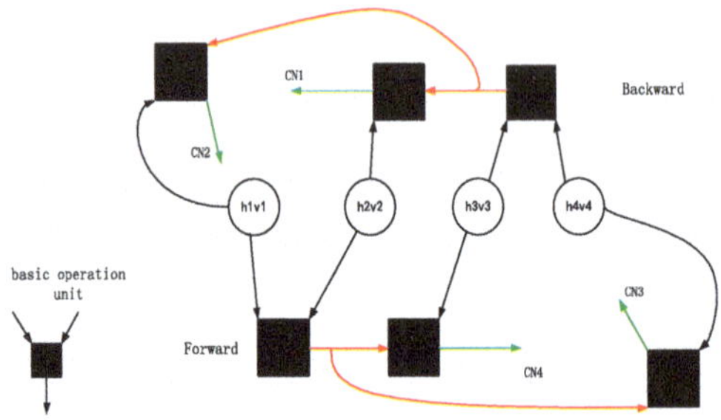

Fig. 2. The recursive structure of check-node

The arrow represents the direction of information transmission, $h_i v_i$, $(i = 1, 2, 3, 4)$ represents the sorted element and $CN_\imath$, $(\imath = 1, 2, 3, 4)$ represents the updated value of the check node. The input vector of the basic operation unit are I and V, the output vector is M, and the right superscript Q represents their associated GF symbols. The elementary steps are as follows:

$$M[k] = \min_{i,j \in [0, n_m - 1]^2} \{V(i) + I(j)\}_{V^Q[i] \oplus I^Q[j] = M^Q[k]} \tag{3}$$

where $\oplus$ represents the addition over GF (64), and the computation complexity of an elementary step can be reduced with a particular strategy in Reference [11].

4. Inverse permutation step

The inverse permutation of the check node is essentially a division over GF (64), which is identical to step 2, but in the reverse order.

5. Variable node update

Defining the basic operation unit I and V as input and Z as output, the operation rules are as follows:

$$\begin{array}{l} Z[i] = V[i] + X,\ Z[n_m + i] = I[i] + \gamma_v, \\ i \in [0, n_m - 1] \end{array} \tag{4}$$

where $X = \begin{cases} I[j], & I^Q[j] = V^Q[i] \\ \gamma_I, & I^Q[j] \notin V^Q \end{cases}$, $j \in [0,\ n_m - 1]$, γ_v and γ_I are the compensate value and are used when the required symbol index is not present in an input message. The output vector is sorted in ascending order and the processing of variable node update is similar to step 3. We note that the degree of the check node $d_c = 2$, thus the variable node update can be simplified as the same step in reference [10].

6. Decision

A decision is taken on each symbol until $\boldsymbol{Hc}^T = \boldsymbol{0}$ or a maximum number of iterations has occurred. Otherwise, it would turn back to step 2 and continue to iterate.

3.2 Proposed Frame Synchronization Method

The data received by the navigation receiver can be expressed as follows:

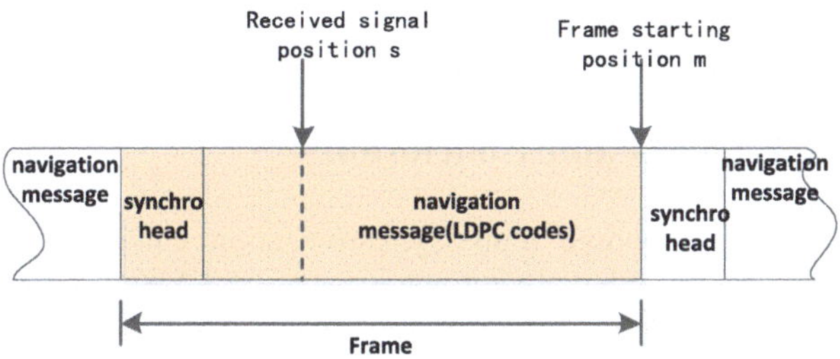

Fig. 3. Frame structure diagram of the received signal

As shown in Fig. 3, the starting position of the received signal is s, Ω_s represents the number that satisfies the parity-check equation after decoding, and m is the starting position of the correct frame. The conventional frame synchronization algorithm firstly calculates the correlation between the known synchronous head sequence and the received symbol sequence, and then compares it with the threshold to determine the starting position of the frame head [12]. However, this method does not make use of the coding gain of LDPC code, thus it cannot work reliably at low SNR. In addition, in order to ensure the reliability of frame synchronization, traditional methods often need to receive multi-frame data for redundancy verification, which undoubtedly increases the time needed to complete frame synchronization. This paper presents a frame synchronization algorithm based on EMS algorithm, which can reliably achieve frame synchronization at low SNR. The proposed algorithm steps are as follows:

1. Storing the received sequence $\boldsymbol{y} = (y_i, y_{i+1}, \cdots\cdots, y_{i+N-1})$ in the shift register;
2. Framing the received sequence, using EMS algorithm to decode and output $\Omega_s (0 \leq \Omega_s \leq 48)$;
3. If $\Omega_s \geq \lambda$, we obtain an estimation of the frame synchronization position, otherwise right shifted the register one bit to get the new sequence $\boldsymbol{y} = (y_{i+1}, y_{i+2}, \cdots\cdots, y_{i+N})$ and jump to step 2.

At this point, it can be deduced that the expression of the frame synchronization error rate is:

$$\begin{aligned} P_{FSE} &= 1 - \frac{1}{N}\sum_{i=0}^{N}(1-P_{fa})^{i}(1-P_{fd}) \\ &= 1 - \frac{1-P_{fd}}{NP_{fa}}\left(1-(1-P_{fa})^{N}\right) \end{aligned} \tag{5}$$

where P_{fa} represents false alarm probability, P_{fd} represents false dismissal probability, and N represents the total length of the search sequence. It is important to note that this implies that the smaller the decision threshold λ is, the larger P_{fa} will be and the smaller P_{fd} will be, and vice versa.

4 Simulation Analysis and Conclusion

The performance of the proposed frame synchronization scheme is estimated using MonteCarlo method. The length of vector messages N_m in EMS algorithm was 4, 8 and 16 respectively. The maximum number of iterations was set as 10 and the following figure shows the distribution of Ω_s under the in-sync and off-syn states at different Eb/N0 (Figs. 4, 5 and 6):

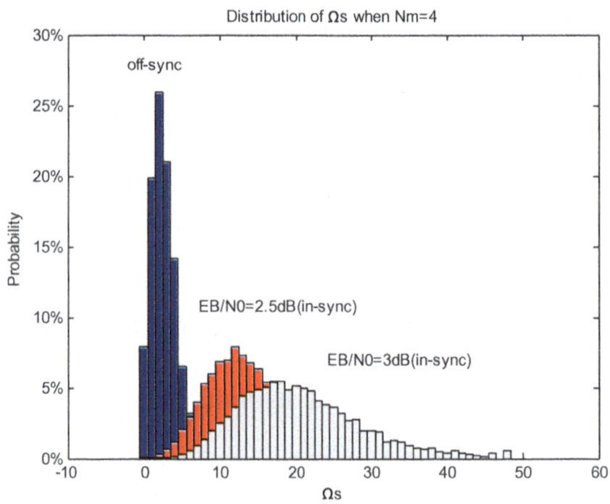

Fig. 4. Histograms of Ω_s when Nm = 4

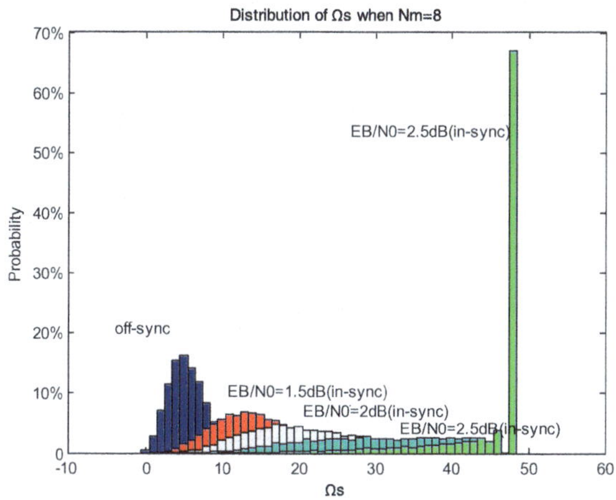

Fig. 5. Histograms of Ω_s when Nm = 8

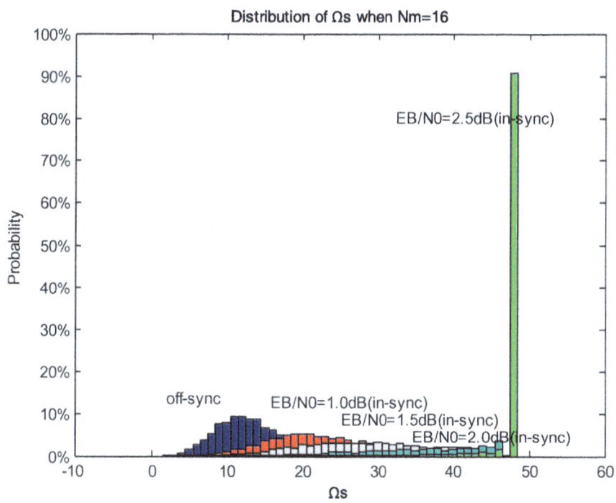

Fig. 6. Histograms of Ω_s when Nm = 16

As can be seen from Fig. 6, in the case of $N_m = 4$, there are overlaps with Eb/N0 = 2.5 which implies performance degradation. At this point, if the decision threshold is given as λ, the corresponding false dismissal probability and false alarm probability can be calculated. Moreover, the frame synchronization algorithm works well when there is no overlap between the histogram [7].

In the case of the same Eb/N0, the longer the vector message N_m is, the less overlapping the image is, and the better the performance of the proposed frame synchronization algorithm is. In the N_m fixed case, the larger Eb/N0 is, the better the performance of the proposed algorithm is.

It can also be seen that the false alarm probability P_{fa} and the false dismissal probability P_{fd} are not only related to the threshold λ, but also depend on the vector messages length N_m of EMS algorithms and Eb/N0.

The comparisons between the proposed frame synchronization algorithm and the conventional one are shown in Fig. 7. The vector messages length of EMS is set as $N_m = 4$, $N_m = 8$ and $N_m = 16$.

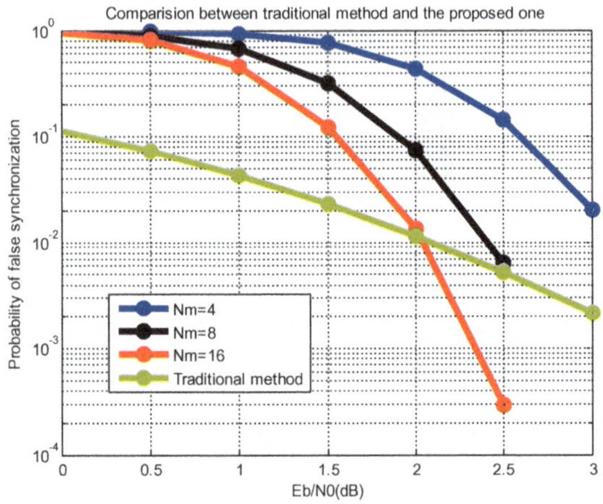

Fig. 7. Probability of false synchronization of different methods

By observing Fig. 7, we can see that under three different N_m values, the performance of the algorithm proposed in this contribution also improves with the increase of Eb/N0, which is consistent with the above analysis results.

Notably, in the example, the traditional frame synchronization algorithm performs better than the proposed algorithm in the condition of Eb/N0 < 2 for the case of $N_m = 16$. In this situation, the navigation receiver cannot decode the navigation message reliably (when $N_m = 16$, it can achieve the bit error rate of 10^{-6} only if Eb/N0 > 3 dB) with traditional method. Therefore, the traditional method would make no sense to improve the performs of navigation receiver. When Eb/N0 > 2 dB, the synchronization performance of the proposed algorithm is obviously better than that of the convenient one. Hence the performance of the proposed algorithm in this paper is effective and it naturally outperforms the convenient frame synchronization in the situation of navigation application.

5 Conclusion

Based on EMS algorithm for BDS B2a signal, we propose a novel frame synchronization algorithm which uses the number of parity-check equations that is satisfied with the LDPC code constraints to determine the frame starting point. The simulation results show that the frame synchronization performance of the proposed EMS algorithm gets better with the increasing of vector messages length N_m. The algorithm proposed in this paper can achieve reliable frame synchronization with low SNR on the premise of successful decoding.

References

1. MacKay, D.J., Neal, R.M.: Near Shannon limit performance of low density parity check codes. Electron. Lett. **32**(18), 1645–1646 (1996)
2. Pfletschinger, S., Mourad, A., Lopez, E., Declercq, D., Bacci, G.: Performance evaluation of non-binary LDPC codes on wireless channels. In: Proceedings of ICT Mobile Summit, pp. 1–8, Santander, June 2009
3. Kim, H., Jones, C.R., Villasenor, J.D.: Pilotless frame synchronization for LDPC-coded transmission systems. IEEE Trans. Signal Process. **56**(7), 2865–2874 (2008)
4. Lee, D.U., Kim, H., Jones, C.R., Villasenor, J.D.: Pilotless frame synchronization via LDPC code constraint feedback. IEEE Commun. Lett. **11**(8), 683–685 (2007)
5. Wymeersch, H., Moeneclaey, M.: ML frame synchronization for turbo and LDPC codes. In: Proceedings of International Symposium on DSP and Communications Systems, pp. 9–14, December 2003
6. Matsumoto, W., Imai, H.: Blind synchronization with enhanced sum-product algorithm for low-density parity-check codes. In: The 5th International Symposium on Wireless Personal Multimedia Communications, vol. 3, pp. 966–970. IEEE, October 2002
7. Sodha, J.: Check node LDPC decoder synchronisation. Int. J. Electron. Lett. **4**(3), 287–295 (2016)
8. Stefanovic, C., Vukobratovic, D., Bajic, D.: Low-complexity list-based frame synchronization for LDPC coded transmission. In: 2009 IEEE International Conference on Communications, pp. 1–5. IEEE, June 2009
9. Imad, R., Poulliat, C., Houcke, S.: Frame synchronization techniques for non-binary LDPC codes over GF (q). In: 2010 IEEE Global Telecommunications Conference GLOBECOM 2010, pp. 1–6. IEEE, December 2010
10. Declercq, D., Fossorier, M.: Decoding algorithms for nonbinary LDPC codes over GF (q). IEEE Trans. Commun. **55**(4), 633–643 (2007)
11. Boutillon, E., Conde-Canencia, L., Al Ghouwayel, A.: Design of a GF (64)-LDPC decoder based on the EMS algorithm. IEEE Trans. Circuits Syst. I: Regul. Pap. **60**(10), 2644–2656 (2013)
12. Li, Z.X., Wu, N., Wang, H., Zhao, H.J., Kuang, J.M.: Code-aided frame synchronization based on LDPC code constraints. Beijing Ligong Daxue Xuebao/Trans. Beijing Inst. Technol. **35**(3), 294–298 (2015)

Quality Analysis of Signal for BDS-3 Basic System

Yilei He(✉)

China Railway Design Corporation, Tianjin 300308, China
heyilei@cumt.edu.cn

Abstract. The BeiDou-3 basic system consisting of 19 new-generation BeiDou global satellites (BDS-3) was officially put into operation on December 27, 2018. In addition to B1I and B3I signals, the BDS-3 also transmit several new navigation signals, namely B1C, B2a and B2b, the performance of basic navigation services has been greatly improved. It is necessary to make a comprehensive evaluation for the quality of its observation data. In this paper, the real multi-GNSS data of 8 iGMAS tracking stations with 4 receiver types from 18 March to 31 March, 2019 are selected for quality analysis in terms of data integrity, SNR, multipath effect, ionospheric delay and cycle slip. By comparing with overlap frequencies from BDS-2 MEO B1I/B3I signals, GPS L1/L5 signals and Galileo E1/E5a signals, the performance of the current BDS-3 satellite signal and the receiving capability of the iGMAS stations are obtained. The results show that, the data integrity rate of each signal of BDS-3 satellite is slightly lower than that of the overlap frequency of BDS-2/GPS/Galileo satellite, but it can meet the daily positioning needs. The pseudorange multipath error and ionospheric delay error are comparable to the overlap frequency of BDS-2/GPS/Galileo satellite. The SNR and CSR are better than the overlap frequency of BDS-2/GPS/Galileo satellite. Therefore, BDS-3 basic system satellite maintains the excellent performance of BDS-2 B1I/B3I signals, and the observation data quality of B1C and B2a signals is comparable to that of GPS L1/L5 and Galileo E1/E5a signals, which can meet the normal working requirements of BDS, lay a solid foundation to establish the BDS-3 global system.

Keywords: BDS-3 basic system · iGMAS · Evaluation indicator · Quality analysis

1 Introduction

Beidou navigation satellite system (BDS) is a global navigation satellite system independently operated by China, which needs the "three-step" to realize the global system (Yang 2010). It completed the deployment of experimental system (BDS-1) and expanded regional navigation system (BDS-2) in 2003 and 2012, respectively, providing positioning, navigation, and timing (PNT) services for the Asia Pacific Region (Li 2019). It plans to complete the construction of global navigation system (BDS-3) in 2020, providing services to global users. Before the formal construction of BDS-3, two inclined geosynchronous orbit (IGSO) and three medium earth orbit (MEO) Beidou-3 test satellites (BDS-3e) were launched in 2015–2016, fully verifying the new-generation

J. Sun et al. (Eds.): CSNC 2020, LNEE 650, pp. 506–516, 2020.
https://doi.org/10.1007/978-981-15-3707-3_48

navigation signal. Then, the networking will be implemented in three steps: the simplest system, the basic system and the global system, the simplest system consisting of 8 BDS-3 satellites was built in March 2018, and the basic system consisting of 19 BDS-3 satellites (18 MEO+1 geostationary orbit (GEO)) was built at the end of 2018 to provide PNT services for "One Belt and One Road" countries and regions (CSNO 2018). The only one GEO satellite is in the in-orbit test state, this paper will not consider. The launch information of 18 BDS-3 MEO satellites is shown in Table 1.

Table 1. The status of BDS-3 basic system satellites

PRN	Launch date	PRN	Launch date	PRN	Launch date	PRN	Launch date
C19	2017/11/05	C22	2018/02/12	C25	2018/08/24	C34	2018/10/15
C20	2017/11/05	C29	2018/03/29	C26	2018/08/24	C35	2018/10/15
C27	2018/01/11	C30	2018/03/29	C32	2018/09/19	C36	2018/11/18
C28	2018/01/11	C23	2018/07/29	C33	2018/09/19	C37	2018/11/18
C21	2018/02/12	C24	2018/07/29				

BDS-3 satellite has applied new satellite signals, satellite platform, new atomic clock and other new technologies. Zhang et al. (2017) and Yang et al. (2018) evaluated five signals of BDS-3e satellite (C32-C34) from the aspects of signal-to-noise ratio (SNR) and pseudorange multipath error, and the results showed that the SNR of BDS-3e MEO satellite was larger than that of IGSO satellite, the pseudorange multipath error was significantly weaker than that of BDS-2, the observation data quality of BDS-3e satellite was equivalent to that of GPS L1/L2/L5 and Galileo E1/E5a/E5b signals. He (2019) analyzed eight BDS-3 simplest system satellites' observation quality, the results showed that BDS-3 satellite B1C and B2a signals quality were better than B1I and B3I signals, and reached the comparable level with GPS and Galileo satellites' overlap frequency signal. At present, the BDS-3 basic system has not long been built, there are relatively few studies on it. Yang et al. (2020) analyzed the orbit and clock offset precision of 18 BDS-3 MEO satellite, the precision of post-processing orbit can reach centimeter level, the average satellite clock offset uncertainty is 1.55 ns. The observation data quality is another important indicator to evaluate the performance of BDS-3 satellite, so it is necessary to study and analyze it deeply. Therefore, this paper will comprehensively and systematically analyze the quality of BDS-3 basic system satellite observation data from data integrity, SNR, multipath effect, ionospheric delay and cycle slip to verify the performance of the current Beidou global system.

2 Data Collection

China began to build the International GNSS Monitoring & Assessment System (iGMAS) in 2011 with the aim of establishing a global multi-mode GNSS tracking network for collecting, storing and processing data. Now, 24 continuous operation tracking stations have been built, including four receiver types, and all of them can receive BDS-3 basic system satellite signals. In order to better study the performance of the BDS-3 basic system satellite signals, this paper selects eight tracking stations with better observation data quality from four receivers. Figure 1 shows the distribution of eight tracking stations.

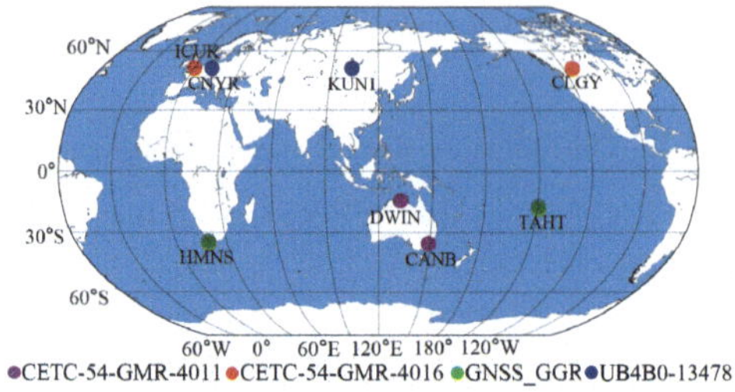

Fig. 1. Distribution of eight selected tracking stations from iGMAS

At present, the BDS-3 basic system MEO Satellite (BDS-3M) provides four public service signals B1I, B3I, B1C and B2a (CSNO 2018). It retains the B1I/B3I signals of the BDS-2 satellite, and adds the overlapping signals B1C/B2a with GPS L1/L5 and Galileo E1/E5a signals, achieving compatibility and interoperability with BDS-2, GPS and Galileo. The frequency allocation of the signals of each satellite navigation system is shown in Table 2 (Zhang 2017).

Table 2. Frequency of different satellite signals

Frequency (MHz)	1575.420	1561.098	1268.520	1176.450
BDS-2		B1I	B3I	
BDS-3	B1C	B1I	B3I	B2a
GPS	L1			L5
Galileo	E1			E5a

In this paper, the multi-mode GNSS observation data for eight iGMAS tracking stations with four receiver types from March 18 to March 31, 2019 (doy: 077-090) are selected. The multi-system GNSS observation data quality analysis software, which is independently developed by author was used to analyzes the observation data quality of 18 BDS-3M satellites based on the GNSS dual-frequency observation data from five aspects: data integrity, SNR, multipath effect, ionospheric delay and cycle slip (He 2017). By comparing with BDS-2 MEO satellite B1I/B3I signals, GPS satellite L1/L5 signals and Galileo satellite E1/E5a signals, the performance indexes and improvement levels of BDS-3M satellite B1I, B3I, B1C and B2a signals are obtained.

3 Data Quality Evaluation Indexes

3.1 Data Integrity Rate

Data integrity refers to the availability and integrity of the data during the observation period. It mainly tests the data loss condition and reflects the influence of satellite signal performance level and surrounding environment (Zhu 2015). The larger the value, the better the data quality, and the data integrity I is defined as follows:

$$I_{r,i}^{s} = \frac{H_{r,i}^{s}}{E_{r,i}^{s}} \tag{3.1}$$

Among them, H and E are the number of complete observations and theoretical observations of the i frequency of s satellite, respectively. In this paper, the complete observation value is defined as the observation value with pseudorange, carrier phase, doppler and SNR at the same epoch. The theoretical observation value means that the receiver can receive the satellite signal in theory.

Figure 2 shows the data integrity statistics results of 18 BDS-3M satellites at B1C and B2a signals for 14 consecutive days at eight tracking stations. As shown in Fig. 2, there are minor differences in data integrity between different BDS-3M satellites. For B1C signal, the data integrity is between 83.53%–99.5%. It is different for different stations. The UB4B0-113478 receiver (UB4B0) is the best, the CNYR and KUN1 station are greater than 99% and 95%, respectively. CETC-54-GMR-4016 (CETC-4016) and GNSS_GGR receivers are second, between 85%–99%; CETC-54-GMR-4011 receiver (CETC-4011) is the worst, between 85%–90%. For B2a signal, the data

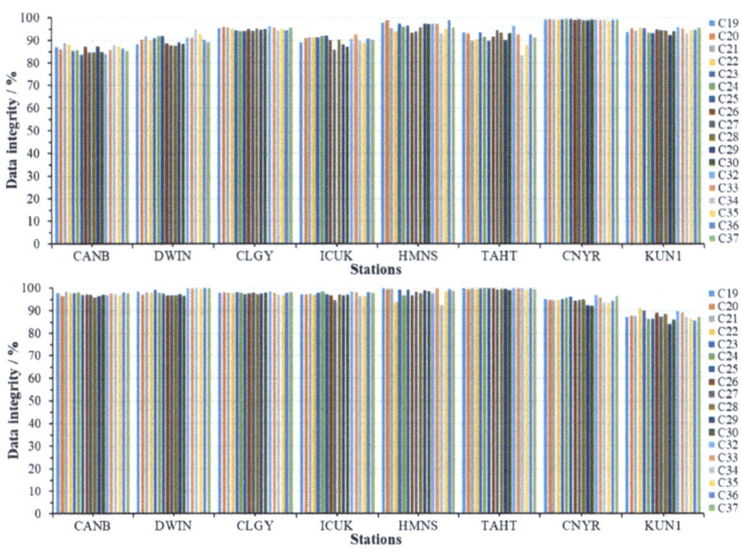

Fig. 2. Data integrity for B1C (up) and B2a (down) signals of selected stations

integrity is between 84.1%–99.97%. CETC-4011, CETC-4016 and GNSS_GGR receivers can all reach more than 95%, while UB4B0 receivers are between 85%–95%. In general, the data integrity rate of the B2a signal is better than the B1C signal, but the UB4B0 receiver shows obvious differences.

The average of the integrity rates of GPS/Galileo/BDS-2/BDS-3 satellites at different signals for the eight tracking stations are shown in Table 3. The BDS-3M satellite B1I and B3I signals data integrity rate is less than BDS-2 MEO satellite (BDS-2M) about 1.5%, the B1C signal is less than L1/E1 signal about 6%, and the B2a signal is less than L5/E5a about 3%. The data integrity rate of BDS-3 satellites at each signal is slightly inferior to that of BDS-2/GPS/Galileo overlap frequency. But they all greater than 95% except for B1C signal, which can meet daily positioning needs.

Table 3. Data integrity of different satellite signals (%)

Frequency (MHz)	1575.420	1561.098	1268.520	1176.450
BDS-2		99.11	98.59	
BDS-3	92.99	97.84	96.96	96.42
GPS	98.90			99.26
Galileo	97.11			97.48

3.2 Signal-to-Noise Ratio

The Signal-to-Noise Ratio (SNR) is the ratio of the receiver carrier signal power to the noise power, the unit is dB-Hz. It can better reflect the quality of the received satellite signal and the performance of the receiver. Normally, the SNR can be directly obtained from the RINEX observation file. The higher the value, the better the signal quality and the higher the observation accuracy (Zhu 2015).

Figure 3 shows the SNR statistics results of 18 BDS-3 M satellites at B1C and B2a signals for 14 consecutive days at eight tracking stations. As shown in Fig. 3, the SNR difference between B1C and B2a signals of different BDS-3M satellites is small, and different type receivers show obvious consistency. The SNR of the B1C and B2a signals are between 44.25–45.64 dB-Hz and 44.32–46.67 dB-Hz, respectively. The B2a signal is better than B1C signal.

The average of the SNR of GPS/Galileo/BDS-2/BDS-3 satellites at different signals for the eight tracking stations are shown in Table 4. It can be seen that the SNR of BDS-3M satellites at B1C and B2a signals are smaller than that of B1I and B3I signals, and the SNR of B1I and B3I signals are greater than BDS-2M satellite about 0.2–0.3 dB-Hz. The B1C signal is greater than that of L1/E1 about 0.7 and 2.2 dB-Hz, respectively, and the B2a signal is greater than that of L5/E5a about 0.8 and 1.1 dB-Hz, respectively. Therefore, the SNR of each signal of BDS-3 satellite is better than that of the overlap frequency of BDS-2/GPS/Galileo.

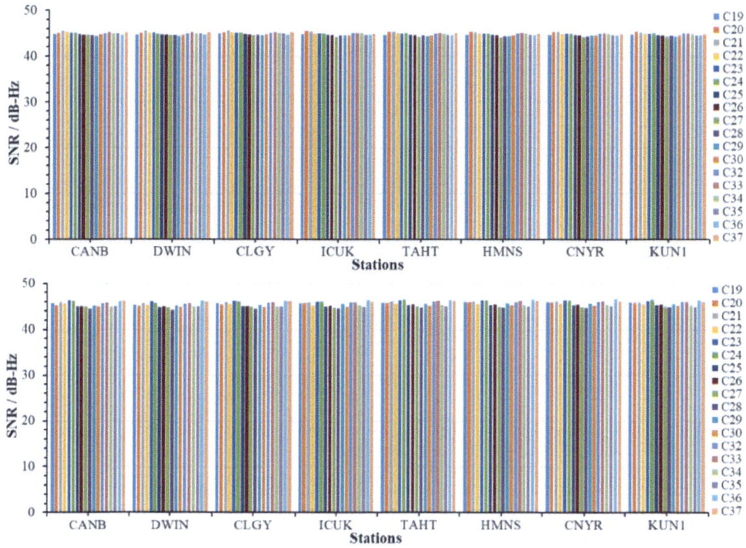

Fig. 3. SNR for B1C (up) and B2a (down) signals of selected stations

Table 4. SNR of different satellite signals (dB-Hz)

Frequency (MHz)	1575.420	1561.098	1268.520	1176.450
BDS-2		44.92	46.21	
BDS-3	44.91	47.33	46.49	45.66
GPS	44.26			44.86
Galileo	42.71			44.60

3.3 Multipath Effect

Multipath effect is mainly affected by the surrounding environment of GNSS receiver, with the characteristics of periodicity and random noise, and cannot be completely separated from noise (He 2018). Therefore, the multipath effect (MP) described in this paper includes pseudorange multipath error and noise, which can be represented by a linear combination of dual-frequency pseudorange and carrier phase observations (Zhang 2013).

Figure 4 shows the pseudorange multipath error statistics results of 18 BDS-3M satellites at B1C and B2a signals for 14 consecutive days at eight tracking stations. As shown in Fig. 4, the MP difference between B1C and B2a signals of different BDS-3M satellites is significantly. The MP of HMNS and CNYR stations are smaller, indicating that the surrounding environment of their stations is great. The MP of the B1C and B2a signals are between 0.33–1.46 m and 0.22–1.69 m, respectively. In general, the pseudorange multipath error at the B2a signal is smaller than that of B1C signal.

Figure 5 shows the pseudorange multipath error time series of each GPS/BDS-2/BDS-3/Galileo satellites' signal for 14 consecutive days at the CNYR station, and the

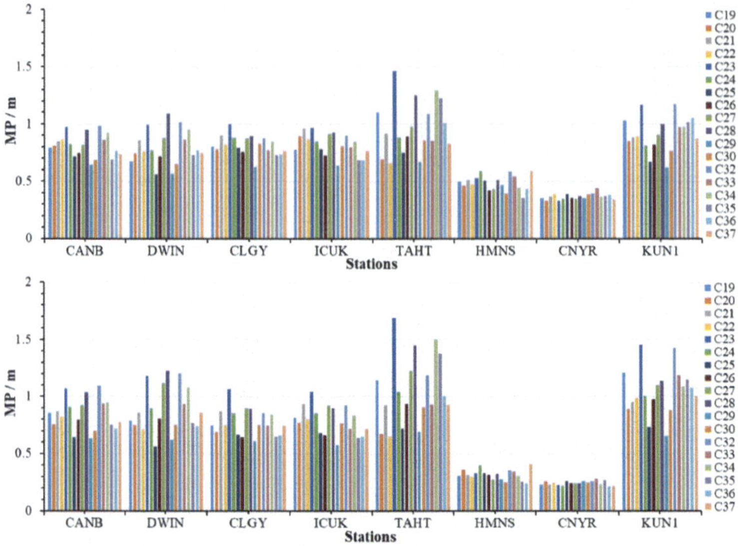

Fig. 4. Pseudorange multipath error for B1C (up) and B2a signals (down) of selected stations

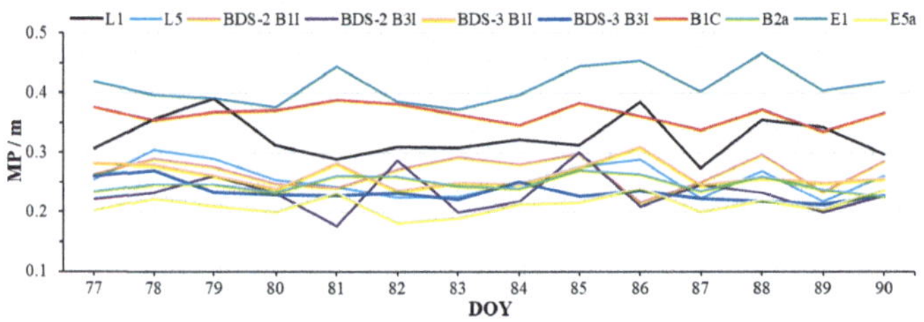

Fig. 5. Pseudorange multipath error for GPS/BDS-2/BDS-3/Galileo of CNYR station

BDS-2 satellite only selects three MEO satellites (C11/C12/C14). As shown in Fig. 5, the MP value of each satellite fluctuates significantly. The MP values of BDS-2 B1I and BDS-3 B1I signals are comparable. The average MP values of 14 days are 0.27 m and 0.26 m, respectively. The BDS-2 B3I and BDS-3 B3I signal have similar MP values, and the MP averages of both are 0.23 m. The MP value of BDS-3 B1C signal is slightly worse than that of L1 signal, but it is better than E1 signal. The average MP values of those are 0.36 m, 0.33 m, and 0.41 m, respectively. The BDS-3 B2a and L5 signal have similar MP values, but both are slightly worse than E5a signal. The average MP values of those are 0.25 m, 0.26 m, and 0.21 m, respectively. Therefore, in terms of suppressing pseudorange multipath, each signal of BDS-3 has reached the level which is comparable to that of the overlap frequency of BDS-2/GPS/Galileo.

3.4 Ionospheric Delay and Change Rate

When the satellite signal passes through the ionosphere, the propagation speed of the carrier phase and pseudorange will change, and then generate ionospheric delay error. The ionospheric delay rate (IOD) can reflect the activity of the ionosphere. When the IOD $\geq$ 4 m/min, that the ionospheric slip (Li 2006).

Figure 6 shows the ionospheric delay change rate RMS (IOD-RMS) statistical results of 18 BDS-3M satellites at B1C and B2a signals for 14 consecutive days at eight tracking stations. As shown in Fig. 6, the IOD-RMS difference between B1C and B2a signals of different BDS-3M satellites is obvious, and the receivers show obvious consistency. The IOD-RMS of the B1C signal is between 0.42–1.60 m/min and the IOD-RMS of the B2a signal is between 0.75–2.24 m/min. It can be seen that the IOD-RMS of the B1C signal is better than that of B2a signal.

Figure 7 shows the IOD-RMS time series of each GPS/BDS-2/BDS-3/Galileo satellites' signal for 14 consecutive days at the CNYR station. It can be seen that the IOD-RMS changes of each satellite signal are gentle, and there are obvious slips in doy 80, 86 and 89, indicating that the ionospheric changes in these three days are obvious. The IOD-RMS value of BDS-2 B1I signal is slightly lower than BDS-3 B1I signal, and the average IOD-RMS values for 14 days are 0.97 m/min and 1.13 m/min, respectively. The IOD-RMS value of BDS-2 B3I signal is slightly lower than BDS-3 B3I signal, and the average IOD-RMS values are 1.54 m/min and 1.67 m/min, respectively. The IOD-RMS values of BDS-3 B1C and E1 signal are comparable, and both are better than L1 signal. The average IOD-RMS values of those are 0.70 m/min, 0.68 m/min, and 0.93 m/min, respectively. The IOD-RMS value of BDS-3 B2a and E5a signal are comparable, and both are better than L5 signal. The average IOD-RMS values of those are 1.14 m/min, 1.14 m/min, and 1.29 m/min, respectively. Therefore, in terms of suppressing ionospheric delay, BDS-3

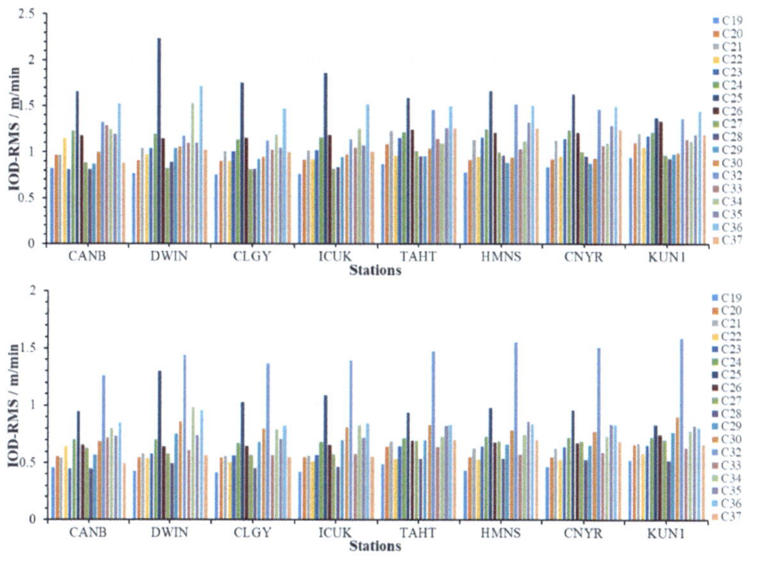

Fig. 6. IOD-RMS for B1C (up) and B2a (down) signals of selected stations

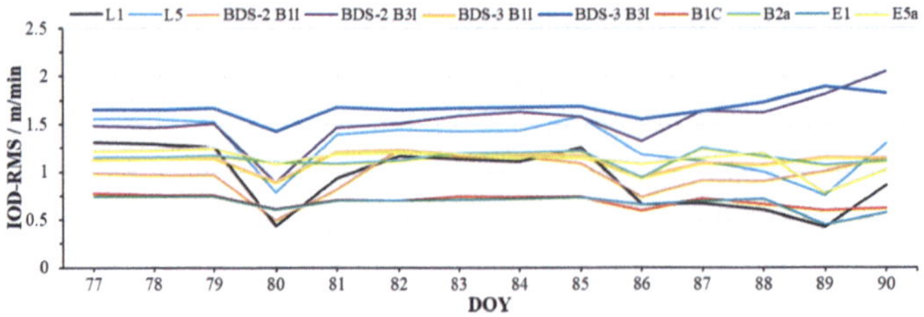

Fig. 7. IOD-RMS for GPS/BDS-2/BDS-3/Galileo of CNYR station

B1I/B3I signals are slightly lower than BDS-2 B1I/B3I signals. BDS-3 B1C/B2a signals and E1/E5a signals are better than L1/L5 signals.

3.5 Cycle Slip Ratio

Cycle slips is an important indicator for evaluating the quality of the observation data. In this paper, the combination of ionospheric residuals combination and Melbourne-Wübbena combination is used to detect the cycle slip (Fang 2009). The Cycle Slip Ratio (CSR) is usually used to measure the cycle slip of the observation data. The larger the value, the more severe the cycle slip (Wu 2015).

$$C = \frac{1000}{O/S} \tag{3.2}$$

Where C represents CSR, O/S represents the ratio of the number of observations to the number of cycle slip.

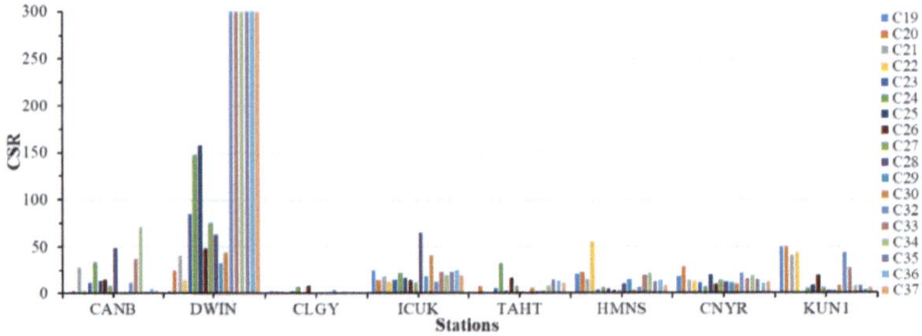

Fig. 8. SNR for B1C/B2a signals combination of selected stations

Figure 8 shows the CSR statistics results of 18 BDS-3M satellites at B1C and B2a signals for 14 consecutive days at eight tracking stations. As shown in Fig. 8, the CSR different of B1C/B2a signals of different BDS-3M satellites is obvious, and each receiver exhibits obvious differences. Among them, CLGY is the best, CSR is below 8, but the cycle slip phenomenon of DWIN station is serious, the maximum CSR can exceed 300, after excluding DWIN station, nearly 77.8% of BDS-3M satellite CSR is less than 20.

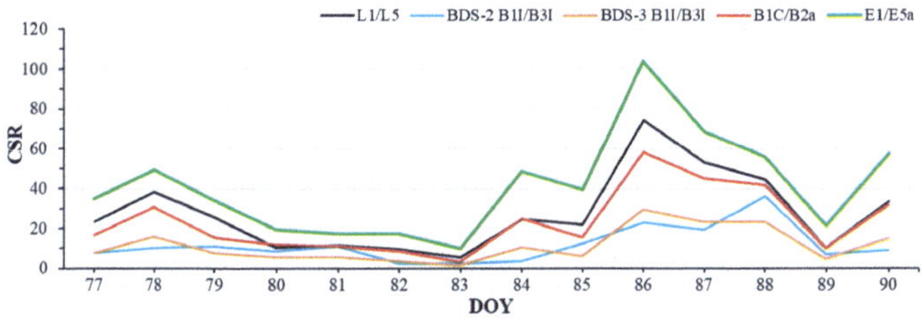

Fig. 9. CSR for GPS/BDS-2/BDS-3/Galileo of CNYR station

Figure 9 shows the CSR time series of GPS/BDS-2/BDS-3/Galileo satellites frequency combination for 14 consecutive days at the CNYR station. It can be seen that the CSR value of each satellite fluctuates significantly, but is consistent, among which the cycle slip phenomenon of doy 86-88 is extremely serious. The CSR values of BDS-2 B1I/B3I and BDS-3 B1I/B3I signals combinations are comparable, and the average CSR values for 14 days are 11.79 and 11.64, respectively. The CSR of BDS-3 B1C/B2a signals combination is better than that of L1/L5 and E1/E5a signals combination with the average CSR of those are 23.42, 27.60 and 41.54, respectively. Therefore, in terms of suppressing cycle slip, the B1I/B3I signals combination of BDS-3 satellite is comparable to that of BDS-2 satellite, and the B1C/B2a signals combination of BDS-3 satellite is better than GPS and Galileo satellites.

4 Conclusions

This paper selected the real multi-GNSS observations of eight iGMAS stations to analyze observation data quality of 18 BDS-3 basic system satellites' B1C/B2a/B1I/B3I/signals from five evaluation indicators, including data integrity rate, SNR, multipath effect, ionospheric delay and cycle slip. The results were compared with BDS-2 MEO B1I/B3I, GPS L1/L5 and Galileo E1/E5a signals of the homologous stations, and then the satellite signal performance of BDS-3 basic system was obtained. The experimental results show that:

(1) The data integrity rate of BDS-3M B1I and B3I signals are slightly larger than that of B1C and B2a signals, but smaller than that of BDS-2M B1I and B3I signals. The signal of B1C and B2a are smaller than those of GPS L1/L5 and Galileo E1/E5a, respectively. (2) The SNR of B1I/B3I signals are better than B1C/B2a signals and BDS-2M B1I/B3I signals, respectively, and the SNR of B1C/B2a signals are better than GPS L1/L5 and Galileo E1/E5a signals, respectively. (3) The pseudorange multipath error of B1I/B3I signals is comparable to B2a signal and BDS-2M B1I/B3I signals, which is better than that of B1C signal. The MP value of B1C signal is slightly less than GPS L1 signal, but better than Galileo E1 signal. The MP value of B2a and GPS L5 signals is comparable, but slightly less than Galileo E5a signal. (4) The IOD-RMS of B1I/B3I signals are less than B1C/B2a signals and BDS-2M B1I/B3I signals. The IOD-RMS of B1C/B2a signals are comparable to Galileo E1/E5a signals, respectively, both are better than that of GPS L1/L5 signals. (5) The CSR of B1I/B3I

combination is comparable to BDS-2M B1I/B3I combination, and smaller than B1C/B2a combination, and the CSR of B1C/B2a combination is smaller than that of GPS L1/L5 and Galileo E1/E5a combination.

In general, the BDS-3 basic system satellite keeps the good performance of BDS-2, and the quality of the new signal has reached the comparable level to GPS and Galileo satellites. It not only improves the performance of Beidou satellites, but also increases compatibility and interoperability with GPS and Galileo, laying a solid foundation for the Beidou global satellite navigation system.

Acknowledgements. This paper was supported by the National Science Foundation of China (Grant No. 41874039). Thanks iGMAS for the data and products.

References

Yang, Y.: The progress, contribution and challenge of the Beidou satellite navigation system. J. Surv. Mapp. **39**(01), 1–6 (2010)

Li, X., Xie, W., Huang, J., et al.: Estimation and analysis of differential code biases for BDS3/BDS2 using iGMAS and MGEX observations. J. Geodesy **93**, 419–435 (2019)

CSNO: Development of the BeiDou Navigation Satellite System (Version 3.0). China Satellite Navigation Office (2018)

Yang, Y., Mao, Y., Sun, B.: Basic performance and future developments of BeiDou global navigation satellite system. Satell. Navig. **1**, 1–8 (2020)

Zhang, X., Wu, M., Liu, W., et al.: Initial assessment of the COMPASS/BeiDou-3: new-generation navigation signals. J. Geodesy **1**, 1–16 (2017)

Yang, Y., Xu, Y., Li, J., et al.: Development and performance prediction of Beidou-3—system-test verification data analysis. Sci. Sinica Terrae **48**(5), 584–594 (2018)

He, Y., Wang, Q., Wang, Z., et al.: Quality analysis of observation data of BeiDou-3 experimental satellites. In: China Satellite Navigation Conference, pp. 275–294 (2018)

He, Y.: A Method for Multi-Satellite System's ISB Modeling and Prediction. China University of Mining and Technology, Xuzhou (2019)

He, Y., Wang, Q.: Quality analysis and preprocess software for multi-GNSS observation data V1.0 (2017). Software copyright: 2017SR426408

Li, Z., Huang, J.: GPS Measurement and Data Processing. Wuhan University Press, Wuhan (2005)

Zhu, J.: A Multi System GNSS Real-Time Data Quality Analysis and Software Implementation. Southeast University, Dhaka (2015)

Zhang, X., Ding, L.: Quality analysis of the second generation compass observables and stochastic model refining. Geomat. Inf. Sci. Wuhan Univ. **38**(7), 832–836 (2013)

Wu, D., Wang, L., Zhang, Q., et al.: The implementation and verification analysis of GNSS data quality assessment software. J. Surv. Mapp. Sci. Technol. **32**(04), 344–348 (2015)

Li, J., Wang, J., Xiong, X.: Quality checking and analysis on GPS data in Northeast Asia. Geomat. Inf. Sci. Wuhan Univ. **31**(3), P209–P212 (2006)

Fang, R., Shi, C., Wei, N., et al.: Research on real-time cycle slip detection in GPS data quality control. Geomat. Inf. Sci. Wuhan Univ. **34**(9), 1094–1097 (2009)

Subcarrier Periodic Shifting BOC Modulations

Xin Zhao, Xinming Huang, Jingyuan Li(✉), Ke Zhang, and Guangfu Sun

College of Electronic Science and Technology, National University of Defense Technology, Changsha, China
Jingyuanli82@sina.com

Abstract. Designing the modern satellite navigation signals is essential for modernizing global navigation satellite systems (GNSS). For satisfying the diverse user requirements, more superior receiving performance is one of the urgent problems solved for GNSS. This paper proposes an improved BOC modulation technique, called subcarrier periodic shifting BOC (SPS-BOC) modulations. The spreading code delay is used as the time interval to form the SPS-BOC modulations by periodically shifting the phase of the subcarrier. The simulation shows that the high-frequency components of the SPS-BOC modulations are larger than traditional BOC modulations. By comparing with traditional BOC signals, the receiving performances of low-order SPS-BOC signal are superior to BOCs signals. Especially, the quality factor of matching interference is 50% more than BOCs. When the filter bandwidth is 10 MHz, Gabor bandwidth of low-order SPS-BOC signal is 26% more than BOCs, and the multipath error envelop is reduced by 34%. Therefore, the SPS-BOC modulation proposed in the paper has superior receiving performance and can be used as a new option for low-order components in MBOC signals and designing next-generation navigation signals.

Keywords: Designing navigation signal · Subcarrier periodic shifting BOC modulations · Receiving performance

1 Introduction

For improving receiving and service ability of global satellite navigation systems (GNSS), the satellite navigation systems have evolved into modernization, which designing the modern navigation signals is important to the GNSS modernization [1]. In 2001, Betz proposed the binary offset carrier (BOC) modulations [2, 3] at the ION conference, which has the split spectrum and meet the requirement of spectrum sharing. In addition, BOC modulation has more advantages in tracking accuracy and multipath performance than traditional BPSK modulation [4]. The ranging performance of satellite navigation systems is a big step forward because of BOC modulations. With the diverse requirements, the higher receiving performance has become one of the urgent problems solved for satellite navigation systems. And it provides a good opportunity for designing the new navigation signal.

Now, the improved modulations based on BOC modulations maintain the structure of BOC modulations and proposes the new design in terms of modulation symbols, belonged to discontinuous phase modulations, where the step code symbols are adopted

J. Sun et al. (Eds.): CSNC 2020, LNEE 650, pp. 517–526, 2020.
https://doi.org/10.1007/978-981-15-3707-3_49

and modulated in the spreading code. The improved modulations take different modulation symbols in the spreading code, which can be divided into binary code symbol (BCS) modulations [5, 7, 8] and multilevel code symbol (MCS) modulations [5, 6] according to quantity coding of symbol amplitude. Meanwhile, MCS modulations are consisted of multiple BCS modulated signals, which can form more flexible modulations by different multiplexed methods in the spreading code. And MBOC modulation is a robust design in MCS modulations [9]. Therefore, BCS modulations are the basis of the improved modulations. BCS modulations have more high-frequency components and more superior in receiving performance than traditional BOC modulations. However, it has brought more complicated discrimination curves, higher receiving complexity and worse spectrum separation ability than BOC modulations [8], which can be improved and reconstruct. Expect adopting different modulation symbols, there is another flexible improved modulations based on BOC modulations by adjust the subcarrier phase periodically and dynamically. And it will also obtain excellent receiving performance and be a new option for the evolution of navigation signals in the future.

In order to improve the signal receiving performance, this paper proposes the subcarrier periodic shifting BOC modulations. The second part of the paper focuses on describing the principle of the subcarrier periodic shifting BOC modulations, and derives the autocorrelation functions and power spectral density. The third part is the simulation results, and the receiving performance is discussed. Finally, the conclusion is presented.

2 Subcarrier Periodic Shifting BOC Modulations

The traditional BOC modulations are formed by multiplying a square subcarrier with a spreading code signal, so that the original BPSK spectrum is moved to both sides of the center frequency. The initial subcarrier phase of the BOCs modulation is 0, and the initial subcarrier phase of the BOCc modulation is pi/2. This part presents a new BOC modulation where the subcarrier phase is not fixed, called the subcarrier periodic shifting binary offset carrier modulation. The phase of the subcarrier is periodically adjusted with the time interval of the spreading code delay to form the subcarrier periodic shifting binary offset carrier modulation, which is denoted as $\mathrm{SPS\text{-}BOC(m, n, \Delta\varphi)}$. The expression of SPS-BOC modulation is:

$$signal_{SPS-BOC(m,n,\Delta\varphi)} = c(t) * sc(t) = \sum_{i=-\infty}^{+\infty} c_i(-1)^{Mi} p_{SPS-BOC}(t - iT_c) \tag{1}$$

The modulation symbols in each chip are as follows:

$$p_{SPS-BOC}(t) = sign(\sin(2\pi f_s t + i * \Delta\varphi)), \quad 0 \le t \le T_c \tag{2}$$

Where c_i is the spreading code, i is the sequence index of the spreading code, M is the modulation order, the subcarrier frequency is f_s, $f_s = m * 1.023$ MHz, and the subcarrier symbol delay is T_s, $T_s = 1/2/f_s$, the spreading code rate is f_c, $f_c = n * 1.023$ MHz, spreading chip delay is T_c, $T_c = 1/f_c$, $\Delta\varphi$ represents the adjustment in the sub-carrier initial phase, and the period is $2 * k$, $k = \mathrm{pi}/\Delta\varphi$, and when the spreading code alternates, the initial subcarrier phase changes by $\Delta\varphi$. According to the modulation principle, the subcarrier signal in the time domain is shown in Fig. 1.

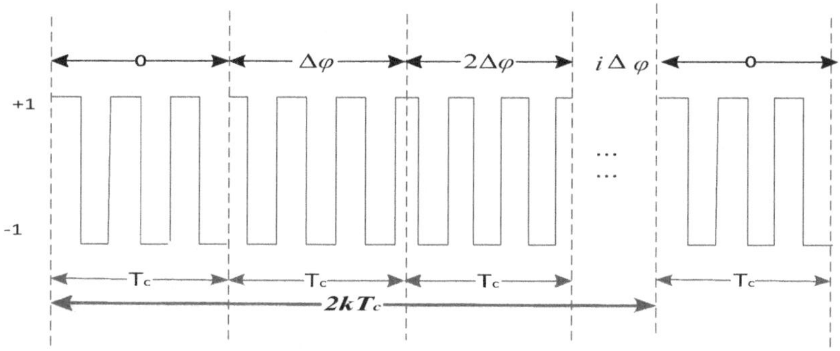

Fig. 1. Subcarrier signal of SPS-BOC in time domain

In Fig. 1, the subcarrier of SPS-BOC modulations in the time domain are combined with different initial phase subcarriers in the time division method, and the spreading code chip is used as an interval. In each spreading code chip, the initial phase of its subcarrier signal is $i * \Delta\varphi$, and its basic modulation symbol waveform is as follows:

$$p_2(t) = \begin{cases} 1, & 0 \leq t \leq (1 - \frac{i*\Delta\varphi}{\pi})T_s \\ -1, & (1 - \frac{i*\Delta\varphi}{\pi})T_s \leq t \leq (2 - \frac{i*\Delta\varphi}{\pi})T_s \\ 1, & (2 - \frac{i*\Delta\varphi}{\pi})T_s \leq t \leq 2T_s \end{cases} \tag{3}$$

In order to characterize the performance of the subcarrier periodic shifting BOC modulation, the theoretical autocorrelation functions and power spectral density are presented in the paper. Since the SPS-BOC modulation is constructed by BOC signals with different initial phase in the code delay T_c, its autocorrelation function is:

$$R(\tau) = \frac{1}{2k}\sum_{i=0}^{2k-1} R_i(\tau) \tag{4}$$

Where $R_i(\tau)$ is the autocorrelation function of the corresponding BOC signal with initial subcarrier phase $i * \Delta\varphi$, the expression is shown as:

$$R_i(\tau) = (-1)^l \begin{cases} \frac{(2l+1)(M-l)+l}{M} - \frac{2M-2l+1}{T_c}|\tau| & lT_s \leq |\tau| \leq lT_s + \frac{i}{k}T_s \\ \frac{(2l+1)(M-l)-l}{M} - \frac{2i}{kM} - \frac{2M-2l-1}{T_c}|\tau| & lT_s + \frac{i}{k}T_s \leq |\tau| \leq (l+1)T_s - \frac{i}{k}T_s \\ \frac{(2l+1)(M-l)-3l-2}{M} - \frac{2M-2l-3}{T_c}|\tau| & (l+1)T_s - \frac{i}{k}T_s \leq |\tau| \leq (l+1)T_s \end{cases} \tag{5}$$

Considering that the autocorrelation functions are the same when the subcarrier phases are $i * \Delta\varphi$ or $\pi + i * \Delta\varphi$, the above autocorrelation function can simplified as:

$$R(\tau) = \frac{1}{k}\sum_{i=0}^{k-1} R_i(\tau) \tag{6}$$

Finally, the autocorrelation function is as:
When the half period k is even:

$$R(\tau) = (-1)^l \begin{cases} \frac{(2l+1)(M-l) + \frac{k-2-4i}{k}l - \frac{2i(i+1)}{k^2}}{M} - \frac{2M - 2l + \frac{k-2-4i}{k}}{T_c}|\tau| & (l+\frac{i}{k})T_s \le |\tau| \le (l+\frac{i+1}{k})T_s \\ \frac{(2l+1)(M-l) - \frac{k-2}{k}l - \frac{(k-2)}{2M}}{M} - \frac{2M - 2l - \frac{k-2}{k}}{T_c}|\tau| & (l+\frac{k-2}{2k})T_s < |\tau| \le (l+0.5)T_s \\ \frac{(2l+1)(M-l) - \frac{k+2}{k}l - \frac{(k+2)}{2k}}{M} - \frac{2M - 2l - \frac{k+2}{k}}{T_c}|\tau| & (l+0.5)T_s \le |\tau| \le (l+1-\frac{k-2}{2k})T_s \\ \frac{(2l+1)(M-l) - \frac{3k-2-4i}{k}l - \frac{2k-2-4i}{k} - \frac{2i(i+1)}{k^2}}{M} - \frac{2M - 2l - \frac{3k-2-4i}{k}}{T_c}|\tau| & (l+1-\frac{i+1}{k})T_s \le |\tau| \le (l+1-\frac{i}{k})T_s \end{cases} \tag{7}$$

Where $i = 0, 1\ldots (k-1)/2 - 1$, $l = 0, 1\ldots M - 1$.
When the half period k is odd:

$$R(\tau) = (-1)^l \begin{cases} \frac{(2l+1)(M-l) + \frac{k-2-4i}{k}l - \frac{2i(i+1)}{k^2}}{M} - \frac{2M - 2l + \frac{k-2-4i}{k}}{T_c}|\tau| & (l+\frac{i}{k})T_s \le |\tau| \le (l+\frac{i+1}{k})T_s \\ \frac{(2l+1)(M-l) - l - \frac{(k-1)(k+1)}{2k^2}}{M} - \frac{2M - 2l - 1}{T_c}|\tau| & (l+\frac{k-1}{2k})T_s < |\tau| \le (l+1-\frac{k-1}{2k})T_s \\ \frac{(2l+1)(M-l) - \frac{3k-2-4i}{k}l - \frac{2k-2-4i}{k} - \frac{2i(i+1)}{k^2}}{M} - \frac{2M - 2l - \frac{3k-2-4i}{k}}{T_c}|\tau| & (l+1-\frac{i+1}{k})T_s \le |\tau| \le (l+1-\frac{i}{k})T_s \end{cases} \tag{8}$$

Where $i = 0, 1\ldots (k-1)/2 - 1$, $l = 0, 1\ldots M - 1$.

Similarly, the normalized power spectral density of SPS-BOC modulation is shown as:

$$G(f) = \frac{1}{k}\left[\sum_{i=0}^{k-1}\left|\sin(\pi f T_s(1-\frac{i}{k})) - \sin(\pi f T_s \frac{i}{k})e^{-j\pi f T_s}\right|^2\right]\frac{\sin^2(\pi f T_c)}{T_c(\pi f \cos(\pi f T_s))^2} \tag{9}$$

3 Simulation and Analysis

In order to analyze the receiving performance of the SPS-BOC modulation proposed in the paper, autocorrelation function (ACF), power spectral density (PSD), Gabor bandwidth, multipath error envelop (MEE) and anti-jamming capability are used for evaluation and comparison.

3.1 Autocorrelation Functions and Power Spectral Density

Assuming that the sampling rate is 50 MHz, the phase offset adjustments $\Delta\varphi$ of the SPS-BOC signal are pi/2, pi/3, pi/4 and pi/12, the modulation order M is 2 and 12, the autocorrelation functions and power spectral density of SPS-BOC are compared with the corresponding BOCs and BOCc signals. And $\Delta\varphi - S$ represents the simulation curve, $\Delta\varphi - T$ represents the theoretical curve. The curves are shown in the figure below.

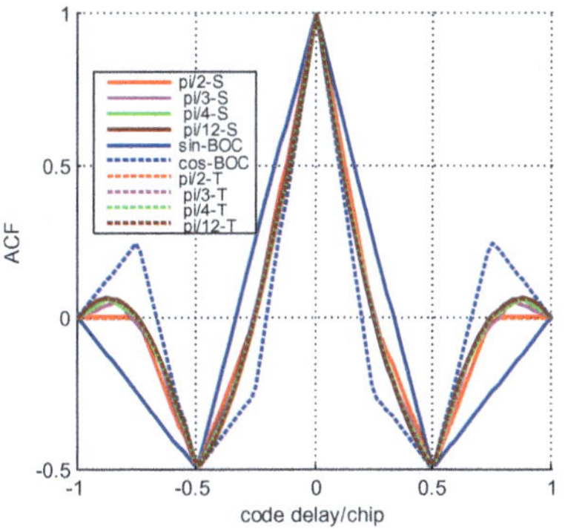

Fig. 2. ACFs of SPS-BOC(1,1,$\Delta\varphi$)

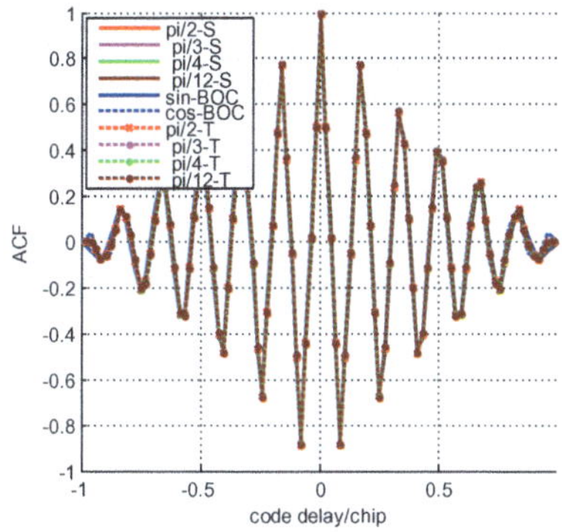

Fig. 3. ACFs of SPS-BOC(6, 1, $\Delta\varphi$)

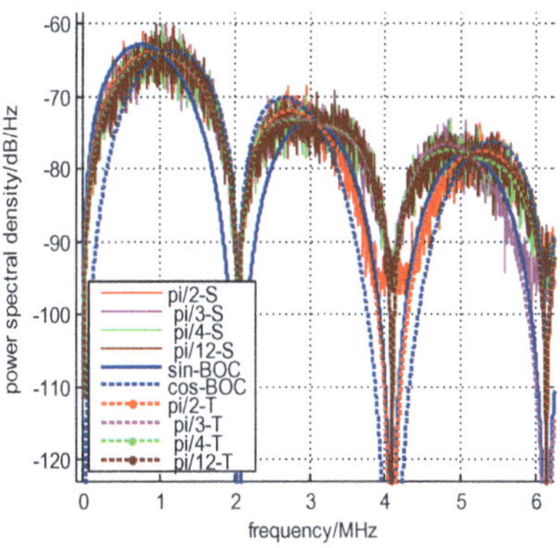

Fig. 4. PSDs of SPS-BOC(1, 1, $\Delta\varphi$)

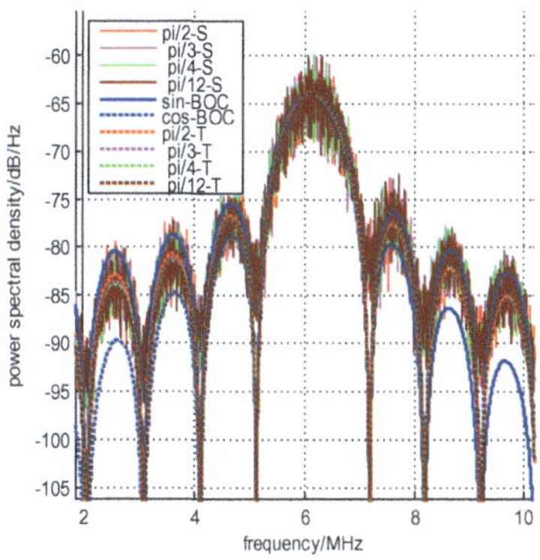

Fig. 5. PSDs of SPS-BOC(6, 1, $\Delta\varphi$)

It can be known from Fig. 2 that the theoretical ACF curve of SPS-BOC modulation is consistent with the actual simulation result, and the theoretical derivation is proved. The main peak of the SPS-BOC modulation is sharper than the BOCs(1,1) signal and close to the BOCc(1,1) signal, which means that the SPS-BOC signal has more potential in tracking performance than the BOCs signal. In addition, the secondary positive peak amplitude of SPS-BOC$(1, 1, \Delta\varphi)$ is 1/4 of BOCc(1,1), so its reception robustness may increase significantly. However, the high-order SPS-BOC signal in Fig. 3 is basically similar as the corresponding BOCs and BOCc signals. It cannot reflect the advantages of SPS-BOC modulations.

Figures 4 and 5 presents the simulation and theoretical power spectral density curves of SPS-BOC signals. And the simulation curves are consistent with the theoretical curve. In Fig. 4, the sideband energy of SPS-BOC$(1, 1, \Delta\varphi)$ is higher than the BOCs(1,1) signal, which is similar to BOCc(1,1). However, the sideband energy of SPS-BOC$(6, 1, \Delta\varphi)$ in Fig. 5 is close to BOCs(6,1) and BOCc(6,1). Therefore, the high-frequency component energy is equivalent to BOCs and BOCc signals. Meanwhile, it can be seen that the SPS-BOC modulations have the similar spectrum separation feature with BOC modulations.

3.2 Anti-interference Ability

In order to analyze the anti-interference ability of SPS-BOC modulation, the quality factors of band-limited white noise interference and the matching spectrum interference are used for comparing its characterization, and the expressions of quality factors are as follows:

$$Q_{band-noise} = \frac{1}{\frac{f_c}{2\beta} \int_{f_l-\beta}^{f_l+\beta} G(f)df} \tag{10}$$

$$Q_{match} = \frac{1}{f_c \int_{-\beta}^{\beta} (G(f))^2 df} \tag{11}$$

Where f_c is the spreading code rate, G(f) is the power spectral density, β is the single sideband bandwidth, and f_l is the center frequency.

Assuming that the sampling rate is 50 MHz, the phase offset adjustments $\Delta\varphi$ of SPS-BOC modulations are pi/2, pi/3, pi/4 and pi/12, the modulation orders M are 2 and 12, the quality factors of SPS-BOC modulations are compared with the corresponding BOCs and BOCc signals. The quality factors of the band-limited white noise is shown in Table 1 (the center frequency is the subcarrier frequency f_s, and the single sideband bandwidth β is the spreading code rate f_c). Meanwhile, the quality factors of matching spectrum interference under different bandwidths are shown in Figs. 6 and 7.

Table 1. Quality factors of band-limited noise interference of SPS-BOC modulations

Signal	Jamming frequency	Quality factors of the band-limited white noise
SPS-BOC$(1, 1, \Delta\varphi)$	1.023 MHz	5.02(pi/2)
		5.14(pi/3)
		5.19(pi/4)
		5.24(pi/12)
BOCs(1,1)	1.023 MHz	4.67
BOCc(1,1)	1.023 MHz	5.43
SPS-BOC$(6, 1, \Delta\varphi)$	6 * 1.023 MHz	5.45 (pi/2)
		5.46 (pi/3)
		5.46 (pi/4)
		5.46 (pi/12)
BOCs(6,1)	6 * 1.023 MHz	5.44
BOCc(6,1)	6 * 1.023 MHz	5.47

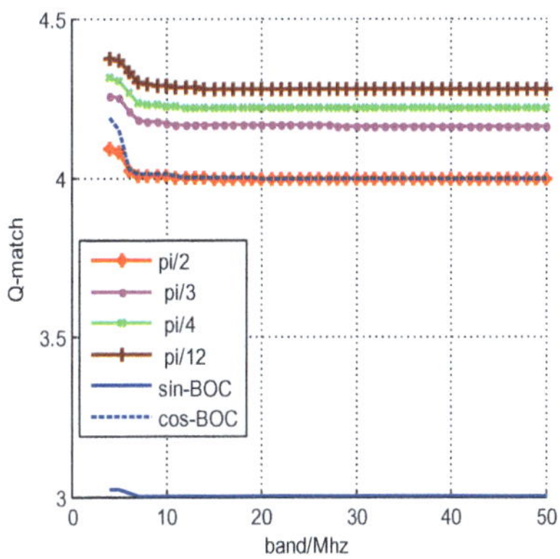

Fig. 6. Quality factors of matched interference of SPS-BOC$(1, 1, \Delta\varphi)$

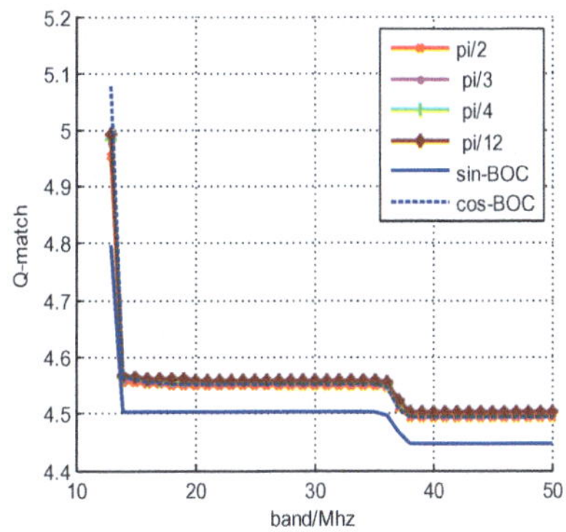

Fig. 7. Quality factors of matching spectrum interference of SPS-BOC$(6, 1, \Delta\varphi)$

Table 1 shows the quality factors of band-limited white noise of low-order and high-order SPS-BOC signals. As can be seen from Table 1, the SPS-BOC modulation has better anti-noise performance than BOCs signals in lower-order signal, and is slightly worse than BOCc signals. However, the band-limited white noise performance of SPS-BOC$(6, 1, \Delta\varphi)$ is equivalent to the corresponding BOCs, BOCc signals in the high-order signal.

In Figs. 6 and 7, the curves of the quality factors of anti-matching spectrum interference is plotted in the paper. It can be seen from the figure that anti-matching spectrum interference performance of the low-order SPS-BOC$(1, 1, \Delta\varphi)$ is slightly better than BOCc signals, significantly 50% better than BOCs signals. But, the anti-matching interference performance of high-order SPS-BOC$(6, 1, \Delta\varphi)$ is basically the same as the corresponding BOCs and BOCc signals, without obvious advantages.

3.3 Gabor Bandwidth

Gabor bandwidth is usually used to characterize the tracking performance of the signal. In this part, the phase offset adjustments Δφ of SPS-BOC modulations are pi/2, pi/3, pi/4 and pi/12, the modulation orders *M* are 2 or 12, the Gabor bandwidth of SPS-BOC modulations are compared with the corresponding BOCs and BOCc signals. The Gabor bandwidths are shown in Figs. 8 and 9, and the expression of Gabor bandwidth is shown:

$$Band_{Gabor} = \sqrt{\int_{-\beta}^{\beta} f^2 G(f) df} \quad (12)$$

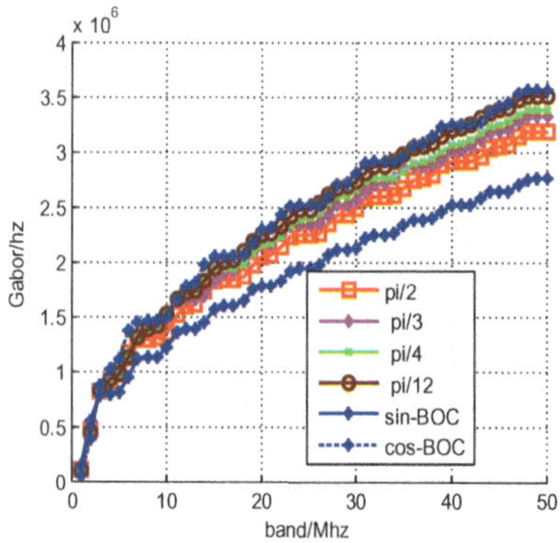

Fig. 8. Gabor bandwidths of SPS-BOC (1, 1, Δφ)

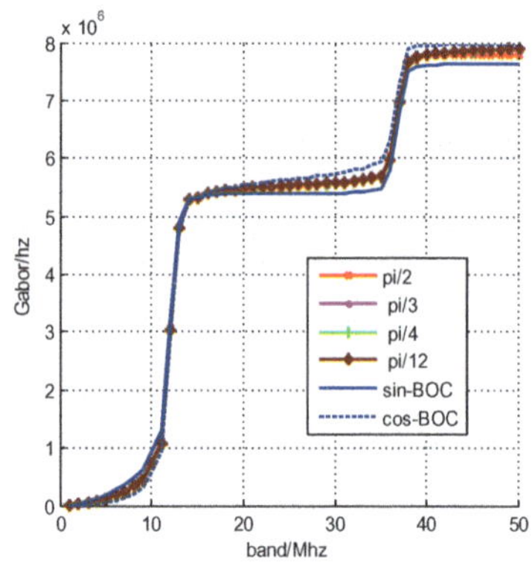

Fig. 9. Gabor bandwidths of SPS-BOC (6, 1, Δφ)

In Figs. 8 and 9, the simulation results of Gabor bandwidth for high-order and low-order SPS-BOC signals are presented. As can be seen from Fig. 8, the Gabor bandwidth of the low-order SPS-BOC(1,1,$\Delta\varphi$) is larger than the BOCs(1,1) signal, which increases by at least 26%. Meanwhile, it is slightly worse than the BOCc signal. It means that low-order SPS-BOC modulation has advantages and potential in tracking performance compared with traditional BOCs signals. When the phase offset period is large enough, the Gabor bandwidth of the SPS-BOC modulation is basically the same as the BOCc signal. In Fig. 9, the Gabor bandwidth of high-order SPS-BOC(6,1,$\Delta\varphi$) signal is similar with the corresponding BOCs and BOCc signals.

3.4 Multipath Error Envelop

The multipath error envelops (MEE) are generally used to evaluate the anti-multipath performance. Assume that the front-end bandwidth is 10 MHz for low-order

SPS-BOC$(1, 1, \Delta\varphi)$ signals, and the front-end bandwidth is 20 MHz for high-order SPS-BOC$(6, 1, \Delta\varphi)$ signals, and the amplitude ratio of the one multipath signal and the direct signal (MDR) is −6 dB, the phase offset adjustments $\Delta\varphi$ of SPS-BOC modulations are pi/2, pi/3, pi/4 and pi/12. The multipath error envelops of SPS-BOC modulations are compared with the corresponding BOCs and BOCc signals in Figs. 10 and 11, and the MEE expression is as follows:

$$\Gamma = (\max\varepsilon(\alpha, u, \Phi), \min\varepsilon(\alpha, u, \Phi)) \tag{13}$$

Where α is the amplitude ratio of multipath and direct signals, Φ is the multipath phase, ε is the code tracking error, and u is the multipath delay.

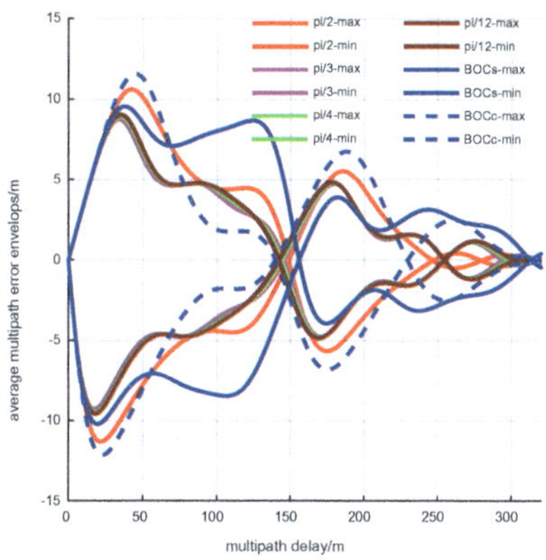

Fig. 10. MEEs of SPS-BOC$(1, 1, \Delta\varphi)$

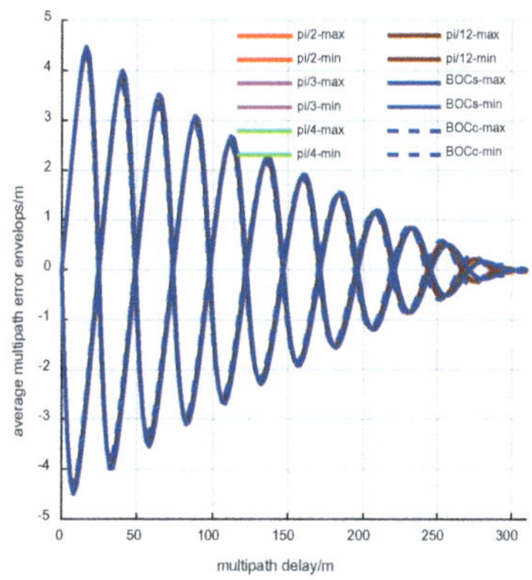

Fig. 11. MEEs of SPS-BOC$(6, 1, \Delta\varphi)$

According to the average multipath error envelope curves in Fig. 10, it can be known that the multipath performance of the low-order SPS-BOC$(1, 1, \Delta\varphi)$ signal is far better than BOCs(1,1) when the front-end filter bandwidth is 10 MHz. Even, The signal is slightly better than BOCc(1,1). Comparing with BOCs and BOCc signals, the overall multipath error envelop areas are respectively reduced by 34% and 17%. And when the phase offset adjustment $\Delta\varphi$ is pi/2, the multipath performance is equivalent to BOCc(1,1). At the same time, it can be known from Fig. 11 that the multipath performance of the high-order SPS-BOC$(6, 1, \Delta\varphi)$ is equivalent to the corresponding high-order BOCs and BOCc signals, and there is no obvious advantage.

4 Conclusions

In order to improve the receiving performance, the paper proposes the subcarrier period shifting BOC (SPS-BOC) modulation. The modulation is formed to adjust the subcarrier phase periodically based on the subcarrier shifting BOC modulations at the interval of spreading code delay. The SPS-BOC modulation is analyzed and evaluated through autocorrelation function, power spectral density, anti-interference ability, tracking performance, and multipath performance.

The receiving performance of low-order SPS-BOC signals is generally more superior than BOCs signals, similar with BOCc signals, even better than BOCc signals in multipath performance. The Gabor bandwidth is at least 26% larger than that of BOCs signals. Meanwhile, it is slightly worse than BOCc signals. Specially, the multipath mitigation performance of low-order SPS-BOC signal is better than BOCs and BOCc signals when the filter bandwidth is 10 MHz, and the envelope areas is respectively reduced by 34% and 17%. However, the receiving performance of high-order SPS-BOC signals is basically the same as that of BOC signals with the increasing modulation orders.

Therefore, the low-order SPS-BOC signals have more receiving advantages than BOCs signals, and can be used as an optimization choice for the low-order components of future MBOC signals, laying the foundation for the design of next-generation navigation signals.

Acknowledgments. This work is supported by National Science Foundation of China (41604016, 61601485).

References

1. Betz, J.W.: The offset carrier modulation for GPS modernization. In: Proceedings of the 1999 National Technical Meeting of the Institute of Navigation, pp. 639–648 (1999)
2. Spilker, J.J.: A family of split spectrum GPS civil signals. In: Proceedings of International Technical Meeting of the Satellite Division of the Institute of Navigation, pp. 1905–1914 (1998)
3. Betz, J.W.: Binary offset carrier modulations for radionavigation. Navigation **48**(4), 227–246 (2001)
4. Kaplan, E.D., Hegarty, C.J.: Understanding GPS: Principles and Applications, 2nd edn. Artech House, Norwood (2006)
5. Avila-Rodríguez, J.A.: On generalized signal waveforms for satellite navigation, Ph.D. thesis, University of FAF Munich, Munich, Germany (2008)
6. De Gaudenzi, R., Hoult, N., Batchelor, A., Burden, G.: Galileo Signal Development. Wiley, New York (2000)
7. Hegarty, C.J., Betz, J.W., Saidi, A.: Binary coded symbol modulations for GNSS. In: Proceedings of ION NTM, pp. 55–64 (2005)
8. Liu, W., Hu, Y., Zhan, X.Q.: Generalised binary offset carrier modulations for global navigation satellite systems. Electron. Lett. **48**(5), 284–285 (2012)
9. Hein, G.W., Avila-Rodriguez, J.A., Wallner, S., Pratt, A.R., Owen, J., Issler, J., et al.: MBOC: the new optimized spreading modulation recommended for GALILEO L1 OS and GPS L1C. In: 2006 IEEE/ION Position, Location, and Navigation Symposium, pp. 883–892. IEEE, Coronado (2006)

Analysis of Multipath Error Characteristics of BeiDou Navigation Signal

Yi Lu, Zhibin Xiao, Yaoding Wang, and Shaojie Ni(✉)

Satellite Navigation R&D Center, School of Electronic Science and Engineering, National University of Defense Technology, Changsha 410073, China
nishaojie123@126.com

Abstract. Pseudorange multipath error is one of the main error sources of BeiDou Navigation Satellite System (BDS), all precise applications which use code measurements are severely affected. In view of the current situation that BDS lacks comprehensive assessment of multipath error characteristics, this paper analyzes time and frequency domain characteristics of multipath error and its relationship with elevation angle, following with its performance under different signal modulation systems, correlator intervals and RF bandwidths. By way of tri-frequency pseudorange and carrier phase combination method, the results demonstrate that three types (MEO, GEO, IGSO) of satellites' multipath error all reflect low frequency characteristics, and the multipath error elevation-dependence of BDS-3 MEO satellites is significantly lower than that of BDS-2. Compared with signal B1I with 4 M modulating bandwidth, the multipath error fluctuation of B2I, B3I and B2a (single sideband reception) with 10 M bandwidth are smaller. For the new system broadband modulation system, signal using TMBOC(6, 1, 4/33) modulating method has better anti-multipath performance than BOC(1, 1). In addition, increasing RF bandwidth or reducing correlator interval can both impair multipath error. The research results would have certain guiding significance for the performance evaluation of the new Beidou navigation signal system and the system design for Beidou navigation receiver.

Keywords: BeiDou · Pseudorange multipath · BOC · RF bandwidth · Correlator interval

1 Introduction

Beidou Satellite Navigation System (BDS) is established and operated independently by China. At present, the construction of BDS-3 system is gradually promoting, its scope of service has achieved coverage of countries and regions along the "Belt and Road" in 2018, then the global coverage will be completed in 2020 [1, 2]. Additionally, the new system B1C (1575.420 MHz) and B2a (1191.795 MHz) signals have also been broadcast on previously launched MEO and IGSO satellites, providing open services [3].

With the continuous development of satellite navigation system, systematic errors derive from ionosphere and troposphere have been effectively suppressed, and multipath error has become one of the main error sources restricting system performance,

J. Sun et al. (Eds.): CSNC 2020, LNEE 650, pp. 527–536, 2020.
https://doi.org/10.1007/978-981-15-3707-3_50

especially for areas with complex observation environments. The concept of code phase divergence is first proposed in literature [4] and was proved in subsequent experiments and operations. The long-term multipath error of BDS IGSO and MEO satellites are analyzed experimentally [5, 6]. Results show that pseudorange multipath error (MP) of BDS satellites are dependent on elevation angle and frequency, causing code-phase divergences of more than 1 m, all precise applications which use code measurements are severely affected [7, 8]. Hauschild [9] points out that this kind of pseudorange error can be eliminated by receiver multipath weakening technique. The pseudorange multipath error caused by satellite is greatly reduced after various correction models are adopted. Literature [10, 11] analyzed the characteristics of satellite error of BDS-3. The experiment showed that the effect of multipath error was not significantly reflected in the new generation satellites. However, none of the above studies analyzed the multipath error characteristics of BDS satellite signals fully. Their experiments only involved MEO, IGSO satellites or B1 and B3 frequency points.

In this paper, an in-depth study is made on problems existing in BDS navigation signals described. Firstly, the extracting method of MP sequence is briefly introduced. On this basis, the actual received data is comprehensively analyzed from three aspects including satellite, signal frequency band and receiver, so as to provide references for theoretical evaluation of signal system performance and system design of BDS navigation receiver.

2 Pseudorange Multipath Extracting Method

Previous studies show that pseudorange measurement error caused by multipath can reach meter lever, while the error of carrier phase will not exceed 1/4 wavelength. Therefore, linear combination of carrier phase and pseudorange can be used to obtain multipath sequence.

The multi-frequency pseudorange and carrier phase observation equations of GNSS system are:

$$\rho_i = R + c(dt_r + dt^s) + T + Ion_i + MP_i + e_i \tag{1}$$

$$\Phi_i = \lambda_i \varphi_i = R + c(dt_r + dt^s) + T - Ion_i + \lambda_i N_i + M\phi_i + \varepsilon_i \tag{2}$$

with ρ, φ the pseudorange (unit: m) and carrier-phase (unit: cycle) measurement, respectively; i the frequency point mark; λ_i, N the wavelength and integer ambiguity; R, Ion, T the geometric distance between satellite and receiver antenna, ionospheric and tropospheric delays, respectively; c the speed of light in vacuum; dt_r, dt^s the satellite and receiver clock offset (unit: s), respectively; $MP, M\phi$ the pseudorange and carrier-phase multipath error; e, ε the measurement noise for range and phase. Among these, the ionospheric delay can be expressed in polynomial form as follow:

$$Ion_i = \sum_{k=1}^{K} \frac{A_k}{f_i^{k+1}} \tag{3}$$

Where f_i, A_k represent carrier wavelength and k-order coefficient of ionosphere, K is the maximum term of the ionosphere. Studies have shown that the first-order term of ionosphere accounts for more than 99% of total delay, the error introduced by the second-order term is on centimeter-level, while the third-order term is generally three orders of magnitude smaller than the first-order [12]. Thereby, third order and higher terms are generally ignored in engineering.

The geometric distance between satellite and earth, tropospheric delay, satellite and receiver clock offset can be eliminated by forming linear combination of Eqs. (1) and (2):

$$MP_i = \rho_i - \sum_{j=1}^{J} b_{i,j}\lambda_i\varphi_j - e_i - \sum_{j=1}^{J} b_{i,j}(M\phi_j + \varepsilon_j) - \sum_{j=1}^{J} b_{i,j}N_j \tag{4}$$

Where the first and second items on the right of the equal sign are measured values; third and fifth terms are error terms; fourth and sixth terms are integer ambiguity of pseudo-code device delay and carrier-phase, which can be treated as a constant [13–16] and eliminated by removing the mean value of the MP sequence of a complete arc.

Taking J = 3, K = 1, consider the first order model of ionosphere:

$$Ion_i = \frac{A_1}{f_i^2}(i = 1, 2, 3) \tag{5}$$

Assuming that the precision of tri-frequency carrier-phase measurements are equal, and then the linear combination coefficients of carrier phase can be obtained by solving the following optimal problem $(i = 1, 2, 3)$, the specific derivation process is not explained here, details can be checked in literature [17]:

$$Min\{b_{i,1}^2 + b_{i,2}^2 + b_{i,3}^2 / b_{i,1} + b_{i,2} + b_{i,3} = 1, \frac{b_{i,1}}{f_1^2} + \frac{b_{i,2}}{f_2^2} + \frac{b_{i,3}}{f_3^2} = -\frac{1}{f_i^2}\} \tag{6}$$

3 Analysis of BDS Signal Multipath Observation Data

The observation data used for analysis were collected during three periods of time: 2019.10.22 11:00-10.23 11:00, 2019.10.25 19:00-10.26 19:00, 2019.11.25 9:00-10.26 9:00, respectively. Each segment of collection time is 24 h, and the sampling interval is 1 s. The observation site is located on the rooftop of Satellite Navigation R&D Center of National University of Defence Technology, equipped with iGMAS antenna, connecting to three GNSS receivers, forming the condition of zero-baseline. What needs to be explained here is that the surrounding environment along with different receiver hardware characteristics would have influence on multipath effect [18]. Therefore, this paper selects three receivers of the same type for analysis to make the results with more reference value (Fig. 1).

Fig. 1. Observation environment

3.1 Analysis of BDS Signal Multipath Observation Data

3.1.1 Spectral Characteristics of GEO, MEO, IGSO Satellite Pseudorange Multipath Error

Carrying out fast Fourier transformation (FFT) on B1 signal MP time series of GEO (C01), MEO(C13) and IGSO(C10) satellites, the results are shown in Fig. 2. The spectrum characteristics of satellites are correspondingly identical, all of which reflect the low-frequency characteristics, indicating that low-frequency components occupy the main part of the signal. This is consistent with the conclusion in literature [19]. However, the acquisition time is not long enough to analyze the periodic characteristics based on satellite orbits by spectral characteristics.

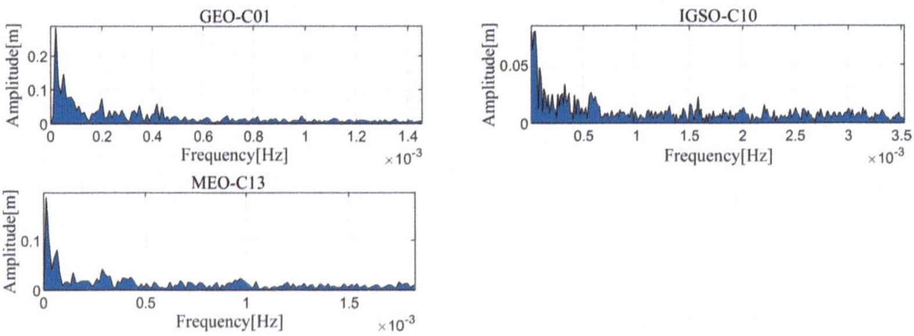

Fig. 2. Spectrum analysis of BDS-2 GEO, MEO, IGSO satellite multipath

3.1.2 Multipath Characteristics Comparison Based on Elevation Variation

Based on MP time series of individual ambiguity blocks, we estimated elevation-dependent polynomial models of the satellite-induced code-pseudorange bias. The function of MP combination values and elevation angles is set as [20, 21]:

$$MP_i(j) = B_0 + B_1E(j) + B_2E(j)^2 + B_3E(j)^3 \tag{7}$$

Where $MP_i(j)$ refers to MP values with unknown ambiguity and hardware delay removed; $E(j)$ denotes the satellite elevation angle in degree and B_0, B_1, B_2, B_3 stands for the coefficients of polynomial models.

The pseudorange multipath of BDS-2 C01, C10, and C13 satellites as a function of elevation angle is shown in Fig. 3 and the RMS of BDS-3 C32, C33 satellites is shown in Table 1. The elevation variation range of GEO satellite is much smaller than that of

MEO and IGSO, within 5° of the observed arc, making it difficult to determine the specific correlation between pseudorange observations and multipath deviation of elevation angle. It can be seen that the pseudorange error of BDS-2 MEO and IGSO satellites is intuitively dependent on elevation, which decreases with the increase of elevation angle, the main fluctuation is within 2 m. Table 1 shows that the multipath-elevation fitting parameters of BDS-3 satellite (data were collected at the same time) are nearly zero, indicating that the elevation-dependence of pseudorange multipath error fluctuation of BDS-3 satellite is basically eliminated. However, further verification is needed to reach a general conclusion.

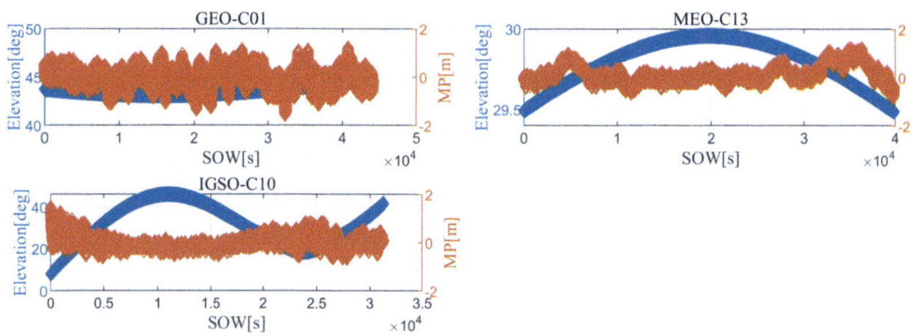

Fig. 3. MP error series over elevation angles for BDS-2 GEO, MEO, IGSO satellites

Table 1. Cubic polynomial fitting parameters of MP series for BDS-3 MEO satellite

Satellite	B_0	B_1	B_3	B_4
C32	0.0087	−0.0021	−0.0003	1.98×10^{-6}
C33	0.0276	0.0039	−0.0001	1.12×10^{-6}

3.2 Comparison of Pseudorange Multipath Error Characteristics of Signals on Different Working Bandwidth

3.2.1 BPSK (2) and BPSK (10)

Figure 4 shows BDS-2 GEO, MEO, and IGSO satellites' multipath error sequences, it can be seen that the B1I frequency point error of GEO and MEO satellites is relatively large, with about 2 m, while the multipath error of each frequency point for IGSO satellite is around 1 m. As can be seen from the figure, multipath error of GEO is more significant than that of MEO and IGSO, showing an obvious periodic changing trend additionally. Besides, the amplitude of multipath error fluctuation for B1, B2, B3 frequency point decreases successively. Thus, the multipath error is bandwidth-dependent, and this phenomenon is most prominently visible to GEO satellites. This is because signal B1I work on 4 M bandwidth while signal B2I, B3I extend to 10 M. For different types of satellites, B1 frequency point signal has the weakest multipath resistance, while B3 frequency point is the strongest.

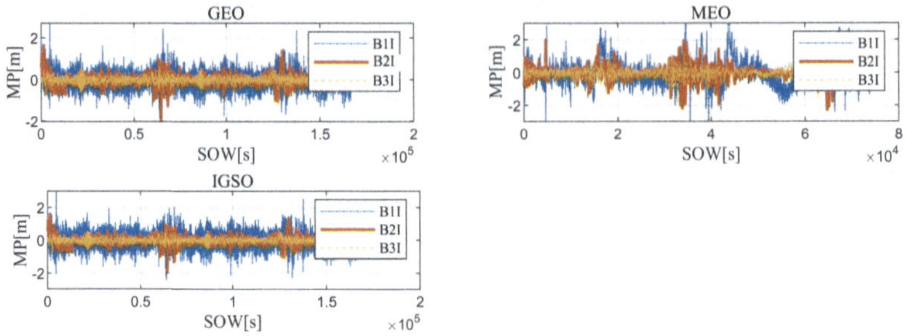

Fig. 4. MP error series for BDS-2 GEO, MEO, IGSO satellites on B1, B2, B3 frequency

3.2.2 TMBOC (6, 1, 4/33) and BOC (1, 1)

B1C signals are transmitted in TMBOC (6, 1) modulation with a center frequency of 1575.42 MHz, which are compatible with GPS/Galileo (L1/E1) signals. Its data channel B1Cd uses the BOC (1, 1) modulation method, and the pilot channel B1Cp uses the TMBOC (6, 1, 4/33) modulation method (Fig. 5).

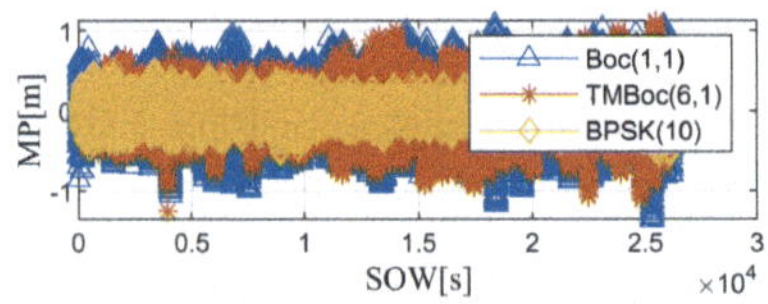

Fig. 5. MP error series under different modulation methods

The results show that the anti-multipath capability of signal B1C is weaker compared with signal B2a. The multipath error of B1C data channel signal is around 1 m, while the pilot channel is around 0.6–0.7 m, and the B2a signal error is reduced to less than 0.5 m. It may be caused by the small bit rate of B1I. The robustness to multipath of the B1C two channels can be clearly seen as follows: TMBOC (6, 1, 4/33)> BOC (1, 1).

3.3 Comparative Analysis of Pseudorange Multipath Error Characteristics Under Different Receiver Parameters

3.3.1 Effects of Different RF Bandwidths

Figure 6 shows the pseudorange multipath error sequence under RF bandwidths of 3 MHz and 28 MHz. Receiver correlator interval is set to 1/8 chip. As can be seen: when the RF bandwidth is relatively large, the multipath error is small; when the RF bandwidth is 3 MHz, the maximum value of the multipath error reaches 8 m; when it is added up to 28 MHz, the maximum value of the multipath error is only 2 m. However, when the receiver bandwidth is larger, the capability of preventing multipath interference of the broadband BOC signal is better than that of the narrowband modulation signal. Increasing the receiver bandwidth significantly improves the multipath error performance of the BOC signal, while the improvement of BPSK modulation signal is limited. This is because the BOC signal is mostly filtered out when the bandwidth is 3 MHz, and the advantage is only apparent when the signal bandwidth is increased. In combination with the above considerations, the receiver should try to ensure a larger receiver bandwidth in the aspect of combating multipath, but at the same time, it should also balance the resulting noise performance degradation (Table 2).

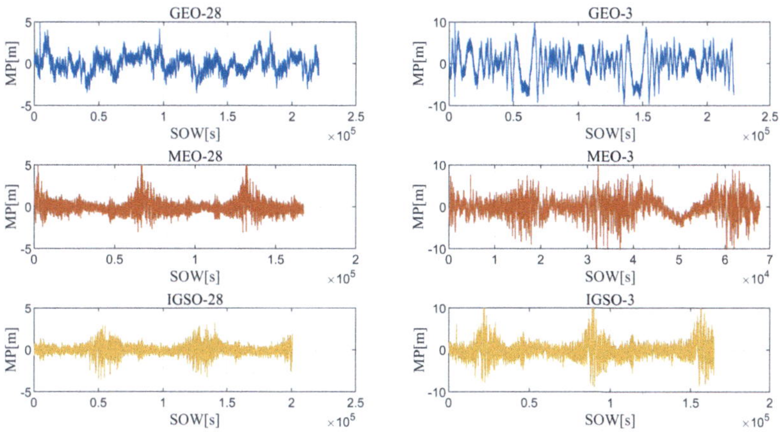

Fig. 6. MP error series for BDS GEO, MEO, IGSO satellites under different RF bandwidths

Table 2. RMS of MP error for BDS GEO, MEO, IGSO satellites under different RF bandwidths

RF bandwidth/MHz	GEO	MEO	IGSO
3	6.4561	3.9526	2.1323
28	0.8156	0.5882	0.3121

3.3.2 Effects of Different Correlator Intervals

Figure 7 shows the multipath error sequence under different correlator intervals when RF bandwidth is fixed at 28 MHz. It can be seen that correlator intervals have a significant impact on multipath errors. When the correlator interval decreases from 1/2 chip to 1/4 chip, the range of satellite multipath error decreases gradually but the envelope morphology remains unchanged, and the variation trend is the same for different satellite types. When the correlator interval is further reduced to 1/8 chip, the multipath error of the satellite presents a more gentle state, which basically eliminates the redundant jitter. During actual measuring of the receiver, narrow correlator interval is generally adopted to improve the tracking accuracy and performance (Table 3).

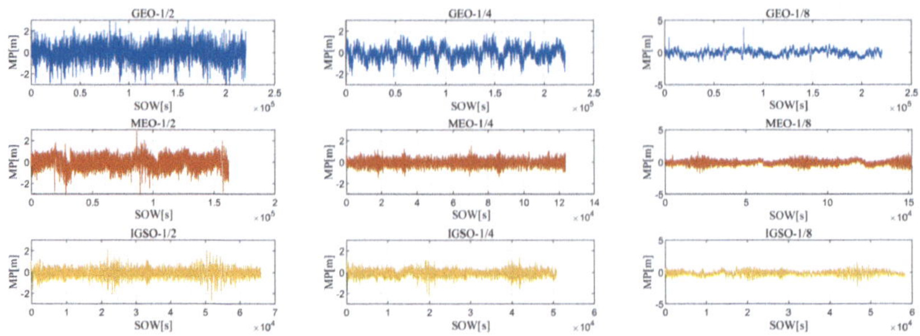

Fig. 7. The MP error series for BDS-2 GEO, MEO, IGSO satellites under different correlation intervals

Table 3. RMS of MP error for BDS GEO, MEO, IGSO satellites under different correlation intervals

Correlation interval/chip	GEO	MEO	IGSO
1/2	0.3743	0.4333	0.4407
1/4	0.2901	0.2620	0.3148
1/8	0.2298	0.1455	0.1732

4 Conclusions

Multipath error has become the main error source of high-precision GNSS positioning. Due to the complexity of multipath effect mechanism, time-varying characteristics and environmental influence, multipath error is difficult to effectively detect and eliminate. In this paper, the combined observation values of carrier phase and pseudorange are used to compare and analyze the pseudorange multipath error characteristics of satellite navigation signals. The main conclusions are as follows:

(1) In time and frequency domain, the pseudorange MP error of each signal shows low frequency characteristics and the error is related to the change of satellite elevation angle.
(2) Under different frequency point, the MP error of B1, B2 and B3 signals decreases successively, from 1 m to 0.5 m, and B3 has the best anti-multipath capability. For the new system of BOC modulation, the anti-multipath capability of TMBOC (6, 1, 4/33) is better than BOC (1, 1).
(3) On the level of receiver parameter setting, MP error can be reduced by decreasing the correlator interval and increasing the RF bandwidth. Reducing correlator interval can reduce multipath RMS from 0.4 to about 0.1.

It should be pointed out that as the data used is only single-station data, relevant test results can only reflect the data quality of a certain type of receiver, and more data support is needed for the comprehensive assessment of the data quality of BDS global system satellite navigation signals.

Acknowledgements. All data in this paper are from the navigation center of National University of Defense Technology and IGS open data. Thank the engineers, teachers and students for their help in data collecting.

References

1. Liu, M.K.: Pseudo noise code tracking performance analysis of Beidou B2 frequency band navigation signal. Telecommun. Eng. **58**(1), 25–29 (2018)
2. Yang, Y.X., Mao, Y., Sun, B.J.: Basic performance and future developments of BeiDou global navigation satellite system. Satell. Navig. **1**, 1–8 (2020). https://doi.org/10.1186/s43020-019-0006-0
3. China Satellite Navigation Office: BeiDou Navigation Satellite Signal in Space Interface Control Document Open Service Signal B1C (Version 1.0) (2017)
4. Hauschild, A., Montenbruck, O., Sleewaegen, J., et al.: Characterization of compass M-1 signals. GPS Solut. **16**(1), 117–126 (2012)
5. Wanninger, L., Beer, S.: BeiDou satellite-induced code pseudorange variations: diagnosis and therapy. GPS Solut. **19**(4), 639–648 (2015)
6. Montenbruck, O., Steigenberger, P., Khachikyan, R., et al.: IGS-MGEX: preparing the ground for multi-constellation GNSS science. Inside GNSS **9**(1), 42–49 (2014)
7. Montenbruck, O., Hauschild, A., Steigenberger, P., et al.: A COMPASS for Asia: first experience with the BeiDou-2 regional navigation system. Notes **36287**(001A), C01 (2010)
8. Xu, H., Cui, X., Lu, M.: Satellite-induced multipath analysis on the cause of BeiDou code pseudorange bia. In: China Satellite Navigation Conference. Springer, Singapore, pp. 11–21 (2017)
9. Hauschild, A., Montenbruck, O., Thölert, S., et al.: A multi-technique approach for characterizing the SVN49 signal anomaly, part 1: receiver tracking and IQ constellation. GPS Solut. **16**(1), 19–28 (2012)
10. Zhang, X., He, X., Liu, W.: Characteristics of systematic errors in the BDS Hatch–Melbourne–Wübbena combination and its influence on wide-lane ambiguity resolution. GPS Solut. **21**(1), 265–277 (2017)
11. Zhang, X., Wu, M., Liu, W., et al.: Initial assessment of the COMPASS/BeiDou-3: new-generation navigation signals. J. Geod. **91**(10), 1225–1240 (2017)
12. Lachapelle, G., Falkenberg, W., Neufeldt, D., et al.: Marine DGPS using code and carrier in a multipath environment. In: Proceedings of ION GPS-89, Second International Technical Meeting of the Satellite Division of the Institute of Navigation, Colorado Springs, vol. 2729, pp. 343–347 (1989)
13. Yuan, Y., Ou, J.: Impact of instrumental deviations in GPS observation data on determining ionospheric delay and processing methods. J. Surv. Mapp. **28**(2), 110–114 (1999)
14. Xu, C.J., Chen, Y.Q., Liu, Y.X.: A new method for determining satellite and receiver signal delay deviation and its application. J. Surv. Mapp. **28**(2), 153–161 (1999)
15. Chang, Q., Zhang, D., Xiao, Z., et al.: GPS hardware delay correction method. Sci. Bull. **45**(15), 1676–1680 (2000)
16. Sardón, E., Zarraoa, N.: Estimation of total electron content using GPS data: how stable are the differential satellite and receiver instrumental biases. Radio Sci. **32**(5), 1899–1910 (1997)
17. Liu, W.X.: Study on techniques of accuracy augmenting and integrity monitoring for satellite based navigation system. National University of Defence Technology (2011)

18. He, C.Y., Lu, X.C., Guo, J., et al.: Initial analysis for characterizing and mitigating the pseudorange biases of BeiDou navigation satellite system. Satell. Navig. **1**, 1–10 (2020). https://doi.org/10.1186/s43020-019-0003-3
19. Ma, X., Shen, Y.: Multipath error analysis of COMPASS triple frequency observations. Positioning **5**(1), 12–21 (2014)
20. Cazals, F., Pouget, M.: Estimating differential quantities using polynomial fitting of osculating jets. Comput. Aided Geom. Des. **22**(2), 121–146 (2005)
21. Guo, F., Li, X., Liu, W.: Mitigating BeiDou satellite-induced code bias: taking into account the stochastic model of corrections. Sensors **16**(6), 909–925 (2016)

A Coherent Processing Technique with High Precision for BDS B1I and B1C Signals

Yang Gao[1,2], Zheng Yao[1,2](✉), and Mingquan Lu[1,2]

[1] Department of Electronic Engineering, Tsinghua University, Beijing, China
yaozheng@tsinghua.edu.cn

[2] Beijing National Research Center for Information Science and Technology, Beijing, China

Abstract. The smooth transition constraint and constant envelope restriction of the Beidou Navigation Satellite System (BDS) make the legacy regional system B1I signal and new global system B1C signal, which are located at different center frequencies, coexist in the BDS-3 B1 band. Therefore, new modulation and multiplexing techniques need to be adopted to meet these system constraints. More specifically, single-sideband complex binary offset carrier (SCBOC) modulation and constant envelope multiplexing via intermodulation construction (CEMIC) techniques are used in B1 band to construct a multicarrier constant-envelope composite navigation signal. In particular, the SCBOC modulation technique is introduced into the B1I signal to move its main energy from the global system B1 frequency to the regional system B1 frequency. However, as of now, the SCBOC modulation has only been regarded as a means to achieve the smooth update of the BDS, whose high-precision ranging potential has not been fully understood and utilized. To solve this problem, this paper proposes a high-precision coherent processing technique, which makes full use of the coherence between B1I and B1C signals. The proposed algorithm can not only exploit the ranging performance brought by high-frequency complex subcarriers without loss, but also greatly simplify the implementation complexity of the receiver. The experimental results of live BDS-3 signals verify the effectiveness and correctness of the proposed method. This paper provides a new solution for the high-precision application of the BDS B1 composite signal and has great reference value for receiver designers.

Keywords: BOC signal · Complex subcarrier · Coherent processing technique · B1 composite signal

1 Introduction

In recent years, the BDS is transitioning from the regional system (BDS-2) to the global system (BDS-3). According to the development plan of the BDS, the BDS-3 satellites not only need to broadcast a new interoperable signal B1C and a new authorized service signal B1A, but also have to be backward compatible with legacy BDS-2 B1I signal. Since the central frequencies of the regional system B1I signal and the global system B1 signal are not the same, the new modulation technique needs to be applied in the BDS B1 band to achieve multi-frequency multiplexing and backward compatibility.

J. Sun et al. (Eds.): CSNC 2020, LNEE 650, pp. 537–550, 2020.
https://doi.org/10.1007/978-981-15-3707-3_51

In addition, due to the constant envelope constraint and the asymmetric spectrum in the B1 band, the new multicarrier constant-envelope multiplexing technique also needs to be introduced.

These system constraints have promoted the emergence of new modulation and multiplexing techniques. More specifically, the SCBOC modulation [1] and CEMIC multiplexing [2, 3] techniques are adopted in the BDS-3 B1 band to construct a multicarrier constant-envelope composite navigation signal. In particular, the SCBOC modulation technique is introduced into the B1I signal to move its main energy from the global system B1 frequency to the regional system B1 frequency. In addition, the CEMIC multiplexing technique combines all the useful signals in the B1 band into a wideband constant-envelope composite signal by introducing additional intermodulation terms. However, as of now, the SCBOC modulation has only been regarded as a means to achieve the smooth update of the BDS, whose high-precision ranging potential has not been fully understood and utilized. Since the single-sideband complex subcarrier introduced in the BDS-3 B1I signal has a high frequency, it is not difficult to predict that the SCBOC modulation will bring great ranging performance potential. However, at the same time, the SCBOC modulated signal also introduces many new tracking challenges.

The first challenge is the well-known tracking ambiguity threat. Similar to other BOC-class modulated signals, the auto-correlation function (ACF) of the SCBOC modulated signal has multiple side peaks, which means that it is possible to track the side peaks instead of the main peak. Since the modulation order of the global system B1I signal is large, the ambiguity threat in this situation is very serious. In addition, different from the binary phase shift keying (BPSK) and other BOC-class modulated signals, the ACF of the SCBOC modulated signal is a complex function. Such a complex correlation peak cannot be tracked by traditional receivers [4]. When the subcarrier phase of the local replica is not aligned with that of the received signal, the energy moves from the real part to the imaginary part, which will cause the estimation deviation of the carrier phase, and vice versa. The coupling relationship between the carrier phase and the subcarrier phase makes it difficult to process the SCBOC modulated signal. One possible solution is the BPSK-Like method [5], but it ignores the ranging performance potential brought by the high-frequency subcarrier. Other existing unambiguous tracking methods [6–10] can only deal with the case where the correlation function is a real function, and cannot be applied to the SCBOC modulated signals. These complex tracking challenges and difficulties prevent the exploiting of the ranging potential of the SCBOC modulated signal.

To solve these problems, this paper proposes a high-precision coherent processing technique, which makes full use of the coherence between B1I and B1C signals. The proposed algorithm can not only exploit the ranging performance brought by high-frequency complex subcarriers without loss, but also greatly simplify the implementation complexity of the receiver. The stable carrier estimation result of the B1C signal is used to help recover the carrier of the B1I signal, while the high-precision delay estimation result of the B1I signal is used to help track the delay of the code and subcarrier of the B1C signal. The experimental results of live BDS-3 signals verify the effectiveness and correctness of the proposed algorithm. This paper provides a new solution for the high-precision application of the BDS B1 composite signal and has great reference value for receiver designers.

2 Problem Description

This section is mainly divided into two parts. The first one introduces the system constraints in the BDS B1 band and the B1 composite signal model. The second part analyses the ranging performance potential and tracking challenges brought by the single-sideband complex subcarrier.

2.1 System Constraints and Signal Model

There exist many system constraints in the signal design of the BDS-3 B1 band. Figure 1 shows the spectrum constraints of the BDS-3 B1 band. As can be seen from Fig. 1, on the one hand, the BDS-3 satellites have to broadcast the legacy BDS-2 B1I signal at the frequency of $f_{B1I} = 1561.098$ MHz to meet the constraint of backward compatibility. On the other hand, the BDS-3 satellites need to broadcast a new interoperable signal B1C at the frequency of $f_{B1} = 1575.42$ MHz to meet the constraints of compatibility and interoperability between GNSSs. In addition, the new authorized service signal B1A is also broadcast at f_{B1} to meet the performance growth needs of advanced users. Finally, these useful baseband signals in the same band need to be multiplexed into a wideband constant-envelope composite navigation signal to meet the constraint of the constant envelope.

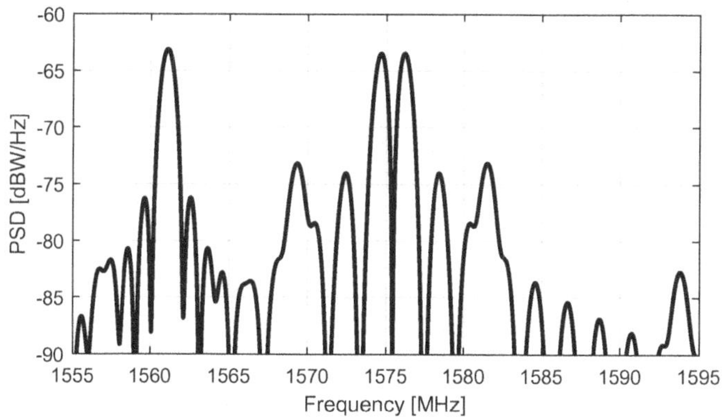

Fig. 1. The spectrum constraint of BDS-3 B1 band

These system constraints have promoted the emergence of new modulation and multiplexing techniques. More specifically, the SCBOC modulation and CEMIC multiplexing techniques are adopted in the BDS-3 B1 band to construct a multicarrier constant-envelope composite navigation signal. In particular, the SCBOC modulation technique is introduced into the B1I signal to move its main energy from the global system B1 frequency to the regional system B1 frequency. In addition, the CEMIC multiplexing technique combines all the useful signals in the B1 band into a wideband constant-envelope composite signal by introducing additional intermodulation terms. Therefore, the B1 wideband composite signal broadcast by BDS-3 satellites can be modeled as

$$S_{\mathrm{B1}}(t) = \mathrm{Re}\left\{\left(\sqrt{P_{\mathrm{B1I}}}\mathrm{e}^{\mathrm{j}\phi_{\mathrm{B1I}}} s_{\mathrm{B1I}}(t) + \sqrt{P_{B1C}}\mathrm{e}^{\mathrm{j}\phi_{B1C}} s_{B1C}(t) + \sqrt{P_{B1A}}\mathrm{e}^{\mathrm{j}\phi_{B1A}} s_{B1A}(t) + I_{IM}(t)\right)\mathrm{e}^{\mathrm{j}2\pi f_{B1}t}\right\} \tag{1}$$

where P_i, ϕ_i, and $s_i(t)$ are the nominal power, initial phase, and baseband complex envelope of the corresponding signal component $i = B1I, B1C, B1A$, respectively. The initial phases ϕ_{B1I} and ϕ_{B1C} are equal, and the baseband complex envelope can be further given by

$$s_{B1I}(t) = d_{B1I}(t)c_{B1I}(t)\gamma_{sc}(t) \tag{2}$$

$$s_{B1C}(t) = \frac{1}{2}d_{B1C}(t)c_{B1C-d}(t)sc_a(t) + c_{B1C-p}(t)\left[\sqrt{\frac{1}{11}}sc_b(t) + \mathrm{j}\cdot\sqrt{\frac{29}{33}}sc_a(t)\right] \tag{3}$$

where $d_{B1I}(t)$ and $d_{B1C}(t)$ are the navigation messages of B1I and B1C signal, respectively. $c_{B1I}(t)$, $c_{B1C-d}(t)$, and $c_{B1C-p}(t)$ are the ranging codes of the B1I signal, data channel, and pilot channel of B1C signal, respectively.

$sc_a(t) = \mathrm{sign}(\sin(2\pi f_{sc,a}t))$ and $sc_b(t) = \mathrm{sign}(\sin(2\pi f_{sc,b}t))$ are the sine-phased square wave subcarriers with subcarrier frequency $f_{sc,a} = 1f_0$ and $f_{sc,a} = 6f_0$ used by the narrowband BOC(1,1) component and the wideband BOC(6,1) component, respectively, where $f_0 = 1.023$ MHz is the GNSS baseline frequency. In addition,

$$\gamma_{sc}(t) = \mathrm{sign}(\cos(2\pi f_{sc,B1I}t)) - \mathrm{j}\cdot\mathrm{sign}(\sin(2\pi f_{sc,B1I}t)) \tag{4}$$

is the single-sideband complex subcarrier with the frequency of $f_{sc,B1I} = f_{B1} - f_{B1I} = 14f_0$, where $\mathrm{sign}(x)$ is the sign function which takes value of 1 for $x \geq 0$ and -1 for $x < 0$. $I_{IM}(t)$ is the additional intermodulation term to maintain the constancy of the composite signal envelope.

Considering the effect of the bandlimiting filter, the single-sideband complex subcarrier $\gamma_{sc}(t)$ can be approximated as $\gamma_{sc}(t) \approx \mathrm{e}^{-\mathrm{j}2\pi f_{sc,B1I}t}$. Ignoring the influences of the authorized service signal and the introduced intermodulation terms, Using (1)–(4), the received B1 wideband composite signal at the antenna of the GNSS receiver can be represented as

$$
\begin{aligned}
r_{B1}(t) = & \underbrace{\sqrt{P_{B1I}} d_{B1I}(t-\tau) c_{B1I}(t-\tau) \cos\left(2\pi\left(f_{B1} - f_{sc,B1I} + f_D\right)t + \varphi + 2\pi f_{sc,B1I}\tau\right)}_{r_{B1I}(t)} \\
& + \underbrace{\frac{1}{2}\sqrt{P_{B1C}} d_{B1C}(t-\tau) c_{B1C-d}(t-\tau) sc_a(t-\tau) \cos(2\pi(f_{B1} + f_D)t + \varphi)}_{r_{B1C-d}(t)} \\
& + \underbrace{\sqrt{\frac{P_{B1C}}{11}} c_{B1C-p}(t-\tau) sc_b(t-\tau) \cos(2\pi(f_{B1} + f_D)t + \varphi)}_{r_{B1C-pb}(t)} \\
& - \underbrace{\sqrt{\frac{29 P_{B1C}}{33}} c_{B1C-p}(t-\tau) sc_a(t-\tau) \sin(2\pi(f_{B1} + f_D)t + \varphi)}_{r_{B1C-pa}(t)} \\
& + n(t)
\end{aligned}
\tag{5}
$$

where τ is the signal propagation delay, f_D is the Doppler shift, φ is the carrier phase, $n(t)$ is the zero-mean Gaussian white noise with PSD N_0. The Eq. (5) is the B1I and B1C signal model to be processed.

2.2 Ranging Potential and Tracking Challenges

The previous subsection gives the model of the BDS B1 composite signal. It can be seen from (5) that the BDS-3 B1I signal using SCBOC(14,2) modulation can be easily processed by the traditional BDS-2 receiver as a legacy BPSK(2) modulated signal, which meets the requirement of backward compatibility. However, as of now, the SCBOC modulation has only been regarded as a means to achieve the smooth update of the BDS, whose high-precision ranging potential has not been fully understood and utilized. Therefore, this paper, for the first time, points out that the $r_{B1I}(t)$ in Eq. (5) can not only meet the constraint of the backward compatibility but also contain great ranging performance potential. To fully use it will probably improve the ranging and positioning precision of the BDS-3 B1 band.

Figure 2 shows the Gabor bandwidth comparison of all civil signals in GNSS community. Generally, the wider the Gabor bandwidth, the better the thermal noise performance, which means the potential of signal in ranging and positioning performance is greater. As can be seen from Fig. 2, the SCBOC(14,2) modulated signal and the AltBOC(15,10) modulated signal have the first-class Gabor bandwidths of all civil signals, which means that they have great ranging performance potential. In addition, compared with the AltBOC(15,10) modualted signal, which has similar Gabor bandwidth, the SCBOC(14,2) modulated signal requires a narrower front-end bandwidth to receive the entire main lobe, which means receiving and processing SCBOC(14,2) modulated signal has a better cost performance. To sum up, the SCBOC(14,2) modulated signal has great ranging potential, and its effective use is worthy of attention. However, at the same time, the SCBOC modulated signal also introduces many new tracking challenges.

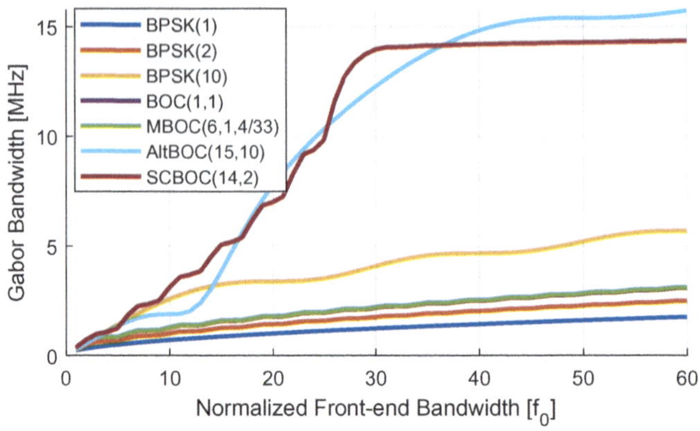

Fig. 2. The Gabor bandwidth comparison of all civil signals in GNSS community

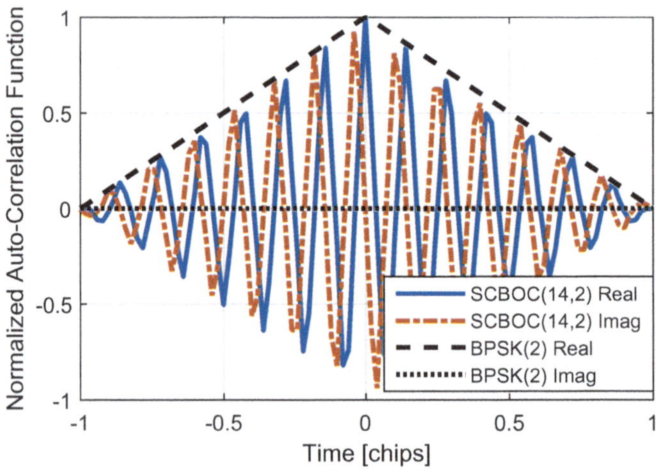

Fig. 3. Normalized auto-correlation function of the SCBOC(14,2) modulated signal

Figure 3 shows the normalized ACF of the SCBOC(14,2) modulated signal. For comparison, the ACF of the BPSK(2) modulated signal is also given. As can be seen from Fig. 3, on the one hand, the ACF of the SCBOC (14,2) modulated signal has multiple side peaks, which means that it is possible to track the side peaks instead of the main peak. This is the well-known ambiguity threat in BOC-class modulated signals. Since the modulation order of the SCBOC(14,2) is large, the ambiguity threat in this situation is very serious. On the other hand, different from the ACF of BPSK and BOC-class modulated signals, the ACF of SCBOC(14,2) modulation signal is a complex function. Such a complex correlation peak cannot be tracked by traditional receivers. When the subcarrier phase of the local replica is not aligned with that of the received signal, the energy moves from the real part to the imaginary part, which will cause the estimation deviation of the carrier phase, and vice versa. The coupling relationship between the carrier phase and the subcarrier phase makes it difficult to process the SCBOC modulated signal. These complex tracking challenges and difficulties prevent the exploiting of the ranging potential of the SCBOC modulated signal.

3 Proposed Algorithm

This section makes full use of the coherence between B1I and B1C signals to solve these tracking challenges while pursing the ranging potential brought by the high-frequency complex subcarrier. More specifically, we propose a high-precision coherent processing technique. The proposed algorithm can not only exploit the ranging performance brought by high-frequency complex subcarriers without loss, but also greatly simplify the implementation complexity of the receiver. The stable carrier estimation result of the B1C signal is used to help recover the carrier of the B1I signal, while the high-precision delay estimation result of the B1I signal is used to help track the delay of the code and subcarrier of the B1C signal. Therefore, using the coherence between B1I and B1C signals, only the prompt correlators are required in the tracking phase of the B1C signal, thus greatly simplifying the implementation complexity of the receiver while exploiting the high-precision ranging performance.

Figure 4 shows the schematic representation of the coherent processing technique for BDS B1I and B1C signals. It should be noted that a coherent discriminator is adopted in subcarrier branch. Colors help identify different loops. The carrier loop is in red, the code loop is in green, and the subcarrier branch is in blue. For the sake of simplicity, all signals in Fig. 4 are complex signals.

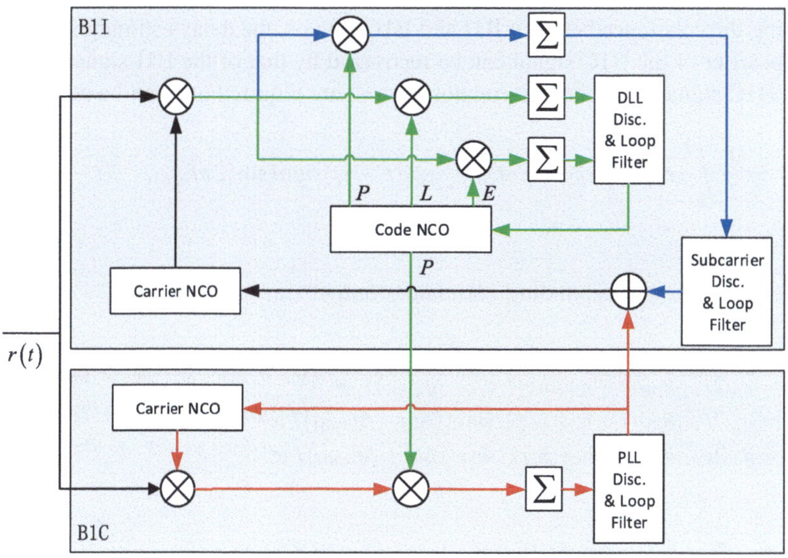

Fig. 4. Schematic representation of the coherent processing technique for BDS B1I and B1C signals (coherent discriminator used by subcarrier branch), carrier loop (red), code loop (green), subcarrier branch (blue)

At first, the carrier of the B1C signal can be removed from the received B1 composite signal by multiplying locally generated complex carrier $y_{\mathrm{B1C}}(t) = \mathrm{e}^{-\mathrm{j}\left(2\pi\left(f_{B1}+\hat{f}_D\right)t+\hat{\varphi}+\pi/2\right)}$, where $\hat{f}_D$ and $\hat{\varphi}$ are the estimations of Doppler shift and carrier phase, respectively. Note that only B1C pilot narrowband BOC(1,1) component is considered here. $\pi/2$ represents that the B1I and B1C pilot BOC(1,1) component are phase-orthogonal. Similarly, the carrier of the global system B1I signal can be removed from the received B1 composite signal by multiplying locally generated complex carrier $y_{SCBOC}(t) = \mathrm{e}^{-\mathrm{j}\left(2\pi\left(f_{B1}-\hat{f}_{sc,B1I}+\hat{f}_D\right)t+\hat{\varphi}+\hat{\varphi}_{sc,B1I}\right)}$, where $\hat{f}_{sc,B1I}$ is the estimation of the subcarrier frequency, and $\hat{\varphi}_{sc,B1I} = 2\pi\hat{f}_{sc,B1I}\hat{\tau}_s$ is the estimation of the subcarrier phase, and $\hat{\tau}_s$ is the phase delay estimation in the subcarrier dimension.

Then, the product of the locally generated code with the received signal is subjected to the coherent integration to calculate the correlators output results. For the SCBOC (14,2) modulated signal, the early (E), prompt (P), and late (L) correlators output results are needed and can be represented as

$$
\begin{aligned}
P_{SCBOC} &= \frac{1}{T}\int_0^T r_{\mathrm{B1}}(t)\cdot y_{\mathrm{SCBOC}}(t)\cdot c_{\mathrm{B1I}}(t-\hat{\tau}_c)\cdot \mathrm{d}t \\
E_{SCBOC} &= \frac{1}{T}\int_0^T r_{\mathrm{B1}}(t)\cdot y_{\mathrm{SCBOC}}(t)\cdot c_{\mathrm{B1I}}(t-\hat{\tau}_c-\Delta_c/2)\cdot \mathrm{d}t \\
L_{SCBOC} &= \frac{1}{T}\int_0^T r_{\mathrm{B1}}(t)\cdot y_{\mathrm{SCBOC}}(t)\cdot c_{\mathrm{B1I}}(t-\hat{\tau}_c+\Delta_c/2)\cdot \mathrm{d}t
\end{aligned}
\tag{6}
$$

where T is the coherent integration time, $\hat{\tau}_c$ is the phase delay estimation in code dimension, Δ_c is the spacing between E and L replicas for the code discriminator.

Using the coherence between B1I and B1C signals, the delay estimation of the code and subcarrier of the B1C signal can be recovered by that of the B1I signal. Therefore, for the B1C signal, only the P correlators output are required and can be represented as

$$
P_{\mathrm{B1C}} = \frac{1}{T}\int_0^T r_{\mathrm{B1}}(t)\cdot y_{\mathrm{B1C}}(t)\cdot C_{B1C\_pilot}(t-\hat{\tau}_c)\mathrm{sign}\left(\sin\left(2\pi f_{sc,B1C\_a}(t-\hat{\tau}_c)\right)\right)\cdot dt
\tag{7}
$$

Therefore, the corresponding correlators output can be given by

$$
\begin{aligned}
P_{B1C} &= \sqrt{2P_{B1C}}R_{B1C}(\tau-\hat{\tau}_c)\sin\mathrm{c}(\Delta f_D T)\mathrm{e}^{\mathrm{j}(\pi\Delta f_D T+\Delta\varphi)} \\
P_{SCBOC} &= d\sqrt{2P_{B1I}}R_{B1I}(\tau-\hat{\tau}_c)\sin\mathrm{c}\left(\left(\Delta f_D-\Delta f_{sc,B1I}\right)T\right)\mathrm{e}^{\mathrm{j}\left(\pi\left(\Delta f_D-\Delta f_{sc,B1I}\right)T+\Delta\varphi+2\pi f_{sc,B1I}\tau-2\pi\hat{f}_{sc,B1I}\hat{\tau}_s\right)} \\
E_{SCBOC} &= d\sqrt{2P_{B1I}}R_{B1I}(\tau-\hat{\tau}_c-\Delta_c/2)\sin\mathrm{c}\left(\left(\Delta f_D-\Delta f_{sc,B1I}\right)T\right)\mathrm{e}^{\mathrm{j}\left(\pi\left(\Delta f_D-\Delta f_{sc,B1I}\right)T+\Delta\varphi+2\pi f_{sc,B1I}\tau-2\pi\hat{f}_{sc,B1I}\hat{\tau}_s\right)} \\
L_{SCBOC} &= d\sqrt{2P_{B1I}}R_{B1I}(\tau-\hat{\tau}_c+\Delta_c/2)\sin\mathrm{c}\left(\left(\Delta f_D-\Delta f_{sc,B1I}\right)T\right)\mathrm{e}^{\mathrm{j}\left(\pi\left(\Delta f_D-\Delta f_{sc,B1I}\right)T+\Delta\varphi+2\pi f_{sc,B1I}\tau-2\pi\hat{f}_{sc,B1I}\hat{\tau}_s\right)}
\end{aligned}
\tag{8}
$$

where $R_{B1C}(\tau-\hat{\tau}_c)$ is the ACF of the B1C pilot BOC(1,1) component, $R_{B1I}(\tau-\hat{\tau}_c)$ is the code dimension ACF of the SCBOC(14,2) modulated signal, $\Delta f_D = f_D - \hat{f}_D$ and $\Delta\varphi = \varphi - \hat{\varphi}$ are the corresponding estimation errors, respectively. $\Delta f_{sc,B1I} = f_{sc,B1I} -$

$\hat{f}_{sc,B1I}$ is the subcarrier frequency estimation error, $2\pi\left(f_{sc.B1I}\tau - \hat{f}_{sc,B1I}\hat{\tau}_s\right) = \Delta\varphi_{sc,B1I}$ is the subcarrier phase estimation error.

By assuming the perfect Doppler shift and subcarrier frequency synchronization, that is, $\Delta f_D \approx 0$ and $\Delta f_{sc,B1I} \approx 0$, the correlators outputs (8) can be further simplified as

$$\begin{aligned} P_{B1C} &= \sqrt{2P_{B1C}}R_{B1C}(\tau - \hat{\tau}_c)\mathrm{e}^{\mathrm{j}(\Delta\varphi)} \\ P_{SCBOC} &= d\sqrt{2P_{B1I}}R_{B1I}(\tau - \hat{\tau}_c)\mathrm{e}^{\mathrm{j}\left(\Delta\varphi + 2\pi f_{sc,B1I}\tau - 2\pi\hat{f}_{sc,B1I}\hat{\tau}_s\right)} \\ E_{SCBOC} &= d\sqrt{2P_{B1I}}R_{B1I}(\tau - \hat{\tau}_c - \Delta_c/2)\mathrm{e}^{\mathrm{j}\left(\Delta\varphi + 2\pi f_{sc,B1I}\tau - 2\pi\hat{f}_{sc,B1I}\hat{\tau}_s\right)} \\ L_{SCBOC} &= d\sqrt{2P_{B1I}}R_{B1I}(\tau - \hat{\tau}_c + \Delta_c/2)\mathrm{e}^{\mathrm{j}\left(\Delta\varphi + 2\pi f_{sc,B1I}\tau - 2\pi\hat{f}_{sc,B1I}\hat{\tau}_s\right)} \end{aligned} \tag{9}$$

Then, the correlator outputs are input to the discriminators to get the estimation error. Considering the code loop uses a standard DLL to process E_{SCBOC} and L_{SCBOC} correlator outputs, detailed DLL operations are not discussed further. Similarly, the carrier loop uses a standard PLL to process P_{B1C} correlator outputs, so the PLL operations are not discussed further. The following discussion focuses on the operation of the subcarrier branch.

From (9), it is possible to design discriminators of subcarrier branch. Generally, there are two classes of discriminators can be employed. The first class is coherent discriminators, in which the effect of the residual carrier phase is neglected ($\Delta\varphi = 0$), while the other class is non-coherent discriminators, which are designed to operate even in the presence of the residual carrier phase errors ($\Delta\varphi \neq 0$). For the coherent discriminators, an example can be

$$\phi_{c,SCBOC}\left(\Delta\varphi_{sc,B1I}\right) = \arctan\left(\frac{\mathrm{Re}(P_{SCBOC})}{\mathrm{Im}(P_{SCBOC})}\right) \tag{10}$$

when the residual carrier phase is small enough and can be approximated as $\Delta\varphi = 0$, the P_{SCBOC} correlator results can be directly used to calculate the subcarrier phase estimation error. For the non-coherent discriminators, an example can be

$$\phi_{nc,SCBOC}\left(\Delta\varphi_{sc,B1I}\right) = \arctan\left(\frac{\mathrm{Re}\left(P_{SCBOC}P^*_{B1C}\right)}{\mathrm{Im}\left(P_{SCBOC}P^*_{B1C}\right)}\right) \tag{11}$$

where $*$ represents conjugate operator. The working principle of (11) can be verified by trigonometric identity.

Next, the filtered B1I subcarrier estimation error and the filtered B1C carrier estimation error are both input to the carrier NCO of the B1I to update the locally generated complex carrier $y_{\mathrm{SCBOC}}(t)$, thus closing the tracking loop.

When the unambiguous but low-precision code dimension delay estimation $\hat{\tau}_c$ and the high-precision but ambiguous subcarrier dimension delay estimation $\hat{\tau}_s$ are obtained, the final propagation delay estimation $\hat{\tau}$ can be represented as

$$\hat{\tau} = \hat{\tau}_s + \text{round}\left(\frac{\hat{\tau}_c - \hat{\tau}_s}{T_s}\right) \times T_s \tag{12}$$

where $T_s = 1/2f_{sc,B1I}$ is the subcarrier chip width.

4 Experimental Results

In this section, the live BDS-3 signals are used to verify the correctness and effectiveness of the proposed algorithm and further investigate the performance of the method. It should be noted that the data set used in this paper was collected on December 9, 2019, which contains valid B1 wideband composite signals since there are already 28 BDS-3 satellites in orbit.

Table 1 shows parameters used in the receiver for real data processing. The front-end central frequency is 1575.42 MHz. The front-end filter bandwidth is 40 MHz. Therefore, zero-IF complex digital signals can be obtained. The sampling frequency is 40 MHz, which is wide enough to contain the entire main lobe of the B1 composite signal. The carrier loop is achieved using a second-order frequency lock loop (FLL) aided three-order PLL. The code loop is implemented using a second-order DLL, and the subcarrier discriminator uses a coherent discriminator. The coherent integration time is constant at 1 ms, which is the length of one code period. It should be noted that the loop parameters used are typical values for land applications.

Table 1. Parameters used in the receiver for real data processing

Parameter	Value
Front-end central frequency	1575.42 MHz
Front-end filter bandwidth	40 MHz
Sampling frequency	40 MHz
Sampling type	Complex I and Q
FLL order	2
FLL bandwidth	2 Hz
PLL order	3
PLL bandwidth	20 Hz
DLL order	2
DLL bandwidth	5 Hz
DLL early-minus-late spacing	0.2 chips
Subcarrier branch order	2
Subcarrier branch bandwidth	5 Hz
Integration time	1 ms

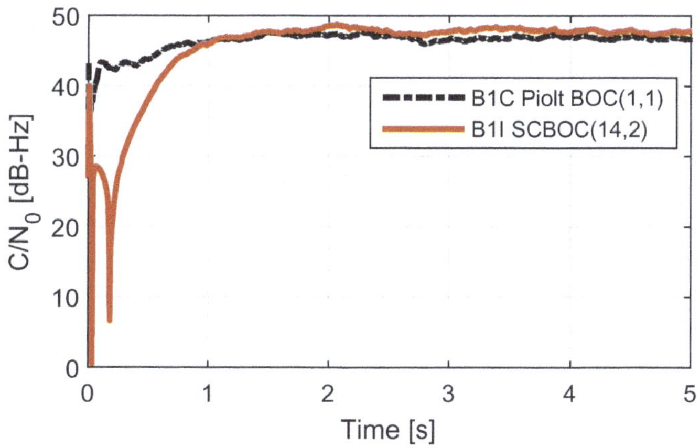

Fig. 5. C/N_0 estimation for the B1I and B1C signals

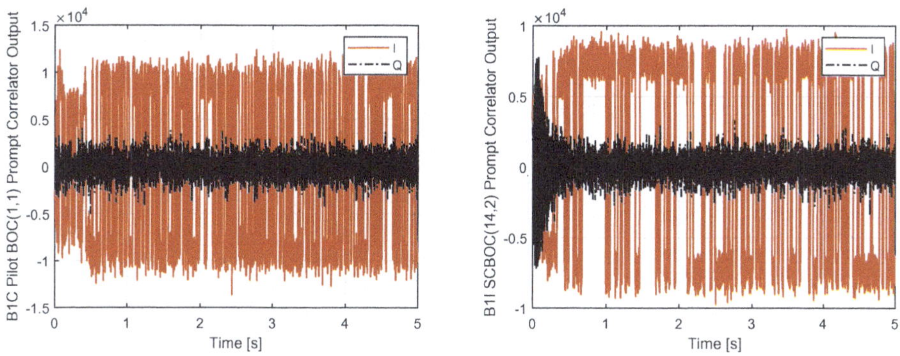

Fig. 6. Prompt correlator output results (Left: B1C pilot BOC(1,1) component. Right: B1I SCBOC(14,2) component)

Figure 5 shows the C/N_0 estimation for the B1I and B1C signals. It can be seen from Fig. 5 that after about 1 s, the C/N_0 of the proposed method tends to be stable, which means the proposed method can stably track the SCBOC(14,2) modulated signal. In addition, the proposed method takes a longer time to enter the stable tracking phase, which is because the carrier needs to be tracked stable at first and then the complex subcarrier starts to effectively demodulate.

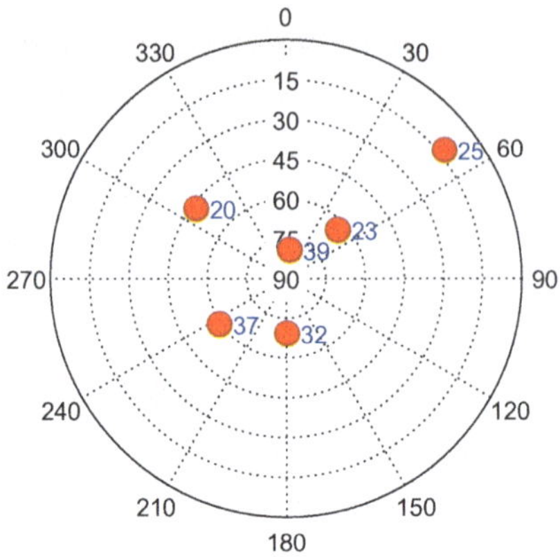

Fig. 7. The skyplot of the BDS-3 satellites

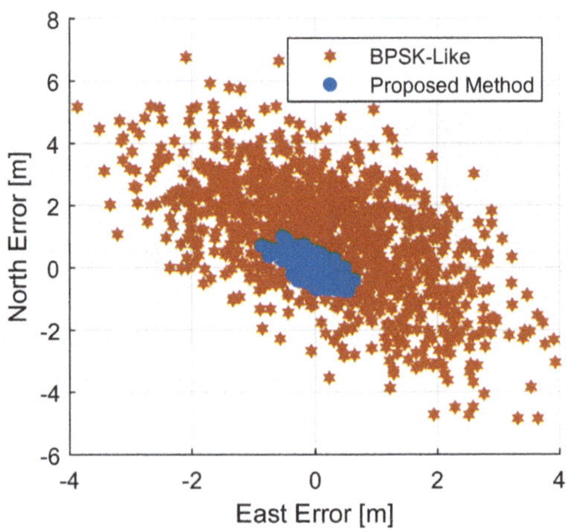

Fig. 8. The comparison of the horizontal positioning errors using proposed method and BPSK-Like technique

Figure 6 shows the prompt correlator outputs of the B1 composite signal. The left is the prompt correlator outputs of the B1C pilot BOC(1,1) component, including the in-phase (I) and quadrature (Q) parts. Comparing the amplitudes of the I and Q parts, it can be clearly seen that the main energy of the prompt correlator output is maintained in the I branch rather than the Q branch, which means the carrier of the composite signal has been perfectly tracked. In addition, it can be seen that the overlay code modulated on the B1C pilot component is also parsed in the in-phase branch. The right is the prompt correlator output of the SCBOC(14,2) component, including the I and Q parts. Similarly, it can be seen that the main energy of the prompt correlator output is retained in the I branch and the B1I navigation message can be clearly demodulated.

The above results show the measurement performance of a single channel, and the following parts analyze the positioning performance of the proposed method. The experimental time was selected at 15:14 on December 9, 2019, Beijing time. Figure 7 shows the skyplot of the BDS-3 at that time. It can be seen from Fig. 7 that there are six available BDS-3 satellites, and their PRN numbers are 20, 23, 25, 32, 37, 39, respectively. The position dilution of precision (PDOP) value is 2.73, which is enough to complete stable single-point positioning.

Figure 8 shows the comparison of the horizontal positioning errors using proposed method and BPSK-like techniques. It can be clearly seen from Fig. 8 that the proposed method has a significant positioning precision improvement due to the utilization of the high-frequency subcarrier ranging performance. In this experiment, compared with the traditional BSPK-Like reception mode, the proposed method reduces the east error from about ± 4 m to ± 1 m, and the north error is reduced from around ± 6 m to ± 1 m. It is not difficult to predict that when the number of available satellites increases, the distribution of visible satellites can be improved. At that time, the positioning performance using proposed method can be further improved, perhaps even to sub-meter positioning applications.

These results verify the correctness and effectiveness of the proposed algorithm. The experimental results show that the proposed algorithm can achieve better ranging performance and higher positioning precision. In the future, the effects of the ionosphere and the multipath on the proposed algorithm should be further analyzed.

5 Conclusions

For the single-sideband complex subcarrier introduced by the BDS-3 B1I signal to achieve the system smooth update, this paper, for the first time, points out that the single-sideband complex subcarrier can not only meet the constraint of the backward compatibility but also contain great ranging performance potential. To fully use it will greatly improve the ranging and positioning precision of the BDS-3 B1 band. However, the single-sideband complex subcarrier also introduces the serious ambiguity threat and complex coupling relationship in correlation function. These complicated tracking challenges and difficulties prevent the exploiting of the ranging potential brought by the single-sideband complex subcarrier.

To solve this problem, this paper proposes a high-precision coherent processing technique, which makes full use of the coherence between B1I and B1C signals. The proposed algorithm can not only exploit the ranging performance brought by high-frequency complex subcarriers without loss, but also greatly simplify the implementation complexity of the receiver. The experimental results of live BDS-3 signals verify the effectiveness and correctness of the proposed method. This paper provides a new solution for the high-precision application of the BDS B1 composite signal and has great reference value for receiver designers.

References

1. Yao, Z., Lu, M.: Constant envelope combination for components on different carrier frequencies with unequal power allocation. Proc. ION ITM 629–637 (2013)
2. Yao, Z., Guo, F., Ma, J., Lu, M.: Orthogonality-based generalized multicarrier constant envelope multiplexing for DSSS signals. IEEE Trans. Aerosp. Electron. Syst. **53**(4), 1685–1698 (2017)
3. Yao, Z., Lu, M.: Signal multiplexing techniques for GNSS: the principle, progress, and challenges within a uniform framework. IEEE Signal Process. Mag. **34**(5), 16–26 (2017)
4. Sleewaegen, J.-M., De Wilde, W., Hollreiser, M. (eds.): Galileo ALTBOC receiver. Proc. GNSS **2**, 3 (2004)
5. Martin, N., Leblond, V., Guillotel, G., Heiries, V. (eds.): BOC (x, y) signal acquisition techniques and performances. In: Proceedings of the 16th International Technical Meeting of the Satellite Division of the Institute of Navigation (ION GPS/GNSS 2003) (2001)
6. Fine, P. (ed.): Tracking algorithm for GPS offset carrier signals. In: 1999 Proceeding of ION National Technical Meeting, 1 (1999)
7. Ward, P.W. (ed.): A design technique to remove the correlation ambiguity in binary offset carrier (BOC) spread spectrum signals. In: 2003 Proceedings of ION 59th Annual Meeting, 6 (2003)
8. Julien, O., Macabiau, C., Cannon, M.E., Lachapelle, G.: ASPeCT: unambiguous sine-BOC (n, n) acquisition/tracking technique for navigation applications. IEEE Trans. Aerosp. Electron. Syst. **43**(1), 150–162 (2007)
9. Hodgart, M., Blunt, P., Unwin, M. (eds.): The optimal dual estimate solution for robust tracking of binary offset carrier (BOC) modulation. In: Proceeding of ION GNSS (2007)
10. Borio, D.: Double phase estimator: new unambiguous binary offset carrier tracking algorithm. IET Radar Sonar Navig. **8**(7), 729–741 (2014)

Preliminary Analysis of BDS-3 and Galileo Compatible Interoperable Positioning Performance

Wenjun Zhao[1] and Wei Wang[2(✉)]

[1] Beijing Satellite Navigation Center, Beijing, China
luckyzhaowenjun@163.com
[2] Beijing Unistrong Science & Technology Co., Ltd., Beijing, China

Abstract. In this paper, using the BDS-3 and Galileo observation data collected by compatible interoperable receivers, the signal-to-noise ratio (SNR), multipath, pseudorange and phase noise, single point positioning performance of B1C, B2a, E1, E5a and BDS-3 SBAS positioning accuracy are compared and analyzed. The results show that the SNR of B1C/B2a are 2–3 dB-Hz higher than E1/E5a. The average of pseudorange multipath of 4 frequencies are between from 0.4 m to 0.8 m. The pseudorange noises of 4 frequencies are all better than 8 cm and phase noises are better than 0.006 cycle. The interoperable positioning accuracy is improved by 11.5%–59.8% compared to BDS-3 single system. Compared with B1C + B2a dual-frequency positioning, the accuracy of BDS-3 dual-frequency SBAS positioning is improved by 5.6%–15.2%.

Keywords: BDS-3 · Galileo · Compatible interoperability · Positioning accuracy

1 Introduction

Compatible interoperability is the development trend of GNSS. Compatibility refers to the use of multiple GNSS and enhanced systems separately or in combination without causing unacceptable interference. Interoperability refers to the comprehensive use of multiple GNSS and enhanced systems, which can obtain better services at the user level than a single system, and minimize the use burden and cost of receiver manufacturers and users [1]. Therefore, the signals of different systems should be as similar as possible. In order to adapt to this trend, in addition to continuing to broadcast B1I and B3I signals to ensure a smooth transition from BDS-2 to BDS-3, BDS-3 also added B1C and B2a compatible interoperable signals. The new signal uses a center frequency point and bandwidth that are highly similar to GPS L1C/L5 and Galileo E1/E5a, thereby achieving interoperability with GPS and Galileo. Compared with B1I and B3I, the new signal has a wider bandwidth and higher ranging accuracy. A pilot channel has been added to improve the reception sensitivity under weak signals, and a new CNAV navigation message has been modulated to use more accurate orbit parameters and the global ionosphere model BDGIM has higher ionosphere correction accuracy [2]. So far, BDS-3 has launched 26 satellites. By 2020, a space constellation consisting of 24

J. Sun et al. (Eds.): CSNC 2020, LNEE 650, pp. 551–561, 2020.
https://doi.org/10.1007/978-981-15-3707-3_52

MEO satellites, 3 GEO satellites, and 3 IGSO satellites (orbital backup satellites deployed as appropriate) will be completed. Therefore, the analysis and evaluation of BDS-3's compatible and interoperable positioning performance with other systems is particularly necessary. This paper analyzes the quality of B1C, B2a, E1, and E5a observation from the aspects of SNR, multipath effects, pseudorange, and phase noise, and evaluates the positioning performance of multiple positioning modes.

2 Analysis Indicators

It mainly includes SNR, multipath effects, pseudorange and phase noises, pseudorange single point positioning accuracy and BDS-3 SBAS positioning accuracy.

2.1 SNR

SNR can be used to measure the quality of ranging signals and is defined as the ratio between signal power and noise power. The SNR is the result of the gain and loss of the entire transmit and receive link signals. It is affected by antenna gain parameters, receiver correlator status, and multipath effects. The higher the signal-to-noise ratio, the better the quality of the observed signal.

2.2 Multipath

If the satellite signal reflected by the reflector near the station enters the receiver antenna, it will interfere with the signal directly from the satellite. This interference delay effect caused by the signal propagation of multiple paths is called multipath effect. Due to the short wavelength of phase observations, measurement noise and multipath effects are much smaller than that of pseudorange. Therefore, this paper only considers the effect of pseudorange multipath. Using the multipath combination of pseudorange and phase observations to compute pseudorange multipath:

$$\begin{cases} MP_1 = p_1 - \frac{f_1^2+f_2^2}{f_1^2-f_2^2}\Phi_1 + \frac{2f_2^2}{f_1^2-f_2^2}\Phi_2 \\ MP_2 = p_2 - \frac{2f_1^2}{f_1^2-f_2^2}\Phi_1 + \frac{f_1^2+f_2^2}{f_1^2-f_2^2}\Phi_2 \end{cases}$$

where MP_1, MP_2 respectively represent the pseudo-range multipath effect corresponding to the frequency 1 and 2. f_1,f_2 represent the frequency values of frequency 1 and 2 respectively. p_1,p_2 represent pseudorange observations at frequency 1 and 2 respectively. Φ_1, Φ_2 represent phase observations at frequency 1 and 2, respectively, in cycles.

2.3 Pseudorange and Phase Noise

The internal noise of the receiver mainly refers to the signal interference error caused by the receiver signal channel, phase-locked loop, and code tracking loop deviation. It is an important indicator for evaluating the performance of the receiver [3]. The zero baseline method is a common method for detecting internal noise in a receiver. For the zero baseline, the double difference pseudorange and phase residuals eliminate the effects of a series of errors such as ephemeris error, satellite and receiver clock errors, ionospheric and tropospheric delay, and multipath effects, which can truly reflect the quality level of the receiver.

2.4 Single Point Positioning Accuracy

Pseudorange single point positioning accuracy directly reflects the basic service performance of a navigation system. The geometric distribution of satellites, the accuracy of orbit and clock parameters, and the accuracy of the broadcast ionosphere model all determine the performance of single-point positioning.

2.5 BDS-3 SBAS Positioning Accuracy

SBAS (Satellite Based Augmentation System) refers to a wide area differential system that uses a geostationary satellite as a communication medium and can provide GNSS differential correction values and integrity data. SBAS plays an increasingly important role in user use. Currently, BDS-3 has broadcast SBAS corrections on the B1C and B2a frequencies through the C59 satellite. B1C SBAS provides services for B1C and B2a single-frequency positioning, and B2a SBAS provides services for B1C + B2a dual-frequency positioning. Since the correction of the BDS-3 satellite is only broadcast in the B2a SBAS message for the time being, only B1C + B2a SBAS positioning accuracy evaluation is performed in this paper. The dual-frequency SBAS positioning mainly uses the orbit and clock correction in the SBAS message type 32. The calculation method of the orbit correction amount $\delta x(t), \delta y(t), \delta z(t)$ is as follows:

$$\begin{bmatrix} \delta x(t) \\ \delta y(t) \\ \delta z(t) \end{bmatrix} = \begin{bmatrix} \delta x \\ \delta y \\ \delta z \end{bmatrix} + \begin{bmatrix} \dot{\delta x} \\ \dot{\delta y} \\ \dot{\delta z} \end{bmatrix} \times (t - t_D)$$

In the formula, $[\delta x, \delta y, \delta z]$ indicates the broadcast orbit corrections, $\left[\dot{\delta x}, \dot{\delta y}, \dot{\delta z}\right]$ indicates the change rate of the orbit corrections, t is the observation time, and t_D is the reference time of the correction.

The amount of clock correction $\delta \Delta t_{SV}$ can be calculated by:

$$\delta \Delta t_{SV} = \frac{\delta B + \dot{\delta B} \cdot (t - t_D)}{c}$$

δB represents the correction of the broadcast clock difference, $\dot{\delta B}$ represents the clock drift correction, t is the observation time, t_D is the reference time of the correction, c is the speed of vacuum light.

3 Experimental Results

Compatible and interoperable receivers developed by Unistrong were respectively used in Beijing and Xi'an to collect two sets of zero baseline data (4 receivers, each represented by BJ01, BJ02, XA01, XA02), with a data interval of 10 s and an observation period of 0: 00-24: 00, November 27, 2019. The cut-off elevation angle is set to 10°. The B1C, B2a, E1, and E5a were subjected to observation quality analysis and single point positioning. The processing results are as follows:

3.1 SNR

The SNR of the receiver is directly extracted from the observation data for analysis. Figures 1 and 2 show the SNR and elevation time series of B1C/B2a and E1/E5a for satellite C21 and E09 respectively. As can be seen from the figures, the SNR of B1C/B2a are distributed between 41–51 dB-Hz. The E1/E5a SNR are significantly lower than that of B1C/B2a, with a distribution between 36 and 49 dB-Hz, and the E1 SNR is lower than E5a. When the elevation is height, the SNR is also large and vice versa. Figure 3 shows the average of the SNR of the four signals from each receiver. The difference between the SNR of the four receivers at the same frequency are not large, and the difference between the average SNR of B1C and B2a is small, the SNR of E1/E5a are 2–3 dB-Hz lower than that of B1C/B2a, and SNR of E1 is lower than E5a by about 1 dB-Hz.

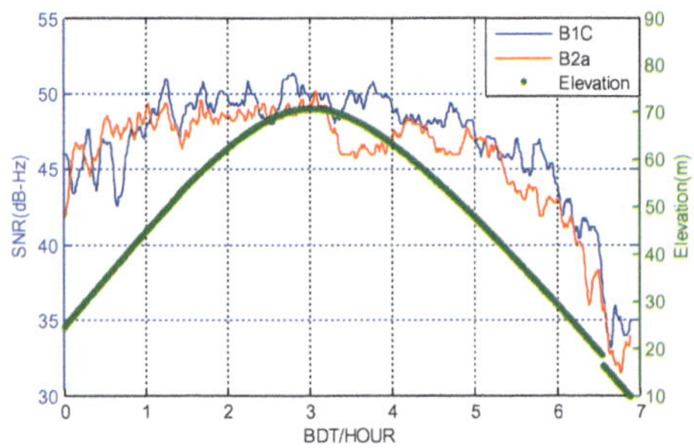

Fig. 1. Time series of SNR and elevation of B1C and B2a for C21

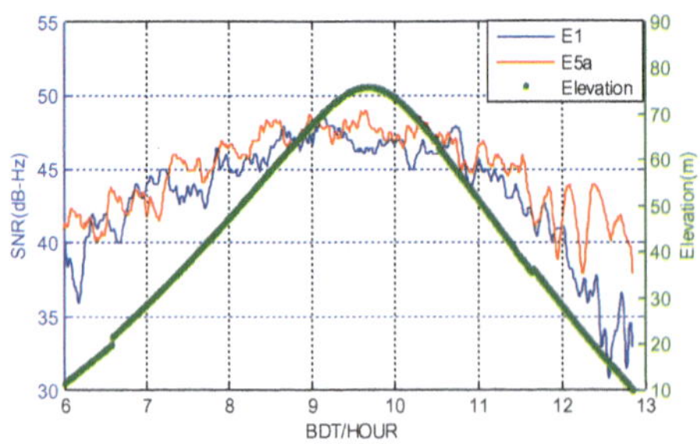

Fig. 2. Time series of SNR and elevation of E1 and E5a for E09

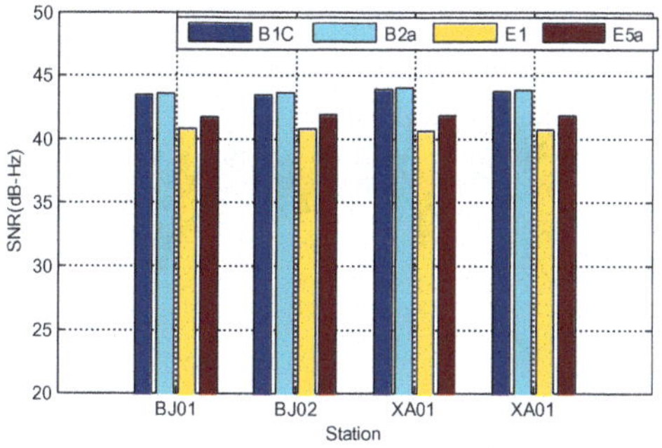

Fig. 3. Average SNR of each frequency for 4 receivers

3.2 Pseudorange Multipath

Figures 4 and 5 show the time series of the receiver BJ02's C21 and E09 satellite multipath effects as a function of elevation angle. It can be seen from the figure that when the elevation angle is low, the pseudorange multipath is large, especially when the elevation angle is below 15°, the multipath effect is even close to 3 m, which will seriously affect the positioning accuracy of the receiver. When the elevation angle is high, the pseudorange multipath is small, and is basically distributed at ±1 m. And satellite C21 does not appear the systematic deviation of pseudo-range multipath caused by inter-satellite multipath error on BDS-2 satellite [4]. It can be inferred that the BDS-3 satellite has eliminated this system deviation. Figures 6 and 7 show the average of the B1C/B2a pseudorange multipath of each satellite of the receiver. It can be seen that the B2a pseudorange multipath effect is about 0.1 m larger than that of B1C. The distribution of B1C pseudorange multipath is about 0.5 m, and B2a is about 0.6 m. The E1 and E5a pseudorange multipaths are distributed in the range of 0.5–0.8 m, which is a normal range. It also shows that the station has a wider field of view and a better observation environment.

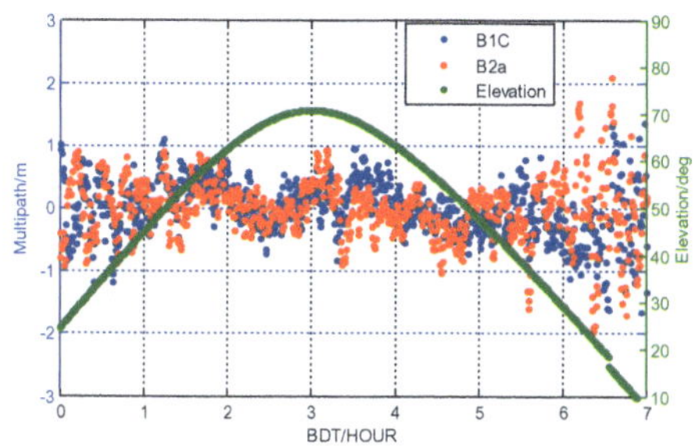

Fig. 4. Pseudorange multipath and elevation of C21

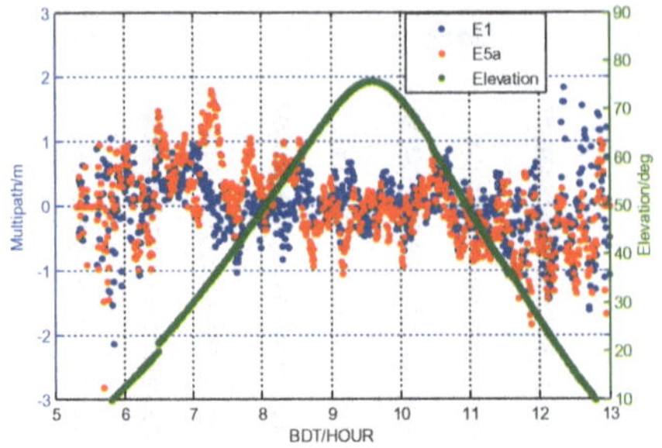

Fig. 5. Pseudorange multipath and elevation of E09

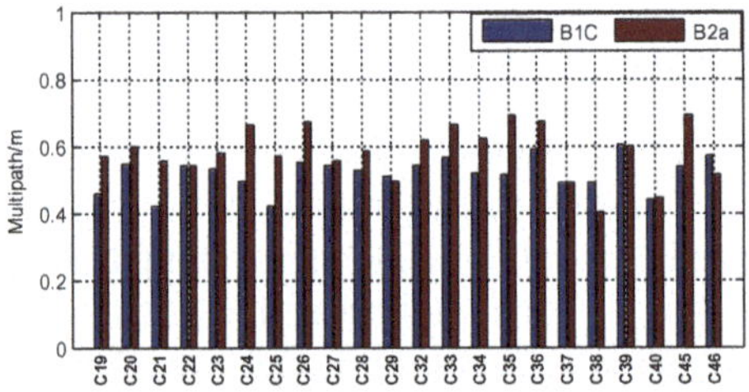

Fig. 6. Average of pseudorange multipath for each BDS-3 satellite

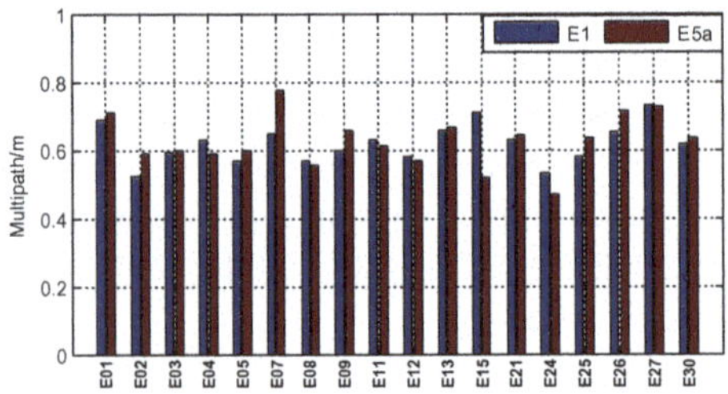

Fig. 7. Average of pseudorange multipath for each Galileo satellite

3.3 Noise of Pseudorange and Phase

Zero baseline BJ01-BJ02 was used for double difference. Figures 8, 9, 10, and 11 show the double-difference pseudorange and phase residuals of B1C, B2a, E1, and E5a, respectively. Blue line indicates pseudorange residuals and green line indicates phase residuals. It can be seen from the figure that the pseudo-range residuals of the four frequencies are basically fluctuated within ±2 m. Except for the phase residuals of B2a at individual time, the phase residuals of the other three frequencies are distributed within ±0.1 cycle. According to the double difference residual error RMS and the law of error propagation, the original pseudorange and phase noise (RMS/2) can be calculated, as shown in Table 1. It can be seen that the pseudorange noises at 4 frequencies are better than 8 cm, and the phase noises are better than 0.006 cycle, and the difference between the frequency is not large, which indicates that the accuracy of this group of data is high.

Table 1. Noise of pseudorange and phase for each frequency

Baseline	BJ01-BJ02				XA01-XA02			
Frequency	B1C	B2a	E1	E5a	B1C	B2a	E1	E5a
Pseudorange noise (m)	0.061	0.054	0.058	0.061	0.074	0.062	0.056	0.063
Phase noise (cycle)	0.005	0.004	0.003	0.003	0.006	0.005	0.004	0.003

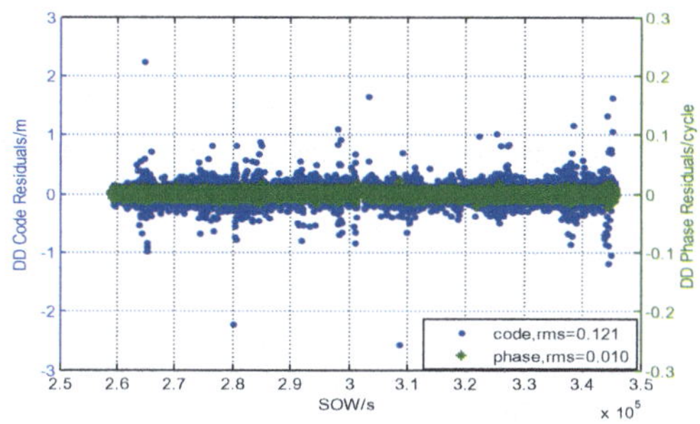

Fig. 8. Double difference residuals of pseudorange and phase for B1C

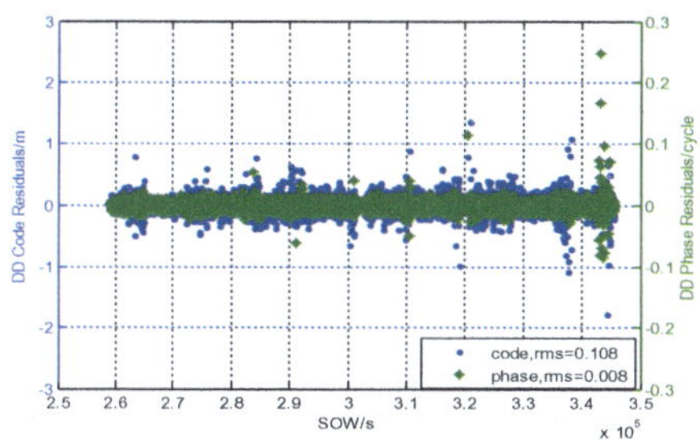

Fig. 9. Double difference residuals of pseudorange and phase for B2a

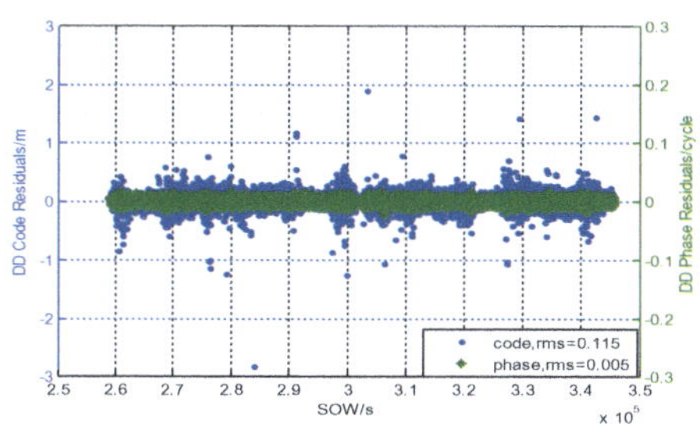

Fig. 10. Double difference residuals of pseudorange and phase for E1

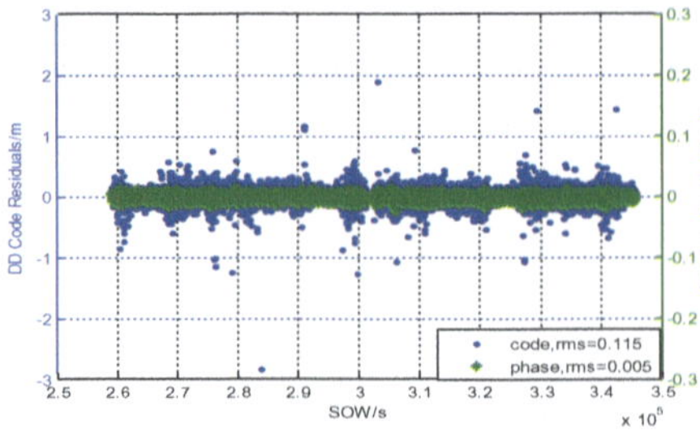

Fig. 11. Double difference residuals of pseudorange and phase for E5a

3.4 Pseudorange Single Point Positioning

Six positioning modes (B1C, B2a, E1, E5a, B1C + E1, B2a + E5a) are used to evaluate the positioning performance of BDS-3, Galilleo and BDS-3 and Galileo compatible and interoperable positioning. Considering the health status of the satellites during positioning, the BDS-3 C39, C45, C46 satellites are not healthy, so they do not participate in the calculation. Figure 12 shows the pseudorange single point positioning results of receiver BJ01. Due to the small number of BDS-3 and Galileo visible satellites in a certain period of time, a large deviation occurred in the positioning. Therefore, the paper excludes data with positioning errors exceeding 20 m. As can be seen from the figure, both B1C and B2a can achieve independent positioning within a day. Except for fewer satellites participating in positioning during a certain period of time, which results in a large positioning error, the positioning errors in the other three directions are basically not more than 6 m. The positioning error of E1 at 0–3 h points is too large, and the positioning accuracy is good in the other periods. The positioning errors of E5a at 0–4 h and 16–19 h exceeded 20 m, and the positioning is normal in the other periods. In compatible interoperable positioning modes of B1C + E1 and B2a + E5a, the positioning accuracies are high throughout the day, and no positioning errors exceeds 20 m. The number of participating positioning satellites in the compatible interoperable mode is from 9 to 13, and the PDOP is below 1.8. Compared with the single system, the geometric distribution in the compatible interoperable mode is better, so the positioning accuracy is higher. Table 2 gives the positioning accuracy statistics of the four receivers under different positioning modes. As can be seen in Table 2, the positioning accuracies of the B1C + E1, B2a + E5a combination are significantly better than that of single system, and the positioning errors in three directions are not more than 4 m. Compared with B1C, the positioning accuracy of B1C + E1 is improved by 11.5%–35.6%, and the positioning accuracy of B2a + E5a is improved by 31.8%–59.8% compared to B2a.

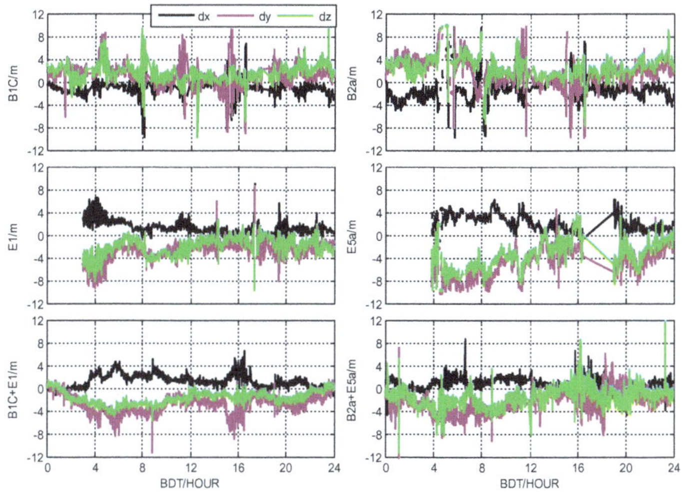

Fig. 12. Time series of positioning errors at different mode for receiver BJ01

Table 2. RMS of positioning errors at different mode for 4 receivers (unit: m)

Mode	BJ01			BJ02			XA01			XA02		
	dx	dy	dz	dx	dy	dz	dx	dy	dz	dx	dy	dz
B1C	2.125	3.766	2.567	2.367	3.848	2.891	3.021	3.866	3.015	2.876	3.452	3.580
B2a	4.715	5.002	4.362	4.768	5.291	4.424	4.914	5.887	4.978	4.836	5.689	4.803
E1	4.469	6.007	5.884	4.256	5.291	5.639	4.058	5.367	5.926	4.167	5.509	5.988
E5a	2.633	5.505	4.344	2.412	5.416	4.123	3.236	5.702	4.344	3.458	5.813	4.579
B1C + E1	1.667	3.143	1.984	1.723	3.267	2.093	2.027	3.336	2.346	1.982	3.055	2.296
B2a + E5a	1.895	2.902	2.271	1.958	2.981	2.515	2.156	3.452	2.940	2.207	3.570	3.274

3.5 Position Accuracy of BDS-3 SBAS

Figure 13 shows the positioning results of one compatible interoperable receiver at Beijing ZPark. The positioning time is from 2019/11/27 13:00 to 2019/11/28 01:00 (GPST). It can be seen from the figure that when the number of satellites is small, the positioning error of both positioning modes is very large, even exceeding 20 m. Overall, the SBAS positioning error is smaller. According to the statistical results, the RMS errors in dx, dy, dz three directions of SBAS positioning are 3.243 m, 4.823 m, and 3.901 m, while the B1C + B2a dual-frequency positioning errors RMS are 3.425 m, 5.731 m, and 4.493 m respectively. The positioning accuracy of the three directions of SBAS has been improved by 5.6%, 18.8%, and 15.2% respectively.

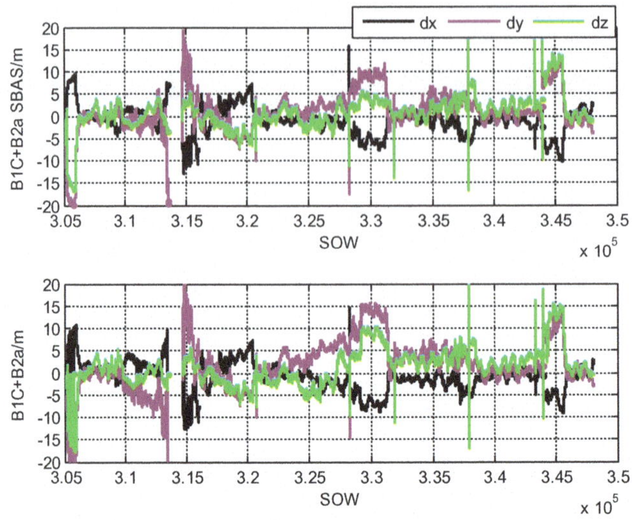

Fig. 13. Positioning errors of B1C + B2a SBAS

4 Conclusion

This paper uses two sets of zero baseline data to analyze the quality of the B1C, B2a, E1, and E5a observation data from the aspects of SNR, multipath, pseudorange and phase noise. Different single point positioning modes accuracy and BDS-3 dual-frequency SBAS positioning accuracy are also analyzed. The conclusions are as follows: The SNR difference between B1C and B2a is very small. The SNR of E1 and E5a is 2–3 dB-Hz lower than that of B1C/B2a, and the SNR of E1 is lower than E5a by about 1 dB-Hz. The distribution of B1C pseudorange multipath is about 0.5 m, and B2a is about 0.6 m. Both E1 and E5a pseudorange multipaths are distributed in 0.5–0.8 m. The pseudorange noises of B1C, B2a, E1 and E5a are all better than 8 cm, phase noises are also all better than 0.006 cycle, and there is not much difference between each frequency. In the compatible interoperable positioning mode, the number of satellites is larger, and the geometric distribution is better. The positioning accuracy of B1C + E1 is 11.5%–35.6% higher than that of B1C, and the positioning accuracy of B2a + E5a is 31.8%–59.8% higher than B2a. The positioning accuracy of the BDS-3 dual-frequency SBAS has been improved by 5.6% to 15.2% compared to the B1C + B2a dual-frequency. With the construction of BDS-3, the compatibility and interoperability of BDS and Galileo will be further improved in the future.

References

1. Yang, Y., Lu, M., Han, C.: Some notes on interoperability of GNSS. Acta Geodaetica Cartogr. Sin. **45**(3), 253–259 (2016)
2. Guo, S., Cai, C., Meng, Y., et al.: BDS-3 RNSS technical characteristics and service performance. Acta Geodaetica Cartogr. Sin. **48**(7), 810–821 (2019)
3. Gao, X., Dai, W., Li, S.: Interior performance test of high precision GPS/BDS compatible receivers. Geomat. Inf. Sci. Wuhan Univ. **40**(6), 795–799 (2015)
4. Wanninger, L., Beer, S.: BeiDou satellite-induced code pseudorange variations: diagnosis and therapy. GPS Solut. **19**, 639–664 (2015)

Robust GNSS Triple-Carrier Joint Estimations Under Strong Ionosphere Scintillation

Rong Yang, Xingqun Zhan(✉), and Jihong Huang

School of Aeronautics and Astronautics, Shanghai Jiao Tong University, Shanghai, China
{rongyang,xqzhan,jihong.huang}@sjtu.edu.cn

Abstract. In this paper, we proposed a tri-frequency joint estimation algorithm to address the loss of lock issues during strong ionospheric scintillations. This algorithm utilizes the frequency dependency to assist the GNSS tri-frequency signal tracking. The tri-frequency measurements are combined in a joint estimator for Doppler estimations so as to perform inter-frequency aiding from uncompensated signals to the interfered signals and mitigate the strong amplitude attenuations. However, as the tri-frequency estimations are coupled together, the bad measurements will deteriorate the strong signal tracking performance. We upgrade the tri-frequency joint estimator with a real time weighting adjustment according to the observed signal strengths on three frequency bands to address the issues. In addition, the algorithm adopts independent tracking on carrier phase, while joint tracking on carrier frequency. Therefore, the phase fluctuation features on each frequency can be preserved for the ionospheric scintillation characterizations and detections. The performance of the proposed algorithm is verified by the simulations. The results show that the tri-frequency joint estimator can effectively suppress tracking errors and recover carrier parameter estimations during strong ionospheric scintillations.

Keywords: Ionosphere scintillation · Triple frequency · GNSS carrier tracking

1 Introduction

Ionospheric scintillation typically refers to the phenomenon of electromagnetic wave signal's refraction, reflection and scattering caused by the abnormal changes of electrons and plasmas [1–3]. Ionospheric scintillation changes the signal speed, polarization characteristics and energy distribution during propagation, leading to strong fluctuations on receiver signal phase, amplitude and arrival angle [4]. Strong signal fluctuations will deteriorate the range measurement accuracy, affect the positioning, navigation, and timing (PNT) performances, and even destroy receiver functioning in some extremely cases. It is the most essential issue to be solved in navigation receiver designs [5].

For Global Navigation Satellite System (GNSS), ionosphere scintillation includes the amplitude scintillation and phase scintillation. They are characterized by the index S_4 and σ_ϕ, respectively. Typically, in the equatorial region, the amplitude scintillation is a dominated effect, while in the high latitude region, phase scintillation is more

J. Sun et al. (Eds.): CSNC 2020, LNEE 650, pp. 562–575, 2020.
https://doi.org/10.1007/978-981-15-3707-3_53

obvious [6]. GPS L2 and L5 bands are more susceptible to ionospheric scintillation than L1 bands [6]. Various signal processing algorithms have been investigated to address the loss of lock issues during strong ionosphere scintillations [7]. For example, reference [8] provided a detailed performance analysis of a phase lock loop (PLL), including tracking error, cycle slip, bit error in the presence of a canonical scintillation with an over half cycle phase change and a 15 dB amplitude fading. Reference [9] firstly built an auto-regression model with consideration of ionospheric phase and amplitude scintillations for tracking loop design. Reference [10] utilized a frequency lock loop (FLL) to assist PLL to increase the pull-in range to improve the robustness when scintillation occurs. Reference [11] proposed an adaptive method that based on the scintillation index to tune the Kalman filter parameters. Reference [12] designed a semi-open loop architecture to further improve tracking accuracy and robustness with a moving measurement window. Since there is a frequency divergence across signal bands, the receiver can use this feature to realize inter-frequency aiding. Reference [13] implemented an inter-frequency tracking loop that uses strong signal measurements to assist weak signal channel based on the signal strength variations for ionosphere scintillation mitigation. Reference [14] extended the single-frequency autoregressive ionospheric scintillation model [9] to a triple-frequency scenario, and achieved joint tracking of triple-frequency signals. Reference [15] established an unified triple frequency signal model based on the characteristics of frequency diversity. Given the estimations of signal strength on each frequency, an optimal combination of triple-frequency signals is designed to minimize the tracking errors induced by the frequency selective fading. However, [15] only provided a general joint tracking method which is not a specific design for ionosphere scintillation scenario. Its performance can be further improved with consideration of ionospheric scintillation characteristics. This paper will focus on this issue. We will expand the work in [15] and propose a better implementation of triple frequency tracking in a strong ionospheric scintillation environment. We will use the simulator to verify the performance of the proposed method.

The rest paper is organized as follows: Sect. 2 introduces the triple frequency signal models including dynamic model as well as measurement model; Sect. 3 describes the design and implementation of the triple-carrier joint estimator; Sect. 4 summarizes the set-up of the ionospheric scintillation simulator and also investigates the triple frequency tracking performance. Finally, Sect. 5 concludes the paper.

2 Tri-Frequency GPS Carrier Model

Modernized GNSS is capable of broadcasting multi-frequency signals to improve signal availability and anti-interference performance, e.g., Beidou B1, B2, B3, and GPS L1, L2, L5. Since the multi-frequency signals are generated by the same satellite and received by the same receiver, they share the same propagation distance, velocity, and acceleration on the line-of-sight (LOS) direction [15]. Therefore, the multi-carrier states satisfy the following linear relation:

$$\left[\varphi_{Li}; f_{di}; \dot{f}_{di}\right]_k = \eta_i \left[\varphi_0; f_{d0}; \dot{f}_{d0}\right]_k \quad (1)$$

where $\varphi_{Li}, f_{di}, \dot{f}_{di}$ represent the carrier phase, Doppler, Doppler rate on the ith band and i = 1, 2, 5 for GPS. $\varphi_0, f_{d0}, \dot{f}_{d0}$ are the carrier phase, Doppler, Doppler rate on the fundamental carrier at 10.23 MHz for GPS [15]. η_i is the frequency ratio relative to the fundamental carrier on the ith frequency, and $\eta_i = 154, 120, 115$ for GPS L1, L2, L5 carriers.

Also taking GPS for example, the triple-carrier state vector $\mathbf{x}_k$ can be constructed as [15]:

$$\mathbf{x}_k = \left[\varphi_{L1}\ \varphi_{L2}\ \varphi_{L5}\ 2\pi f_{d0}\ 2\pi\dot{f}_{d0}\right]_k^{\mathrm{T}} \quad (2)$$

The reason of building independent phase for each carrier is that the phases on each carrier will be affected by the ionosphere, troposphere multipath effects. The hardware delays also introduce phase biases. All these facts bring the phase divergence so that the phase does not follow the linear relationship in (1). Therefore, we reserve each individual carrier phase term to absorb the environmental information. We assume that those effects on the carrier Doppler and Doppler rate are negligible small so that the fundamental Doppler and Doppler rate can be used to link LOS triple-carrier dynamics. Detailed signal models are discussed in [15].

2.1 Dynamic Model

The dynamic model for $\mathbf{x}_k$ can be expressed as [15–17]:

$$\mathbf{x}_{k+1} = \mathbf{A}\mathbf{x}_k + \mathbf{n}_k \quad (3)$$

where $\mathbf{A}$ is the system transition matrix with expression of (4), T is the coherent integration time. $\mathbf{n}_k$ in (3) represents the system noise which typically refers to the oscillator noise in the receiver front end and the random dynamic noise on the LOS direction. The covariance matric of $\mathbf{n}_k$ can be expressed as (5), where f_0 is the 10.23 MHz fundamental carrier frequency, q_φ and q_ω are the oscillator noise power spectral densities (PSD) which is determined by the oscillator characteristics [16, 17], q_a represents the random walk noise PSD which is related to the maximum LOS jerk [16, 17]. q_{i1}, q_{i2}, q_{i5} are the ionosphere scintillation noise PSD on GPS L1, L2, and L5 bands, their values can be estimated from the carrier tracking outputs [11].

$$\mathbf{A} = \begin{bmatrix} 1 & 0 & 0 & \eta_1 T & \eta_1 \frac{T^2}{2} \\ 0 & 1 & 0 & \eta_2 T & \eta_2 \frac{T^2}{2} \\ 0 & 0 & 1 & \eta_5 T & \eta_5 \frac{T^2}{2} \\ 0 & 0 & 0 & 1 & T \\ 0 & 0 & 0 & 0 & 1 \end{bmatrix} \quad (4)$$

$$\mathbf{Q}_k = (2\pi f_0)^2 \begin{bmatrix} \eta_1^2\left(Tq_\varphi + \frac{T^3 q_\omega}{3} + \frac{T^5 q_a}{20c^2}\right) + q_{i1} & 0 & 0 & \eta_1\left(\frac{T^2 q_\omega}{2} + \frac{T^4 q_a}{8c^2}\right) & \eta_1\left(\frac{T^3 q_a}{6c^2}\right) \\ 0 & \eta_2^2\left(Tq_\varphi + \frac{T^3 q_\omega}{3} + \frac{T^5 q_a}{20c^2}\right) + q_{i2} & 0 & \eta_2\left(\frac{T^2 q_\omega}{2} + \frac{T^4 q_a}{8c^2}\right) & \eta_2\left(\frac{T^3 q_a}{6c^2}\right) \\ 0 & 0 & \eta_5^2\left(Tq_\varphi + \frac{T^3 q_\omega}{3} + \frac{T^5 q_a}{20c^2}\right) + q_{i5} & \eta_5\left(\frac{T^2 q_\omega}{2} + \frac{T^4 q_a}{8c^2}\right) & \eta_5\left(\frac{T^3 q_a}{6c^2}\right) \\ \eta_1\left(\frac{T^2 q_\omega}{2} + \frac{T^4 q_a}{8c^2}\right) & \eta_2\left(\frac{T^2 q_\omega}{2} + \frac{T^4 q_a}{8c^2}\right) & \eta_5\left(\frac{T^2 q_\omega}{2} + \frac{T^4 q_a}{8c^2}\right) & \left(Tq_\omega + \frac{T^3 q_a}{3c^2}\right) & \frac{T^2 q_a}{2c^2} \\ \eta_1\left(\frac{T^3 q_a}{6c^2}\right) & \eta_2\left(\frac{T^3 q_a}{6c^2}\right) & \eta_5\left(\frac{T^3 q_a}{6c^2}\right) & \frac{T^2 q_a}{2c^2} & \frac{Tq_a}{c^2} \end{bmatrix}_k \tag{5}$$

2.2 Measurement Model

Typically, the discriminator outputs are selected as the carrier measurements, therefore the measurement vector $\mathbf{z}_k$ is comprise of triple-carrier phase errors $\Delta\theta_{L1}$, $\Delta\theta_{L2}$, $\Delta\theta_{L5}$:

$$\mathbf{z}_k = \begin{bmatrix} \Delta\theta_{L1} & \Delta\theta_{L2} & \Delta\theta_{L5} \end{bmatrix}_k^{\mathrm{T}} \tag{6}$$

$\mathbf{z}_k$ can be linearized as the following form [15]:

$$\mathbf{z}_k = \mathbf{H}\Delta\mathbf{x}_k + \mathbf{v}_k \tag{7}$$

where $\Delta\mathbf{x}_k = \begin{bmatrix} \Delta\varphi_{L1} & \Delta\varphi_{L2} & \Delta\varphi_{L5} & \Delta f_{d0} & \Delta\dot{f}_{d0} \end{bmatrix}_k^{\mathrm{T}}$ represent the state estimation error, $\mathbf{H}$ is the measurement matrix:

$$\mathbf{H} = \begin{bmatrix} 1 & 0 & 0 & \eta_1 \frac{T}{2} & \eta_1 \frac{T^2}{6} \\ 0 & 1 & 0 & \eta_2 \frac{T}{2} & \eta_2 \frac{T^2}{6} \\ 0 & 0 & 1 & \eta_5 \frac{T}{2} & \eta_5 \frac{T^2}{6} \end{bmatrix} \tag{8}$$

$\mathbf{v}_k$ represents the measurement noise term, which is assumed as a white Gaussian noise from the discriminator output. $\mathbf{v}_k$ on each frequency bands are uncorrelated, therefore its covariance matrix can be written as:

$$\mathbf{R}_k = \begin{bmatrix} \sigma_{L1,k}^2 & 0 & 0 \\ 0 & \sigma_{L2,k}^2 & 0 \\ 0 & 0 & \sigma_{L5,k}^2 \end{bmatrix} \tag{9}$$

$\sigma_{L1,k}^2$, $\sigma_{L2,k}^2$, $\sigma_{L5,k}^2$ represents the noise variances on L1, L2, and L5 bands, which can be estimated via:

$$\sigma_{Li,k}^2 = \frac{1}{2T \cdot C/N_{0Li,k}} \left(1 + \frac{1}{2T \cdot C/N_{0Li,k}}\right) \tag{10}$$

$C/N_{0Li,k}$ is the carrier to noise ratio on the ith frequency. It is a measure of received signal quality. Since the ionosphere, troposphere, and multipath scintillation cause fluctuations on signal amplitude and phase, $C/N_{0Li,k}$ can be chosen as an indicator to associate with different scintillation levels. The tracking loop can adaptively tune the filter parameters according to the variations of $C/N_{0Li,k}$ to mitigate the scintillation errors.

3 Tri-frequency GPS Carrier Model

The theoretical triple frequency joint tracking has been discussed in [15], however, [15] only provided a general triple frequency combination model applicable to various frequency selective fading scenarios. In this paper, we will specified it for anti-ionospheric scintillation designs. The tri-frequency measurements are combined in a joint estimator for Doppler estimations so as to perform inter-frequency aiding from uncompensated signals to the interfered signals and mitigates the strong amplitude attenuations. However, as the tri-frequency estimations are coupled together, there is a problem that the bad measurements will deteriorate the strong signal tracking performance. We up-grade the tri-frequency joint estimator with a real time weighting adjustment according to the observed signal strengths on three frequency bands to eliminate the mutual influence when the strong and weak signal co-exit. In addition, the algorithm adopts independent tracking on carrier phase and frequency, that is, the phase on each frequency is estimated independently, only the frequency tracking is realized in the joint estimator.

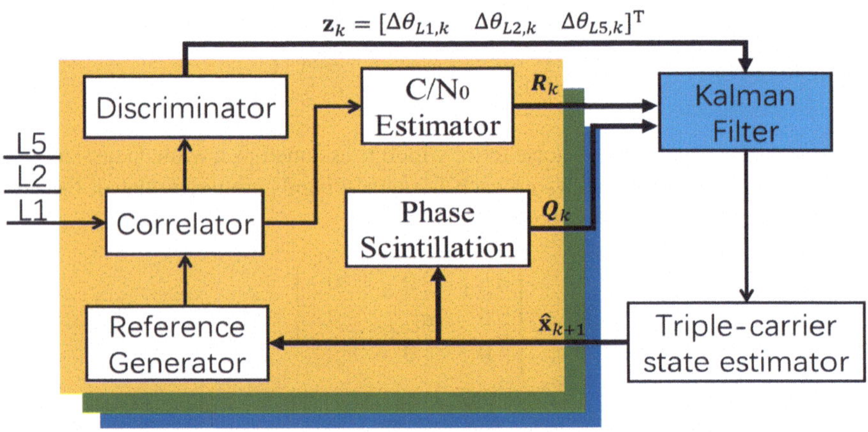

Fig. 1. The structure of the GNSS triple-carrier joint estimator

Figure 1 shows the structure of the three-frequency carrier joint estimation in this paper. L1, L2, and L5 have their own independent tracking channels. Each channel has a Correlator, a Discriminator, and a Local Reference Signal Generator, C/N_0 estimator, and phase scintillation estimation module; the three channels share a common Kalman filter and a triple-carrier state estimator.

The specific implementation method is as follows: L1, L2, and L5 digital IF signals are sent to a correlator to perform a correlation operation with a locally generated carrier signal to generate the accumulation results for in-phase, I, and quadrature, Q branches. The output interval is determined by the coherent integration time T. The phase detector obtains the phase errors on L1, L2, and L5 according to the I and Q results of each channel. Generally, two-quadrant and four-quadrant arctangent are more commonly used for phase detection in the data channel and pilot channel, respectively. In addition, the I and Q results of each channel can also be input to the C/N_0 estimator for real-time C/N_0 estimation to evaluate the signal quality and calculate the observation noise covariance matrix. The phase scintillation module is used to estimate the phase scintillation noise PSD at each frequency. It is used to calculate the values of q_{i1}, q_{i2}, q_{i5} and thus adjust the system noise covariance matrix $\mathbf{Q}_k$. According to the Kalman filter principle, its filtering **K** gain can be determined by the state estimation error covariance matrix $\mathbf{P}_k$, observation matrix $\mathbf{H}_k$, and observation noise covariance matrix $\mathbf{R}_k$:

$$\mathbf{K} = \mathbf{P}_k\mathbf{H}^{\mathrm{T}}\left(\mathbf{H}\mathbf{P}_k\mathbf{H}^{\mathrm{T}} + \mathbf{R}_k\right)^{-1} \tag{11}$$

where $\mathbf{P}_k$ can be obtained by solving the following Riccati equation:

$$\mathbf{A}\mathbf{P}\mathbf{A}^{\mathrm{T}} - \mathbf{A}\mathbf{P}\mathbf{H}^{\mathrm{T}}\left(\mathbf{H}\mathbf{P}\mathbf{H}^{\mathrm{T}} + \mathbf{R}_k\right)^{-1}\mathbf{H}\mathbf{P}\mathbf{A}^{\mathrm{T}} + \mathbf{Q}_k - \mathbf{P} = 0. \tag{12}$$

The state estimation can then be obtained as [15–17]:

$$\hat{\mathbf{x}}_{k+1} = \mathbf{A}\hat{\mathbf{x}}_k + \mathbf{A}\mathbf{K}\mathbf{z}_k \tag{13}$$

It is worth noting that the signal amplitude change and phase change in the strong ionospheric scintillation scene can be adjusted and reflected in $\mathbf{Q}_k$ and $\mathbf{R}_k$, and the phase filter coefficients and frequency filter coefficients in **K** on L1, L2, and L5 can be adjusted separately to achieve stable carrier tracking and preserve characteristics of phase scintillation in each band.

4 Strong Ionosphere Scintillation Simulation and Performance Analysis

The measured results of tri-frequency tracking under strong ionospheric scintillation conditions have been discussed in the paper [15]. In order to better quantitatively verify the effectiveness of the algorithm in this paper, the open source ionospheric scintillation signal simulator provided by the University of Colorado Boulder SenSe Lab will be used. The simulator [18, 19] generates GPS digital intermediate frequency signals for tracking and performance analysis. This section will give the ionospheric simulator setup and tracking results, respectively.

4.1 Ionosphere Scintillation Simulator

In the current research, the ionospheric scintillation simulator mainly includes two parts: modeling of ionospheric irregularity and signal propagation [18–20]. Ionospheric irregularity is simulate by setting parameters such as height, thickness, intensity, and spectral density of the ionosphere [18–20]. The signal propagation model is generally implemented using the phase screen model. The simulator used in this paper is a multi-frequency strong ionospheric scintillation simulator is developed on the basis of a two-component power-law phase screen model and a numerical mapping-driven scintillation physical model [18, 19]. The simulator only needs to input the amplitude scintillation index S_4 and the decorrelation time τ_0 estimated from the measured data to generate a time series of scintillation amplitude and phase for GPS intermediate frequency signal simulation. In this paper, we set $S_4 = 0.9$ and $\tau_0 = 0.75$. This is the default setting of the simulator in [18]. Based on this setting 300-s GPS L1, L2, and L5 intermediate frequency data with 0 MHz intermediate frequency and 4 bit quantization were generated.

The signal strength variations of GPS L1, L2, and L5 caused by ionospheric scintillation are shown in Fig. 2. After initial 30 s, the GPS tri-frequency signals begin to have different levels of attenuation. Most of the attenuation on each frequency signal is more than 20 dB and the highest can be more than 40 dB. The attenuation of L2 signal is deeper than that of L1 and L5. Severe amplitude attenuation is accompanied by rapid phase changes. As shown in Fig. 3, the phase change produced by the ionosphere can change from −6 cycles to 10 cycles in the entire 300 s. In general, the phase change trends of L1, L2, and L5 are similar. In terms of the degree of phase change, the relatively less affected in the last 200 s is L1, followed by L5, and L2 has the largest phase change.

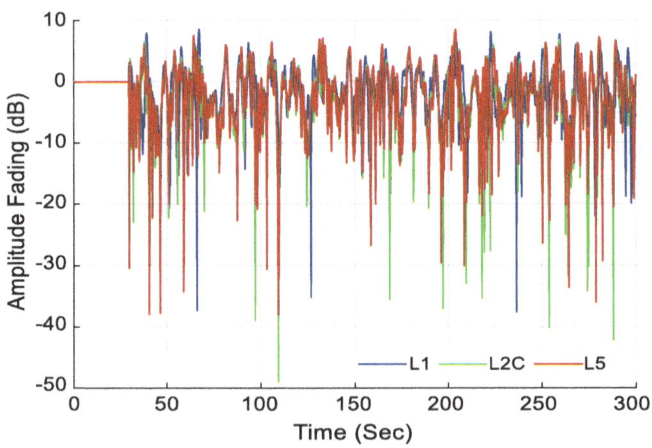

Fig. 2. GPS triple frequency amplitude fading generated by the ionosphere scintillation simulator

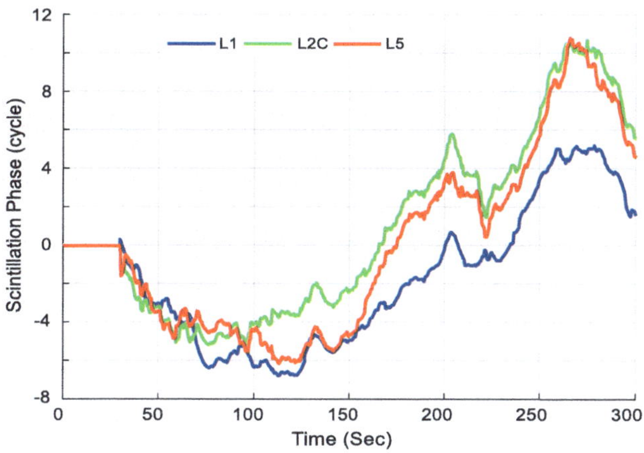

Fig. 3. GPS triple frequency phase sequence generated by the ionosphere scintillation simulator

4.2 Triple-Carrier Tracking Results

The tri-frequency joint estimation algorithm described in Sect. 3 is used to process the IF signal of the ionospheric scintillation, where $h_0 = 6.3 \times 10^{-26}$ (s^2/Hz) and $h_{-2} = 3.4 \times 10^{-26}$ (1/Hz), $q_a = 10^{-4}$ (m^2/s^6)/Hz is used to initialize $\mathbf{Q}_k$ in (5), details can be refer to [15–17]. We tested the traditional PLL single-frequency tracking and triple-frequency joint tracking methods based on Kalman filtering, in which the loop integration time T is uniformly selected for 10 ms, and the L1C/A, L2CM, and L5Q signals are tracked accordingly. The navigation message is not modulated, so each signal tracking uses a four-quadrant arctangent phase detector. Figure 4 shows the C/N_0 estimation results of the GPS L1, L2, and L5 signals at 1 Hz. It can be seen that the signal attenuation caused by the ionosphere scintillation reaches more than 20 dB after 30 s. The general trend is similar to the amplitude scintillation of the simulator in Fig. 2.

Traditional single-frequency PLL tracking results are shown in Figs. 5 and 6, where Fig. 5 shows the estimated error of the Doppler frequency tracking results relative to the true geometric Doppler in the simulator, and Fig. 6 shows the phase tracking results compared with the estimation error of the total phase including ionospheric scintillation changes. It can be seen that under the strong ionospheric scintillation, there will be a lot of errors in the traditional PLL Doppler tracking. In addition, phase scintillation on each frequency signal can also cause phase tracking errors. Figure 6 shows the PLL phase estimation error. It can be seen that L1, L2, and L5 all have different cycle slip occurrences. Among them, L1 showed a full cycle slip at 230 s, L2 and L5 signals started cycle slip at the 30 s, and cycle slip occurred frequently in the entire 300 s interval. L2 has 10 cycle slips, L5 has 9 cycle slips with the largest phase error reaching to 4 cycles.

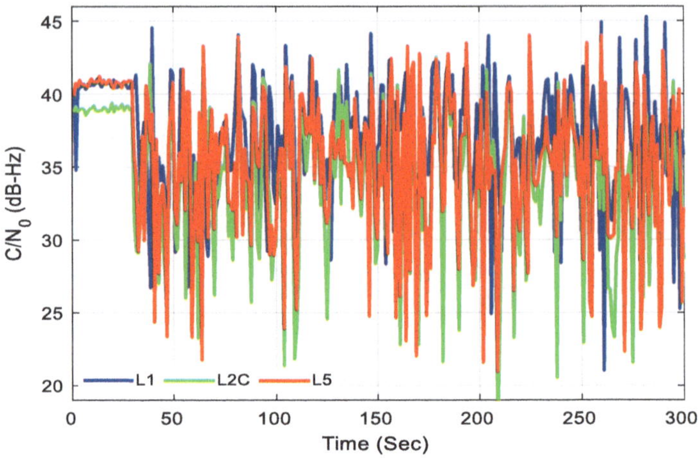

Fig. 4. GPS L1, L2, L5 signal C/N_0 estimation

The three-frequency joint estimation algorithm using frequency diversity can effectively improve the carrier tracking accuracy of each frequency signal. The results of the Doppler estimation error and the carrier phase estimation error are given in Figs. 7 and 8, respectively. Compared with the Doppler error of the PLL in Fig. 5, the three-frequency joint estimation algorithm reduces the amount and magnitude of errors. Most of the Doppler errors in Fig. 7 are in the range of −0.5 Hz to 0.5 Hz, and a small amount exceeds the level of ±1 Hz. In addition, comparing Fig. 8 with Fig. 6, it can be found that the use of the three-frequency joint tracking algorithm can also greatly reduce the phase error. The L1 signal has no cycle slip, and the characteristic information of the ionosphere scintillation at this frequency is completely retained. L2 and L5 have 5 and 3 cycle slips respectively, compared with PLL results, the cycle slips are reduced by more than 50%. The maximum cycle slip of L2 and L5 is at the level of 2 cycles, which reduces the phase error by 1 cycle as compared to PLL.

The three-frequency joint estimation algorithm can also estimate the phase change trend of ionospheric scintillation. The idea is that the geometric phase can be subtracted from each phase estimate so that the residual term contains the phase information of the ionospheric scintillation. However, since the ionospheric scintillation phase trends are similar at each frequency, the common phase component that satisfies the linear relationship in formula (1) will be absorbed by the Doppler, and the estimation of each frequency phase only retains the difference term of the ionospheric scintillation phase.

This paper estimates the true difference based on the ionospheric phase change sequence generated by the simulator in Fig. 3. The results are shown in Fig. 9. It can be seen that after subtracting the common trend in Fig. 3, the phase difference terms of each frequency have their own characteristics. The difference term of L1 is opposite to that of L2 and L5. The difference term of L2 is similar but slightly larger than that of to L5.

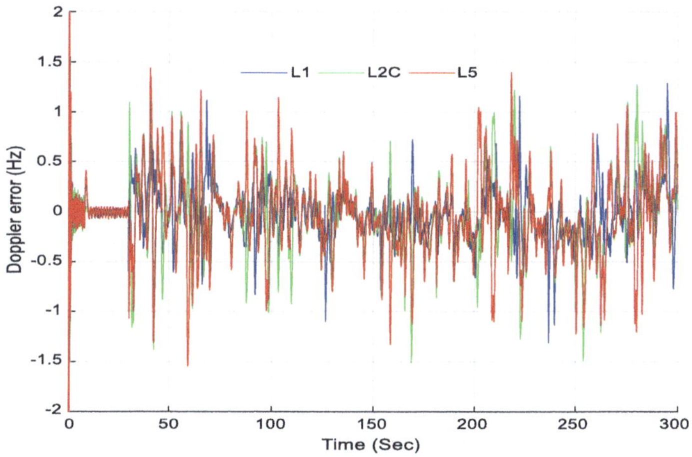

Fig. 5. Doppler estimation errors in traditional PLL

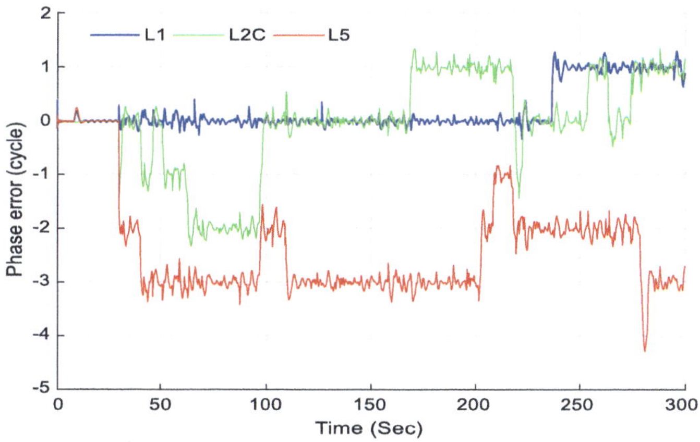

Fig. 6. Carrier phase estimation errors in traditional PLL

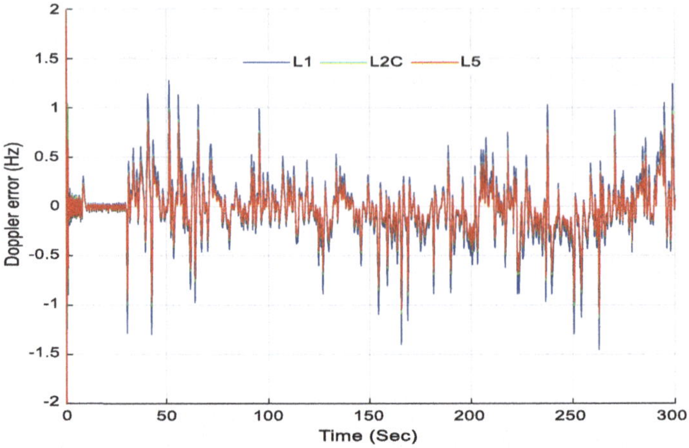

Fig. 7. Doppler estimation errors in triple frequency tracking

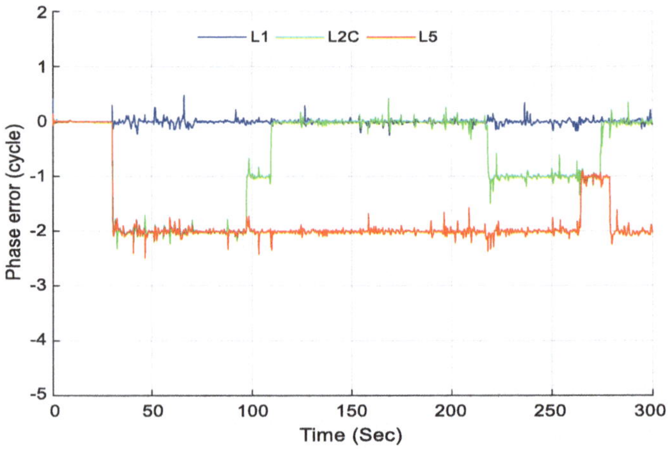

Fig. 8. Carrier phase estimation errors in triple frequency tracking

Figure 10 shows the ionospheric scintillation phase estimation at each frequency extracted from the carrier phase result and Doppler result in the three-frequency joint estimation algorithm. Comparing with Fig. 9, it can be seen that the estimated phase change is consistent with the true value trend. L1 changes from 2 to 6 cycles after 200 s, which is slightly larger than the true value. The phase change of L2 is slightly smaller than that of L5 after 150 s, which is contrary to truth in Fig. 9. The reason may be that there is cycle slip occurred in carrier phase estimation which may makes the phase estimation inaccurate and deviate from the truth.

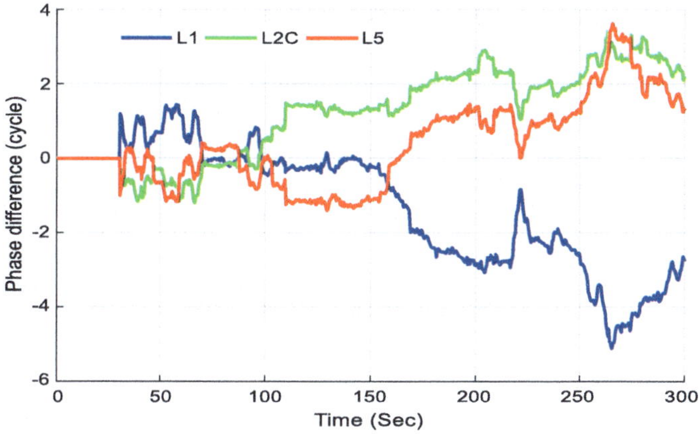

Fig. 9. The simulated scintillation phase residuals that removed common component from the original scintillation phase on GPS L1, L2, L5.

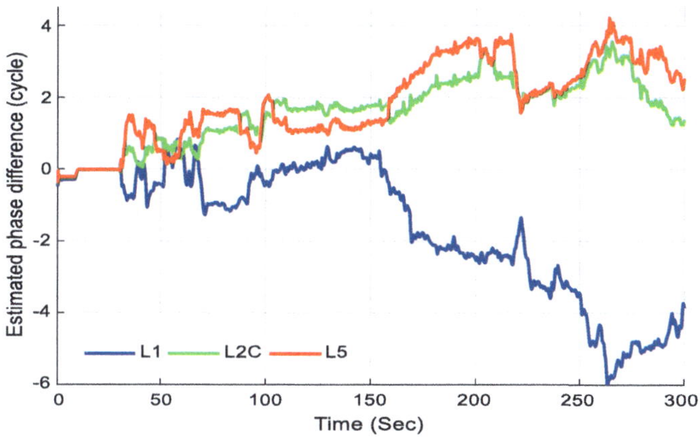

Fig. 10. The estimated phase differences induced by ionosphere scintillation on GPS L1, L2, L5 in triple frequency tracking.

5 Conclusion

This paper proposes a three-frequency joint estimation carrier tracking algorithm under strong ionospheric scintillation environment. An ionospheric scintillation simulator was used to generate IF signal to verify the proposed algorithm. Compared with the traditional PLL, the algorithm can effectively reduce the Doppler estimation error and phase estimation error, and the number of cycle slips is reduced by more than 50% compared with the PLL. The maximum cycle slip of L2 and L5 is at the level of 2 cycles, which is 1 cycle less than that of PLL. In addition, the algorithm in this paper can also estimate the phase change trend of ionospheric scintillation, which is helpful for the extraction of phase scintillation characteristic parameters and the modeling of ionospheric scintillation monitoring.

The algorithm in this paper is only a preliminary algorithm design. There is still cycle slip existed in phase tracking. The future research is to optimize the design and conduct detailed theoretical analysis of the three-frequency joint tracking algorithm to further improve the tracking accuracy. In addition, since most parts of the Chinese southern areas are located in the middle and low latitudes which are likely to have ionosphere scintillations. Therefore, future work intends to conduct study on the Chinese ionospheric scintillation characteristics, especially in low latitudes of China, and to develop anti-scintillation receiver algorithms suitable for Chinese ionospheric activity, to facilitate Chinese high-precision positioning applications.

Acknowledgement. The ionosphere scintillation simulator used in the paper is an open source simulator developed by SenSe Lab in Colorado University at Boulder. We appreciate Professor Y. T. Jade Morton and Professor C. R. Rino for providing this valuable resource which helps us to finish the research. In addition, we also appreciate the suggestions from Dr. Dongyang Xu in GPS Solutions Corporation. Finally, we appreciate the financial support from Shanghai Jiao Tong University (WF220541306) and Key laboratory of space microwave technology (6142411193113).

References

1. Wen, D.: Investigation of GPS-based ionospheric tomographic algorithms and their applications. Institute of Geodesy and Geophysics Chinese Academy of Sciences (2007)
2. Zhong, J.: Investigation on the variations of the topside ionosphere using low earth orbit satellite based TEC. University of Science and Technology of China (2017)
3. Rino, C.L.: The Theory of Scintillation with Applications in Remote Sensing. Wiley, New York (2011)
4. Kintner, P.M., Humphreys, T., Hinks, J.: GNSS and ionospheric scintillation. Inside GNSS **4** (4), 22–30 (2009)
5. Li, Q., Yin, P.: The characteristic study of ionospheric scintillations over China based on GNSS data. In: The 9th Chinese Satellite Navigation Conference (2018)
6. Jiao, Y., Morton, Y.T.: Comparison of the effect of high-latitude and equatorial ionospheric scintillation on GPS signals during the maximum of solar cycle 24. Radio Sci. **50**(9), 886–903 (2015)
7. Lee, J., Morton, Y.T.J., Lee, J., et al.: Monitoring and mitigation of ionospheric anomalies for GNSS-based safety critical systems: a review of up-to-date signal processing techniques. IEEE Sig. Process. Mag. **34**(5), 96–110 (2017)
8. Humphreys, T.E., Psiaki, M.L., Kintner, P.M.: Modeling the effects of ionospheric scintillation on GPS carrier phase tracking. IEEE Trans. Aerosp. Electron. Syst. **46**(4), 1624–1637 (2010)
9. Vila-Valls, J., Closas, P., Fernández-Prades, C.: Advanced KF-based methods for GNSS carrier tracking and ionospheric scintillation mitigation. In: 2015 IEEE Aerospace Conference, pp. 1–10. IEEE (2015)
10. Xu, R., Liu, Z., Chen, W.: Improved FLL-assisted PLL with in-phase pre-filtering to mitigate amplitude scintillation effects. GPS Solut. **19**(2), 263–276 (2015)

11. Susi, M., Andreotti, M., Aquino, M., et al.: Tuning a Kalman filter carrier tracking algorithm in the presence of ionospheric scintillation. GPS Solut. **21**(3), 1149–1160 (2017)
12. Xu, D., Morton, Y.: A semi-open loop GNSS carrier tracking algorithm for monitoring strong equatorial scintillation. IEEE Trans. Aerosp. Electron. Syst. **54**(2), 722–738 (2017)
13. Yin, H., Morton, Y., Carroll, M., et al.: Implementation and performance analysis of a multi-frequency GPS signal tracking algorithm. In: Proceedings of the ION GNSS+, Tempa, Florida (2014)
14. Vilà-Valls, J., Closas, P., Curran, J.T.: Multi-frequency GNSS robust carrier tracking for ionospheric scintillation mitigation. J. Space Weather Space Clim. **7**, A26 (2017)
15. Yang, R., Xu, D., Morton, Y.T.: Generalized multi-frequency GPS carrier tracking architecture: design and performance analysis. IEEE Trans. Aerosp. Electron. Syst. (2019). https://doi.org/10.1109/TAES.2019.2948535
16. Yang, R., Ling, K.V., Poh, E.K., et al.: Generalized GNSS signal carrier tracking: part I—modeling and analysis. IEEE Trans. Aerosp. Electron. Syst. **53**(4), 1781–1797 (2017)
17. Yang, R., Morton, Y., Ling, K.V., et al.: Generalized GNSS signal carrier tracking—part II: optimization and implementation. IEEE Trans. Aerosp. Electron. Syst. **53**(4), 1798–1811 (2017)
18. Xu, D., Morton, Y.T., Rino, C.L., Carrano, C.S., Jiao, Y.: A two-parameter multi-frequency GPS signal simulator for strong equatorial ionospheric scintillation: modeling and parameter characterization. Navigation (2019)
19. Jiao, Y., Xu, D., Rino, C.L., et al.: A multifrequency GPS signal strong equatorial ionospheric scintillation simulator: algorithm, performance, and characterization. IEEE Trans. Aerosp. Electron. Syst. **54**(4), 1947–1965 (2018)
20. Liu, D., Yu, X., Feng, J., et al.: Simulating the impacts of ionospheric scintillation on GNSS signals with phase screen method. Chin. J. Radio Sci. **31**(4), 632–638 (2016). https://doi.org/10.13443/j.cjors.2015083102. (in Chinese)

Research on Enhancement Scheme of GPS Occultation Open-Loop Tracking Strategy

Lu Zhang(✉), Xiaojiang Yang, Qijia Dong, Juanjuan Dong, GenJin, Xianyang Liu, YanCheng, and Lijing Pan

Space Star Technology Co., Ltd., Beijing, China
zhanglucandi@163.com

Abstract. GPS radio occultation detection technology is a new application technology, mainly using GPS for edge detection of the atmosphere and ionosphere. In the low troposphere, it is difficult to track the low tropospheric part (about below 10 km) of the rising occultation using the closed-loop mode. The open-loop mode can effectively overcome the limitations of the closed-loop mode, improving the quality and quantity of low tropospheric occultation observations. The traditional open loop tracking is L1C guided L1P capture, and then L2P capture. The overall design is a serial design, and L2P tracking takes too long. In order to improve the available duration of dual-frequency, this paper optimizes the L2 capture strategy. Through the open-loop prediction, the model pseudo-range placement channel is obtained to get code phase, which directly guides L2P code capture and tracking. At the same time, L1P and L1CA are made into a synchronous code loop, and the L1P code acquisition and tracking process is eliminated to shorten the L2P code acquisition and tracking time, thereby solving the problem of long time to track L2. This article compares the availability of in-orbit data and inversion results of the two tracking schemes, and concludes that the new tracking strategy has increased the available time of dual-frequency data by about 20 s, while increasing the atmospheric dry temperature and dry pressure index profiles below 10 km, which provides effective data support for the low tropospheric atmosphere detection, and at the same time provides verification of tracking algorithms for the development of GNSS occultation receivers.

Keywords: GPS occultation · Open loop · Dual frequency · Atmosphere · Capture and tracking

1 Introduction

GNSS occultation exploration has the advantages of all-weather observation, high vertical resolution and global coverage. The complexity of the atmospheric environment requires that the occultation receiver must be real-time fast and stable continuous tracking. The GPS/LEO occultation observation diagram is shown in Fig. 1. In the low troposphere, the atmosphere becomes complicated, and it is difficult to track the low troposphere (below about 10 km) of the rising occultation using closed-loop mode. Using the open-loop mode can effectively overcome the limitations of the closed-loop mode. In the low troposphere, the closed-loop mode can only track the descending

J. Sun et al. (Eds.): CSNC 2020, LNEE 650, pp. 576–584, 2020.
https://doi.org/10.1007/978-981-15-3707-3_54

occultation, not the rising occultation. The open-loop mode not only accurately recovers the phase of the signal, but also processes rising and falling occultations, thereby improving the quality and quantity of low tropospheric occultation observations [1].

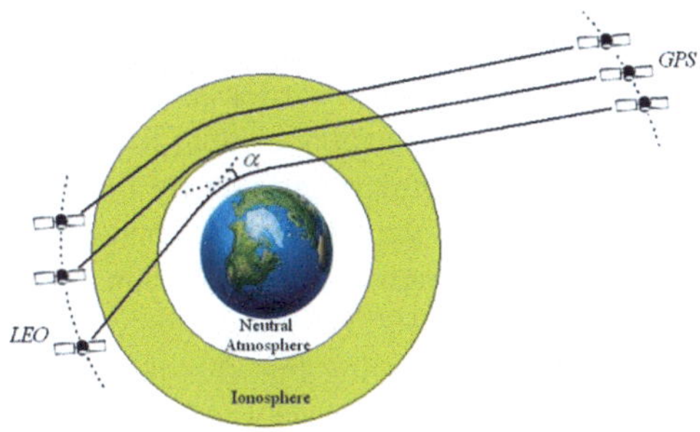

Fig. 1. GPS/LEO occultation observation diagram

Atmospheric open-loop tracking mainly uses LEO satellite position velocity and prior atmospheric model (atmospheric bending angle model) and prior ionosphere model to estimate Doppler observations and pseudorange observations. The duration of the entire atmospheric occultation event is about 1 min, the atmospheric occultation signal needs to be captured as soon as possible. This article analyzes the traditional serial design open-loop tracking strategy, that is L1C guides L1P capture and then L2P capture. This capture strategy makes the available arc segments of dual-frequency data too short, which affects the inversion of atmospheric parameters. This paper further proposes an optimized atmospheric open-loop tracking strategy, optimized L2 acquisition strategy, and obtained the model pseudo-range insertion channel through open-loop prediction to obtain accurate code phases, directly guiding L2P code acquisition and tracking, and simultaneously making L1P and L1CA into The synchronous code processing loop removes the L1P code capture and tracking process and shortens the L2P code capture and tracking time. By simulating and verifying the optimized tracking strategy, it is concluded that the optimization strategy improves the available time of the dual-frequency data by about 20 s, and increases the atmospheric dry temperature and dry pressure index profiles below 10 km, increasing the availability of atmospheric occultation data, and the accuracy of the inversion results.

2 Tracking Strategy Program Analysis

2.1 Defects of Traditional Tracking Strategies

The traditional receiver software design tracking strategy is: The receiver is designed with two atmospheric channels to handle the same atmospheric event, one channel adopts open-loop processing, and one channel uses closed-loop processing. Atmospheric occultation events begin, closed-loop channels begin to enter L1CA capture state, and L1CA tracking stable L1 carriers are available. After the L1CA code bits are synchronized with the frame, the accurate Z count and CodeCycle count are obtained to guide the L1P code capture tracking. After the L1P code tracking is stable, the L2P code capture tracking is guided. After the L2P code capture, the L2 carrier loop begins to converge, and a few seconds later L2 carrier is available. From L1CA to the L2 carrier stable tracking, it took more than 40 s, that is, there is no L2 carrier data in the first 40-odd seconds of the occultation event. Generally, an atmospheric occultation event is about one minute, and there is no L2 carrier data in the first 40-odd seconds of the occultation event, resulting in less available dual-frequency data for the entire occultation event, and the accuracy of atmospheric retrieval is affected. Figure 2 below shows the capture tracking strategy in the traditional scheme.

Now randomly select 6 sets of traditional occultation receivers of atmospheric occultation satellites receiving Class 0 100 Hz simulation data to analyze the tracking status of each frequency point. The following Fig. 3 is the analysis of each frequency point tracking. It is obtained that the open-loop/closed-loop L1CA carrier-to-noise ratio increases when the atmospheric occultation event reaches about 10 s (1000 epochs). When the occultation event is about 30 s (3000 epochs), as showing in Fig. 4(c), the L1P carrier-to-noise ratio increases. When it is about 40 s (4000 epochs), as showing in Fig. 3(d), the L2P carrier-to-noise ratio increases, and tracking starts. It takes about 40 s for the receiver to capture the L2 carrier from L1. The dual-frequency stable tracking time during the entire occultation event is only about 20 s to 30 s. It cannot meet the tracking time required for atmospheric occultation inversion. The acquisition and tracking strategy needs to be optimized. To increase the arc duration available for occultation data.

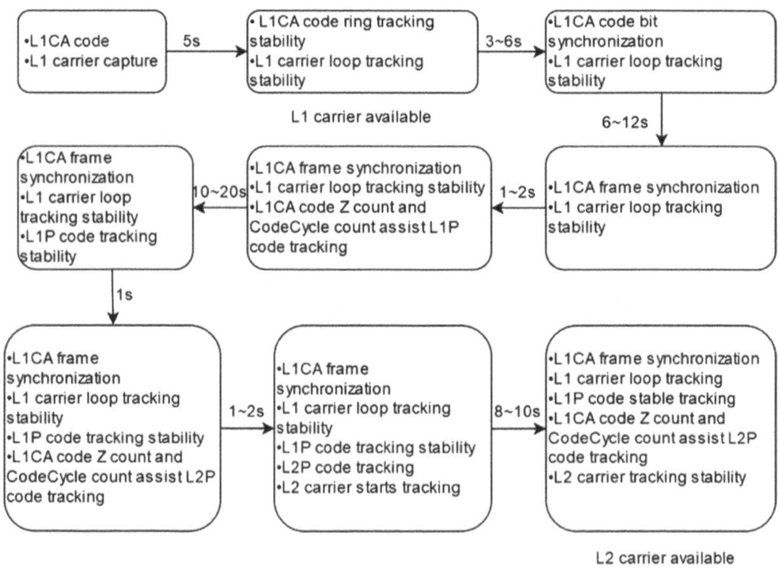

Fig. 2. Flowchart of the traditional plan capture tracking strategy

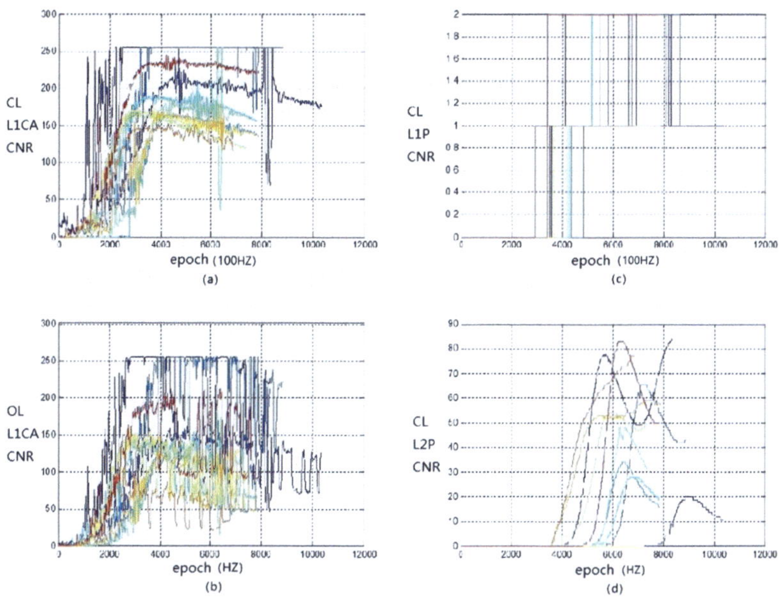

Fig. 3. Flowchart of the traditional plan capture tracking strategy

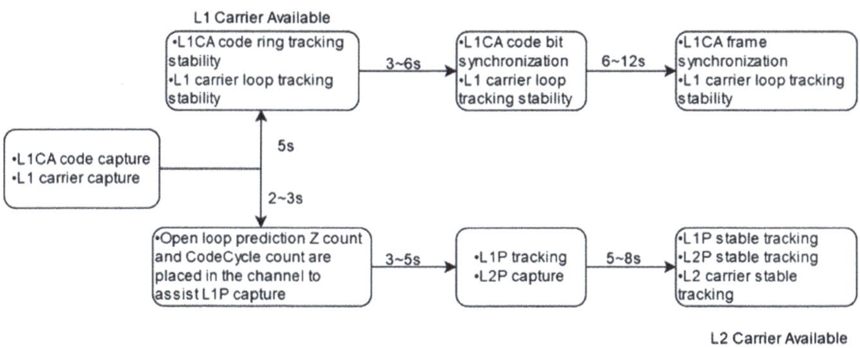

Fig. 4. L1/L2 acquisition and tracking optimization strategy

2.2 Optimize Tracking Strategy Plan and Effect

Analyze the problem of long L2 carrier acquisition and tracking time. L2 carrier tracking is stable, the premise is that the L2P code capture and tracking is stable. L2P code capture and tracking needs to obtain accurate Z count and CodeCycle count. Obtaining accurate Z count and CodeCycle count depends on L1CA code capture and tracking and achieve frame synchronization status. At the same time, the L2P code capture tracking depends on the tracking status of the L1P code. The L1P code is not captured successfully, and the L2P code ring tracking is affected. To improve the tracking strategy, consider combining open-loop forecasting and closed-loop tracking to reduce the L2 carrier acquisition time. The optimized tracking strategy is shown in

Fig. 4. Open-loop prediction can obtain model pseudorange and model Doppler, and obtain accurate Z count and CodeCycle count through the open-loop model pseudorange insertion channel, which directly guides L2P code capture and tracking. At the same time, L1P and L1CA are made into a synchronous code processing loop, and the L1P code capture and tracking process is eliminated, which can shorten the L2P code capture and tracking time, thereby solving the problem of L2 capture and tracking time.

The open loop tracking of the software receiver developed by our company is optimized, and the accurate Z count and CodeCycle count are obtained through putting the pseudo-range into the channel of the open loop model, which directly guides the L2P code capture and tracking. Run the occultation data simulation system and receive the atmospheric occultation simulation data to analyze the available time arc of the dual-frequency data. Randomly select 6 sets of occultation receiver 0-level 100 Hz simulation data received by the occultation receiver with optimized tracking strategy, and analyze the tracking status of each frequency point. Figure 5 below shows the tracking analysis chart of each frequency point. Figure 3(a, b) It is concluded that when the atmospheric occultation event progresses to about 10 s (1000 epochs), the open-loop/ closed-loop L1CA carrier-to-noise ratio increases. When the occultation event progressed to about 15 s (1500 epochs), as shown in Fig. 4(c), the L1P carrier-to-noise ratio increased. When it is about 20 s (2000 epochs), as shown in Fig. 3(d), the L2P carrier-to-noise ratio increases, and tracking starts. It takes only about 20 s for the receiver to capture the L2 carrier from L1. The dual-frequency stable tracking time is about 40 s to 60 s during the entire occultation event, which meets the tracking time required for atmospheric occultation inversion, and greatly improves the availability of occultation data.

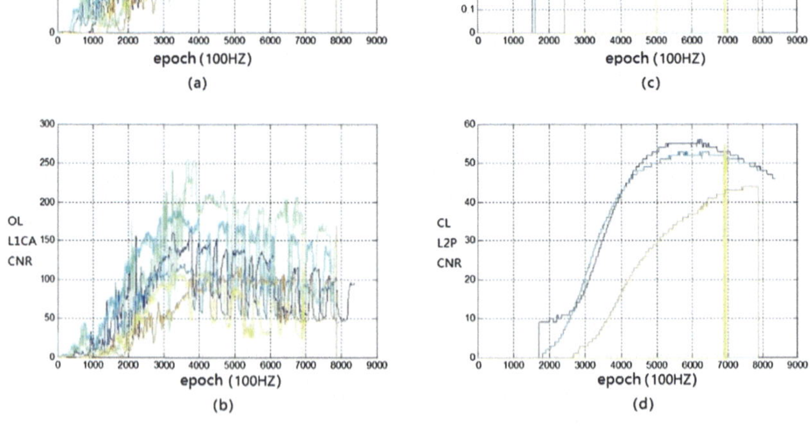

Fig. 5. Analysis of the available time arc of the dual-frequency data of the optimization scheme

3 Verification Analysis

3.1 Data Sources

In this article, the GNSS occultation signal source and occultation receiver developed by Space Star Technology Co., Ltd, are used to simulate the GPS/LEO occultation event data. The self-developed GNSS occultation signal source can simulate forward or backward atmospheric occultation events of LEO, and supports GPS L1/L2 BD2 B1/B2. The composition of its data simulation system is shown in Fig. 6 below.

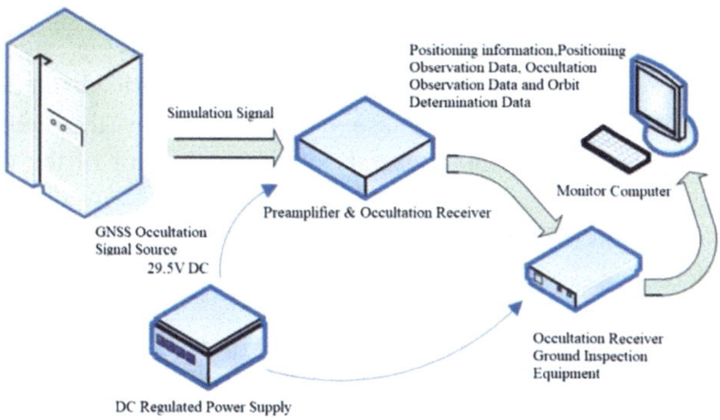

Fig. 6. Data simulation system composition

The traditional open-loop tracking strategy and the optimized open-loop tracking strategy optimized in this paper are used to design different occultation receiver software. The simulation system is run to receive atmospheric occultation observation data of different tracking strategies, which provides reliable data source for this paper.

3.2 GPS Atmospheric Occultation Inversion Algorithm

Neutral atmospheric parameter inversion technology based on GNSS neutral atmospheric occultation observation data can pre-process the occultation observation data, use precision ephemeris and single-difference technology to eliminate receiver clock error, and generate additional atmospheric phase delay data; The atmospheric Doppler is derived under the assumption of local symmetry, and the atmospheric bending angle is further deduced; the atmospheric refractive index profile is obtained by inverse Abel inverse transformation of the atmospheric bending angle [2]; The occultation receiver observes the occultation GPS satellite and another non-occultation reference GPS satellite signal at the same time, ignoring the integer ambiguity and measurement noise [5]. The GPS observation phase can be expressed as

$$L_K = -\frac{C}{f_k}\varphi_k = \rho + dA_k + dI_k + c(t - T) \tag{1}$$

Using single differential technology, the difference between the receiver and the occultation satellite observation L_k^0 and the receiver and the reference GPS satellite observation L_k^r eliminates the receiver's clock error.

$$L_k^0 - L_k^r = \rho_k^0 - \rho_k^r + c(t^0 - t^r) + dI_k^0 - dI_k^r + dA_k^0 - dA_k^r \quad (2)$$

The additional phase delay term required for its occultation observation can be written as

$$dA_k^0 + dI_k^0 = (L_k^0 - L_k^r) - \rho_k^0 + \rho_k^r - c(t^0 - t^r) + dI_k^r + dA_k^r \quad (3)$$

The subscripts K = 1 and 2: refer to L1 and L2 respectively; φ_k signal observation phase; C is the speed of light in vacuum, ρ is the geometric distance between the transmitter and receiver; dA_k is the phase delay caused by the neutral layer; dI_k Generate a phase delay for the ionosphere; L is the clock error of the transmitter and receiver, respectively.

Under the assumption of large balloon symmetry [3], according to snell's law, the collision parameters

$$a = n_2 r_2 \sin\varphi_2 = n_1 r_1 \sin\varphi_1 \quad (4)$$

n_1, n_2 is the atmospheric refractive index of the receiver and transmitter, r_1, r_2 is the distance from the refraction point to the center of the earth, and the bending angle is α. Through Abel integral transformation, part of the bending angle profile can be converted into the atmospheric refractive index profile [4]. The Smith-Weintraub equation, the ideal gas equation, and the hydrostatic equation were used to gradually retrieve the dry temperature, dry pressure, and electron density profiles of the neutral atmosphere.

$$n(a) = n_1 exp\left[\frac{1}{\pi}\int_a^{n_1 r_1} \frac{\tilde{a}(x)}{\sqrt{x^2 - a^2}} dx\right] \quad (5)$$

3.3 Analysis of Inversion Results

The inversion of the atmospheric occultation data before and after the optimization of the tracking strategy are respectively performed. Figures 7 and 8 show the refractive index, dry temperature, and atmospheric pressure profiles of the atmospheric simulation data of the traditional tracking strategy. The height range of the inversion altitude of the dry temperature and atmospheric pressure profiles is 10–60 km; Figs. 9 and 10 are the refractive index, dry temperature and atmospheric pressure profiles retrieved from the atmospheric simulation data optimized by the L2P tracking strategy. The height range of profile inversion is 5–60 km, and the profile trend and numerical magnitude are normal. Comparing Fig. 7 with Fig. 9, the L2P tracking strategy is optimized, and the atmospheric inversion results have increased by less than 10 km (as shown in the red frame in the figure); Comparing Fig. 8 with Fig. 10, it can be concluded that compared with the serial tracking strategy, the tropospheric atmospheric dry temperature and dry pressure index profile are increased by less than 10 km (shown in the red box in the figure), greatly improving the depth of atmospheric occultation detection Provide more effective and reliable atmospheric parameters for exploring the atmosphere.

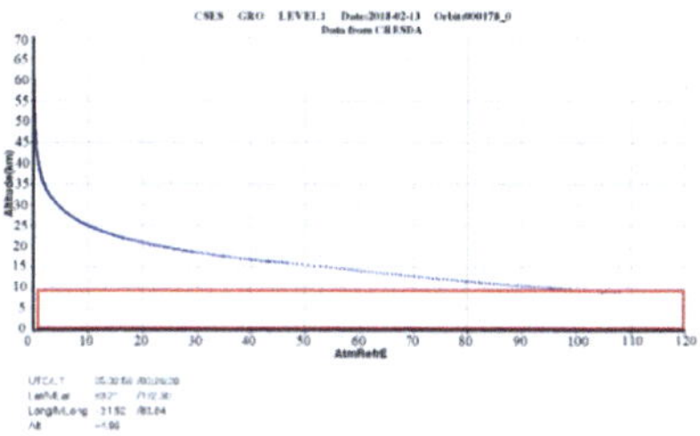

Fig. 7. Atmospheric refractive index profile retrieved from original scheme data

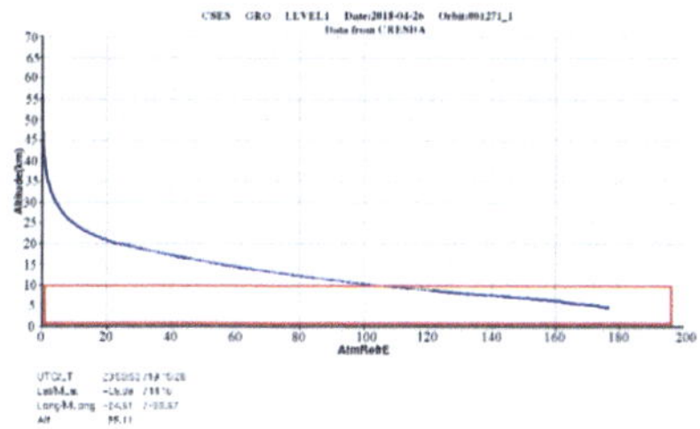

Fig. 9. Atmospheric refractive index profile retrieved from optimization scheme data

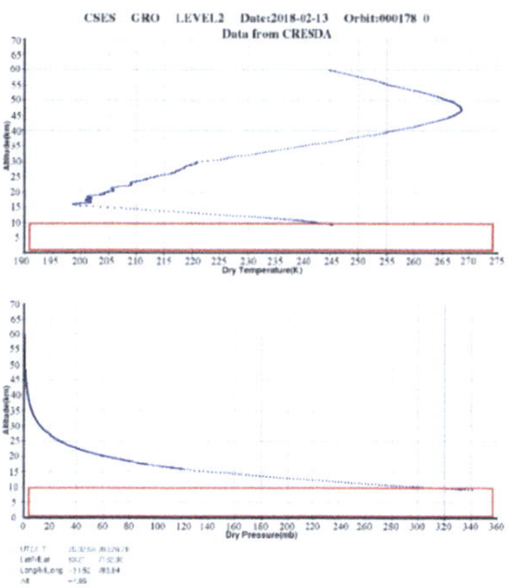

Fig. 8. Dry temperature and atmospheric pressure profiles retrieved from original scheme data

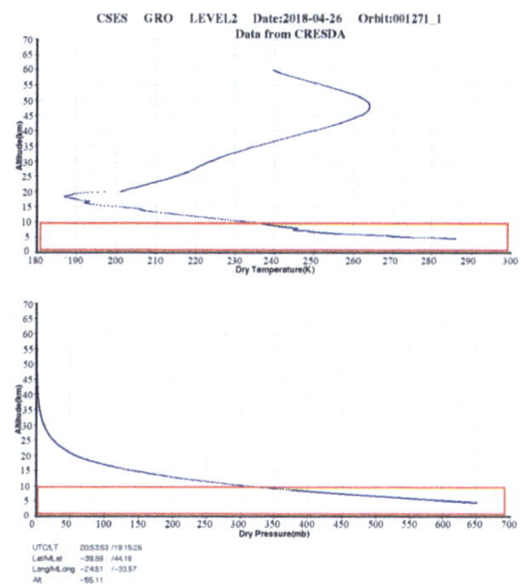

Fig. 10. Dry temperature and atmospheric pressure profiles retrieved from optimization scheme data

4 Summary

This paper introduces an optimization scheme based on the traditional open-loop tracking strategy, and proposes a method for occultation tracking loops to track L1p and L2p carriers simultaneously. Using open-loop prediction, the model pseudo-range is placed in the channel to obtain accurate code phase, which directly guides L2P Code capture tracking, changing L1 and L2 dual-frequency capture tracking from the original serial design to parallel design. By comparing and analyzing the availability of in-orbit data and inversion results of the two schemes, the available time of open-loop capture and tracking dual-frequency in the low troposphere has been increased from 20 s to

40 s, and the atmospheric wetness index profile below 10 km has been increased. The low tropospheric atmosphere detection provided effective data support and verified the reliability of the optimized tracking scheme. The research in this paper provides a more optimized and reliable tracking algorithm verification for the development of GNSS occultation receiver.

Thanks. Thanks to the project team for their support and assistance in the article demonstration process.

References

1. Xu, X.H.: Research on observing the earth's atmosphere with the GNSS radio occultation technique. Wuhan University (2003)
2. Hu, X., Zeng, Z., Zhang, X., et al.: Atmospheric inversion methods of GPS occultation observation. J. Geophys. **48**(4), 768–774 (2005)
3. Mungufeni, P., Rabiu, B.A., Okoh, D., Jurua, E.: Characterisation of total electron content over african region using radio occultation observations of COSMIC satellites. Adv. Space Res. **65**, 19–29 (2019)
4. Stolle, C., Jakowski, N., Schlegel, K., et al.: Comparison of high latitude electron density profiles obtained with the GPS radio occultation technique and EISCAT measurements. Ann. Geophys. **22**(6), 2015–2028 (2004)
5. Steiner, A.K., Kirchengast, G., Ladreiter, H.P.: Inversion, error analysis, and validation of GPS/MET occultation data. Ann. Geophys. **17**, 122–138 (1999)

Policies, Regulations, Standards and Intellectual Properties

Patent Analysis Reveals the Development Route of the Indoor High-Accuracy Positioning Technology

Huiying Li[1], Jinping Yu[1(✉)], and Qingyi Gao[2]

[1] China Industrial Control Systems Cyber Emergence Response Team, Beijing 100040, China
yujinping@infoip.org
[2] School of Reliability and Systems Engineering, Beihang University, Beijing 100191, China

Abstract. As an important part of PNT system (Positioning, Navigation and Timing), indoor positioning technology can solve the vulnerability problem of space-based navigation system in the presence of occlusion and interference and provide users with completed services such as indoor positioning, navigation, query, identification and event inspection. The analysis of a large number of patents related to indoor positioning technology indicates that the relevant patents are relatively concentrated in China, the United States, Korea, Japan and other countries and a large-scale patent layout has been carried out all around the world. It is of great significance for us to track the development route of PNT technology to understand the technological development trend of foreign countries, to optimize the technological development route, and to make full use of the supporting role of patents in technological forecasting.

Keywords: PNT · Patent · Indoor navigation

In outdoor, space-based satellite navigation system, such as GPS, Beidou, etc., can carry out high-accuracy navigation and positioning. As an important part of PNT system, indoor positioning technology can solve the vulnerability problem of space-based navigation system in the presence of occlusion and interference and provide users with complete services such as indoor positioning, navigation, query, identification and event inspection.

1 General Situation of Indoor High-Precision Positioning Technology Patent

The patents related to indoor positioning technology are searched and analyzed and the patent application trend is shown in Fig. 1. The total number of patent applications of indoor positioning technology is increasing year by year and it has developed even rapidly in recent years. Up to August 20, 2019, the total number of patent applications for global indoor positioning technology is 15665, including 8701 Chinese patents and 6964 foreign patents.

J. Sun et al. (Eds.): CSNC 2020, LNEE 650, pp. 587–594, 2020.
https://doi.org/10.1007/978-981-15-3707-3_55

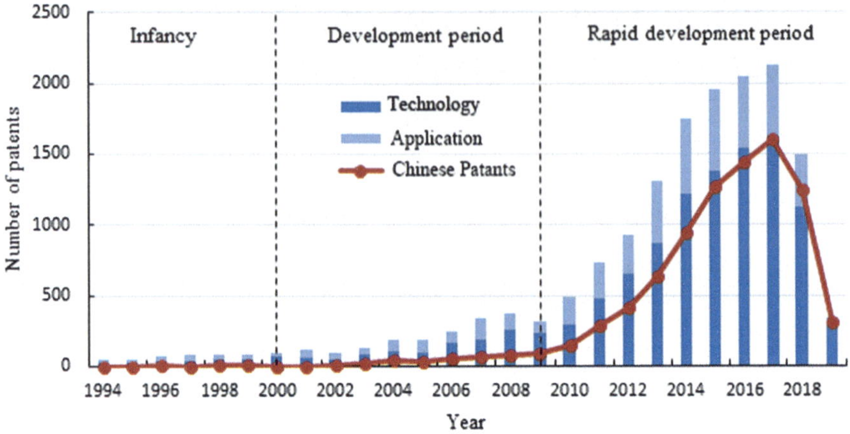

Fig. 1. Global patents application trend of indoor positioning technology

Patent applications are mainly concentrated in the United States, South Korea, Japan and China, as shown in Fig. 2.

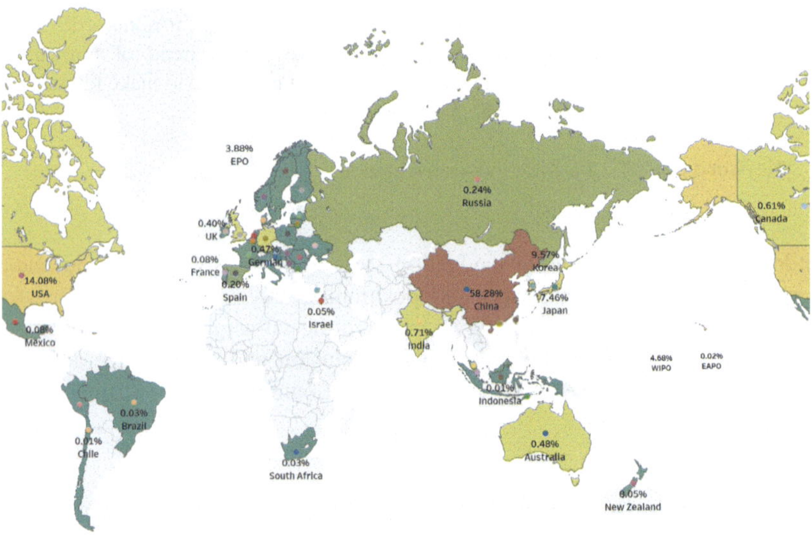

Fig. 2. The geographical distribution of patent applications

The patent related to indoor high precision mainly covers indoor positioning technology of mobile communication network, indoor wireless positioning technology (including WiFi, Bluetooth, ZigBee, infrared, ultrasonic, radio frequency and ultra-bandwidth), pseudo satellite indoor positioning technology, geomagnetic indoor positioning technology, visual positioning technology and all kinds of hybrid positioning technology. There are 2202 patent applications, accounting for 21% of the total indoor positioning technology patents. As of August 20, 2019, the proportion of patent applications in each branch of indoor positioning technology is shown in Fig. 3.

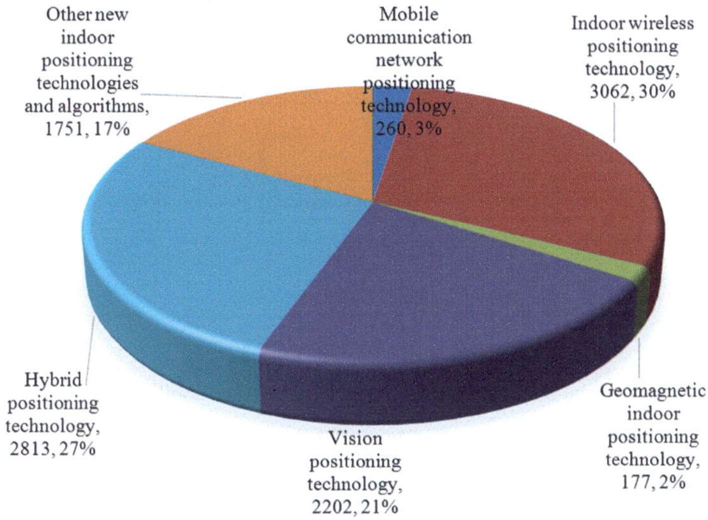

Fig. 3. The proportion of patent applications in each branch of indoor positioning technology

2 Patent Technology Route Analysis of Indoor High-Precision Positioning

Through the analysis of these patents, it is found that the development routes of indoor positioning technology are as follows:

Technical route 1: improving the accuracy and reliability of indoor positioning by improving technology, optimizing layout and algorithm.

Indoor positioning technology is based on mobile communication, indoor wireless signal transmission, image recognition, geomagnetism, pseudo satellite and other related technologies. In recent years, the rapid progress of these technologies has promoted the progress of indoor positioning technology.

Taking the application of mobile communication network in indoor positioning as an example, patent retrieval and analysis are carried out and the technical development route is shown in Fig. 4. Since the positioning was realized through 2G, mobile communication network technology has experienced the development of 3G/4G and the issuance of 5G license. The number of base stations has increased dramatically and the data transmission speed has been greatly improved, which greatly improve the mobile communication network based indoor and outdoor positioning accuracy and positioning service level.

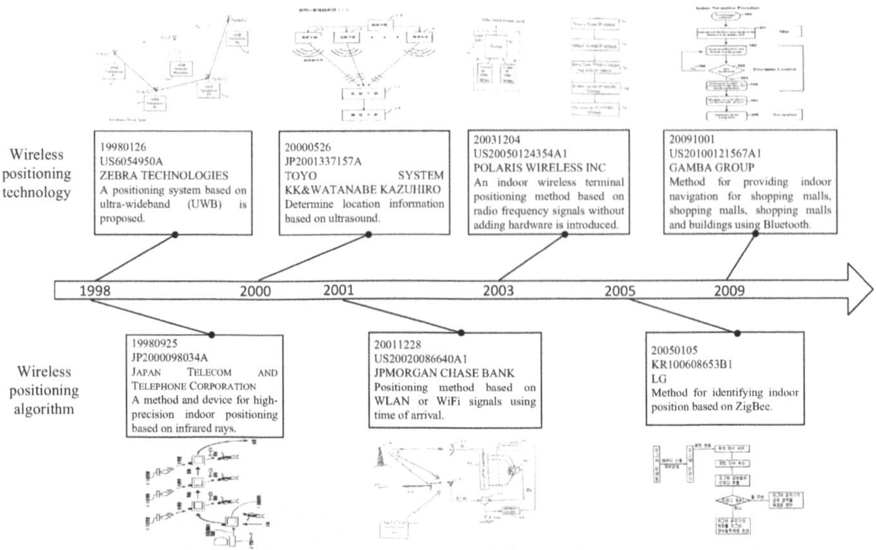

Fig. 4. Development route of indoor positioning technology of mobile communication network

In addition to the progress of mobile communication network technology, the location algorithm has also been optimized, which has gone through the stages of arrival time method (US5317323A), arrival time and angle of arrival method (US5945948A), multi-receiver method (EP964265A2), fingerprint method (US20080167049A1), forward link method (US8369872B2), multi-source location (US7847734B2) and hybrid algorithm (US12786429), etc. These methods have improved continuously the anti-interference ability, accuracy and reliability of positioning.

Technical route 2: exploring the application of new technology in the field of high-accuracy indoor positioning.

With the development of science and technology, the emergence of new technology also provides a new way for indoor positioning technology. For example, with the development of artificial intelligence, image recognition and big data analysis have been widely used. The visual positioning technology, as a carrier of the application of these technologies, has high positioning accuracy and reliability and expanded the scope of indoor positioning technology application. Besides, its patent application number has increased greatly in recent years.

In the visual positioning system, the road sign images in indoor scene, including natural scenery and artificial scenery, are collected. The features of edge, texture and other features of the image are extracted and then matched with the image features from the road sign database to enable image classification and recognition. As a result, the location information of the target can be determined.

The patent application in recent years reveals the technical development route of visual positioning method, as shown in Fig. 5. For example, the patent US200801 47730A1 selects the image with specific features such as buildings, and matches the location of the image from the database to locate it; the patent US20130045751A1 extracts the matching between the visual features and the specific visual signatures of one or more brands rom one or more images to estimate the location of the mobile device; The patent US20170138740A1 recognizes the transformation between image frames in camera to estimate the position change of mobile devices. At the same time, the patent US9693198B2 determines the position and navigates according to the distance of letters at the end of one or more text information sources. Furthermore, the patent CN108957504A uses image recognition to locate the user's geographic coordinates through satellites or base stations, then the image information around the user is collected and the user's indoor position is located through that image recognition. Afterwards, an indoor and outdoor continuous positioning can be realized.

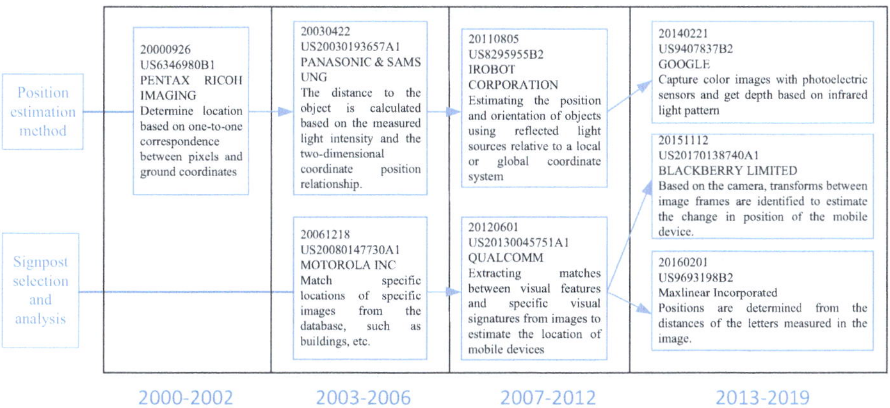

Fig. 5. The development route of visual positioning technology

Technical route 3: adopting the heterogeneous collaborative positioning technology, complementing each other's advantages, and forming a systematic and multi-level hybrid indoor positioning technology framework to increase the accuracy and stability of indoor positioning.

Due to the advantages, disadvantages, the most appropriate application scenarios of various navigation and positioning technologies, as well as the relatively complex indoor environment, it is difficult to achieve high reliability and high accuracy positioning by using a single positioning method. At the same time, the continuous indoor and outdoor positioning in the modern positioning frame also needs the cooperation of various positioning technologies. In recent years, hybrid positioning technology has attracted more attention by combining multiple positioning technologies to realize navigation and positioning as well as improve the positioning accuracy. According to the summary and analysis of the indoor positioning technology patents, there are at least 2813 patents, 27% of the indoor positioning technology patents, adopting the integration of more than two positioning technologies, as shown in Fig. 4.

The hybrid positioning technology includes: (1) multi algorithm cooperative positioning, such as the wireless positioning technology using signal fingerprint and other position estimators in US11739097; (2) multi sensor combined positioning technology, such as the method published in US12014092 using multiple position estimators together to locate positions; (3) opportunity signal supplementary positioning technology, such as US12107457, which provides a navigation system and method for obtaining precise navigation information in a signal challenging environment; (4) multi system complementary positioning technology, such as A-GPS technology used in US10313586. The application of hybrid positioning technology has become an important direction in the development of various indoor positioning technology.

3 Conclusion

According to the statistic analysis of patent data, indoor high-accuracy positioning technology patents are relatively concentrated in China, the United States, South Korea, Japan and other countries, which have carried out large-scale patent portfolio. Compared with other countries, although the number of patents applied for indoor positioning technology in China is large, they are mainly concentrated in scientific research institutions such as universities and institutes. The PNT technology powerhouses represented by the United States have a relatively early patent portfolio with a wide range of coverage and a strong systematicness. Companies such as Qualcomm, Motorola, and Google have made use of the industrial advantages to carry out systematic patent portfolio in the indoor positioning fields such as A-GPS, mobile communication network positioning, and positioning services, mastering large number of core patents and have important influence. In terms of product research and development, it is recommended that related companies start from the following aspects:

(1) Seize the opportunity of technological upgrading and carry out patent layout and algorithm innovation

In the current era of mobile communication network upgrading, 5G technology is the latest generation of cellular mobile communication technology, characterized by high data transmission efficiency, low latency, and large system capacity, etc., which is conducive to improving the positioning accuracy and response speed. It is important to seize the development opportunities of indoor positioning technology in the 5G era, actively carry out research on 5G-based high-accuracy indoor positioning technology, innovate algorithms, systematically carry out patent portfolio to form new indoor positioning leading technologies, and focus on technology advanced research, and actively cooperate with 6G and other technologies to explore the future development mode of indoor positioning technology. In addition, it is necessary to continue to pay attention to the research progress of wireless signals (including Wifi, Bluetooth,

Zigbee, infrared, ultrasonic, radio frequency and ultra-bandwidth, etc.), pseudo satellite technology, geomagnetic measurement, image recognition and other technologies to improve indoor positioning technology.

(2) Continue to make efforts in the field of new technologies and pay attention to industry-university-research cooperation and intellectual property protection to form a new indoor positioning technology patent group

As a new indoor positioning method, visual positioning, not affected by the indoor complex environment, relies on visual sensors to obtain image features, and then uses feature matching to estimate the position and displacement information of the target, thus obtaining the position information of the target to be measured. New patent technologies continue to emerge in the fields of landmark recognition, image understanding and analysis, depth estimation, and so on. Moreover, image recognition technology, which is relied on by visual positioning, is also a key research direction in the hot field - artificial intelligence technology, which is also a research hotspot in various countries around the world. Relevant institutions and enterprises needs to keep abreast of research trends, make technological breakthroughs, and pay attention to industry-university-research cooperation and intellectual property protection to form a patent protection group in emerging fields.

(3) Combined with the characteristics of China's industrial layout to develop hybrid positioning technology, improve the accuracy, reliability and continuity of indoor positioning

So far, no any technology can independently and perfectly implement PNT services, and different positioning technologies have their own advantages and disadvantages. Hybrid positioning technology should also be combined with the layout characteristics of China's industry and technology to reduce costs, improve applicability, and form regional characteristics. The global networking of China's BeiDou Navigation Satellite System (BDS), the booming development of the mobile Internet, the extensive deployment of wireless local Internet, and the formation of a good intelligent ecosystem by a large number of domestic smartphones and other conditions can all become the basis for the formation of China's unique indoor positioning technology.

(4) Continuously strengthen the tracking and analysis of patents to avoid intellectual property risks

The technical solutions of analyzing patents and tracking related patent display are of great significance for sorting out indoor positioning technology, clarifying the development direction of indoor positioning technology, and avoiding intellectual property risk. According to a large number of analysis of patents related to indoor positioning technology, the patents carry the latest technical intelligence information, and the technical solutions are full and accurate. By tracking and analyzing indoor high-accuracy patents, on the one hand, it helps to understand the technological development trends in related fields around the world, helps to optimize the technological development route of the subject of innovation, enhance the speed of technological development, and play a supporting role of patents in tackle key technological problems; on the other hand, it helps to better understand the development path of PNT technology, helps to formulate relevant policies for the development of indoor positioning technology, avoid intellectual property risks, and guide the direction of industry.

Reference

1. Deng, Z., Yin, L., Tang, S., et al.: Overview of key technologies of indoor positioning. Navigation Positioning and Timing, Beijing (2018)

Research on Legal Protection Mechanism of BeiDou Related Names and Marks

Lin Su[1] and Jingfan Yang[2(✉)]

[1] Defense Intellectual Property Office, Beijing 100009, China
18647097@qq.com
[2] Beihang University, Beijing 100083, China
yangjingfan@buaa.edu.cn

Abstract. The development of BeiDou system contains the hard work and unremitting struggle of several generations of scientific researchers, has accumulated a strong social attraction and appeal, and is a heavyweight national business card for China's high-tech exploration. The names and marks of "BeiDou" have high social recognition, wide influence and great commercial value. However, due to the lack of relevant regulations, there are some confusion in them use. In view of the practical need to standardize the use of BeiDou's related names and marks, this paper systematically analyzes the multiple mechanisms of trademark law, anti-unfair competition law and special legal protection under the current legal framework, and puts forward suggestions for the improvement of legal protection in order to promote the construction of "BeiDou ruled by law".

Keywords: Right of name · Mark · Legal protection mechanism of BeiDou

1 Introduction

Although the word of "BeiDou" has existed since ancient times, with the development of the construction, promotion and application, globalization of "BeiDou satellite navigation system", BeiDou related industries has developed rapidly. Now people refer to the word "BeiDou" and the first thought is China's BeiDou satellite navigation system. In the news media and social media, people talk about BeiDou, they mostly refer to BeiDou satellite navigation system. Therefore, "BeiDou" refers specifically to "BeiDou satellite navigation system" which is widely known to the relevant public in China. As an important space-time infrastructure of the country, BeiDou's development contains the hard work and unremitting struggle of several generations of scientific researchers. It has accumulated its unique attraction and appeal to the public in the whole society and has become a heavyweight national business card of our country. At the same time, the promotion and application of BeiDou system is not only related to the country's major core interests, but also related to major industrial interests and the interests of hundreds of millions of users. After long-term use, BeiDou related names have won wide recognition in society and have high social popularity. BeiDou related marks have also become an important carrier of wealth, reputation and market

J. Sun et al. (Eds.): CSNC 2020, LNEE 650, pp. 595–602, 2020.
https://doi.org/10.1007/978-981-15-3707-3_56

competitiveness. At present, it is urgent to explore and construct a legal mechanism to protect BeiDou's relevant names and marks by means of legal thinking and legal methods.

2 The Status Quo of Legal Protection of BeiDou Related Names and Marks

Incompatible with the strategic status of the BeiDou national business cards and important national space infrastructure are the many illegal acts carried out in the name of the BeiDou system in the online media There have been rumors of "BeiDou map navigation accurate to within 1 m", "in the next three years, get rid of the dependence on U.S. GPS", "BeiDou phone calls don't cost money to get online" and so on There are even "BeiDou Office" wechat public accounts, operating as wine "BeiDou Office" of the trademark, these will undoubtedly bring some negative impact to BeiDou Directly or indirectly damaged the reputation and image of the BeiDou navigation system. Judging from the current situation, there are mainly the following ways to infringe BeiDou's relevant names and marks:

2.1 Malicious Cybersquatting

From China Trademark Network using "BeiDou" as a search term, 3,652 trademark registrations involving the word "BeiDou" or graphic signs can be retrieved, covering all trademark categories. In addition to the normal registration or use of BeiDou related industrial companies, there are also some companies that apply for registration of the company name or trademark with the "BeiDou", but the industry or business scope they are engaged in has nothing to do with the satellite navigation industry.

2.2 False Propaganda of Confusing BeiDou's Relevant Names and Marks

There are also individual companies or individuals who use consumers' national sentiments towards BeiDou to make false propaganda, confuse audiovisual, and attract attention from bloggers. There have been reports that the BeiDou Maps App is expected to be launched on May 1st, supplemented by titles such as "Beijing's navigation is used in May." This behavior of equating the BeiDou Map APP with BeiDou satellite navigation system is actually a commercial hype based on the name of BeiDou. The unauthorized use of a unique name related to BeiDou as an enterprise name, the use of trademarks similar to BeiDou related marks, and registered trademarks have led people to mistakenly believe that they have a specific connection with BeiDou officials and pass this information on to consumers. Use the good reputation carried by the BeiDou name and related signs to promote your own products in order to obtain improper benefits or gain some kind of competitive advantage obtained without effort.

2.3 Abuse Leads to Dilution of BeiDou's Relevant Name and Mark

BeiDou has huge industry development space, so it has won the favor of many companies, including many successful experiences, but there is also the phenomenon of using the name of the BeiDou for other purposes. The registration of names, marks, and registered trademarks in products that are not similar or related has destroyed consumers' specific perceptions of BeiDou-related products, distorted the value of their symbol guidance, and made the BeiDou name and mark special and proprietary. Sex is weakened. Affects the national honor carried by the "Bei Dou" mark, affecting the authority of the mark.

BeiDou related names and marks are protected by multiple laws based on their own special attributes. By reviewing the current violations, it can be seen that BeiDou related names and marks can be protected through trademark law when they are registered as trademarks according to law, and when they are not registered as trademarks, they can choose anti-unfair competition law, general rules of civil law and special signs. Management regulations and other protections serve as the basis for rights protection. Different rights bases determine the choice of different rights protection paths.

3 The Current Legal Protection and Path Selection of BeiDou Related Names and Marks

3.1 Protection of Trademark Law

The related names and marks of BeiDou are trademark and meet the basic requirements of trademark law for trademark identification. Registering a trademark is the key way to actively protect the related names and marks of BeiDou. As a commercial mark, a trademark must bear the task of distinguishing the goods or services it uses from those using other marks. The distinctiveness is the prerequisite and basis for the trademark to be protected by law. The distinctiveness of a trademark is divided into inherent distinctiveness and acquired distinctiveness. Intrinsic distinctiveness refers to the fact that, at the beginning of the design, the mark itself has distinctive features due to its falsification and uniqueness, which complies with the constituent elements of the trademark mark. The specificity and artistic creativity of BeiDou Engineering and BeiDou related signs mean that they are likely to have inherent distinctiveness and can be registered directly as trademarks. "BeiDou", although it is a common term, is not inherently significant. But it has produced de facto distinctive features through its widespread use in society, allowing the public to link it to specific technologies and national projects, thereby achieving saliency. The so-called distinctiveness refers to a mark that does not have inherent distinctiveness itself and is known by the relevant public after use, thereby enabling the relevant public to establish a connection between the mark and a specific provider regarding the relevant goods or services. Therefore, the relevant authorities managing the BeiDou project should formulate clear regulatory requirements, and the relevant legal entity undertaking the task of BeiDou construction shall adopt the trademark registration method in a timely manner to obtain the trademark rights of the relevant name and mark.

At the same time, the Trademark Law also provides exclusion rules that exclude BeiDou related names and marks as registered trademarks under special circumstances, thereby protecting some government official BeiDou related names and marks from being used in the commercial sector. Article 10 of China's "Trademark Law" stipulates that "the following signs shall not be used as trademarks: (1) the same or similar to the national name, national flag, national emblem, national anthem, military flag, military emblem, military song, medal, etc. of the People's Republic of China, and The same as the name and mark of the central state organ, the name of a specific place or the name and graphics of a landmark building; … (8) those that are harmful to socialist morality or have other adverse effects." Therefore, when the BeiDou related names and marks, especially the same or similar to the characteristic names and official marks of the management agencies, are likely to cause ambiguity to the relevant public and will adversely affect the functions and credibility of state agencies, they shall not be used as registered trademarks use.

After the relevant names and marks of BeiDou have been approved for registration, the relevant units shall enjoy trademark rights over their names. On the one hand, BeiDou related units enjoy the exclusive use rights of trademarks for rights holders, and can realize their rights without the cooperation of others, using their approved registered trademarks on the products they have approved for use. By actively registering trademarks, we seek the protection of BeiDou related names and marks under the Trademark Law. As a prior right, these registered trademarks can effectively prevent subsequent squatting behaviors of others and prevent relevant units from losing the right to use BeiDou related names and marks. On the other hand, BeiDou related units have also enjoyed the right of prohibition, thereby preventing and restricting the abuse or unauthorized use of BeiDou related names and marks. At the same time, the protection of BeiDou related names and marks through the Trademark Law can also bring future trademark benefits to BeiDou related units and better stimulate their endogenous motivation to protect BeiDou related units.

3.2 Protection of Anti-unfair Competition Law

However, as BeiDou related names need to be protected, they include not only the names and marks of the central state organs, but also many names containing national interests and public interest, such as "BeiDou satellite navigation system", "BeiDou system" and "BDS". In the case where the subject fails to register its registered trademark in a timely manner, there is still a risk of malicious cybersquatting, and in social practice, in addition to the typical trademark infringement that uses a trademark mark in the sense of trademark, there are still a large number of Circumstances such as non-trademark use of trademarks that cannot be protected by trademark rights. Therefore, relying solely on trademark law is still not enough to achieve full protection. Uncertainty and flexibility in the definition of unfair competition acts provide appropriate protection beyond trademark law for the control of these acts.

With regard to the protection of BeiDou related names and marks, the unfair competition acts involved in the typed provisions of the anti-unfair competition law are mainly concentrated in two categories: counterfeiting and false publicity:

Regarding counterfeiting, the protection of BeiDou related names and marks by the Anti-Unfair Competition Law is not based on granting rights of the proprietors, but is based on the protection of a kind of rights. The protection of the identification function of the identification subject to avoid market confusion. Article 6 of the Anti-Unfair Competition Law stipulates that "Business operators shall not engage in the following confusion, which may be mistakenly considered to be another person's goods or have a specific connection with others: (1) unauthorized use of product names and packaging that have a certain influence on others. The same or similar marks, decoration, etc.; (2) unauthorized use of company names (including short names, font sizes, etc.), social organization names (including short names, etc.), names (including pen names, stage names, translated names, etc.) that have an influence on others; (3) unauthorized use of the main body of the domain name, website name, web page, etc.; (4) other confounding behaviors that are misleadingly considered to be other people's products or have specific connections with others."

In judicial practice, the feasibility of similar protection methods involving special signs in the aerospace field has also been confirmed. From 2008 to 2012, because the launch image of the CZ-2F (Long March II F) launch vehicle was used without permission in the advertisement, China Academy of Launch Vehicle Technology Dairy (Group) Co., Ltd. and BMW have filed a lawsuit in court, arguing that their advertisements used the graphic mark of the Chinese Long March carrier rocket and the popularity of the Chinese Long March carrier rocket to attract the attention of relevant audiences, thereby increasing the social visibility and market of their products. Competitiveness constitutes unfair competition and violates the legitimate rights and interests of the Launch Vehicle Research Institute.

In terms of false publicity, when BeiDou related names and marks are used on products and services and do not play a role of identifying sources, they are only used by "near name" to promote product quality and functions, etc, to deceive and mislead consumers. Legal protection is achieved through regulations regulating false propaganda. Article 8 of the "Anti-Unfair Competition Law" states that "operators shall not make false or misleading business propaganda, deceit, or mislead consumers about the performance, function, quality, sales status, user evaluation, or honors of their products. Operators shall not help other operators to carry out false or misleading business propaganda by organizing false transactions, etc."

However, the limitations of the protection function of anti-unfair competition law cannot be ignored. On the one hand, the protection of the name and mark by the anti-unfair competition law depends more on the measurement of compliance with specific conditions, such as the distinctiveness of the mark, the popularity and influence it gained in use, and the subjective intentions of the actor. Judging by factors, the protection of such interests has characteristics such as uncertainty and instability. On the other hand, the commonweal attributes of BeiDou related names and marks are greater than commercial attributes to a certain extent, and their management scale is not entirely a market behavior. Relying on relevant entities to protect relevant rights and interests through the anti-unfair competition system, there are still Lack.

3.3 Other Special Legal Protections

Article in addition to the trademark law and anti-unfair competition law in the general sense, there are some related laws and regulations that may also form a protection mechanism for BeiDou related names and marks. Such as special signs, official signs protection and business registration system.

China's "Regulations on the Management of Special Signs" provides a basis for incorporating BeiDou related names and marks into the management of special signs. According to Article 2 of the Regulations, "Special signs referred to in these regulations refer to national and international approvals organized by the State Council. Names, abbreviations, emblems, mascots, etc. that are used in cultural, sports, scientific research, and other social welfare activities. Obviously, BeiDou related names and marks meet the requirements of special mark protection at the event level. Similar to the BeiDou related names and marks, the Ministry of Railways has formulated the "Administrative Measures for Official Symbols of the Ministry of Railways" at the level of departmental regulatory documents, and the State Administration for Industry and Commerce has also formulated the "12315" official symbol protection rules. At present, the State Intellectual Property Office is also formulating the "Measures for the Protection of Official Signs" to further expand the scope of protection of official signs. In December 2019, the registration of the official signs for geographical indications and the official signs of the Audit Office of the People's Republic of China have been completed. Registration. This provides a new legal path for protecting BeiDou related names and signs with special signs and official signs.

In addition, when the BeiDou related name is used in the name of the relevant company, it may cause deception or misunderstanding to the public, or be confused with the name and abbreviation of the relevant authority, according to law, it will not be protected by law. For example, Article 9 of the "Administrative Regulations on the Registration of Enterprise Names" stipulates that "the name of an enterprise shall not contain the following contents and words: (1) those that are detrimental to the national and social public interests; (2) those that may cause deception or misunderstanding to the public; (3) Names of foreign countries (regions) and names of international organizations; (4) Names of political parties, party, government, and military organs, names of mass organizations, names of social groups, and army numbers; (5) Hanyu Pinyin (except those used in foreign names), Figures; (6) prohibited by other laws and administrative regulations".

4 Legal Improvement of BeiDou Related Name and Mark Protection

4.1 Clarify the Ownership of BeiDou Related Names and Marks

It is recommended that BeiDou related names and marks be treated as special types of special names and marks, and incorporated into the relevant national satellite regulations to protect them and clarify the state's ultimate control over them. According to the three situations of commercial use, public welfare use, and official use, the trademark rights, trade name rights, name rights, and official use rights of BeiDou related names and marks are reasonably allocated, and the trademark rights, trade name rights, and names related to BeiDou related names and marks are formulated. The special rules and management restrictions on the use and disposal of rights, clarify the authority of BeiDou authorities in the official use of names, relevant procedures for official use, and legal responsibilities.

4.2 Strengthen the Name Management and Law Enforcement Inspection of the BeiDou Industry Sector

It Strengthen the administrative supervision of the registration of institutions and organizations in the field of BeiDou. Without approval, do not use the word "BeiDou" in the name of an institution or unit or other similar expressions that are likely to cause confusion. Strengthen the supervision of the registration of relevant enterprise names, strictly regulate the management of enterprise names and trademarks containing the word "BeiDou", and do not use enterprise names and trademarks that mislead the public, harm public interests and disrupt social order. Strengthen the cleanup of domain names, Weibo, WeChat public accounts and other online platforms related to the expression "BeiDou". Except for the official websites, Weibo, and WeChat public accounts in the BeiDou field, no organization or individual may use names similar to the official websites, Weibos, or official public accounts of countries, military agencies, and state-owned enterprises and institutions.

Relevant departments should conduct a comprehensive inventory and focus on verifying registration, permitting, or filing procedures on websites, Weibo, and WeChat public accounts that release BeiDou industry development information as their main content, and investigate and correct violations of laws and regulations in accordance with the law. Strictly standardize the names of various BeiDou industrial parks and demonstration parks.

4.3 Improve the Rights Protection Mechanism to Protect BeiDou's Development Rights

Clarify the law enforcement and rights coordination agency for the infringement of BeiDou related names and marks, establish a joint law enforcement inspection mechanism with market supervision, intellectual property, public security and other government departments, strengthen ties with procuratorial organs, courts and other departments, and make it easier for BeiDou related names and signs to defend rights Green channel to timely resolve and deal with various types of infringements and abuses.

5 Conclusion

BeiDou satellite navigation system has developed rapidly, and the research and development work has come to an end. The development of satellite navigation has entered a new stage. Next, the application service and product promotion of "BeiDou" will be the focus of our work. The soundness of the BeiDou legal system is the fundamental guarantee for “BeiDou” to enter the world. In this process, the legal protection of BeiDou related names and marks will help BeiDou to further An important factor for growth.

References

1. Fu, D., Du, Y.: Regulating the use of the BeiDou name is imperative, Legal Daily, 31 May 2018
2. Xin, J.: The name of BeiDou urgently needs legislative protection. China Bus. Times **14**(003), 06 (2018)
3. Bo, Y.: Public property and goodwill of trademark symbols: a Perspective from the carrier rocket unfair competition case. Chin. Trademark **09**, 80 (2012)
4. Yu, X.-Y., Qi, D.-J., Zhang, X.: The US defense trademark system construction and its enlightenment. Mil. Econ. Res. **11**, 73 (2011)

Quantitative Research of Satellite Navigation Industry Policy Based on Text Analysis

Xinran Peng, Xiaosong Li(✉), and Mingxing Yuan

Military Science Information Research Center, PLA Military Science,
Beijing 100142, China
164832732@qq.com

Abstract. Satellite navigation industry policy modernization depend upon systematic, standardized satellite navigation industry policy. Applied with the method of policy text analysis, this paper construct three-dimensional framework based on object including satellite navigation industry policy tools, satellite navigation industry process, satellite navigation industry policy targets. This paper analyzed 25 satellite navigation industry policy samples, analyzed satellite navigation industry policy by three-dimensional framework, as well as conclusion: policy tool genres appear uneven, policy targets appear even. Then carry out quantitative analysis research on satellite navigation industry policy hot spot, collaborating following aspects: policy text hotspot, policy publisher, policy publish date and policy text key words, thus obtain satellite navigation industry policy key words: reinforce infrastructure, develop core technologies edge, broaden application range. The research result is meant to provide reference for optimize satellite navigation industry policy.

Keywords: Satellite navigation industry policy · Policy text · Quantitative analysis · Policy tool

1 Introduction

The Beidou satellite navigation system is an independently developed global satellite navigation system. The goal is to build a stable global satellite navigation system, promote the development of satellite navigation related industries, and promote the large-scale application of satellite navigation in various industries of the national economy and society [1]. In November 2017 and January 2018, the Xichang Satellite Launch Center successfully launched the first batch of four networked satellites of BeiDou-3 in two ways with a "single arrow and double star", which marked the gradual improvement of Beidou navigation infrastructure construction. With the global satellite navigation system becoming more mature and the scope of applications expanding, competition in the satellite navigation industry market is intensifying. In order to ensure the stable and healthy development of China's satellite navigation industry, maintain the safety of China's geographic information and other industries, and meet the future development strategy of China's space industry, demand not only requires continuous breakthroughs in innovation in technology, but also the protection and guidance of policies and regulations play a decisive role. In 2013, the National Satellite Navigation

J. Sun et al. (Eds.): CSNC 2020, LNEE 650, pp. 603–611, 2020.
https://doi.org/10.1007/978-981-15-3707-3_57

Industry Medium- and Long-Term Development Plan focused on navigation positioning and applications, and required the further development of China's global navigation positioning system to enhance the industrialization level of satellite navigation positioning applications.

Based on the analysis of domestic and foreign scholars' policy text analysis methods, this paper conducts quantitative research on satellite navigation industry policy. Wang [2] used the quantitative analysis method of policy texts to code and classify the policy texts of China's ex-situ poverty alleviation and relocation policy, and extracted seven policy dimensions including fiscal and financial policies. Sun [3] et al. Carried out a textual quantitative research on 40 strategic emerging industry policies by constructing a two-dimensional analysis framework such as industrial development dimensions and policy support dimensions, and using content analysis. Bai [4] and others used content analysis and quantitative analysis to select the policy text of entrepreneurship to promote employment, based on the perspective of policy tools, combined the entrepreneurial cycle dimension and policy object dimension to build a three-dimensional policy analysis framework, and proposed improvements Paths and methods of entrepreneurship driving employment policy system. Li [5] learned that the number of higher-level policies is closely related to the higher stage of popularization through text measurement and content analysis of higher policy texts. Zeng [6] collected 233 Pan-Pearl River Delta regional cooperation policies. Yang [7] selected the textual data of international climate policy, adopted a thematic model perspective, integrated word frequency, and comprehensively compared and analyzed the climate policies of China and the US and EU Happening.

At present, relevant scholars at home and abroad have used policy text analysis methods to compare and analyze policy texts in fields such as industry, technological innovation, and the natural environment from multiple perspectives, dimensions, and levels, and provide policy and institutional reforms in related fields. A combination of qualitative and quantitative methods. This paper draws on the methods of policy text analysis and co-word network diagrams, and uses satellite navigation industry policy as its main research object. The satellite navigation industry policy tools, satellite navigation industry chain links, and satellite navigation industry policy role objects are used to construct satellites. The three-dimensional analysis framework of the navigation industry policy texts, using the database of policies and regulations to retrieve the relevant policy documents of the satellite navigation industry issued by various agencies. Quantitative analysis of industrial policies, analysis of problems existing in the satellite navigation industry, and suggestions for countermeasures are provided, which provides a reference for the reform of the policy system of the satellite navigation industry.

2 Three-Dimensional Analysis Framework of Satellite Navigation Industry Policy

The satellite navigation industry involves many subjects, covers a wide range of fields, and has complex and multi-process processes. As a result, the satellite navigation industry's policies are complex and diverse, and policy analysis can be conducted from multiple dimensions. The three-dimensional framework of the satellite navigation industry policy, including the role of the policy object, is a three-dimensional framework for the quantitative analysis of the satellite navigation industry policy.

2.1 X Dimension: Dimensions of Policy Tools for Satellite Navigation

At present, scholars at home and abroad have conducted in-depth research on the classification of policy tools from different perspectives in accordance with the characteristics of the field. For example, Howlett et al. [8] classify policy tools into voluntary tools, hybrid tools, and compulsory measures based on the direct government participation Tools; Rothwell et al. [9] divided policies into three types: supply, environment, and demand according to different levels of policies and their impact. Ingram et al. [10] divided policy tools into four types: persuasion, command, institutional change, and capacity building, in accordance with the objectives to be achieved by the policy.

Based on the characteristics of the satellite navigation industry, this article uses ROTHWELL ideas, and divides the satellite navigation industry policy into categories such as supply, environment, and demand.

Supply-oriented policy tools for the satellite navigation industry, optimizing the supply environment for the elements of the satellite navigation system, and providing a driving force for the development of the satellite navigation system. The market role is relatively small. The policy-oriented tool is to stimulate the growth of the satellite navigation industry through indirect measures.

2.2 Y Dimension: Dimensions of the Satellite Navigation Industry Chain

Generally, the satellite navigation industry policy needs to be formulated differently in accordance with the characteristics of different links in the satellite navigation industry chain, to ensure that the policy system and related text content can fully reflect the special characteristics of different links in the satellite navigation industry chain, and to ensure the self-sustainability and development ability of the internal system Promotion. According to the general characteristics of the high-tech industry, this article divides the satellite navigation industry policy into four types of policies: investment, research and development, production, and application according to each link of the satellite navigation industry chain.

2.3 Z Dimension: Dimension of the Satellite Navigation Industry Policy Objects

The role of the satellite navigation industry policy can be understood as the various subjects involved in the content of the policy text, including the communications department, transportation department, power department and other departments, as well as enterprises, universities and scientific research units that undertake research and development and production of satellite navigation systems. This article divides the role of satellite navigation industry policy into two categories: enterprises and scientific research units.

2.4 3D Frame Construction

According to the analysis above, a three-dimensional framework for the quantitative analysis of the satellite navigation industry policy is established from the dimensions of the X satellite navigation industry policy tool dimension, the Y satellite navigation industry chain link dimension, and the Z satellite navigation industry policy role object dimension, as shown in Fig. 1.

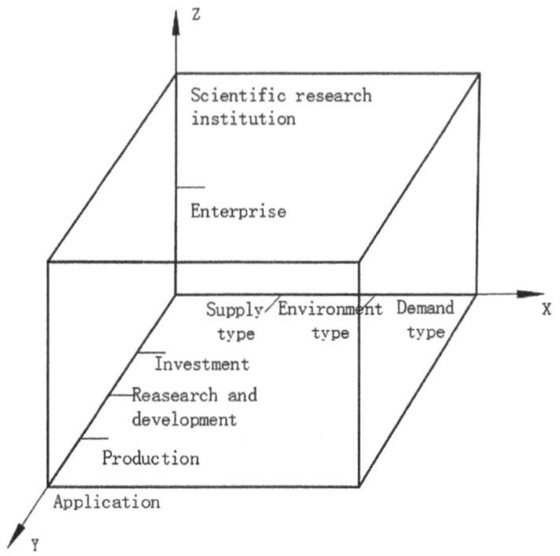

Fig. 1. Satellite navigation industry policy three-dimensional framework

3 3D Quantitative Analysis of Satellite Navigation Industry Policy

First collect and organize the policy text of the satellite navigation industry, and then select and code the policy text. On this basis, according to the three-dimensional analysis framework, a three-dimensional quantitative analysis of the satellite navigation industry policy is carried out.

3.1 Satellite Navigation Industry Policy Text Selection and Coding

From the website of the State Council, the Beidou satellite website, and the online legal literature retrieval database, this article collects 25 different policies.

According to the three-dimensional framework for the quantitative analysis of Beidou satellite navigation industry policies constructed above, 25 policies are coded, of which the X policy tool dimension coding method is: "Policy Sequence Number-X-S (Supply Type)", "Policy Sequence Number-X-E (Environment) Type) "," Policy sequence number-X-R (demand type) ". Y satellite navigation industry chain link dimension coding methods are: "Policy Sequence Number-Y-1 (Investment)", "Policy Sequence Number-Y-2 (R & D)", "Policy Sequence Number-Y-3 (Production)", "Policy Sequence Number-Y-4 (Application)". The coding method of the

Z-dimensional policy object is: "Policy Sequence Number-Z-1 (Scientific Research Institution)", "Policy Sequence Number-Z-2 (Enterprise)". e.g. The code for "Accelerate the development and industrialization of Beidou navigation core technology" is 1-X-E\1-Y-1\1-Z-1.

3.2 Three-Dimensional Quantitative Analysis of Satellite Navigation Industry Policy Text

This section conducts a three-dimensional quantitative analysis of the satellite navigation industry policy text from the dimensions of the satellite navigation industry policy tool dimension, the satellite navigation industry chain link dimension, and the satellite navigation industry policy role object dimension, and provides prerequisites for the subsequent policy text analysis.

3.2.1 Quantitative Analysis of X Dimension: Satellite Navigation Industry Policy Tool

Based on the dimensions of the X policy tool, the 25 Beidou satellite navigation industry policies collected and analyzed were analyzed. Among them, demand-based policy tools accounted for 18%, environmental-based policy tools accounted for 68.4%, and supply-based policy tools accounted for 23.6%. It can be seen that there are more environmental-based policy tools in the satellite navigation industry, while fewer demand- and supply-based policy tools. Therefore, the next step of Beidou satellite navigation industry policy reform should focus more on providing supply-oriented policy tools such as technical support and financial support that help form core competitive advantages, as well as demand-based policies such as expanding domestic and foreign markets and increasing the scale of satellite navigation applications On the tool.

3.2.2 Quantitative Analysis Based on the Y Dimension: The Satellite Navigation Industry Chain

Based on the dimensions of the Y industry chain, this article analyzes 25 satellite navigation industry policies collected and sorted out. Among them, the investment link accounts for 9%, the research and development link accounts for 17%, the production link accounts for 36%, and the application link accounts for 59% (because some sub-policies involve multiple stages, the total percentage is not 1). It can be seen that there are more policies and measures in the two links of production and application, while fewer policies and measures are involved in the investment and research and development links. At this stage, China's satellite navigation industry chain is concentrated in production and application links, and the front-end development of industrial chains such as research and development and investment is not mature enough. At this stage, policies on investment and research and development in the satellite navigation industry should be moderately increased, the market should be encouraged to play a role in the development and innovation of satellite navigation technology, and guide developers in technological breakthroughs and innovations.

3.2.3 Quantitative Analysis of the Z Dimension: Satellite Navigation Industry Policy Objects

According to the dimension of the role of Z policy, the 25 satellite navigation industry policies collected and analyzed are analyzed. Among them, 73.5% are policies that affect enterprises and 26.5% are policies that affect scientific research institutions. Insufficient policies issued by scientific research institutions may cause many weaknesses in the introduction and training of talents in the satellite navigation industry.

4 Quantitative Analysis of the Satellite Navigation Industry Policy Text Hot Spots

This article quantitatively analyzes the hotspots of the satellite navigation industry's policy texts in terms of the types of policy release agencies, the time of policy release, and key phrases of policy texts.

4.1 Quantitative Analysis of the Types of Satellite Navigation Industry Policy Issuers

The construction, operation, application and service of the BeiDou system involve a large number of departments and agencies. According to the types of policy promulgators, the satellite navigation industry policies can be divided into: policies issued by the State Council, policies issued by ministries and commissions under the State Council, and policies issued by local governments. This article collects and organizes 25 satellite navigation industry policies, 13 items promulgated by the State Council, 7 items promulgated by various ministries and commissions under the State Council, and 5 items promulgated by local governments.

The policies issued by the State Council mainly propose forward-looking guiding principles for industrial development, development principles, and specific goals. However, from the perspective of the existing policies and systems, due to the current construction status and management system of the Beidou navigation system in China, existing policies have problems such as incomplete overall planning, inconsistencies between regions and departments, etc. The next step should be to improve China. The management system of Beidou satellite navigation system strengthens the communication links between the main bodies of each policy promulgation and strengthens departmental communication and coordination in policy formulation.

4.2 Quantitative Analysis of Satellite Navigation Industry Policy Release Time

China's satellite navigation policy has entered a stage of rapid development since 2013, and 2016 was the peak year for policy release, reaching seven. The number of laws and regulations issued and modified during this period increased and the coverage was more comprehensive. In addition to the guiding program documents, the corresponding supporting laws and regulations were issued at a much higher frequency than in the previous stage. However, since the National Medium- and Long-Term Development

Plan for the National Satellite Navigation Industry was promulgated in 2013, the state has not yet introduced a satellite navigation policy that includes both the general thinking and goals of the development of the satellite navigation industry and specific safeguard measures. Standards related to satellite navigation in a certain industry field, and lack of overall regulations related to various industries.

4.3 Quantitative Analysis of Keyword Sentences in the Satellite Navigation Industry Policy Text

This article summarizes and refines the keyword sentences of 25 satellite navigation industry policy texts, analyzes the frequency of occurrence of keyword sentences, and statistics the keyword sentences that appear more frequently than 2 times. Keyword sentences are to improve infrastructure construction, increase the application scale, strengthen policy support, encourage product innovation, develop core technologies, introduce professional talents, strengthen international cooperation, strengthen financial support, expand markets, encourage dual-use military and civilians, and Tax reduction, scientific and technological information support, etc.

This article uses the co-word analysis method to carry out a hotspot analysis of the satellite navigation industry policy text. Co-word analysis refers to judging the situation and content structure of research and development in a certain field based on the situation that two different words in a certain number of documents appear in the same document together. Co-word network of hot spots in the text of the satellite navigation industry policy, as shown in Fig. 2.

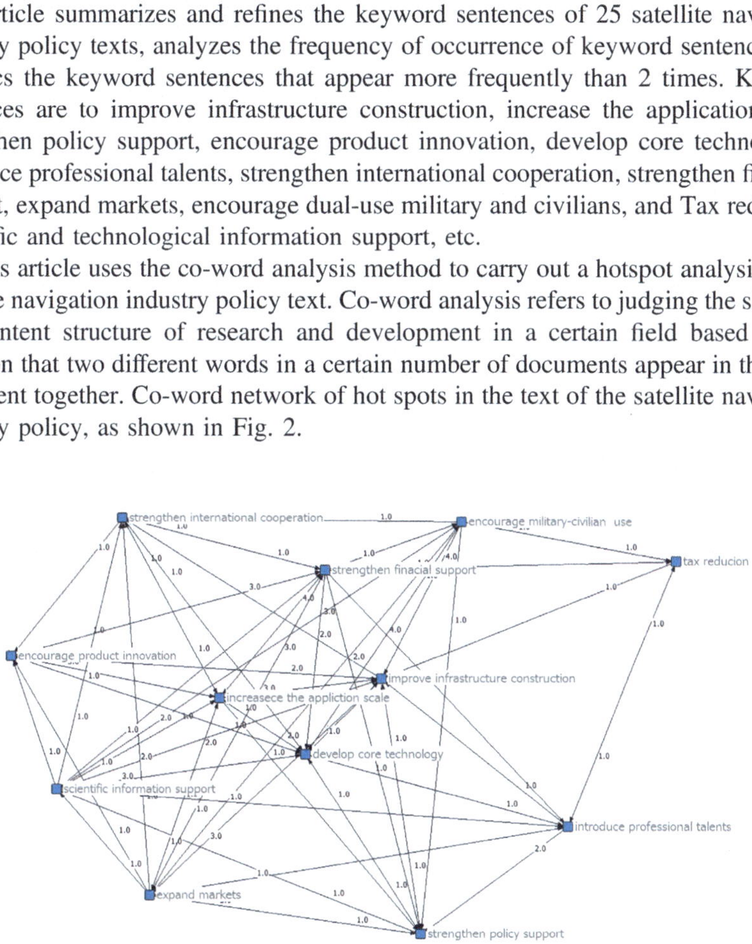

Fig. 2. Satellite navigation industry policy text hotspot co-word network

According to the connection density in Fig. 2, it can be seen that improving infrastructure construction, developing core technologies, and increasing the scale of application of results have a greater influence in the co-word network, indicating that the above keyword sentences are the focus, hotspots and priorities of policy systems. Analyze key words and phrases of policies and systems with high influence.

4.3.1 Improve Infrastructure Construction

Infrastructure construction is an environmental policy tool, and it belongs to the investment link in the industrial chain. The main role of policy is enterprises. The construction of satellite navigation infrastructure includes hardware facilities, such as the equipment and facilities required for satellite navigation, but also software, such as the construction of industry standard systems. Backward infrastructure construction has greatly restricted the development of Beidou satellite navigation application services on a large-scale market. As a result, the main competitiveness is not strong. Should respond to national strategic needs and application needs in key related fields, accelerate the construction of a coordinated, open and complete satellite navigation infrastructure system, promote data sharing, improve resource utilization efficiency, innovate service models, consolidate the foundation for industrial development, and enhance the ability of sustainable industrial development.

4.3.2 Develop the Core Technologies

The development of core technologies belongs to the supply-oriented policy tool, which belongs to the research and development link in the industrial chain. The main policy object is scientific research institutes. China's satellite navigation application theory research and technology research and development have developed rapidly, and key technologies such as navigation chips and antennas have made major breakthroughs and achieved commercialization. However, China's satellite navigation field still has insufficient independent innovation capabilities, lack of core technologies, significant gaps between important products and solutions and the international advanced level, and the Beidou satellite navigation system application market space has been severely squeezed. The level of compatible application technologies for satellite navigation chips, time-sharing general-purpose products, Beidou satellite navigation systems will be further improved, breakthroughs will be made in the integration of satellite navigation and mobile communications, Internet, remote sensing and other fields, and core products will be upgraded and scaled up.

4.3.3 Increase the Application Scale

Increasing the scale of application of results is an environmental policy, which belongs to the application chain in the industrial chain. The object of the policy involves enterprises and scientific research institutes. The Beidou satellite navigation system service has a wide range of applications and has played a role that cannot be ignored in transportation, agriculture, forestry and water conservancy. However, the current policy lacks mandatory requirements for the application and promotion of Beidou system and related technologies and products, which has led to the backward application in energy, communications and other fields. Promote the formation of comprehensive industry application solutions, improve the efficiency of industry operations, and promote the transformation and upgrading of related industries. Support key enterprises and scientific research institutions to build innovation capabilities, speed up the construction of engineering experiments and results transformation platforms, and form a technology innovation system that combines production, education, and research.

5 Conclusion

In the three-dimensional quantitative analysis of the satellite navigation industry policy text, through the classification and statistics of policy tools, the division of the industrial chain, and the analysis of the main role of the policy, we have obtained the need to increase demand and supply policies, investment and research and development policies, and targeted Suggestions on scientific research institutions' policies, etc. In the quantitative analysis of the satellite navigation industry policy texts, through quantitative analysis of the types of policy release agencies, the time of policy release, and the key words of policy texts, the coordination of inter-agency policy planning has been strengthened, and the national level Policy recommendations. In the next step, the author will combine the key points of the policy reform of the satellite navigation industry and use a combination of qualitative and quantitative methods to refine the key points and key points of the policy text.

References

1. Gao, G.: Policy research on Beidou satellite navigation industry. J. Beijing Univ. Aeronaut. Astronaut. (Soc. Sci. Ed.). **30**(04), 48–55 (2017)
2. Wang, H., Fu, T., Zhang, W.: The evolution characteristics of China's ex-situ poverty alleviation and relocation policy: based on quantitative analysis of policy text. J. Natl. Adm. Inst. (03), 48–53+129 (2017)
3. Sun, R., Wu, J.: Quantitative research on China's strategic emerging industry policy text. Sci. Sci. Technol. Manag. **36**(02), 3–9 (2015)
4. Bai, B., Zhang, Z.: Analysis of entrepreneurship-driven employment policy from the perspective of policy tools: content analysis and quantitative analysis based on policy text. Sci. Sci. Technol. Manag. **37**(12), 92–100 (2016)
5. Li, K., Liang, L.: Quantitative analysis of China's higher education policy texts: from the perspective of policy tools. China High. Educ. Res. (08), 50–56 (2015)
6. Zeng, J.: Quantitative analysis of the pan-pearl river delta regional cooperation policy text: 2004–2014. China Adm. (07), 110–116 (2015)
7. Yang, H., Yang, J.: Quantitative analysis of policy texts based on LDA model: based on empirical studies in the international climate field. Mod. Inf. **36**(05), 71–81 (2016)
8. Howlett, M., Perl, A., Ramesh, M.: Studying Public Policy: Policy Cycles and Policy Subsystems, p. 163. Oxford University Press, Oxford (1995)
9. Rothwell, R., Zegveld, W.: Reindustrialization and Technology. Longman, Harlow (1985)
10. Schneider, A., Ingram, H.: Behavioral assumptions of policy tools. J. Polit. **55**(2), 513–522 (1990)

Development Strategy of Chinese Satellite Navigation Technology: A Research Based on SWOT Method

Wenbo Chen(✉) and Xiaole Li

School of Public Administration, Beihang University, Beijing 100191, China
wenbo.chen@buaa.edu.cn

Abstract. In recent years, western countries have been strengthening their containment of China's technology development in an all-round way, and the scope of technical restrictions has involved satellite navigation and other strategic emerging industries. As an important support for national security, China's science and technology security in the field of satellite navigation is facing major challenges. The purpose of this paper is to make a comprehensive analysis of the advantages and disadvantages of China's satellite navigation technology security and what opportunities and challenges it will face, and to make a qualitative analysis of all combined strategies, then to put forward suggestions for the development of China's satellite navigation. The analysis based on the SWOT shows that, in order to prevent from major risks in the field of science and technology, and break through the West's "encirclement" of science and technology, the optimal choices are the "SO" strategies of distinctive advantages, independent innovation, international cooperation, and industrialization. Also, a torsion strategy requires an in-depth analysis of the challenges and weaknesses. To accelerate the construction of an monitoring and pre-alarming system for science and technology security in the field of satellite navigation is a feasible solution.

Keywords: China Satellite Navigation · Technology security · Development strategy · SWOT analysis

1 Introduction

The topic of national security have been discussing since its connotations and focuses are expanding over time. After the evolution of concepts of political security, military security, and economic security, as the status of science and technology rises, a new security concept, which hinges on technology security, has gradually formed. In this concept, technology security and other security elements interpenetrate with each other. Lian and Ma [1] define the concept of scientific and technological security for the first time, including scientific and technological security into the category of national security. The narrow sense of science and technology security concerns the security of the science and technology system itself while the broad sense of science and technology security includes efforts to resist attempts and actions by external science and

J. Sun et al. (Eds.): CSNC 2020, LNEE 650, pp. 612–623, 2020.
https://doi.org/10.1007/978-981-15-3707-3_58

technology to harm national interests [2]. The security of science and technology has the dual significance of national security and national development interests. With the connotation of national security thoughts extending in the new era, satellite navigation has become a focal point of science and technology security, and its status of has become increasingly important.

As a global satellite navigation system independently developed by China, the Beidou actively promotes technological innovation with a gradually expanding application range. The satellite navigation industry has also become a strategic emerging industry in China, which undertakes the vital mission of the modernization of national defense technology. It has become the fastest-growing electronic information industry after mobile communications and the Internet. However, the development of China's satellite navigation is not as smooth as it seems, in fact, there are many difficulties behind the glory. Therefore, it is necessary to carry out a comprehensive review of the safety of satellite navigation technology, and to establish a monitoring and pre-alarming mechanism.

2 Research Methods

The SWOT (Strengths Weaknesses Opportunities Threats) is a qualitative research method widely used in the field of strategic management in recent years. It combines the evaluation of the advantages, disadvantages, opportunities, and challenges of the strategic environment and the evaluation of the influence of each subdivision strategy. By the SWOT analysis method, this research intends to analyze the development environment of China's satellite navigation, and propose appropriate development suggestions.

3 SWOT Analysis of China Satellite Navigation Development

3.1 Identification of Key Factors

A review of existing literature reveals that the factors affecting China's science and technology security include six dimensions: science and technology environment, national awareness, science and technology resources, science and technology strength, science and technology system, policies and regulations, and safety management level (Table 1).

Table 1. Main factors affecting science and technology security

Technological environment	Globalization [3], External environment [4], Internal environment [5]
National awareness	Strategic positioning [6], Demands of science and technology security [7], Technology safety culture [8]
Technology resources	Human resources, Information resource [9], Technology funding, Information communication channels [10]
Technological strength	Basic R&D capabilities, Technological output, Self-innovation ability [11]
Science and technology system	Organization structure of science and technology system, Operating mechanism, Management model and institutional system, Science and technology security pre-alarming system [12], Reform of science and technology system
Policies and regulations	Science and technology regulations, Existing policies on technology security, Science and technology security planning, Technology safety regulations, Regulations on science and technology import and export [13]
Safety management level	Emergency plan, Crisis management capabilities, Information management capabilities

3.2 Internal Factor Analysis

3.2.1 Dominant Factor Analysis

S1. Policy Support Advantages

The development of China's satellite navigation system has drawn on the experience of developed countries, and it has shown certain advantages. As a matter of fact, there have already been over 170 publicized documents concerning satellite navigation policies in the past decade. Their content has involved implementing the national security strategy, promoting the application of Beidou satellite navigation system in the transportation industry, clarifying the core position of Beidou satellite and navigation system in national information infrastructure and strategic emerging industries from the top-level planning, and formulating the strategic goals of the construction of Beidou satellite navigation system in specific stages.

S2: Technical Advantages

1. Property rights' autonomy

Satellite navigation systems have important defense and military significance and are one of the basic strategic infrastructures. The Beidou satellite navigation system is independently developed and constructed by China. The system, which is safe, stable and reliable, has independent property rights, uses high-strength encryption design, and is suitable for key sector applications. It can not only provide basic technical support for military defense but also lay a good foundation for the healthy development of China's economy.

2. Dual functions, compatibility and openness

The Beidou satellite navigation system not only has the ability to determine the geographic latitude, longitude and altitude of the user at any time and place, but also contains multiple innovations in positioning performance. It has the dual functions of positioning and communication, and is especially suitable for places that need navigation and mobile data communication. In addition, the compatibility of Beidou satellite navigation system with other global satellite navigation systems is also at the forefront.

3. The performance of the third generations of Beidou improved, and global deployment continued to advance

In 2018, China launched the Beidou-3 global networking. In 2019, it is expected to complete all mid-Earth orbit satellite launch missions and continue to promote global coverage. China and the United States have signed the Joint Declaration on Compatibility and Interoperability of the Beidou and the GPS Signals, reflecting the globalization of Beidou satellite navigation systems, which is compatible and interoperable with other GNSS navigation systems. This can make better use of resources and provide better services in advancing satellite applications.

S3: Military-Civilian Integration Development Advantage

The satellite navigation system was originally established for military use, and is mainly used for land, sea, and air high-precision positioning and navigation, targeted bombing, and ship-based missile guidance. The biggest feature of the Beidou system is that the precise position of the moving target can be displayed on the electronic map in real time, which is convenient for commanders to grasp the moving track of target in time. Therefore, it has performed exceptionally in military command, troop operations, and target reconnaissance.

1. High-Precision Application Development

Beidou system with high-precision provides lane-level positioning accuracy within meters to the car. Unmanned driving is inseparable from 5G + BeiDou high precision, and the low latency of 5G provides a reliable network environment for unmanned driving. In addition, 5G + Beidou is expected to have a broad range of application scenarios in various fields such as airport dispatch, robot inspection, drones, building monitoring, vehicle monitoring, and logistics management.

2. Military-Civilian Integration Advantage

The military-civilian integration industry is the foundation of military equipments development. From the perspective of innovation economics, the sign of the integration of military and civilian science and technology is the commercialization of innovation results [14]. The Beidou satellite navigation industry is a national strategic emerging industry and also a typical representative of the military-civilian integration industry.

S4: Multi-channel Fund Raising Advantage

The *Notice on Accelerating the Application of Beidou Navigation System* defines the construction of the Beidou navigation system as a national infrastructure plan, and also resolves the problem of insufficient funding channels by overcoming the disadvantages of large system investment, slow market cultivation, and long return cycles in the process of industrial development. From a practical point of view, Beidou Civil drew on the experience of Russia's GLONASS and Galileo to absorb private funds. For example, the private enterprise Beidou Star Navigation Technology Co., Ltd. has the operating right of "Beidou No. 1", and raised funds through the issuance of stocks.

S5: Industry-Leading Advantages
The Beidou satellite navigation system is suitable for large-scale monitoring and management and data collection of group users. It can be authorized to system operators with industry background to form an industry advantage. Beidou Xingtong, Divine Tianhong and China Satcom, which have obtained the civil operation rights of "BeiDou No. 1", have opened up the civilian market in marine fishery, water conservancy and hydropower, transportation mobilization, port management and other fields, and have played a leading role in the industry.

S6: Broad Market Advantage
The satellite navigation industry has a wide application field. All information related to location, speed, and time is related to satellite navigation. It is used in various areas of the national economy. It is limited only by people's imagination. With the development of the navigation industry, the personal consumption field has grown. It has broad prospects and huge market capacity from professional fields to popular applications.

3.2.2 Analysis of Disadvantages

W1: Departmental Coordination and Regional Restrictions in Existing Policies
Constrained by China's current Beidou navigation system construction and management system that lacks of consideration of the differences in military and civilian application, the existing management system has poor communication, low efficiency and lack of responsibilities in terms of inter-ministerial coordination and guidance to promote industrial development [15]. Problems of the management system have led to the relatively fragmented Beidou policy at a national level. Government departments have different tempo in formulating policies and lack effective communication and coordination. Although local governments have also formulated some industrial promotion policies, they are only effective in their regions, and some reasonable measures and opinions are not universally applicable.

W2: Lack of Relevant Legislation Affects Policy Implementation
China currently lacks laws and regulations for the construction, operation, application and industrial promotion for the Beidou system. Especially, in the civilian field, there are legal gaps in promoting the Beidou system and expanding its application. As is known to all, policies provide a reference for the formulation of laws, and laws provide a guarantee for the implementation of policies. If the operation and application of the Beidou system is clearly defined in the legislation, it will lay a solid foundation for the implementation of relevant policies.

W3: Insufficient Professional Talents
Cao Chong, chief scientist of the China Satellite Navigation and Positioning Association, believes that talent is a relatively large gap in the Beidou industry. In the process

of promoting the development of the Beidou industry, we must also make up for the insufficiency of talents [16]. The satellite navigation industry involves a variety of professional technologies. The urgent need for professionals is a big problem for the satellite navigation industry. At the moment, professionals are selected from students majored in either surveying, mapping or navigation. However, the professional content settings of these majors need much improvement and integrating to help cultivate professionals. Since surveying and mapping students are relatively familiar with positioning theory and algorithms, but they are not familiar with the composition of large systems, receiver hardware, and working principles; while navigation students have less knowledge about positioning algorithms.

W4: Technical Gaps and Technical Difficulties
During the construction of Beidou satellite navigation system, it faced a series of technical problems such as navigation signal system and high-precision space-borne atomic clock technology. With the accelerating upgrading of the global satellite navigation system, researches on key and basic technologies in the field of satellite navigation are advancing rapidly, hence the competition is extremely fierce. Compared with the United States and Russia, China's satellite navigation research started relatively late, and the technical gaps and technical problems of the research are prominent.

W5: Relatively Weak Industrial Foundation
The construction of Beidou satellite navigation system requires high precision and high reliability while the operation requires continuity and stability. This means that autonomous manufacturing and mass production of key components must be achieved. For a country that has a relatively weak industrial foundation, without mastering core technologies, it could be easily turned into a disadvantage of relying on imports for the production of important parts.

W6: The industry Has Not Yet Formed a Scale Effect
Despite the rapid development of satellite navigation in China, the development of the Beidou system is still under great pressure of the development of the GPS, which is relatively mature and has locked a first-mover position in the Chinese market. Firstly, the lack of high-tech content and high value-added products makes it difficult to promote applications for the Beidou system. Secondly, since the product is in its growth stage, the development ability of application market is still weak. Consequently, an industrial scale is hard to form, resulting in a navigation terminal cost premium, high market prices, and weak competitiveness.

3.3 Analysis of External Factors

3.3.1 Opportunity Analysis

O1: Global Expansion of Technological Revolution Drives Technological Innovation
Since the new period, a large number of new scientific and technological fields such as biomedical engineering, aerospace, and information technology have developed

rapidly. As General Secretary Xi said, "Science and technology have never so profoundly affected the country's future and destiny, and have never so profoundly affected people's life and well-being like today." The linkage of scientific and technological progress and industrial development has brought human civilization progress. And the crossover and specialization promote new technological innovations and breakthroughs. The iterations and technological advancements of the scientific and technological revolution around the world provide a strong technical impetus for the innovative development of satellite navigation systems.

O2: Meeting the Needs of National Interest Expansion in the Context of Globalization

The construction and application of autonomous satellite navigation are on the front line of national security and the development of national economy. It is an inevitable choice for the expansion of national interests under the background of globalization. It is in line with the need to promote independent innovation in aerospace science and technology and the establishment of a national satellite navigation industry, indicating that a new generation of strategic emerging industries is on the rise.

O3: Its Development Has a Broad International Market

The international development pattern of multi-polarization of global satellite navigation provides a broad space for the development of China's Beidou system. The development of satellite navigation systems other than GPS has encountered bottlenecks, and rules and standards for international satellite navigation are still being formed. As an active builder of the international satellite navigation system, China is working hard to participate in the formulation of international rules and standards. This has created favorable conditions for Beidou's strategic rise.

O4: A Period of Historical Opportunities in Which Competition and Cooperation Coexist

The development of global satellite navigation has experienced a period of precipitation, and we have gradually entered into a period of opportunity in which multiple systems coexist and competition and cooperation coexist. In attempts to seize the commanding heights of technology and formulating international cooperation rules and standards, competitions promote the progress of countries, and cooperations reflect the spillover effect brought by the proliferation of technology. As a late-joined party in the development of satellite navigation, through cooperations and competitions, China can effectively improve the quality of Beidou satellite navigation development and elevate its discourse power at the global stage.

3.3.2 Challenge Analysis

T1: The Challenge of International Political Instability

At present, international relations are changing, and the international order is also facing transformation and adjustment. The trend of multi-polarity in the world is becoming increasingly apparent. With the rise of unilateralism, global governance has also begun to move towards national governance. The field of satellite navigation will definitely become a competition upland for international forces.

T2: The Challenge of Policy and Standard Setting in Line with International Standards
Public policy-making inevitably exhibits delay. In the early stage of the construction of Beidou satellite navigation system in China, national policies and standards were in lack to some extent. Different countries have their own conditions in terms of the basis of scientific and technological development, legislative concepts and procedures. Coupled with the stance biases held by western countries, they all pose a challenge to Beidou's internationalization of policies and standards formulation.

T3: The Challenge of Space Resources
In an age of information technology, major powers in space have stepped up their capacity building. Up to now, in addition to the GPS, GLONASS, Galileo, and Beidou satellite navigation systems determined by the Global Satellite Navigation System Committee, India and Japan will also be included with their independent satellite navigation systems on the agenda, making the international space environment increasingly complex. Overall, the exploitation of space resources and the environment are becoming new challenges.

4 Suggestions

4.1 Strategic Choices to Cope with China's Satellite Navigation Technology Security Risks

4.1.1 Strength-Opportunity Strategy

According to SWOT analysis results, deepening advantages and seizing opportunities are the optimal strategic choices for the development of satellite navigation in China. Specifically, the government must continue to attach importance to scientific and technological innovation, leverage policy support, and strengthen guidance and support for the construction of satellite navigation systems. Combined with the background of global technology and market penetration, China will need to strengthen cooperation with international satellite navigation systems and enhance existing navigation and communication technology advantages. At the same time, China can give full play to the advantages of military-civilian integration that focus on the civilian market, and promote China's satellite navigation industry to improve market competitiveness.

4.1.2 Independent Innovation Strategy

The development experience of global satellite navigation and the practice of China's satellite navigation system construction show that independent innovation is the only way for China's development. The nature of satellite navigation system determines the continuity of its development. This puts forward requirements for the independent and stable supply of key components in China. As high-tech products at the core, these components condense many core key technologies. With the western technology blockade against China intensifying, China is facing multiple difficulties in introducing core satellite navigation technology. Aiming at the sustainability of satellite navigation development, China must adhere to an independent innovation strategy and achieve independent research and production of key single machines and components.

4.2 Speeding up the Construction of Satellite Navigation Technology Safety Pre-alarming Monitoring System

According to the SWOT analysis, weak technological security management capabilities are one of the disadvantages of China's satellite navigation development. The strategic coupling attempts to reverse this disadvantage, and it is necessary to strengthen the construction of satellite navigation technology safety monitoring and pre-alarming system. On January 21, 2019, at the provincial and ministerial level leading cadres' seminar, General Secretary Xi Jinping pointed out "The security in the field of science and technology is a vital part of national security. We should speed up the construction of an monitoring and pre-alarming system for science and technology security." Science and technology security carry the dual tasks of building a strong country with science and technology and building national security. The establishment of a science and technology security pre-alarming mechanism has important practical significance for ensuring national security interests [17].

4.2.1 The Mechanism of Satellite Navigation Technology Safety Warning Monitoring System

From planning to the feedback loop, a reasonable monitoring and pre-alarming system can form a reasonable system. It contains the following parts:

1. The construction of the basic system: the planning of the science and technology safety monitoring and pre-alarming system. First of all, according to the overall goal of the current development of satellite navigation, the goal of science and technology security pre-alarming must be clarified. Secondly, we must define the content of the scientific and technological safety pre-alarming management of functional departments. Since the functioning of the monitoring and pre-alarming system is based on timely response, it is necessary to ensure that the monitoring and pre-alarming system collects information in a timely and accurate manner, and information communication channels are unblocked.
2. Contents of system construction: The operation of the pre-alarming mechanism is based on a comparative idea, and its content includes top-down targets for science and technology security pre-alarming, identification of science and technology security risk elements, calculation and adjustment of security thresholds, evaluation of science and technology security status, logical judgment, and results feedback, state clustering, and strategy adjustment. Information gathering runs throughout the process.

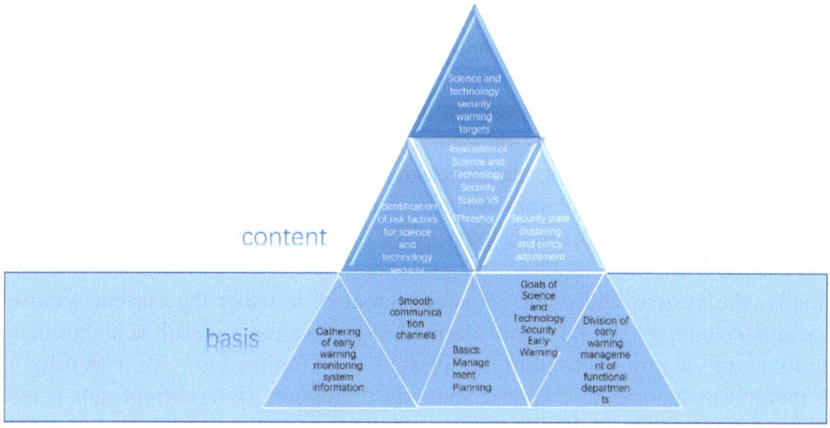

Fig. 1. Construction mechanism of science and technology security pre-alarming monitoring mechanism

4.2.2 Operation Logic of Satellite Navigation Technology Safety Pre-alarming System

The entire process of the science and technology security pre-alarming system is inseparable from the flow of information. Therefore, the operation of the science and technology security pre-alarming system as shown in Fig. 1 can be divided into three major links: key information input, science and technology security status evaluation and pre-alarming and decision response.

The key information input link mainly uses the Delphi method to identify the risk factors affecting the safety of satellite navigation technology and determine the risk threshold based on the identification; the technology security status evaluation link is based on the latest information input to evaluate and compare the technology security risk; the early-warning and decision-making response link is based on the compare results to respond to and control whether pre-alarming should be called and subsequent strategy should be adjusted.

The identification of the factors that affect the development risk of national satellite navigation systems is part of the pre-alarming monitoring mechanism. Its mission is to continuously identify current objective potential project risks. The results are updated at all time and incorporated into China's satellite navigation technology safety early-warning monitoring system library with timely supplementation.

The steps of determining the risk threshold include: forming an expert group; determining early-warning indicators; determining "homogeneous" benchmarking groups; measuring early-warning indicator parameter values; risk weighting; determining deviation domains; re-consulting experts in the deviation domain; experts repeating the consultation on the basis of communication until all experts' threshold differences on this indicator fall within an acceptable deviation range; averaging the thresholds determined by each expert to obtain the final pre-alarming indicator threshold.

The second step is to carry out a science and technology security risk assessment of the current project. According to literature review, current risks will be assessed in four dimensions: the integrity and effectiveness of the national science and technology system, the degree of independent control of core technologies in key areas, the degree to which national security meets its own development needs, the status of national development interests. Besides, the assessment will incorporate the secondary and tertiary indicators of the four dimensions. Combined with the Delphi method, this system will determine weights of the possibility and impact of risk events, and establish its safety level.

Finally, the system will come to a judgement of whether the current security of science and technology should be pre alarmed. If the evaluation result is higher than the threshold, an pre-alarming program is supposed to be initiated to issue a security alert, which requires instantaneous response and risks regulation. If the current state is judged as safe, monitoring will be continued (Fig. 2).

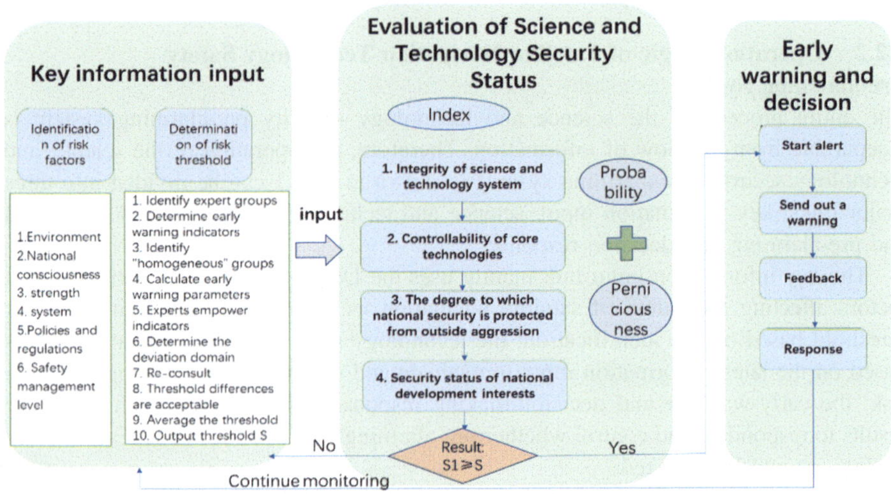

Fig. 2. Operation logic of science and technology security pre-alarming monitoring system

5 Conclusion

According to the analysis above, China's satellite navigation technology security development strategy faces a complex strategic environment. Along with many internal and external advantages, there are also many shortages in technology, industry, and professional talents. In order to cope with the complex and changing strategic environment and meet the need of maintaining national scientific and technological security in the field of satellite navigation in China, scientific strategic evaluation and accurate strategic selection must be performed first in line. Then, it is vital to speed up the construction of a science and technology safety monitoring and pre-alarming system.

References

1. Lian, Y., Ma, W.: Science and technology security: a new concept of national security. Sci. Technol. Manag. (11), 20–22 (1998)
2. Ma, W.: Science and technology security: definition, connotation and extension. Int. Technol. Econ. Res. (02), 14–18 (1999)
3. Yang, M.: Inquiry into socialist science and technology security under the background of economic globalization. Acad. Exchange (10), 86–89 (2009)
4. Ma, W.: Thinking about the main challenges and countermeasures of science and technology security in China in the early 21st century. Sci. Technol. Rev. (7), 28–30 (2003)
5. Lin, C., Li, Z.: Theoretical thoughts on science and technology security. Sci. Technol. Manag. Res. (12), 68–70 (2007)
6. Pan, Z., Yang, Y.: Globalization and national science and technology security. China Sci. Technol. Forum (5), 19–23 (2007)
7. Yang, M.: On the difficult situation and way out of national science and technology security demands. Acad. Exchange (9), 95–98 (2011)
8. Li, D.: National Security. China Yanshi Press, p. 223 (2016)
9. Anderson, J.O.L.: Security on the cheap: creating a cost-effective information security program. Indianapolis 23–27 (2013)
10. Wang, L.: Study on the construction of science and technology security system from the perspective of dissipative structure theory. Enterprise Econ. (2), 87–91 (2014)
11. Zhang, J., Ma, F.: National science and technology security information system and construction. J. Inf. **35**(05), 483–491 (2016)
12. Liu, Z.: Science and technology of people's livelihood from the perspective of science of science—recommend Jia Pinrong's monograph "technology of people's livelihood: innovation model and evaluation system". Sci. Res. **34**(12), 1916–1917 (2016)
13. National Research Council: Beyond 'fortress America': National Security Controls on Science and Technology in a Globalized World. National Research (2009)
14. You, G., Zhao, L.: Theory and Practice of Military-civilian Technology Integration Development, p. 159. National Defense Industry Press (2017)
15. Gao, G.: Policy research on Beidou satellite navigation industry. J. Beijing Univ. Aeronaut. Astronaut. (Soc. Sci. Edn.) (04), 51–58 (2017)
16. Liang, S., Cao, C.: Chief scientist of China satellite navigation and positioning association: talent is a relatively large gap in the Beidou industry [EB/OL]. http://field.10jqka.com.cn/20171229/c602255795.Shtml. Accessed 29 Dec 2017
17. Li, L., Liao, J., Zhang, Y.: Establishment and improvement of science and technology security pre-alarming mechanism. Sci. Technol. Rev. **37**(19), 26–32 (2019)

Technologies for Navigation of Autonomous Systems

Integrated Error Compensation Method for Three-Axis Magnetometer in Geomagnetic Navigation

Binfeng Yang(✉), Run Wang, and Huan Sun

School of Information and Navigation,
Air Force Engineering University, Xi'an, China
bf_yang@163.com

Abstract. In the process of geomagnetic field measurement, the three-axis magnetometer is affected by various interference factors and produces various errors, which have an impact on the measurement accuracy. In this paper, the joint estimation iterative algorithm is used to calibrate the three-axis magnetometer, and the performance of the method is compared with the EKF algorithm and the nonlinear least squares algorithm. The simulation results show that in the three methods, the error average and the standard deviation of the joint estimation iterative algorithm are the least and the convergence rate is the fastest, which proves that the method is effective and usable. The experimental results show that the joint estimation iterative algorithm reduces the mean average of the measured value of the three-axis magnetometer from 69.5211nT to 9.241nT, the standard deviation is reduced from 122.0014nT to 19.941nT, and the suppression ratio reaches 86.71%. The result suggest an effective way for the calibration of three-axis fluxgate magnetometers.

Keywords: Three-axis magnetometers · Joint estimation iterative algorithm · EKF algorithm · Nonlinear Least Square algorithm · Nonmagnetic rotation equipment

1 Introduction

In recent years, geomagnetic navigation has developed rapidly, and its features of strong concealment, high autonomy and strong anti-interference ability make geomagnetic navigation have obvious advantages and broad development prospects in underwater and terrible electromagnetic environment [1–3].

The observation accuracy of the geomagnetic sensor will directly affect the navigation accuracy. However, in the actual measurement process, the geomagnetic signal and various error signals are coupled to each other to form the geomagnetic sensor output, which seriously affects the geomagnetic field measurement accuracy and inhibits the potential performance of the magnetometer. Therefore, Please note that the first line of text that follows a heading is not indented ("p1a" style) compensation of geomagnetic sensors are of great significance for improving the accuracy of geomagnetic matching navigation [4–7].

J. Sun et al. (Eds.): CSNC 2020, LNEE 650, pp. 627–640, 2020.
https://doi.org/10.1007/978-981-15-3707-3_59

In response to the error compensation problem of geomagnetic sensors, many scholars have conducted research. Renaudin et al. studied the correction problem of AMR in the magnetic field, and analyzed the sensor in detail. The source and generation mechanism of self-meter error, soft magnetic and hard magnetic error are fully considered in modeling [8]. In [9, 10], the geomagnetic sensor measurement error and the soft and hard magnetic errors are integrated. The former uses the ellipsoid fitting calibration method to compensate. The latter considers that both the observation vector and the data matrix have errors, so the overall minimum is used. The second ride was compensated. For the above vector modeling method, the true state of the magnetic field in the carrier cannot be accurately represented. The eddy current magnetic field and the associated low frequency magnetic field are generated by the attitude transformation of the carrier. The eddy current magnetic field does not depend on the ferromagnetism of the carrier, but only depends on its conductivity [11]; the literature [12] uses the COMSOL Multiphysics simulation software to establish the carrier. The simulation model of the eddy current interference field is analyzed qualitatively. Literature [13] put forward the idea of integrated modeling, and studied the characteristics of rotation speed and so on.

For the magnetic field compensation algorithm, the current mainstream algorithms include aeromagnetic compensation method, two-step estimation method, ellipse fitting method and unscented Kalman filter algorithm (UKF), and aeromagnetic compensation method [11, 14] is based on Tolles-Lawson equation. The three components of the measured value of the earth's magnetic field are projected in the direction of the earth's magnetic field, but the error between the true value of the earth's magnetic field and the direction of the measured value is neglected, so it is only applicable to the case where the interference field is small; in the two-step estimation algorithm [5, 10], due to the existence of intermediate variables, the correlation of various parameters may appear singularity of the coefficient matrix, which cannot be solved correctly; the ellipse fitting method [15, 16] requires the measured values to fit an ellipse, so that the modeling method is compared. Simple, difficult to adapt to complex measurement environments; UKF algorithm [6, 17] can achieve high-precision real-time compensation, but high sensitivity to initial values, improper selection may cause filter divergence, and can not get correct results.

Through analysis, previous studies have insufficient research on eddy current magnetic field and double noise in the carrier, lacking an integrated model and efficient compensation algorithm. In this paper, an integrated model is proposed for this problem, and an iterative algorithm based on joint estimation is proposed. The performance is compared between EKF and nonlinear least squares algorithm. The method has good simulation and experimental performance and can be used to improve the measurement accuracy of the three-axis magnetometer.

2 Compensation Model

According to previous research, the geomagnetic sensor error model can be expressed as:

$$B = CH + b' + \varepsilon \tag{1}$$

among them, $C = C_M C_{NO} C_{SF} C_{SI}$, $b' = C_M C_{NO} C_{SF} C_{SI} B_h + w + b_s$, C_M is an error matrix generated by the geomagnetic sensor being installed in the carrier, and the three axes of the sensor and the carrier are completely not coincident; C_{NO} is an error matrix caused by the three internal axes of the sensor being not completely orthogonal; C_{SF} is the scale factor error produced by the manufacturing process; C_{SI} is a matrix of soft magnetic error coefficients in the carrier; B_h is the interference generated by the hard magnetic material in the carrier; w is the internal zero-scale drift error of the sensor; b_s is the residual magnetic error of the sensor; ε is the observation noise during the measurement process. Previous studies have shown that ε has a Gaussian distribution [6].

For the carrier with high speed and frequent attitude change, eddy current will be generated inside, which will generate eddy current magnetic field, which will affect the accurate measurement of the sensor. Taking it into account, the model can be obtained:

$$B_k = CH_k + b' + P'(\Delta B/\Delta t) + \varepsilon \tag{2}$$

The expression that transforms the true value of the earth's magnetic field is:

$$\hat{H}_k = QB_k + b + P(\Delta B/\Delta t) + \varepsilon \tag{3}$$

among them, $Q = C^{-1}$, $b = -C^{-1}b'$, $P = -C^{-1}P'$, Gaussian noise ε is still Gaussian after being transformed, so we're not doing the transformation. In the above formula, the parameter space dimension is:

$$\begin{aligned} \dim \Theta &= \text{elements}(Q) + \text{elements}(b) + \text{elements}(P) \\ &= 21 \end{aligned} \tag{4}$$

It can be seen from the above analysis that Q and P are non-singular matrices in the environment set in this paper, that is, the instrument error and the eddy current magnetic field error must exist (even if it is small), so through QR decomposition, the parameter matrices Q and P are expressed as follows:

$$Q = Q_A R_A,\ P = Q_B R_B \tag{5}$$

After QR decomposition, the Q and P parameter matrices are decomposed into orthogonal matrix Q_A, Q_B and upper triangular matrix R_A, R_B. Since R_A, R_B is unique, on the basis of guaranteeing the uniqueness of the parameter matrix, the parameter space is reduced from 21 to 15 dimensions. Get the model:

$$\hat{H}_k = R_A B_k + b + R_B(\Delta B/\Delta t) + \varepsilon \tag{6}$$

The parameter matrix expressions are as follows:

$$R_A = \begin{bmatrix} q11 & q12 & q13 \\ 0 & q22 & q23 \\ 0 & 0 & q33 \end{bmatrix}, b = \begin{bmatrix} b1 \\ b2 \\ b3 \end{bmatrix}, R_B = \begin{bmatrix} p11 & p12 & p13 \\ 0 & p22 & p23 \\ 0 & 0 & p33 \end{bmatrix}$$

When the parameters of each system are clear, the geomagnetic field measurement value can be compensated, so the parameter vector X_k is set to be determined:

$$\begin{aligned} X_k = [&q11, q12, q13, q22, q23, q33, b1, b2, \\ &b3, p11, p12, p13, p22, p23, p33]^T \end{aligned} \tag{7}$$

In the actual measurement process, the sensor is affected by various factors, resulting in unstable measurement results, which are reflected in small changes in system parameters, that is $\tilde{X}_k = X_k + \eta_k$, and η_k are system noise. It is assumed that the system noise and the measurement noise are Gaussian noise and are independent of each other.

2.1 EKF Algorithm

The scalar square of magnetic field true value can be expressed as Eq. 8:

Then the equation of state and the observation equation for the problem can be expressed as:

$$X_k = f_k(X_{k-1}) + \eta_k \tag{8}$$

$$Z_k = H_k^2 = h_k(X_k) + v_k \tag{9}$$

$$f_k(X_k) = X_k \tag{10}$$

$$\begin{aligned} h_k(X_k) = B_k^T[Q^2B + Q^TB + Q^TP(\Delta B/\Delta t)] + b^T[QB + P(\Delta B/\Delta t)] \\ + (\Delta B/\Delta t)^T[P^TQB + P^Tb + P^2(\Delta B/\Delta t)] + b^2 \end{aligned} \tag{11}$$

$$v_k = (2Q_k(B_k + b_k + P_k(\Delta B/\Delta t)))^T \varepsilon_k + \|\varepsilon_k\|^2 \tag{12}$$

Among them, X_k is the parameter vector to be estimated, η_k is the system noise, v_k is the observation noise, and the system noise η_k is the Gaussian noise with the mean value of 0 and the covariance matrix of Q.

From Eq. 12, the mean and variance of the observation noise can be expressed as:

$$\mu_k = E[v_k] = -tr(\Sigma) \tag{13}$$

$$\begin{aligned} R &= \sigma_k^2 = E\left[v_k^2\right] - \mu_k^2 \\ &= 4(Q(B - b - P(\Delta B/\Delta t)))^T\Sigma(Q(B - b - P(\Delta B/\Delta t))) + 2\left(tr\Sigma^2\right) \end{aligned} \tag{14}$$

$$\Sigma = \left\{ \varepsilon\varepsilon^T \right\} \tag{15}$$

X_k is the amount to be estimated, $\hat{X}_k$ is the predicted value of the k step. According to the EKF algorithm, the compensation model can be expressed as:

$$\begin{aligned}
\hat{X}_{k|k-1} &= f_k(\hat{X}_{k-1}) \\
P_{k|k-1} &= F_{k-1} P_{k-1} F_{k-1}^T + Q_{k-1} \\
G_k &= P_{k|k-1} H_k^T (H_k P_{k|k-1} H_k^T + R_k)^{-1} \\
\hat{X}_k &= \hat{X}_{k|k-1} + G_k (Z_k - h_k(\hat{X}_{k|k-1})) \\
P_k &= (I - G_k H_k) P_{k|k-1}
\end{aligned} \tag{16}$$

Among them, the Jacobian matrix $F_{k-1} = \left.\frac{\partial f_k(\hat{X}_{k-1})}{\partial X}\right|_{X=\hat{X}_{k-1}}$, $H_k = \left.\frac{\partial h_k(\hat{X}_k)}{\partial X}\right|_{X=\hat{X}_{k|k-1}}$.

In the initial value setting of EKF, X_0 takes the empirical value, P_0 can take larger, because the amount of datas are sufficient, after k times, the final result is X_k is the required parameter value.

2.2 Nonlinear Least Square Algorithm

In the case of the nonlinear least squares algorithm, since the average value can be taken after multiple measurements, the system noise η_k and the observed noise ε_k can be ignored. According to the above, the geomagnetic compensation model becomes:

$$\hat{H}_k = R_A B_k + b + R_B(\Delta B / \Delta t) \tag{17}$$

Then the scalar square of magnetic field true can be calculated by Eq. (19):

The difference between the measurement value square and the true value square can be expressed as Eq. (20):

After k measurements, the total difference expression obtained by summing the differences is:

$$E = \sum_{i=1}^{k} D_k = \sum_{i=1}^{k} (B_k^2 - H_k^2) \tag{18}$$

According to the principle of the nonlinear least squares algorithm, when E obtains the minimum value, X obtains the optimal solution, that is $X = \arg\min(E)$, and then obtains various system parameters in the compensation model.

2.3 Joint Estimation Iterative Algorithm

According to the above, the geomagnetic compensation model is:

$$z = H_k^2 = H_k^T H_k = f(\tilde{X}_k, w) + v_k \tag{19}$$

Considering that both system noise η_k and observation noise v_k exist in the model, the covariance matrix is Q, R, and when the systematic error exists, the traditional Gauss-Newton iteration method will no longer be applicable. This paper proposes to use the Taylor-B iterative method to jointly estimate the evaluation X_k and the parameter w. The basic idea is to use the observation z and the observation system parameter w as joint observations for Taylor expansion and iteration, assuming that the result of the k iteration is $[\hat{X}_k^b, \hat{w}_k^b]^T$, the first-order Taylor expansion is performed here to obtain the result shown in Eq. 20.

Among them, $F_1(\hat{X}_k^b, \hat{w}_k^b) = \frac{\partial f(\tilde{X}_k, w)}{\partial \hat{X}_k^{b\,T}}$, $F_2(\hat{X}_k^b, \hat{w}_k^b) = \frac{\partial f(\tilde{X}_k, w)}{\partial \hat{w}_k^{b\,T}}$, Since η_k and v_k are statistically independent, the weighting matrix is $Q' = blkdiag[Q^{-1} R^{-1}]$, so the k + 1 iteration result is solved Eq. 21.

This is also a secondary optimization problem for x and w, so there is a closed solution, as shown in Eq. 22.

After several iterations, the convergence value is reached, which is recorded as: $\hat{X}_{Nlwls}^b = \lim_{k \to +\infty} \hat{X}_k^b$, $\hat{w}_{Nlwls}^b = \lim_{k \to +\infty} \hat{w}_k^b$, then, the joint estimated variance matrix of $\hat{X}_{Nlwls}^b$ and $\hat{w}_{Nlwls}^b$ in the presence of systematic error is obtained as shown in Eq. 23.

$$\begin{aligned} H_k^2 &= H_k^T H_k = (R_{Ak} B_k + b_k + R_{Bk}(\Delta B/\Delta t) + \varepsilon_k)^T (R_{Ak} B_k + b_k + R_{Bk}(\Delta B/\Delta t) + \varepsilon_k) \\ &= B_k^T R_{Ak}^T R_{Ak} B_k + B_k^T R_{Ak}^T b_k + B_k^T R_{Ak}^T R_{Bk}(\Delta B/\Delta t) + b_k R_{Ak} B_k + b_k^T b_k + b_k R_{Bk}(\Delta B/\Delta t) \\ &+ (\Delta B/\Delta t)^T R_{Bk}^T R_{Ak} B_k + (\Delta B/\Delta t)^T R_{Bk}^T \mathrm{b_k} + (\Delta B/\Delta t)^T R_{Bk}^T R_{Bk}(\Delta B/\Delta t) + v_k \\ &= B_k^T [R_{Ak}^2 B_k + R_{Ak}^T B_k + R_{Ak}^T R_{Bk}(\Delta B/\Delta t)] + b_k^T [R_{Ak} B_k + R_{Bk}(\Delta B/\Delta t)] \\ &+ (\Delta B/\Delta t)^T [R_{Bk}^T R_{Ak} B_k + R_{Bk}^T b_k + R_{Bk}^2(\Delta B/\Delta t)] + b_k^2 + v_k \end{aligned} \tag{20}$$

$$\begin{aligned} H_k^2 &= H_k^T H_k = [R_A B_k + b + R_B(\Delta B/\Delta t)]^T [R_A B_k + b + R_B(\Delta B/\Delta t)] \\ &= B_k^T [R_A^2 B_k + R_A^T B_k + R_A^T R_B(\Delta B/\Delta t)] + b^T [R_A B_k + R_B(\Delta B/\Delta t)] \\ &\quad + (\Delta B/\Delta t)^T [R_B^T R_A B_k + R_B^T b + R_B^2(\Delta B/\Delta t)] + b^2 \end{aligned} \tag{21}$$

$$\begin{aligned} D_k = B_k^2 - H_k^2 &= B_k^2 - B_k^T [R_A^2 B_k + R_A^T B_k + R_A^T R_B(\Delta B/\Delta t)] \\ &- b^T [R_A B_k + R_B(\Delta B/\Delta t)] - (\Delta B/\Delta t)^T [R_B^T R_A B_k + R_B^T b + R_B^2(\Delta B/\Delta t)] - b^2 \end{aligned} \tag{22}$$

$$\begin{bmatrix} f(X, w) \\ X \end{bmatrix} \approx \begin{bmatrix} f(\hat{X}_k^b, \hat{w}_k^b) \\ \hat{X}_k^b \end{bmatrix} + \begin{bmatrix} F_2(X_k^b, \hat{w}_k^b) & F_1(\hat{X}_k^b, \hat{w}_k^b) \\ O_{q \times r} & I_q \end{bmatrix} \cdot \begin{bmatrix} w - \hat{w}_k^b \\ X - \hat{X}_k^b \end{bmatrix} \tag{23}$$

$$\begin{bmatrix} \hat{w}_{k+1}^b \\ \hat{X}_{k+1}^b \end{bmatrix} = \arg\min \left(\left[\begin{array}{c} \left(\begin{bmatrix} F_2(\hat{X}_k^b, \hat{w}_k^b) & F_1(\hat{X}_k^b, \hat{w}_k^b) \\ O_{q\times r} & I_q \end{bmatrix} \cdot \begin{bmatrix} w - \hat{w}_k^b \\ X - \hat{X}_k^b \end{bmatrix} - \begin{bmatrix} z - f(\hat{X}_k^b, \hat{w}_k^b) \\ X - \hat{X}_k^b \end{bmatrix} \right)^T \\ \cdot Q' \cdot \left(\begin{bmatrix} F_2(\hat{X}_k^b, \hat{w}_k^b) & F_1(\hat{X}_k^b, \hat{w}_k^b) \\ O_{q\times r} & I_q \end{bmatrix} \cdot \begin{bmatrix} w - \hat{w}_k^b \\ X - \hat{X}_k^b \end{bmatrix} - \begin{bmatrix} z - f(\hat{X}_k^b, \hat{w}_k^b) \\ X - \hat{X}_k^b \end{bmatrix} \right) \end{array} \right] \right) \tag{24}$$

$$\begin{aligned} \begin{bmatrix} \hat{w}_{k+1}^b \\ \hat{X}_{k+1}^b \end{bmatrix} = \begin{bmatrix} \hat{w}_k^b \\ \hat{X}_k^b \end{bmatrix} + & \begin{bmatrix} F_1^T(\hat{X}_k^b, \hat{w}_k^b)Q^{-1}F_1(\hat{X}_k^b, \hat{w}_k^b) & F_1^T(\hat{X}_k^b, \hat{w}_k^b)Q^{-1}F_2(\hat{X}_k^b, \hat{w}_k^b) \\ F_2^T(\hat{X}_k^b, \hat{w}_k^b)Q^{-1}F_1(\hat{X}_k^b, \hat{w}_k^b) & F_2^T(\hat{X}_k^b, \hat{w}_k^b)Q^{-1}F_2(\hat{X}_k^b, \hat{w}_k^b) + R \end{bmatrix}^{-1} \\ & \cdot \begin{bmatrix} F_1^T(\hat{X}_k^b, \hat{w}_k^b)Q^{-1} & O_{q\times r} \\ F_2^T(\hat{X}_k^b, \hat{w}_k^b)Q^{-1} & R^{-1} \end{bmatrix} \cdot \begin{bmatrix} z - f(\hat{X}_k^b, \hat{w}_k^b) \\ X - \hat{X}_k^b \end{bmatrix} \end{aligned} \tag{25}$$

$$\begin{aligned} \operatorname{cov}\left(\begin{bmatrix} \hat{w}_{Nlwls}^b \\ \hat{X}_{Nlwls}^b \end{bmatrix} \right) &= E\left[\begin{bmatrix} \hat{w}_{Nlwls}^b - w \\ \hat{X}_{Nlwls}^b - X \end{bmatrix} \cdot \begin{bmatrix} \hat{w}_{Nlwls}^b - w \\ \hat{X}_{Nlwls}^b - X \end{bmatrix}^T \right] \\ &= \begin{bmatrix} F_1^T(\hat{X}_{Nlwls}^b, \hat{w}_{Nlwls}^b)Q^{-1}F_1(\hat{X}_{Nlwls}^b, \hat{w}_{Nlwls}^b) & F_1^T(\hat{X}_{Nlwls}^b, \hat{w}_{Nlwls}^b)Q^{-1}F_2(\hat{X}_{Nlwls}^b, \hat{w}_{Nlwls}^b) \\ F_2^T(\hat{X}_{Nlwls}^b, \hat{w}_{Nlwls}^b)Q^{-1}F_1(\hat{X}_{Nlwls}^b, \hat{w}_{Nlwls}^b) & F_2^T(\hat{X}_{Nlwls}^b, \hat{w}_{Nlwls}^b)Q^{-1}F_2(\hat{X}_{Nlwls}^b, \hat{w}_{Nlwls}^b) + R \end{bmatrix}^{-1} \end{aligned} \tag{26}$$

3 Simulation

The simulation datas are used to verify the above algorithms. According to the above-mentioned geomagnetic field observation model, the true value of the geomagnetic field vectors were obtained with the orthogonal three-axis magnetometer model:

$$\begin{cases} H_x = R \sin\theta \cos\gamma \\ H_x = R \sin\theta \sin\gamma \\ H_z = R \cos\theta \end{cases} \tag{27}$$

Where R denotes the scalar of magnetic field. The magnetic field vector will vary with the change of θ and γ when the orthogonal three-axis magnetometer rotates. The simulation measured vector of the three-axis magnetometer were obtained by combining (27) with prearranged error parameters. In this simulation. Assume the true scalar value of magnetic field is 50000nT.

The system noise and observation noise in the observation process obey the zero-mean Gaussian distribution and are statistically independent of each other. In order to better explain the compensation effect, this paper uses error mean $E(n)$ and standard deviation δ for analysis:

$$E(n) = \frac{1}{n}\sum_{i=1}^{n}(B_i - B_0) \tag{28}$$

$$\delta = \sqrt{\frac{1}{n-1}\sum\nolimits_{i=1}^{n}(B_i - B_m)^2} \tag{29}$$

Where n is the number of measurement points, B_i is the scalar measurements of three-axis magnetometer, B_m is the average of scalar measurements, and B_0 is the true scalar value of magnetic field.

In the simulation, in order to compare the compensation effects of the three algorithms, a set of data is generated, and the simulation data is processed by the EKF algorithm, the nonlinear least squares algorithm and the joint estimation iterative algorithm respectively. After calculation, the prearranged parameters and estimated parameters of each system parameter are as shown in Table 1:

Table 1. Prearranged and estimated parameters by different algorithm

Par	q11	q12	q13	q22	q23	q33	b1	b2	b3	p11	p12	p13	p22	p23	p33
X_P	1.03	0.97	0.85	1.52	1.30	0.99	28.0	35.0	10.0	1.12	0.58	0.85	1.68	0.02	1.77
X_E	1.11	0.89	0.93	1.60	1.36	1.32	25.3	27.7	12.2	1.47	0.52	1.07	1.71	0.00	2.01
X_N	2.03	1.56	1.20	2.35	1.09	1.24	87.2	65.1	95.4	0.98	0.02	0.99	1.62	0.54	2.71
X_J	1.09	1.21	0.78	1.32	1.57	1.60	32.2	33.0	9.01	1.20	0.09	0.74	1.74	0.00	1.80

The meaning of the symbols:

Par: parameters X_P: prearranged parameters X_E: estimated parameters by EKF algorithm
X_N: estimated parameters by nonlinear Least Square algorithm
X_J: estimated parameters by Joint Estimation Iterative algorithm

It can be seen from Table 1 that all three algorithms can compensate for the system parameters. The nonlinear least squares algorithm is less effective than the other two algorithms because it does not consider the noise. The compensation effects of the three algorithms are shown in Figs. 1, 2, and 3, and Table 2:

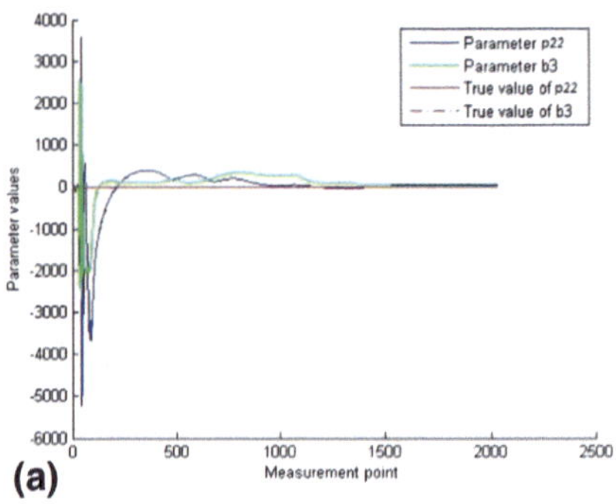

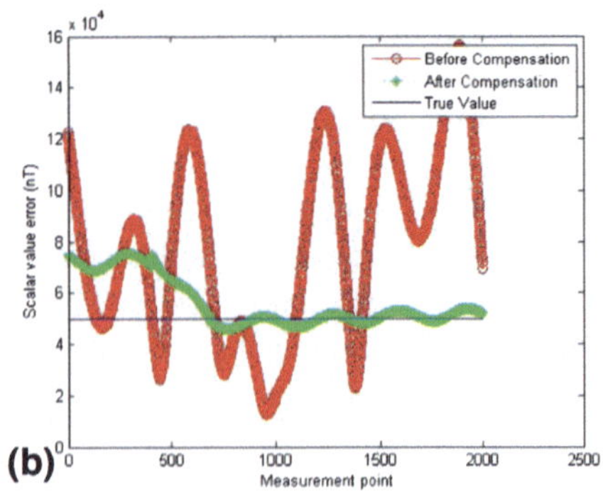

Fig. 1. a. EKF algorithm partial parameter convergence b. Compensation performance with EKF (Simulation datas)

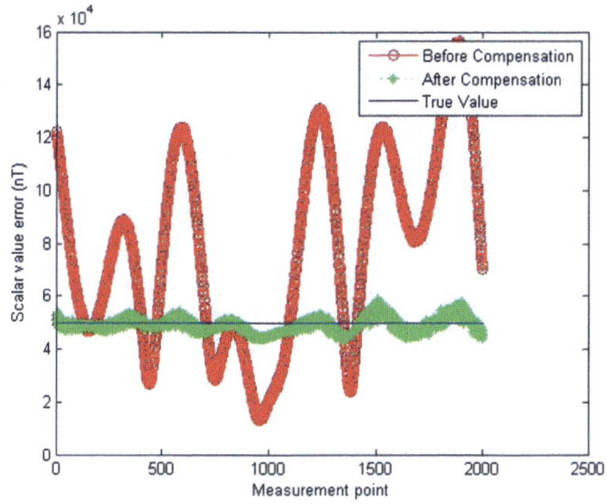

Fig. 2. Compensation performance with nonlinear Least Square

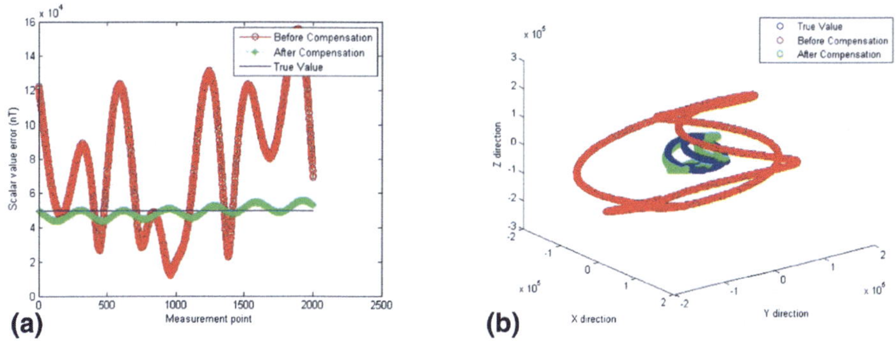

Fig. 3. a. Compensation performance of scalar value with Joint Estimation Iterative (Simulation datas), b. Compensation performance of vector with Joint Estimation Iterative (Simulation datas)

When the EKF algorithm compensates, the two values in any of the parameters are analyzed for convergence. As shown in Fig. 1a, after about 800 sets of data operations, the parameters converge and tend to be stable, so in the initial stage of compensation, the overall compensation effect is not ideal; the nonlinear least squares algorithm has large error due to no noise reduction processing (Fig. 2); the joint estimation iterative algorithm can achieve better compensation whether it is the geomagnetic component or the total amount (Fig. 3a, b), the compensation result data of each algorithm is shown in Table 2:

The simulation results show that the nonlinear least squares algorithm has large error in compensation result, and the other two algorithms can achieve higher precision compensation, but the compensation precision of EKF algorithm has a large dependence on data volume (In this simulation, high-precision compensation cannot be achieved when the amount of data is less than 800 sets). Compared with the other two algorithms, the joint estimation iterative algorithm can achieve the highest precision and the suppression ratio can reach 95.4%.

Table 2. Compensation results of different algorithms (Simulation data)

Scalar error	Average (nT)	Standard deviation (nT)
Before compensation	529.7809	2676.6033
EKF	30.4121	92.3430
Nonlinear least square	57.8438	118.0087
Joint estimation iterative	24.2051	58.9042

4 Experimental System

In the experimental design, a Bartington company's Mag-690-FL100 (measurement accuracy of 0.1nT) three-axis magnetometer was used for compensation, and it was placed above the 1:48 aircraft scale model to simulate the real situation inside the flight carrier. In order to simulate the attitude change of the flying carrier, a 3FHT30C non-magnetic turntable is used to provide attitude information. The turntable has a residual magnetism of 7nT. The measurement system uses 12 V DC power to supply power to isolate the adverse effects of unstable mains voltage. The data acquisition software used National Instruments, the test system is shown in Fig. 4, and later Matlab was used for data processing. Another DM050 proton magnetometer is used to provide the true value of the geomagnetic field.

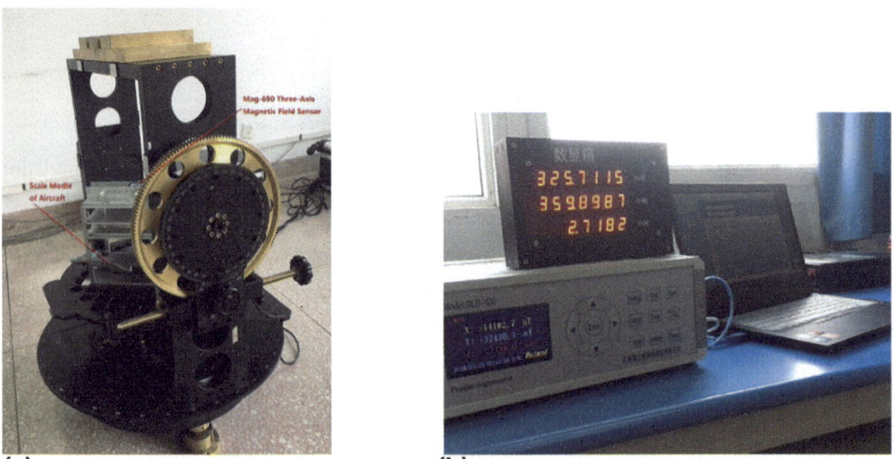

(a) (b)

Fig. 4. a. Geomagnetic field measurement system, b. Geomagnetic data acquisition system

(1) The experimental site is selected in an open place away from strong magnetic interference. The non-magnetic turntable is adjusted with a spirit level. After the three-axis magnetometer is fixed, a scale model of the aircraft is placed below to

simulate the softness generated inside the flight carrier. Hard magnetic errors and other interference.

(2) Connect the equipment, connect the DC power supply, set the data acquisition software acquisition frequency to 50 Hz, perform continuous data acquisition, and quickly rotate the non-magnetic turntable along the axis to perform all-round attitude information measurement, simulating the full range of the flight carrier during the flight The attitude is rotated, and the acquired data is processed by Matlab for post-processing.

5 Experimental Result

In the experiment, the geomagnetic field value of the experimental site was measured by proton magnetometer to be 46157nT. The experimental data was compensated by EKF algorithm, nonlinear least squares algorithm and joint estimation algorithm. The initial value of EKF algorithm was set as:

$$X_0 = [0,0,0,0,0,0,0,0,0,0,0,0,0,0,0]^T, P_0 = \begin{bmatrix} 100I_{6\times6} & 0_{6\times3} & 0_{6\times6} \\ 0_{3\times3} & 10^4 I_{3\times3} & 0_{3\times3} \\ 0_{6\times6} & 0_{6\times3} & 100I_{6\times6} \end{bmatrix}$$

After three kinds of algorithm compensation, the compensation effect diagram is shown in Figs. 5, 6 and 7:

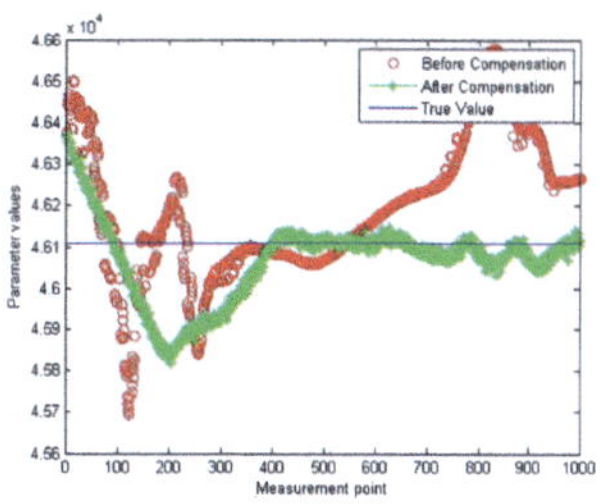

Fig. 5. Compensation performance with EKF (Expermental datas)

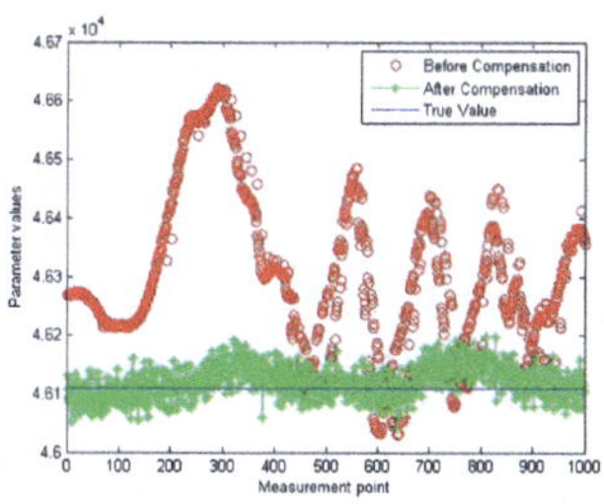

Fig. 6. Compensation performance with Nonlinear Least Square (Expermental datas)

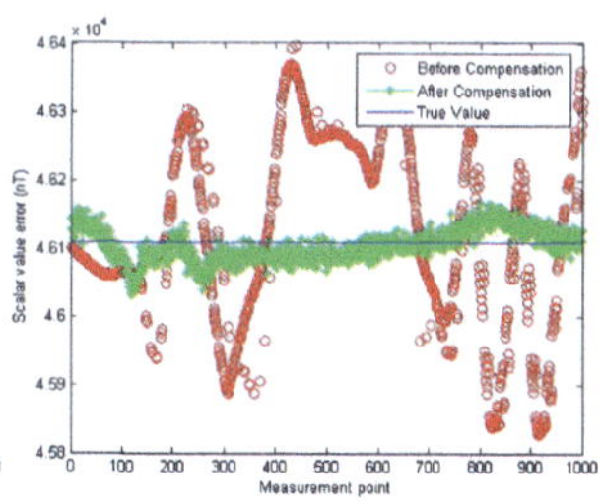

Fig. 7. Compensation performance with Joint Estimation Iterative (Expermental datas)

It can be seen that the experimental results are basically consistent with the simulation results. The 1000 sets of data used in the experiment, because the initial value is set to 0 vector, the EKF algorithm tends to converge after about 420 sets of data, and the compensation effect in the early stage is second to the other two algorithms; The compensation result of the nonlinear least squares algorithm still has large noise interference; the joint estimation iterative algorithm method has fast convergence speed and good compensation effect.

The mean and variance of the compensation results of each algorithm are shown in Table 3:

It can be seen from Table 3 that the joint estimation iterative algorithm has obvious advantages compared with the other two algorithms, and t The compensation error average can be reduced from 69.5211nT to 9.2410nT, the standard deviation is reduced from 122.0014nT to 19.941nT, and the suppression ratio is 86.71%, considering that the residual magnetic error of the non-magnetic turntable reaches 7nT, and the experimental environment is around other interferences, so the results of this experiment have verified the theory proposed in this paper, which has practical significance.

Table 3. Compensation results of different algorithms (Expermental data)

Scalar error	Average (nT)	Standard deviation (nT)
Before compensation	69.5211	122.0014
EKF	18.6232	30.7500
Nonlinear least square	16.9814	47.5022
Joint estimation iterative	9.2410	19.9410

6 Discussion

(1) The EKF algorithm has a simple principle and can achieve higher precision compensation in a shorter time, and real-time compensation makes its timeliness better than other types of algorithms, but the Kalman filter algorithm is highly dependent on the initial value. In this paper, because the prior knowledge of the system parameters is insufficient, the initial value is taken as 0 vector, and the covariance matrix is relatively large. Because of the large amount of datas, a better compensation effect can be achieved in the later stage. For the first-order EKF algorithm, the compensation accuracy is not enough. Because the model studied in this paper is weakly nonlinear, its compensation accuracy is within the acceptable range.

(2) During the experiment, the rapid rotation of the non-magnetic turret makes the eddy current magnetic field inside the aircraft contraction model, which has an influence on the measurement of the three-axis magnetometer, which is reflected in the large fluctuation of the measured value, and the integrated model containing the eddy current magnetic field. The eddy current magnetic field information is taken into account in the modeling process, so the fluctuation can be better filtered than the model without the eddy current magnetic field, and the compensation result is more stable.

(3) The problem of instability of the observing system is mathematically reflected in the variation of the parameters of each system. In this experiment, the new experimental site, the leveling of the non-magnetic turret and the full attitude measurement process will make the system parameters minor changes occur, which is bound to produce systematic errors. For this error, this paper simplifies it

to zero-mean Gaussian noise and obtains a better compensation effect. In order to further improve the compensation accuracy, the error should be further studied.

(4) When using the nonlinear least squares algorithm, the method of averaging multiple sets of data is used for the noise, but the essence of the mean is still noisy. The simulation and experimental results show that the limitations of the proposed algorithm for this problem lead to a larger error than the other two algorithms.

(5) The proton magnetometer has high measurement accuracy, so the measured geomagnetic field value is taken as the true value of the experimental site. In this paper, an integrated model with eddy current magnetic field and double noise is proposed, which contains more complete information types, more complete models, and more advantages in compensation. In the joint estimation iterative algorithm, the observed system noise and measurement noise are jointly estimated and processed, and the compensation effect indicates this method has higher compensation precision than the nonlinear least squares algorithm and converges faster than the compensation result of EKF algorithm, which proves the development potential of the compensation model and compensation method.

7 Conclusion

Geomagnetic field measurement error compensation has always been a key issue in geomagnetic navigation. Aiming at the problem of eddy current magnetic field and double noise in the carrier, this paper proposes a joint estimation iterative algorithm for double noise to calibrate the Mag-690-FL100 geomagnetic sensor, and compare it with the EKF algorithm and the nonlinear least squares algorithm. The simulation and experimental results show that the joint estimation iterative algorithm has better compensation effect. The method has simple principle, small dependence on data volume and strong practicability, which is beneficial to improve the accuracy of geomagnetic navigation system.

References

1. Alonso, R., Shuster, M.D.: Complete linear attitude-independent magnetometer calibration. J. Astronaut. Sci. **50**(4), 477–490 (2002)
2. Foster, C.C., Elkaim, G.H.: Extension of a two-step calibration methodology to include nonorthogonal sensor axes. IEEE Trans. Aerosp. Electron. Syst. **44**(3), 1070–1078 (2008)
3. Siddharth, S., Ali, A.S., Sheimy, N.E.: A game-theoretic approach for calibration of low-cost magnetometers under noise uncertainty. Meas. Sci. Technol. **23**, 1–12 (2012)
4. Ghanbarpour, H.A., Pourtakdoust, S.H., Samani, M.: A new non-linear algorithm for complete pre-flight calibration of magnetometers in the geomagnetic field domain. J. Aerosp. Eng. **223**, 729–739 (2009)
5. Gebre-Egziabher, D., Elkaim, G.H., Powell, J.D., et al.: A non-linear, two-step estimation algorithm for calibrating solid-state strapdown magnetometers. In: 8th International St. Petersburg Conference on Navigation Systems, pp. 28–30 (2001)

6. Crassidis, J.L., Lai, K.L.: Real-time attitude-independent three-axis magnetometer calibration. J. Guidance Control Dyn. **28**(1), 115–120 (2005)
7. Wang, Y., Zhang, J.Y., Zhang, D.W.: Error analysis and algorithm validation of geomagnetic sensor. Appl. Mech. Mater. **742**(2), 21–26 (2015)
8. Renaudin, V., Afzal, M.H., Lachapelle, G.: Complete triaxis magnetometer calibration in the magnetic domain. J. Sens. **2010**, 1–10 (2010)
9. Huang, X., Wang, J.: Error analysis and compensation methods for geomagnetic signal detection system. Acta Armamentarii **32**(1), 33–36 (2011)
10. Wu, Y., Wang, T., Liang, J.: In-suit error calibration of three-axis magnetometer for unmanned aerial vehicle. Acta Aeronautica et Astronautica Sinica **32**(2), 330–336 (2011). (in Chinese)
11. Bickel, S.H.: Small signal compensation of magnetic fields resulting from aircraft maneuvers. IEEE Trans. Aerosp. Electron. Syst. **15**, 518–525 (1979)
12. Li, C., Yang, B., Fan, B., et al.: An analysis of eddy current interference field characteristics of carrier in geomagnetic navigation. J. Air Force Eng. Univ. (Natl. Sci. Edn.) **19**(4), 72–78 (2018). (in Chinese)
13. Hong, F.P., Qi, Z., Ji, L., et al.: Improvement of vector compensation method for vehicle magnetic distortion field. J. Magn. Magn. Mater. **353**, 1–5 (2014)
14. Pang, X., Ling, C., Zhang, N.: Parameter estimation of airplane magnetic model based on TSVD. J. Detect. Control **31**(5), 48–51 (2009)
15. Zhang, X., Zhao, Y.: New auto-calibration and compensation method for vehicle magnetic field based on ellipse restriction. Chin. J. Sci. Instrum. **30**(11), 2438–2443 (2009). (in Chinese)
16. Qingde, L., John, G.G.: Least squares ellipsoid specific fitting. In: Proceeding of Geometric Modeling and Processing, pp. 335–340 (2004)
17. Li, J., Pan, M.C., Luo, F.L., et al.: Vehicle magnetic field compensation method using UKF. In: Proceeding of 10th International Conference on Electronic Measurement & Instruments, vol. IV, pp. 25–28 (2011)

Azimuth Error Suppression Method Based on the Rotation Modulation and Acoustic Navigation Assistance for Polar Grid SINS

Yingyao Kang, Lin Zhao(✉), Jianhua Cheng(✉), and Mouyan Wu

College of Automation, Harbin Engineering University, Harbin, China
{zhaolin, chengjianhua}@hrbeu.edu.cn

Abstract. The grid strapdown inertial navigation system (SINS) can conquer the orientation problem caused by the meridian convergence in the polar region, but azimuth errors of the grid SINS still exist. The navigation error characteristic is analyzed based on the grid SINS mechanization. According to the error analysis, the azimuth error of the grid SINS will accumulate with time and it is hard to be estimated in polar regions because of the low observability. To improve the orientation accuracy, an azimuth error suppression method based on the rotation modulation and acoustic navigation assistance is proposed in this paper. The grid SINS mechanization is modified to update without longitude and avoid the influence from the longitude error amp. As the latitude increased, the compass effect decreased, which leads to the reduction of azimuth errors observability. So the rotation modulation, which chooses the azimuth axis as rotation axis, is employed to restrain the azimuth errors caused by the horizontal inertial measurement devices, and simultaneously improve the observability of azimuth errors. Then the azimuth errors accumulate with time and will increase the nonlinearity of system, and together with the velocity and position navigation information from acoustic measurement units, the nonlinear filter model is designed to further estimate and suppress the azimuth errors. Finally the simulation results show that with the rotation modulation and acoustic navigation assistance the proposed method in this paper can suppress the accumulated azimuth errors of the grid SINS effectively when working in the polar regions, and prove that the scheme has practical engineering application value.

Keywords: Grid SINS · Polar grid azimuth · Error suppression · Rotation modulation · Information fusion

1 Introduction

With the exploration and development of polar ocean, more and more kinds of navigation vehicles are working in polar regions. In order to ensure the normal work of polar vehicles, it is particularly important to achieve safe and reliable navigation in polar regions [1–3]. The polar regions have special geographical location but less available navigable equipment. The strapdown inertial navigation system (SINS) has been one of the optimal choices for polar vehicles because of its highly autonomy,

J. Sun et al. (Eds.): CSNC 2020, LNEE 650, pp. 641–655, 2020.
https://doi.org/10.1007/978-981-15-3707-3_60

stealthiness and information comprehensiveness [4–6]. But the SINS still has some limitations about the azimuth datum selection and azimuth accuracy [7].

The traditional north-oriented SINS cannot be employed by the vehicle in polar regions because it loses the geographic north. Therefore, the wander azimuth inertial navigation systems, such as the grid SINS and the transversal SINS, are employed to solve the problem of azimuth datum in the polar regions [8]. The transversal earth-fixed frame is obtained by revolving the geographic earth-fixed frame, and the north of transversal frame is also changed with the Greenwich meridian and equatorial plane. Together with the transverse Mercator chart the transversal SINS can solve the azimuth datum problem in polar regions [9]. The grid SINS chooses the grid north which is parallel to the Greenwich meridian as the azimuth datum together with polar stereographic map projection chart to solve the problem [10].

The wander azimuth frame can solve the azimuth datum problem, but the accumulated errors and three kinds of oscillation errors of SINS still exist. To improve the navigation accuracy, some error suppression method, such as the rotation modulation [11], the damping technology [12] and the integration algorithms [13], are proposed for the wander azimuth SINS and proved effective. The damping technology for azimuth SINS is designed based on the control network or filter algorithm [11]. In [12] the availability of rotating modulation in polar region is proposed and verified. In [13, 14] some acoustics navigation system is employed as external measurements to restrain the SINS errors. However, the traditional error suppression method treat the azimuth errors as little angle. The azimuth accuracy and error propagation characteristic have not been analysed comprehensively. And the special effect to observability of azimuth misalignment angle has not been considered.

Compared to the grid SINS, the transversal SINS designs a novel earth model, which brings more computational overhead to the navigation update. What is more, when the integrated filter is designed, there may be compatibility issues with other navigation devices about earth models. Therefore, in this paper the grid frame is employed as the navigation frame. Based on the grid SINS, the error propagation function is derived and the error propagation characteristic is analysed. Then according to the azimuth error characteristics of the grid SINS in polar region, the grid SINS mechanization without longitude is proposed to eliminate the effects of longitude error amp caused by the meridian convergence. Based on the novel grid SINS mechanization, the nonlinear filtering model is designed with a large azimuth misalignment angle. Then the rotation modulation is employed to restrain the navigation error caused by inertial measurement unit (IMU) and improve the observability of azimuth error. Meanwhile, considering the convenience of acoustic navigation sensors, the DVL and USBL output information is introduced in to construct the integrated filter to suppress the azimuth misalignment angle and improve the navigation performance.

2 The Azimuth Error Analysis of the Grid SINS

2.1 The Grid Frame

Some different frames are used in this paper, i.e. the inertial frame i, the earth centred earth fixed frame $ECEF(\mathrm{e})$, the geographic frame g, the body frame b and the grid frame G.

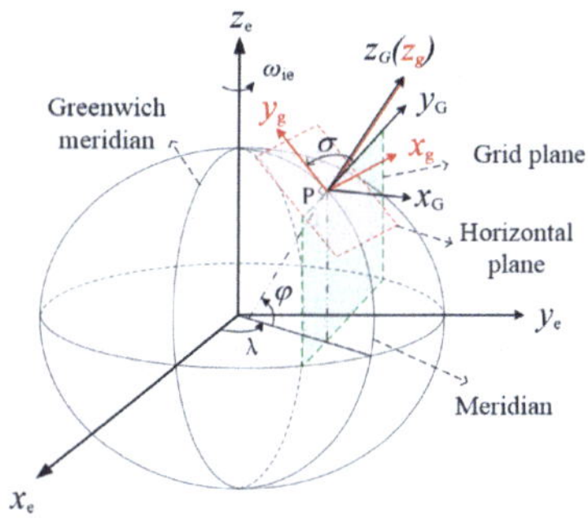

Fig. 1. The description of the grid frame

The grid frame is shown as Fig. 1. The location of vehicle mass center is the grid frame origin and the angle between the geographic north and grid north axis is the grid angle σ.

2.2 The Azimuth Error Propagation Characteristic of the Grid SINS

The error propagation differential function has been derived in [10, 13]. Supposing the earth is a sphere, on stationary base the error propagation function of grid SINS can be simplified as:

$$\dot{\phi}_x = \omega_{iez}^G \phi_y - \omega_{iey}^G \phi_z - \frac{1}{R}\delta v_y^G + C_{11}\delta\varphi + C_{12}\delta\lambda - \delta\tilde{\omega}_{ibx}^G \tag{1}$$

$$\dot{\phi}_y = -\omega_{iez}^G \phi_x + \omega_{iex}^G \phi_z + \frac{1}{R}\delta v_x^G + C_{21}\delta\varphi + C_{22}\delta\lambda - \delta\tilde{\omega}_{iby}^G \tag{2}$$

$$\dot{\phi}_z = \omega_{iey}^G \phi_x - \omega_{iex}^G \phi_y - \frac{\cot L \sin\sigma}{R}\delta v_y^G + C_{31}\delta\varphi - \delta\tilde{\omega}_{ibz}^G \tag{3}$$

$$\delta\dot{v}_x^G = -g\phi_y + 2\omega_{iez}^G \delta v_y^G - 2\omega_{iey}^G \delta v_z^G + \delta f_x^G \tag{4}$$

$$\delta\dot{v}_y^G = g\phi_x - 2\omega_{iez}^G \delta v_x^G + 2\omega_{iex}^G \delta v_z^G + \delta f_y^G \tag{5}$$

$$\delta\dot{v}_z^G = 2\omega_{iey}^G \delta v_x^G - 2\omega_{iex}^G \delta v_y^G + \delta f_z^G \tag{6}$$

$$\delta\dot{\varphi} = \frac{\sin\sigma}{R}\delta v_x^G + \frac{\cos\sigma}{R}\delta v_y^G \tag{7}$$

$$\delta\dot{\lambda} = \frac{\cos\sigma}{R}\sec\varphi\delta v_x^G - \frac{\sin\sigma}{R}\sec\varphi\delta v_y^G \tag{8}$$

where $\boldsymbol{\phi} = \begin{bmatrix} \phi_x & \phi_y & \phi_z \end{bmatrix}^T$ is the attitude error of grid SINS in the grid frame; $\delta\boldsymbol{V}^G = \begin{bmatrix} \delta v_x^G & \delta v_y^G & \delta v_z^G \end{bmatrix}^T$ is the velocity error in grid frame; $\delta\varphi$ is the latitude error and $\delta\lambda$ is the longitude error. $\boldsymbol{\omega}_{ie}^G = \begin{bmatrix} \omega_{iex}^G & \omega_{iey}^G & \omega_{iez}^G \end{bmatrix}^T$ is the the angular velocity of earth in G frame; $\delta\boldsymbol{\omega}_{ib}^G = \begin{bmatrix} \delta\omega_{ibx}^G & \delta\omega_{iby}^G & \delta\omega_{ibz}^G \end{bmatrix}^T$ is the measurement error of gyroscopes and $\delta\boldsymbol{f}^G = \begin{bmatrix} \delta f_x^G & \delta f_y^G & \delta f_z^G \end{bmatrix}^T$ is the measurement error of accelerometer s, and some parameters are [13]:

$$C_{11} = \sin L\sin\sigma - \cot L\cos L\cos^2\sigma\sin\sigma, C_{31} = \cos L$$

$$C_{12} = -\frac{\cot L\cos\sigma\sin^2\sigma}{\sin^2\lambda}, C_{22} = -\frac{\cot L\sin^3\sigma}{\sin^2\lambda}$$

$$C_{21} = -\sin L\cos\sigma - \cot L\cos L\sin^2\sigma\cos\sigma$$

According to the Eqs. (1) to (8), the error propagation diagram can be designed and shown as Fig. 2:

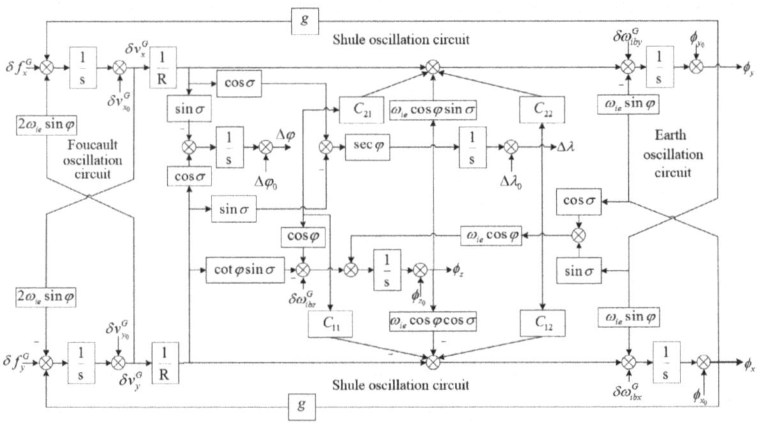

Fig. 2. The error propagation diagram of the grid SINS

According to the definition of grid frame, the conversion relation between the grid azimuth Yaw^G and geographic azimuth Yaw^g can be expressed as:

$$\mathrm{Yaw}^g = \mathrm{Yaw}^G + \sigma \tag{9}$$

therefore,

$$\phi_z^G = \phi_z^g - \delta\sigma \tag{10}$$

where $\delta\sigma = \begin{bmatrix} \frac{\sin\lambda\cos\lambda\cos\varphi}{1-\cos^2\varphi\sin^2\lambda} & \frac{\sin\varphi}{1-\cos^2\varphi\sin^2\lambda} \end{bmatrix} \begin{bmatrix} \delta\varphi \\ \delta\lambda \end{bmatrix}$ [13]. Define the parameter $\Lambda = 1 - \cos^2\varphi\sin^2\lambda$ and $\Omega = \omega_s^2 - \omega_{ie}^2$. To simplify the problems, the harmful acceleration is ignored, and the gyro constant drift is considered as the main error source. Then the azimuth error caused by the gyroscope drift on stationary base can be expressed as:

$$\begin{aligned}
\phi_z^G &= \phi_z^g + \frac{\sin\lambda\cos\lambda\cos\varphi}{1-\cos^2\varphi\sin^2\lambda}\delta\varphi + \frac{\sin\varphi}{1-\cos^2\varphi\sin^2\lambda}\delta\lambda \\
&= \{\frac{\cos^2\lambda\cos\varphi}{\omega_{ie}\Lambda} + [\frac{(\omega_s^2 - \omega_{ie}^2\cos^2\varphi)}{\omega_{ie}\Omega\cos\varphi} - \frac{\omega_s^2\sin^2\varphi}{\omega_{ie}\Omega\cos\varphi\Lambda}]\cos\omega_{ie}t \\
&+ \frac{\omega_s^2\sin\lambda\cos\lambda\cos\varphi}{\omega_{ie}\Omega\Lambda}\sin\omega_{ie}t + \\
&[\frac{\omega_{ie}\sin^2\varphi}{\Omega\cos\varphi} + \frac{\omega_{ie}\sin^2\varphi}{\Omega\cos\varphi\Lambda}]\cos\omega_s t - \frac{\omega_s\sin\lambda\cos\lambda\cos\varphi}{\Omega\Lambda}\sin\omega_s t\}\varepsilon_x \\
&+ \{-\frac{\cos\varphi\sin\varphi}{\Lambda} * t + \frac{\sin\lambda\cos\lambda\, sin\,\varphi\cos\varphi}{\omega_{ie}\Lambda} \\
&+ [\frac{\omega_{ie}^2\sin\varphi\cos^2\varphi - \omega_s^2\sin\varphi}{\omega_{ie}\Omega\cos\varphi} - \frac{\omega_s^2\sin^3\varphi}{\omega_{ie}\Omega\cos\varphi\Lambda}]\sin\omega_{ie}t \\
&- \frac{\omega_s^2\sin\lambda\cos\lambda\sin\varphi\cos\varphi}{\omega_{ie}\Omega\Lambda}\cos\omega_{ie}t \\
&- [\frac{\omega_{ie}^2\sin\varphi\cos^2\varphi - \omega_s^2\sin\varphi}{\omega_s\Omega\cos\varphi} - \frac{\omega_s^2\sin\varphi - \omega_{ie}^2\cos^2\varphi\sin\varphi}{\omega_s\Omega\cos\varphi\Lambda}]\sin\omega_s t \\
&+ \frac{\omega_{ie}\sin\lambda\cos\lambda\sin\varphi\cos\varphi}{\Omega\Lambda}\cos\omega_s t\}\varepsilon_y \\
&+ \{-\frac{\sin^2\varphi}{\Lambda} * t - \frac{\sin\lambda\cos\lambda\cos^2\varphi}{\omega_{ie}\Lambda} \\
&+ [\frac{\omega_s^2 - \omega_{ie}^2\cos^2\varphi}{\omega_{ie}\Omega} + \frac{\omega_s^2\sin^2\varphi}{\omega_{ie}\Omega\Lambda}]\sin\omega_{ie}t + \frac{\omega_s^2\sin\lambda\cos\lambda\cos^2\varphi}{\omega_{ie}\Omega\Lambda}\cos\omega_{ie}t \\
&- [\frac{\omega_{ie}^2\sin^2\varphi}{\omega_s\Omega} + \frac{\omega_{ie}^2\sin^2\varphi}{\omega_s\Omega\Lambda}]\sin\omega_s t - \frac{\omega_{ie}\sin\lambda\cos\lambda\cos^2\varphi}{\Omega\Lambda}\cos\omega_s t\}\varepsilon_z
\end{aligned} \tag{11}$$

where $[\varepsilon_x \quad \varepsilon_y \quad \varepsilon_z]$ is the gyroscope drift. And ignore all the oscillation errors, the azimuth misalignment can be rewritten as:

$$\begin{aligned}\phi_z^G = &\frac{\cos^2\lambda\cos\varphi}{\omega_{ie}\Lambda}\varepsilon_x - \frac{\sin\lambda\cos\lambda\sin\varphi\cos\varphi}{\omega_{ie}\Lambda}\varepsilon_y + \frac{\sin\lambda\cos\lambda\cos^2\varphi}{\omega_{ie}\Lambda}\varepsilon_z \\ &+ (\frac{\cos\varphi\sin\varphi}{\Lambda}\varepsilon_y + \frac{\sin^2\varphi}{\Lambda}\varepsilon_z) * t\end{aligned} \tag{12}$$

According to the error propagation diagram and azimuth error propagation function, the longitude error is no longer open-loop and still relative to $\sec\varphi$. As the latitude gets higher, the longitude error and the error amp of longitude will be more and more serious. According to the Eq. (12), because of the grid angle, the $\sec\varphi$ or $\tan\varphi$ has not been contained in the grid azimuth error function, which can avoid the error amp of azimuth. However, there are also some new problems about the grid azimuth. The grid azimuth will accumulate with time, and some technologies should be employed to suppress the azimuth errors.

A simulation is conducted to verify the propagation characteristics of the grid SINS errors. The latitude and longitude of the initial position P is set as $(75°\text{ N},\ 126°\text{ E})$. The gyroscopes drifts are set as 0.05 degrees per hour and the accelerometer biases are set as 6×10^{-5} g. The initial pitch and roll error are set as $3'$ and the azimuth error is set as $6'$. The simulation results are shown as Fig. 3.

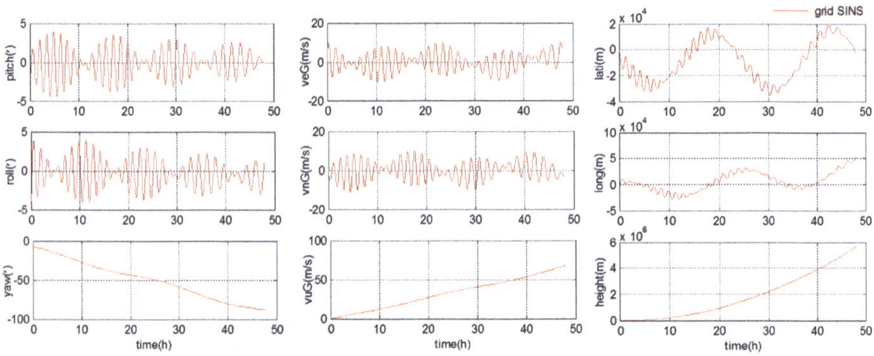

Fig. 3. The navigation errors of the grid SINS

As shown in the Fig. 3, the grid SINS has three kind of oscillation errors and accumulated azimuth and position errors, which are consistent with the theoretical analysis. By analyzing the traditional algorithm about the grid SINS in polar regions, the main factors, which have main impacts to the azimuth accuracy, are shown as follows:

Firstly, the meridian convergence in polar regions leads to the error amp of longitude. The traditional grid SINS mechanization choose the position coordinates in ECEF frame to be updated, but the longitude still participate in the update of main navigation parameters, such as the grid angle σ, the angular velocity ω_{eG}^{G} and transition matrix C_e^g. To improve the accuracy of grid SINS, a novel grid SINS mechanization without longitude is proposed. And based on the novel mechanization, the azimuth will be suppressed with the help of acoustic navigation information.

Secondly, compared with the traditional north-oriented SINS, the grid azimuth errors will accumulate with time. If operating without correction for a long time, the grid SINS may contain a large azimuth misalignment angle, which may lead to a serious nonlinear of the grid SINS. So a nonlinear filter with large misalignment angle is designed for the grid SINS.

Besides, the azimuth error establish relations with the velocity and position observed information by compass effect. With the increase of latitude, the decrease of compass effect leads to the decrease of observability of azimuth errors and it will be more difficult to estimate azimuth errors. Therefore, the rotation modulation is employed by the grid SINS in this paper to restrain the error of IMU and improve the observation and estimation accuracy of azimuth errors, which may also improve the performance of navigation system in polar regions.

3 The Novel Grid SINS Mechanization Without Longitude

3.1 The Grid SINS Mechanization

The vectors $\boldsymbol{e}_{Gx}\boldsymbol{e}_{Gy}\boldsymbol{e}_{Gz}$ is set as the unit vectors along the axes of the G frame. The vectors $\boldsymbol{e}_{gx}\boldsymbol{e}_{gy}\boldsymbol{e}_{gz}$ is set as the unit vectors along the axes of the g frame. The vectors $\boldsymbol{e}_{ex}\boldsymbol{e}_{ey}\boldsymbol{e}_{ez}$ is set as the unit vectors along the axes of the e frame. The vectors $\boldsymbol{e}_{Gy}$ and $\boldsymbol{e}_{ey}$ can be expressed as:

$$\boldsymbol{e}_{Gy} = \sin\sigma\boldsymbol{e}_{ex} + \cos\sigma\boldsymbol{e}_{ey} \tag{13}$$

$$\boldsymbol{e}_{ey} = \cos\lambda\boldsymbol{e}_{ex} - \sin\varphi\sin\lambda\boldsymbol{e}_{ey} + \cos\varphi\sin\lambda\boldsymbol{e}_{ez} \tag{14}$$

where φ is the latitude and λ is the longitude.

The vectors $\boldsymbol{e}_{Gy}$ and $\boldsymbol{e}_{ey}$ are perpendicular to one another:

$$\boldsymbol{e}_{Gy} \cdot \boldsymbol{e}_{ey} = \sin\sigma\cos\lambda - \cos\sigma\sin\varphi\sin\lambda = 0 \tag{15}$$

$$\sin^2\sigma + \cos^2\sigma = 1 \tag{16}$$

The position is represented as the coordinate $\boldsymbol{R}^e = \begin{bmatrix} x & y & z \end{bmatrix}^T$ in e frame.

$$\begin{cases} x = R_{Nh}\cos\varphi\cos\lambda \\ y = R_{Nh}\cos\varphi\sin\lambda \\ z = [R_N(1 - e^2) + h]\sin\varphi \end{cases} \tag{17}$$

The trigonometric functions of σ can be obtained as:

$$\cos\sigma = \frac{[R_N(1-e^2)+h]x}{\sqrt{[R_N(1-e^2)+h]^2x^2+y^2z^2}} \qquad \sin\sigma = \frac{yz}{\sqrt{[R_N(1-e^2)+h]^2x^2+y^2z^2}} \tag{18}$$

The transformation matrix between the g frame and the G frame can be obtained as:

$$\boldsymbol{C}_g^G = \begin{bmatrix} \cos\sigma & -\sin\sigma & 0 \\ \sin\sigma & \cos\sigma & 0 \\ 0 & 0 & 1 \end{bmatrix} \tag{19}$$

and the transformation matrix between the e frame and the g frame can be obtained as:

$$\boldsymbol{C}_e^g = \begin{bmatrix} -\frac{x}{\sqrt{x^2+y^2}} & \frac{y}{\sqrt{x^2+y^2}} & 0 \\ -\sin\varphi\frac{y}{\sqrt{x^2+y^2}} & -\sin\varphi\frac{x}{\sqrt{x^2+y^2}} & \cos\varphi \\ \cos\varphi\frac{y}{\sqrt{x^2+y^2}} & \cos\varphi\frac{x}{\sqrt{x^2+y^2}} & \sin\varphi \end{bmatrix} \tag{20}$$

The differential equation of grid SINS attitude is:

$$\dot{\boldsymbol{C}}_b^G = \boldsymbol{C}_b^G(\boldsymbol{\omega}_{Gb}^b \times) \tag{21}$$

where $\boldsymbol{\omega}_{Gb}^G$ is the angular velocity of G frame relative to b frame in G frame;

$$\boldsymbol{\omega}_{Gb}^b = \boldsymbol{\omega}_{ib}^b - \boldsymbol{C}_G^b(\boldsymbol{\omega}_{ie}^G + \boldsymbol{\omega}_{eG}^G) \tag{22}$$

$$\boldsymbol{\omega}_{ie}^G = \boldsymbol{C}_g^G\boldsymbol{\omega}_{ie}^g = \begin{bmatrix} \cos\sigma & -\sin\sigma & 0 \\ \sin\sigma & \cos\sigma & 0 \\ 0 & 0 & 1 \end{bmatrix}\begin{bmatrix} 0 \\ \omega_{ie}\cos\varphi \\ \omega_{ie}\sin\varphi \end{bmatrix} \tag{23}$$

The horizontal component of the angular velocity $\boldsymbol{\omega}_{eG}^G$ can be obtained from the angular velocity $\boldsymbol{\omega}_{eg}^g = \begin{bmatrix} \omega_{egx}^g & \omega_{egy}^g & \omega_{egz}^g \end{bmatrix}^T$ as:

$$\begin{bmatrix} \omega_{eGx}^G \\ \omega_{eGy}^G \end{bmatrix} = \begin{bmatrix} \cos\sigma & -\sin\sigma \\ \sin\sigma & \cos\sigma \end{bmatrix}\begin{bmatrix} \omega_{egx}^g \\ \omega_{egx}^g \end{bmatrix} = \begin{bmatrix} (\frac{1}{R_{Mh}} - \frac{1}{R_{Nh}})\sin\sigma\cos\sigma & -(\frac{\cos^2\sigma}{R_{Mh}} + \frac{\sin^2\sigma}{R_{Nh}}) \\ (\frac{\sin^2\sigma}{R_{Mh}} + \frac{\cos^2\sigma}{R_{Nh}}) & -(\frac{1}{R_{Mh}} - \frac{1}{R_{Nh}})\sin\sigma\cos\sigma \end{bmatrix}\begin{bmatrix} v_x^G \\ v_y^G \end{bmatrix} \tag{24}$$

And the vertical component of $\boldsymbol{\omega}_{eG}^{G}$ can be obtained as:

$$\omega_{eGz}^{G} = \omega_{egz}^{g} - \dot{\sigma} = \frac{x\cos\varphi}{\sqrt{x^2\sin^2\varphi + y^2}}[(\frac{1}{R_{Nh}} - \frac{1}{R_{Mh}})\sin\sigma\cos\sigma v_x^G + \frac{\cos^2\sigma}{R_{Mh}} + \frac{\sin^2\sigma}{R_{Nh}} v_y^G] \tag{25}$$

Define the parameters:

$$\begin{cases} \frac{1}{R_{xG}} = \frac{\sin^2\sigma}{R_{Mh}} + \frac{\cos^2\sigma}{R_{Nh}} \\ \frac{1}{R_{yG}} = \frac{\cos^2\sigma}{R_{Mh}} + \frac{\sin^2\sigma}{R_{Nh}} \end{cases} \quad \begin{cases} \frac{1}{\tau_{fG}} = (\frac{1}{R_{Mh}} - \frac{1}{R_{Nh}})\sin\sigma\cos\sigma \\ \kappa = \frac{x\cos\varphi}{\sqrt{x^2\sin^2\varphi + y^2}} \end{cases} \tag{26}$$

The angular velocity $\boldsymbol{\omega}_{eG}^{G}$ can be expressed as:

$$\boldsymbol{\omega}_{eG}^{G} = \begin{bmatrix} \omega_{eGx}^{G} \\ \omega_{eGy}^{G} \\ \omega_{eGz}^{G} \end{bmatrix} = \begin{bmatrix} \frac{1}{\tau_{fG}} & -\frac{1}{R_{yG}} \\ \frac{1}{R_{xG}} & -\frac{1}{\tau_{fG}} \\ \frac{\kappa}{\tau_{fG}} & \frac{\kappa}{R_{yG}} \end{bmatrix} \begin{bmatrix} v_x^G \\ v_y^G \end{bmatrix} \tag{27}$$

The differential equation of grid SINS velocity is:

$$\dot{\boldsymbol{V}}^G = \boldsymbol{C}_b^G \boldsymbol{f}^b - \left(2\boldsymbol{\omega}_{ie}^G + \boldsymbol{\omega}_{eG}^G\right) \times \boldsymbol{V}^G + \boldsymbol{g}^G \tag{28}$$

The differential equation of grid SINS position is:

$$\dot{\boldsymbol{R}}^{\mathrm{e}} = \boldsymbol{C}_{\mathrm{G}}^{\mathrm{e}} \boldsymbol{V}^{\mathrm{G}} = (\boldsymbol{C}_{\mathrm{e}}^{\mathrm{g}})^T (\boldsymbol{C}_{\mathrm{g}}^{\mathrm{G}})^T \boldsymbol{V}^{\mathrm{G}} \tag{29}$$

The latitude can be updated by:

$$\sin\lambda = \frac{z}{R_N(1 - e^2) + h} \tag{30}$$

As shown in the Eqs. (18) to (30), in the novel grid SINS mechanization the longitude information no longer participates in the navigation update and the error amp of longitude caused by meridian convergence will not be introduced into the system and influence the navigation performance. As a convenience to the display, the position coordinate in e frame can be converted to longitude as:

$$\cos\lambda = \frac{x}{\sqrt{x^2 + y^2}} \quad \sin\lambda = \frac{y}{\sqrt{x^2 + y^2}} \tag{31}$$

3.2 The Single-Axis Rotation Modulation for Grid SINS

To restrain the measurement errors of the IMU and improve the observation of the azimuth misalignment, z axis is chosen as the rotation modulation axis and the IMU rotates at an angular speed ω_s.

The rotation frame, s frame, is coincide exactly with the b frame at the inertial moment. During the rotation modulation, the transformation matrix between the s frame and b frame can be obtained as:

$$\boldsymbol{C}_b^s = \begin{bmatrix} \cos\omega t & \sin\omega t & 0 \\ -\sin\omega t & \cos\omega t & 0 \\ 0 & 0 & 1 \end{bmatrix} \tag{32}$$

The outputs of IMU with rotation modulation are $\boldsymbol{\omega}_{is}^s$ and $\boldsymbol{f}_{is}^s$. After the following transformation, $\boldsymbol{\omega}_{ib}^b$ and $\boldsymbol{f}_{ib}^b$ can be employed to update the navigation information:

$$\begin{aligned} \boldsymbol{\omega}_{ib}^b &= (\boldsymbol{C}_b^s)^T \boldsymbol{\omega}_{is}^s + \boldsymbol{\omega}_{sb}^b \\ \boldsymbol{f}_{ib}^b &= (\boldsymbol{C}_b^s)^T \boldsymbol{f}_{is}^s + \boldsymbol{f}_{sb}^b \end{aligned} \tag{33}$$

where $\boldsymbol{\omega}_{sb}^b = [0 \quad 0 \quad \omega_s]^T$ and $\boldsymbol{f}_{sb}^b = 0$.

4 The Nonlinear Grid SINS/Acoustic Navigation Integration Filter Model

4.1 The Error Function of the Grid SINS with Large Misalignment Angle

The calculated navigation frame G^c can be obtained from the navigation frame G after three rotation as follows:

$$Ox_G y_G z_G \xrightarrow[\phi_z]{\text{绕 } -z_G \text{ 轴}} Ox_G^c y_G^1 z_G \xrightarrow[\phi_x]{\text{绕 } x_G^c \text{ 轴}} Ox_G^c y_G^2 z_G^c \xrightarrow[\phi_y]{\text{绕 } y_G^2 \text{ 轴}} Ox_G^c y_G^c z_G^c$$

The transformation can be expressed as:

$$\boldsymbol{C}_G^{G^c} = \begin{bmatrix} c\phi_y c\phi_z + s\phi_x s\phi_y s\phi_z & -c\phi_y s\phi_z + s\phi_x s\phi_y c\phi_z & -c\phi_x s\phi_y \\ c\phi_x s\phi_z & c\phi_x c\phi_z & s\phi_x \\ -s\phi_x c\phi_y s\phi_z + s\phi_y c\phi_z & -s\phi_y s\phi_z - s\phi_x c\phi_y c\phi_z & c\phi_x c\phi_y \end{bmatrix} \tag{34}$$

where s represent the sine function and c represent the cosine function.

The angle velocity between the G^c and G frame can be expressed as:

$$\omega_{GG^c}^{G^c} = \begin{bmatrix} c\phi_y & 0 & -c\phi_x s\phi_y \\ 0 & 1 & s\phi_x \\ s\phi_y & 0 & c\phi_x c\phi_y \end{bmatrix} \dot{\boldsymbol{\phi}} = \boldsymbol{C}_\phi \dot{\boldsymbol{\phi}} \tag{35}$$

and

$$\omega_{GG^c}^{G^c} = (\boldsymbol{I} - \boldsymbol{C}_G^{G^c})\omega_{iG}^G + \boldsymbol{C}_G^{G^c}\delta\omega_{iG}^G - \boldsymbol{C}_{\mathrm{b}}^{G^c}\delta\omega_{ib}^b \tag{36}$$

The following relationships exist between some navigation parameters [13]:

$$\delta\sigma = \boldsymbol{C}_1\delta\boldsymbol{P}\ \delta\sigma = \boldsymbol{C}_{R2\sigma}\delta\boldsymbol{R}^e\quad \delta\boldsymbol{P} = \boldsymbol{C}_{R2P}\delta\boldsymbol{R}^e$$
$$\delta\omega_{iG}^G = \delta\omega_{ie}^G + \delta\omega_{eG}^G\quad \delta\omega_{ie}^G = \boldsymbol{C}_5\delta\sigma + \boldsymbol{C}_6\delta\boldsymbol{P}\ \ \delta\omega_{eG}^G = \boldsymbol{C}_2\delta\boldsymbol{P} + \boldsymbol{C}_3\delta\boldsymbol{V} + \boldsymbol{C}_4\delta\sigma$$

Therefore, $\delta\omega_{iG}^G$ can be expressed as:

$$\delta\omega_{iG}^G = \boldsymbol{C}_{R\_\omega}\delta\boldsymbol{R}^e + \boldsymbol{C}_{V\_\omega}\delta\boldsymbol{V} \tag{37}$$

where $\boldsymbol{C}_{R\_\omega} = \boldsymbol{C}_5\boldsymbol{C}_{R2\sigma} + \boldsymbol{C}_6\boldsymbol{C}_{R2P} + \boldsymbol{C}_2\boldsymbol{C}_{R2P} + \boldsymbol{C}_4\boldsymbol{C}_{R2\sigma}, \boldsymbol{C}_{V\_\omega} = \boldsymbol{C}_3$.

The propagation characteristic function of the grid azimuth with large misalignment angle can be obtained as:

$$\begin{aligned}\dot{\boldsymbol{\phi}} &= \boldsymbol{C}_\phi^{-1}[(\boldsymbol{I} - \boldsymbol{C}_G^{G^c})\hat{\omega}_{iG}^G + \boldsymbol{C}_G^{G^c}(\boldsymbol{C}_{R\_\phi}\delta\boldsymbol{R}^e + \boldsymbol{C}_{V\_\phi}\delta\boldsymbol{V}) - \boldsymbol{C}_{\mathrm{b}}^{G^c}\delta\omega_{ib}^b]\\ &= \boldsymbol{C}_\phi^{-1}(\boldsymbol{I} - \boldsymbol{C}_G^{G^c})\hat{\omega}_{iG}^G + \boldsymbol{C}_{V\_\phi}\delta\boldsymbol{V} + \boldsymbol{C}_{R\_\phi}\delta\boldsymbol{R}^e - \boldsymbol{C}_\phi^{-1}\boldsymbol{C}_{\mathrm{b}}^{G^c}\delta\omega_{ib}^b\end{aligned} \tag{38}$$

where $\boldsymbol{C}_{R\_\phi} = \boldsymbol{C}_\phi^{-1}\boldsymbol{C}_G^{G^c}\boldsymbol{C}_{R\_\phi}$, $\boldsymbol{C}_{V\_\phi} = \boldsymbol{C}_\phi^{-1}\boldsymbol{C}_G^{G^c}\boldsymbol{C}_{V\_\phi}$.

The velocity error equation can be calculated as [13]:

$$\delta\dot{\boldsymbol{V}}^G = [\boldsymbol{I} - (\boldsymbol{C}_G^{G^c})^T]\boldsymbol{C}_b^{G^c}\tilde{\boldsymbol{f}}^b + \boldsymbol{C}_{V\_V}\delta\boldsymbol{V}^G + \boldsymbol{C}_{R\_V}\delta\boldsymbol{R}^e + \boldsymbol{C}_b^{G^c}\delta\boldsymbol{f}^b \tag{39}$$

where $\boldsymbol{C}_{R\_V} = (\boldsymbol{V}^G\times)(2\boldsymbol{C}_6 + \boldsymbol{C}_2)\mathrm{C}_{R2P} + (\boldsymbol{V}^G\times)(2\boldsymbol{C}_5 + \boldsymbol{C}_4)\boldsymbol{C}_{R2\sigma}$, $\boldsymbol{C}_{V\_V} = (\boldsymbol{V}^G\times)\boldsymbol{C}_3 - (2\omega_{ie}^G + \omega_{eG}^G)\times$.

The position error equation can be calculated as [13]:

$$\delta\dot{\mathbf{R}}^e = \boldsymbol{C}_{V\_R}\delta\boldsymbol{V}^G + \boldsymbol{C}_{R\_R}\delta\boldsymbol{R}^e \tag{40}$$

where $\boldsymbol{C}_{V\_R} = \boldsymbol{C}_G^e$, $\boldsymbol{C}_{R\_R} = -\boldsymbol{C}_G^e(\boldsymbol{V}^G\times)\boldsymbol{C}_{R2\delta\theta}$.

4.2 The Integration Filter Model

The attitude error, velocity error and position error are chosen as the main states of integration filter to be estimated. The Doppler velocity log (DVL) provides velocity and the ultra-short base line position system (USBL) provides position measurements. According to the derived error equations of grid SINS in Sect. 4.1, the filter's dynamic model can be described as:

$$\begin{bmatrix} \dot{\phi} \\ \delta\dot{V}^G \\ \delta\dot{\mathbf{R}}^e \\ \dot{\varepsilon}_c^b \\ \dot{\nabla}_c^b \\ \delta\dot{V}_{DVL}^m \end{bmatrix} = \begin{bmatrix} C_\phi^{-1}(I - C_G^{G^c})\hat{\omega}_{iG}^G + C_{V\_\phi}\delta V + C_{R\_\phi}\delta R^e - C_\phi^{-1}C_b^{G^c}\delta\omega_{ib}^b \\ [I - (C_G^{G^c})^T]C_b^{G^c}\hat{f}^b + C_{V\_V}\delta V^G + C_{R\_V}\delta R^e + C_b^G\delta f_{ib}^b \\ C_{V\_R}\delta V^G + C_{R\_R}\delta R^e \\ 0 \\ 0 \\ -\delta V_{DVL}^m/\tau_V + \mathbf{w}_V \end{bmatrix} \tag{41}$$

where $\dot{\varepsilon}_c^b$ and $\dot{\nabla}_c^b$ are the constant errors of the gyroscopes and accelerometers, respectively and $\delta\dot{V}_{DVL}^m$ is the measurement error of the DVL; τ_V is the correlation time of Markov process, and $\mathbf{w}_V$ is the zero-mean Gaussian white noise.

The filter's observation model can be described as:

$$Z = \begin{bmatrix} V_{SINS}^G - V_{DVL}^G \\ R_{SINS}^e - R_{USBL}^e \end{bmatrix} = Hx + V \tag{42}$$

where $H = \begin{bmatrix} 0_{3\times3} & I_{3\times3} & 0_{3\times3} & 0_{3\times9} \\ 0_{3\times3} & 0_{3\times3} & I_{3\times3} & 0_{3\times9} \end{bmatrix}$ and V is the observation noise.

5 Experiments and Performance Analysis

To validate the performance of the method which suppresses the azimuth errors of the grid SINS in polar regions, some simulation experiments are designed. The simulation time is 4 h. Gyroscopes drifts are 0.05°/h and the accelerometer bias are 6×10^{-5} g. The grid SINS update frequency is 100 Hz. The DVL update frequency is 1 Hz and the velocity measurement error is less than 0.3 m/s. The USBL update frequency is 1 Hz. The range measurement error is less than 5 m and the angle measurement error is less than 1.5°.

Two integrated methods are discussed in this section. The algorithm 1 is the grid SINS/DVL/USBL integrated method without the rotation modulation and the algorithm 2 is the grid SINS/DVL/USBL integrated method with the rotation.

Simulation 1: the initial pitch and roll error is set as 3′ and the azimuth error is set as 6′. The algorithm 1 and algorithm 2 are employed to restrain the SINS errors respectively, and the navigation errors are depicted as Figs. 4 and 5.

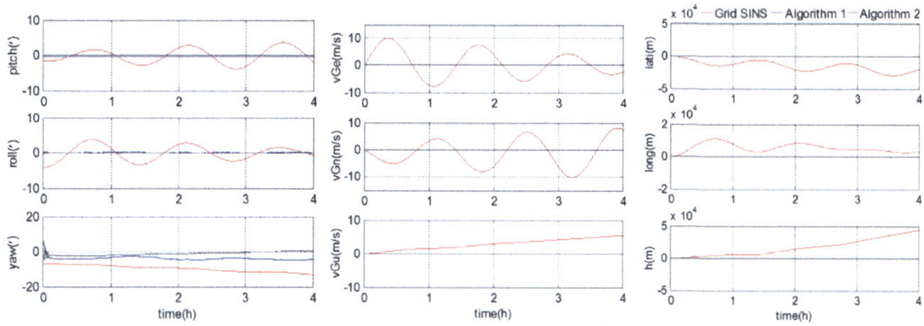

Fig. 4. The attitude errors of the grid SINS with little misalignment angle

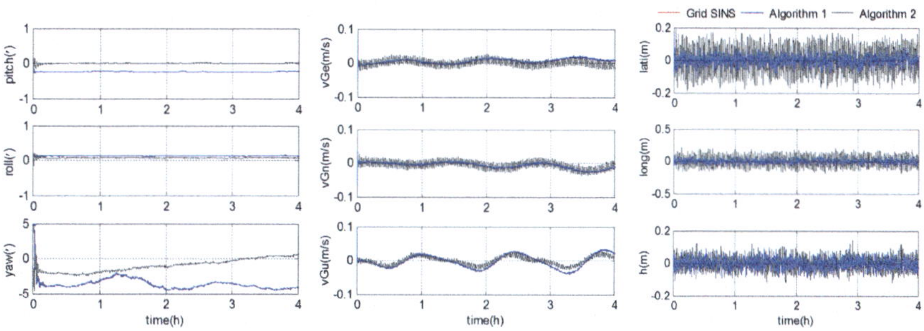

Fig. 5. The velocity errors of the grid SINS with little misalignment angle

Simulation 2: the initial pitch and roll error is set as 3′ and the azimuth error is set as 60′. The algorithm 1 and algorithm 2 are employed to restrain the SINS errors respectively, and the navigation errors are depicted as Fig. 6.

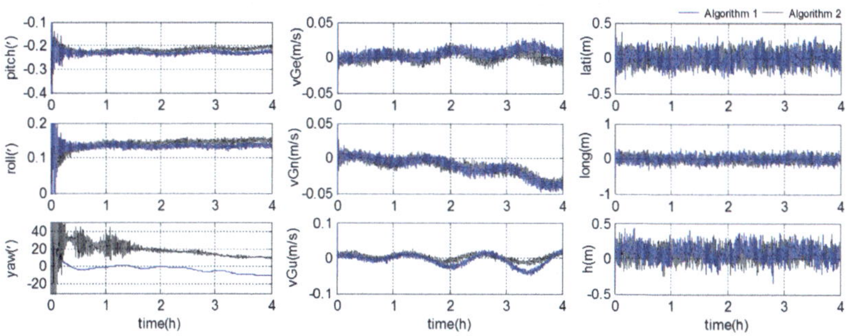

Fig. 6. The navigation errors of the grid SINS with large misalignment angle

As shown in the Figs. 4 and 5, with the little inertial azimuth error, the introduction of acoustic navigation information and the integrated filter can restrain the navigation errors effectively. Algorithm 1 and algorithm 2 has almost the same accuracy of pitch, roll, velocity and position. However, compared to the algorithm 1, the algorithm 2 with rotation modulation has higher azimuth accuracy with a lower state error. The reason is that the rotation modulation can restrain the IMU low-frequency errors and improve the observation of azimuth error, which can improve the accuracy of filter estimation.

As shown in the Fig. 6, when the SINS has a large azimuth error, the nonlinear integrated filter with acoustic navigation information can estimate and suppress the azimuth errors in some degree. And the algorithm 2 has a higher accuracy than the algorithm 1 because of the rotation modulation. The filter without rotation modulation convergences more slowly and has larger state error than the filter with rotation modulation. The filter with rotation modulation can restrain the azimuth errors more effectively. And the pitch, roll, velocity and position error of the grid SINS can also be restrained.

As the simulation results shown, the proposed integrated technology with rotation modulation method can suppress the azimuth errors of the grid SINS in polar regions.

6 Conclusion

In this paper, the error propagation characteristic of the grid azimuth is analyzed and the azimuth error suppression method for the grid SINS in polar regions is proposed based on the rotation modulation and acoustic navigation assistance. Firstly, the longitude error of grid SINS is influenced by meridian convergence and no longer open-loop, and the grid azimuth is influenced by the longitude errors. Therefore, a novel grid SINS mechanization without longitude is proposed. Secondly, the grid azimuth errors accumulate with time, which may lead to the large azimuth error and nonlinearity of SINS. And the nonlinear integrated filter model is designed with the acoustic navigation reference to estimate and correct the navigation error, in which the azimuth error is included. Thirdly, the observation of polar azimuth errors will decrease with latitude escalating. The rotation modulation is employed to improve the observation and estimation accuracy of the integrated filter, which can help suppress the azimuth errors effectively. Finally, the simulation experiments are conducted to validate the performance of the proposed method. The simulation results shown that the azimuth and other navigation errors can be estimated and suppressed by the nonlinear filter effectively.

References

1. Anthony, S., Ayanna, M.H., Britney, S., Matthew, M.: Design and development of an under-ice autonomous underwater vehicle for use in Polar regions. In: Oceans, pp. 1–6. IEEE (2014). https://doi.org/10.1109/oceans.2014.7002991
2. Cheng, J.H., Wang, T.D., Guan, D.X., Li, M.L.: Polar transfer alignment of shipborne SINS with a large misalignment angle. Measur. Sci. Technol. **27** (2016). https://doi.org/10.1088/0957-0233/27/3/035101
3. Marco, B., Antonio, P., Pere, R., Enrica, Z.: Introduction to the special section on navigation, control, and sensing in the marine environment. Ann. Rev. Control **40**, 127–128 (2015). https://doi.org/10.1016/j.arcontrol.2015.09.007

4. Zhou, J., Nie, X.M., Lin, J.: A novel laser Doppler velocimeter and its integrated navigation system with strapdown inertial navigation. Opt. Laser Technol. **64**, 319–323 (2014). https://doi.org/10.1016/j.optlastec.2014.06.001
5. Chen, G., Li, K., Wang, W., Li, P.: A novel redundant INS based on triple rotary inertial measurement units. Measur. Sci. Technol. **27** (2016). https://doi.org/10.1088/0957-0233/27/10/105102
6. Naser, E., Ahmed, Y.: Inertial sensors technologies for navigation applications: state of the art and future trends. Satell. Navig. **1**, 1–21 (2020). https://doi.org/10.1186/s43020-019-0001-5
7. Zhou, Q., Yue, Y.Z., Zhang, X.D., Tian, Y.: Indirect grid inertial navigation mechanization for transpolar aircraft. J. Chinese Inertial Technol. **22**(1), 18–17 + 66 (2014). https://doi.org/10.13695/j.cnki.12-1222/o3.2014.01.005
8. Wu, Y.X., He, C., Liu, G.: On inertial navigation and attitude initialization in polar areas. Satell. Navig. **1**, 1–6 (2020). https://doi.org/10.1186/s43020-019-0002-4
9. Li, Q., Ben, Y.Y., Yu, F., Sun, F.: Transversal strapdown INS based on reference ellipsoid for vehicle in the Polar Region. IEEE Trans. Veh. Technol. **65**(9), 7791–7795 (2016). https://doi.org/10.1109/TVT.2015.2497713
10. Zhou, Q., Qin, Y.Y., Fu, Q.W., Yue, Y.Z.: Grid mechanization in inertial navigation systems for transpolar aircraft. J. Northwestern Polytech. Univ. **31**(2), 210–217 (2013). CNKI:SUN:XBGD.0.2013-02-013
11. Li, Q., Ben, Y.Y., Sun, F., Huo, L.: Transversal strapdown INS and damping technology for marine in polar region. In: IEEE/ION Position, Location and Navigation Symposium – PLANS (2014). https://doi.org/10.1109/plans.2014.6851511
12. Zhao, C., Wu, W., Lian, J.: Research on rotating modulation inertial navigation system error characteristics simulation method in polar area. In: Proceedings of 2014 IEEE Chinese Guidance, Navigation and Control Conference (2014). https://doi.org/10.1109/cgncc.2014.7007605
13. Kang, Y.Y., Zhao, L., Cheng, J.H., Wu, M.Y., Fan, X.L.: A novel grid SINS/DVL integrated navigation algorithm for marine application. Sensors **18**(2) (2018). https://doi.org/10.3390/s18020364
14. Zhao, L., Kang, Y.Y., Cheng, J.H., Wu, M.Y.: A fault-tolerant polar grid SINS/DVL/USBL integrated navigation algorithm based on the centralized filter and relative position measurement. Sensors **19**(18) (2019). https://doi.org/10.3390/s19183899

X-Ray Pulsar-Based Navigation Method Verification by Insight-HXMT Satellite Data

Dapeng Zhang[1(✉)], Yidi Wang[2], Wei Zheng[2], and Mingyu Ge[3]

[1] State Key Laboratory of Astronautic Dynamics, Xi'an 710043, China
zhangdapeng12@163.com
[2] College of Aerospace Science and Engineering, National University of Defense Technology, Changsha 410073, China
[3] Institute of High Energy Physics, CAS, Beijing 100049, China

Abstract. X-ray pulsar-based navigation is a novel spacecraft autonomous navigation method. At present, pulsar-based navigation technology has entered the stage of space experiment verification. In recent years, many countries are carrying out X-ray pulsar-based navigation experiments. China's first large-scale X-ray astronomical observation satellite—Insight-HXMT also regards the X-ray pulsar-based navigation technology verification as one of its scientific research tasks. Its X-ray sensors' total effective area is near 1 m^2 in an energy range from 1 to 250 keV. Therefore it could be used to verify X-ray pulsar-based navigation methods. The Insight-HXMT satellite has been sent into space for 2 years, and part of observation data has been open to researchers. This paper processed the raw observation data from the satellite's high-energy X-ray telescope and the navigation results are present. At the same time, the results are compared with the external measurement values, so the accuracy of the experiment is analyzed.

Keywords: X-ray pulsar-based navigation · Experiment verification · Insight-HXMT satellite · Autonomous navigation

1 Introduction

Since Downs put forward the deep space navigation concept based on pulsar radio signal observation in 1974, the researches for pulsar-based navigation technology has been carried out for more than 40 years [1]. The aerospace powerful countries represented by the United States conducted space experiments for pulsar-based navigation technology for a long time [2]. Especially, In June 2017, the SEXTANT project entered the engineering stage, and its pulsar observation device NICER was launched and successfully installed on the International Space Station. NICER is composed of 56 groups of Wolter-I grazing incidence lens and silicon drift sensors (SDD). The geometric area of the detector is as high as 6400 cm^2 [3]. In a flight test in November 2017, by observing a series of millisecond pulsars, the navigation accuracy of the test is better than 10 km [4]. This is the first published results of X-ray pulsar-based navigation in the world using the measured data. It marks the basic integrity of the X-ray pulsar-based navigation theories, and the X-ray pulsar-based navigation technology has entered into a new stage of space experiments.

J. Sun et al. (Eds.): CSNC 2020, LNEE 650, pp. 656–665, 2020.
https://doi.org/10.1007/978-981-15-3707-3_61

Keeping up with international development trends, China also actively conducted X-ray pulsar-based navigation experiments. In September 2016, China's Tiangong-2 space laboratory was launched, and it was equipped with a payload which is gamma-ray burst polarization detector (POLAR). By observed Crab pulsars data from POLAR detector, Zheng conducted a pulsar navigation space experiment, and they preliminarily explored the feasibility of pulsar navigation [5]. In November 2016, the XPNAV-1 satellite developed by China Academy of Space Technology was successfully sent into the orbit. It mainly researched the high-flow pulsar detection technology [6].

On June 15, 2017, China's first large astronomy satellite "Insight" hard X-ray modulation telescope (Insight-HXMT) was successfully launched. It uses a direct demodulation imaging method to achieve wide-band X-ray imaging surveys, and carries out high-precision fixed-point celestial bodies observation such as black hole binaries, and studies their multi-band X-ray fast optical changes [7]. In addition, the Insight-HXMT satellite's scientific research team also regards pulsar-based navigation as one of its important research projects.

2 Profile for Insight-HXMT

2.1 Insight-HXMT's Payloads

The Insight-HXMT satellite is a sub-chamber design. The payload is located at the top and the service cabin is located at the bottom. The satellite's total mass is about 2.7 tons and it operates in a near-Earth orbit with an altitude of 550 km and an inclination of 43°. The designed life is about 4 years. Insight-HXMT's pointing accuracy is 0.1° (3σ), attitude measurement accuracy is 0.01° (3σ), and attitude stability is 0.005° per second [7].

Insight-HXMT satellite carries 4 kinds of payloads: high-energy X-ray telescope (HE, photon energy range: 20–250 keV), medium-energy X-ray telescope (ME, photon energy range: 5–30 keV), and low-energy X-ray telescope (LE, photon Energy range: 1–15 keV) and Space Environment Monitor (SEM).

Among them, the HE detector is a detector array composed of 18 groups of NaI/CsI composite crystals. Each group's diameter is 19 cm, and the area is 283.5 cm^2. So the whole geometric area of HE detector is about 5100 cm^2. After a special arrangement design, the field of view of the HE detector is about $5.7° \times 5.7°$ [7].

2.2 Insight-HXMT's Scientific Data

At present, the Insight-HXMT satellite has operated in orbit for more than two years, and plentiful observation data is open access to scholars or engineers around the world. It could be downloaded on the Insight-HXMT's official website.

In the downloaded files, two list files are provided in the root directory. "FileList.fits" contains all file information of archived observations (name, path, occupied space, type, md5 check code, etc.). The "ExpoList.xml" file gives the exposure list for the observation. The root directory archives an exposure data folder that stores photon arrival events and engineering data during the exposure cycle. In addition, two auxiliary data files "ACS" and "AUX" are provided in the directory. The "ACS" directory stores satellite attitude data and orbit data. The "AUX" folder is used to archive auxiliary data generated by ground stations, such as data quality reports, good time interval, etc., which may be used in future data analysis. More than 20 types of observation data are provided under the exposure data folder which is related to three main payloads (HE, ME, and LE). Considering the HE detector's large effective area, In this paper, we select to use "HE-Evt" type data for X-ray pulsar-based navigation experiments [8].

3 Methods for X-Ray Pulsar-Based Navigation

3.1 Principle

X-ray pulsar is a kind of neutron star that rotates at a certain period. A periodic statistical regularity of the X-ray photon flux is existed when the X-ray radiation beam sweeps through the spacecraft. Making use of the periodic characteristics of X-ray pulsar signals and calculating the signal phase difference between the spacecraft and the solar system's centroid (SSB), a basic measurement value of pulsar-based navigation is obtained. Converting the phase difference into distance, it can be known that the basic measurement value of X-ray pulsar-based navigation is the projection of the spacecraft's distance between spacecraft and SSB in the direction of the pulsar.

More measurement value vectors can be obtained by observing one or more pulsars in different orientations sequentially or parallelly. Combined with the spacecraft's precise orbit dynamics model, the best estimate of the spacecraft's position and velocity can be obtained. A schematic diagram of this principle is shown in Fig. 1.

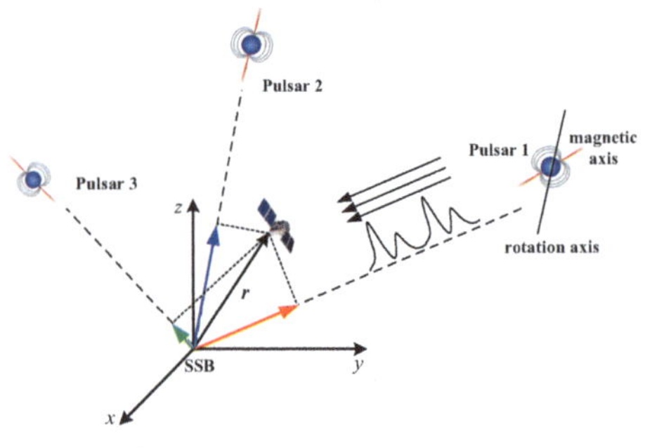

Fig. 1. Principle of X-ray pulsar-based navigation

3.2 Process of Measurement Value

The spacecraft moves relative to the inertial system in orbit, so the pulsar signal received at the spacecraft contains Doppler frequency shift. However, the pulsar X-ray signal is extremely weak, and generally cannot obtain the waveform of the pulsar signal instantaneously. This Doppler frequency shift characteristic is implicit in the statistical characteristics of the photon flux. The procedure of achieving navigation measurement value from this statistical characteristic is called "dynamic pulsar signal processing", and it is one of the focuses and difficulties of pulsar-based navigation technology.

The pulse arrival time conversion equation from the spacecraft to the SSB can be expressed as [9]

$$t_{SSB} = t_{sc} + \Delta_{R\odot} + \Delta_{other} \tag{1}$$

where t_{SSB} is the time when pulse arrives at SSB. t_{SC} is the time when pulse arrives at the detector. $\Delta_{R\odot}$ is Roemer delay. It reflects the geometric distance between the spacecraft and SSB. Δ_{other} includes Einstein delay, Solar system objects Shapiro delay, clock error correction, etc. They are small delay terms.

Roemer delay is related with the spacecraft position. If the position vector of the spacecraft relative to the SSB is $\vec{r}_{SC}$, the pulsar direction unit vector is $\vec{n}$, and the speed of light is expressed as c, then the Roemer delay is

$$\Delta_{R\odot} = \frac{\vec{n} \cdot \vec{r}_{SC}}{c} \tag{2}$$

The signal phase at SSB can be predicted using a pulsar timing model, which is shown in Eq. (3). Due to the limitation of data accuracy and observation interval, the higher order terms of the pulsar frequency derivative are difficult to be estimated. For high flux pulsars, usually only the second derivative of the signal frequency can be obtained within a month. Phase prediction truncation errors and all unmodeled errors are called timing noise.

$$\phi(t_{SSB}) = \phi_0 + \sum_{n=1}^{3} \frac{\upsilon^{(n-1)}}{n!} \cdot (t_{SSB} - t_0)^n + o \tag{3}$$

where, $\upsilon^{(n)}$ presents the derivatives of the pulsar signal frequency usually from pulsar ephemeris. t_0 is the epoch of the parameters in the pulsar ephemeris, and ϕ_0 is the phase of the pulsar signal at time t_0.

Substituting Eqs. (1) and (2) into Eq. (3), and ignoring higher-order truncation errors, the phase of the pulsar signal at the spacecraft t_{sc} is

$$\phi_{SC} = \phi_0 + \sum_{n=1}^{3} \frac{\upsilon^{(n-1)}}{n!} \left(t_{sc} + \frac{\vec{n} \cdot \vec{r}_{SC}}{c} + \Delta_{other} - t_0 \right)^n \tag{4}$$

Divided $\vec{r}_{SC}$ into two parts, the estimated position and the estimated position error, namely $\vec{r}_{SC} = \vec{r}_{est} + \delta\vec{r}_{SC}$. So the phase ϕ_{SC} at the spacecraft can also be expressed as

$$\begin{gathered} \phi_{SC} = \phi_0 + \phi_{est} + \delta\phi \\ \phi_{est} = \sum_{n=1}^{3} \frac{\upsilon^{(n-1)}}{n!} \left(t_{sc} + \frac{\vec{n} \cdot \vec{r}_{est}}{c} + \Delta_{other} - t_0 \right)^n \end{gathered} \tag{5}$$

where $\delta\phi$ is the phase error to be estimated caused by the position error $\delta\vec{r}_{SC}$, and it can be considered to use a one-variable polynomial approximation fitting.

$$\delta\phi = \delta f_0 + \delta f_1 \cdot (t_{sc} - t_{sc0}) + \cdots \tag{6}$$

According to the literature [3, 10], the parameters can be estimated using the maximum likelihood method. If only the constant and first-order terms are fitted, the maximum likelihood estimator can be expressed as

$$\left(\delta\hat{f}_0, \delta\hat{f}_1\right) = \underset{\delta f_0, \delta f_1}{\arg\ \max} \sum_{k=1}^{N} \log \lambda[\phi_0 + \phi_{est} + \delta f_0 + \delta f_1 \cdot (t_{sc} - t_{sc0})] \tag{7}$$

where $\lambda(\phi), \phi \in [0, 1)$ denotes the empirical profile of pulsar signal.

Substituting $\delta\hat{f}_0$ and $\delta\hat{f}_1$ into Eqs. (5) and (6), we can get the estimated phase $\hat{\phi}_{SC}$ at the spacecraft at time t_{sc}. According to Eq. (4), the phase at SSB is $\phi(t_{SC})$. Then the final navigation measurement value is

$$z = \left[\hat{\phi}_{SC} - \phi(t_{SC})\right] \cdot c/f \tag{8}$$

where f is the frequency of the pulsar signal at time t_{sc}, which can be propagated by the pulsar ephemeris.

3.3 Navigation Filter

Unscented Kalman Filter (UKF) transmits state mean and variance by sigma points. Compared with the EKF method, its accuracy is higher in theory and does not need to calculate the complicated Jacobian matrix. Therefore UKF is more convenient for practical use. Suppose the pulsar navigation nonlinear system is

$$\begin{cases} X_{k+1} = f(X_k) + w_k \\ Z_k = h(X_k) + v_k \end{cases} \tag{9}$$

where X_k is system state variables. w_k is system noise. Z_k is measurement value. v_k is measurement noise.

The system equation is

$$f(X) = \begin{bmatrix} \vec{v} \\ \vec{a} \end{bmatrix} \qquad \vec{a} = \vec{a}_{TB} + \vec{a}_{NS} + \vec{a}_{ST} + \vec{a}_{SP} + \vec{a}_{DR} + \vec{a}_{T} \tag{10}$$

where $\vec{a}_{TB}$ is the two-body gravitational acceleration. $\vec{a}_{NS}$ is the nonspherical perturbation acceleration. $\vec{a}_{ST}$ is the perturbation acceleration due to the solid tide of the sun and the moon. $\vec{a}_{SP}$ is the perturbation acceleration of the solar pressure. $\vec{a}_{DR}$ is the perturbation acceleration of atmosphere drag. $\vec{a}_T$ is third body gravitational perturbation acceleration.

The observation equation is

$$h(X) = \vec{n} \cdot \vec{r}_{SC} + c \cdot \Delta_{other} \tag{11}$$

4 X-Ray Pulsar-Based Navigation Verification

4.1 The Data Used for Experiments

The information of Insight-HXMT's measured data which is used for X-ray pulsar-based navigation experiments is shown in Table 1.

Table 1. The data that is used for the experiments

Observation ID		P0101299003
Proposal ID		P0101299
Target		Crab
Direction	Right ascension [°]	83.633
	Declination [°]	22.0145
Start observation time [UTC]		2017-08-31 10:00:17.0
Interval [s]		230121.0

Decompress the downloaded original data package, and find out the "HE-Evt" type data. "HE-Evt" is a standard fits format file. In it, the data space named "Events" is the time event of X-ray photons. The information in "Events" includes "Time" (Arrival time), "Det_ID" (Detector ID), "Channel" (Photon energy channel), "Pulse_Width" (Electronic pulse width) and "ACD" (Anti-Compliance Detector Information).

"Time" is the time of photon arrival to HE detector, which is expressed in the style of accumulated seconds relative to a reference time. The reference time is available in the header of the fits file. The reference time is represented by two key words: "MJDREFI" and "MJDREFF". Their values are MJDREFI = 55927, MJDREFF = 7.6601852E-4, and the time system is TT.

The "ACS" folder contains a "Orbit" type data with fits format, which contains the orbital information of the satellite in the observation interval, which can be used to estimate the accuracy of the navigation experiments' precision.

4.2 The Procedure of the Experiments

The basic process of pulsar navigation test verification is shown in Fig. 2.

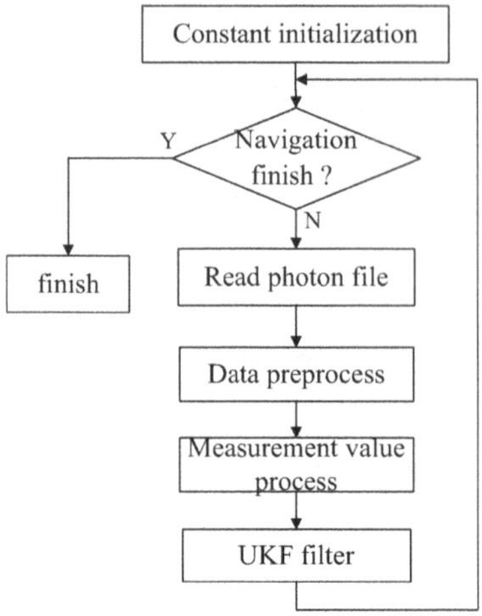

Fig. 2. Procedure for X-ray pulsar-based navigation

"Constant initialization" mainly accomplishes reading the orbit dynamics model constants, the pulsar ephemeris, the pulsar empirical profile, initial values of the UKF filter and so on. "Photon data preprocessing" mainly removes invalid photon arrival events by judging whether the satellite is entry into the Earth and Moon shadow or not. "Measurement processing" and "UKF filter" can be referred in Sects. 3.2 and 3.3 in detail.

In this paper only Crab pulsar is observed, and its ephemeris is shown in Table 2. The initial value of the UKF filter is shown in Table 3.

Table 2. Ephemeris of Crab

Name	Value
Epoch MJD [TDB]	57992.165481
Right ascension [°]	83.6332167
Declination [°]	22.0144639
Frequency F0 [Hz]	29.639022542
First derivative of frequency F1 [Hz s^{-1}]	−3.6867E-10
Second derivative of frequency F2 [Hz s^{-2}]	0

Table 3. Initial value for UKF filter

Filtering step	1800 s
System noise	Position: 1.0E-6 m Velocity: 1.0E-9 m/s
Measurement noise	5 km
Initial covariance	Position: 20 km Velocity: 20 m/s
Initial epoch	178797663 s
Initial orbital elements (J2000)	1905608.64032341 m, −5115172.24880413 m, 4243757.37600491 m, 6009.590021381 m/s, 4103.661265044 m/s, 2298.571735840 m/s

where, the initial orbit error in Table 3 is consistent with the covariance matrix, that is, the position error is 20 km and the Velocity error is 20 m/s. The epoch reference time of the initial orbital value is described in Sect. 4.1.

4.3 Results of the Experiments

The initial position error of the orbit is 20 km and the velocity error is 20 m/s. Without any measurement, by the orbit propagation, the error will quickly diverge with time. By pulsar observation, the orbital divergence will be effectively controlled, and the results are shown in Fig. 3, in which y axis is logarithmic axis.

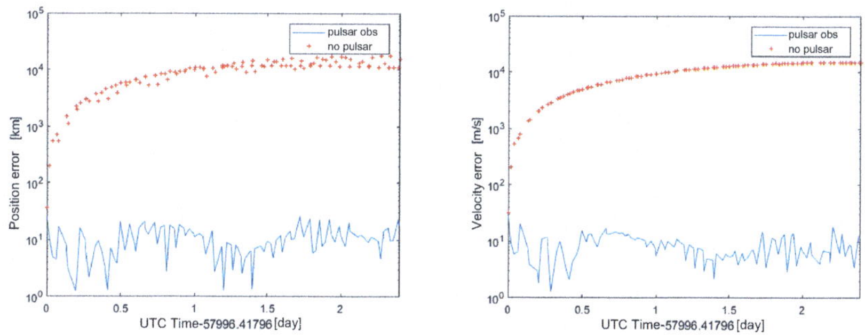

(a) (b)

Fig. 3. The error keeping by X-ray pulsar observation

From Fig. 3, by pulsar observation, the orbit error can keep with about 10.88 km and the velocity error can keep with about 8.62 m/s. However, with no pulsar observation, the orbit error diverges to more than 10,000 km.

With a data interval about 2.5 days, when only Crab pulsars are observed, the accuracy of the spacecraft's autonomous navigation is shown in Fig. 4.

Figure 4(a) shows the results of the 3-axis position error, and Fig. 4(b) shows the results of the 3-axis velocity error. The solid blue line in Fig. 4 represents the difference between the optimal estimated value of the UKF filter and the true orbit of the Insight-HXMT satellite. The red dashed line represents the 3 times sigma limit of the navigation error. The accuracy is also shown in Table 4.

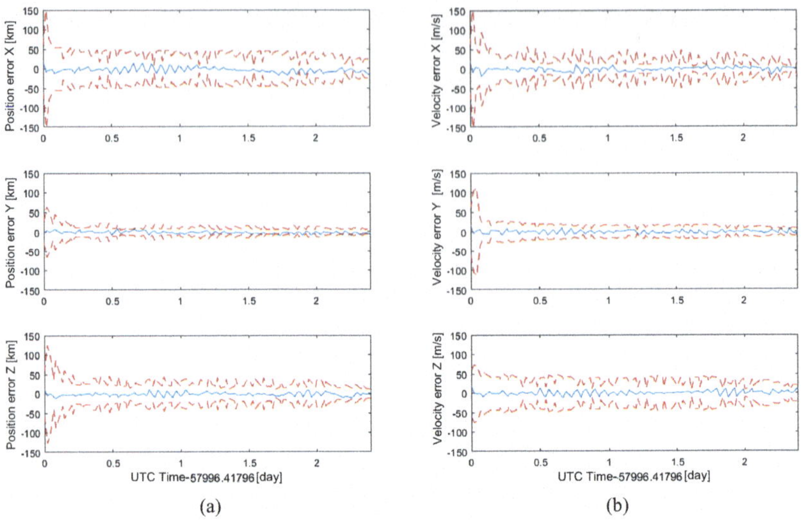

Fig. 4. The navigation error in the three axes

Table 4. Navigation error in the experiment

Error		Max	Min	Mean	3δ
Position [km]	X	18.1	0.0997	6.789	12.713
	Y	10.2	0.0676	3.623	4.218
	Z	11.6	0.00206	4.437	9.988
Velocity [m/s]	X	19.6	0.00869	4.326	9.264
	Y	16.9	0.0805	3.760	5.955
	Z	20.6	0.0232	4.863	11.075

In the table, 3δ is the average of the 3 times sigma error limit corresponding to the estimated variance in UKF filter. It can be seen that the navigation accuracy of the pulsar-based navigation experiment is about 10 km.

5 Discussion

This paper analyzes the principles of X-ray pulsar navigation, navigation measurement value processing methods, and uses the measured data of China's Insight-HXMT astronomical satellite to carry out X-ray pulsar navigation experiments. The results showed that:

(1) By Crab pulsar observations the divergence of orbital errors can be effectively suppressed. The measured data experiments prove that when observation single pulsar for navigation the Kalman filter still has a good convergence performance.

(2) The navigation accuracy in this experiment is about 10 km, which is comparable to the accuracy of the pulsar navigation test carried out by NISER on the International Space Station.

However there are still some issues need to be resolved in the follow-up works.

(1) The onboard high-stable clock in the Insight-HXMT satellite has been corrected by the GPS second pulse, so the clock error is not included in the published data.
(2) Limited by current X-ray detector technology, the accuracy of X-ray pulsar-based navigation is about ten kilometers level. This accuracy has little advantage in near-Earth space applications. In deep space applications, the convergence performance of the filter, especially when fewer pulsars can be observed, requires further research.

Acknowledgements. This work made use of the data from the HXMT mission, a project funded by China National Space Administration (CNSA) and the Chinese Academy of Sciences (CAS).

References

1. Downs, G.S.: Interplanetary Navigation Using Pulsation Radio Sources, Washington (1974)
2. Keith, G.: NICER: neutron star interior composition explorer and SEXTANT. NASA Technical report, GSFC.CPR.6621.2012, 1–32 (2012)
3. Winternitz, L.M.B., Hassouneh, M.A., Mitchell, J.W., et al.: X-ray pulsar navigation algorithms and testbed for SEXTANT. In: Aerospace Conference (2015)
4. Mitchell, J.W., Winternitz, L.B., et al.: SEXTANT X-ray pulsar navigation demonstration: initial on-orbit results. NASA Technical report, AAS 18-155, pp. 1–12 (2018)
5. Zheng, S., Ge, M., Han, D., et al.: Experiment of pulsar navigation based on Tiangong-2 POLAR. Sci. Sinica **9**(47), 0995051–0995059 (2017)
6. Huang, L., Shuai, P., Zhang, X., et al.: XPNAV-1 Satellite timing data analysis and pulse proile recovery. Chinese Space Sci. Technol. **37**(3), 1–10 (2017)
7. HXMT Core Scientific Proposal Call for White Paper. Institute of High Energy Physics, CAS
8. The HXMT Data Reduction Guide v2.01. Institute of High Energy Physics, Chinese Academy of Sciences
9. Lorimer, D.R., Kramer, M.: Handbook of Pulsar Astronomy. Cambridge Uiversity Press, Cambridge (2005)
10. Emadzadeh, A.A., Speyer, J.L.: Navigation in Space by X-ray Pulsars. Springer, London (2011)

High Precision Integrated Navigation Algorithms for Weak Observation of Quasi-One-Dimensional Application and on Track Test

Peng Li(✉)

Institute of Spacecraft System Engineering, Beijing, People's Republic of China
Pengli.richard@foxmail.com

Abstract. The intelligent and safe operation requirements of rail application put forward higher requirements to navigation and positioning technology. For example, strong multipath, weak signal and fast scene-switching, tunnel scene, occlusion scene in the station and so on. In the application of integrated navigation, rail application belongs to the issue of weak observation (slow acceleration and deceleration), rapid environment change (need adaptive modelling), quasi one-dimensional motion. At the same time, the application of rail application requires strict reliability and maintenance free. The above factors have blocked the application of integrated navigation system in rail application for a long time. In this paper, an integrated navigation algorithm based on adaptive modelling is proposed for the application under the weak observation and quasi one-dimensional constraints. By modelling the IMU signal, combined with the quasi one-dimensional motion constraints, the automatic initial alignment under the condition of weak observation is realized. Through the real application of train on operation, the model parameters are trained and tested. From the results of on track test, under the condition that the hardware platform of commercial MEMS devices with gyro stability less than 20°/h and accelerometer stability less than 30 mg without wheel speedometer and other auxiliary signals, the algorithm can achieve better than 3% * distance error accumulation under the conditions of insufficient visible satellites, no satellites and weak signals. Furthermore, the algorithm is expected to achieve 0.1% * distance error accumulation when the subsequent high-precision MEMS devices are used. The research results of this paper can be popularized in China's rail application, including railway, subway, rail mine transportation, rail port transportation and other fields.

Keywords: Integrated navigation · Railway navigation · IMU

1 Introduction

Chinese rail transportation has made considerable progress. The operating mileage of railways and subways has exceeded 120,000 km and 30,000 km, respectively. There is also a large number of rail needs such as automatic mine transportation and automatic rail port transportation. With the completion of BDS construction in the year of 2020,

J. Sun et al. (Eds.): CSNC 2020, LNEE 650, pp. 666–678, 2020.
https://doi.org/10.1007/978-981-15-3707-3_62

BDS can support the users in various industries in system level. The unique characteristics of the BDS, such as the integration of communication and navigation, full coverage, and satellite-based enhancement, create a broad imagination space for the future BDS support for rail transportation [1].

The demand for intelligent and safe operation of rail transportation places higher requirements on navigation and positioning technology under complicated operating environment, such as various strong multipath, weak signal scenes, fast switching scenes, tunnel scenes, occlusion in the stations, etc. The excellent anti-interference ability of the inertial navigation system make it possible that the integrated navigation technology should get great application in rail navigation.

However, the rail vehicles are different from ordinary road vehicles. They are confronted with special problems such as slow start, blocked in the station, locomotive non-turning, and are subject to weak signal or no satellites such as canopies, buildings, and tunnels on the way. In integrated navigation applications, rail transportation belongs to weak observations (slow acceleration and deceleration), rapid environmental changes (requires adaptive modeling), and quasi-one-dimensional motion problems. At the same time, rail applications have strict requirements for reliability and maintenance-free performance. The above factors have long restricted the application of integrated navigation systems in rail transportation.

MEMS (Micro Electro Mechanical Systems) is made of Micro-electro-machined inertial sensors and has extremely low size, weight and power consumption [5]. The integrated systems including MEMS have the advantages of convenient installation and maintenance free, and is much suitable for the rail application after inertial error modeling [5].

Aiming at high-precision applications under weak observations and quasi-one-dimensional constraints, this paper proposes an integrated navigation algorithm based on adaptive modeling. Modeling and adaptive filtering for severe train occlusion, tunnel and other scenarios are processed. At the same time, combined with the quasi-one-dimensional motion constraint, automatic initial alignment under weak observation conditions is achieved. This is very important for the commercialization and maintenance-free operation of the locomotive integrated navigation system. Through real locomotive on track test, the model parameters were trained and tested.

2 System Algorithm Description

2.1 Algorithm Architecture

As far as the actual on track test is concerned, the proportion of pure GNSS that cannot be solved normally in southern regions railway is as high as 50%. For rail applications, a tightly coupled integrated navigation structure is used, and its block diagram is shown in Fig. 1 [2]. In the tightly coupled structure, the observed values of GNSS such as pseudorange ρ, pseudorange rate $\dot{\rho}$, carrier phase ϕ and carrier phase rate $\dot{\phi}$ are sent to the integrated filter as filter observations.

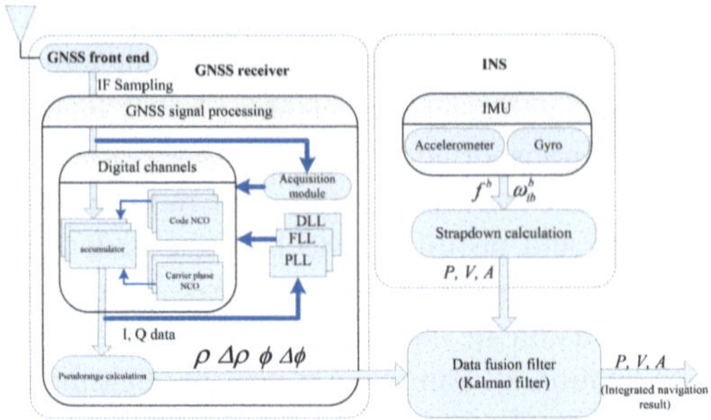

Fig. 1. GNSS/IMU tight-integrated algorithm architecture

2.2 INS State Vector Construction

The INS system state vector $\boldsymbol{X}_{INS}$ contains 24 error components [2],

$$\begin{aligned}\boldsymbol{X}_{INS} = [\,&\delta r_N \quad \delta r_E \quad \delta r_D \quad \delta v_N \quad \delta v_E \quad \delta v_D \quad \delta\psi_N \quad \delta\psi_E \quad \delta\psi_D \quad \nabla_{bx} \quad \nabla_{by} \quad \nabla_{bz} \\ &\nabla_{fx} \quad \nabla_{fy} \quad \nabla_{fz} \quad \varepsilon_{bx} \quad \varepsilon_{by} \quad \varepsilon_{bz} \quad \delta g_N \quad \delta g_E \quad \delta g_D \quad \delta A_N \quad \delta A_E \quad \delta A_D\,]^T\end{aligned} \tag{1}$$

These include 4 types as follows: 9 navigation parameters, where N, E, D represents three directions of the north east and down, position error: δr_N, δr_E, δr_D (unit: m), speed error: δv_N, δv_E, δv_D (unit: m/s), and attitude error: $\delta\psi_N$, $\delta\psi_E$, $\delta\psi_D$ (unit: rad).

9 IMU error parameters, where x, y, z represent the three axes of the IMU: accelerometer deviation: ∇_{bx}, ∇_{by}, ∇_{bz}, accelerometer scale factor: ∇_{fx}, ∇_{fy}, ∇_{fz}, gyro deviation: ε_{bx}, ε_{by}, ε_{bz}. 3 gravity model errors: δg_N, δg_E, δg_D, 3 antenna position errors: δA_N, δA_E, δA_D.

The INS system state vector determines that $\boldsymbol{\Phi}_{INS}$ is a 24×24 matrix. To facilitate the setting of these 576 parameters, the 24 state parameters are divided into 8 groups of 3 each. So the form of $\boldsymbol{\Phi}_{INS}$ can be written as 64 3×3 matrix as follows,

$$\boldsymbol{\Phi}_{INS} = I_{24\times 24} + F_{24\times 24} \tag{2}$$

Where,

$$F_{24\times 24} = \begin{bmatrix} f_{1\ 1} & f_{1\ 2} & \cdots & f_{1\ 24} \\ f_{2\ 1} & f_{2\ 2} & \cdots & f_{2\ 24} \\ \vdots & \vdots & \ddots & \vdots \\ f_{24\ 1} & f_{24\ 2} & \cdots & f_{24\ 24} \end{bmatrix} = \begin{bmatrix} F_{11} & F_{12} & \cdots & F_{18} \\ F_{21} & F_{22} & \cdots & F_{28} \\ \vdots & \vdots & \ddots & \vdots \\ F_{81} & F_{82} & \cdots & F_{88} \end{bmatrix} \tag{3}$$

$$F_{11} = \begin{bmatrix} f_{11} & f_{12} & f_{13} \\ f_{21} & f_{22} & f_{23} \\ f_{31} & f_{32} & f_{33} \end{bmatrix}, \cdots, F_{88} = \begin{bmatrix} f_{22\ 22} & f_{22\ 23} & f_{22\ 24} \\ f_{23\ 22} & f_{23\ 23} & f_{23\ 24} \\ f_{24\ 22} & f_{24\ 23} & f_{24\ 24} \end{bmatrix} \tag{4}$$

3 Observation Model and Preprocessing

3.1 Observation Model

Considering that rail transportation is often in a state without a fixed GNSS solution, the GNSS pseudorange is selected, and the pseudorange rate and the inertial navigation system estimate the pseudorange, and the difference between the pseudorange rate is used as the observation measurement.

$$Z_{Tight} = \begin{bmatrix} Z_\rho \\ Z_{\dot{\rho}} \end{bmatrix} \tag{5}$$

Where,

$$Z_\rho = \begin{bmatrix} \delta\rho_1 \\ \vdots \\ \delta\rho_n \end{bmatrix} = \begin{bmatrix} \rho_{1(GNSS)} - \rho_{1(INS)} \\ \vdots \\ \rho_{n(GNSS)} - \rho_{n(INS)} \end{bmatrix} \tag{6}$$

$$Z_{\dot{\rho}} = \begin{bmatrix} \delta\dot{\rho}_1 \\ \vdots \\ \delta\dot{\rho}_n \end{bmatrix} = \begin{bmatrix} \dot{\rho}_{1(GNSS)} - \dot{\rho}_{1(INS)} \\ \vdots \\ \dot{\rho}_{n(GNSS)} - \dot{\rho}_{n(INS)} \end{bmatrix} \tag{7}$$

ρ_{INS} and $\dot{\rho}_{INS}$ can be obtained as,

$$\rho_{INS}^k = \left\| r_s^k - r \right\| = \sqrt{\left(r_{s,x}^k - r_x\right)^2 + \left(r_{s,y}^k - r_y\right)^2 + \left(r_{s,x}^k - r_y\right)^2} \tag{8}$$

$$\dot{\rho}_{INS}^k = \frac{\left(r_{s,x}^k - r_x\right)\left(v_{s,x}^k - v_x\right) + \left(r_{s,y}^k - r_y\right)\left(v_{s,y}^k - v_y\right) + \left(r_{s,z}^k - r_z\right)\left(v_{s,z}^k - v_z\right)}{\rho_{INS}^k} \tag{9}$$

Where r_s^k and v_s^k are the position and velocity of the k satellite in the Earth-fixed system. The observation equation is as follows,

$$H_{Tight} = \begin{bmatrix} H_\rho \\ H_{\dot{\rho}} \end{bmatrix} \tag{10}$$

Where,

$$H_\rho = \begin{bmatrix} \underbrace{\begin{matrix} \frac{\partial\rho^1}{\partial r_x} & \frac{\partial\rho^1}{\partial r_y} & \frac{\partial\rho^1}{\partial r_z} & 0 & \cdots & 0 \\ \vdots & \vdots & \vdots & 0 & \ddots & 0 \\ \frac{\partial\rho^n}{\partial r_x} & \frac{\partial\rho^n}{\partial r_y} & \frac{\partial\rho^n}{\partial r_z} & 0 & \cdots & 0 \end{matrix}}_{24INS\ States} & \underbrace{\begin{matrix} 1 & 0 \\ \vdots & \vdots \\ \vdots & \vdots \\ 1 & 0 \end{matrix}}_{2GNSS\ States} \end{bmatrix} \tag{11}$$

$$H_{\dot{\rho}} = \left[\underbrace{\begin{matrix} \frac{\partial \dot{\rho}^1}{\partial r_x} & \frac{\partial \dot{\rho}^1}{\partial r_y} & \frac{\partial \dot{\rho}^1}{\partial r_z} & \frac{\partial \dot{\rho}^1}{\partial v_x} & \frac{\partial \dot{\rho}^1}{\partial v_y} & \frac{\partial \dot{\rho}^1}{\partial v_z} & 0 & \cdots & 0 \\ \vdots & \vdots & \vdots & \vdots & \vdots & \vdots & 0 & \ddots & 0 \\ \frac{\partial \dot{\rho}^n}{\partial r_x} & \frac{\partial \dot{\rho}^n}{\partial r_y} & \frac{\partial \dot{\rho}^n}{\partial r_z} & \frac{\partial \dot{\rho}^n}{\partial v_x} & \frac{\partial \dot{\rho}^n}{\partial v_y} & \frac{\partial \dot{\rho}^n}{\partial v_z} & 0 & \cdots & 0 \end{matrix}}_{24INS\ States} \underbrace{\begin{matrix} 1 & 0 \\ \vdots & \vdots \\ 1 & 0 \end{matrix}}_{2GNSS\ States} \right] \tag{12}$$

Under tight coupling, the systematic error clock error and frequency error of GNSS can also affect the correction of pseudorange and pseudorange rate. Differentiating the position and speed from Eqs. (8) and (9) gives:

$$\left.\frac{\partial \rho^k}{\partial r_e}\right|_{e=x,y,z} = \frac{r_e - r_{s,e}^k}{\rho_{INS}^k} \tag{13}$$

$$\left.\frac{\partial \dot{\rho}^k}{\partial r_e}\right|_{e=x,y,z} = \frac{v_e - v_{s,e}^k}{\rho_{INS}^k} + \frac{\left(r_e - r_{s,e}^k\right)\left(v_e - v_{s,e}^k\right)}{\left(\rho_{INS}^k\right)^3} \tag{14}$$

$$\left.\frac{\partial \dot{\rho}^k}{\partial v_e}\right|_{e=x,y,z} = \frac{r_e - r_{s,e}^k}{\rho_{INS}^k} \tag{15}$$

In rail transportation navigation, since the speed basically exists only in the direction of the rolling axis of the carrier system, the speed observation can only select the speed along the pitch axis and yaw axis. It can be assumed that the speed of the GNSS receiver in these two directions is 0, and the speed observation measurement can be easily obtained (16),

$$\begin{bmatrix} \delta v_{\phi}^b \\ \delta v_{\psi}^b \end{bmatrix} = \begin{bmatrix} 0 \\ 0 \end{bmatrix} - \begin{bmatrix} v_{\phi(INS)}^b \\ v_{\psi(INS)}^b \end{bmatrix} \tag{16}$$

3.2 IMU Weak Observation State Processing

In the application of rail transit, although the speed of running vehicles is very high, there is no obvious acceleration, deceleration, turning, fluctuation and other dynamic conditions, which is manifested as obvious weak observation. As a result, the three-dimensional observation on the body coordinate system is not obviously above the noise level, and the observation is weak, which is formed after coupling with the gravity field in the inertial navigation error equation. If the conventional modeling method is used, it will cause the underestimation on the covariance matrix, and the output is only not smooth and ignore the actual movement. Aiming at the problem of weak observation peculiar to rail transportation, this paper proposes to build a window which is often 1 s wide to accumulate the sensor measurement in the body coordinate system, so as to better remove the noise.

As mentioned before, the system noise matrix of the combined filter is adaptively adjusted according to the window accumulation result. Assuming that Q_0 is the noise matrix used in system initialization, the adaptive noise matrix Q in the case of IMU weak dynamic measurement will change with the accumulated peak value of the window, as follows:

$$Q_{arw} = |(W_{\max} - W_{\min})|^2 \times T^{1/2} \times Q_{0,arw} \tag{17}$$

$$Q_{yrw} = |(f_{\max} - f_{\min})|^2 \times T^{1/2} \times Q_{0,yrw} \tag{18}$$

Where Q_{arw} and Q_{yrw} are the angle random walk and the velocity random walk noise matrix of the filter respectively along the three directions of the body coordinate system. $|(W_{\max} - W_{\min})|$ and $|(f_{\max} - f_{\min})|$ represent the difference between the peak-to-peak of gyroscope and accelerometer after window accumulation. T is the length of the time window.

4 Adaptive Scene Modeling Filtering

The traditional Kalman filter (KF) is an estimator for a linear dynamic system. However, rail transportation integrated navigation systems are non-linear systems. For non-linear systems, non-linear filter design methods can be adopted. For rail transit scene modeling, multi-model adaptive estimation (MMAE) is used for modeling and model switching. For each model, an innovation-sequence adaptive estimation (IAE) algorithm can be adopted, and the actual on track data is used to train the parameters of each scene model.

4.1 Multi-model Adaptive Estimation (MMAE)

In the MMAE algorithm, a filter bank containing multiple Kalman filters is processed in parallel. These filters can use different observation noise matrix R or system noise matrix Q. They independently estimate the system state vector $\hat{x}_k$ according to their preset R or Q values, and the total estimated value of the system is obtained by linearly weighting these components. The structural block diagram of MMAE is shown as in Figs. 2 [2].

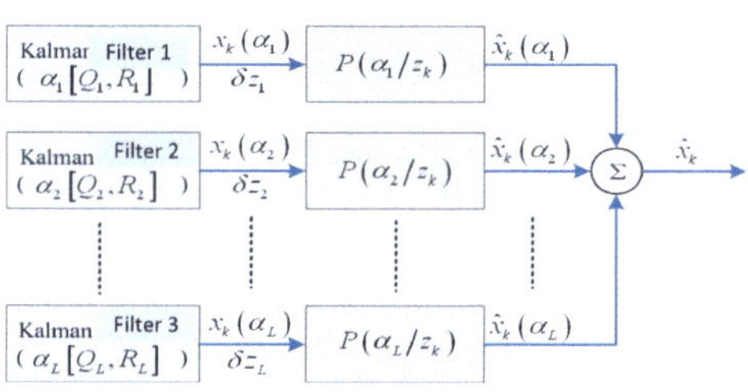

Fig. 2. GNSS/IMU tight-integrated algorithm architecture

Each filter can calculate the posterior probability for each preset hypothesis based on the innovation sequence obtained after receiving the observations. The posterior probability can then be used as a weight for each filter estimator. In this way, the total estimated value of the system can be obtained from the following eqn. [3],

$$\hat{x}_k = \sum_{i=1}^{L} \hat{x}_k(\alpha_i) P(\alpha_i / z_k) \tag{19}$$

Where L is the total number of filters, z_k is the measurement, and α_i is the posterior probability. The filter bank can gradually identify which filter is the best, and the weight of other filters will gradually approach 0.

4.2 Innovation-Sequences Adaptive Estimation (IAE)

Based on the innovation-sequence adaptive estimation (IAE), the parameters R_k and Q_k in each orbit scene are obtained through adaptive estimation of R or Q, as shown in Fig. 3 [2].

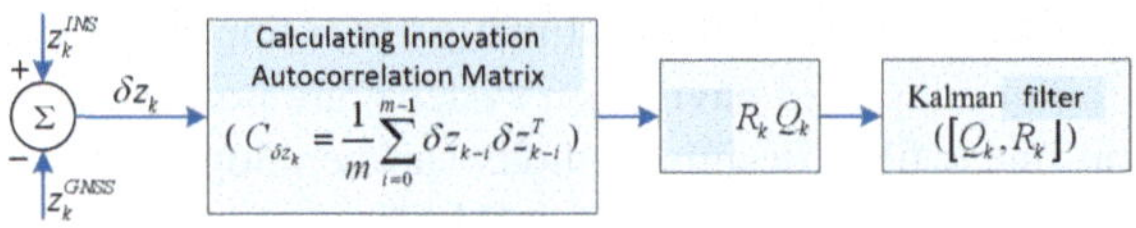

Fig. 3. Innovation-based Adaptive Estimation (IAE) algorithm architecture

The calculation of R_k and Q_k is based on the maximum likelihood (ML) principle. Let the adaptive parameter be α, then the probability density function measured under this adaptive parameter satisfies [6],

$$P(z/\alpha) = \frac{1}{\sqrt{(2\pi)^m |C_{\delta z_k}|}} e^{-\frac{1}{2}\delta z_k^T C_{\delta z_k}^{-1} \delta z_k} \tag{20}$$

The maximum likelihood solution satisfies,

$$\partial P / \partial \alpha = 0 \tag{21}$$

Get,

$$\sum_{j=j_0}^{k} \left[trace\left\{ C_{\delta z_j}^{-1} \frac{\partial C_{\delta z_j}}{\partial \alpha_k} \right\} - \delta z_j^T C_{\delta z_j}^{-1} \frac{\partial C_{\delta z_j}}{\partial \alpha_k} C_{\delta z_j}^{-1} \delta z_j \right] = 0 \tag{22}$$

Then by,

$$C_{\delta z_k} = H_k P_k^- H_k^T + R_k \tag{23}$$

Get,

$$\frac{\partial C_{\delta z_j}}{\partial \alpha_k} = \frac{\partial R_k}{\partial \alpha_k} + H_k \frac{\partial P_k(-)}{\partial \alpha_k} H_k^T \tag{24}$$

Bring in,

$$\frac{\partial P_k(-)}{\partial \alpha_k} = \frac{\partial Q_k}{\partial \alpha_k} \tag{25}$$

Get,

$$\frac{\partial C_{\delta z_j}}{\partial \alpha_k} = \frac{\partial R_k}{\partial \alpha_k} + H_k \frac{\partial Q_{k-1}}{\partial \alpha_k} H_k^T \tag{26}$$

By bringing (24) into (20), the maximum likelihood equation of the adaptive Kalman filter can be obtained.

$$\sum_{j=j_0}^{k} trace\left\{ \left[C_{\delta z_j}^{-1} - C_{\delta z_j}^{-1} \delta z_j \delta z_j^T C_{\delta z_j}^{-1} \right] \left[\frac{\partial R_k}{\partial \alpha_k} + H_k \frac{\partial Q_{k-1}}{\partial \alpha_k} H_k^T \right] \right\} = 0 \tag{27}$$

For self-adaptive R, setting Q is known, $\alpha_i = R_{ii}$, bring into (25), we can get,

$$\begin{aligned} \sum_{j=j_0}^{k} trace\left\{ \left[C_{\delta z_j}^{-1} - C_{\delta z_j}^{-1} \delta z_j \delta z_j^T C_{\delta z_j}^{-1} \right] [I + 0] \right\} = 0 \\ \sum_{j=j_0}^{k} trace\left\{ C_{\delta z_j}^{-1} \left[C_{\delta z_j} - \delta z_j \delta z_j^T \right] C_{\delta z_j}^{-1} \right\} = 0 \end{aligned} \tag{28}$$

Bring into (25), we can get the adaptive estimation equation [4]

$$\hat{R}_k = \hat{C}_{\delta z_k} - H_k P_k^- H_k^T \tag{29}$$

where,

$$\hat{C}_{\delta z_k} = \frac{1}{m} \sum_{i=0}^{m-1} \delta z_{k-i} \delta z_{k-i}^T \tag{30}$$

Where m is the length of self-adaptive window.

Similarly, for self-adaptative Q, we can get the self-adaptive Q estimation equation [4],

$$\hat{Q}_k = \frac{1}{m}\sum_{i=0}^{m-1} \delta x_{k-i}\delta x_{k-i}^T + P_k^+ - \varphi_k P_{k-1}^+ H_k^T \tag{31}$$

5 On Track Testing and Results Analysis

5.1 Experiment Configuration

In order to test the validity of the algorithm, experiments were carried out by means of on-rail locomotive experiments. The experimental vehicle is a K express train in operation, and the main operating area is Fujian, Guangdong with multiple mountains, multiple tunnels and multiple obstructions. After the equipment is installed, the operating frequency of the locomotive is about 5 days a week. The experimental train platform is shown in Fig. 4.

Fig. 4. Train platform carrying out experiment

The integrated navigation hardware system equipped with a dual-frequency RTK GNSS module. The MEMS device uses commercial-grade devices with gyroscope stability less than 20°/h and accelerometer stability less than 30 mg. The integrated navigation algorithm and model filtering processing algorithm are used in embedded ARM of the integrated navigation system. The system obtains commercial CORS network data through the 3G mobile communication module of the train communication equipment, and can achieve RTK centimeter positioning in open areas. The hardware platform is shown in Fig. 5.

a. b.

Fig. 5. Hardware system a. Train communication and navigation system b. Integrated navigation module

5.2 Experiment Result on Track

After training with on track data, the device performed IAE-based scene modeling and parameter extraction, as well as MMEA scene mode recognition and handover processing for each scene. After three months of data training, the algorithm was solidified, and data from three consecutive on track tasks in March were noted. The results are shown in Figs. 6, 7 and 8.

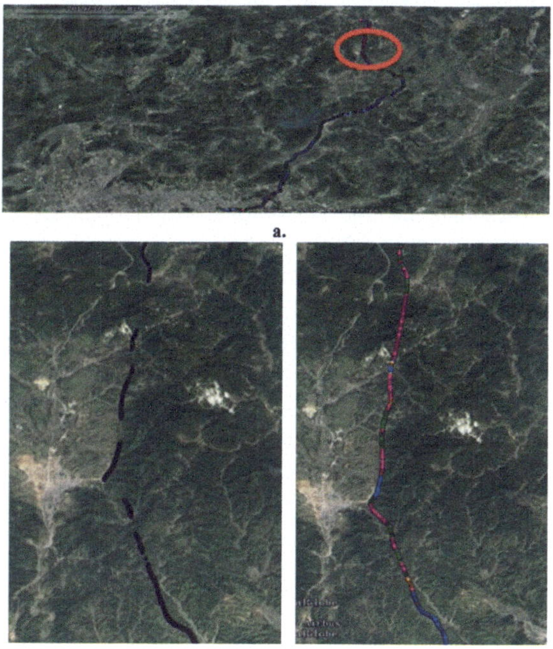

a.

b. c.

Fig. 6. Experimental results on March 17th a. Whole journey results b. GNSS results in occluded scene c. Integrated navigation results in occluded scene

Figure 6 shows the results of the K88 train from Dongguan to Heyuan on March 17th, 2019. Figure 6a shows that the line has a total length of about 3.5 h, of which the severely blocked area has about 15 min and a continuous tunnel section of about 30 km. In this area, Fig. 6b shows that the GNSS cannot be resolved for about 40% of the time, with the longest tunnel being 2 min and nearly 3000 km. Figure 6c shows that the integrated navigation system performs well and achieves uninterrupted positioning under the conditions of few stars, no stars, and weak signals.

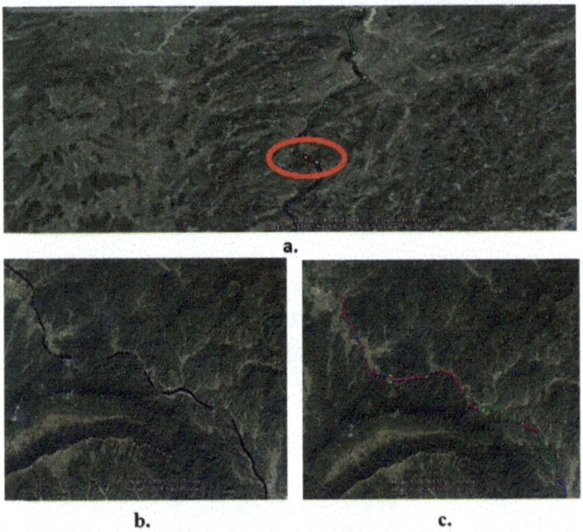

Fig. 7. Experimental results on March 18th a. Whole journey results b. GNSS results in occluded scene c. Integrated navigation results in occluded scene

Figure 7 shows the results of the K88 train from Heyuan to Jiujiangon March 18th, 2019. Figure 7a shows that the line has a total length of about 8 h, of which the severely blocked area has about 22 min and a continuous tunnel section of about 30 km. In this area, Fig. 7b shows that the GNSS cannot be resolved for about 30% of the time, with the longest tunnel being 2 min and nearly 3000 km. Figure 7c shows that the integrated navigation system performs well and achieves uninterrupted positioning under the conditions of few stars, no stars, and weak signals.

Figure 8 shows the results of the K8725 train from Jiujiangon to Xinguo March 20th, 2019. Figure 8a shows that the line has a total length of about 8 h, of which the severely blocked area has about 12 min and a continuous tunnel section of about 20 km. In this area, Fig. 8b shows that the GNSS cannot be resolved for about 50% of the time, with the longest tunnel being 1 min and nearly 2000 km. Figure 8c shows that the integrated navigation system performs well and achieves uninterrupted positioning under the conditions of few stars, no stars, and weak signals.

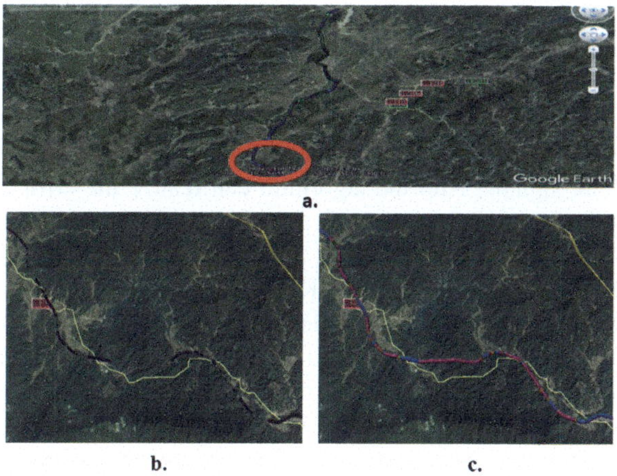

Fig. 8. Experimental results on March 20th a. Whole journey results b. GNSS results in occluded scene c. Integrated navigation results in occluded scene

5.3 Data Analysis

Based on the measured data in the open orbit RTK environment, the performance of the integrated navigation rail transportation algorithm is quantitatively analyzed by constructing the states of bad visibility and even no signal. From the results shown in Table 1, the integrated navigation system achieves an error performance of 3% * mileage under an occlusion state for 3 min.

Table 1. Positioning error in set blocked scene (speed per hour)

No signal period (s)	10	30	60	120	180
First test error (m, %)	1.51, 0.6%	18.08, 2.4%	58.15, 3.9%	70.15, 2.3%	69.07, 1.5%
Second test error (m, %)	0.56, 0.2%	8.55, 1.1%	28.82, 2%	69.07, 2.3%	104.07, 2.3%
Third test error (m, %)	0.93, 0.4%	5.12, 0.7%	49.40, 3.3%	59.46, 2%	89.02, 2.0%

6 Conclusion and Suggestion

Rail transportation has higher requirements for navigation and positioning technology. Such as various strong multipath, weak signal scenes, fast switching scenes, tunnel scenes, stop scenes, etc. Aiming at high-precision applications under weak observations and quasi-one-dimensional constraints, this paper proposes a integrated navigation algorithm based on adaptive modeling. For severe train occlusion, tunnel and other scenarios, modeling and model adaptive filtering are processed. At the same time, through the modeling and processing of the inertial navigation signal, combined with the quasi-one-dimensional motion constraint, automatic initial alignment under weak observation conditions is achieved.

Through real train on track testing, the model parameters were trained and tested. From the results of on track tests, commercial-grade MEMS with gyro stability below 20°/h and accelerometer stability below 30 mg were used. The hardware platform of the device does not have auxiliary signals such as a wheel speed meter, which can realize the uninterrupted operation of the vehicle under the conditions of bad visibility, occlusion, and weak signals. And the algorithm can realize the error accumulation of less than 3% * mileage.

In this experiment, a commercial RTK network is used to achieve high-precision GNSS. In terms of actual operation, the RTK network's available time ratio is less than 50%. Subsequent improvements were made in the following three areas. Firstly, the ground-based RTK network was replaced by BDS-3 satellite-based enhancement. Secondly, it is connected to sensors such as the wheel speedometer of the train platform to strengthen the data fusion capability of the integrated system. Finally, the performance of MEMS commercial devices was enhanced, and replacement experiments of domestic high-performance MEMS devices were performed. The algorithm is expected to achieve 0.1% * distance error accumulation when the subsequent high-precision MEMS devices are used. The research results of this paper can be promoted in China's rail application, including railway, subway, mine transportation, rail port transportation and other fields.

Acknowledgements. The on track train experiment described in this article was strongly supported by CRRC Zhuzhou Institute Times Communication Co., Ltd. A lot of help such as device installation and data acquisition were get form them. At the same time, Beijing Sandcanyon Technology Co., Ltd. provided the INS hardware platform to enable data collection and processing. Thanks also here.

References

1. Xie, J., Wang, H., Li, P., Meng, Y.: Satellite Navigation System and Technology, pp. 74–78. Bejing Institute of Technology Press, Beijing (2018)
2. Li, P.: Data fusion for GNSS/IMU integrated navigation, Thesis, Tsinghua University (2010)
3. Li, P., Lu, M., Feng, Z.: Positioning accuracy analysis for adaptive Kalman filtering in measurements degradation. J. Syst. Eng. Electron. **32**(7), 1489–1492 (2010)
4. Li, P., Li, C., Wu, X., Chen, Z.: A modified IAE algorithm for GNSS and IMU integration. In: Proceedings of the 26th Conference of Spacecraft TT&C Technology in China, pp. 427–437. Springer (2012)
5. El–Sheimy, N., Youssef, A.: Inertial sensors technologies for navigation applications: state of the art and future trends. Satell. Navig. (2020). https://doi.org/10.1186/s43020-019-0001-5
6. Mohamed, A.H., Schwarz, K.P.: Adaptive Kalman filtering for INS/GPS. J. Geodesy **73**, 193–203 (2003)

Design and Verification of Long-Term Reliable Autonomous Navigation System of Navigation Satellite

Weisong Jia(✉), Qiuli Chen, Ying Wu, and Haihong Wang

Beijing Institute of Spacecraft System Engineering, Beijing, China
kingarthurjs@163.com, cql_hong@163.com,
wy630628@163.com, 5585737@qq.com

Abstract. Beidou-3 global navigation satellite system (GNSS) has the capability of autonomous navigation. Without the support of the ground system for a long time, it uses the pre-recorded ground auxiliary data, with the help of intersatellite/satellite-ground measurement, intersatellite/satellite-ground data exchange and the processing of the satellite autonomous navigation software to ensure the autonomous and stable operation of the navigation system. Autonomous navigation service is built on the complex dynamic network system of satellite and ground station, involving diverse system equipments and complex information process. In order to ensure the long-term, high-precision and reliable service capability of autonomous navigation, it is necessary to adopt various means in the design of satellite autonomous navigation to improve the reliability and robustness. In this paper, the profile analysis of autonomous navigation long-term operation scenario is carried out, and the on-board realization of autonomous navigation satellite system is designed by combining the working mode, adaptive data processing, intelligent start synchronization and other technologies, the design of the autonomous navigation system on the satellite was researched. Also the autonomous navigation reliable operation ability was validated based on long-term tasks, abnormal interactions between satellites, constellation refactoring conditions.

Keywords: Long-term reliable · Design · Verification · Autonomous navigation

1 Introduction

Autonomous navigation can enhance the viability of the satellite navigation system [1], improve system performance and improve the accuracy of navigation and positioning. At the same time, autonomous navigation can also decrease the amount of ground station, reduce the times of information injections to satellites and reduce the cost of system maintenance. The ability of autonomous navigation has become an important characteristic of security and advancement in the satellite navigation system.

Autonomous navigation service of Beidou global system is based on the complex large system which integrated the satellites and earth. The process of service is based on high precision measurement and communication between satellites, as well as the

J. Sun et al. (Eds.): CSNC 2020, LNEE 650, pp. 679–690, 2020.
https://doi.org/10.1007/978-981-15-3707-3_63

data exchange between each, completion of autonomous orbit determination and time synchronization. Autonomous navigation generates and broadcasts auto-generated ephemeris, guarantees the stability of autonomous operation of the Global Navigation Satellite System, and continuously provides high precision location and time product for the user services. The above process has the dynamic characteristics of multi-stage, multi-task and multi-mode. At the same time, multi-task is concurrent in the networking stage of the constellation. System configuration, failure criterion, failure rate of bottom event, etc. at system level, component level may change with the task transformation. The accuracy and stability of the autonomous navigation service are affected by a variety of factors, including operational control interaction strategy, inter-satellite link transceiver communication, inter-satellite network topology and route management, inter-satellite link antenna pointing, and equipment temporary failure caused by the space environment.

In order to improve the reliability and robustness of satellite autonomous navigation system, it is necessary to carry out a variety of design methods according to the characteristics of diverse equipment, complex information flow and dynamic configuration change.

This paper carries out profile analysis on the scenario of autonomous navigation for long-term operation, and designs the realization of autonomous navigation system on the satellite, with long-term reliable autonomous operation as the core goal. This paper also verifies the reliable autonomous navigation capability based on long-term missions, inter-satellite interaction anomalies, constellation reconstruction and other working conditions.

2 The System Design

2.1 Scene Profile Analysis

Constellation autonomous navigation requires each satellite to establish inter-satellite links with as many satellites as possible to obtain better geometric precision factor values. Therefore, space time division multiple access (STDMA) system based on Ka phased array can provide more measurement data support by inter-satellite high-frequency ranging and communication [2].

When a navigation satellite carries a phased array transceiver, a half-duplex intersatellite link is established between satellites within the planned time slot [3], and the distance measured by the other satellite cannot be obtained immediately, so it needs to be exchanged with the autonomous navigation message and covariance through the network, and then matched respectively [4].

Based on phased array antenna and phased array to send and receive messages built in the network, with the forecasting ephemeris and precise ephemeris data injected by the control station [5], on the basis of two-way ranging and information exchange, information processing of antenna direction pointing, form the input of autonomous navigation algorithm. After the completion of range data preprocessing, filtering correction of orbit and clock parameters [6], fitting calculation, autonomous navigation message is generated. The navigation task processing units broadcasted the ephemeris and clock products, achieve long-term reliable autonomous navigation service.

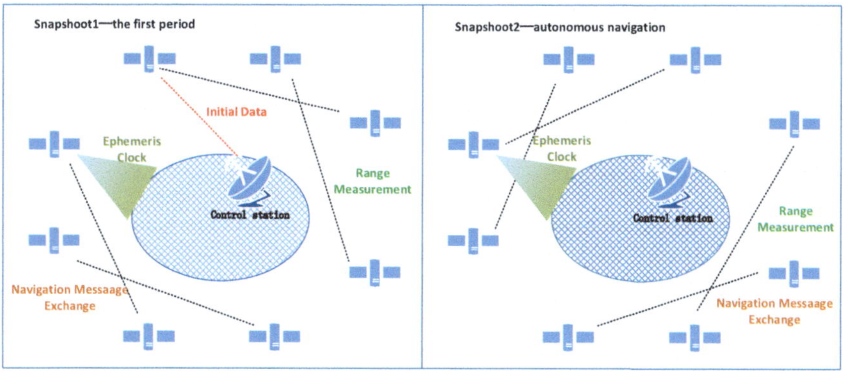

Fig. 1. Time-division network constellation autonomous navigation work snapshot

When autonomous navigation completes the long-term operation of measurement, filtering and fitting in a cycle of 5 min to 60 min, the factors affecting the stable operation of autonomous navigation of the constellation include the following typical problem scenarios:

1. When the calculation of one satellite or multiple satellites stops due to a failure, the message generated by the satellite will stop update, and the last message will become unavailable with passage of time. At this point, the interaction between the failed satellite and other satellites should be terminated.
2. When the phased array transceiver or other intersatellite link equipment fails due to the space environment, the local satellite cannot obtain the measurement information and intersatellite interaction information for distributed autonomous navigation calculation. Before the intersatellite link equipment recovers autonomously, the autonomous navigation system should be able to maintain the services of ephemeris and clock.
3. When the network planning is not reasonable, some satellites will establish a small number of links, which will affect the orbit filtering effect and result in a decrease in the single-star accuracy.
4. In the non-continuous link network, when the inter-satellite interaction information cannot arrive on time due to network problems, the measurement data in the current working cycle will be reduced, and the non-real-time measurement data input will arrive in the next working cycle.
5. When the accuracy of calculation results of one or more satellites decreases, the autonomous navigation message generated by it will be used by other satellites, which will cause the autonomous navigation accuracy of other satellites to decrease simultaneously.
6. When the device running the autonomous navigation algorithm is reset due to single event effect (SEE) or other space environment problems, the autonomous navigation should recover from the fault quickly and resume normal operation without ground intervention, and the high accuracy of autonomous navigation should be maintained during device reset.

Based on the analysis of factors affecting the stable operation of constellation autonomous navigation, the autonomous navigation system focuses on inter-star network anomalies, algorithm accuracy anomalies and equipment failures, and designs flexible working modes, adaptive data processing and intelligent startup synchronization strategies.

2.2 Working Mode Design

In order to solve the problem that partial satellite faults pollute the whole constellation in the scenario 1mentioned above, the distributed autonomous navigation system designed the isolation mode in addition to the normal operation mode.

In normal mode, the measurement information and autonomous navigation message data exchange between the satellites in the autonomous navigation system, autonomous navigation software generated autonomous navigation message and covariance. The message and its integrity provided to navigation service equipment, according to the current satellite navigation service mode, was decided whether adopted and transmitted to the user.

In isolation mode, the local satellite receives the measurement information of other satellites but does not share the measurement information of the local satellite. It receives autonomous navigation messages and covariance from other satellites, but does not send local autonomous navigation messages and covariance. The autonomous navigation software calculates the results, but does not provide the message and integrity information to the navigation service equipment.

That is, satellites in isolation mode only receive input and complete their own calculations, but other satellites in the constellation do not take data from the isolated satellite as input.

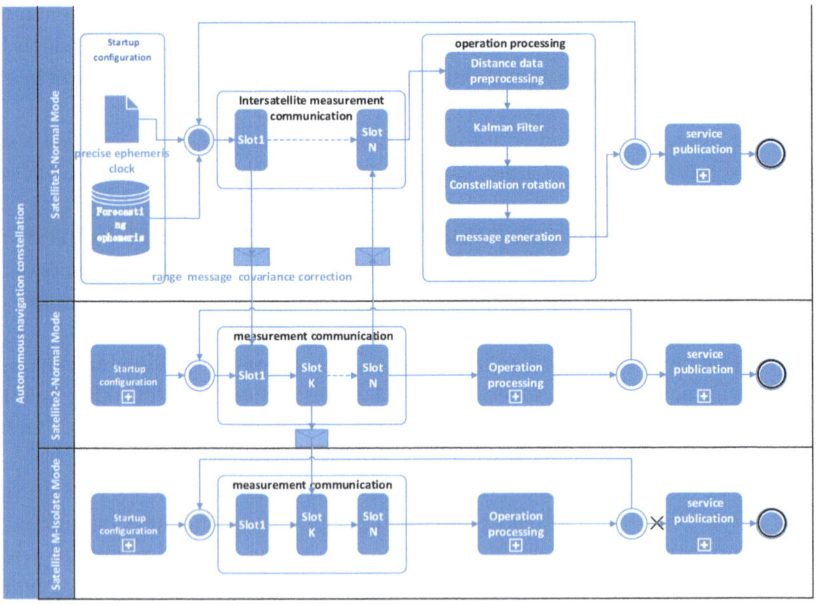

Fig. 2. Autonomous navigation workmode

In the problem scenarios 2 and 3, when the number of satellite links for distance measuring is small, the effect of kalman filter will be reduced.

If there is an error in kalman filtering and message fitting based on the measured data of the current autonomous navigation operating period, or the number of measurement links is sufficient, then the ephemeris and clock difference will maintain the last correct result, so as to ensure that the new measured data can be used for another attempt in the next working period.

If the fitting is consistently wrong over several duty cycles, it will lead to the decrease of the availability of the results broadcasted by the satellite. As time goes on, when the URE (User Range Error) of the last correctly updated ephemeris and clock error information has dropped to an unacceptable level, the autonomous navigation automatically switches to isolation mode to avoid pollution diffusion and wait for ground treatment.

2.3 Adaptive Data Processing Design

The information foundation of the autonomous navigation system is the inter-satellite network equipment and navigation payload equipment, as well as the antenna pointing software, inter-satellite network software and navigation information processing software running on it.

The autonomous navigation algorithm obtains the information through the infrastructure before starting each duty cycle to form a complete input packet.

Input data packets originate from local satellite generation, operational control injection and inter-satellite interaction, including bidirectional ranging information R, measurement correction from phase center of phased array antenna to satellite centroids $C_{correctKa}$, measurement correction from phase center of navigation L antenna to satellite centroids $C_{correctL}$, and receiving and transmitting delay of phased array transmitter C_{delay}.

$$C = R + C_{correctL} + C_{correctKa} + C_{delay}$$

For an certain inter-satellite link, only when all the relevant delay and distance information is collected, and the message and covariance information of satellites at both ends of the link are also exchanged, can the link be used as an effective input link of the autonomous navigation algorithm.

In the problem scenarios 4, antenna pointing and any anomaly in the network will cause the link information to be unavailable.

Therefore, in inter-satellite network transmission, the priority of autonomous navigation service information is set as the highest priority. At the same time, strict inspection is carried out in the matching of link information, including the following key points:

1. The measurement information of the repetitive links in the same working cycle is screened out, and only the bidirectional measurement information that is closest to the calculation time of kalman filter is retained.
2. For the satellite with repetitive links, make sure that the measurement time of local ranging and the measurement time of other satellites are in the same time slot, so as to avoid the large deviation of two-way measurement time, which lead to the decrease of calculation accuracy.

To avoid problem scenarios 4, when the autonomous navigation algorithm obtains the bidirectional measurement information and messages from other satellites, it also needs to judge the quality of the ranging information and the accuracy of the ephemeris and clock differences, so as to prevent the error of the other satellite spreading to the local.

Here, the difference between the measured data and the theoretical data of the interstellar distance is used as a criterion: First, delay correction, relativity correction, and phase center-to-centroid correction are performed on the bidirectional measurement information, and the corrected two-way measurement information is calculated to C, which represents the measured distance at kalman filtering time; at the same time, the theoretical distance between the local star and other stars, O is calculated by using the local satellite's ephemeris and the other satellite's ephemeris.

Combined together with the inter-satellite measurement noise, ephemeris dynamic noise, and message covariance, the difference between the theoretical distance and the measured distance $O - C$ is evaluated, to determine whether the measurement information of the link can be used as the input of the Kalman filtering algorithm.

2.4 Intelligent Startup Synchronization Design

Affected by the space environment, in order to cope with the failure caused by a single particle, the satellite equipment occasionally performs an autonomous reset operation. In order to ensure that the autonomous navigation can operate autonomously for 60 days, the autonomous navigation software is required to be able to recover quickly after a stand-alone reset in the problem scenario 6. when autonomous navigation is interrupted during the algorithm, intelligent autonomous start synchronization is required.

The autonomous navigation software stores the precise ephemeris, precise clock parameters and covariance parameters of the input data packet of the current working cycle into the non-volatile memory every hour. The non-volatile memory device will not lose data due to machine reset, and can recover these data from the non-volatile memory device after machine reset.

It should be noted that when storing the input data packet of the current working cycle, all parameters in the input data packet must belong to the same calculation cycle, otherwise it will lead to an error in the autonomous navigation calculation after machine reset.

Since important data is written into FLASH as pages, the data that changes with the cycle in the algorithm input packet is deployed on the same page, and the binding parameters and long-term ephemeris parameters injected by operation control are deployed on other pages.

When the machine is reset due to fault, the autonomous navigation first checks whether the important data has been restored successfully. If the recovery fails, the operational control injection parameters such as the maximum fitting times are initialized as default values, and the autonomous navigation software enters the shutdown mode.

The autonomous navigation software detects the current time on the satellite in real time. If the time on the satellite is an integer multiple of the duty cycle and important data is successfully restored, the autonomous navigation automatically enters the synchronized state. The parameters restored from the non-volatile memory device FLASH to the input data packet take effect, and the normal operation of this cycle begins.

When the time on the satellite deviates, the ground station will perform phase modulation or time calibration operation as soon as possible to recover. In order to avoid restarting the autonomous navigation operation manually, autonomous navigation system has the function of automatic resynchronization after the time deviation: at the beginning of each working cycle, it is checked whether the time on the satellite is an integer multiple of the working cycle. If it is not an integer multiple of the duty cycle, the autonomous navigation synchronization status becomes out of sync. In the out-of-synchronization state, the autonomous navigation interface and algorithm stop working; the autonomous navigation detects the current time every second, if the onboard time is an integer multiple of the working cycle again, and ephemeris is successfully fitted in the last normal cycle of operation, while the time on the satellite after calibration is less than one hour compared with the reference epoch of ephemeris, the autonomous navigation automatically switches to the synchronized state.

3 Testing and Validation

In December 2018, China completed the global network of 18 beidou-3 satellites, and the beidou-3 navigation system officially provided basic navigation services to the world. In this paper, 18 beidou satellites that constitute the basic navigation constellation are taken as the analysis objects. Based on the inter-satellite measurement data and accurate real-time ephemeris from January to February 2019 as references, a verification platform for the satellite-borne autonomous navigation system of beidou 3 is built in the laboratory to evaluate the capability of long-term reliable autonomous operation of navigation satellites. The basic parameters of the beidou 3 basic navigation constellation are as follows:

Table 1. Beidou 3 basic navigation constellation configuration

Satellite	Orbital location	Launch time (UTC)
BeiDou-3 M1	MEO ~ 21500kmB07	2017.11.5
BeiDou-3 M2	MEO ~ 21500kmB08	2017.11.5
BeiDou-3 M3	MEO ~ 21500kmB06	2018.2.12
BeiDou-3 M4	MEO ~ 21500kmB05	2018.2.12
BeiDou-3 M5	MEO ~ 21500kmC07	2018.7.29
BeiDou-3 M6	MEO ~ 21500kmC01	2018.7.29
BeiDou-3 M7	MEO ~ 21500kmA04	2018.1.11
BeiDou-3 M8	MEO ~ 21500kmA05	2018.1.11
BeiDou-3 M9	MEO ~ 21500kmA02	2018.3.29
BeiDou-3 M10	MEO ~ 21500kmA03	2018.3.29
BeiDou-3 M11	MEO ~ 21500kmC02	2018.8.24
BeiDou-3 M12	MEO ~ 21500kmC08	2018.8.24
BeiDou-3 M13	MEO ~ 21500kmB01	2018.9.19
BeiDou-3 M14	MEO ~ 21500kmB03	2018.9.19
BeiDou-3 M15	MEO ~ 21500kmA01	2018.10.15
BeiDou-3 M16	MEO ~ 21500kmA07	2018.10.15
BeiDou-3 M17	MEO ~ 21500kmC04	2018.11.18
BeiDou-3 M18	MEO ~ 21500kmC06	2018.11.18

Phased-array satellite measurement information and phased-array antenna pointing information are obtained through Beidou-3 satellite telemetry. Long-term forecast ephemeris parameters, ephemeris parameters, clock parameters, and covariance parameters for full constellation are initialized. Routing table, judgement threshold of measurement information, process noise of ephemeris and clock, channel delay of inter-satellite links, channel noise parameters and measurement noise parameters [7] of inter-satellite links are configured. By adjusting the autonomous navigation duty cycle and the fit thresholds, the parameters of the autonomous navigation system are fully consistent with the on-orbit operating environment. The key configuration parameters are as follows:

Table 2. Key parameters of autonomous navigation system

Parameter	Parameter settings
Simulation time (Beijing time)	0:00 10 January 2019-23:00 28 February 2019
Simulation time (beidou hour)	349,200 s for 679 weeks - 431,700 s for 686 weeks
Ephemeris measure the discard threshold	20 m
The star clock measures the discarding threshold	20 m
Intersatellite link bandwidth	50 frames/slot
Working cycle	5 min
Intersatellite measurement noise	0.04 m

Based on the configuration parameters above, the accuracy and reliability indexes of the autonomous navigation system during the long-term autonomous operation are concerned.

Since beidou-3 has not yet formed a complete MEO constellation with 24 satellites in January 2019, the number of non-repeat links will be less than 4 in some time intervals during the autonomous navigation working period.

Figure 3 shows the number of effective ranging links between 18 satellites, and the number of effective links is between 0 and 13.

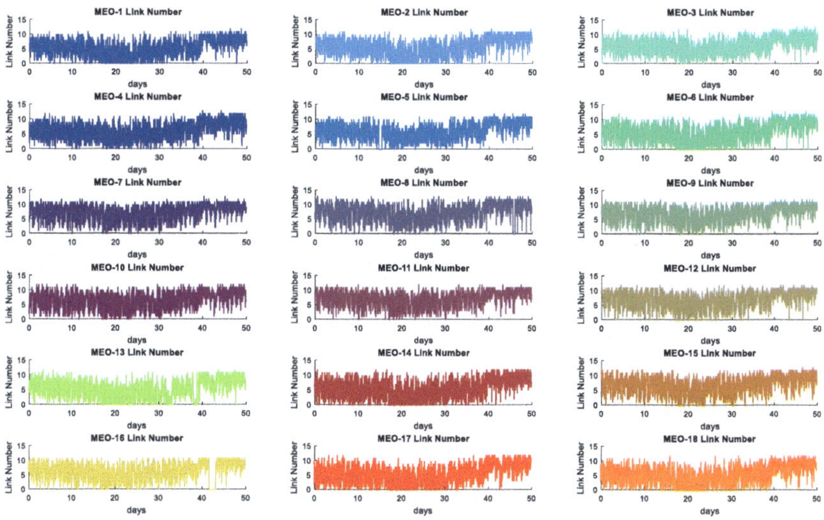

Fig. 3. Number of effective inter-sat links

In order to verify the effect of long-term and reliable operation of the autonomous navigation system, faults were injected during the autonomous operation.

1. Noise is randomly added to intersatellite measurement data and intersatellite messages are tempered;
2. Faults are injected into the satellite of meo-8, reduce the accuracy of the algorithm running in the satellite;
3. The number of inter-satellite ranging links of meo-16 reduces to 0. And it is re-synchronized after a period of time to rejoin the constellation.

The satellite autonomous navigation system gets the theoretical inter-satellite distance calculated by the autonomous navigation messages, and compares it with the actual measured distance between the satellites to obtain the difference between the theoretical value and the actual measurements, which is identified by OC. Figure 4 shows the maximum value of the ephemeris OC of 18 satellites. Ignoring the jump-value, Fig. 5 shows the normal value of the ephemeris OC of 18 satellites. Under the condition of random failure injection in full constellation, the OC deviation of the

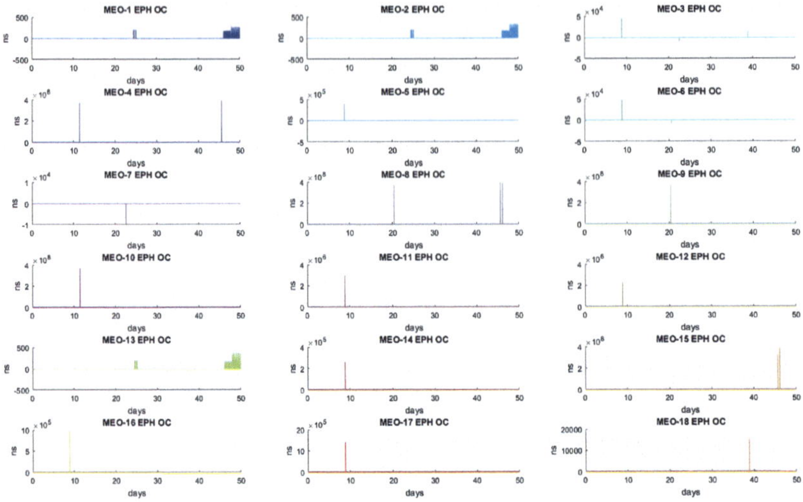

Fig. 4. Maximum OC Value of satellite ephemeris

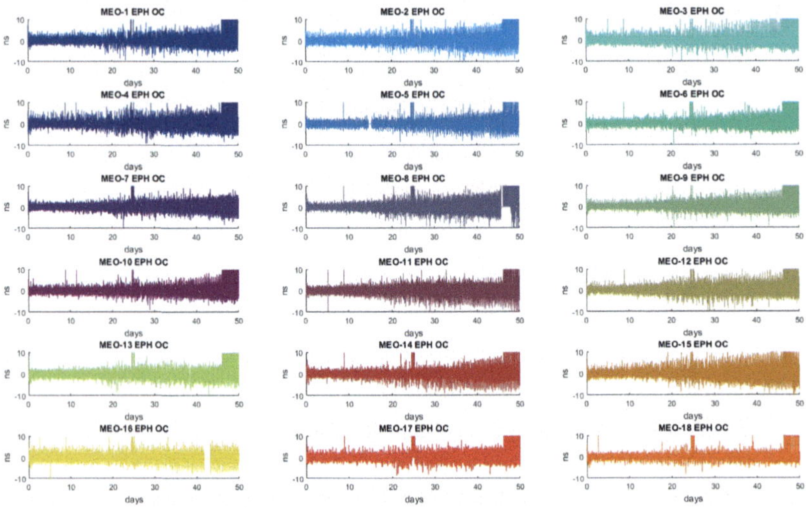

Fig. 5. Normal OC Value of satellite ephemeris

autonomous navigation system occasionally appears with a large value, the maximum deviation is greater than 3.5×10^8 ns. When the failure injection of the meo-8 satellite on the 46th day reduced the precision of its ephemeris, the OC values of other satellites who build cross links with meo-8 at different time slots deviated from the average.

On the basis of fault injection, the User Range Error (URE) of the satellite under test was used as an indicator to evaluate the long-term stability of the autonomous navigation system, as shown in the Fig. 6.

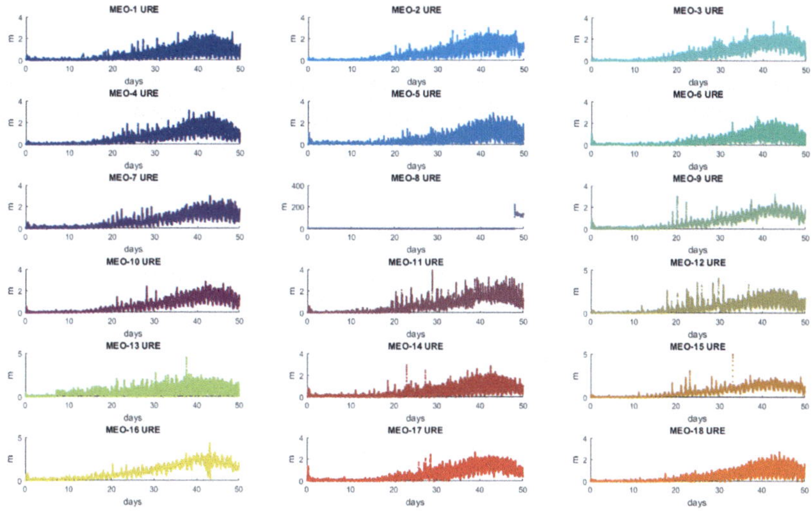

Fig. 6. URE of Autonomous orbit determination

The URE of 18 satellites during long-term operation was observed. On the 46th day, a fault was injected into the MEO-8 satellite. The error of the message broadcasted from the MEO-8 did not affect the normal autonomous operation of the other 17 satellites and did not cause pollution to the full constellation. The accuracy of autonomous orbit determination for all satellites stopped decreasing after the 40th day, and entered a stable convergence state. Due to the limited input data in the experimental part of this paper, a 50-days evaluation was performed instead of 60-days. The URE of the autonomous navigation system was less than 5 meters during the 50-days autonomous operation.

Verification shows that the navigation satellite autonomous navigation system always maintains long-term stable operation under fault injection conditions such as intersatellite interaction anomalies. Failure of one satellite will not cause constellation-level autonomous navigation failure. The orbital accuracy of the system is always maintained at a high level.

4 Conclusion

In order to ensure that the autonomous navigation system provides long-term high-precision and high-reliability services, this article analyzes the long-term operation scenarios of autonomous navigation, and introduces the system design method of autonomous navigation in combination with key technologies such as network information exchange, intelligent startup synchronization, data pre-processing, and algorithm adaptive adjustment. Based on the basic navigation constellation, the reliable operation capability of autonomous navigation under abnormal working conditions are verified. The evaluation results prove that the autonomous navigation service can continuously provide users with high-precision location and time services. At present, Beidou-3 autonomous navigation on-orbit tests are in progress. Subsequent research will focus on the calibration of the constellation system difference, the correction strategy of earth orientation parameter based on anchor stations, etc, to further improve the long-term accuracy and stability of autonomous operation of navigation satellites.

References

1. Shuai, P., Qu, G.: Time synchronization techniques of the autonomous navigation of navigation constellation. J. Astronaut. **26**(6), 768–772 (2005)
2. Chen, Y., Hu, X.: A new autonomous orbit determination algorithm based on inter-satellite ranging measurements. SCIENTIA SINICA Physica Mech. Astronom. **45**(7), 079511 (2015)
3. Jia, W., Wang, Q.: Optimization of contact design plan of time-division navigation satellite network based on load balancing. Spacecraft Eng
4. Jia, W., Wang, H.: Design of ground simulation and test system of Beidou autonomous navigation. Spacecraft Eng. **26**(3), P195–P197 (2017)
5. Wang, H., Chen, Z.: On-board autonomous orbit prediction algorithm for navigation satellites. J. Astronaut. **33**(8), P1019–P1026 (2012)
6. Lin, Y., Qin, Z.: A satellite cross link-based GNSS distributed autonomous orbit determination algorithm. J. Astronaut. **31**(9), 2088–2094 (2010)
7. Yang, Y., Yang, Y.: Comparison and analysis of two orbit determination methods for BDS-3 satellites. Acta Geod. et Cartographica Sin. **48**(7), 831–839 (2019)

Research on the Verification of Autonomous Navigation Technology Based on Inter-satellite Link of BDS Satellite

Qiuli Chen(✉), Ying Wu, Haihong Wang, and Weisong Jia

Beijing Institute of Spacecraft System Engineering, Beijing, China
cql_hong@163.com, wy630628@163.com,
5585737@qq.com, kingarthrjs@163.com

Abstract. Up to now, the new generation of BDS-3system has nearly completed the global networking. Different from the BDS-2 satellite, the BDS-3 satellite has the inter-satellite ranging function supported by the inter-satellite link. The introduce of inter-satellite link and inter-satellite ranging enables that BDS-3 system has autonomous navigation function by autonomous orbit and clock difference determination, which has less or no dependence on ground systems. Taking the BDS-3 satellite as an example, this paper evaluates the autonomous navigation results based on inter-satellite and satellite to ground ranging. the influence of the different ranging data to satellites orbits and clock errors was analyzed. The precision of autonomous navigation of BDS-3 system based on inter-satellite ranging has been given.

Keywords: BDS satellite · Inter-satellite link · Autonomous navigation

1 Introduction

As of now, the new generation of BDS-3 system is nearing completion of global networking. The new generation BDS system with full constellation configuration includes 24MEO + 3IGSO + 3GEO satellites, referred to as BDS-3. Affected by various factors, the ground monitoring station of BDS control system has not achieved global deployment. In order to make up for the limitation of regional ground station observation, the BDS-3 satellite has the inter-satellite ranging function supported by the inter-satellite link. The introduction of inter-satellite links/inter-satellite ranging enables BDS-3 satellites to autonomously complete the determination of orbits and clock differences without relying on the ground at all, thereby achieving system autonomous navigation services. Autonomous navigation refers to the use of pre-annotated ground-assisted data, inter-satellite/satellite-ground survey, inter-satellite/satellite-to-ground data exchange, and satellite-borne autonomous navigation software for processing when the satellite does not receive normal support from the ground system for a long time. The ability and process ensure the constellation's independent and stable provision of navigation services.

During the BDS-3 test satellite phase, autonomous navigation functions and performance under a few links were verified in orbit. For the BDS-3 network satellite,

J. Sun et al. (Eds.): CSNC 2020, LNEE 650, pp. 691–701, 2020.
https://doi.org/10.1007/978-981-15-3707-3_64

various verification tests were carried out based on the ground simulation system. Currently, in-orbit validation of autonomous navigation function BDS-3 satellite network is fully operational. In this paper, the BDS-3 satellite was used as an example to perform ground test and test verification base on in-orbit data of the full-constellation satellite's autonomous navigation function based on inter-satellite ranging. The results were evaluated. The influence of different data on satellite orbit and clock error was analyzed. The analysis results of autonomous navigation accuracy of BDS-3 system with inter-satellite link were given.

2 Autonomous Navigation Technology of BDS System

Autonomous satellite navigation refers to a satellite navigation technology that does not completely rely on the support of ground systems and can operate autonomously. Spaceborne data processing technology focuses on solving the problem of autonomously updating the navigation ephemeris using interstellar measurements. The autonomously generated navigation message on the satellite is an indispensable part of the realization of autonomous navigation of the navigation constellation. The navigation satellite ephemeris in the autonomously generated navigation message on the satellite is calculated by the onboard computer. It is released to users through satellites to achieve real-time navigation and positioning [1]. Autonomous navigation system design, autonomous orbit determination and time synchronization algorithms, directional parameter constraints and corrections, and broadcast ephemeris fitting are several key technologies for achieving autonomous navigation [2–4].

Two major error sources of satellite navigation system autonomous navigation. The first is the position error of constellation satellites in Earth Center Inertial (ECI) space maintained using interstellar ranging information. The other is the long-term forecast error of Earth Orientation Parameter (EOP), which is no longer updated from the ground during autonomous navigation [5]. Aiming at the first source of error that can be improved by algorithm design, BDS autonomous navigation adopts a distributed algorithm improvement based on long-term ephemeris parameters on the ground. The core algorithm of calculation based on long-term ephemeris parameters is the on-board autonomous orbit prediction technology and forecasting the results of autonomous orbit determination filtering forward for a specific time period. Orbit determination is to calculate the orbit parameters at the corresponding moment based on the measurement data of a segment of orbit, including the initial orbit determination and the orbit improvement. In the traditional sense, the initial orbit determination is based on six independent observations to determine the number of six orbits, corresponding to a non-perturbed two-body problem. The orbit improvement is to refine the orbital number of corresponding epochs by using more observations based on the initial orbit [6–8].

2.1 Autonomous Navigation Operation Mode

The BDS system autonomous navigation task is designed to work in semi-autonomous navigation mode and fully autonomous navigation mode. The semi-autonomous navigation working mode refers to a working mode in which the system relies on the anchoring station to continue to provide the space-time reference information of the navigation constellation when the ground operation control system cannot work normally. In the semi-autonomous navigation mode, the anchor station is equipped with satellite-ground bidirectional measurement and data transmission equipment. It is also equipped with equipment with strong data processing capabilities. It adopts a centralized data processing mode similar to ground operation control. Anchor stations and in orbit satellites generate clock difference and ephemeris data close to the ground's operational control level. They also complete data transmission and distribution. The operating mode is shown in Fig. 1.

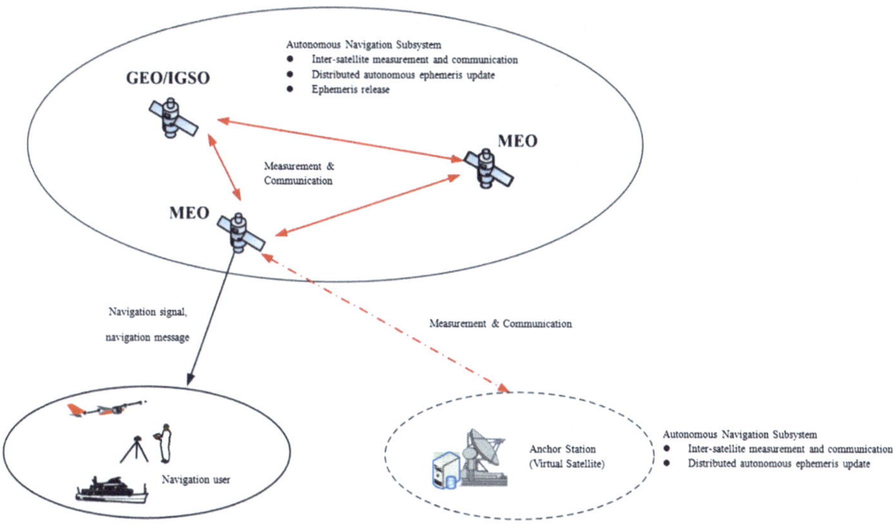

Fig. 1. Semi-autonomous navigation mode with ground anchor support

The fully autonomous navigation operating mode refers to the situation where the ground operation control system, measurement and control system, and anchoring station are unable to inject the space-time reference information required for the operation of the navigation constellation. The system only relies on a working mode in which satellites automatically generate full constellation clock difference and ephemeris data. The operating mode is shown in Fig. 2.

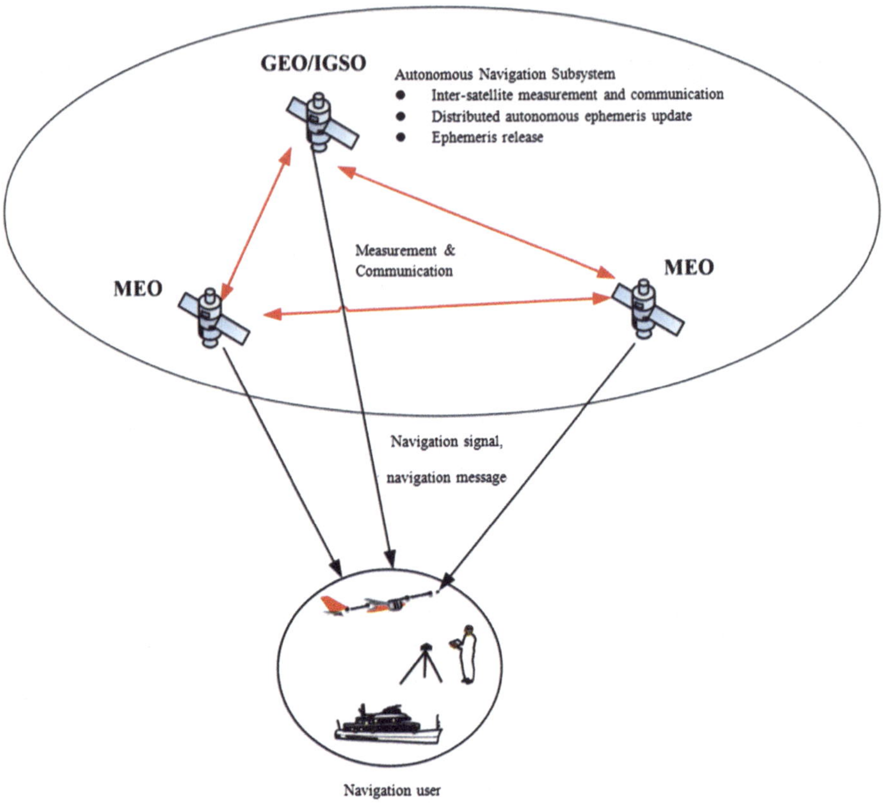

Fig. 2. Full autonomous navigation working mode

2.2 Key Algorithms for Autonomous Navigation

There are two ways to solve the navigation constellation autonomous navigation. One is distributed solution. Each satellite calculates ephemeris clock offset and deviation parameters. The other is a centralized solution, where all satellite broadcast ephemeris and system deviations are calculated by a specified satellite. At present, BDS system adopts distributed solution, that is, all in-orbit satellites are equipped with the same autonomous navigation calculation unit, participate in the calculation of orbit and clock difference, and generate a broadcast ephemeris and transmit it to the broadcast business unit. The autonomous navigation algorithm consists of different functional modules, including data preprocessing, epoch reduction, orbit prediction, orbital clock error parameter filtering and correction, and broadcast ephemeris fitting. This section mainly introduces the filtering correction of orbital clock difference parameters and the fitting of broadcast ephemeris.

2.2.1 Key Algorithms for Autonomous Navigation [9, 10]

For a non-linear system with continuous time and discrete observation, after omitting the linearization process, the state quantity $X^*(t_{k-1})$ at time t_{k-1} is known. And the reference track $X^*(t_k)$ and the state transition matrix $\Phi(t_k, t_{k-1})$ are obtained at time t_k.

$$A(t) = [\frac{\partial F(X,t)}{\partial X}]^* \tag{1}$$

$$\dot{X}^* = F(X^*(t), t) \tag{2}$$

$$\dot{\Phi}(t, t_{k-1}) = A(t)\Phi(t, t_{k-1}) \tag{3}$$

$$\bar{P}_k = \Phi(t_k, t_{k-1})P_{i-1}\Phi^T(t_k, t_{k-1}) \tag{4}$$

Then the observation bias, observation transition matrix, and gain matrix are calculated.

$$y_k = Y_k - G(X_k^*, t_k) \tag{5}$$

$$\tilde{H} = [\frac{\partial G(X, t_k)}{\partial X}]^* \tag{6}$$

$$K_k = \bar{P}_k\tilde{H}_k^T(\tilde{H}_k\bar{P}_k\tilde{H}_k^T + R_k)^{-1} \tag{7}$$

Finally, the orbit and clock difference parameters are updated.

$$\hat{x}_k = K_k y_k \tag{8}$$

$$X_k^* = X_k^* + \hat{x}_k = X_k^* + K_k y_k \tag{9}$$

$$P_k = (I - K_k\tilde{H}_k)\bar{P}_k \tag{10}$$

2.2.2 Broadcast Ephemeris Fitting Algorithm [11]

The Beidou system broadcast ephemeris includes an ephemeris reference time toe and 17 broadcast ephemeris parameters $M_0, \Delta n, \Delta\dot{n}, e, \Delta A, \dot{A}, \Omega_0, \dot{\Omega}, i_0, \dot{i}, w, C_{rs}, C_{rc}, C_{us}, C_{uc}, C_{is}, C_{ic}$. The broadcast ephemeris user algorithm can be found in the user's public file, which is abbreviated as:

$$R_k^{'} = h(\xi, t_k) \tag{11}$$

The coordinate residuals at each sampling point are:

$$\Delta R_k = R_k - R_k^{'} = R_k - h(\xi, t_k) \tag{12}$$

The correction amount of the broadcast ephemeris parameter is:

$$\Delta\xi = (H^T H)^{-1} H^T \Delta R \tag{13}$$

H is the partial derivative matrix of the position coordinate R (XYZ) on the broadcast ephemeris parameters.

$$H = \frac{\partial h}{\partial \xi} = \begin{bmatrix} \frac{\partial X}{\partial \xi_1} & \cdots & \frac{\partial X}{\partial \xi_{17}} \\ \frac{\partial Y}{\partial \xi_1} & \cdots & \frac{\partial Y}{\partial \xi_{17}} \\ \frac{\partial Z}{\partial \xi_1} & \cdots & \frac{\partial Z}{\partial \xi_{17}} \end{bmatrix} \tag{14}$$

There are more than one condition for the convergence of broadcast ephemeris fitting. It is set here that the root of the sum of the squares of the position errors is less than a certain threshold Δ_{lim}:

$$\Delta = \sqrt{\sigma_{\Delta x}^2 + \sigma_{\Delta y}^2 + \sigma_{\Delta z}^2} \le \Delta_{\text{lim}} \tag{15}$$

$$\sigma_{\Delta x}^2 = \frac{\sum_{k=1}^{N} \Delta x_k^2}{N-1}, \sigma_{\Delta y}^2 = \frac{\sum_{k=1}^{N} \Delta y_k^2}{N-1}, \sigma_{\Delta z}^2 = \frac{\sum_{k=1}^{N} \Delta z_k^2}{N-1} \tag{16}$$

3 Autonomous Navigation Simulation and Ion-Orbit Test Results

During the BDS-3 test satellite phase, in-orbit verification and technical tests were carried out. A short-term (less than 7 days) on-orbit test of autonomous navigation services was conducted using 5 test satellites. The ground transportation control system injected autonomous navigation service parameters and command information into the test satellite uplink. The inter-satellite link equipment of 5 test satellites was established normally. After receiving the autonomous navigation service parameters and command information, the satellite enabled the corresponding autonomous navigation service parameters. Autonomous navigation instructions were executed. The satellite launched the on-board autonomous navigation software. The ground operation control system judged the correctness of the satellite enabling autonomous navigation service parameters and executing the autonomous navigation instructions from the S operation control service information. The results showed that the satellite has the capability of autonomous navigation calculation. And the autonomous navigation capability based on the inter-satellite link was verified.

The satellites of the BDS-3 system are equipped with new inter-satellite link equipment. According to the system time slot allocation plan, it can establish links, distance measurement and data interaction with ground stations or other satellites in the constellation. Among them, the duration of each link is 3 s. That is, according to the link establishment time slot configuration table, there is a maximum of 20 link establishment data every 1 min. In this chapter, the 18 satellites basic systems simulated on the ground and 14 satellites in orbit were allocated corresponding chain-building time slots to verify and evaluate the results of autonomous navigation.

3.1 Key Algorithms for Autonomous Navigation

18 MEO satellite simulation basic systems were composed of ground simulation test equipment and ground test data. This section used a simulated basic system for simulation testing. The composition of the basic navigation system is shown in Fig. 3.

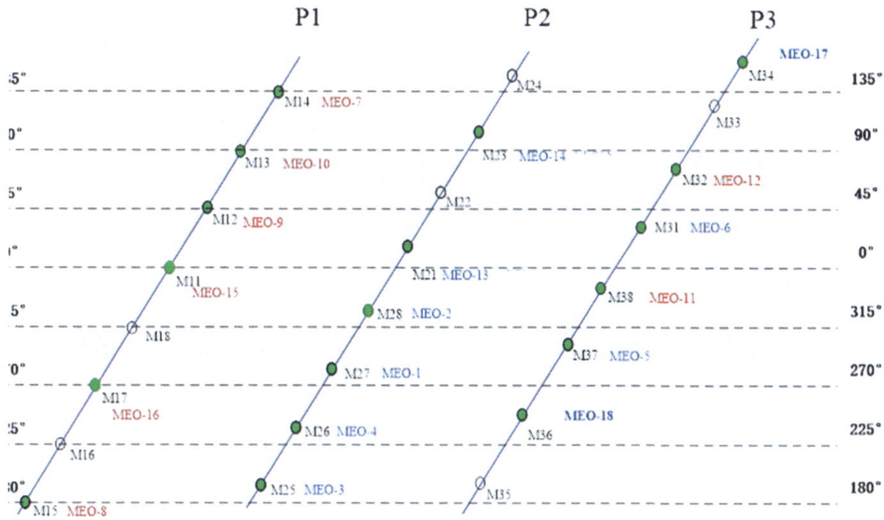

Fig. 3. Composition of Beidou-3 in-orbit basic system

The satellite simulators were loaded with spaceborne autonomous navigation software. The basic operating data included ephemeris parameters, satellite clock parameters, long-term forecast ephemeris parameters, and autonomous navigation service parameters and instructions. Supporting information such as reference ephemeris, clock difference, EOP parameters, ionospheric parameters necessary for the above-mentioned autonomous navigation test were generated by ground simulation.

The 18 MEO satellites were injected with startup parameters of 676 weeks 439,200 s. The satellite simulator's autonomous navigation software was turned on. Synchronously, S telemetry received autonomous navigation messages. The 18 satellites autonomous navigation calculation results were processed by the ground data processing terminal. Compared with the theoretical precision ephemeris of the evaluation module, the correctness of the autonomous navigation message was checked. Whether the difference of orbit and clock between the autonomous navigation message and the ground transportation control system meets the requirements of the index was evaluated.

Figure 4 shows the change of the average URE accuracy during the long-term autonomous navigation of a basic system consisting of 18 MEO satellites.In the figure, R, T, and N are radial, along the track direction, and normal.

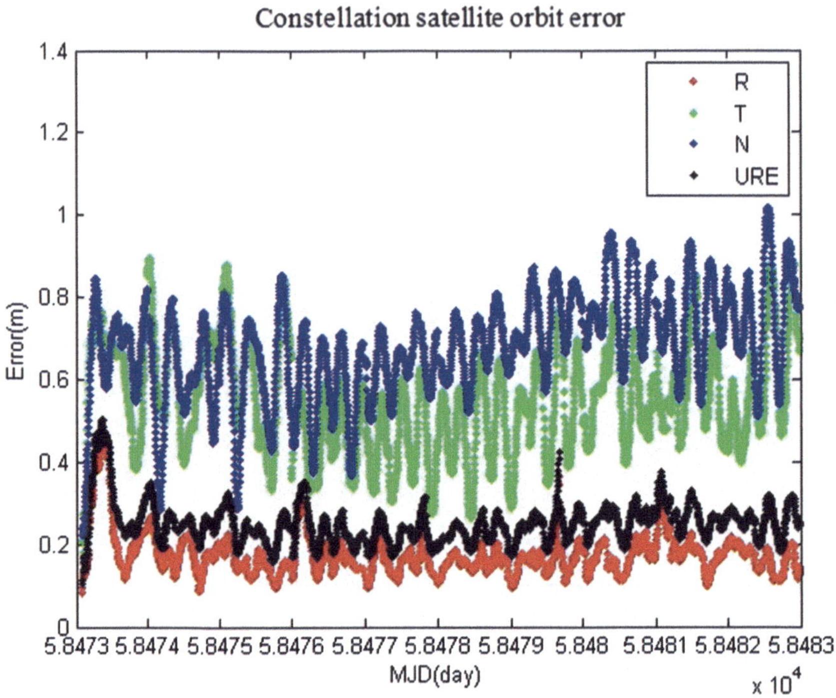

Fig. 4. Orbital URE changes of autonomous constellation for all constellations

The above simulation results showed that the ground simulation module completely consistent with the spaceborne autonomous navigation software runs continuously for 240 h. The average URE for autonomous orbit determination in the whole constellation was significantly better than 2.85 m.

3.2 Verification Results of Autonomous Navigation Based on Orbit Data

The BDS-3 system satellite, which has been launched into orbit, generated autonomous navigation broadcast ephemeris and clock offset parameters based on satellite-borne autonomous navigation software and inter-satellite ranging data. In this section, the precise ephemeris and clock offset parameters provided by the operation control were used as standards to evaluate the accuracy of the ephemeris and clock offset parameters of the autonomous navigation. The performance of spaceborne autonomous navigation algorithms and software was examined.

The basic process of data processing was to take the flight version of the spaceborne autonomous navigation software as the core. Satellite inter-satellite Ka ranging was processed. Autonomous navigation broadcast ephemeris and clock offset parameters were generated. In addition to the core spaceborne autonomous navigation software, the data analysis software also included an inter-satellite interaction module, an operation control simulation injection module, and an evaluation module. The inter-satellite interaction module completed the inter-satellite interaction of autonomous navigation messages, variances, and directional parameter corrections. The link-building relationship and time slot table were restored based on the telemetry data. The mode of inter-satellite interaction was consistent with the on-board network protocol software. The operation control simulation injection module completed the simulation of betting on autonomous navigation support information. The software parameter setting was consistent with the on-orbit test.

The interstellar measurement data file, from 671 weeks 34,600 s to 675 weeks 601200 s (November 15, 2018–December 15, 2018), was a total of 30 days. The data of 14 satellites were analyzed and processed, namely M1–M14 (PRN is 36, 37, 38, 39, 40, 41, 25, 26, 27, 28, 29, 30, 42, 43).

Figure 5 shows the processing results of time synchronization. The figure above is the 14 satellite autonomous clock difference parameter minus the precision clock difference data, which represents the absolute timing accuracy. The following is the 14 satellite time synchronization error. Figure 6 is the URE evaluation result of ephemeris filtering state quantity, and the URE evaluation result of autonomous navigation broadcast ephemeris at the end of two hours. And the URE evaluation results of the two in the three aspects of RTN.

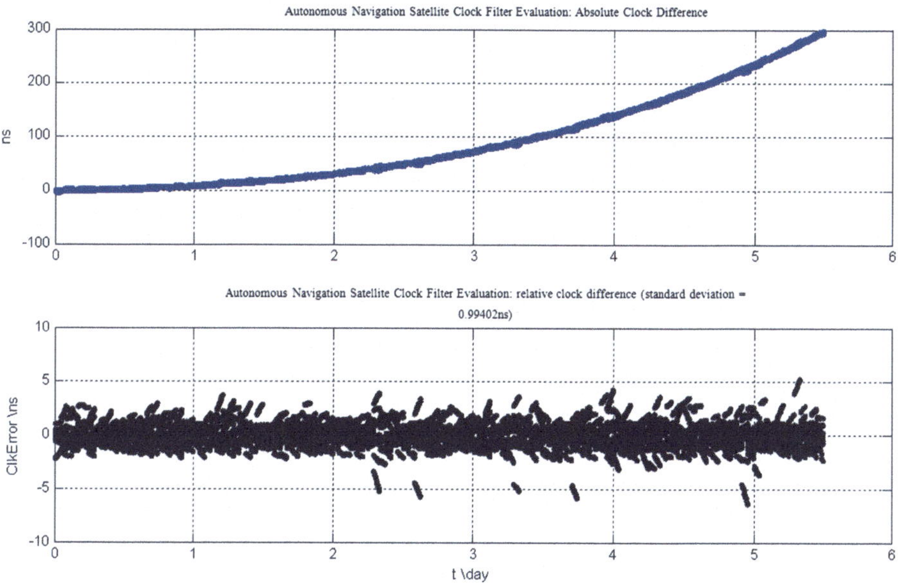

Fig. 5. Autonomous time synchronization results of 14 satellites in orbit

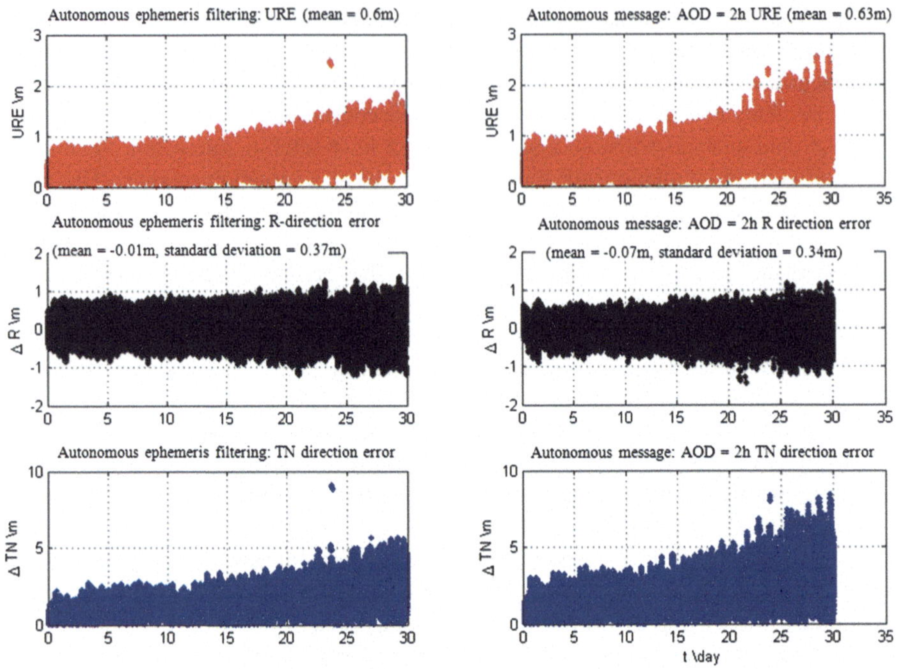

Fig. 6. Results of autonomous orbit determination by 14 in orbit satellites

The above test results based on in-orbit data show that the time synchronization of the 14 constellations was about 1 ns. Compared with the ephemeris precision ephemeris of the ephemeris autonomous filtering state, the error variance in the R direction was 0.37 m. The error in the plane direction was basically stable in the first 10 days, not exceeding 3 m. And there was a slight divergence trend after 10 days, which increased to not more than 6 m. During the 30-day period, the mean URE of the filtered state quantities was about 0.6 m. The maximum was no more than 2 m. For broadcast ephemeris, the accuracy of each autonomous text at the end of two hours was evaluated. Compared with the operation control broadcast ephemeris, the error variance in the R direction was 0.34 m. The error in the plane direction was basically stable in the first 10 days, not exceeding 4 m. After 10 days, there was a slight divergence trend, which increased to no more than 8 m. During the 30-day period, the mean URE of the broadcast ephemeris was about 0.63 m. The maximum was no more than 3 m.

4 Conclusion

The new generation of BDS satellite navigation system includes 24MEO + 3IGSO + 3GEO satellite, referred to as BDS-3. The constellation has the inter-satellite ranging function supported by the inter-satellite link. The introduction of inter-satellite links/ inter-satellite ranging enables BDS-3 satellites to autonomously complete the determination of orbits and clock differences without relying on the ground at all, thereby achieving system autonomous navigation services. Autonomous navigation technology can effectively reduce the number of ground stations, reduce the number of times of information injection from ground stations to satellites, and improve system survivability.

This article detailed the autonomous navigation technology of Beidou system. The long-term results of autonomous navigation of a basic navigation system composed of 18 MEO satellites were simulated. The average URE of autonomous orbit determination for more than 10 days of continuous operation was less than 2.85 m. The in-orbit results of the 14 satellite autonomous navigation launched into the network were evaluated. The in-orbit test results showed that during 30 days, the time of the 14 constellations was synchronized with a variance of about 1 ns. The mean value of the URE of the filtering state was about 0.6 m, and the maximum was not more than 2 m. The average URE of the broadcast ephemeris was about 0.63 m, and the maximum didn't exceed 3 m.

References

1. Chen, Z., Liu, G., Liao, Y., Wen, Y.: Autonomously updated broadcast ephemeris algorithm. J. Natl. Univ. Defense Technol. **33**(3), 1–4 (2011)
2. Liu, J., Zeng, X., Xia, L., Zhao, Q.: Algorithm and simulation of autonomous orbit determination for navigation satellites. Geomatics Inf. Sci. Wuhan Univ. **29**(12), 1040–1044 (2004)
3. Zhang, Y., Zhang, Y.: Design and implementation of autonomous navigation of constellation. J. Astronaut. **24**(5), 525–528 (2003)
4. Song, X., Mao, Y., Jia, X., Wu, X.: The distributed processing algorithm for autonomously updating the ephemeris of navigation satellite by inter-satellite links. Geomatics Inf. Sci. Wuhan Univ. **35**(10), 1161–1164 (2010)
5. Wang, H., Chen, Z., Chu, H., Wu, X., Zheng, J.: On-board autonomous orbit prediction algorithm for navigation satellites. J. Astronaut. **33**(8), 1019–1023 (2012)
6. Liu, L., Wang, H., Hu, S.: Summary on satellite orbit determination. J. Spacecraft TT&C Technol. **24**(2), 28–31 (2005)
7. Schutz, B.Z., Tapley, B.D., Born, G.H.: Statistical Orbit Determination, pp. 1–11. Elsevier Academic Press, Burlington (2004)
8. Guo, H., Guo, J., Zhao, Q., Wang, C.: Precise orbit determination for Beidou-3 experimental satellites and its impact on PPP. J. Geodesy Geodyn. **37**(12), 1263–1266 (2017)
9. Wang, J., Wang, J., Chen, J.: BDS broadcast ephemeris fitting based on satellite's position velocity. J. Tongji Univ. (Nat. Sci.) **44**(1), 155–159 (2016)
10. Gu, Y., Chen, Z., Shuai, P.: Autonomous Time Synchronization Algorithm on Different Crosslink Structures among Navigation Satellites, CSNC 2010 (2010)
11. Wang, H., Wu, X., Chen, Q.: New design for GPS broadcast ephemeris with 24 parameters. Spacecraft Eng. **23**(4), 27–32 (2014)

Research on Inter-satellite Link Network Routing Algorithm Based on Multi-objective Optimization

Sixin Wang[1(✉)], Qi Wang[2], Hao Yin[1], and Yu Zhou[1]

[1] The 6th Research Institute of China Electronics Corporation, Beijing 10083, China
sixin_wang@qq.com, yinhao@ncse.com.cn, zhouyu0909@bupt.edu.cn

[2] Beijing Institute of Tracking and Telecommunications Technology, Beijing 10083, China
softfiona@163.com

Abstract. In view of the problem that the inter-satellite link network routing planning algorithm adapts to the single business scenario and has poor generalization ability, this paper proposes a multi-objective optimized inter-satellite link routing model based on the breadth-first routing algorithm. In this model, the optimization objects are the data transmission performance and connectivity of the time-varying inter-satellite network, the inputs are the point-to-point path hop threshold, the preferred path number threshold, and the node load weight in the route planning algorithm, and the evaluating indicators are data transmission delay, satellite node load, and network connectivity. In this paper, the fast non-dominated sorting genetic algorithm with elite strategy (NSGA-II) is used to solve the model. In addition, this paper establishes a data transmission model based on the STDMA communication protocol of the satellite navigation system to obtain the data transmission delay in each iteration, which is composed of 24 MEO, 3 GEO and 3 IGSO satellites. Finally, the Pareto solution set of point-to-point path hop threshold, the preferred path number threshold, and the node load weight is determined when the data transmission performance and network connectivity performance of inter satellite link are optimal. According to the Pareto solution set obtained in this paper, it can effectively modify the key parameter settings of the routing planning algorithm and improve the generalization ability of the algorithm in multi service scenarios.

Keywords: Inter-satellite link · Routing planning · Multi-objective optimization · Data transmission · Network connectivity

1 Introduction

The inter-satellite link network is a wireless network with precision measurement and data transmission functions established between navigation satellites. The satellite navigation system based on inter-satellite links can achieve autonomous constellation orbit determination by inter-satellite ranging. On the other hand, the system can realize

J. Sun et al. (Eds.): CSNC 2020, LNEE 650, pp. 702–712, 2020.
https://doi.org/10.1007/978-981-15-3707-3_65

the information transmission between satellites and satellites, and support the system to operate independently of the ground station for a period of time. From the perspective of engineering practice, the establishment of a highly connected inter-satellite network can reduce the dependence on the ground and enhance the satellite's measurement performance and fault tolerance, but earth-satellite-station data interaction performance will be significantly reduced correspondingly [1, 2].

At present, the research on routing planning algorithms for inter-satellite link networks is mainly aimed at a single service scenario, and the performance of the algorithm is mainly evaluated from the satellite-ground data transmission time delay [3–5]. But in terms of the original intention and development direction of the inter-satellite link, data transmission is only one of the business scenarios. In addition, building a robust, reliable, and highly fault-tolerant inter-satellite link network is also an important business requirement. The current routing planning algorithms do not consider this problem enough, and the generated routing planning results must not meet the connectivity requirements of the inter-satellite network.

Aiming at this problem, this paper will propose an inter-satellite link routing planning model based on multi-objective optimization based on the breadth-first [6] routing planning algorithm of the inter-satellite link. The model planning goal is to improve the data transmission performance of the inter-satellite link. And time-varying network connectivity. The model inputs are the "point-to-point" path hop threshold, the optimal path number threshold, and the node load weight in the routing planning algorithm. The model evaluation indicators are data transmission delay, satellite node load, network connectivity. By solving the model, the optimal input conditions of the routing planning algorithm are determined.

2 Routing Planning Algorithm Design

Directional antennas are usually are used to construct inter-satellite link networks under the space-time division multiple access (STDMA), and directional antenna can only establish at most one link in the same time period, the link with different satellites by adjusting antenna direction to meet the communication and measurement requirements of inter-satellite links. Therefore, it is necessary to reasonably plan the sequence and duration of satellite chain building to build an efficient inter-satellite link. Based on the STDMA, the planning period can be divided into multiple time slots (minimum time units), and the inter-satellite link is established at any time slot in the planning cycle considering the inter-satellite visual relationship and various business constraints, that is, the dynamic change of the inter-satellite link network topology. The above process is called slot planning.

From the perspective of data transmission, the network topology at any timeslot can not have full connectivity due to the constraints of link resources, and the data transmission between two satellites that can not build the link directly must be completed with the help of the network topology relationship under multiple time-slots. The routing planning mainly determines the jump paths between any two satellites through path search based on the slot planning results, and select the actual paths between satellites from jump paths based on path evaluation factors such as the number of hops, transmission time delay, and load of the jump path. Forwarding path. At the same time, the routing table is generated according to the selected point-to-point paths, which is used to data transmission between the inter-satellite link network nodes.

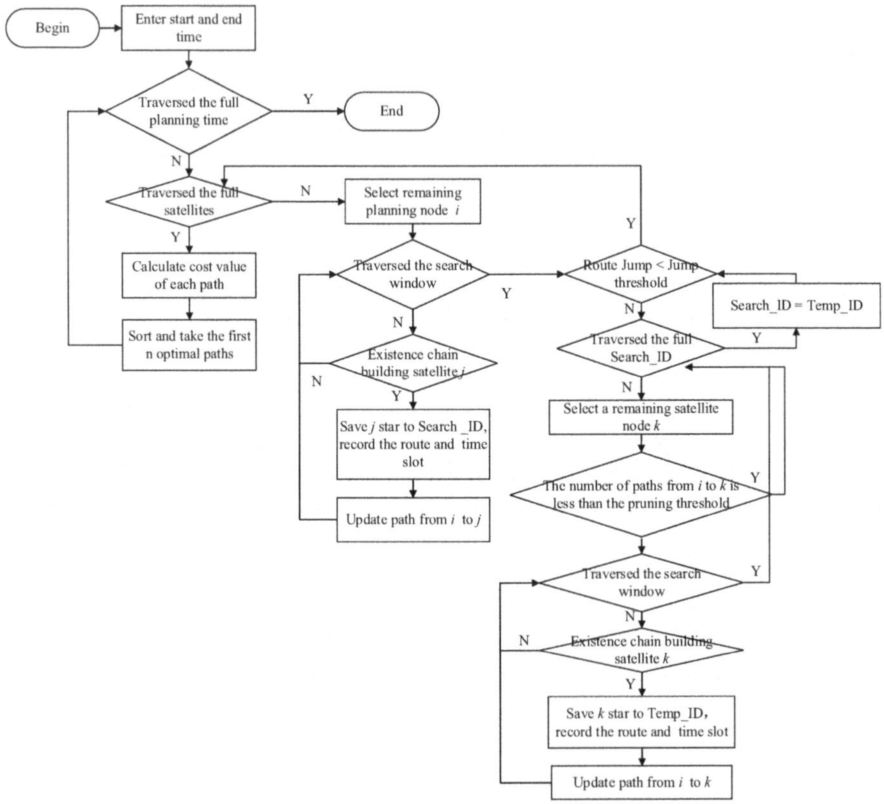

Fig. 1. Flow chart of inter satellite link network routing planning algorithm

The routing planning algorithm flow designed in this paper is shown in Fig. 1. The planning algorithm takes 1 min as the planning unit, and there are three basic process modules in each planning unit, which are: searching paths → sorting paths → generating routing table.

2.1 Path Search

This paper uses breadth-first (BFS) algorithm for path search. As shown in Fig. 2, the search starts from a certain network node A_0, the node set A_1 and the corresponding time slot $Slot_1$ that the node A_0 link directly can be obtained by searching the time slot planning table, which is called one-hop path search from A_0. The next search starts from one node of set A_1, and the node set A_2 and the corresponding time slot $Slot_2$ that the node A_1 link directly can be obtained after $Slot_2$. The two-hop path search is

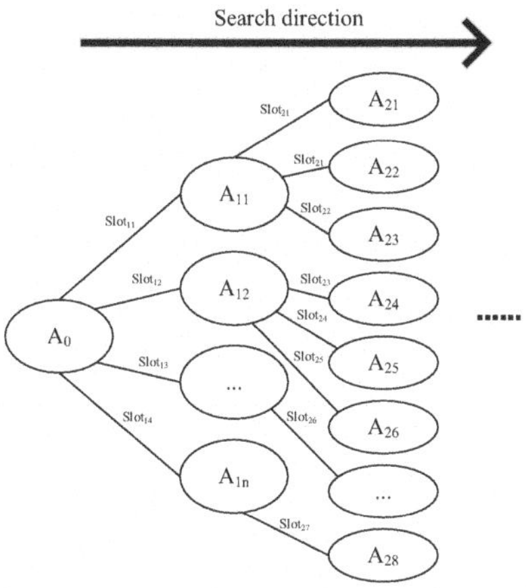

Fig. 2. Schematic diagram of path breadth first search

completed until all nodes of set A_1, are searched. Repeating the above process until the jump threshold is reached.

Considering that the BFS has high overhead and affects the operation efficiency of the algorithm, the path search algorithm adopts path pruning strategy and time slot sliding window strategy to reduce the overhead of the path search algorithm.

- Pruning strategy: Pruning strategy mainly includes two aspects. One is to delete invalid paths in time and stop searching them, such as backtracking paths, paths including two ground stations; If the number of paths between them exceeds the set pruning threshold, the search will stop searching for deeper direction.
- Time slot sliding window strategy: The time slot sliding window strategy is mainly considered from the perspective of path time delay. In path search, there are some paths that exceed the time delay threshold. For example, the node a_1 links the node a_2 at the first time slot, and the node a_2 links a_3 at nineteenth time slot. The above paths will be eliminated in path selection. Setting a time window can delete paths that exceed the time threshold in time to avoid the time overhead caused by this type of path search.

2.2 Path Sorting and Selection

After the path search is over, sorting multiple paths with the same start and end points according to the comprehensive evaluation, and selecting the top n optimal paths instead of reserve all paths.

For the evaluation of the paths, the path time delay and load balancing are considered comprehensively, and the path with low path delay and less load is preferentially selected. For the i-th path, the cost function can be expressed as:

$$C(i) = w_1 * D(i) + w_2 * L(i) \quad (1)$$

Where:

$D(i)$ is the transmission time delay of path i, and its mathematical expression is:

$$D(i) = Slot_{end}(i) - Slot_{begin}(i) + 1 \quad (2)$$

$Slot_{begin}(i)$, $Slot_{end}(i)$ are the start time slot and the end time slot of path i, respectively.

$L(i)$ is the load of path i, and its mathematical expression is:

$$L(i) = L_1(i) + L_2(i) + L_3(i) + \ldots + L_m(i) \quad (3)$$

Among them, $L_1(i), L_2(i), L_3(i), \ldots, L_m(i)$ are the real-time loads of the m intermediate nodes of path i in the corresponding time slots.

w_1 and w_2 are the weights of transmission time delay and load, respectively.

2.3 Generation of Route Planning Table

According to the data transmission system of the inter-satellite link network, a network node communicates with at most another network node in a certain time slot, so the dynamic routing planning results can be saved in the form of a static 0–1 table. Assume that a three-dimensional array $R[N][N][T]$ is used to store the routing plan table, where N is the number of network nodes and T is the number of time slots. When $R[i][j][k]$ is set to 1. It means that when node i is in the current time slot k, the data transmitted to the destination node j should be sent out; otherwise, the value is 0, it means that the data transmitted to the destination node j should wait.

3 Building a Multi-objective Optimization Model for Routing Planning

The route planning determines whether the data arriving at a destination node is forwarded by the current satellite according to the node link relationship. Therefore, there must be a link establishment without forwarding within a certain time in the inter-satellite link network, that is, the actual network topology is finally determined by the routing planning.

The evaluation of the effect of routing planning is mainly considered from two aspects: on the one hand, it can make the data transmission delay small, and on the other hand, it makes the connectivity high to improve the robustness and fault tolerance of the network.

Aiming at the two optimization goals of delay and network connectivity, this paper establishes the following multi-objective optimization model to determine the optimal value of the variable parameters of the routing planning algorithm.

3.1 Objective Function

1. Transmission time delay

Transmission delay is an important indicator to measure the pros and cons of routing planning algorithms. Therefore, the following optimization goals exist in the routing planning model:

$$\min TimeDelay \tag{4}$$

The goal is to minimize the data transmission delay. There are generally two ways to calculate the transmission time delay. One is to calculate the transmission delay of each path and then calculate the average value. The other is to establish a data transmission model and calculate the average transmission delay of the data from the time slot span of the starting node to the destination node by simulating the process of data generation, forwarding, and receiving processes. From the comparison of effect, the time delay obtained by data transmission simulation is generally higher than the calculation delay of the path, the main reason is that the overlap of the time and space of the path node easily leads to the phenomenon that the actual path is not forwarded according to the planned path. This phenomenon is also the route planning is unreasonable, and the problem cannot be found through the path calculation delay. Therefore, this paper uses a simulation method to calculate the transmission delay. In this paper, a navigation system consisting of 24 medium-orbit satellites (MEO), 3 geostationary orbit satellites (GEO), and 3 inclined geosynchronous orbit satellites (IGSO) and 8 ground stations is establish ed (24 MEO satellites use Walker 24/3/1 Constellation structure) and simulate data transmission [7, 8].

2. Network connectivity

Another important indicator of route planning is to establish a high connectivity inter-satellite network to improve the fault tolerance of the network. Therefore, the route planning model also has the following optimization goals:

$$\max Connectivity \tag{5}$$

For the calculation of network connectivity, this paper calculates the network degree (Degree) value per minute and then takes the average value to characterize the connectivity of the entire network in a period.

$$Connectivity = \left(\sum_{t=1}^{T} \sum_{n=1}^{Slot} \sum_{i=1}^{N} \sum_{j=1}^{N} R(i, j, (t-1) * N + n) \right) / T \tag{6}$$

Where T is the planning time (minutes), s is the number of timeslots per minute, and N is the number of nodes, $R(i, j, (t-1) * S + s)$ indicating the routing relationship between i and j in the timeslot of $(t-1) * S + s$.

In order to unify with the transmission delay, the network connectivity is reciprocal, that is, the optimization goal of maximizing the network connectivity is transformed into a minimization problem, namely:

$$\min \frac{1}{Connectivity} \tag{7}$$

3.2 Input Variables and Value Constraints

According to the design of routing planning algorithm, the optimizable variables in the routing planning model include the number of hops (*Jump*), the number of point-to-point reserved paths (*RouteNum*), and the load weight (w_2).

The number of hops and the number of reserved paths are integers in the range of 2–7, and the load weight is real in the range of 0–1.

4 Model Solving Based on NSGA-II

The inter-satellite link network routing planning model proposed in this paper has two different dimensions of optimization goals: minimizing transmission delay and maximizing network complexity, which belongs to multi-objective optimization. Its optimal solution is a set of non-dominated solutions. Since the general genetic algorithm (GA) is not suitable for multi-objective optimization problems, this paper uses a fast non-dominated sorting genetic algorithm with an elite strategy (NSGA-II) to solve it.

4.1 Pareto Optimal Solution and Frontier

For multi-objective optimization model, it is usually difficult to determine the only optimal solution, but to obtain a solution set. It is impossible to compare the advantages and disadvantages of the solutions in the solution set through the objective function, because improving any objective value will lead to the worse of other objective values. The solution set is called the Pareto solution set. The surface formed by the Pareto solution set corresponding to the target value is called the Pareto optimal front [9].

4.2 Characteristics of NSGA-II Algorithm

Compared with conventional genetic algorithms, the advantages of NSGA-II are:

- Fast non-dominated sorting: NSGA-II can sort the dominating relationships of the individual populations before executing selection operator, which can determine the optimal Pareto solution set and ensure these solution sets have a greater probability of inheriting to the next generation. Therefore, it is suitable for multi-objective optimization problems, and reduces the connectivity of sorting algorithm from mN^3 to mN^2 [10] compared with the first generation NSGA algorithm.
- Elite retention strategy: The parent population and its offspring population are used to generate a new generation of population through competition to ensure that the

elite population is retained during the evolution process, which reduce the cost of finding the optimal solution and improving the algorithm's convergence speed.

- Ensuring the diversity of the population based on the degree of crowding: The degree of crowding represents the density of the surrounding individuals of the individuals in the population. It is generally expressed by calculating the sum of the differences in the distances of multiple targets between adjacent individuals [11]. As shown in Fig. 3, the calculation formula for the congestion degree of the *i*-node is:

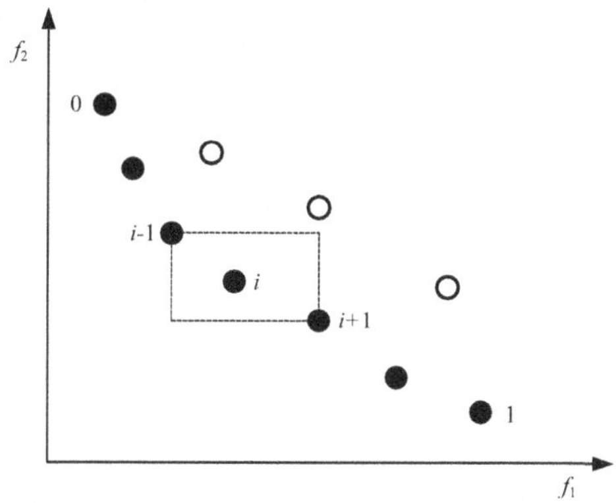

Fig. 3. Schematic diagram of individual crowding degree of population

$$Di = |f_{i+1,1} - f_{i+1,1}| + |f_{i+1,2} - f_{i+1,2}| \quad (8)$$

According to the above analysis, during the evolution process, each population individual will calculate the non-dominated level and crowding degree. For the non-dominated solutions belonging to different levels, priority is given to the solution with a lower rank number, and for the non-dominated solutions of the same level, crowding is preferred Degree of solution.

4.3 Chromosome Coding

Coding is the basis for NSGA-II to solve optimization problems. This paper uses a real number coding scheme to code the optimization problems. The length of the chromosome is taken as the total number of input variables 3, and its values represent the number of hops, the number of point-to-point reserved paths, and the weight of the load.

4.4 Model Solving Process

- Generating an initial population P_0 of a specific size according to the chromosome coding method and the range of each genetic locus, and performing fast non-dominated sorting on all initial population individuals;

- Specifying one of the target values as the evolution fitness value, and generating a new generation of population C_0 through evolution operators including selection, crossover, and mutation;
- Combining the offspring population C_0 and the parent population P_0 into N_0, performing a fast non-dominated sorting on N_0, determining the non-dominated level of each individual, and calculating the degree of congestion;
- Putting the non-dominated rank-first solution set into the next generation population P_1, where the number of individuals in P_1 should be equal to the population size. When selecting the next generation of individuals, non-dominated level is preferred. When the non-dominated levels are equal, the solution with a large congestion degree is selected.
- Iterating the above process until the maximum number of iterations is reached.

5 Simulation Analysis

5.1 Simulation Scenarios and Parameter Settings

In this paper, an inter-satellite and inter-satellite link network consisting of 24 medium-orbit satellites (MEO), 3 geostationary orbit satellites (GEO), 3 inclined geosynchronous orbit satellites(IGSO), and 8 ground stations is established. Among them, 24 MEO satellites adopt Walker 24/3/1 constellation structure. For this network, the slot planning table is used as the model input, the route table is planned for 7-days based on routing planning algorithm, the NSGA-II algorithm is used to solve pareto solution of input variables when the time delay and complexity are optimal. The parameter settings of the NSGA-II algorithm are shown in Table 1.

Table 1. NSGA-II algorithm parameter setting

Parameter name	Parameter value
Population size	100
Population dimension	3
Number of evolutions	200
Number of targets	2
Cross probability	0.8
Mutation probability	0.05

5.2 Results and Analysis

According to the simulation scenario described in Sect. 5.1, the NSGA-II algorithm was used to solve the multi-objective optimization model of inter-satellite link network routing planning. The Pareto optimal solution is shown in Table 2. The Pareto optimal frontier was obtained. As shown in Fig. 4.

Table 2. Pareto optimal solution

Num.	Input variable			Target value	
	Jump	*RouteNum*	w_2	*1/Connectivity*	*TimeDelay*
1	7	7	1	0.236081	11.06856
2	6	7	1	0.248113	10.63843
3	7	6	1	0.260191	10.27942
4	6	5	1	0.293938	9.47555
5	5	6	0.964274	0.311365	9.315984
6	5	5	0.980855	0.329308	8.965413
7	7	4	0.323812	0.350775	8.856639
8	5	3	0.384792	0.438116	7.809732
9	3	2	0.258896	2.98112	4.92023
10	2	2	0.255812	3.02267	3.702435

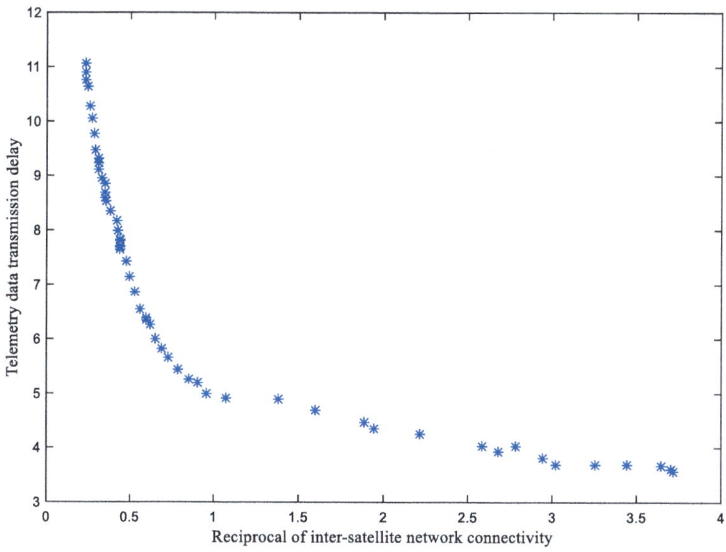

Fig. 4. Pareto optimal frontier

According to the Pareto optimal front, it can be determined that the connectivity inverse of the interstellar network is high, namely when the inter-satellite network connectivity is low, the data transmission delay is relatively low, so the two indicators are mutually restricted. The Pareto optimal solution has important guiding significance for the value of the routing planning algorithm under different engineering requirements. When the network connectivity needs to be met first, the hop planning and optimal path number of the routing planning algorithm should be above 5. The node load weight should be above 0.9; conversely, when the data transmission delay requirement is preferentially satisfied, the hop count and optimal path number of the routing planning algorithm should be below 3, and the node load weight should be below 0.3.

References

1. Shi, L.G.: Research on Key Technologies of Inter Satellite Links for Distributed Satellites System. National Center for Space Science, Chinese Academy of Sciences (2016)
2. Gao, H.: Research on Inter-satellite Link Allocation Method in Beidou Navigation System. Hunan University, Changsha (2018)
3. Li, R.N.: Research on Inter-satellite Link Cross-layer Routing Algorithm. Beijing University of Posts and Telecommunications, Beijing (2018)
4. Li, Z.X.: Research on Inter-satellite Link Assignment and Algorithm for Navigation Constellation Network. Nanjing University, Nanjing (2019)
5. Teng, Y., Wang, Y.K., Chen, J.Y., et al.: Inter-satellite link directivity algorithm research and performance validation. Chinese Journal of Scientific Instrument (2014)
6. Xu, Q.Z., Han, W.Y., Chen, J.S.H., et al.: Optimization of breadth-first search algorithm based on many-core platform. Comput. Sci. **46**(01), 314–319 (2019)
7. Shao, F.W., Gong, W.B., Jiang, X.L.: General simulation model of navigation satellite system based on OPNET. Electron. Des. Eng. **25**(14), 105–110 (2017)
8. Mo, Y.: Multi-Objective Optimization Design of LEO Satellite Constellations for Communication. National University of Defense Technology (2016)
9. Cao, H.T.: Research on Multi-objective Flexible Job Shop Scheduling Problem Based on Improved NSGA-II. Zhejiang University of Technology (2019)
10. Han, K.Q., Ding, D.J., Qian, K.J., et al.: Multi-objective charging network planning for electric vehicles based on NSGA-II (2017). https://doi.org/10.3969/j.issn.1009-1831.2017.S1.026
11. Wang, W.W.: Research on Location Model of Emergency Distribution Center based on Improved NSGA-II Algorithm. Wuhan University of Technology (2018)

Pedestrian Autonomous Positioning System Based on Inertial Navigation

Bowen Xing[1,2(✉)], Haonan Jia[1,2], Pengfei Liu[1,2], Guoju Ma[1,2], and Xiaonan Li[1,2]

[1] The 54th Research Institute of CETC, Shijiazhuang 050081, China
xingbowenstu@163.com
[2] State Key Laboratory of Satellite Navigation System and Equipment Technology, Shijiazhuang, China

Abstract. Under the background of mobile Internet era, the application requirements of indoor positioning technology is increasingly urgent, and accurate positioning information has become an important prerequisite and guarantee for the successful completion of various tasks. Since GNSS faces problems such as wireless signal attenuation and positioning accuracy degradation in a complex indoor environment, an autonomous positioning system based on the strapdown inertial navigation principle has become an important way to obtain position information in complex environments. Aiming at the requirements of wearable, miniaturization and accurate positioning of the autonomous positioning system, this paper carried out the structure and autonomous positioning algorithm design. An algorithm structure combining complementary filtering attitude solution, zero-speed detection and zero-speed update was proposed, the system miniaturization structure design and low-power power management strategy were also studied. The system can complete all-weather, full-area, high-concealment three-dimensional real time positioning. Experimental results show that the system's autonomous positioning accuracy is within 6‰ of the walking distance.

Keywords: Inertial navigation · Autonomous positioning · Algorithm design · Hardware design

1 Introduction

Pedestrian positioning and navigation has a growing demand in the field of military single combat system, underground venue location service, large-scale venue guidance and so on [1, 2]. Positioning technologies based on WIFI [3], pseudo-satellite [4], radio frequency identification [5] and other methods are constantly emerging, but these technologies all need external auxiliary equipment, and the scope of use is limited. This paper focuses on the autonomous positioning system based on the principle of inertial navigation [6, 7]. The algorithm based on complementary filtering is used in attitude calculation. This algorithm structure has been verified to provide effective fusion performance with a relatively less calculation [8]. In addition, in order to improve positioning accuracy, zero-speed detection and update algorithms are used to eliminate

J. Sun et al. (Eds.): CSNC 2020, LNEE 650, pp. 713–722, 2020.
https://doi.org/10.1007/978-981-15-3707-3_66

accumulated errors. In hardware design, it mainly aims at miniaturization and low power consumption.

2 Fundamental Principles

The overall structure of the algorithm of pedestrian autonomous positioning system is shown in Fig. 1. Firstly, the output values of gyroscope and accelerometer are calibrated and low-pass filtered. Then, the complementary filtering algorithm is used to calculate the attitude, and the position change of pedestrian can be obtained by integrating the speed. Finally, in order to get more accurate position information, the zero-speed update algorithm is used to further suppress the integration error.

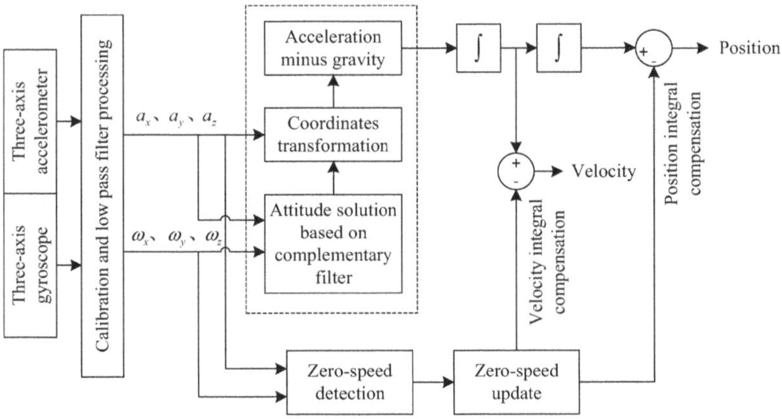

Fig. 1. Overall structure diagram of pedestrian autonomous positioning system algorithm

3 Attitude Solution Based on Complementary Filtering

Normally, indoor pedestrians will not have severe unidirectional vibrations during walking, running and ascending and descending stairs, and the frequency of shaking is relatively low. Therefore, the characteristics of higher accuracy of the accelerometer at lower frequencies can be effectively used to continuously modify the angular rate measured by the gyroscope, and output more accurate roll and pitch angles. By using the complementary characteristics of the two sensors in the frequency domain [9], a complementary filter is designed to fuse the data of the two sensors to improve the accuracy.

The value of the gravity vector measured by the three-axis accelerometer in the body coordinate frame is expressed as $a = [a_x \quad a_y \quad a_z]^{\mathrm{T}}$, the unit is m/s^2, and the angular rate measured by the gyroscope is denoted as $\omega = [\omega_x \quad \omega_y \quad \omega_z]^{\mathrm{T}}$, the unit is rad/s. Let the attitude transformation matrix of the geodetic coordinate system

(E system) relative to the body coordinate system (B system) be C_e^b, define the initial quaternion $q = [q_0 \quad q_1 \quad q_2 \quad q_3] = [1 \quad 0 \quad 0 \quad 0]$, and then:

$$C_e^b = \begin{bmatrix} 1-2(q_2^2+q_3^2) & 2(q_1q_2+q_0q_3) & 2(q_1q_3-q_0q_2) \\ 2(q_1q_2-q_0q_3) & 1-2(q_1^2+q_3^2) & 2(q_2q_3+q_0q_1) \\ 2(q_1q_3+q_0q_2) & 2(q_2q_3-q_0q_1) & 1-2(q_1^2+q_2^2) \end{bmatrix} \tag{1}$$

The gravity component V in the body coordinate system can be obtained from the attitude transformation matrix,

$$V = \begin{bmatrix} v_x \\ v_y \\ v_z \end{bmatrix} = C_e^b \begin{bmatrix} 0 \\ 0 \\ 1 \end{bmatrix} \tag{2}$$

The measurement error vector ω_{error} of the gyroscope is calculated by the following formula, where, "$\wedge$" represents the normalized vector.

$$\omega_{error} = \hat{a} \times V \tag{3}$$

Input the error into the PI controller, adjust the parameters of the filter, and then correct the angular velocity value measured by the gyroscope:

$$\begin{aligned} \Delta\omega &= K_p\omega_{error} + K_i dt \sum \omega_{error} \\ \omega &= \omega + \Delta\omega \end{aligned} \tag{4}$$

Extend the angular rate measured by the three-axis gyroscope to the quaternion q_ω,

$$q_\omega = [0 \quad \omega_x \quad \omega_y \quad \omega_z]. \tag{5}$$

The attitude quaternion of the E system relative to the B system at time t is described as Q_t, which can be obtained by integrating the attitude update rate $\frac{dQ}{dt}$, as shown in Eq. (6):

$$\begin{aligned} \frac{dQ_t}{dt} &= \frac{1}{2}\hat{Q}_{mix,t-1} \otimes q_{\omega,t} \\ Q_t &= \hat{Q}_{mix,t-1} + \frac{dQ_t}{dt}\Delta t \end{aligned} \tag{6}$$

where, "$\otimes$" represents the product between quaternions, $q_{\omega,t}$ is the angular rate measured by gyroscope at time t, $\hat{Q}_{mix,t-1}$ is the attitude information of the previous process, and Δt is the sampling period of the sensor.

From the normalized quaternion $\hat{Q}_t$, the attitude transformation matrix C_b^e of the E system relative to the B system can be obtained:

$$C_b^e = \begin{bmatrix} q_0^2 + q_1^2 - q_2^2 - q_3^2 & 2(q_1 q_2 - q_0 q_3) & 2(q_1 q_3 + q_0 q_2) \\ 2(q_1 q_2 + q_0 q_3) & q_0^2 - q_1^2 + q_2^2 - q_3^2 & 2(q_2 q_3 - q_0 q_1) \\ 2(q_1 q_3 - q_0 q_2) & 2(q_2 q_3 + q_0 q_1) & q_0^2 - q_1^2 - q_2^2 + q_3^2 \end{bmatrix} \tag{7}$$

4 Zero-Speed Detection and Update

4.1 Zero-Speed Detection

The movement cycle of a person's foot during walking can be divided into four stages as shown in Fig. 2: S1 stage, stationary stage, S2 stage, and stepping stage. Among them, S1 stage represents the process from heel to toe touching the ground; The stationary stage is a state that the foot completely touches the ground. In this stage, the output of the gyroscope is approximately zero, and the output of the accelerometer is approximately equal to the acceleration of gravity, this stage is also called zero-speed interval; The S2 stage represents the process of lifting the toes off the ground from the heel. The gait characteristics of running and walking are basically similar, except that the frequency of the feet is different. The gait characteristics of going up and down stairs and walking are slightly different, but there is also a zero-speed interval in each gait cycle.

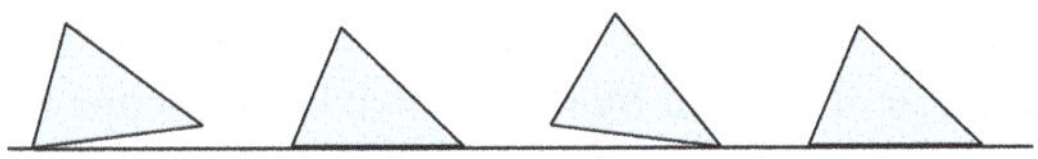

Fig. 2. Schematic diagram of foot movement period

In this paper, the generalized likelihood ratio tests (GLRT) method is adopted to identify the zero-speed interval, which can meet the requirements of the autonomous positioning system for pedestrians to judge the zero-speed interval in real time and stably. The specific discrimination conditions are as follows [10]:

$$\frac{1}{N}\sum_{k \in \Omega_n}\left(\frac{1}{\sigma_a^2}\left\| y_k^a - g\frac{\bar{y}_n^a}{\|\bar{y}_n^a\|}\right\|^2 + \frac{1}{\sigma_\omega^2}\|y_k^\omega\|^2\right) < -\frac{2}{N}\ln\gamma \tag{8}$$

Where, N is the size of the sliding window, $y_k^a \in R^3$ and $y_k^\omega \in R^3$ represent the measurement results of accelerometer and gyroscope respectively, σ_a^2 and σ_ω^2 represent noise variances of accelerometer and gyroscope respectively, $\|\bar{y}_n^a\|$ is the mean value of the acceleration output in the sliding window, γ is the detection threshold, whose value can be determined according to the actual situation.

4.2 Zero-Speed Update

Due to the drift characteristics of the inertial device, there will be a large integral drift deviation in the speed and position information obtained directly from the complementary filtering solution process. The zero-speed update algorithm can eliminate this deviation to a large extent. The flowchart is shown in Fig. 3.

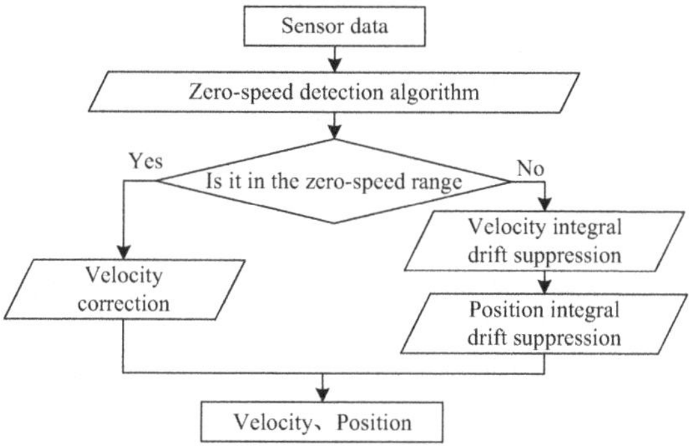

Fig. 3. Zero-speed update algorithm flowchart

At the end of each step period, the drift error (only occurs only during the non-stationary step) is expressed as the error between the integral and the theoretical value. Based on the assumption that the error will accumulate at the end of each step, the following model is applied to eliminate the velocity drift:

$$V_{corrected}(t) = V(t) - \frac{V(T)}{T}t \tag{9}$$

where, $V(t)$ is the velocity obtained by integrating the output value on the x, y and z axes of the accelerometer, $V(T)$ is the calculated velocity at the end of the step interval, t is the instantaneous time of the current step interval, T is the duration of the step interval, and $V_{corrected}(t)$ is the entire step speed after interval correction.

The position of the pedestrian can be obtained by integrating the corrected speed, but a large drift error will occur in the height value, so the second correction of the vertical displacement is performed here. Taking advantage of the strong regularity of vertical displacement changes of pedestrians during indoor walking, the movement state of pedestrians is divided into two types: walking on flat ground and going up and down stairs. Due to the good stability of the drift of the inertial devices, the slope of the integral drift is approximately replaced by the slope of the integral drift when walking on the ground at the latest time. Therefore, the following model is used for correction:

$$D_{s\_corrected}(t) = D_s(t) - \frac{D_{s1}(T)}{T}t \tag{10}$$

where $D_s(t)$ is the vertical displacement of the foot obtained through the trapezoid integral of the corrected velocity $V_{corrected}(t)$, $D_{s1}(T)$ is the vertical displacement when walking on flat ground at the latest moment, and $D_{s\_corrected}(t)$ is the corrected vertical displacement.

5 System Hardware Design

5.1 System Structure

The indoor pedestrian autonomous positioning system adopts the foot wearing mode, and the hardware structure is shown in Fig. 4. Functionally, the pedestrian autonomous positioning system is mainly composed of sensors, power supply and communication modules; In terms of hardware, it is mainly composed of wearable structure, battery, processor, memory, clock and micro-inertial sensor, etc. among which the inertial measurement sensor includes three-axis gyro and three-axis accelerometer. The physical picture of the whole system is shown in Fig. 5.

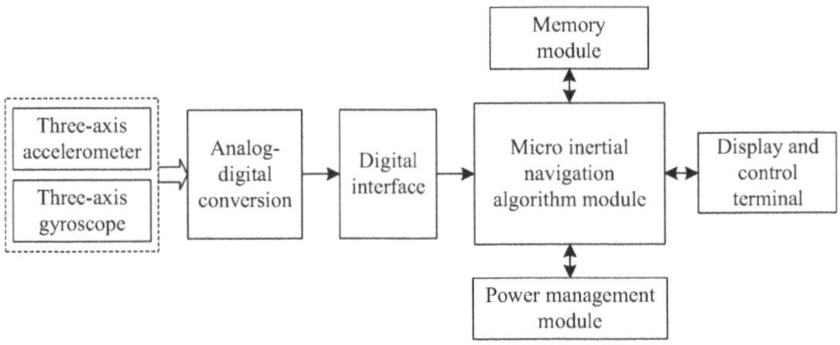

Fig. 4. Navigation microsystem composition and interface diagram

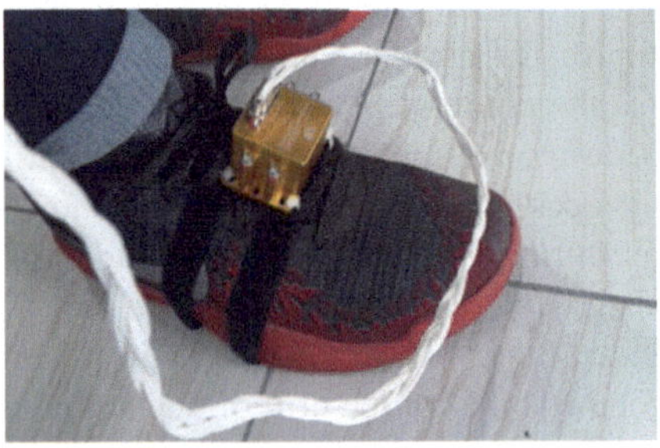

Fig. 5. Physical picture of indoor pedestrian navigation microsystem

5.2 Low-Power Power Management Strategies

Low-power power management is essential for wearable navigation micro-system applications. The dormancy strategy based on motion detection is used to achieve low-power power management. The specific scheme is as follows:

- The microprocessor processes acceleration data to determine whether it is currently in motion or stationary.
- When stationary state over a period of time, the navigation micro-system enters the sleep mode. In the sleep mode, the motion detection accelerometer, microprocessor, crystal oscillator, and micro-inertial resonance mechanism are in operation, and all other chips and mechanisms are in the off state.
- When moving again, enable the power management chip to power the chip and the mechanism in the off state.

6 Experiment

In order to verify the performance and positioning accuracy of the autonomous positioning system, a three-dimensional walking experiment was performed. The distance difference between the end point and the starting point was used to measure the accuracy of the pedestrian positioning. In addition, the experiment is space relative positioning, so it does not involve the initial alignment process. The performance parameters of the inertial devices used in the experiment are shown in Table 1.

Table 1. Performance parameters of inertial devices

Inertial device	Bias stability (1 σ)	Bias repeatability (1 σ)	Measurement range	Unit
Accelerometer	$\leq 5 \times 10^{-4}$	$\leq 5 \times 10^{-4}$	± 15	g
Gyroscope	≤ 10	≤ 10	$\pm 1.08 \times 10^{6}$	$^{\circ}/h$

Testing process: The inertial measurement unit is installed on the back of the pedestrian's feet. The experimental site was chosen as a three-story building. In order to facilitate the determination of walking baseline, the path is planned as a regular route.

Figure 6(a) shows the three-dimensional trajectory obtained by the wearable autonomous positioning system in the above experiment. Figure 6(b) shows the position estimates of x, y and z axes, respectively. The total walking distance is about 250 m, and the positioning error is less than 6‰D (D is the walking distance). Calculated as follows:

$$error = \frac{\sqrt{1.299^2 + 0.1032^2 + 0.3901^2}}{250} = 0.0054$$

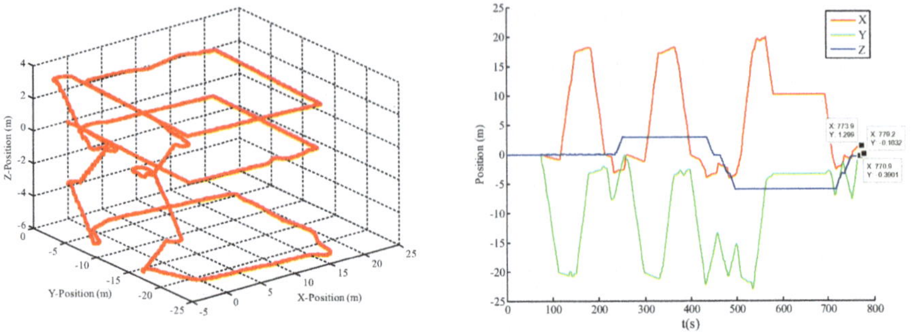

a) b)

Fig. 6. (a) Three-dimensional trajectory diagram of pedestrian autonomous positioning. (b) x, y, z position coordinate estimation diagram

In order to verify the effect of zero-speed detection algorithm, the outputs of accelerometer and gyroscope between 122 s–129 s in the above experiment were intercepted. Figure 7(c) is the detection result of zero-speed detection algorithm. The logical "0" in the figure represents the zero-speed interval, while the logical "1" represents the non-zero speed interval. It can be seen that the algorithm can accurately detect the zero speed interval, and can determine the entry time and end time of the zero-speed interval.

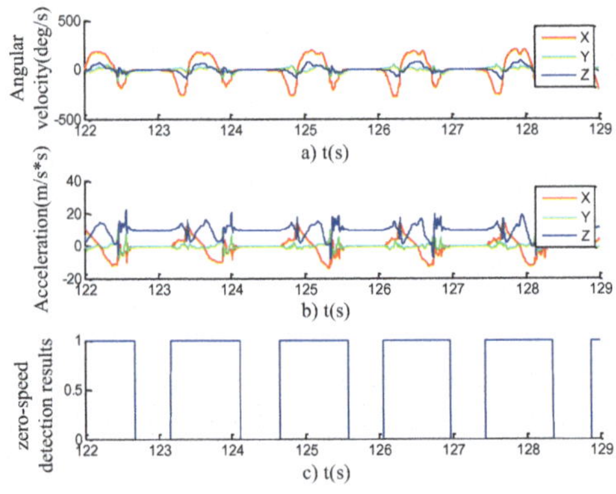

Fig. 7. Comparison of gyroscope, accelerometer output value and zero speed detection result

7 Conclusion

This paper designs an autonomous positioning system based on inertial navigation, which does not need any external reference, and can achieve autonomous and accurate positioning function for a long time indoors. This paper proposes an algorithm structure that uses complementary filtering algorithms to perform attitude resolution, and uses zero-speed detection and update algorithms to perform integral error correction. The system hardware structure design and low-power power management strategies are also studied. It is verified by experiments that the expected target is satisfied and the algorithm is stable, which has a good application prospect.

Fund Projects. National Defense Science and Technology Pre-Research Fund (41418040102, 31512020205-2, 30102120201, 315060203), National Key Research and Development Program (2016YFB0502100).

Appendices

Statement on Ethical Approval

I am the author of this article, Bowen Xing. Now I hereby solemnly declare that the experiments involved in this article have received ethical approval from the Satellite Navigation Department of the 54th Research Institute of China Electronics Technology Group Corporation, which confirms that we have appropriate qualifications to conduct the proposed research projects and be able to conduct the research in a safe and ethical manner.

Informed Consent Form

Dear Haonan Jia,

I am Bowen Xing at the Satellite Navigation Department of the 54th Research Institute of China Electronics Technology Group Corporation. I will conduct a research project on pedestrian autonomous positioning system based on inertial navigation and would like to invite you to participate.

In the experiment, you need to wear our equipment to walk on a predetermined path (it takes about 30 min). We will take photos of your feet, and you have the right to view and delete these photos.

Please complete the reply slip below to indicate whether you do decide to participate in this research. All information obtained will be used for research purposes only. Participant will not be identified by name in any report of the completed study.

Participation is entirely voluntary. This means that you can choose to stop at any time without negative consequences. If you have any questions about the research, please feel free to contact Bowen Xing (email address: xingbowenstu@163.com).

If you understand the contents described above and agree to participate in this research, please sign below. Your help is very much appreciated.

Yours sincerely,

Bowen Xing
The Satellite Navigation Department

Reply Slip

Name of Participant: Haonan Jia

I will participate in the research.

I agree to take photos during the procedure.

Signature: Haonan Jia
Date: 2019.11.26

References

1. Wang, Q.: Research on the Individual Soldier Navigation Technology in Complex Environment Based on MEMS/GPS Integration. Harbin Engineering University (2019)
2. El-Sheimy, N., Youssef, A.: Inertial sensors technologies for navigation applications: state of the art and future trends. Satell. Navig. **1**, 1–21 (2020). https://doi.org/10.1186/s43020-019-0001-5
3. Huang, F., Xiao, S., et al.: Comparison of common algorithm of Wi-Fi indoor location based on RSSI. Inf. Technol. (12) (2017)
4. Xia, Y., Pan, S., Yu, B., et al.: Asynchronous pseudolite indoor positioning method based on C/No weighted fusion. J. Chin. Inertial Technol. (2), 154–159 (2019)
5. Sue, K.-L., Tsai, C.-H., Lin, M.-H.: FLEXOR: a flexible localization scheme based on RFID. In: Information Networking, Advances in Data Communications and Wireless Networks, International Conference, ICOIN 2006, Sendai, Japan, 16–19 January 2006, Revised Selected Papers. DBLP (2006)
6. Qin, Y.: Inertial Navigation, 2nd edn. The Science Publishing Company (2014)
7. Wu, Y., He, C., Liu, G.: On inertial navigation and attitude initialization in polar areas. Satell. Navig. **1**, 1–6 (2020)
8. Mahony, R., Hamel, T., Pflimlin, J.M.: Nonlinear complementary filters on the special orthogonal group. IEEE Trans. Autom. Control **53**(5), 1203–1218 (2008)
9. Mahony, R., Hamel, T., Pflimlin, J.M.: Complementary filter design on the special orthogonal group SO (3). In: Proceedings of the IEEE Conference on Decision and Control and European Control Conference, pp. 1477–1484 (2005)
10. Li, L., Xu, X.: Pedestrian zero velocity detection algorithm based on inertial navigation. Transducer Micrasystem Technol. **38**(03), 154–156+160 (2019)

Integrated Precise Positioning System for Autonomous Level II Driving Offering Lane Level Accuracy

Haiyu Lan(✉), Hongzhou Yang, Yashar Balazadegan Sarvrood, Fei Liu, and Ahmed Wahdan

Profound Positioning Inc., Calgary, AB T2P 3G3, Canada
{hlan, hyang, ysarvrood, fliu, awahdan}@profoundpositioning.com

Abstract. Precise Point Positioning (PPP) has become an attractive topic as consumer low-cost receivers are able to provide raw observations and signals with multiple frequencies and support multiple GNSS systems. Generally, PPP can provide sub-meter level accuracy in favorable environments. However, such an accuracy degrades in challenging environments where the GNSS observations are attenuated or blocked. In order to reach seamless and reliable navigation performance, in this research, PPP is designed to integrate with a low-cost IMU which runs the Inertial Navigation System (INS) algorithm on it. A low-cost Profound-IP3/DR (dead reckoning) integrated navigation system targeted for precise vehicular navigation is proposed. The integrated system is accessed through various road tests which cover good LOS GNSS environments, sub-urbans, short-term GNSS outages, and complete GNSS outages. The results indicate that the Profound-IP3/DR is able to maintain sub-meter RMS accuracy in open-sky and sub-urban areas. For complete GNSS outages and challenging downtown environments with severe multi-path and signal blockages, the Profound-IP3/DR shows robust long-term performance with decimeter to meter-level accuracy. The high positioning accuracy supports Profound-IP3/DR to be a key player in continuous and precise vehicular navigation applications such as asset tracking and car navigation. Moreover, such accurate navigation performance shows the potential and readiness to be integrated with other sources of update such as vision, LiDAR, radar and digital maps to fully benefit the technologies for navigation of autonomous systems.

Keywords: Precise Point Positioning (PPP) · Inertial Navigation System (INS) · Profound-IP3/DR

1 Introduction

Navigation and localization have recently become a hot research topic for autonomous vehicles. A fully autonomous vehicle is capable of navigating itself without human input. Commonly, navigation in autonomous vehicles has five major components [1, 2]. The first component is mobility, which is related to the motion control. Second is the positioning/localization component which provides the vehicle's position, velocity,

J. Sun et al. (Eds.): CSNC 2020, LNEE 650, pp. 723–739, 2020.
https://doi.org/10.1007/978-981-15-3707-3_67

and attitude (PVA). Next comes the guidance component which is used to guide and control the vehicle by utilizing the outputs of the positioning/localization component and perception of the local environment. This component plans paths to avoid collision with other objects. The fourth component is the mission and task planning which provides the waypoints and routes for vehicles. The last is the communication component, which provides a link or network between the vehicle and the road infrastructures or other vehicles.

While the vehicle motion control, trajectory planning algorithms and communication technologies have already been demonstrated successfully, robust vehicle positioning and localization that work well in all-inclusive environments is still a challenging problem. GNSS (Global Navigation Satellite System) receivers, IMUs (Inertial Measurement Units), wheel encoders, and stereo cameras have been widely used to tackle this problem [3].

Recently, the Precise Point Positioning (PPP) has become a very attractive topic since most of the latest consumer low-cost receivers are able to provide raw observations and signals with double-frequency and support multiple GNSS systems. Generally, PPP can provide sub-meter level accuracy in favourable environments [4]. Yet, such an accuracy degrades in challenging environments where the GNSS observations are attenuated or completely down. In order to reach seamless and preferable navigation performance, in this research, PPP is designed to integrate with an IMU which runs the Inertial Navigation System (INS) algorithm on it [5]. A low-cost PPP/INS integrated navigation approach targeted for precise vehicular navigation is proposed. The INS algorithm employs a full MEMS (Micro-Electro-Mechanical System) grade IMU in favour of low-cost applications. In order to solve the issue of inherent inertial sensor error drift, the design of the integrated PPP/INS system relies on the accurate PPP solutions and employs a robust integration algorithm. In general conditions, precise navigation solutions from PPP dominates the overall navigation outputs. Besides, PPP brings faster convergence and more accurate estimation of the INS-derived errors in PVAs as well as error drift in the inertial sensors. The correct estimation of the PVA errors of the INS and IMU sensor biases is a critical factor for an integrated GNSS/INS system in challenging environments [6].

The proposed PPP/INS system is accessed through various road tests which cover good line-of-site GNSS environments, sub-urban, short-term GNSS outages, and complete GNSS outages. The test results indicate that the proposed PPP/INS system is able to maintain sub-meter RMS (Root Mean Square) accuracy in open-sky and sub-urban areas. For complete GNSS outage scenarios (e.g., an underground parking lot) and downtown scenarios with sever multi-path, the proposed PPP/INS system shows robust long-term performance with meter level accuracy. Based on the research and test results, the proposed PPP/INS system is expected to benefit low-cost, continuous and precise land vehicular navigation.

2 System Description

Due to the reason that a single-frequency PPP is able to provide sub-meter accuracy at relative low-cost and faster than a typical double-frequency PPP to reach such a level of accuracy, the single-frequency PPP makes a desirable candidate for ground vehicular navigation. However, a standalone single-frequency PPP cannot continuously output reliable navigation performance for all operating environments, it has to be combined with other sensors or systems for robust navigation.

This section starts by introducing the basics of the single-frequency PPP, namely, the Porfound-IP3 and the mechanization of the INS. Then it describes the Profound-IP3/DR system utilizing a low-cost single frequency GNSS receiver and a MEMS-based IMU that is capable of providing continuous and robust navigation solutions for general land vehicle applications.

2.1 Precise Point Positioning (PPP)

Taking GPS L1 signal as an example, a single-frequency receiver can provide raw code and carrier-phase observations (i.e., P_1 and Φ_1) at the frequency of L1, that is,

$$P_1 = \rho_r^s + c(dt_r - dt^s) + d_{orb} + d_{trop} + d_{ion} + d_{rel} + \epsilon_P \tag{1}$$

$$\Phi_1 = \rho_r^s + c(dt_r - dt^s) + d_{orb} + d_{trop} - d_{ion} + d_{rel} + \lambda_1 N_1 + \epsilon_\Phi \tag{2}$$

in which,

P_1 is the raw code measurements at L1 (m);
Φ_1 is the raw carrier-phase measurement at L1 (m);
ρ_r^s is the geometric distance as between the receiver (r) and the satellite (s);
c is the speed of light (m/s);
dt^s is the satellite clock (s);
dt_r is the receiver clock (s);
d_{orb} is the satellite orbit error (m);
d_{trop} is the tropospheric delay (m);
d_{ion} is the first-order ionospheric delay on frequency L1 (m);
d_{rel} is the error derived from relativistic effects (m);
λ_1 is the wave-length of carrier-phase on frequency L1 (m);
N_1 is the integer ambiguity on frequency L1 (cycle);
ϵ_P is the code error including multipath and noise (m);
ϵ_Φ is the carrier-phase error including multipath and receiver noise (m)

In order to reach sub-meter level accuracy through the single-frequency PPP, the errors due to relativistic effects and the tropospheric delay can be explicitly modeled and resolved [7].

Other than the error sources described in the Eqs. (1) and (2), there are other error components need to be taken into account, those error components include but not limited to the code biases, phase windup, earth tides, satellite antenna phase center offsets. Some errors can be ignored such as ocean and polar tides, antenna phase center variations, given that these small effects cannot affect the sub-meter level accuracy. After applying all the necessary corrections, the remaining unknowns of the Eqs. (1) and (2) are the receiver position, receiver clock error, ionospheric delay, and carrier-phase integer ambiguities. The ionospheric delay can be corrected by either through the ionospheric-free combinations or through corrections from precise products. The Profound-IP3 obtains corrections directly from IGS-RTS for code bias, satellite orbital error, satellite clock error, and ionospheric delay. Meanwhile, the receiver clock error, and the carrier-phase ambiguity (float) are estimated through the Kalman Filter.

2.2 Inertial Navigation System (INS)

The process of INS mechanization is to use the inertial sensor measurements of triaxial accelerometers and triaxial gyroscopes to determine the PVA of the vehicle. The INS process starts with a set of initial PVA information and then adds the changes in the PVA once new sensor measurements are obtained. Below shows the differential equation (continuous model) of the typical INS mechanization [8].

$$\begin{bmatrix} \dot{r}^n \\ \dot{v}^n \\ \dot{T}_b^n \end{bmatrix} = \begin{bmatrix} D^{-1}\, v^n \\ T_b^n f^b - \left(2\Omega_{ie}^n + \Omega_{en}^n\right) v^n + g^n \\ T_b^n\left(\Omega_{ib}^b - \Omega_{in}^b\right) \end{bmatrix} \tag{3}$$

in which

$$D = diag([R_M + h \quad (R_N + h)cos\varphi \quad -1]) \tag{4}$$

The n frame is the navigation frame (local north-east-down); r represents the INS position; v denotes the velocity; T_b^n stands for the 3×3 transition matrix expressing the attitude ε (including pitch, roll and azimuth) of the IMU b frame (body frame) relative to the n frame; ε is then obtained from T_b^n; f^b denotes the accelerometer measurements; ω^b is the angular rate readings measured by the gyroscope. We assume the sampling period of INS is Δt; the non-linear function F denotes the INS mechanism. At time step k, the navigation solution $x_k = [r_k \quad v_k \quad \varepsilon_k]$ can then be updated as follows,

$$x_k = F\left(x_{k-1},\, f_k^b,\, \omega_k^b\right); \quad x_0 = [initial\ value] \tag{5}$$

where x_0 is the initial PVA of INS, more details about the discrete representation of INS mechanization can be referenced in [9].

2.3 PPP/INS Integration

This paper presents the performance of the Profound-IP3/DR system developed by Profound Positioning Inc. targeting land vehicle navigation. Profound-IP3/DR is an integrated navigation library, which provides loosely-coupled integration of Profound-DR and Profound-IP3 solutions. It can provide reliable long-term positioning performance even in the absence of reliable GNSS signal in covered areas and in dense urban regions. Profound-IP3/DR library lies at the core of PPI tethered integrated navigation solution, which has the following capabilities:

- A Plug-and-play capability enabling the system to operate once installed inside the land vehicle with no resection to any special initialization or calibration procedures like any of the competitors.
- An accurate and continuously available PVA solution in GNSS challenging and denied environments using low-cost GNSS receiver and MEMS inertial sensors for low-cost applications where current systems cannot.
- A fast acquisition of PPP solution at the end of GNSS outages/attenuations.

Profound-IP3/DR is capable of providing a superior stand-alone inertial sensor-only navigation performance based on a number of specific designs in the integration engine:

- The state-of-the-art inertial sensor mechanization: Profound-IP3/DR library adopts INS mechanization with precise modeling of sensor errors and higher-order attitude computation algorithm. The navigation libraries feature a special INS mechanization algorithm that eliminates several sources of errors and prevents them to propagate through the INS algorithm. Consequently, a more reliable positioning performance can be obtained. The integrated navigation solution supports various attitude parameterizations including the direction cosine matrix, Euler angles, rotation vector, and quaternion.
- Using several options for initial alignment such as user-defined, leveling-given heading, analytic coarse alignment, fine alignment and in-motion GNSS alignment.
- Extending stand-alone IMU operation using speedometer.
- Applying robust dynamic error modeling: Enabling robust performance over short periods of GNSS outages required the development of a robust dynamic error model that takes into account the propagation of all the navigation error states. While linearizing the model around the true trajectory, the robustness of the model is maintained by including more sophisticated error terms. An EKF is realized for the integration with Profound-IP3 GNSS solutions. The EKF utilizes the robust error model prepared for the navigation error states and feeds back the estimated errors to be removed from future inertial sensor measurements before processing by the mechanization module [10]. The whole flowchart of the Profound-IP3/DR navigation system can be seen in the following Fig. 1.

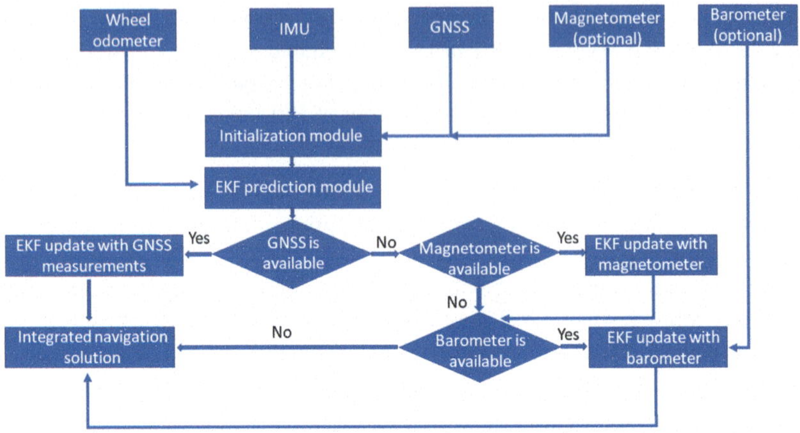

Fig. 1. Flowchart of the Profound IP3/DR integrated navigation system

3 Results and Discussion

The performance of the Profound-IP3/DR integrated navigation system was evaluated by abundant road tests covering various driving environments in a typical North American city - Calgary. Raw observations from GPS, Galileo, and GLONASS were logged and utilized in real-time on a Linux operation system through using a u-blox M8T platform which is a consumer low-cost single-frequency GNSS receiver. The raw observations were also processed in real-time along with the precise corrections from IGS to provide reliable single frequency-PPP solution.

A Bosch BMI088 IMU is used for the INS mechanization, which is a consumer grade 6-axis MEMS IMU. The Profound-IP3/DR integration library is developed using C/C++ language to be desirable for real-time processing. The reference was obtained from a GNSS-RTK/INS integrated solution in which the real-time observations of the receiver are collected using a NovAtel ProPak6 receiver along with a navigational grade KVH 1750 IMU. The IGS-UCAL station was selected as the reference station with baseline less than 15 km during operation. Moreover, the raw data from the reference platform are post-processed through Waypoint Inertial Explorer software provided by NovAtel. Figure 2(a) shows the Profound-IP3/DR system and Fig. 2(b) displays the NovAtel SPAN system with the ProPak6 Triple-Frequency GNSS Receiver and a KVH-1950 IMU inside the testing van and its GPS-703-GGG Antenna on top of the testing van shown in the Fig. 2(c).

To evaluate the general performance of the Profound-IP3/DR system, four different testing scenarios are selected for evaluation, they are

1. Open-sky environment with good LOS GNSS observations for 2.55 min;
2. Passing through several bridges each has a few seconds of GNSS signal blockages;

3. Downtown with severe multipath and signal attenuations for 8.1 min;
4. Complete GNSS outage in an underground parking lot for 4.27 min.

The testing results and discussions to the four testing scenarios are summarized in the following Sects. 3.1–3.4, respectively.

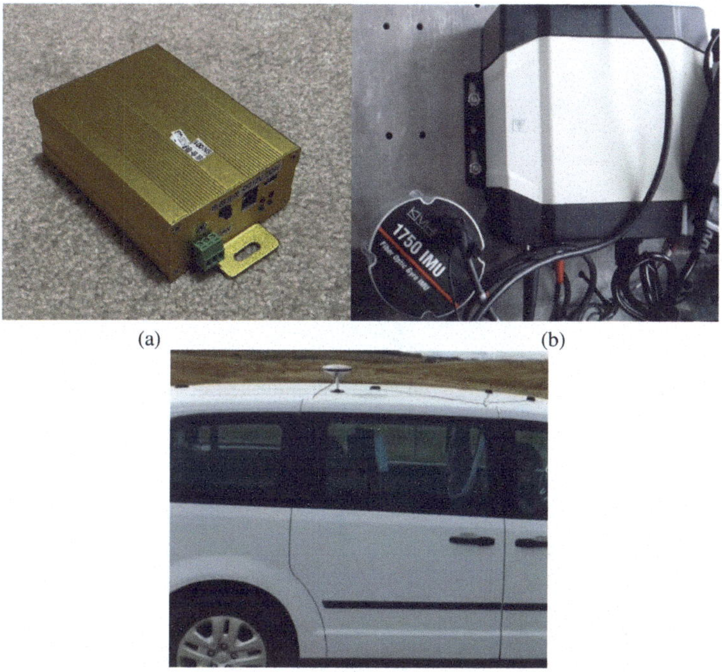

(a) (b) (c)

Fig. 2. Illustration of the (a) Profound-IP3/DR system, (b) NovAtel SPAN KVH-1950 IMU with the ProPak6 triple-frequency GNSS receiver inside the testing van and its GPS-703-GGG Antenna on top of the testing van (c)

3.1 Open-Sky

Figure 3 depicts the open-sky testing environment in this trajectory. Figure 4 shows trajectories drawn from the IP3, IP3/DR, and reference system in this environment. Figure 5 compares the 2D position errors of the IP3 and IP3/DR systems. Benefiting from the favorable GNSS LOS conditions, it is indicated that the IP3/DR solution mostly follows the IP3 solution. Table 1 summarizes the RMS and Maximum errors of both systems. The proposed IP3 and IP3/DR systems are able to achieve decimeter-level positioning accuracy in term of good GNSS LOS environments.

Fig. 3. Open-sky testing environment indicated in Google Earth

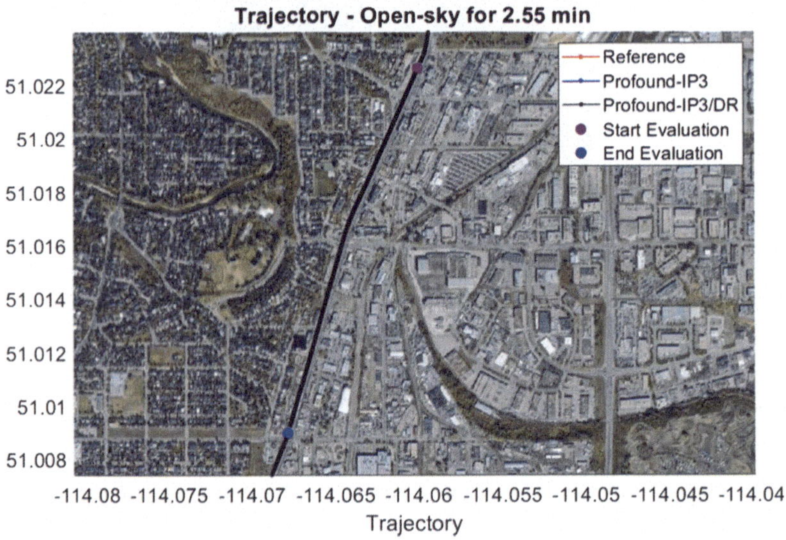

Fig. 4. Trajectories from different systems to the open-sky testing environment

Table 1. The errors statistics for the Profound-IP3 and Profound-IP3/DR during open-sky for 2.55 min

Open-sky (2.55 min)	RMS horizontal position error (m)	Maximum horizontal position error (m)
Profound-IP3	0.259	0.355
Profound-IP3/DR	0.241	0.356

3.2 Passing Bridges

In this testing scenario, the moving speed of the vehicle is around 80 km/h on a highway with several overhead bridges crossed, see the whole trajectory in Fig. 5 and on ridge view instance in Fig. 6. Figures 9 and 10 are two examples of the local zoom-in of the trajectories amid passing bridges. As seen from Fig. 8 and Table 2 that, the integrated IP3/DR solution smooths the spikes/jumps from the IP3 solutions due to limited visible number of satellites and cycle slips when driving under the overpasses. However, the proposed IP3/DR could bridge the transitory GNSS blockages by keeping sub-meter level accuracy and maintains continuous performance. The RMS and Maximum errors of the IP3/DR for the whole trajectory are 0.379 m and 1.02 m, respectively. Moreover, after crossing the overpasses, the fast convergence of the PPP solution of the IP3 contributes to a stronger and more reliable IP3/DR performance.

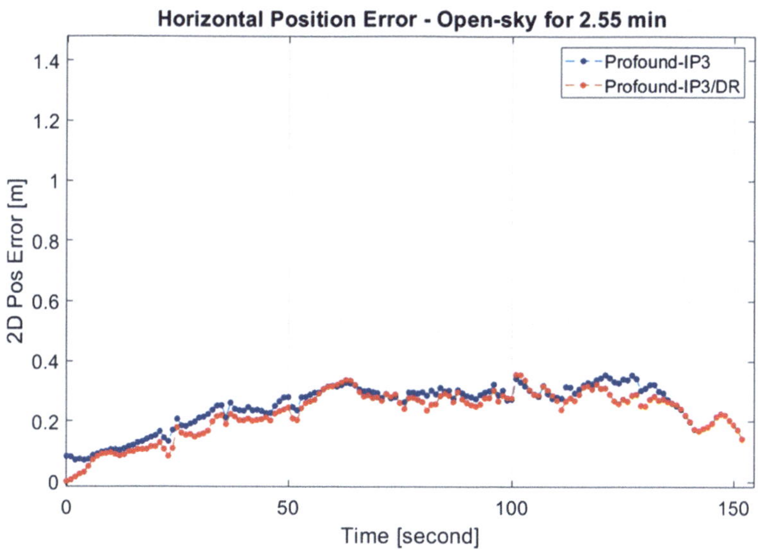

Fig. 5. Performance comparison of the Profound-IP3 and Profound-IP3/DR system in typical open-sky environment

Fig. 6. Test environment of passing several bridges indicated in Google Earth

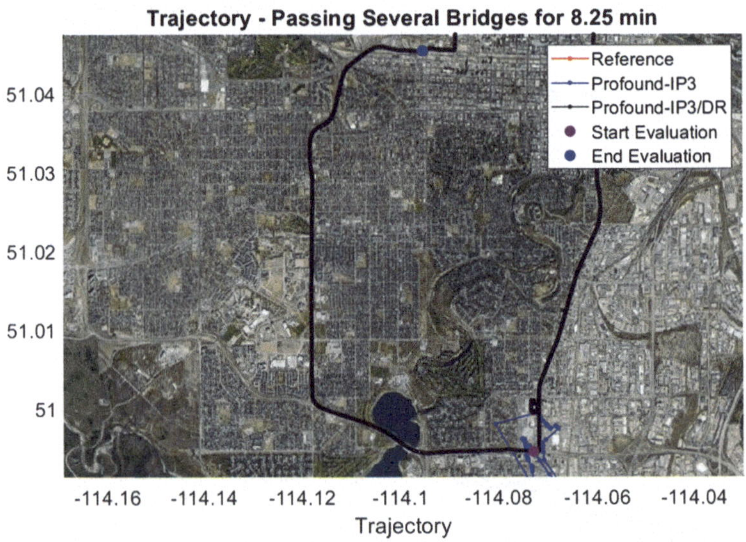

Fig. 7. Trajectories from different systems to the testing environment of passing several bridges

Table 2. The errors statistics for the Profound-IP3 and Profound-IP3/DR in the environment of passing several overhead bridges

Passing bridges (8.25 min)	RMS horizontal position error (m)	Maximum horizontal position error (m)
Profound-IP3	0.784	3.825
Profound-IP3/DR	0.379	1.102

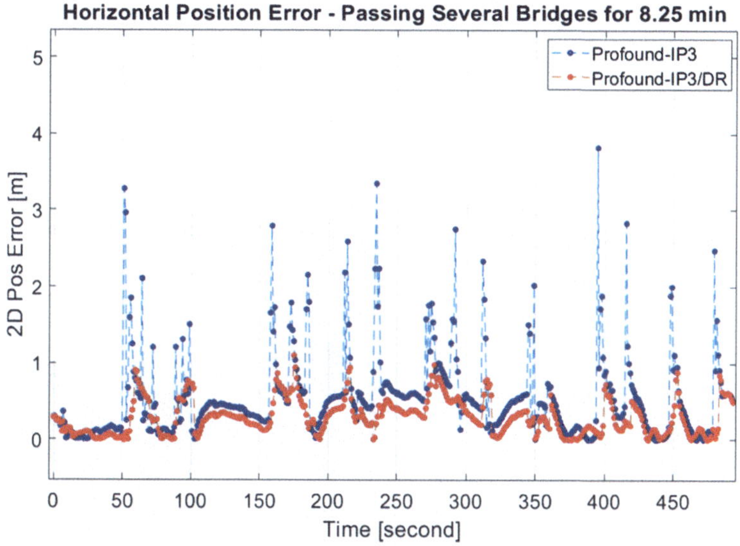

Fig. 8. Performance comparison of the Profound-IP3 and Profound-IP3/DR system in the environment of consecutively passing several overhead bridges

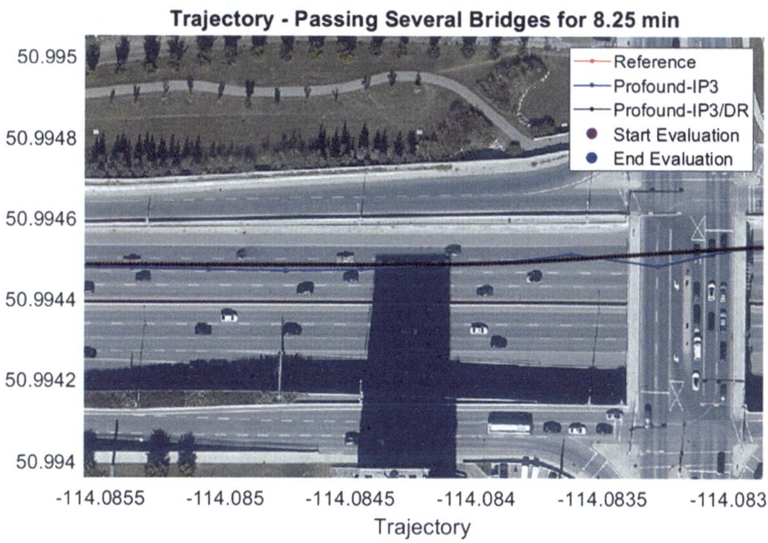

Fig. 9. A typical example of the trajectories comparison when pass a bridge shown in Fig. 7

3.3 Downtown

Figure 11 shows the typical surroundings of the downtown environment in this test. Figure 12 plots trajectories of the IP3, IP3/DR, and reference system in the whole downtown core. The error matrices for IP3 and IP3/DR are shown in Fig. 13 and summarized in Table 3. It is concluded that the IP3/DR solution outperforms the stand-alone IP3 and improves the IP3 performance which was switched into SPP mode instead of PPP mode due to the challenges of severe multipath, poor satellite geometry, and signal blockage, etc. However, the IP3/DR library is able to recognize the GNSS environment and automatically weights the GNSS updates accurately. The smart combination of GNSS updates and inertial sensor measurements result in continuous meter-lever accuracy in such changeling environments.

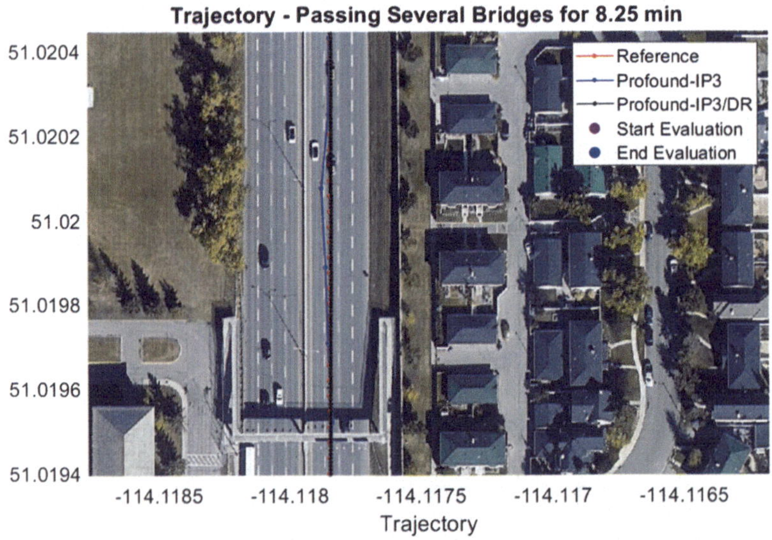

Fig. 10. Second example of the trajectories comparison when passing a bridge shown in Fig. 7

Fig. 11. Test environment of downtown and urban canyon indicated in Google Earth

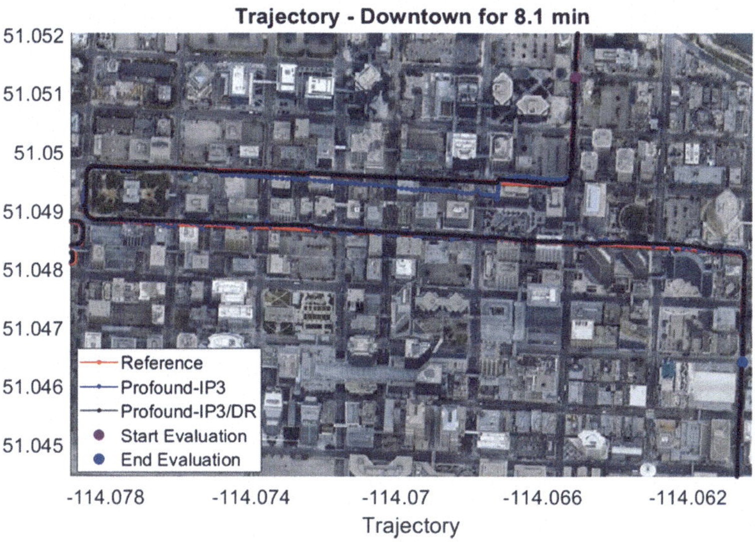

Fig. 12. Trajectories from different systems to the testing environment of downtown and urban canyon

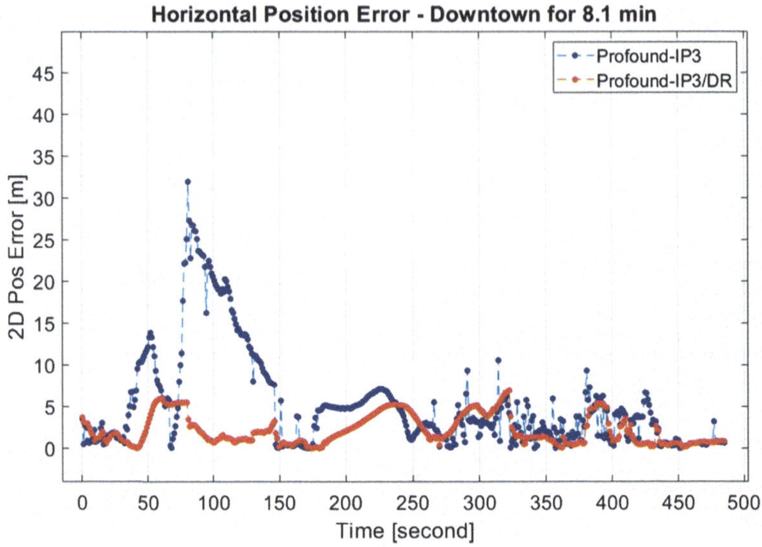

Fig. 13. Performance comparison of the Profound-IP3 and Profound-IP3/DR system in the environment of downtown and urban canyon

Table 3. The error statistics for the Profound-IP3 and Profound-IP3/DR in the environment of downtown and urban canyon

Downtown (8.1 min)	RMS horizontal position error (m)	Maximum horizontal position error (m)
Profound-IP3	7.884	31.953
Profound-IP3/DR	2.819	6.928

3.4 Underground Parking Lot

As seen from Figs. 14, 15 and 16, once entering the underground parking lot, IP3 lost GNSS signal right way with very few solution outputs. However, the IP3/DR can continuously provide reliable navigation solutions in meter-level accuracy. It is summarized in Table 4 that the RMS and Maximum errors of the IP3/DR during GNSS outages are 2.892 m and 5.162 m. The reliable navigation performance during the complete GNSS blockage is a strong indication of the optimally-estimated inertial sensor error and bias drift before the blockage as a result of the precise IP3 navigation solutions and the solid integration algorithm before encountering GNSS outages. Figure 14 also proves the fast acquisition of GNSS satellite signals and the fast convergence of the PPP solutions of the IP3 system at the end of long GNSS outages, which seamlessly contributes to reliable overall navigation performance.

Fig. 14. Test environment of an underground parking lot

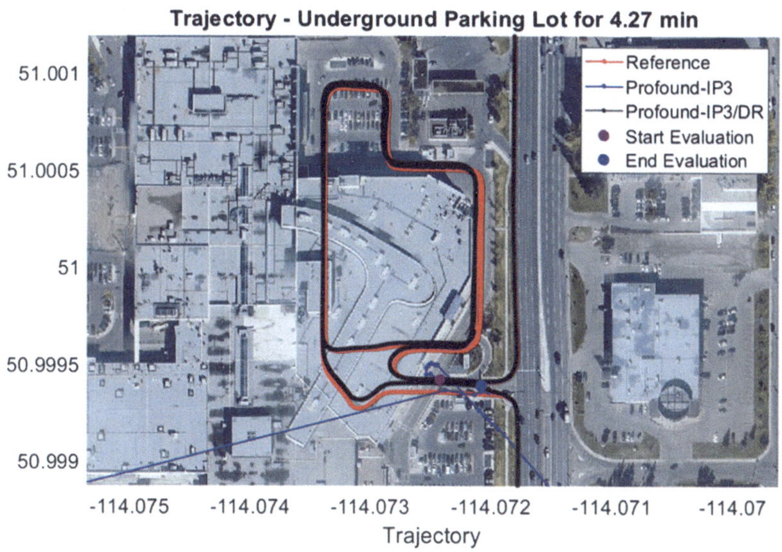

Fig. 15. Trajectories from different systems to the testing environment of an underground parking lot without GNSS availability.

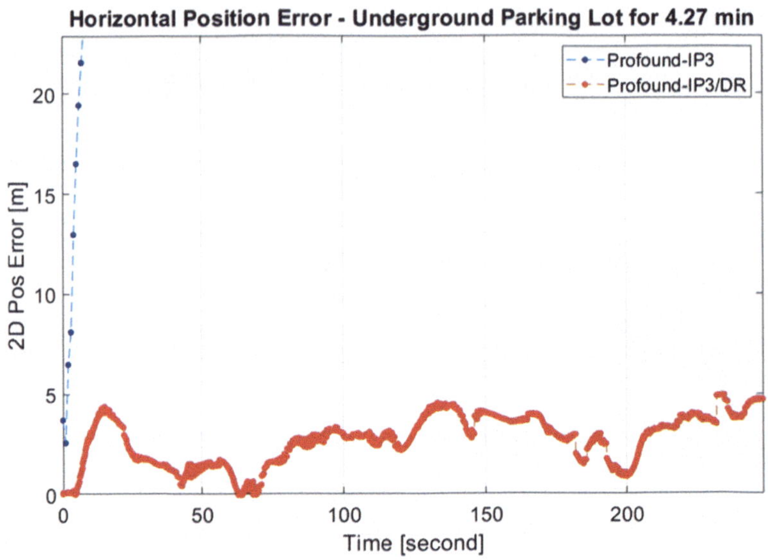

Fig. 16. Performance comparison of the Profound-IP3 and Profound-IP3/DR system in the testing environment of an underground parking lot

Table 4. The error statistics for the Profound-IP3 and Profound-IP3/DR in the environment of downtown and urban canyon

Underground (4.27 min)	RMS horizontal position error (m)	Maximum horizontal position error (m)
Profound-IP3	N/A	N/A
Profound-IP3/DR	2.892	5.162

4 Conclusion

Profound-IP3/DR has the advantage of providing reliable positioning solutions in all-inclusive environments while utilizing low-end consumer-level inertial sensors and GNSS receivers. The high positioning accuracy opens the door for Profound-IP3/DR to be a key player in continuous and precise vehicular navigation applications such as asset tracking and safety assurance. Likewise, such accurate navigation performance shows the potential and readiness to be integrated with other sources of update such as vision, LiDAR, radar and digital maps to fully benefit the industry of autonomous driving and the development of the advanced driver-assistance systems.

References

1. Kurata, K., Morizane, H., Matsuo, S.: Autonomous driving vehicle and autonomous driving system. USA Patents (2019)
2. Paden, B., Čáp, M., Yong, S.Z., Yershov, D., Frazzoli, E.: A survey of motion planning and control techniques for self-driving urban vehicles. IEEE Trans. Intell. Veh. **1**(1), 33–55 (2016)
3. Noureldin, A., Karamat, T.B., Georgy, J.: Fundamentals of Inertial Navigation. Satellite-Based Positioning and Their Integration, Springer, Heidelberg (2012)
4. Kouba, J., Héroux, P.: Precise point positioning using IGS orbit and clock products. GPS Solution **5**(2), 12–28 (2001)
5. Lan, H., Elsheikh, M., Abdelfatah, W., Wahdan, A., El-Sheimy, N.: Integrated RTK/INS navigation for precision agriculture. In: 32nd International Technical Meeting of the Satellite Division of the Institute of Navigation (ION GNSS+ 2019), pp. 4076–4086 (2019)
6. Elsheikh, M., Abdelfatah, W., Wahdan, A., Gao, Y.: Low-cost PPP/INS integration for continuous and precise vehicular navigation. In: 31th International Technical Meeting of the Satellite Division of the Institute of Navigation, ION GNSS 2018, pp. 3169–3178 (2018)
7. Gao, Y., Shen, X.: A new method for carrier-phase-based precise point positioning. Navigation **49**(2), 109–116 (2002)
8. Lan, H., El-Sheimy, N.: A state constraint Kalman filter for pedestrian navigation with low cost MEMS inertial sensors. In: 27th International Technical Meeting of the Satellite Division of the Institute of Navigation, ION GNSS 2014, pp. 579–589 (2014)
9. El-Sheimy, N.: Inertial techniques and INS/DGPS Integration. Engo 623-Course Notes (2004)
10. Shin, E.-H., El-Sheimy, N.: Accuracy improvement of low cost INS/GPS for land applications. University of Calgary (2001)

Reliable Localization Using Multi-sensor Fusion for Automated Valet Parking Applications

Mostafa Sakr[1](✉), Adel Moussa[2], Walid Abdelfatah[1], Mohamed Elsheikh[1], and Naser El-Sheimy[1]

[1] Profound Positioning Inc., Suite 820, 800 6th Avenue SW, Calgary, AB T2P 3G3, Canada
msakr@profoundpositioning.com
[2] University of Calgary, 2500 University Dr NW, Calgary, AB T2N 1N4, Canada

Abstract. Multi-sensor fusion is the key to enabling reliable and accurate localization for automated valet parking (AVP) systems that can autonomously navigate in highly structured indoor environments. The current state-of-the-art in AVP systems depends on light detection and ranging (LiDAR) systems, visible-light cameras, or infrastructure-based solutions. LiDAR systems can provide high positioning accuracy across a variety of operating conditions. However, they are generally more expensive, have moving parts that create more room for error, and the processing of its 3D point cloud is computationally demanding. As for cameras, the performance of any camera-based localization system is susceptible to variations in lighting conditions and fail altogether under degraded visual environments. This paper presents a real-time, radar-based localization library developed by Profound Positioning Inc. (PPI). PPI's localization library utilizes a multi-sensor fusion approach for real-time vehicle localization in structured environments such as underground parking lots. PPI's algorithm integrates the onboard motion sensors, a set of mid-range automotive radars, and two-dimensional (2D) high definition (HD) maps. PPI's radar-based localization library is capable of maintaining a decimeter-level accuracy, and it is validated and evaluated using real-life scenarios in an indoor parking lot.

Keywords: Navigation · Sensor fusion · Automated valet parking · Autonomous system · Real-time localization

1 Introduction

1.1 Background

Positioning services for car navigation have long relied on the global navigation satellite system (GNSS); however, GNSS cannot maintain an accurate position while traveling under bridges, around tall buildings, and under tree canopies due to signal blockage or multipath [1, 2]. Autonomous and connected vehicles (AV/CVs) aim to rely on remote sensing systems (e.g., LiDAR, cameras) to perceive the environment

J. Sun et al. (Eds.): CSNC 2020, LNEE 650, pp. 740–747, 2020.
https://doi.org/10.1007/978-981-15-3707-3_68

and provide accurate positioning [2, 3]. However, these systems suffer from degraded performance in some environments, such as those with low visibility, rain, or snow [4].

Considering availability, accuracy, and cost factors in the production of AV/CVs, candidate positioning and navigation technologies face some limitations for practical implementation [5]. Present land vehicles rely on GNSS receivers for positioning services. For AV/CVs requiring a submeter level of accuracy, GNSS receivers should operate on either differential mode (which requires a local reference station within 40 km) or precise point positioning (PPP) mode [6, 7]. However, the submeter level of accuracy obtained from these modes can be only achieved in an open sky environment when there is a direct line of sight to GNSS satellites. The accuracy may even deteriorate in the open sky when traveling under overpasses or due to short outages and cycle slips [8, 9]. Moreover, GNSS has limitations such as multipath errors, signal blockage in urban canyons, ionospheric delays, and natural or intentional interference/jamming; hence, backup systems are needed [10, 11]. Vision-based positioning relying on cameras can be utilized for pose estimation [12, 13], based on processing camera frames (also known as visual odometry or VO), which performs well in textured and ambient environments. Since VO integrates small incremental motions over time, it is bound to drift due to the variations observed in the scene. The main limitation of VO is its inability to extract features in a degraded visual environment, where low contrast is abundant in the acquired image frames [13, 14].

Despite the fact that state-of-the-art algorithms perform well in urban environments [14, 15], most of the tests were conducted using the KITTI benchmark suite [15, 16] or the Ford campus vision and LiDAR dataset [17], which do not provide insight into the feasibility of these algorithms in challenging scenarios such as downtown cores and in areas where the vehicle travels through a tunnel, or in parking garages where parts of the scenes may suffer from complete absence of light. LiDAR can operate in this degraded vision environment and can provide accurate measurements of range information with respect to the surrounding objects. However, LiDAR is generally more expensive, may cause design restrictions, has moving parts that create more room for error, and the processing of its 3D point cloud is computationally demanding [20]. Machine vision technologies (cameras and LiDAR) may also fail to provide accurate positioning accuracies when operating under challenging weather conditions such as rain, snow, and fog [4]. Leading car manufacturers and pioneers in the self-driving car industry, such as Tesla and Nissan, have decided to exclude LiDAR from their autonomous driving systems and move toward vision and radar [18].

Automotive radar is an all-weather system that has been used for a long time in land vehicles for adaptive cruise control (ACC). It can be used as an alternative to LiDAR to detect objects and to provide range estimation [19]. Nevertheless, most present radar-based odometry and localization methods lack the target submeter level accuracy required by AV/CVs [20, 21]. Another source of positioning information is the inertial navigation system (INS), which relies on the measurements of accelerometers and gyroscopes to detect incremental changes in position, velocity, and orientation, allowing the navigation solution to extend from a previously known position and orientation [22, 23]. However, the INS solution may drift rapidly over time, especially when low-cost Micro-Electro-Mechanical-Systems (MEMS) based sensors are used [24, 25]. Updates from odometers, barometers, and magnetometers can be used in denied GNSS and vision environments to limit the INS position drift [26, 27]. Nevertheless, these updates cannot enable the standalone operation of INS for an extended duration [28, 29]. The continuous advancements in sensor and computing technologies are the driving forces behind the progress of autonomous vehicles. However, the widespread market adoption of full autonomy (level-5 autonomy according to the SAE classification) still faces many challenges, in particular, the assurance of reliable operation across different driving modes under uncertain environments and changing operating conditions. Meanwhile, the current generation of sensors and systems are capable of supporting a high-level of autonomy (level-4), and one of the important applications of this automation category is the automated valet parking (AVP).

1.2 Motivation

Autonomous vehicle (AV) technologies are being developed, tested, and deployed by several private companies and public agencies. AVs promise to enhance safety, reduce emissions, and improve transportation system efficiency and reliability [30]. The growing demand for AVs is shaping the future of the automotive industry by connecting vehicles to one another and to the surrounding infrastructure (V2X), transforming the in-vehicle experience, and paving the way for large-scale implementation of autonomous driving [31]. AV technology requires onboard intelligence relying on a suite of sensors and systems such as global navigation satellite systems (GNSS), including GPS, vehicle motion sensors (gyroscopes, accelerometers, and speedometers), cameras, LiDAR, and radar technologies. AV/CVs will be able to share common perception and awareness of other traffic participants, reduce the risk of accidents, and improve traffic flow [33]. AVs that are capable of sensing the environment and navigating without human input require robust high precision positioning and orientation information at all times in all environments. A key AV/CV system requirement is to have reliable decimeter-level positioning accuracy everywhere under all operational environments [32]. The availability of the above-mentioned sensors and systems in future AV/CVs provides an attractive opportunity to increase the accuracy and reliability of the positioning system.

1.3 Objectives

The objective of this paper is to demonstrate PPI's real-time, radar-based localization library. The objective of PPI's solution is to provide an efficient algorithm capable of running in real-time, providing continuous localization of land vehicles at denied GNSS environment and challenging degraded vision environment with decimeter-level accuracy. The proposed method offers a reliable localization solution that could be used in the context of AVP applications based on the integration of four mid-range automotive radars (mounted at the four vehicle corners) fused with the onboard vehicle's sensors.

2 Technology

This section provides an overview of the AVP systems and provides a top-level description of the PPI's radar-based localization solution. At the core of PPI's solution is a new sensor fusion method using a set of mid-range automotive radars and onboard sensors to provide a reliable localization solution in GNSS denied and visually challenging environments.

Figure 1 shows an overview of the AVP system, highlighting the different components of the system and outlining the major tasks involved in typical AVP operation scenarios. The successful implementation of an AVP system takes into consideration the interaction between the user, the vehicle, and the parking facility. A typical AVP scenario can be initiated when the user reaches a point of interest (POI) such as a building entrance or an underground elevator. The user then disembarks from the vehicle and instructs the vehicle to park. In a managed parking facility, the request to park is handled by a centralized facility management layer that processes the parking request, assigns a free parking spot to the vehicle, and directs the vehicle to the designated spot. The parking facility management layer also disseminates the parking lot floorplan and the high-definition (HD) maps to the vehicle. The objective of the localization algorithm is to determine the ego-vehicle position with respect to the parking floorplan. This location estimate is used by the motion planning and control algorithm to compute a valid trajectory to the assigned parking spot, to control the vehicle motion towards this spot park, and to perform the final parking maneuver. Later, the user can command the vehicle to return to a prearranged meeting spot. Again, the accurate real-time localization of the vehicle ensures a timely and safe journey towards the user pickup region.

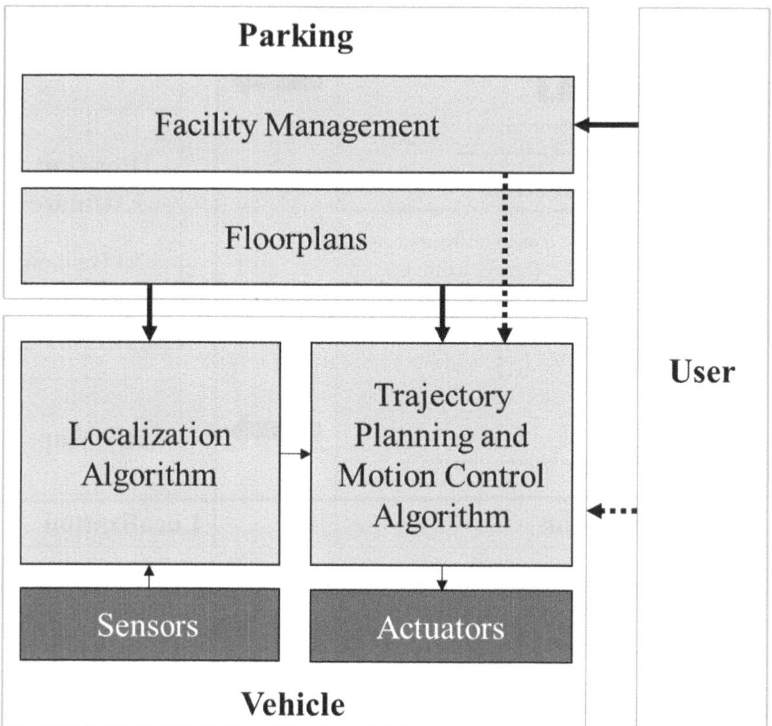

Fig. 1. Automated valet parking: system components

Providing a reliable and accurate localization is essential to the integrity of the AVP system. In PPI's radar-based solution, the localization algorithm depends on two types of sensors available on-board the test vehicle; the mid-range radar and the onboard motion sensors. Figure 2 shows an overview of the radar localization algorithm. Limiting the localization algorithm to the radar and the onboard motion sensors ensures that the system is viable under a wide range of environmental conditions, like degraded vision conditions.

The radars are mounted at the four corners of the vehicle to provide good coverage of the surrounding environment. The radar reflections over the different objects in the parking lot are represented as a sparse point cloud representing the range and the azimuth of the peaks of the reflected signals relative to each radar element. These reflections are processed to form the input to the localization algorithm, which relies on HD maps created from a prearranged and saved LiDAR scan for the environment.

The radar reflections are used to generate a radar point cloud, which is compared to the floorplan of the parking lot to provide position and heading information. The output from the odometry and the map localization blocks are then fused to generate the final local pose, consisting of two-dimensional position and heading, of the vehicle in the target floorplan.

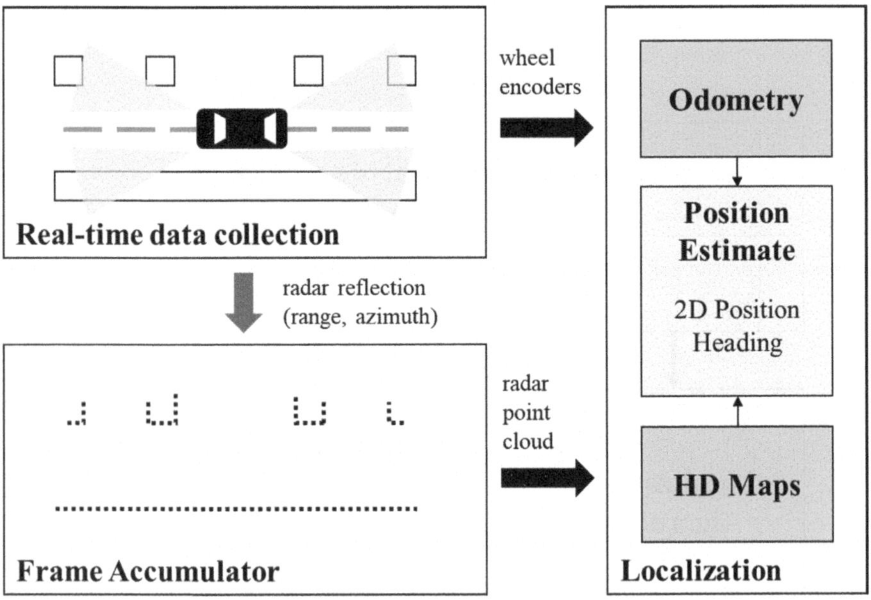

Fig. 2. Radar-based localization system overview

3 Results and Discussion

This section shows the results of a real-time test performed to demonstrate the real-time performance of PPI's radar-based localization library in an underground parking lot, in the same operating conditions for AVP applications.

The real-time test was performed at the B1 level of an underground parking lot. The test vehicle moved at an average speed of 8 km/h inside the parking area, did reverse parking, and then returned close to the starting point. The test vehicle completed one loop in approximately 2.2 min, traveling for more than 300 m. Figure 3 shows, on the B1 map, the performance of PPI's radar-based real-time localization algorithm for AVP, along with the reference trajectory. The reference trajectory is obtained using a 3D LiDAR device mounted onboard the test vehicle to obtain a centimeter-level reference. Figure 3 clearly shows that PPI's radar-based localization library exhibits decimeter-level accuracy with stable pose estimate that does not suffer from drift, guaranteeing robust performance for such safety-critical application.

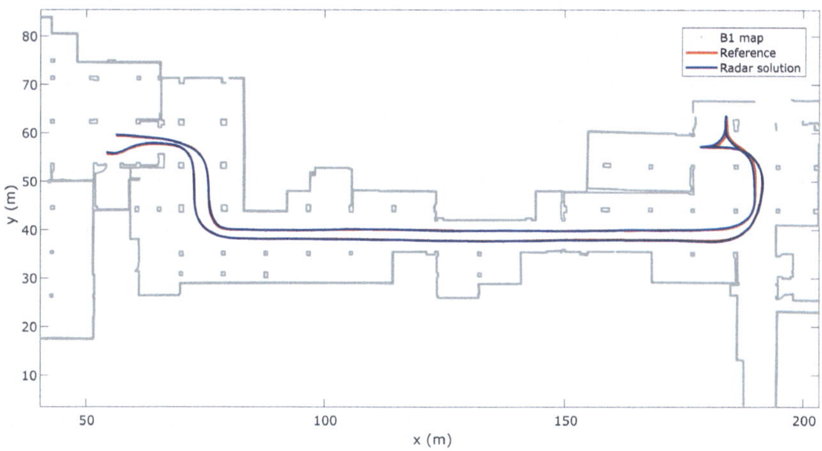

Fig. 3. PPI's localization solution compared to the reference

4 Conclusion

This paper provided an overview of PPI's radar-based real-time localization system. PPI's solution performance is tested and evaluated using a scenario for autonomous valet parking in covered parking garages. The system utilizes data from four mid-range radars mounted at the four corners of the vehicle along with measurements from the onboard motion sensors. The system is designed to operate under challenging conditions and in the absence of an absolute positioning method, such as GNSS, and to provide a reliable localization solution in real-time. PPI's approach exploits sensors already available on most modern vehicles, which ensures a seamless deployment of AVP functionality to a wide range of operating vehicles while keeping down the total cost of the system. Additionally, PPI's approach is based on the vehicle's on-board sensors and does not require installing specific infrastructure in the target area. PPI's system was tested in real-world conditions in an underground parking facility. The real-time test results show that the PPI's radar-based localization library can achieve the localization accuracy required for AVP operation.

References

1. Rabiee, R., Zhong, X., Yan, Y., Tay, W.P.: LaIF: a lane-level self-positioning scheme for vehicles in GNSS-denied environments. IEEE Trans. Intell. Transp. Syst. **20**(8), 2944–2961 (2019)
2. Rose, C., Britt, J., Allen, J., Bevly, D.: An integrated vehicle navigation system utilizing lane-detection and lateral position estimation systems in difficult environments for GPS. IEEE Trans. Intell. Transp. Syst. **15**(6), 2615–2629 (2014)
3. Sun, P., Zhao, X., Xu, Z., Min, H.: A 3D LiDAR data-based dedicated road boundary detection algorithm for autonomous vehicles. IEEE Access **7**, 29623–29638 (2019)
4. Aldibaja, M., Suganuma, N., Yoneda, K.: Improving localization accuracy for autonomous driving in snow-rain environments. In: Proceedings of the 2016 IEEE/SICE International Symposium on System Integration, Sapporo Convention Center, Sapporo, Japan, 13–15 December 2016 (2016)
5. Mohamed, S., Haghbayan, M., Westerlund, T., Tenhunen, H., Plosila, J.: A survey on odometry for autonomous navigation systems. IEEE Access **7**, 97466–97486 (2019)
6. Zhou, P., Yang, H., Xio, G., Du, L., Gao, Y.: Estimation of GPS LNAV based on IGS products for real-time PPP. GPS Solutions **23**(1), 27 (2019)
7. NovAtel. Precise positioning with NovAtel CORRECT including performance analysis, April 2015. http://www.novatel.com/assets/Documents/Papers/NovAtel-CORRECT-PPP.pdf
8. Hadas, T., Bosy, J.: IGS RTS precise orbits and clocks verification and quality degradation over time. GPS Solutions **19**(1), 93–105 (2015)
9. Karaim, M., Elsheikh, M., Noureldin, A.: GNSS error sources. In: Rustamov, R.B., Hashimov, A.M. (eds.) Multifunctional Operation and Application of GPS, Chap. 4, pp. 69–85. IntechOpen (2018)
10. Mannings, R.: Ubiquitous Positioning. Mobile Communication Series. Artech House (2008)
11. Farrell, J.: Aided Navigation, GPS with High-Rate Sensors. McGraw Hill, New York (2008)
12. Wang, S., Clark, R., Wen, H., Trigoni, N.: DeepVO: towards end-to-end visual odometry with deep recurrent convolutional neural networks. In: Proceedings of IEEE International Conference on Robotics and Automation, Singapore, 29 May–3 June 2017 (2017)
13. Ci, W., Huang, Y., Hu, X.: Stereo visual odometry based on motion decoupling and special feature screening for navigation of autonomous vehicles. IEEE Sens. J. **19**(18), 8047–8056 (2019)
14. Zhang, J., Singh, S.: Visual-lidar odometry and mapping: low-drift, robust, and fast. In: Proceedings of the IEEE International Conference on Robotics and Automation, Seattle, WA, 25–30 May 2015 (2015)
15. Yin, H., Berger, C.: When to use what data set for your self-driving car algorithm: an overview of publicly available driving datasets. In: Proceedings of IEEE 20th International Conference on Intelligent Transportation Systems, Yokohama, 16–19 October 2017 (2017)
16. Parmar, Y., Natarajan, S., Sobha, G.: Deeprange: deep-learning-based object detection and ranging in autonomous driving. IET Intel. Transport Syst. **13**(8), 1256–1264 (2019)

17. Pandey, G., McBride, J., Eustice, R.: Ford campus vision and lidar data set. Int. J. Robot. Res. **30**(13), 1543–1552 (2011)
18. Manners, D.: Nissan follows Tesla in ruling out Lidar for autonomous EVs. Electronics Weekly. https://www.electronicsweekly.com/news/business/nissan-follows-tesla-ruling-lidar-autonomous-evs-2019-05/. Accessed 16 May 2019
19. Crowe, S.: Researchers back Tesla's non-LiDAR approach to self-driving cars. The ROBOTREPORT. https://www.therobotreport.com/researchers-back-teslas-non-lidar-approach-to-self-driving-cars/. Accessed 25 April 2019
20. Lee, T.-Y., Skvortsov, V., Kim, M.-S., Han, S.-H., Ka, M.-H.: Application of W-Band FMCW Radar for road curvature estimation in poor visibility conditions. IEEE Sens. J. **18**(13), 5300–5312 (2018)
21. Rapp, M., Barjenbruch, M., Hahn, M., Dickmann, J., Dietmayer, K.: Probabilistic ego-motion estimation using multiple automotive radar sensors. Robot. Auton. Syst. **89**, 136–146 (2017)
22. Titterton, D., Weston, J.: Strapdown Inertial Navigation Technology, 2nd edn. IEE (2004)
23. Noureldin, A., Karamat, T., Georgy, J.: Fundamentals of Inertial Navigation, Satellite-based Positioning and their Integration. Springer, Heidelberg (2012). ISBN 978-3-642-30465-1
24. Quinchia, A., Falco, G., Falletti, E., Dovis, F., Ferrer, C.: A comparison between different error modeling of MEMS applied to GPS/INS integrated systems. Sensors **13**(8), 9549–9588 (2013)
25. Wei, W., Gao, Z., Gao, S., Jia, K.: A SINS/SRS/GNS autonomous integrated navigation system based on spectral redshift velocity measurements. Sensors **18**(4), 1145–1164 (2018)
26. Georgy, J., Noureldin, A.: Vehicle navigator using a mixture particle filter for inertial sensors/odometer/map data/GPS integration. IEEE Trans. Consum. Electron. **58**(2), 544–552 (2012)
27. Abosekeen, A., Noureldin, A., Korenberg, M.: Improving the RISS/GNSS land-vehicles integrated navigation system using magnetic azimuth updates. IEEE Trans. Intell. Transp. Syst. (2019). https://doi.org/10.1109/TITS.2019.2905871. 14 p.
28. Gao, K., Ren, S., Yi, G., Zhong, J., Wand, Z.: An improved ACKF/KF initial alignment method for odometer-aided strapdown inertial navigation system. Sensors **18**(11), 3896–3914 (2018)
29. Lim, J., Yoo, W.-J., Kim, L., Lee, Y., Lee, H.: Augmentation of GNSS by low-cost MEMS IMU, OBD-II, and digital altimeter for improved positioning in urban area. Sensors **18**(11), 3830–3852 (2018)
30. Research and Markets: North America's Autonomous Car Market to 2030: Projecting to Garner $52.3 Billion, Driven by the Evolution in Connected Car Technology. https://www.globenewswire.com/news-release/2019/07/24/1887021/0/en/North-America-s-Autonomous-Car-Market-to-2030-Projecting-to-Garner-52-3-Billion-Driven-by-the-Evolution-in-Connected-Car-Technology.html. Accessed 24 July 2019
31. PPSC Working Group on Automated and Connected Vehicles: Automated and Connected Vehicles Policy Framework for Canada. Developed by the Policy and Planning Support Committee (PPSC), Cat. No. T42-13/2019E-PDF, Presented to the Council of Ministers of Transportation and Highway Safety, January 2019. ISBN 978-0-660-29330-1
32. Welbur, J.: AV/CV Consumer Awareness Literature Review. Automotive and Surface Transportation, National Research Council of Canada (NRC-CNRC), Cat. No. NR16-262/2019E-PDF, March 2019. ISBN 978-0-660-29965-5
33. GAA Automotive Association: An assessment of LTE-V2X (PC5) and 802.11p direct communications technologies for improved road safety in the EU. http://5gaa.org. Accessed Dec 2017

Author Index

J. Sun et al. (Eds.): CSNC 2020, LNEE 650, pp. 749–752, 2020.
https://doi.org/10.1007/978-981-15-3707-3

The manufacturer's authorised representative in the EU is Springer Nature Customer Service Centre GmbH, Europaplatz 3, 69115 Heidelberg, Germany. If you have any concerns regarding our products, please contact ProductSafety@springernature.com

Printed and bound by CPI Group (UK) Ltd, Croydon, CR0 4YY
30/09/2026
02225498-0019